RIVER BASIN SEDIMENT SYSTEMS

River Basin Sediment Systems: Archives of Environmental Change

Edited by

Darrel Maddy
Department of Geography, University of Newcastle, Newcastle upon Tyne, UK

Mark G. Macklin
Institute of Geography and Earth Sciences, University of Wales, Aberystwyth, UK

Jamie C. Woodward
School of Geography, University of Leeds, UK

A.A. BALKEMA PUBLISHERS / LISSE / ABINGDON / EXTON (PA) / TOKYO

Library of Congress Cataloging-in-Publication Data

Applied for

Cover design: Studio Jan de Boer, Amsterdam, The Netherlands.
Typesetting: C&D Composition & Design services, Minsk, Belarus.
Printed by: Grafisch Produktiebedrijf Gorter, Steenwijk, The Netherlands.

Published by
A.A. Balkema Publishers, a member of Swets & Zeitlinger Publishers
www.balkema.nl and www.szp.swets.nl

ISBN 90 5809 342 5

Contents

Preface ix

Part 1: The context 1

1. Fluvial archives of environmental change 3
D. Maddy, M.G. Macklin & J.C. Woodward

2. Alluvial systematics 19
J. Lewin

Part 2: Tectonic forcing: crustal instability and the fluvial record 43

3. The Maas terrace sequence at Maastricht, SE Netherlands: evidence for 200 m of late Neogene and Quaternary surface uplift 45
M.W. van den Berg & Ton van Hoof

4. Flow in the lower continental crust as a mechanism for the Quaternary uplift of the Rhenish Massif, north-west Europe 87
R. Westaway

Part 3: Climate forcing: records of pleistocene and holocene river behaviour 169

5. Early Pleistocene fluvial and estuarine records of climate change in the southern Netherlands and northern Belgium 171
C. Kasse & S. Bohncke

6. Fluvial responses to external forcing: examples from the French Massif Central, the Texas Coastal Plain (USA), the Sahara of Tunisia, and the Lower Mississippi Valley (USA) 195
M. D. Blum & E.C. Straffin

7. River terrace formation in synchrony with long-term climatic fluctuation: supporting mammalian evidence from southern Britain 229
D.R. Bridgland & D.C. Schreve

8. Lateglacial and Holocene environmental change indicated by floodplain deposits of the Hessian Depression (Central Germany) 249
P. Houben, S. Nolte, H. Rittweger & J. Wunderlich

9. Lateglacial and Holocene palaeohydrology of the lower Vychegda river, western Russia 265
A. Sidorchuk, O. Borisova, N. Kovalyukh & A. Panin

10. A 300-year history of flooding in an Andean mountain river system: the Rio Alizos, southern Bolivia 297
G.S. Maas, M.G. Macklin, J. Warburton, J.C. Woodward & E. Meldrum

Part 4: Geoarchaeological perspectives and the human impact 325

11. The Holocene fluvial sedimentary record and alluvial geoarchaeology in the Nile Valley of northern Sudan 327
J.C. Woodward, M.G. Macklin, D.A. Welsby

12. Lateglacial and Holocene fluvial records from the central part of the Paris Basin (France) 357
J.-F. Pastre, N. Limondin-Lozouet, C. Leroyer, M. Fontugne, A. Gebhardt, V. Krier

13. Historic and possible prehistoric human impacts on floodplain sedimentation, North Branch of the Susquehanna River, Pennsylvania, USA 375
D.M. Thieme

14. Archaeological resources, preservation and prospection in The Trent valley: the application of Geographical Information Systems to Holocene fluvial environments 405
A.J. Howard, K. Challis & M.G. Macklin

Part 5: Modelling and monitoring of sediment fluxes and river channel dynamics 421

15. Holocene sediment budgets in an upland gravel bed river: the River South Tyne, northern England 423
D.G. Passmore & M.G. Macklin

16. The potential for high resolution fluvial archives in braided rivers: quantifying historic reach-scale channel and floodplain development in the River Feshie, Scotland 445
B. Rumsby, J. Brasington & R. McVey

Nile

17. Exploring the possibilities and limitations of modelling Quaternary fluvial dynamics: a case study of the River Meuse 469
L.A. Tebbens & A. Veldkamp

18. Modelling the impacts of different flood magnitudes and frequencies on catchment evolution 485
T.J. Coulthard, M.G. Macklin & M.J. Kirkby

Preface

The last decade has seen a burgeoning of research activity concerned with the relationship between river behaviour and environmental change. This research has involved a wide range of disciplines including geomorphology, sedimentology, geophysics, archaeology, palaeoecology, engineering and planning. Central to many studies has been causality and, in particular, establishing the extent to which river behaviour reflects extrinsic or intrinsic controls. This issue is equally important to the earth scientist, or archaeologist, attempting to establish whether episodes of valley floor cut-and-fill result from tectonic activity or climate change, or for the river manager who, confronted by an unstable reach, must establish its cause, including both the temporal and spatial context.

The Fluvial Archives Group (FLAG)
Reflecting this increasing interest in fluvial systems, the Fluvial Archives Group (FLAG) formed in 1996 as a Research Group of the British Quaternary Research Association for a fixed term of 3 years. Since that time its membership has grown to over 160, with members from all over the globe representing 22 countries. FLAG is now an independent, internationally oriented group of researchers (http://qra.org.uk/FLAG) which has set out the following aims:

- To promote the value of investigating fluvial archives through the production of widely available and readily accessible published information;
- To establish a forum for the exchange of information and ideas;
- To identify foci for future research, and, in particular, to identify the gaps in current knowledge;
- To facilitate joint 'focused' initiatives to formulate strategies for addressing problems.

In order to address these aims two major research themes were identified at the first meeting of the Group in Durham, UK in 1996.

Theme 1 addresses *Long Terrestrial Records* spanning the entire Quaternary. Theme 2 addresses *Fluvial Environments and Processes* in relation to external and internal forcing. After the second meeting of the group in Arcen, the Netherlands (1997), it was decided to focus these research themes into more tightly defined research programmes. Four such programmes are presently defined and these are:

Theme 1

- *Global correlation of Late Cenozoic fluvial sequences* (contact Dr David Bridgland [d.r.bridgland@durham.ac.uk]). This programme is now being developed as IGCP449 with a programme of events planned during the first project term 2000-2003.
- *Fluvial activity and crustal instability* (contact Prof Frank Sirocco [sirocko@mail.uni-mainz.de])

Theme 2

These programmes are being developed to link closely with the international Past Global Change programme (PAGES).

- *Fluvial response to rapid environmental change during the last two interglacial-glacial cycles: The north-west European fluvial archive.* This programme is now being developed as an INQUA project (contact Prof Jef Vandenberghe [vanj@geo.vu.nl]; Dr Darrel Maddy [darrel.maddy@ncl.ac.uk]) with a series of meetings planned during the project term 1999-2003.
- *Holocene fluvial system response to frequent and rapid periods of environmental change: identification and modelling of forcing factors* (contacts Prof Mark Macklin [mvm@aber.ac.uk] and Dr David Passmore [d.g.passmore@ncl.ac.uk]).

The third meeting of the Fluvial Archives Group was held in Cheltenham, UK during September 1998. In accordance with the FLAG objectives, this meeting aimed to bring together researchers from a wide range of disciplines with an interest in river dynamics and environmental change, and to provide a forum for the discussion of innovative techniques and methodologies, research gaps, and opportunities for interdisciplinary work. There was no restriction on time period or geographical area of research. The meeting called for papers on topics of interest that included:

- River response to long-term (Milankovitch frequencies e.g. monsoon extent and intensity, ice sheet build up and decay) and short-term (sub-Milankovitch frequencies e.g. Heinrich events, NADW, ENSO) climate change.
- The role of tectonic activity, in controlling river development, sediment supply and preservation of the fluvial sedimentary record.
- The impact of human activity, both inadvertent (e.g. land-use change, mining) and planned change (e.g. channelisation), promoting and preventing river instability.

The two-day symposium was attended by over 60 delegates and generated much debate and a good deal of consensus on these and related issues. This volume brings together a series of papers presented at the Cheltenham meeting and we would like to express our gratitude to all the authors who have contributed papers to this volume.

All the papers in this volume have been referred by at least two specialists and we would like to thank the following for their patience and diligence in this process: Geoff Bailey (Newcastle), Raimo Becker-Haumann (Cologne), Mike Blum (Nebraska), David Bridgland (Durham), Paul Carling (Southampton), Jo Cheesman (Manchester), Phil Collins (Brune), Tom Coulthard (Aberystwyth), Martin Evans (Manchester), Rob Ferguson (Sheffield), Phil Gibbard (Cambridge), Mark Gillings (Leicester), Chris Green (Royal Holloway), Andy Howard (Leeds), Richard Huggett (Manchester), Stuart Lane (Leeds), John Lewin (Aberystwyth), Kees Kasse (Amsterdam), J-F Pastre (Paris), Frank Pazzaglia (Le-High), Barbara Rumsby (Hull), Rob Scaife (Southampton), Meindert Van den Berg

(Dutch Geological Survey), Jef Vandenberghe (Amsterdam), Tom Veldkamp (Wageningen), John Wainwright (London), Jeff Warburton (Durham), Rob Westaway (Open University, UK) and Jan Zalasiewicz (Leicester).

Finally, we would like to thank the Quaternary Research Association (UK) for supporting the FLAG initiative during its formative years and particularly during the period prior to and during the Cheltenham conference.

Darrel Maddy, Newcastle
Mark Macklin, Aberystwyth
Jamie Woodward, Leeds

Part 1: The context

1. Fluvial archives of environmental change

DARREL MADDY
Department of Geography, University of Newcastle, Newcastle upon Tyne, UK

MARK G. MACKLIN
Institute of Geography and Earth Sciences, University of Wales, Aberystwyth, UK

JAMIE C. WOODWARD
School of Geography, University of Leeds, UK

1 INTRODUCTION

It has long been known that rivers respond to a number of stimuli or forcing functions which act over a variety of time and space scales (e.g. Schumm & Lichty, 1965). Over the longer term (i.e. 10^3-10^6 years) geology (including lithological properties and tectonic setting) and climate may play the dominant role in fluvial system behaviour, generating basin-wide sediment-landform assemblages (Leeder, 1993; Leeder & Jackson, 1993; Macklin et al., 1995). Changes in these variables exert control over sediment and water supply and these promote system adjustment. Runoff and sediment supply, however, do vary over the shorter-term (i.e. 10^1-10^2 years) as a result of sediment exhaustion and starvation, and changes in flood frequency and magnitude. In the fluvial sedimentary record, establishing the relative importance of forcing functions which act at different time and spatial scales can be extremely difficult – especially when the changes observed within the sedimentary record are very similar (see papers in Lewin et al., 1995). One approach to resolving such problems, and this is illustrated in a number of papers in this volume (Coulthard et al., Chapter 18; Tebbens & Veldkamp, Chapter 17), is to explicitly model fluvial system behaviour.

Table 1 summarises the time periods and topics of studies reported in this volume, and Figure 1 shows the location of the study areas. Figure 2 illustrates the spatial and temporal scale of some important climate phenomena that influence drainage basin development and river behaviour. Over longer timescales is the importance of the astronomical frequencies (Milankovitch Cycles). These cycles, reflected in the global deep-sea oxygen isotope records (Shackleton & Opdyke, 1973; Imbrie et al., 1984; Shackleton et al., 1990) govern not only the growth and decay of continental ice-sheets (Hays et al., 1976), which in turn influence global sea-level change, but also have a significant effect on river basin hydrology and sediment fluxes. Not surprisingly, therefore, there has been increasing interest in establishing land-ocean correlations using dated fluvial sedimentary sequences (Bridgland, 1994; Fuller et al., 1998), a theme illustrated in papers in this volume (Van den Berg & Hoof, Chapter 3; Bridgland & Schreve, Chapter 7).

Over progressively shorter timescales ocean and atmospheric circulation variability such as the ENSO phenomenon can play a more dominant role (see Mass et al., this volume, Chapter 10).

The human impact on catchment systems can also provide a significant impetus for change, particularly through adjustments in sediment and water budgets following shifts

Table 1. List of authors, summary geographical data for each study area and topics and timescales of investigation for each of the chapters in Parts 2 to 5 in this volume.

Author	Chapter	Location	River basin(s)	Climate	Present tectonic activity	Fluvial setting	Time period of investigation	Topic of investigation
Part 2. Tectonic forcing: Crustal instability and the fluvial record								
Van den Berg, M. & Ton van Hoof	3	Netherlands	Maas	Humid mid-latitude	Moderate	Lowland	Quaternary	Tectonics, Climate change
Westaway, R.	4	Germany, Netherlands	Rhine, Maas	Humid mid-latitude	Moderate	Lowland	Quaternary	Tectonics, Climate change
Part 3. Climate forcing: Records of Pleistocene and Holocene river behaviour								
Kasse, C. & Bohncke, S.	5	Belgium Netherlands	Rhine, Meuse (Maas)	Humid mid-latitude	Moderate	Lowland	Early Pleistocene	Climate change reflected in palaeoecology and sedimentology of fluvial/estuarine sediments
Blum, M.D. & Straffin, E.C.	6	USA, Tunisia, France	Colorado, Mississippi, various Wadis Loire	Humid mid-latitude Tropical Arid Humid mid-latitude	Low, Low High Low	Lowland Mountain Lowland	Middle Pleistocene-Holocene	Climate change, Tectonics
Bridgland, D.R. & Schreve, D.	7	UK	Thames, Avon, Cam, Ouse, Welland, Nene, Solent	Humid mid-latitude	Low	Lowland	Middle-Late Pleistocene	Climate change as manifest in the biostratigraphy of fluvial terrace sequences
Houben, P., Nolte, S., Rittweger, H. & Wunderlich, J.	8	Germany	Main, Wetter, Nidda, Nidder, Ohm	Humid mid-latitude	Moderate	Piedmont	Late Pleistocene-Holocene	Climate change manifest in the fluvial sedimentological and palaeoecological records
Sidorchuk, A., Borisova, O., Kovaliukh, N. & Panin, A.	9	Russia	Vychegda	Periglacial	Low	Piedmont	Late Pleistocene-Holocene	Climate change reflected in changing palaeochannel geometry
Maas, G., Macklin, M.G., Warburton, J., Woodward, J.C. & Meldrum, E.	10	Bolivia	Alizos	Humid tropical	High	Mountain	Late Holocene	ENSO related changes in flood frequency and magnitude

Table 1. Continued.

Author	Chapter	Location	River basin(s)	Climate	Present tectonic activity	Fluvial setting	Time period of investigation	Topic of investigation
Part 4. Geoarchaeological perspectives and the human impact								
Woodward, J.C., Macklin, M.G. & Welsby, D.	11	Sudan	Nile	Tropical Arid	Low	Piedmont	Holocene	Geochronology of palaeo-channel systems and associated archaeological sites
Pastre, J.-F, Limondin-Lozouet, N., Leroyer, C., Fontugne, M., Gebhardt, A. & Krier, V.	12	France	Seine	Humid mid-latitude	Low	Lowland	Late Pleistocene-Holocene	Climate change and human impact
Thieme, D.M.	13	USA	Susquehanna	Humid mid-latitude	Moderate	Piedmont	Holocene	Human impact
Howard, A., Challis, K. & Macklin, M.G.	14	UK	Trent	Humid mid-latitude	Low	Lowland	Holocene	Human impact
Part 5. Modelling and monitoring of sediment fluxes and river channel dynamics								
Passmore, D.G. & Macklin, M.G.	15	UK	South Tyne	Humid mid-latitude	Low	Upland	Holocene	Holocene valley floor sediment budgets
Rumsby, B., Brassington, J. & McVey, R.	16	UK	Feshie	Humid mid-latitude	Low	Mountain	Late Holocene	River channel change and bar development
Tebbens, L.A. & Veldkamp, A.	17	Netherlands	Meuse (Maas)	Humid mid-latitude	Moderate	Lowland	Middle-Late Pleistocene	Modelling fluvial system dynamics in response to long-term climate change and tectonics
Coulthard, T.J., Kirkby, M.J. & Macklin, M.G.	18	UK	Cam Gill Beck	Humid mid-latitude	Low	Upland	Holocene	Modelling fluvial system dynamics in response to short-term climate and land-use change

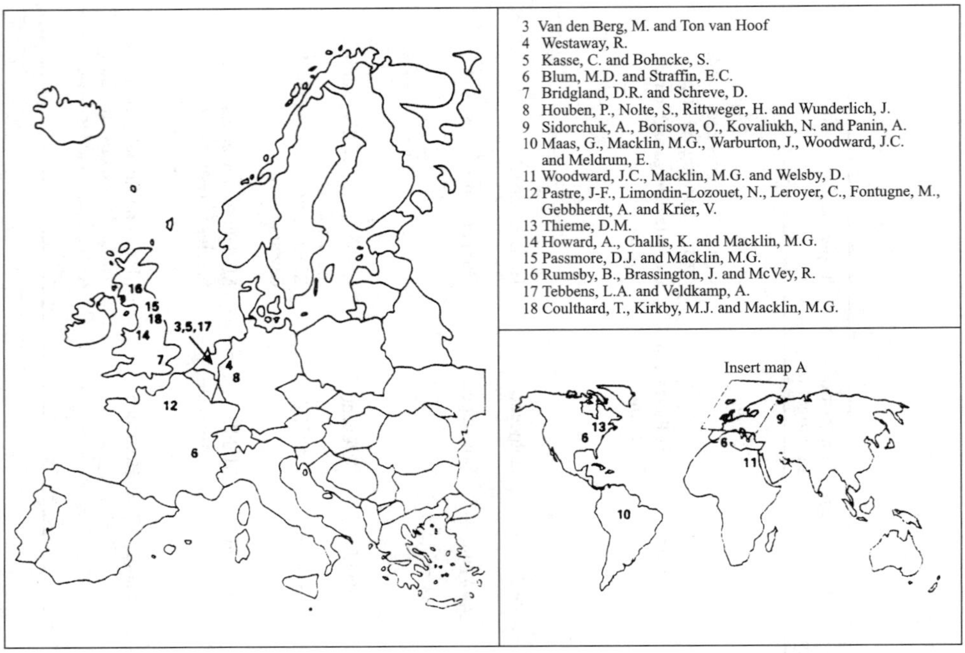

Figure 1. The location of the study areas and river catchments reported in this volume.

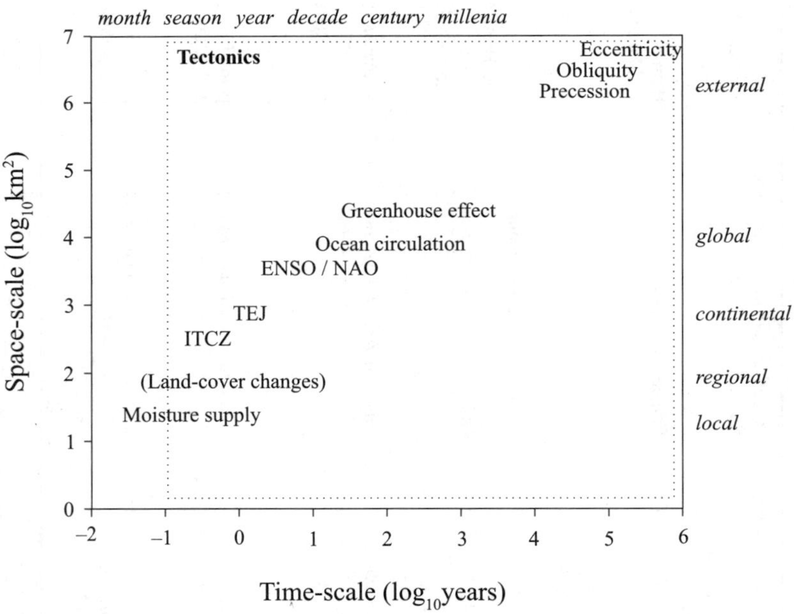

Figure 2. The spatial and temporal expression of some of the environmental changes that can impact on river basin sediment systems (modified from Hulme, 1994).

in land-use (see Trimble, 1983; Macklin & Lewin, 1997; Abbott & Valastro, 1995). The studies reported in this volume span a wide range of temporal scales - from Tertiary to late Holocene times – and document field research in markedly contrasting settings, including present-day periglacial and tropical environments. The volume contains recent research on some of the world's largest river basins including the Nile (2.96×10^6 km^2) and the Mississippi (3.27×10^6 km^2), and several papers report recent work on the great rivers of Northwest Europe including the Rhine, the Maas, the Seine and the Thames. The convergence of these European river systems to form a major fluvial system in the English Channel following remarkable environmental changes during the Middle Pleistocene has been well documented (Gibbard, 1988). At this spatial scale the predominant forcing factors governing fluvial system behaviour are tectonics and climate (including ice sheet dynamics and vegetation and sea level change) and major human impacts are confined to the Holocene. In exploring some of these issues the volume is divided into five parts on a thematic basis.

2 VOLUME STRUCTURE AND CONTENT

Part 1. The context

In the second paper in this section Lewin presents a review of the hierarchical and self-organising structure of river alluvium as a 5-level system from Level 0: Particles to Level IV: Alluvial Complexes (the total preserved fluvial archive of a drainage basin). He suggests that geoscientists have generally focussed on the workings of a limited selection of these levels, while perhaps drawing inferences about others. However, Lewin points out that recent research has extended the application of stochastic and deterministic models to non-linear and chaotic modelling, and a more holistic appreciation of alluvial systems allows these and other appropriate models to be set in context, with the levels providing a process framework for autogenic (intrinsic) system modelling. He acknowledges that the response to allogenic (extrinsic) environmental change may be rather different, with rapid local transformation in some circumstances (such as channel avulsion or the onset of braiding), but with the complexity of higher-level phenomena being liable to dissipative modifications over long time periods. He concludes that, although researchers may eventually achieve a grand 'theory of everything'; present understanding should allow forecasting at and between levels on a limited basis. He suggests that there now needs to be more than a descriptive presence-absence record of environmental happenings in the alluvial archive and this requires both an improved understanding of process and quantitative modelling of the complex multi-level alluvial system itself.

Part 2. Tectonic forcing: crustal instability and the fluvial record

The two papers in this section use river terrace records to make inferences about crustal instability. It is now generally accepted that large-scale terrace flights resulting from progressive valley incision require regional uplift (Vandenberghe, 1995). However, there is considerable debate concerning the issue of how far river incision can be used to quantify both the magnitude and rate of crustal uplift (Van den Berg, 1996; Maddy, 1997, 1998; Kiden & Tornqvist, 1998; Maddy et al., 2000). Many problems exist, not least of which is establishing the difference between surface and crustal uplift. For instance, if erosion rates are high, leading to substantial surface denudation, then river incision may underestimate crustal uplift. Further debate surrounds the extent to which the longitudinal profiles of rivers are in equilibrium with present day uplift and to what extent rivers are still adjusting to Tertiary uplift (see Kuzucuoglu, 1995). Both papers in this section also highlight an important new area of research which links Late Cenozoic crustal instability directly to climate change. The disposition of the continents is a fundamental control on the direction of ocean currents and the uplift of the major mountain belts such as the Himalayas can influence the trajectory of air streams in the upper atmosphere. The interaction of climate and tectonics is now believed to be highly significant and may even have triggered the Quaternary Ice Age (Ruddiman & Raymo, 1988; Raymo & Ruddiman, 1992).

Van den Berg and Ton van Hoof describe a sequence of 31 alluvial terraces of the River Maas on the north-western flank of the Ardennes uplift centre, which form a staircase some 200 m in height, and are considered to have formed during the past 4 million years. The results reported in this paper focus on a 30 × 35 km area of the Dutch part of the Lower Rhine Embayment. This is structurally controlled by the interplay of basement uplift, crustal movement along the Midi-Aachen Thrust zone, as well as along the southern principal displacement zone of the Lower Rhine graben. The authors illustrate that variations in river sediment flux match various frequencies of the cold stage marine isotope chronology. Using changes in sediment characteristics constrained by palaeomagnetism, pollen, palaeosols and thermoluminescence dating the authors attempt to establish a correlation with the 100 ka (Eccentricity), 41 ka (Obliquity) and multiples of the 100 ka Milankovitch cycles (see Figure 2). The terrace record is used as a proxy to reconstruct the regional uplift history. It shows a marked synchrony between the onset of uplift and the change in climate from warm Pliocene conditions to the colder Pleistocene. Interestingly, a cyclic pattern in uplift variations appears to match variations in tilt amplitude.

Uplift-driven incision is not a new concept, but explanations for why areas away from plate boundaries are experiencing uplift have been more elusive. Westaway discusses one potential, if controversial, mechanism in which surface uplift is modelled as a consequence of crustal thickening caused by the flow of lower crust to beneath the Rhenish Massif from beneath its surroundings. This flow is driven by pressure variations induced by ice-loading in surrounding areas and by sea-level fluctuations caused by glaciation. He uses the terrace sequences of the Maas and Rhine rivers adjoining the Rhenish Massif to constrain a model of Quaternary uplift history. His results suggest uplift rates increased by an order of magnitude in the latest Pliocene, since when the land surface has risen by ~150 m in the Maas and up to ~230 m in the Rhine basins. This sudden increase in uplift rates was followed by a reduction, then a renewed increase which peaked in the Middle Pleistocene. Significantly, the Moho temperature increase, which is a consequence of the crustal thickening, is sufficient to cause the local volcanism. The initial increase in uplift rates is interpreted as a consequence of repeated transient loading of surrounding areas by ice following the start of upland glaciation in Europe at approximately 3.1 Ma. It is supplemented by a smaller contribution reflecting the start of lowland glaciation in the Northern Hemisphere around 2.5 Ma, and the associated seawater-loading fluctuations. The second increase in uplift rate is caused by the intensification of glaciation at c. 1.2 Ma, supplemented by a larger contribution from the further intensification at c. 0.9 Ma. Westaway also suggests that this process may have added significantly to the global relief during Neogene and Quaternary times, and may thus have contributed to the progressive intensification of global cooling and glaciation, as well as global volcanism (see Ruddiman & Raymo, 1988).

Part 3. Climate forcing: records of Pleistocene and Holocene river behaviour

For well over a century, fluvial sedimentary sequences have played a highly significant role in the development of models of Quaternary climate change as well as forming important regional lithostratigraphical sequences (e.g. Penck & Bruckner, 1909). Although other climate proxy archives such as ice-cores, ocean core isotope signals and lacustrine sequences generally provide more continuous sequences, fluvial sediments and landforms can still provide valuable archives of palaeoclimatic data for specific events (see Baker et al., 1983) and provide significant terrestrial stratotypes within a formal lithostratigraphy.

In the first paper in Part 3, Kasse and Bohncke describe sediment exposures in northern Belgium and the southern Netherlands, close to the type sections of the Tiglian Stage of the Dutch Early Pleistocene at Tegelen. The Early Pleistocene sediments at these sites are predominantly fluvial sequences of the Rhine and Meuse (Maas). The sequences described include palaeobotanical remains and intraformational periglacial structures. The authors report a new glacial cycle within the Tiglian Stage and demonstrate a more complete Early Pleistocene terrestrial stratigraphy for northwestern Europe than previously recognised. Importantly, this paper illustrates the presence of erosional hiatuses in the terrestrial Early Pleistocene record and discusses some of the mechanisms responsible for this.

Blum and Straffin discuss the morphological, sedimentological, and stratigraphic records of Late Quaternary fluvial systems of varying scale in four contrasting climatic-tectonic settings with particular attention given to the dominant signatures present in each environment (Table 1).

The Loire River in the Massif Central of France and the Colorado River on the Gulf Coastal Plain of Texas are used as examples of unglaciated mid-latitude river basins, one in a continental interior, and the other on a passive continental margin. In these systems, the authors argue that glacial-interglacial climate changes forced significant transformations in channel geometry and depositional style. Cold stage alluvial units dominate the geomorphic and stratigraphic record of these river valleys, and consist of channel-belt sands and gravels. By comparison, the relatively subtle climatic changes of the Holocene have initiated only minor adjustments in channel geometry and the record is dominated by episodic overbank flooding and the deposition of overbank fines, alternating with phases of stability and soil development. Sea level changes impact only on the lowermost reaches of the Colorado system, typically by changing the geometry of stratigraphic units. The Late Pleistocene to middle Holocene units, which were deposited when sea-level was lower, are buried by those deposited contemporaneously with the late Holocene highstand. Drawing on fieldwork in a rather different environment, the paper also considers the impact of environmental change on a number of wadis on the margins of the northern Sahara in Tunisia. According to the authors, here the dominant geomorphic and stratigraphic signature is the transgressive and regressive responses of wadi-terminal fan systems to high-amplitude / low frequency changes in lake level driven by the 40 ka obliquity cycles (Figure 2). They also suggest that lower amplitude and higher frequency climate changes have been of insufficient magnitude to effect significant landscape change.

Finally, the authors consider the Pleistocene fluvial sequence of the Lower Mississippi Valley where the geomorphic and stratigraphic record is dominated by large braided river deposits. These sequences represent relatively short periods of time (c. 10-15 ka) at the end of successive 100 ka (eccentricity-driven) glacial cycles (Figure 2), specifically those time-equivalent to marine (O)xygen (I)sotope (S)tages 6 and 2 when there were tremendous fluctuations in the discharge of water and sediment from the Laurentide ice margin. Deposits from the last interglacial (OIS 5e) and OIS 4-3 are either absent or preserved only in protected settings. The series of depositional and erosional events in much of the Lower Mississippi Valley largely reflect upstream controls on discharge and sediment flux and were independent of sea-level change.

Bridgland and Schreve discuss aggradational river terraces in the Lower Thames Valley, UK. This terrace sequence has been shown to have formed in response to both regional uplift (Maddy et al., 2000) with glacial-interglacial climatic fluctuations (100 ka cycles) governing the timing and magnitude of sediment supply necessary for valley-floor aggradation (Bridgland et al., 1995). The identification of four separate interglacials within the terrace staircase during the post-Anglian Stage (post OIS12) is underpinned by mammalian biostratigraphy, with the terrace lithostratigraphy providing a framework for internal correlation. This scheme is supported by data from the molluscan biostratigraphy and amino-acid geochronology. The authors compare this sequence to mammalian evidence from other British river systems.

Houben, Nolte, Rottweger and Wunderlich describe fluvial system changes in the Hessian Depression, Germany in response to Lateglacial and Holocene environmental changes. Pollen assemblages, together with ^{14}C age estimation and the Laacher See Tephra (c. 12,000 cal BP), are used to constrain the timing of fluvial system changes inferred from sedimentological variations. This multi-proxy study demonstrates a now familiar pattern of changing fluvial sedimentation across this major phase of climatic change (e.g. see Frenzel, 1995). The authors, however, suggest that the reworking of coarse-grained sediment during the early Younger Dryas was not associated with deforestation in response to the cooling, a pattern identified widely elsewhere in Europe (Kasse et al., 1995; Starkel, 1995). Instead, the authors argue that it was controlled by increased snowmelt originating from the headwaters of the catchment and thus emphasising the importance of establishing the local context for palaeohydrological changes, even when those changes are externally driven.

In the penultimate paper in this section, Sidorchuk, Borisova, Kovaliuhk and Panin continue the discussion of Lateglacial and Holocene environmental change focusing on the lower Vychegda River in the extensive taiga zone of European Russia. The authors use palaeochannel morphology to reconstruct palaeoflow depths for various time intervals and compare these with flow depths in the modern river. They demonstrate that palaeochannels were larger than the modern channel during the Lateglacial and early Boreal periods, though during the late Boreal and Atlantic periods, the channels were significantly smaller than at present, increasing to their present size during the Sub-boreal. These reconstructions are significant in that they appear to contradict the generally held view that the Lateglacial in this region was very dry and the Atlantic was a relatively humid period. Analogous present day vegetational and hydrological regions have been identified using palaeobotanical data from the fluvial record. This has allowed mean annual and mean maximum discharges, the ratio of annual and winter-spring flow, and evapotranspiration losses to be estimated. Using the analogue data in a novel and stimulating way the authors are able to suggest solutions to apparent contradictions in the various proxy climate records.

Most of the papers in this volume document the fluvial sedimentary record in the lowland or piedmont zones of river basins (Table 1). In contrast, however, Maas, Macklin, Warburton, Woodward and Meldrum show that mountain river catchments can be sensitive to environmental change brought about by either natural or anthropogenic causes and may preserve high resolution records of river activity. High relief, limited vegetation cover and intense storm activity in these catchments often lead to high rates of geomorphic activity typified by landslides, debris flows and widespread valley floor sedimentation. The rivers draining the eastern Cordillera of southern Bolivia are typical in this respect with steep gradients and cobble- and boulder-bed braided channels that are laterally and vertically unstable as a result of periodic large floods. During the last 50 years or so soil erosion and flooding have been perceived as major hazards in this region, though in the absence of a chronology for river channel change, it was unclear as to whether recent erosion and flooding were dominated by natural or anthropogenic controls.

Geomorphological investigations have been carried out in the Rio Alizos, southern Bolivia to establish the timing, magnitude and impact of major floods and determine the significance of climate change upon river development. Detailed field mapping, clast size measurements and lichenometric dating have been used to establish a 300-year flood history which indicates a link between the timing and nature of floods and the Southern Oscillation. In general terms, alluviation tends to be concentrated during periods dominated by La Niña-type phases of the Southern Oscillation (Figure 2). Analysis of local precipitation and regional environmental proxy data indicate that these phases promote an increase in precipitation in southern Bolivia. Using boulder size data as a proxy for flood magnitude, the record shows that extreme flood events have declined in magnitude from the end of the seventeenth century to the present day. Variations in the calibre and quantity of sediment supplied to the channel over this time period are likely to have altered in response to changes in hillslope vegetation density and weathering rates associated with the climatic deterioration during the Little Ice Age. Extreme flood events occurring approximately every 5 to 15 years have effected considerable modifications to the valley floor of the Rio Alizos over the last three centuries.

Part 4. Geoarchaeological perspectives and the human impact

Earth scientists and archaeologists have enjoyed a long and fruitful relationship and alluvial geoarchaeology in particular has emerged as an important sub-discipline over the last two decades or so (see Needham & Macklin, 1992; Brown, 1997). Long-term river behaviour is of direct relevance to a good deal of archaeological field research for many reasons. Fluvial sediments and landforms can provide stratigraphic and environmental context for the archaeological record and river behaviour can be a major control on the preservation and exposure of cultural materials.

The Nile corridor contains a varied and complex geomorphological and archaeological record and it constitutes one of the most important fluvial archives of environmental change. Nile basin hydrology is intimately linked to the dynamics of the African monsoon and changes in global climate during the Pleistocene and Holocene have lead to major fluctuations in the discharge of water and sediment. Woodward, Macklin and Welsby present the results of a long-term interdisciplinary geoarchaeological investigation in an 80 km length of the Northern Dongola Reach in Northern Sudan. This part of the Nile valley floor contains a series of large and well preserved palaeochannel belts up to 15 km to the east of the modern channel. A four-season survey conducted by the Sudan Archaeological Research Society identified over 350 archaeological sites in this reach. The sites that date to the Kerma Period (c. 2500 to 1500 BC) are closely associated with the margins of these palaeochannel systems while earlier Neolithic sites are more dispersed across the valley floor. The region constitutes an important geoarchaeological archive and a full interpretation of the cultural record would not be possible without an understanding of the Holocene alluvial sequence – including the age and evolution of the palaeochannel belts. Thus, the authors have employed Optically Stimulated Luminescence (OSL) and radiocarbon dating to develop a chronology for Holocene river behaviour to provide geomorphological context for the archaeological record. Ten OSL and two radiocarbon dates cover a 6000 year period from the sixth millennium BC to the first millennium AD. These dates are in good agreement with the independent age control derived from analysis of the Neolithic, Kerma and later pottery collected during survey and excavation. The record of river activity is compared to other proxy climate records in the Nile basin. Interestingly, several large palaeochannels identified on SPOT satellite imagery (see Macklin and Woodward, in press) appear not to be associated with Kerma or later sites, and further work is required to establish the age and significance of these features.

Pastre, Lomondin-Lozouet, Leroyer, Fontugne, Gebhardt and Krier present results from a survey of the floodplains of the Paris Basin (middle Seine River basin) which contain a fluvial record of river response to Lateglacial and Holocene climatic variability as well as the influence of human impact during the late Holocene. The accumulation of sands and silts occurred in a braided fluvial system during the Upper Pleniglacial (c. 30 ka-13 ka) and the beginning of the Bølling interstadial. Towards the end of the interstadial, only one channel remained active, incision ceased and sedimentation was locally characterized by the deposition of organic-rich silts. The cessation in fluvial dynamics was probably linked to vegetation development during the Allerød interstadial. A thin soil appears as a key horizon in the main valleys and peat formation took place in some channels. During the Younger Dryas stadial, calcareous overbank silt deposits covered most of the floodplain, infilling abandoned channels of the warmer Bølling and the Allerød periods. At a few sites, cryogenic involutions are recorded from the beginning of the stadial.

The early postglacial on these floodplains was characterized by marked incision at the beginning of the Pre-boreal period. Lower energy fluvial conditions prevailed during the Boreal and Atlantic with the main river channels being progressively filled by organic-rich sediments as the supply of minerogenic sediment from the upstream catchment diminished significantly. This period was characterized by peat accumulation and tufa deposition in many of the small valleys. During the second half of the Atlantic period, sedimentological and palynological indicators do not indicate any major human modification of the environment. However, by the Middle Neolithic, pollen data show evidence for the development of agriculture and localised opening up of vegetation. The second half of the Subboreal was characterized by increased slope erosion leading to the infilling of the main channels with clayey silts. A concomitant increase in runoff generated numerous channels, which were especially active during the Bronze Age, though many of these were filled during the Second Iron Age at the beginning of the Subatlantic. During Roman times, the main valley floors continued to be filled with fluvial silts while the Middle Ages were generally a period of decreasing sedimentation. Significant erosion marked by thick silty fills in the small valleys, is recorded during the Little Ice Age and the modern period.

Thieme offers a New World perspective to Part 4 and discusses the historic and possible prehistoric human impact on floodplain sedimentation in the North Branch of the Susquehanna River, eastern USA. The paper presents a review of previously documented types of human impact on rivers and floodplains with particular emphasis on the temperate mid-latitudes of North America. Thieme concludes that early Holocene floodplain alluviation was largely controlled by climate. However, around 4500 BC a widespread episode of valley alluviation was associated with deforestation following the destruction of hemlock trees by an insect pest. Later in the Holocene, alluvial terraces appear to result from the introduction of maize agriculture some time after AD 800, by the late prehistoric Clemson Island, Owasco and Wyoming Valley cultures. The impacts of 19th century industrial development, including logging, anthracite mining and direct channel modifications are also discussed.

In the final paper in this part of the volume, Howard, Challis and Macklin describe the construction of a Geographical Information System (GIS) for integrating data derived from the archaeological survey of valley floor environments with geological and geomorphological information. They present data for three reaches of the meandering River Trent in central England, which has a rich and varied cultural record. The authors discuss the potential of this approach and some of the problems associated with data collection and data compatibility. The use of a GIS has allowed the production of maps showing the distribution of archaeological features by period and type. It is also a useful way of documenting the archaeological resource for management and planning purposes. All three of the reaches investigated show evidence of abandoned channels, which may represent a previously braided or anastomosing system. By studying the pattern of cultural features within the landscape, relationships between archaeology and landscape can be established in reaches with rather different fluvial characteristics. At the same time, this approach can be used to identify particular geomorphological and archaeological site associations where further survey and excavation would be beneficial.

Part 5. Modelling and monitoring sediment fluxes and river channel dynamics

The sediment budget concept is a fundamental part of geomorphology and is now established as an important framework for considering the response of river systems to environmental change (see Meade, 1982; Trimble, 1983). Lehre (1982) has outlined three important components that should be addressed in fluvial sediment budget investigations (see Sutherland & Bryan, 1991, p. 383):

1. Identification of erosional processes and understanding their controls and interrelationships;
2. Measurement of the magnitude and frequency of sediment mobilisation by each process, and of sediment outputs from a basin;
3. Identification of sediment storage elements and quantification of the volume, residence time, and changes in storage of sediment in each element.

Many sediment budget investigations have involved intensive field-based monitoring in small catchments – over relatively short timescales – to quantify sediment inputs and changes in storage volumes. Such studies have contributed to important advances in our understanding of catchment sediment dynamics (see Sutherland & Bryan, 1991). Recent work has begun to construct longer term sediment budgets for larger river basins to investigate the relationship between changes in some of the extrinsic and intrinsic variables discussed above and their impact on sediment sources and storage volumes. In this context, recent developments in the use of quantitative sediment fingerprinting techniques to identify suspended sediment sources in larger river basins (see Walling & Woodward, 1995) have also been successfully applied over Pleistocene and Holocene timescales (Hamlin et al., 2000). Part V of this volume contains four papers that adopt contrasting theoretical and field-based approaches to examine parts or all of the sediment budget components identified above by Lehre (1982) across a range of temporal and spatial scales.

Passmore and Macklin have computed a Holocene river sediment budget along a 10.5 km long reach of the South Tyne, a gravel-bed river in northern England. A digitised database, combining the results of geomorphological mapping, topographic survey and geochronological analyses has allowed the age and volume of fluvial sedimentary units to be determined for twenty 0.5 km sub-reaches. Morphological changes in each sub-reach were quantified to 1. Estimate rates of sediment reworking and export during a series of Late Pleistocene and Holocene river incision and deposition cycles, and 2. Examine the relationship between slope-valley floor coupling and long-term patterns of sediment transfer.

Net export of fluvial sediment from valley floor storage over the Holocene is estimated at c. 7.9 million m^3, leaving a total of c. 7.43 million m^3 remaining as Late Pleistocene and Holocene terraced alluvial units, and a further 0.37 million m^3 stored in Holocene alluvial fans. Significant variability in the rate of valley floor reworking and sediment transfer is evident over a Holocene timescale, particularly during the past 1000 years when the South Tyne has become increasingly entrenched and sensitive to changes in flood regime. The highest rates of valley floor reworking are recorded between c. cal AD 1160-1380 and c. AD 1850-1975 and are related to variations in flood frequency and magnitude, and catchment land-use changes.

The paper by Rumsby, Brasington and McVey examines the two- and three-dimensional patterns of channel and bar development on the braided River Feshie, Scotland over the last 100 years and two main approaches are used. The first involves the construction and comparison of geomorphological maps, within a GIS framework, at five timeslices between 1899 and 1997. Analysis of these shows that over the last century a high proportion of the active valley floor of the Upper Feshie has been re-worked laterally, with limited preservation of deposits and landforms associated with the late nineteenth century river system. The best preservation of deposits is associated with a channel abandoned by avulsion during the 1960s, which placed a large volume of gravel in storage. This study demonstrates that the high activity rates of braided and near-braided rivers may constrain their utility as fluvial archives, due to the high probability of valley floor reworking and lower preservation potential of sediments and landforms.

The second approach examines channel change using repeat detailed topographic survey of the contemporary bed and bar morphology using a differential GPS-based survey system. Results are presented here for the high resolution baseline survey conducted in 1998, with resurvey planned on an annual basis. Such surveys may improve our understanding of process dynamics at the reach scale (see Lane et al., 1995) and provide valuable data sets for the calibration of numerical models. At the same time, it is important to appreciate that such process-based research focuses on very short-term or small-scale phenomena and the challenge is how to overcome the obvious scaling problems when time- and spatial scales are extended (see Lewin, Chapter 2).

The last two papers in this volume describe two contrasting modelling strategies that present exciting prospects for improving our understanding of river basin response to environmental change. Tebbens and Veldkamp describe a series of validation and sensitivity tests for the FLUVER2 model set, a series of models, which simulate the development of the 2D longitudinal profile of a river. The model is driven by tectonic uplift rates and climate and sea-level changes, which interact to perturb the balance between sediment supply and sediment transport capacity at the 10^3-10^6 km^2 scale. The simulated fluvial dynamics are visualised in the form of a Profile Evolution Map which illustrates the temporal evolution of river bed height for each 1 km segment, calculated at 20 year intervals but plotted at 1 ka time-steps. The 2D longitudinal model output is subjected to a series of sensitivity analyses. The model output is also used to inform simulations of terrace development using the semi-3D model of Veldkamp and Vermeulen (1989). Sensitivity tests on both models are examined with respect to a 250 ka calibration dataset from the River Meuse (Maas) terrace sequence. Encouragingly, the results demonstrate that the models perform well enough to allow the testing of hypotheses of long-term fluvial system dynamics.

Over much shorter timescales and at a smaller spatial scale, Coulthard, Kirkby and Macklin present results from a high resolution cellular automaton model of a small upland catchment, Cam Gill Beck (4.2 km^2), in the Yorkshire Dales, UK. The combination of hydrological, hydraulic and slope models captures many of the complex interrelationships within the catchment at a scale rarely achieved. The model is designed to examine the relative importance of single large floods as opposed to the cumulative effects of lower magnitude, higher frequency events – a long-standing and controversial topic in fluvial geomorphology. In this paper the model is used specifically to investigate two scenarios: periods of sustained moderate flooding and a single extreme flood event. Patterns of sediment movement and storage have been modelled and their long-term impact on the evolution of the valley floor topography can be assessed. Although the last two papers focus on rather different temporal and spatial scales, they do show how modelling studies are beginning to converge on the study of millennial scale problems as well as adopting a holistic, catchment-based approach.

3 FUTURE RESEARCH NEEDS

The papers presented in this volume clearly identify some common problems that need to be addressed by future research. Three areas emerge as key priorities for future investigations.

1. Perhaps the most pressing problem is that of improving the dating context of fluvial sequences. All of the studies presented in this volume require an accurate timeframe in order to allow inferences to be made with regard to cause and effect relationships. Recent advances in, for example, luminescence dating (Duller, 1996), have significantly enhanced the chronological resolution possible on sequences spanning the last 100 ka, but only a few studies have undertaken comprehensive dating programmes of Quaternary river deposits and landforms (Fuller et al., 1998; Mol, 1997). Advances in amino acid geochronology (Sykes et al., 1995) may also help to resolve sequences spanning the last 0.5 Ma, building on studies such as Bowen et al. (1995) on the River Thames. Most river basins are, however, still dated using isolated samples and very few apply multiple dating techniques that are required for significant confidence to be placed in the chronology.
2. Sediment-based fluvial archives are known to exist in virtually all major river systems across the globe. As such the studies illustrated in this volume represent only a small subset of the available information. However, even when more general consideration is taken of the total available literature, major areas of the globe remain as data vacuums. With relatively sparse coverage across the continents of Africa and South America, it is extremely difficult to determine how far patterns identified in individual basins reflect

the wider regional picture. The answer, of course, is simple – a larger number of studies are required to establish both regional and continental scale variation (see Gregory et al., 1995).

3. A final theme is the emergence of new kinds of modelling studies. The integration of field and process-based modelling studies offers not only the potential for greater understanding of fluvial archives by better targeted fieldwork, but also much greater flexibility in terms of the questions that can be asked of the archive (Coulthard and Macklin, in press).

The papers in this volume show clearly that major advances in our understanding of fluvial archives of environmental change can only be achieved through interdisciplinary and multidisciplinary investigations of fluvial sediments and landforms in a range of environments. Indeed, the study of fluvial archives of environmental change has benefited in particular from:

- The integration of concepts and methods derived from the observation of contemporary fluvial processes in a range of climatic and tectonic settings;
- The incorporation of detailed biostratigraphic analyses of organic materials preserved within fluvial sediments;
- Closer co-operation with archaeological survey and excavation in fluvial settings at a range of scales;
- Technological and conceptual advances in the fields of geochronology, satellite remote sensing, GPS survey and photogrammetry, geophysical methods, and the numerical simulation of catchment processes.

It is the role of the Fluvial Archives Group to ensure that this healthy and productive dialogue is encouraged further.

REFERENCES

Abbott, J.T. & Valastro, S. 1995. The Holocene alluvial records of the chorai of Metapontum, Basilicata, and Croton, Calabria, Italy. In: Lewin, J., Macklin, M.G. & Woodward, J.C. (eds) *Mediterranean Quaternary River Environments*. Rotterdam: Balkema, 195-205.

Baker, V.R., Kochel, R.C., Patton, P.C. & Pickup, G. 1983. Palaeohydrologic analysis of Holocene slackwater sediments. *Special Publication of the International Association of Sedimentology*, 6: 229-239.

Bowen, D.Q., Sykes, G.A., Maddy, D., Bridgland, D.R. & Lewis, S.G. 1995. Aminostratigraphy and amino-acid geochronology of English lowland valleys: the Lower Thames in context. In: Bridgland, D.R., Allen, P. & Haggart, B.A. (eds) *The Quaternary of the Lower reaches of the Thames*. Cambridge: Quaternary Research Association, 61-64.

Bridgland, D.R. 1994. *Quaternary of the Thames*. Geological Conservation Review Series 7, Chapman and Hall, London.

Bridgland, D.R., Allen, P. & Haggart, B.A. (eds) 1995. *The Quaternary of the lower reaches of the Thames*. Quaternary Research Association Field Guide, Durham.

Brown, A.G. 1997. *Alluvial Geoarchaeology: Floodplain Archaeology and Environmental Change*, Cambridge University Press, Cambridge.

Coulthard, T.J. & Macklin, M.G., in press. How sensitive are river systems to climate and land-use changes? A model based evaluation. *Journal of Quaternary Science*.

Duller G.A.T. 1996 Recent developments in luminescence dating of Quaternary sediments. *Progress in Physical Geography*, 20: 133-151.

Frenzel, B. (ed.) 1995. *European river activity and climatic change during the Lateglacial and early Holocene.* (Special Issue: ESF Project European Palaeoclimate and Man 9). Stuttgart: Gustav Fischer Verlag.

Fuller, I., Macklin, M.G., Lewin, J., Passmore, D.G. & Wintle, A.G. 1998. River response to high frequency climate oscillations in southern Europe over the last 200 k.y. *Geology*, 26: 275-278.

Gibbard, P.L. 1988. The history of the great Northwest European rivers during the past three million years. *Philosophical Transactions of the Royal Society*, 318: 559-602. London.

Gregory, K.J., Starkel, L. & Baker, V.R. (eds) 1995. *Global Continental Palaeohydrology*. Chichester, John Wiley & Sons.

Hamlin, R.H.B., Woodward, J.C., Black, S. & Macklin, M.G. 2000. Sediment fingerprinting as a tool for interpreting long-term river activity: the Voidomatis Basin, Northwest Greece. In: I.D.L. Foster (ed.) *Tracers in Geomorphology*, Chichester, John Wiley & Sons, 473-501.

Hays, J.D., Imbrie, J. & Shackleton, N.J. 1976. Variations in the earth's orbit: Pacemaker of the ice ages. *Science*, 194: 1121-1131.

Hulme, M.A. 1994. Global climate change and the Nile basin. In: P.P. Howell & J.A. Allen (eds) *The Nile: Sharing a Scarce Resource*, Cambridge, Cambridge University Press, 139-162.

Imbrie, J., Hays, J.D., Martinson, D.G., McIntyre, A., Mix, A.C., Mauley, J.J., Pisias, N.G., Prell, W.L. & Shackleton, N.J. 1984. The orbital theory of Pleistocene climate: support from a revised chronology of the marine $/^{18}O$ record. In: Berger, A., Imbrie, J., Hays, J., Kukla, G. & Saltzman, B. (eds) *Milankovitch and Climate, Part 1*, Dordrecht, Reidel, 269-306.

Kasse, K., Vandenberghe, J. & Bohncke, S. 1995. Climatic change and fluvial dynamics of the Maas during the Weichselian and early Holocene. In: Frenzel, B. (ed.) *European river activity and climatic change during the Lateglacial and early Holocene* (Special Issue: ESF Project European Palaeoclimate and Man 9), Stuttgart: Gustav Fischer Verlag, 123-150.

Kiden, P. & T.E. Törnqvist 1998. Can river terrace flights be used to quantify Quaternary tectonic uplift rates? *Journal of Quaternary Science*, 13: 573-574.

Kuzucuoglu, C. (1995) River response to Quaternary tectonics with examples from northwestern Anatolia, Turkey. In: Lewin, J., Macklin, M.G. & Woodward, J.C. (eds) *Mediterranean Quaternary River Environments*, Rotterdam, Balkema, 45-53.

Lane, S.N., Richards, K.S. & Chandler, J.H. 1995. Within-reach spatial patterns of process and channel adjustment. In: E.J. Hicken (ed.). *River Geomorphology*, Chichester: John Wiley & Sons, 105-130.

Leeder, M.R. 1993. Tectonic controls upon drainage basin development, river channel migration and alluvial architecture: Implications for hydrocarbon reservoir development and characterisation. *Special Publication of the Geological Society of London*, 73: 7-22.

Leeder, M.R. & Jackson, J.A. 1993. The interaction between normal faulting and drainage in active extensional basins with examples from the Western United States and Greece. *Basin Research*, 5: 79-102.

Lehre, A.K. 1982. *Sediment mobilisation and production from a small mountain catchment: Lone Tree Creek, Marin County, California.* PhD Thesis, University of California, Berkeley.

Lewin, J., Macklin, M.G. & Woodward, J.C. (eds) 1995. *Mediterranean Quaternary River Environments*. Rotterdam, Balkema.

Macklin, M.G. & Lewin, J. 1997. Channel, floodplain and drainage basin response to environmental change. In: Thorne, C.R., Hey, R.D. & Newson, M.D. (eds) *Applied fluvial geomorphology for river engineers and management*, Chichester: John Wiley & Sons, 15-45.

Macklin, M.G. & Woodward, J.C. 2001, in press. Holocene alluvial history and the palaeochannels of the River Nile in the Northern Dongola Reach. In: D.A. Welsby (ed.) *Life on the Edge of the Desert: 5000 Years of Human Settlement in the Northern Dongola Reach of the Nile, Sudan.* London: British Museum Press.

Macklin, M.G., Lewin, J. & Woodward, J.C. 1995. Quaternary fluvial systems in the Mediterranean basin. In: Lewin, J., Macklin, M.G. & Woodward, J.C. (eds) *Mediterranean Quaternary River Environments*, Rotterdam: Balkema, 1-25.

Maddy, D. 1997. Uplift-driven valley incision and river terrace formation in southern England. *Journal of Quaternary Science*, 12: 539-545.

Maddy, D. 1998. Reply: Can river terrace flights be used to quantify Quaternary tectonic uplift rates? *Journal of Quaternary Science*, 13: 574-575.

Maddy, D., Bridgland, D.R. & Green, C.P. 2000. Crustal uplift in southern England: Evidence from the river terrace records. *Geomorphology*, 33: 167-182.

Meade, R.H. 1982. Sources, sinks, and storage of river sediment in the Atlantic drainage of the United States. *Journal of Geology*, 90: 235-252.

Mol, J. 1997. Fluvial response to Weichselian climate changes in the Niederlausitz (Germany). *Journal of Quaternary Science*, 12: 43-60.

Needham, S. & Macklin, M.G. (eds) 1992. *Archaeology Under Alluvium*. Oxford: Oxbow Books.

Penck, A. & Bruckner, E. 1909. *Die Alpen im Eiszeitalter*. Leipzig: Tauchnitz.

Raymo, M.E. & Ruddiman, W.F. 1992. Tectonic forcing of Late Cenozoic climate. *Nature*, 359: 117-122.

Ruddiman, W.F. & Raymo, M.E. 1988. Northern hemisphere climate régimes during the past 3 Ma: Possible tectonic connections. *Philosophical Transactions of the Royal Society of London, Ser. B*, 318: 411-430.

Schumm, S.A. & Lichty, R.W. 1965. Time, space and causality in Geomorphology. *American Journal of Science*, 263: 110-119.

Shackleton, N.J. & Opdyke., N.D. 1973. Oxygen isotope stratigraphy of equitorial Pacific core V28-238: oxygen isotope temperatures and ice volumes on a 10^5 year and 10^6 year scale. *Quaternary Research*, 3: 39-55.

Shackleton, N.J., Berger, A. & Peltier, W.R. 1990. An alternative astronomical calibration of the Lower Pleistocene Timescale based on ODP Site 677. *Transactions of the Royal Society of Edinburgh*, 81: 252-261.

Starkel, L. 1995. The place of the Vistula river valley in the late Vistulin-early Holocene evolution of the European valleys. In: Frenzel, B. (ed.) *European river activity and climatic change during the Lateglacial and early Holocene* (Special Issue: ESF Project European Palaeoclimate and Man 9), Stuttgart: Gustav Fischer Verlag, 75-88.

Sutherland, R.A. & Bryan, R.B. 1991. Sediment budgeting: A case study in the Katiorin drainage basin, Kenya. *Earth Surface Processes and Landforms*, 16: 383-398.

Sykes G.A., Collins M.J. & Walton D.I. 1995. The significance of a geochemically isolated (intracrystalline) organic fraction within biominerals. *Organic Geochemistry*, 23: 1059-1066.

Trimble, S.W. 1983. A sediment budget for Coon Creek basin in the Driftless Area, Wisconsin, 1853-1977. *American Journal of Science*, 283: 454-474.

Van den Berg, M.W. 1996. *Fluvial sequences of the Maas: a 10 Ma record of neotectonics and climate change at various time-scales*. PhD Thesis, University of Wageningen, the Netherlands: 181 pp.

Vandenberghe, J. 1995. Timescales, climate and river development. *Quaternary Science Reviews*, 14: 631-638.

Veldkamp, A. & Vermeulen, S.E.J.W. 1989. River terrace formation, modelling and 3-D graphical simulation. *Earth Surface Processes and Landforms*, 14: 641-654.

Walling, D.E. & Woodward, J.C. 1995. Tracing sources of suspended sediment in river basins: a case study of the River Culm, Devon, UK. *Marine and Freshwater Research*, 46: 327-336.

2. Alluvial systematics

JOHN LEWIN
Institute of Geography and Earth Sciences, University of Wales, Aberystwyth, UK

1 INTRODUCTION

The placing of phenomena within an all-embracing and unifying system has frequently inspired considerable growth and rapid progress in many sciences. Underlying connections have suddenly been appreciated or sought, and underpinning dynamic mechanisms identified. Individual sciences have consequently matured from being disorganised collections of disparate (though often individually valuable) projects at a variety of levels and scales. They have gained a structure to their bodies of knowledge which commonly has gone on to allow greater prediction, better identification of crucial research objectives, and indeed the revelation of sets of relatively simple principles which can be seen to make collective sense of hitherto largely unconnected contemporary variations and temporal changes.

Illustrations of this can be taken from both the physical and biological sciences. In chemistry, the periodic table with its rows and columns first set out by J.A.R. Newlands in 1864 was crucially transformed by D. Mendeleev leaving gaps for new elements whose properties he could predict. In biology, the hierarchical grouping of taxa has underpinned both description and evolutionary hypotheses. In physics, the identification of a scale of entities from atoms and molecules to solar systems and galaxies has been augmented in this century by the postulation of new entities from quarks to black holes.

Quaternary research has similarly both advanced and been constrained by the frameworks and levels of resolution available to it at particular points in time – a 4-fold glacial subdivision, oxygen-isotope stages, Heinrich events and so on. Of course both time and space can be continuously scaled (for us usually in years or metres), but periods or events are invested with process meaning when the impacts of such separable phenomena at different levels in space or time can be qualitatively identified. Thus temporal phases may exert crucial allogenic (external) control on basin sediment systems, including their transformation at periods of system change.

But what of alluvial systems themselves, and their own characteristic spatial and temporal structures involved in autogenic (internal) processes? Where systems intersect there is unlikely to be direct correspondence so that, for example, climatic change achieves immediate fluvial transformation at all scales. Indeed with concepts like 'complex response' and 'geomorphic thresholds' (Schumm, 1977) we do appreciate that this is not the case, even though many older current ideas were framed within an equilibrium and linear response framework at particular levels. The latter have included useful relationships between sediment sizes and river channel dimensions and discharge, as have been used in palaeohydrology computations (e.g. Baker et al., 1988; Williams, 1983).

The nature of alluvial studies has, however, moved on. This has involved a variety of active sciences with contrasting foci. Sedimentologists have mainly geared their attention to the interpretation of what can be seen in section, from small-scale sandy bedforms to the architecture of alluvial basin fills, but stratigraphers now have access to three-dimensional subsurface data and this has led to important developments in sequence stratigraphy. Geomorphologists have supplemented a strong focus on river channel planform patterns and their development with field and laboratory studies of sediment particle dynamics. Engineers have coupled a comparable concern with one of practical design and management objectives for eroding and sedimenting rivers. There is, therefore, a very large and growing literature of work on relevant alluvial studies in many nominally separate research disciplines. Indeed, if research were music, this orchestra would be making a wonderful cacophony with some players in tune and following one score, but others with different or dated score versions. The listener would be hard-put to know what to make of it all – and perhaps the same could be said of the Quaternary scientist concerned with the actual impact of environmental change on alluvial systems.

Progress has also occurred in theoretical understanding, particularly involving dynamical systems (see, for example, Huggett, 1988; Montgomery, 1989; Phillips, 1992; Prigogine, 1989; Williams, 1997). An alluvial river system involves the sediment-storing and sediment-transferring parts of an eroding drainage basin which is becoming progressively reduced in relief as sediment is evacuated. It can exhibit patterns of self-organisation at different hierarchical levels during this dissipative process.

But states of dynamic stability may also be subject to catastrophic change (as when meander forms are terminated by cutoffs), and there is at least the suggestion that some forms exhibit deterministic chaos (Stølum, 1996). Although equilibrium thinking has proved enormously stimulating for many decades, non-linear dynamical modelling should now re-focus attention on additional, and in some cases, neglected aspects of alluvial systems – their 'memory', irreversible changes, self-organisation including chaotic behaviour, and hierarchical structures.

A third impetus to progress lies in available techniques, specifically of laboratory and computer modelling, which have both shed light on alluvial systems behaviour. As we shall see, they require both calibration and field validation and may not yet be developed in such ways as to apply to very many types of, or indeed some levels within, alluvial systems. But they are showing the way in which autogenic processes may work and, potentially at least, how allogenic change may be propagated within alluvial systems in less simplistic ways than our recent levels of functional modelling have allowed.

This paper's purpose is two-fold. First, to take a new look at the full scale-range of hierarchical studies of self-organising structures for alluvial phenomena in the light of recent research, bringing together familiar but hitherto apparently disconnected types and levels of analysis. Second, to ask whether this suggests particularly critical types or levels of explanation for new or enhanced research programmes. The focus is on the alluvial archive, but it could be argued that recent developments have equivalent significance more broadly within the geosciences.

2 ARRIVING AT ALLUVIAL SYSTEMATICS

There are a great many classificatory systems concerned with some aspect or other of alluvial systems. Some present qualitative names for scaled timespans or dimensions (e.g.

short, medium and long timescales; micro- meso- and macro-forms; well-, moderately- or poorly-sorted sediments). These do not involve discrete entitation of phenomena believed to possess underlying difference; there is indeed often the contrary presumption that both forms and time are really on a continuum. Where some difference of level is perceived, some researchers 'order' (1st order, 2nd order – but from high to low, or low to high) whilst others use hierarchical naming, as in an agreed stratigraphic practice (bed, member, formation, group). Finally, some classifications are essentially driven by groupings of previously identified field 'examples' or 'case studies' (as in stratotype or channel pattern), whilst others rely more on formal divergence criteria at various levels, sometimes producing large numbers of 'types' which may or may not have real-world meaning and separate existence. This is not the place to cover classification procedures in general (see, for example, Jardine & Simpson, 1971), but it is important to appreciate that alluvial phenomena have already been much-classified with a variety of perspectives, procedures, and purposes in mind. This occasions conflict in terminology (one person's mega- is another's meso-), and in the nature and number of levels or types perceived. For example, Nikora (1991) distinguishes five hierarchical levels covering about the same phenomenal range as Miall (1993) does in ten. Rust (1978) has four channel patterns, Schumm (1991) illustrates fourteen members of a continuum, and Rosgen (1994) has forty-one.

One might be forgiven for finding all this arbitrary and obfuscating, but it has to be remembered that very many of these classifications have been well-used by researchers with particular objectives in mind; the relative isolation of academic disciplines and their somewhat different needs has meant that they are unwilling to use, or are even unaware of, classifications used by others. And different bases for classification may genuinely be appropriate in different circumstances. But this all does create difficulties for our present purposes – to define the 'real' functioning autogenic structure of alluvial systems and so to see what allogenic change is likely to produce.

Five objectives could characterise a useful scheme for alluvial systematics:

- It should aspire to be an inclusive, hierarchical 'natural ordering' of dynamical entities;
- Whilst convenient for descriptive purposes, it should also be a process framework for system modelling, and a means for the identification of needed research programmes. It should provoke searches for the appropriate level(s) of explanation and forecasting variables;
- System timescales, levels and their components should be defined with concern for autogenic meaning (rather than be, for example, transferred forms of allogenic episodes, phases or cycles);
- Terminology should as far as possible not provoke misunderstanding by giving existing terms different meaning (or selecting one use when several uses are current);
- The minimum number of levels/types/subdivisions consistent with utility should be the aim.

Given such objectives, it is possible to use or re-frame a number of existing schemes and approaches, but also to simplify, extend and re-name components of others, and to avoid some entirely as unsuited for present purposes. Terms like 'element', for example, have been variously used as have the 'micro- to mega-' qualifications, and so they are here avoided. Whilst much influenced by Miall's ideas, I have also compacted his 10 'groups' to five 'levels'. His Groups 2-5 (involving phenomena of significance in sedimentology on the scale of ripples, diurnal dunes, and 'mesoform' dunes) are all form-component strata-types, but their hierarchical subdivision does not seem absolutely essential here, and they have proved somewhat controversial in practice (see Bridge, 1993; Miall, 1995). Again, his higher groups (8-10) are ordered sedimentary cycles responding to allogenic change, and they are also grouped together here. Finally, alluvial systems can be self-similar in structure at a number of scales (which fortunately and usefully permits laboratory modelling) and so levels are not given ranges in size in metres or times of formation in seconds/years as in Brunsden (1996), Church (1996) or Miall (1993) – though Church intriguingly proposes that it is the domains of explanation (stochastic, deterministic, chaotic and contingent) that are spatially and temporally constrained. This insight is equally relevant to the systems-level scaling here used.

Table 1 presents an Alluvial Systematics framework in five levels with brief descriptive notes and examples. Figure 1 presents the framework in graphic form. Each level will now be briefly reviewed, noting the within-level typologies that have previously been proposed (Tables 2 and 3). Since these have been used by different researchers at different times (with contrasted disciplinary allegiances and purposes), they are not surprisingly contrasted in type. Note also that alluvial phenomena have:

Table 1. Alluvial Systematics.

Level 0 Particles

Scales of size, sorting, shape, etc. may be assigned to classes.

e.g. poorly-sorted gravel

Lithofacies coding (Miall, 1977)

Level I Strata sets

Involve sets and cosets of sheets, lenses, wedges, lobes, etc.

Contacts may be concordant and non-erosional, or discordant and erosional, reflecting depositional patterns. In section may involve lithofacies sequences; surface lineations may be mappable from ground survey or remote sensing imagery. Hierarchies of bounding surfaces have been proposed (Allen, 1983; Miall, 1988)

e.g. individual point bar scrolls, growth increments

Level II Form units

Generally relate to sediment supply/energy environments. Unit identification relatively straightforward for surface features, but in section involves interpretation of contact morphology, accordance and internal stratigraphy.

e.g. channels, bars or levees

Level III Architectural ensembles

Complex of unit forms and sediments. External boundaries may be erosional, inset or overlapping. Represents the total local alluvial complex produced by autogenic processes: alloformations, 'genetic floodplains'.

e.g. lateral-migration meander floodplains

Level IV Alluvial complexes

Two or more ensembles which contrast in contact patterns, sedimentation styles, elevation, etc., following allogenic changes (e.g. in discharge, sediment supply, tectonics, base level or human influence) and related to down-valley variation.

e.g. preserved total record of drainage basin alluvial systems

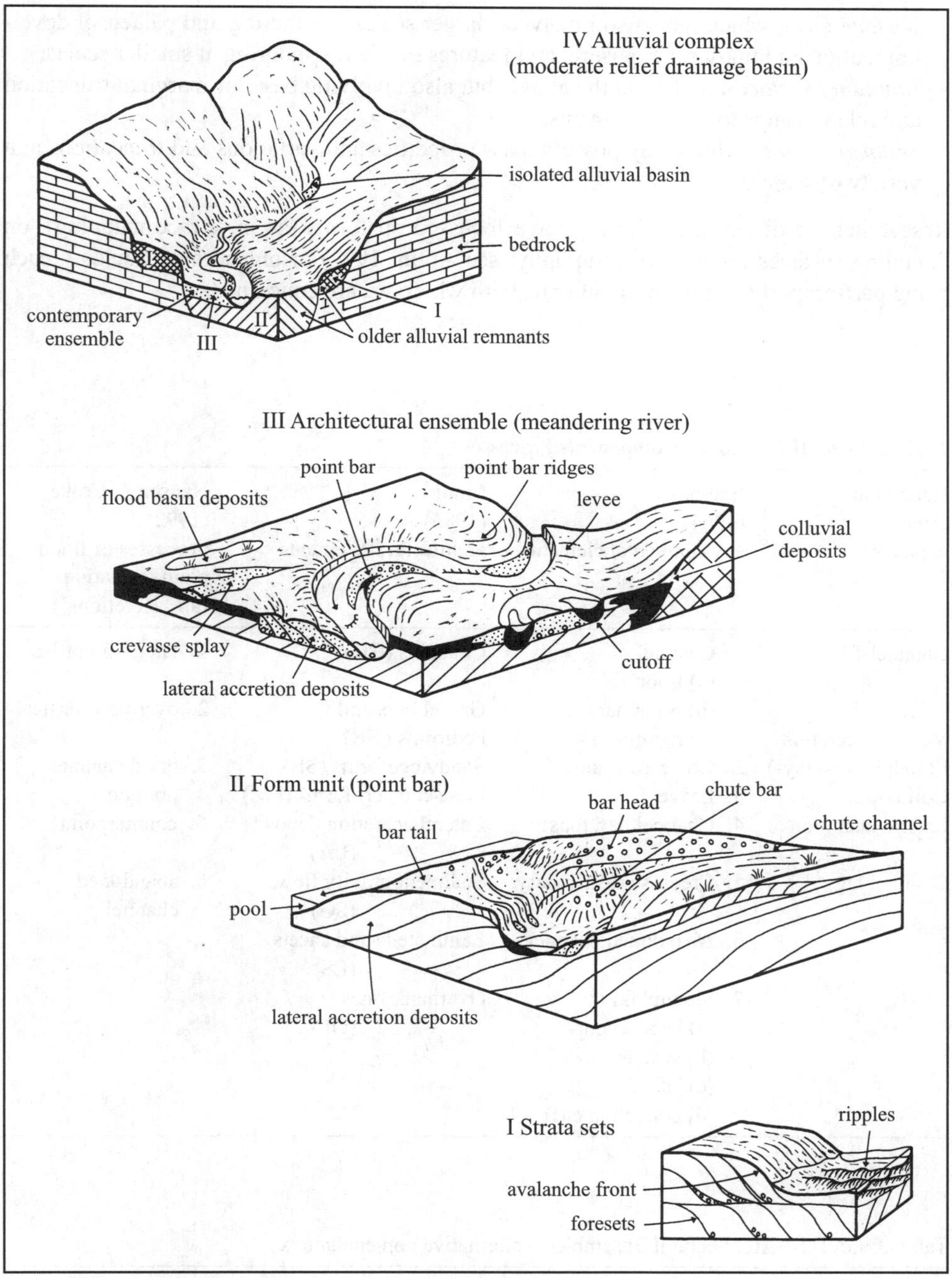

Figure 1. Summary diagram for a five-level alluvial systems model, illustrating the hierarchy of distinct entities that can be found at different levels. It should be appreciated that at higher levels, a range of ensembles and complexes has been distinguished, whilst not all form units will necessarily be present at a single site.

- *Surface form*, which may also involve at larger scales weathering and palaeosol development given long-term exposure, and features like bed armouring at smaller scales;
- *Bounding surfaces*, including the above, but also involving basal or marginal truncation and relationships to prior sediments;
- *Sediment bodies* which may possess mass properties and variations and truncations at a variety of scales.

Researchers in different disciplines have focussed on forms (as in terrace mapping), on bounding surfaces (as in allostratigraphy; see Autin, 1992) or on sediment facies – each being part-properties of the alluvial units with which we are concerned.

Table 2. Level II Floodplain Components/Processes.

Happ et al. (1940) deposits	Beerbower (1964) environmental elements	Miall (1985) architectural elements	Nanson & Croke (1992) processes of floodplain formation (all 'accretions')
Channel-fill	1. Channel	Channels (CH)	1. lateral point-bar
	(a) floor		
	(b) point bar	Gravel bars and	2. overbank vertical
Vertical accretion	(c) margin	bedforms (GB)	
Floodplain (splays)	2. Crevasse channel	Sandy bedforms (SB)	3. braid channel
Colluvium	3. Levee	Foreset macroforms (FM)	4. oblique
Lateral accretion	4. Channel bar, tops	Lateral accretion deposits (LA)	5. counterpoint
Channel lag	5. Crevasse distributary	Sediment gravity flows (SG)	6. abandoned channel
	6. Abandoned channel	Laminated sand sheets (LS)	
	7. Floodplain	Overbank fines (OF)	
	(a) backswamp		
	(b) swamp		
	(c) lake		
	(d) collection channel		

Table 3. Level III Architectural Ensembles – alternative nomenclatures.

Genetic type associations (Happ et al., 1940)
Facies models (Allen, 1963)
Alluvial architecture (Allen, 1978)
Fluvial style (Miall, 1985)
Alloformation (Autin, 1992)
Genetic floodplain (Nanson & Croke, 1992)

3 ALLUVIAL SYSTEM LEVELS

3.1 *Level 0*

This is effectively the lowest level identified; it involves what are for alluvial systems the basic particles (clasts, grains). These may be individually or collectively classified by size, sorting, shape, packing, etc. These are usually measured on continuous scales (with convenient qualitative names like sand and gravel classes), and it is only necessary here to note three points. First, alluvial sediment generally involves aggregates of individuals with a range of attributes, and the general tenor of much recent research is that in itself this is highly significant in process terms. Thus pebble clusters, patches and strips as well as the presence of an armour layer on the bed are important for particle entrainment and deposition (Paola & Seal, 1995); and sediment may also be transported in pulses or sheets (Gomez et al., 1989; Hoey, 1992). All this makes palaeohydrological reconstructions much more difficult, and linear relationships between mean or maximum particle size and average hydraulic transport parameters more dubious.

A second point is that there may be qualitatively different properties in particle-type entities at this level – the distinction between cohesive and non-cohesive sediment that is particularly important for channel bank failure, for example (see Thorne, 1982). Again, the products of high-concentration grain flows, as in debris flows, may be distinctive (see Costa, 1984). Very large boulders may be partially emergent during fluvial transport in small streams where they approach channel depth in dimension and this, too, may lead to qualitatively different particle mixtures on stream beds. Finally, tree roots or fallen trees may make entirely non-organic 'particle' descriptions too simple. This is particularly significant in mountain streams where bed sediments may have complex arrangements of clusters, rocks, berms and bars, as well as log jams (de Jong & Ergenzinger, 1995).

A third point is that it is at this level that archive *evidence* for alluvial system functioning is often derived, for instance where sediment sources are identified for coarser sediments by clast lithological analysis (Bridgland, 1986) or finer sediments by multiparameter 'fingerprinting' (Walling et al., 1993).

3.2 *Level I*

Bedform-scale *strata sets* are fundamental to sedimentology; as previously discussed, they may be classified hierarchically (Fig. 2a, Allen, 1983; see also Miall, 1993) and represent small-scale bedforms or growth increments as in the case of inclined heterolithic stratification (IHS). In addition to being potentially classifiable in terms of cross-cutting contact hierarchies, it is worth noting that internal characteristics may vary within and between classes. Fining-upward sequences were considered a major diagnostic feature for the deposits of meandering rivers (Allen, 1965); the situation has become more complex, with Thomas et al. (1987) noting seven possible fining trends on a hypothetical point bar, with others seeing coarsening trends on some meandering rivers (Bluck, 1971) and fining trends in other river types (Jackson, 1978).

A point worth reiterating is that these units are three dimensional, and whilst most work has been accomplished (by sedimentologists) using vertical sections, it is also possible to see (Fig. 2) planform cross-cutting sequences (see, for example, Hickin, 1974; Baker & Penteado-Orellana, 1978).

(a)
(b)
(c)
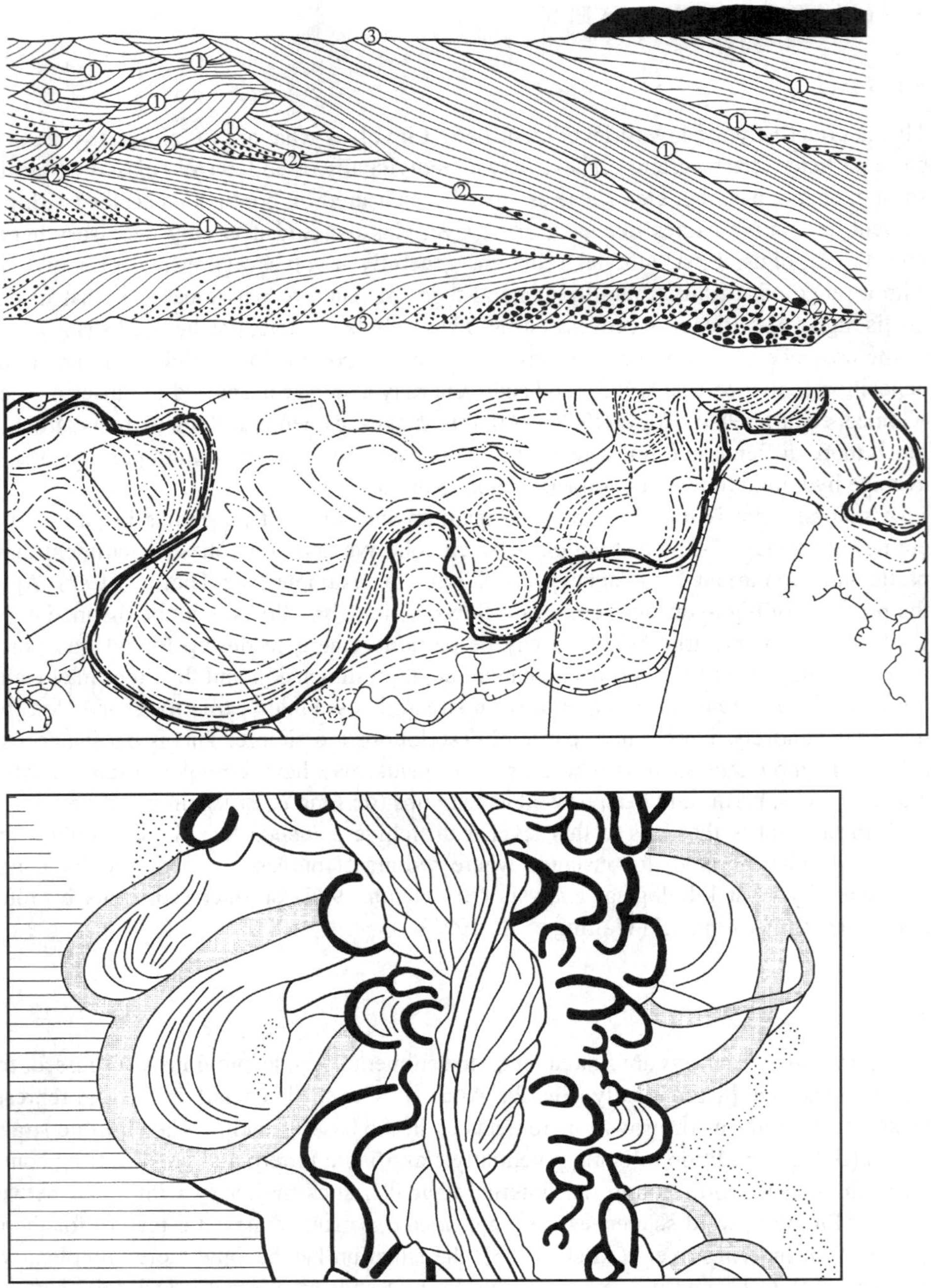

Figure 2. (a) Level I hierarchy of bedding contracts seen in section in sheet sandstones, Devonian Brownstones, Welsh Borderland, UK (after Allen, 1983). (b) Level II/III cross-cutting planform assemblages, Colorado River, Texas, USA (after Baker & Penteado-Orellana, 1978). (c) Level IV alluvial complex of channel traces and related terrace ensembles, of Pleniglacial age through to the present day, San Valley, Poland (after Szumanski, 1972).

All these smaller strata units generally scale with flow depth and are likely to be event-related as well as hydraulics-dependent. As flow depths and energy distributions vary both in time and across channel forms and floodplains, it is possible to get stacking of units sequentially during the rising and falling of a flood, and the formation of contemporary units of highly contrasted type spatially. And an 'event' for a ripple may be much briefer in duration than, say, the duration of an annual or extreme flood producing a distinct lateral aggradation growth increment.

3.3 *Level II*

These *form units* have been variously classified (Happ, 1940; Beerbower, 1964: see Table 2) and are most easily seen in planform. There is a major functional distinction between channel and overbank sedimentary environments (as they have also been called, though 'environment' is also used in the larger sense of marine, glacial, etc.). It is not too difficult to identify forms such as levees on the ground (though see Brierley et al., 1998), but interpretation of assemblages of Level I sets in form unit terms may require some subjective assumptions with evidence available only in sections (see Miall, 1985).

Vigorous research has recently been conducted on overbank environments and this has been comprehensively covered in Anderson et al. (1996). Overbank sedimentation models (James, 1985; Pizzuto, 1987) have been developed, and the modelling of overbank flows and sedimentation has been linked to observed rates of deposition by Nicholas & Walling (1995). But the largest component of recent alluvial research by far has been on channels and their sediments using both theoretical and physical models. Especially significant has been the linking of channel bottom instability with bank erosion models in developing meanders (see review by Rhoads & Welford, 1991; Seminara & Tubino, 1992), whilst the significance of junction scour and avulsion in braided rivers has become much better understood especially from scaled laboratory modelling (Ashmore, 1993; Best, 1988; Leddy et al., 1993).

It is also significant that field studies of active systems are revealing greater complexity, especially as more environments than those commonly found in mid-latitudes in the Northern Hemisphere are investigated. Anastomosing rivers have progressed from being viewed as rather unusual low-energy multichannel systems distinct from braided and meandering types (Smith & Smith, 1980; Rust, 1978) to being one of six types of anabranching system (rivers with islands) with equivalents in straight, meandering or braided types (Knighton & Nanson, 1993; Nanson & Knighton, 1996).

3.4 *Level III*

Such *ensembles* represent a summation of Level II forms: approximately the same concept has been widely recognised in sedimentology, stratigraphy and geomorphology (see Table 3). This is despite the fact that not all Level II components may be preserved for long – either because of autogenic cyclical elimination, or because of subsequent destruction as in the case of the fragmentary preservation of terrace sediments.

Recognition of ensemble types has progressed from Allen's seminal distinctions (Allen, 1965), through the six- and five-fold type distinctions of braided and meandering rivers respectively by Miall (1977) and Jackson (1978), to the alluvial architectural and genetic floodplain classifications of Miall (1985) and Nanson & Croke (1992).

Miall builds his models from eight 'elements' which are largely strata sets (and essentially in the terms of this paper, Level I sets); he exemplifies these with twelve fluvial styles. Nanson and Croke treat floodplains holistically and propose thirteen derivative orders and sub-orders distinguished using energy, sediment-size and geomorphic factors. These types are related to field examples provided by numerous previous researchers. All these authors provide three-dimensional block diagrams so that postulated relationships between features can be appreciated in both plan and section, a diagrammatic tradition that goes back a long way in the earth sciences.

With the exception of Rosgen (1994), the classification of architectural ensembles (or facies models, fluvial styles, genetic floodplains, etc.) that are available are subjective arrangements of known field cases, so that it is clearly possible to add to the types as empirical experience widens. For example, the avulsion process has been important in the development of alluvial architecture models (Bridge & Leeder, 1989) and processes in both modern (Smith et al., 1998) and ancient (Törnqvist, 1994) fluvial systems on a large scale justify this. Even on quite small streams, avulsion during extreme events may be highly significant in re-organising valley floor environments (Lapointe et al., 1998).

3.5 *Level IV*

Valley floors commonly contain both variation along their courses and also survivals from several architectural ensembles: the survivals may be overlapping, inset, or fragments isolated on bedrock slopes or terraced benches. They reflect changing environments. Even at the level of present-day valley bottoms, such remnants which are not equilibrated or adjusting to present environments can be found beyond the reach of active erosion/sedimentation processes (this zone being the present 'genetic floodplains' of Nanson & Croke, 1992) because of lateral distance or vertical elevation or burial. Apparently simple alluvial surfaces may disguise subsurface Level IV complexity, with many valley floors showing the remains of contrasting fluvial regime (see, for example, Starkel, 1982). This is equally true of terrace sediments: simple surfaces may be underlain by fragments of inset channels and overlapping units of contrasting type and date (see Fig. 2c).

Comparatively little published work is available on Level IV typology other than in near-actualistic terms in which field examples are grouped to show some of the possible modes of development (but see Leopold, Wolman & Miller, 1964, pp. 458-465; Starkel, 1981; Green & McGregor, 1986). Terrace formation has been physically modelled (Lewis, 1944; Germanoski & Harvey, 1993), and some field studies of rapidly developing forms or ones which are well-calibrated with dating have shown intrinsically how terracing in particular may develop (Ritter, 1982; Rains & Welch, 1988; Bull, 1991).

Distinction between fragmentarily-preserved Level IV components has been the essence of Quaternary fluvial studies. Nowadays this generally relies on a multifactor approach involving radiometric, luminescence or other dating techniques, lithostratigraphy, pedogenic development, and elevation in relation to reconstructed river profiles (see, for example, Gibbard, 1985; Bridgland, 1994).

Two points may usefully be made concerning all the levels discussed above. The first is that the record is nowhere temporally continuous. Whether it is the relatively short time-period between laminae, or the centuries-to-millennia between higher level units, the hiatuses record the fact that (as in stratigraphy generally) there is probably 'more gap than record' by a long way. Periods of effective non-deposition with surface weathering are augmented by ones of channel incision and sediment removal which are difficult to pinpoint.

Secondly, alluvial systems at all levels are only the storage part of fluvial systems which involve the loading of rivers with rock waste from slopes and its transfer through bedrock or alluvial channels to depositional basins which may well be in marine environments. River systems may have alluvial sedimentation zones that are differentiated down-valley, but these are only those parts of the total system (see Schumm, 1977) which may be found on-lapping or off-lapping in the lower part of drainage basins or perhaps temporarily extant at certain points in largely bedrock upstream systems (e.g. the slackwater sediments produced at extreme, rare events in gorge environments; see Baker et al., 1988). At Level IV, forecasting the location of alluviation is significant – whether or not the detail of Level 0-III phenomena is possible with the techniques used.

4 SIGNIFICANCE OF INTRA-LEVEL STUDIES FOR QUATERNARY STUDIES

A number of recent developments in understanding phenomena at each of the various levels described above are of clear significance to the study of Quaternary alluvial environments. To take one example, the modelling of scour depths at braided-river channel junctions has been supported by field studies (Best & Ashworth, 1997) in showing this may be up to about five times mean channel depth and that such scours may migrate across the full width of braidplains. Thus the 'working depth' of braided rivers may be considerable and perhaps greater than that of meandering rivers. Transferring this conclusion to the gravel deposits of the major rivers in lowland Britain suggests that gravel thicknesses in the order of 5-10 m are unlikely to represent more than the lateral accretion units of a single river level. The term 'aggradation' for such units could be misleading in this context.

Furthermore, Siegenthaler & Huggenberger (1993) have suggested from their analysis of Rhine gravels that preserved deposits represent only deeper pool fills, together with some channel bottom moderate-flow bedload sheets and high-suspension level sheet flows. Bar and braidplain deposits formed at higher topographic levels have been removed, probably by lateral channel shifting. Similar contrasts in meandering river sediments, in which lower point bar sediment responds to rare floods on the scale of decades to centuries (rather than the more frequent cycling of higher elevation units) has been suggested following ground-penetrating radar and coring studies by Bridge et al. (1995). As far as Quaternary sediments are concerned, therefore, it is important to appreciate that it may be *only* Level 0/I evidence that is preserved (especially from topographically low parts of the original depositional system), and that it may not be possible to reconstruct higher elevation and higher level (II-III) phenomena at all.

5 COMBINING LEVEL STUDIES

It has been suggested in passing that different research traditions have tended to focus on different levels, particularly given the evidence they use. Geographers, for example, may focus on Levels II and III given their primarily surface viewpoint. Sedimentologists working with section logs may most easily appreciate Level 0/I phenomena. Nowadays neither group would find such limitations satisfactory. Just as vibration, orbiting motion and waves are widely studied in physics at levels from the sub-atomic to cosmic, so alluvial systems may benefit from the application of concepts across the scales. Bed material sediment pulses and slugs whose form and motion exist at a variety of scales in alluvial systems (Hoey & Sutherland, 1991; Nicholas et al., 1995) are a case in point.

A further question concerns the significance of particular levels for understanding alluvial proceses in general. Some have seen Level I/II 'elements' as the key building blocks (Miall, 1985). Holistic Level III classification has been another focus (Nanson & Croke, 1992). At the other end of the scale, a kind of geoscience atomism might lead to attention being focussed on lower and lower levels, with interactions between flow hydraulics and particles or groups of particles being seen as the subject of the ultimate key research effort. Indeed such work is extremely important, but effects work both ways in the hierarchy of alluvial systems. Level I growth increments may reflect the geometry of the larger forms of which they form part. Meander loop development is affected by meander train evolution in a manner not predictable from single-bend hydraulics. The location of Level III ensembles is constrained by energy and sediment supply factors for river basins as a whole. It is this higher level complexity which has led to the exploration of new ways of analysing non-linear dynamics in addition to better known equilibrium response models. This includes re-examining of basin (Level IV) evolutionary trends, with a number of areas perhaps being ripe for future development:

- The exhaustion of paraglacial sediment stores (Church & Ryder, 1972) and modelling of the dissipation of cold-phase materials;
- The effects of progressive sediment re-working (as Holocene sediments become increasingly sorted, rounded, weathered, etc. over time);
- Progressive pedogenesis (see, for example, Howard et al., 1993; Woodward et al., 1994);
- Confinement during valley/channel incision, so that with decreasing effective floodplain width, flood depths for any given magnitude become greater.

6 MODELLING GROWTH AND CHANGE

A number of studies have taken phenomena at one or more levels and attempted to model developments in space and time. Autogenic behaviour may involve characteristic activity rates which, like the levels distinguished above, have intrinsic meaning. Absolute timescales may vary between laboratory models and 'real world' rivers. The following will be involved:

- Sedimentation rates and events. These have varying magnitude and frequency in different Level II environments. Models may be constructed to run continuously (whether in numerical simulations with their iterations, or physical models with constant flows), but then these need to be calibrated; furthermore, flow variations are the norm for natural rivers even to the extent that forms may be dominated by an alternation between sets of extreme destructional or formative events and intervening constructional or quiescent events (see Baker, 1988; Nanson, 1986). Dryland rivers subjected to ENSO regimes may function in this way (Wells, 1990) in response to well-spaced activity periods.
- Autogenic channel relocation or sediment cycling rates. These may involve progressive change or catastrophic transformation, as in meander development and cutoffs, whilst

there is also an enlarging literature in which change is viewed in terms of input and output from sediment stores/reservoirs (see Dietrich et al., 1982; Nakamura, 1986; Kelsey et al., 1986; Hoey, 1996).
- Reaction and relaxation rates (see Allen, 1974). These may in effect be subsumed within the two factors above.

One of the most interesting and satisfying of what are in practice Level II-III model developments to date is that by Howard (see Howard & Knutson, 1984; Howard, 1992, 1996) for meandering rivers. He used models of flow and bed topography, a bank erosion rate law and an overbank sedimentation rate law to simulate channel change and alluvial sedimentation. Variable-relief floodplains, with sinuous channel patterns and cutoffs, were simulated together with the effects of valley or clay plug confinement. The models bear close resemblance to many properties of full-scale rivers (see Fig. 3) and it is salutary to observe how continuous processes can produce a series of discrete features as their results are preserved. Included in this case are cutoffs, levees (in the models, primarily with bed aggradation), meander belts, 'back-swamps', and point and counterpoint relief. Howard (1996) observes that his model relates to a particular type of meandering (type B3 rivers of Nanson & Croke, 1992); it does not mimic the effects of channel bars in the early stages of development (see Rhoads & Welford, 1991) and meandering is (perhaps for this reason) slow to 'kick-off' from an initial upstream perturbation; and it does not produce compound loops involving bar-related mechanisms (see Whiting & Dietrich, 1993). Gross & Small (1998) present a similarly-based model additionally incorporating crevassing. Entirely different models have been developed for braided streams (Murray & Paola, 1994).

Meander train modelling has been further extended by Sun et al. (1996) and Stølum (1996), the latter showing that high- and low-sinuosity domains develop on his simulated floodplains as a result of cutoff clustering. This self-ordering leads to the achievement of realistic average sinuosity levels, which are thus controlled as much by cutoff as meandering processes. Stølen views this as an example of deterministic chaos achieved by the holistic modelling of meandering systems which would not be forecast by physical processes at a lower level.

A final example of sedimentation modelling is that of Nicholas & Walling (1996). This couples a hydraulic model for inundation and flow with a sediment transport component for dispersion and deposition of suspended sediment. It predicts spatial patterns which compare reasonably with assessed rates for deposition derived from ^{137}Cs dating.

Examples of autogenic modelling at different levels are summarised in Figure 4. Some are stochastic, others deterministic and others chaotic; some link form evolution to hydraulics and sediment transport, whilst others mimic evolving channel traces which can be presumed to have accompanying erosion/sedimentation (such as point bars and cutoff channel fills) although these are not necessarily modelled in themselves. Other models could be added (including several now presented in this volume), but those illustrated are sufficient to underline three points:

- There is no sign (yet) of some grand 'theory of everything' which allows modelling of alluvial systems across all levels;
- Whilst we may subjectively or objectively classify phenomena such as strata sets, channel patterns or architectural ensembles, we cannot predict their occurrence very widely;

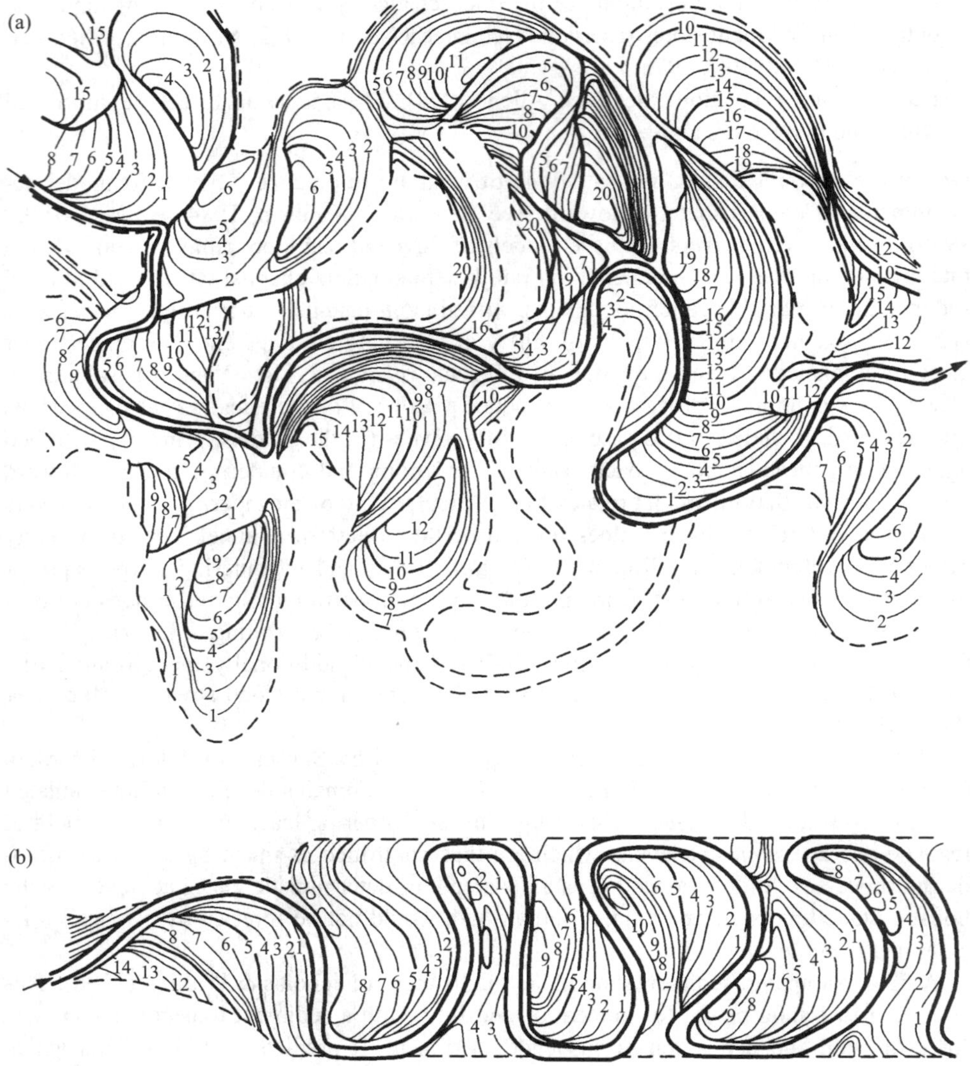

Figure 3. Simulation of meandering river floodplain evolution in free (a) and confined (b) situations. Floodplain age in hundreds of iterations is shown (from Howard, 1992).

- There is no necessary or proven association between types of phenomena at different levels, and conjectural associations involving level 'jumps' between, say, Level 0 sandy gravels and Level II channel patterns (and concomitant alluvial features) are somewhat hypothetical.

This must seem ominous at this stage in knowledge for assessing the impact of external change on alluvial systems – even though such systems commonly record, somewhere, that such change has occurred.

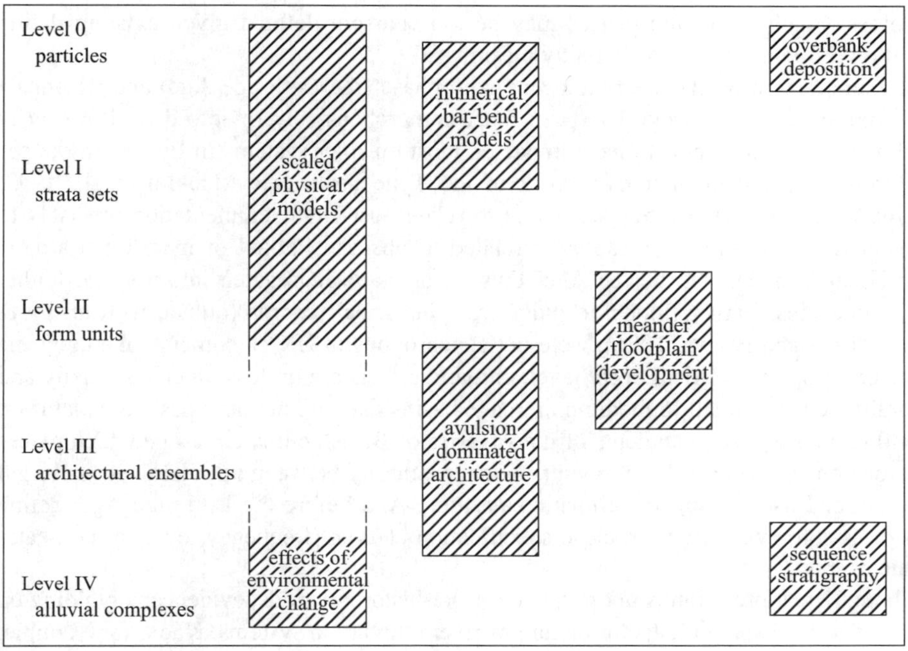

Figure 4. The level range covered by selected autogenic models.

7 FORECASTING THE EFFECTS OF ALLOGENIC CHANGE

Allogenic changes may, at the onset, have a range of potential effects on the functioning of autogenic sedimentation systems. These may be:

- Imperceptible or non-detectable, reflecting the system insensitivity to the change involved (see Brunsden & Thornes, 1979);
- Accommodated by a change in parameters within existing units (e.g. a change in meander dimensions);
- Transformational, in which one form of system is replaced by another (as in channel-type metamorphosis (Schumm, 1969));
- Destructional, as when fluvial systems are terminated by relative sea-level rise, overwhelmed by volcanic eruption, or disrupted by tectonic uplift.

As Brunsden & Thornes (1979) have discussed, allogenic changes themselves may be pulsed or ramped. Reaction and relaxation times (Allen, 1974) are also likely to vary with the systems level concerned.

Our knowledge of several major transformational or destructional happenings has been much extended, but in general allogenic change is seldom so precisely dated or scaled as we might wish. Forest clearance and its catchment-wide dating is commonly difficult to piece together from the site studies available (especially if one is trying not to use as evidence the actual alluvial sediments whose derivation is to be assigned to human or hydroclimatic causes). Tectonic activity is again often inferred from alluvial sediments themselves. Finally, climatic phases may be too coarsely defined given existing dating calibration refinement to be sufficiently meaningful.

For example, the most recent Holocene cold phase (the Little Ice Age) and its impact on alluvial systems in Europe has been most comprehensively reviewed by Rumsby & Macklin (1996). This seems to have impacted most on river systems in Europe in the period 1750-1900, and in their own work in the Tyne basin (see Macklin et al., 1992; Rumsby & Macklin, 1994) they were able to relate post-1890 sedimentation precisely to documented floods. In turn, these were related to phases of zonal or meridional atmospheric circulation. The 'Little Ice Age' thus becomes disaggregated into a set of higher flood frequencies. Larger floods, to judge from historical records (quoted in Rumsby & Macklin, 1994 and Brown, 1996), were not synchronous in UK catchments like the Tyne and Severn. So at the critical *event* level, climatic effects are no less diverse spatially and temporally between small to medium drainage basins than are human ones – a conclusion that will be no surprise to students of the vagaries of British climate but one which places limitations on the argument that country-wide synchrony between alluvial deposits might be present and used to suggest climatic causation. And before the Little Ice Age, dating precision is some way off from close specification at event frequency level for non-catastrophic events.

Although great precision is not possible for pre-historic periods, evidence is building up for a more refined allogenic factor picture of river alluviation systems. Knox, for example, has noted the significance of relatively small climatic changes in the Holocene Upper Mississippi valley sediment record (Knox, 1985, 1993). Groups of extreme palaeofloods in the southwestern USA have been described by Ely et al. (1993). Alluvial variability in the Holocene has also been widely documented in Europe (Starkel, 1982; Becker & Schirmer, 1977; Vandenburghe, 1995b; Schirmer, 1995; Starkel, 1995). Phases of soil erosion have similarly been pinpointed even to particular medieval decades (Bourke, 1989). Vandenberghe (1993, 1995) has noted that phases of high fluvial activity lasted only a relatively short time, mainly at times of climatic change, with incision at the beginning and at the end of cold cycles. In bulk terms, colder stages of the Quaternary appear the most significant though we know rather less about their environments than warm stages.

So many allogenic changes have been no less complex than our previous discussion of autogenic systems has shown these to be. Merely in climatic terms these range from flood- and drought-dominated regimes in southeastern Australia (Warner, 1977), to ENSO oscillations (Wells, 1990), to Holocene palaeoclimatic fluctuations (Rumsby & Macklin, 1994) and to glacial stages and sub-stages (Starkel, 1994; Vandenburghe, 1995). Some of these complexities relate to non-alluvial factors (such as slope sediment storage and the coupling/decoupling of slope-alluvial systems). Sediment storage of various kinds has been demonstrated to be significant in producing alluvial diachronism (Brown, 1987, 1990).

Autogenic system levels and their functioning suggest that routine responses may further be refined as follows:

7.1 *Level 0*

Fine sediment units may result from rapid catchment-long dissipation in suspension in single events. These are not usually competence limited. Brown and Kennett (1998) record deposits from one megaflood on the Mississippi which could have travelled many hundreds of kilometres effectively in one event. However, with increasing sediment size, particle travel distances are much more confined for single flood events. Differential movement within the gravel size range is not highly significant, however (Church & Hassan, 1992), and we are commonly looking at bank/bar to bar distances.

7.2 *Level I*

Rapid hydraulic response is possible, but at a *local* level using sediments available on-site with sets representing discharge events on a scale of minutes to decades. Reformation of units from local sediment sources is going to be much more rapid than the downstream propagation of coarse materials for long distances.

7.3 *Level II*

Again, local adjustment of, say, channel cross-section to hydraulic variables may be rapid, but wholesale change involving the need for new coarser sediment, or pattern transformation, will take much longer. Thus meander development commonly takes some hundreds of years following an initial rapid phase. Comparison of field and modelled patterns is, however, still required to validate transformation patterns and rates.

7.4 *Level III*

To build up a well-developed alluvial ensemble, incorporating sedimentation rates on a variety of scales (the slowest being in areas remote from channels in low-energy and low sediment-supply environments) may be on a scale of hundreds to thousands of years unless allogenic change is so extremely large as to destroy or transform the alluvial system. Even then down-valley change dissipation could be involved. Thus glaciated British Colombia is effectively still responding to the last glaciation (Church & Ryder, 1972; Church & Slaymaker, 1989). In Church & Slaymaker's phrase, 'the natural landscape of British Colombia is imprisoned in its history'.

In modelling terms, it therefore appears important to appreciate (following Church, 1996) that whilst deterministic approaches may serve local transformation and lower levels well in terms of forecasting allogenic changes, for higher levels chaotic and contingent models are more appropriate. Whilst competent flows may transform bedform or channel dimensions on a local scale quite rapidly, higher level phenomena may require sediment transfer over long distances, including periods of storage in a variety of niches, and this may involve an extended timescale.

8 CONCLUSIONS

In Britain, relatively few studies of pre-Holocene fluvial sediments have been undertaken from a sedimentological point of view (see Bryant, 1991), but Holocene sediments have demonstrated contrasting records from upland and lowland Britain. Thus lowland studies have pointed to the invasion of Devensian gravel rivers (often anabranching) or valley systems by fine sediment (Burrin & Scaife, 1994; Robinson & Lambrick, 1984; Robinson, 1992; Brown et al., 1994). In the uplands, progressive incision into Devensian glacial and periglacial sediments has been common (e.g. Macklin & Lewin, 1986; Harvey & Renwick, 1987; Hooke et al., 1990; Macklin et al., 1992; Tipping, 1992).

For alluvial systems and their more complete interpretation, we should also now consider the following:

(a) System response to allogenic change must be mediated through autogenic processes which can be viewed as operating within a hierarchical system with variable activity rates and response times;
(b) Research on autogenic processes at various levels has transformed our understanding in numerous ways. Equilibrium thinking is under challenge. Alluvial modelling of new kinds has produced fresh insights at different levels but using different approaches – the behaviour of sediment aggregates, the development of partially-preserved alluvial strata sets, and the development of meander trains for example. Higher-level modelling must now be important for Quaternary scientists interested in the impact of environmental changes;
(c) Some palaeosystem reconstructions now seem less reliable – palaeodischarge estimates or the reconstruction of Level III and especially Level IV environments on the basis of very limited Level I exposures for instance;
(d) But given the dynamic modelling developments of recent years it should soon be possible to forecast dissipative responses to allogenic changes in much more meaningful ways. It is thus opportune to take a new look at Quaternary alluvial fills, bearing 'level' considerations in mind, to see if new ways of response forecasting suggest themselves.

We may finally revert to the concept of alluvial systematics. An hierarchical structure of levels is not simply a matter of descriptive convenience; it is important to recognise the fundamental process-based nesting of entities, which includes all those scales of concern for geoscientists with a variety of traditions, but who operate within one over-arching field of study, in order to allow proper and unconstrained progress in understanding. No particular level has as yet proved demonstrably more significant for overall modelling than any other, but in confronting the challenge of predicting change in alluvial systems it is surely important to see what it is that evidence at particular levels can tell us, and for particular purposes to decide what modelling on different bases and at different levels can reliably demonstrate.

REFERENCES

Allen, J.R.L. 1965. A review of the origin and characteristics of recent alluvial sediments. *Sedimentology,* 5: 89-191.

Allen, J.R.L. 1974. Reaction, relaxation and lag in natural sedimentary systems: general principles, examples and lessons. *Earth-Science Reviews,* 10: 263-342.

Allen, J.R.L. 1983. Studies in fluviatile sedimentation: bars, bar-complexes and sandstone sheets (low sinuosity braided streams) in the Brownstones (L. Devonian), Welsh Borders. *Sedimentary Geology,* 33: 237-293.

Anderson, M.G., Walling, D.E. & Bates, P.D. 1996. *Floodplain process.* Chichester: John Wiley.

Ashmore, P. 1993. Anabranch confluence kinetics and sedimentation processes in gravel-braided streams. In: Best, J.L. & Bristow, C.S. (eds) *Braided Rivers*, 129-146. London: The Geological Society.

Autin, W.J. 1992. Use of alloformations for definition of Holocene meander belts in the middle Amite River, southeastern Louisiana. *Geological Society of America Bulletin,* 104: 233-241.

Baker, V.R. 1988. Cataclysmic processes in geomorphological systems. *Zeitschrift fur Geomorphologie*, 67: 25-32.

Baker, V.R., Kockel, R.C. & Patton, P.C. 1988. *Flood Geomorphology*. Chichester: John Wiley.

Baker, V.R. & Penteado-Orellana, M.M. 1978. Fluvial sedimentation conditioned by Quaternary climatic change in central Texas. *Journal of Sedimentary Petrology,* 48: 433-51.

Becker, B. & Schirmer, W. 1977. Palaeoecological study on the Holocene valley development of the River Main, southern Germany. *Boreas,* 21: 303-21.

Beerbower, J.R. 1964. Cyclothems and cyclic depositional mechanisms in alluvial plain sedimentation. *Bulletin of Kansas University Geological Society,* 169: 31-42.

Best, J.L. 1988. Sediment transport and bed morphology at river channel confluences. *Sedimentology,* 35: 481-98.

Best, J.L. & Ashworth, P.J. 1997. Scour in large braided rivers and the recognition of sequence stratigraphic boundaries. *Nature,* 387: 275-277.

Bluck, B.J. 1971. Sedimentation in the meandering river Endrick. *Scottish Journal of Geology,* 7: 93-138.

Borke, H.-R. 1989. Soil erosion during the past millennium in central Europe and its significance within the geomorphodynamics of the Holocene. In: Ahnert, F. (ed.) *Landforms and landform evolution in West Germany*, Catena Supplement 15, Cremlingen: Catena Verlag, 121-131.

Bridge, J.S. 1993. Description and interpretation of fluvial deposits: a critical perspective. *Sedimentology,* 40: 801-810.

Bridge, J.S., Alexander, J., Collier, R.Ll., Gawthorpe, R.L. & Jarvis, J. 1995. Ground-penetrating radar and coring used to study the large-scale structure of point-bar deposits in three dimensions. *Sedimentology,* 42: 839-852.

Bridge, J.S. & Leeder, M.R. 1989. A simulation model of alluvial stratigraphy. *Sedimentology,* 26: 617-44.

Bridgland, D.R. (ed.) 1986. *Clast lithological analysis*. London: Quaternary Research Association.

Bridgland, D.R. (ed.) 1994. *Quaternary of the Thames*, Geological Conservation Review Series, 7. London: Chapman and Hall.

Brierley, G.J., Ferguson, R.J. & Woolfe, K.J. 1997. What is a fluvial levee? *Sedimentary Geology,* 114: 1-9.

Brown, A.G. 1987. Holocene floodplain sedimentation and channel response of the lower River Severn, UK. *Zeitschrift fur Geomorphologie,* 31: 293-310.

Brown, A.G. 1990. Holocene floodplain diachronism and inherited downstream variations in fluvial processes: a study of the River Perry, Shropshire, England. *Journal of Quaternary Science,* 5: 39-51.

Brown, A.G. 1996. Human dimensions of palaeohydrological change. In: Branson, J., Brown, A.G. & Gregory, K.J. (eds) *Global continental changes: the context of Palaeohydrology*, London: The Geological Society, 57-72.

Brown, A.G., Keough, M.K. & Rice, R.J. 1994. Floodplain evolution in the East Midlands, United Kingdom: the Lateglacial and Flandrian alluvial record of the Scour and Nene valleys. *Philosophical Transactions of the Royal Society,* 348A: 261-293.

Brown, P.A. & Kennett, J.P. 1998. Megaflood erosion and meltwater plumbing changes during the last North American deglaciation recorded in Gulf of Mexico sediments. *Geology,* 26: 599-602.

Brunsden, D. 1996. Geomorphological events and landform change. *Zeitschrift fur Geomorphologie,* 40: 273-288.

Brunsden, D. & Thornes, J.B. 1979. Landscape sensitivity and change. *Transactions of the Institute of British Geographers*, New Series 4: 463-84.

Bryant, I.D. 1991. Sedimentology of glaciofluvial deposits. In: Ehlers, J., Gibbard, P.L. & Rose, J. (eds) *Glacial deposits in Great Britain and Ireland*, Rotterdam: Balkema, 437-442.

Bull, W.B. 1991. *Geomorphic responses to climatic change*. New York: Oxford University Press.

Burrin, P.J. & Scaife, R.G. 1984. Aspects of Holocene valley sedimentology and floodplain development in southern England. *Proceedings of the Geologists' Association,* 95: 81-86.

Church, M. 1996. Space, time and the mountain – how do we order what we see? In: B.L. Rhoads & C.E. Thorne (eds) *The Scientific Nature of Geomorphology*, Chichester: John Wiley, 147-170.

Church, M. & Hassan, M.A. 1992. Size and distance of travel of unconstrained clasts in a streambed. *Water Resources Research,* 28: 299-303.

Church, M. & Ryder, J.M. 1972. Paraglacial sedimentation: a consideration of fluvial processes conditioned by glaciation. *Geological Society of America Bulletin,* 83: 3059-3072.

Church, M. & Slaymaker, O. 1989. Disequilibrium of Holocene sediment yield in glaciated British Colombia. *Nature,* 337: 452-454.

Costa, J.E. 1984. Physical geomorphology of debris flows. In: J.E. Costa & P.J. Fleisher (eds) *Developments in Applications of Geomorphology*, 268-317. Berlin: Springer-Verlag.

Dietrich, W.E., Dunne, T., Humphrey, N.F. & Reid, L.M. 1982. Construction of sediment budgets for drainage basins. In: Swanson, F.J., Janda, R.J., Dunne, T. & Swanson, D.N. (eds) *Sediment budgets and routing in forested drainage basins*, U.S. Dept. Agriculture, Forest Services General Technical Report PNW-141, 5-23.

Ely, L.L., Enzel, Y., Baker, V.R. & Cayan, D.R. 1993. A 5000 year record of extreme floods and climate change in the south western United States. *Science,* 262: 410-444.

Germanoski, D. & Harvey, M.D. 1993. Asynchronous terrace development in degrading braided channels. *Physical Geography,* 14: 16-38.

Gibbard, P.L. 1985. *Pleistocene history of the Middle Thames Valley*. Cambridge: Cambridge University Press.

Gomez, B., Naff, R.L. & Hubbell, D.W. 1989. Temporal variations in bedload transport rates associated with the migration of bedforms. *Earth Surface Processes and Landforms,* 14: 135-156.

Green, C.P. & McGregor, D.F.M. 1986. River terraces: a stratigraphic record of environmental change. In: V. Gardiner (ed.) *International Geomorphology 1986*, Chichester: John Wiley, I: 977-987.

Gross, L.J. & Small, M.J. 1998. River and floodplain process simulation for subsurface characterization. *Water Resources Research,* 34: 2365-2376.

Happ, S.C., Rittenhouse, G. & Dobson, G.C. 1940. Some principles of accelerated stream and valley erosion. *Technical Bulletin 695*. U.S. Department of Agriculture, Washington D.C.

Harvey, A.M. & Renwick, W.H. 1987. Holocene alluvial fan and terrace formation in the Bowland Fells, northwest England. *Earth Surface Processes and Landforms,* 12: 249-257.

Hickin, E.J. 1974. The development of river meanders in natural river channels. *American Journal of Science,* 274: 414-442.

Hoey, T.B. 1992. Temporal variations in bed load transport rates and sediment storage in gravel-bed rivers. *Progress in Physical Geography,* 16: 319-338.

Hoey, T.B. 1996. Sediment dispersion and duration of storage in a model braided river. *Journal of Hydrology* (NZ), 35: 213-237.

Hoey, T.B. & Sutherland, A.J. 1991. Channel morphology and bedload pulses in braided rivers: a laboratory study. *Earth Surface Processes and Landforms,* 16: 447-62.

Hooke, J.H., Harvey, A.M., Miller, S.Y. & Redmond, C.E. 1990. The chronology and stratigraphy of the alluvial terraces of the River Dane valley, Cheshire, NW England. *Earth Surface Processes and Landforms,* 15: 717-37.

Howard, A.D. 1992. Modelling channel migration and floodplain sedimentation in meandering streams. In: Carling, P.A. & Petts, G.E. (eds) *Lowland Floodplain Rivers,* Chichester: John Wiley, 1-41.

Howard, A.D. 1996. Modelling channel evolution and floodplain morphology. In: Anderson, M.G., Walling, D.E. & Bates, P.D. (eds) *Floodplain Processes*, Chichester: John Wiley, 15-62.

Howard, A.D. & Knutson, T.R. 1984. Sufficient conditions for river meandering: a simulation approach. *Water Resources Research,* 20: 1659-1667.

Howard, J.L., Amos, D.F. & Daniels, W.L. 1993. Alluvial soil chronosequence in the inner coastal plain, Central Virginia. *Quaternary Research,* 39: 201-213.

Huggett, R.J. 1988. Dissipative systems: implications for geomorphology. *Earth Surface Processes and Landforms,* 13: 45-49.

Jackson, R.G. II 1978. Preliminary evaluation of lithofacies models for meandering alluvial streams. In: Miall, A.D. (ed.) *Fluvial Sedimentology*, Calgary: Canadian Society of Petroleum Geologists, 543-576.

James, C.S. 1985. Sediment transfer to overbank sections. *Journal of Hydraulics Research,* 23: 435-452.

Jardine, N. & Simpson, R. 1971. *Mathematical taxonomy*. London: Wiley.

de Jong, C. & Ergenzinger, P. 1995. The interrelations between mountain valley forms and river-bed arrangement. In: Hickin, E.J. (ed.) *River Geomorphology*, Chichester: John Wiley, 55-91.

Kelsey, H.M., Lamberson, R. & Madej, M.A. 1986. Modelling the transport of stored sediment in a gravel bed river, northwestern California. *IASH Publication 159*, 367-391.

Knighton, A.D. & Nanson, G.C. 1993. Anastomosis and the continuum of channel pattern. *Earth Surface Processes and Landforms,* 18: 613-25.

Knox, J.C. 1985. Responses of floods to Holocene climatic change in the Upper Mississippi valley. *Quaternary Research,* 23: 287-300.

Knox, J.C. 1993. Large increases in flood magnitude in response to modest changes in climate. *Nature,* 361: 430-432.

Lapointe, M.F., Secretan, Y., Driscoll, S.N., Bergeron, N. & Leclerc, M. 1998. Response of the Ha! Ha! River to the flood of July 1996 in the Sagueney Region of Quebec: large-scale avulsion in a glaciated valley. *Water Resources Research,* 34: 2383-2392.

Leddy, J.O., Ashworth, P.J. & Best, J.L. 1993. Mechanisms of anabranch avulsion within gravel-bed braided rivers: observations from a scaled physical model. In: Best, J.L. & Bristow, C.S. (eds) *Braided Rivers*. London: The Geological Society, 119-127.

Leopold, L.B., Wolman, M.G. & Miller, J.P. 1964. *Fluvial Processes in Geomorphology*. San Francisco: W.H. Freeman.

Lewis, W.V. 1944. Stream trough experiments and terrace formation. *Geological Magazine,* 81: 241-253.

Macklin, M.G., Rumsby, B.T. & Newson, M.D. 1992. Historical floods and vertical accretion of fine grained alluvium in the Lower Tyne Valley, North East England. In: Billi, P., Hey, R.D., Thorne, C.R. & Tacconi, P. (eds) *Dynamics of Gravel Bed Rivers*, Chichester: John Wiley, 573-589.

Macklin, M.G. & Lewin, J. 1986. Terraced fills of Pleistocene and Holocene age in the Rheidol Valley, Wales. *Journal of Quaternary Science,* 6: 225-232.

Macklin, M.G., Passmore, D.G. & Rumsby, B.T. 1992. Climate and cultural signals in Holocene alluvial sequences: the Tyne Basin, Northern England. In: Needham, S. & Macklin, M.G. (eds) *Alluvial Archaeology in Britain*, Oxford: Oxbow Press, 123-139.

Miall, A.D. 1977. A review of the braided river depositional environment. *Earth-Science Reviews,* 13: 1-62.

Miall, A.D. 1985. Architectural element analysis: a new method of facies analysis applied to fluvial deposits. *Earth-Science Reviews,* 22: 261-308.

Miall, A.D. 1993. The architecture of fluvial-deltaic sequences in the Upper Mesaverde Group (Upper Cretaceous) Book Cliffs, Utah. In: Best, J.L. & Bristow, C.S. (eds), *Braided Rivers*, London: The Geological Society, 305-332.

Miall, A.D. 1995. Description and interpretations of fluvial deposits: a critical perspective. *Sedimentology,* 42: 379-389.

Montgomery, K. 1996. Sinuosity and fractal dimension of meadering rivers. *Area,* 28: 491-500.

Murray, A.B., Paola, C. 1994. A cellular model of braided rivers. *Nature,* 371: 54-57.

Nakamura, F., Araya, T., & Higashi, S. 1987. Influence of river channel morphology and sediment production on residence time and transport distance. *IASH Publication,* 165: 355-364.

Nanson, G.C. 1986. Episodes of vertical accretion and catastrophic stripping: a model of disequilibrium flood-plain development. *Geological Society of America Bulletin,* 97: 1467-1475.
Nanson, G.C. & Croke, J.C. 1992. A genetic classification of floodplains. *Geomorphology,* 4: 459-486.
Nanson, G.C. & Knighton, A.D. 1996. Anabranching rivers: their cause, character and classification. *Earth Surface Processes and Landforms,* 21: 217-239.
Nicholas, A.P. & Walling, D.E. 1995. Modelling contemporary overbank sedimentation in floodplains: some preliminary results. In: Hicken, E.J. (ed.) *River Geomorphology*. Chichester: John Wiley.
Nicholas, A.P., Ashworth, P.J., Kirkby, M.J., Macklin, M.G. & Murray, T. 1995. Sediment slugs: large-scale fluctuations in fluvial sediment transport rates and storage volumes. *Progress in Physical Geography,* 19: 500-519.
Nikora, V.I. 1991. Fractal structure of river plan forms. *Water Resources Research,* 27: 1327-1333.
Paola, C. & Seal, R. 1995. Grain size patchiness as a cause of selective deposition and downstream fining. *Water Resources Research,* 31: 1395-1407.
Phillips, J.D. 1992. Nonlinear dynamical systems in geomorphology: revolution or evolution. *Geomorphology,* 5: 219-229.
Pizzuto, J.E. 1987. Sediment diffusion during overbank flow. *Sedimentology,* 34: 301-317.
Prigogine, I. 1980. *From being to becoming: time and complexity in the physical sciences*. San Francisco: W.H. Freeman.
Rains, B. & Welch, J. 1988. Out-of-phase Holocene terraces in part of the North Saskatchewan river basin, Alberta. *Canadian Journal of Earth Sciences,* 25: 454-464.
Ritter, D,.F. 1982. Complex river terrace development in the Nenana Valley near Healy, Alaska. *Geological Society of America Bulletin,* 93: 346-356.
Robinson, M. 1992. Environment, archaeology and alluvium on the river gravels of the South Midlands. In: Needham, S. & Macklin, M.G. (eds) *Alluvial archaeology in Britain*, Oxford: Oxbow Books, 197-208.
Robinson, M. & Lambrick, G.H. 1984. Holocene alluviation and hydrology in the Upper Thames Basin. *Nature,* 308: 809-14.
Rosgen, D.L. 1994. A classification of natural rivers. *Catena,* 22: 169-199.
Rumsby, B.T. & Macklin, M.G. 1994. Channel and floodplain response to recent abrupt climate change: the Tyne basin, northern England. *Earth Surface Processes and Landforms,* 19: 499-515.
Rumsby, B.T. & Macklin, M.G. 1996. River response to the last neoglacial (the 'Little Ice Age') in northern, western and central Europe. In: Branson, J., Brown, A.G. & Gregory, K.J. (eds) *Global Continental Changes: the Context of Palaeohydrology*, London: The Geological Society, 217-233.
Rust, B.R. 1978. Depositional models for braided alluvium. In: A.D. Miall (ed.) *Fluvial Sedimentology*, Calgary: Canadian Society of Petroleum Geologists, 605-625.
Schirmer, W. 1995. Valley bottoms in the late Quaternary. *Zeitschrift fur Geomorphologie*, 100: 27-51.
Schumm, S.A. 1969. River metamorphosis. *Journal of the Hydraulics Divisions, Proceedings of the American Society of Civil Engineers*, 95 HYI: 255-273.
Schumm, S.A. 1977. *The Fluvial System*. New York: John Wiley.
Schumm, S.A. 1981. Evolution and response of the fluvial system, sedimentological implications. *SEPM Special Publication,* 31: 19-29.
Seminara, G. & Tubino, M. 1992. Weakly non-linear theory of regular meanders. *Journal of Fluvial Mechanics,* 244: 257-88.
Siegenthaler, C. & Huggenberger, P. 1993. Pleistocene Rhine gravel: deposits of a braided river system with dominant pool preservation. In: Best, J.L. & Bristow, C.S. (eds) *Braided Rivers*, London: The Geological Society, 147-162.
Smith, D.G. & Smith, N.D. 1980. Sedimentation in anatomosed river systems: examples from alluvial valleys near Banff, Alberta. *Journal of Sedimentary Petrology,* 50: 157-164.

Smith, N.D., Slingerland, R.L., Pérez-Arlucea, M. & Morozova, G.S. 1998. The 1870s avulsion of the Saskatchewan River. *Canadian Journal of Earth Sciences,* 35: 453-466.

Starkel, L. & J.B. Thornes (eds) 1981. Palaeohydrology of river basins. *Technical Bulletin 28, British Geomorphology Research Group*. Norwich: Geobooks.

Starkel, L. (ed.) 1982. *Evolution of the Vistula River Valley during the last 15000 years*. Wrøclaw: Polish Academy of Sciences.

Starkel, L. (ed.) 1994. Reflection of the glacial-interglacial cycle in the evolution of the Vistula river basin, Poland. *Terra Nova,* 6: 486-494.

Stølum, H.H. 1996. River meandering as a self-organisation process. *Science* 271: 1710-1713.

Sun, T., Meakin, P., Jøssang, T. & Schwarz, K. 1996. A simulation model for meandering rivers. *Water Resources Research,* 32: 2937-2954.

Szumanski, A. 1972. The valley of the Lower San River in the Sandomierz Basin. *Excursion guide book, Symposium of the INQUA Commission on Studies of the Holocene*, 58-68.

Thomas, R.G. and five others 1987. Inclined heterolithic stratification – terminology, description, interpretation and significance. *Sedimentary Geology,* 53: 123-179.

Thorne, C.R. 1982. Processes and mechanisms of bank erosion. In: Hey, R.D., Bathurst, J.C. & Thorne, C.R. (eds) *Gravel Bed Rivers*, Chichester: John Wiley, 227-221.

Tipping, R. 1992. The determination of cause in the generation of major prehistoric valley fills in the Cheviot Hills, Anglo-Scottish border. In: Needham, S. & Macklin, M.G. (eds) *Alluvial Archaeology in Britain*, Oxford: Oxbow Books, 111-121.

Törnqvist, T.E. 1994. Middle and Late Holocene avulsion history of the River Rhine (Rhine-Meuse delta, Netherlands). *Geology,* 22: 711-714.

Vandenberghe, J. 1993. Changing fluvial processes under changing periglacial conditions. *Zeitschrift fur Geomorphologie*, 88: 17-28.

Vandenberghe, J. 1995. European river activity and climatic change during the Lateglacial and early Holocene. *Palaoklimaforschung,* 14: 11-19.

Walling, D.E., Woodward, J.C. & Nicholas, A.P. 1993. A multi-parameter approach to fingerprinting suspended-sediment sources. *IAHS Publication,* 215: 329-38.

Warner, R.F. 1997. Floodplain stripping: another form of adjustment to secular hydrologic regime changes in Southeast Australia. *Catena,* 30: 263-282.

Wells, L.E. 1990. Holocene history of the El Niño phenomenon as recorded in flood sediments of northern coastal Peru. *Geology,* 18: 1134-1137.

Whiting, P.J. & Dietrich, W.E. 1993. Experimental constraints on bar migration through bends: implications for meander wavelength selection. *Water Resources Research,* 29: 1091-1102.

Williams, G.P. 1983. Palaeohydrological methods and some examples from Swedish fluvial environments, I Cobble & Boulder deposits. *Geografiska Annaler,* 65A: 227-243.

Williams, G.P. 1997. *Chaos Theory Tamed.* London: Taylor and Francis.

Woodward, J.C., Macklin, M.G. & Lewin, J. 1994. Pedogenic weathering and relative age dating of Quaternary alluvial sediments in the Pindus Mountains of Northwest Greece. In: Robinson, D.A. & Williams, R.B.G. (eds) *Rock weathering and landform evolution*, Chichester: John Wiley, 259-283.

Part 2: Tectonic forcing: crustal instability and the fluvial record

3. The Maas terrace sequence at Maastricht, SE Netherlands: evidence for 200 m of late Neogene and Quaternary surface uplift

MEINDERT W. VAN DEN BERG
Netherlands Institute of Applied Geosciences TNO-National Geological Survey, Zwolle, the Netherlands

TON VAN HOOF
University of Utrecht, Faculty of Earth Sciences Department of Geophysics, Utrecht, the Netherlands

1 INTRODUCTION

Rivers are continuously present in areas with humid climatic conditions. Their evolution is recorded in sediment-landform assemblages. Over the years, several river-terrace flights have been reported from European drainage basins outside areas affected by glacial isostasy. Their geomorphology is typically well established but their chronological framework is typically poor. From the process-response point of view these records can be split up into two different groups: Late Glacial terrace records with a well established age framework and older Pleistocene records with a poorly established age framework. They differ fundamentally in the sense that Late Glacial terraces are primarily responding to climate change, in particular to millennial-scale climate cycles (van den Berg, 1996). The uplift rates in most of NW Europe are so low that this factor plays a minor role only in this type of terrace. This is in contrast to the longer-term Pleistocene terrace staircases. Interpretation of the later proxies can potentially provide insight into the main controlling factors of these records: climate change and crustal deformation.

Both river records and the marine record have climate – change as a common controlling factor. This provides the potential for the correlation of long river records with the much better dated marine records. These correlations are of great importance. They will provide insight into temporal dynamics of rivers and thus of crustal dynamics, in particular on the landward sides of the hinges of sedimentary basins. This is a fundamental aspect of the understanding of crustal dynamics.

In this paper we focus on a terraced area in NW Europe at 51°N, shaped by the north flowing, predominantly rain-fed, River Maas (Meuse). The discussed area is 30-x35 km and is here referred to as the Maastricht section (M, in Fig. 1). The area is enclosed by two main structural elements: in the south by the frontal thrust zone of the Ardennes Massif (the Midi-Aachen thrust which is part of the Variscan Front); and in the northeast by the Feldbiss Fault, the southern principle displacement zone of the Roer Valley rift system. Data concerning the river evolution within the Roer valley rift system will also be considered as boundary conditions in the age approximations.

The geology and geomorphology of the area have been mapped in great detail (e.g. Van Straaten, 1946; Zonneveld, 1949; Breuren, 1945; Krook, 1961; Felder & Bosch, 1989; Ruyters et al., 1995; Van den Berg, 1989). This large dataset provides a detailed overview on the distribution, lithology, sedimentology and morphology of the sediments. The terrace staircase contains 31 levels, including a Weichselian Late Glacial level and the Holocene floodplain. Terraces are vertically separated by 4 to 9 m. Conventional age data on

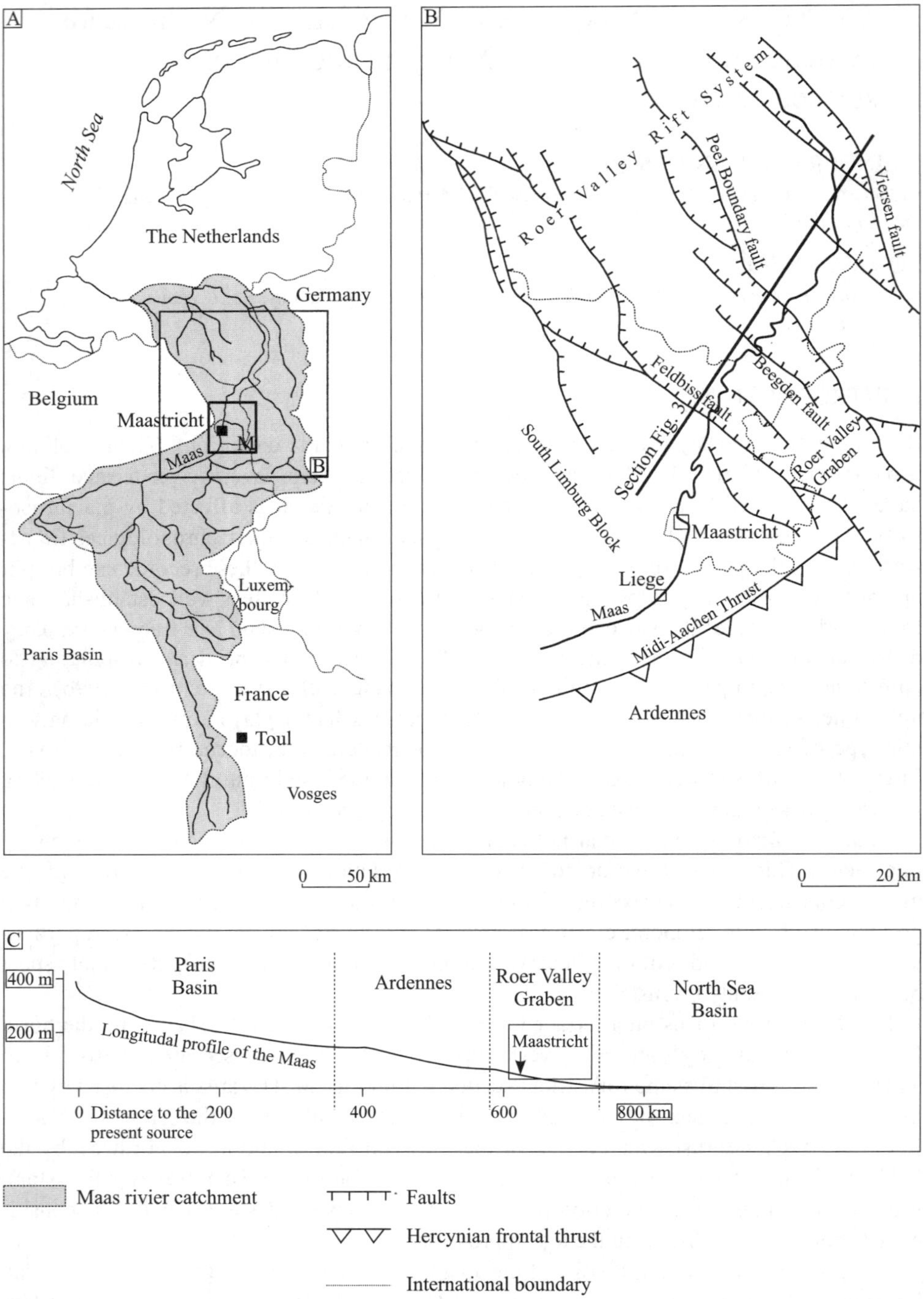

Figure 1. Setting of the Maastricht section: (M in: 1 A) within the catchment of the Maas river and in the structural domain of the Roer Valley rift System.

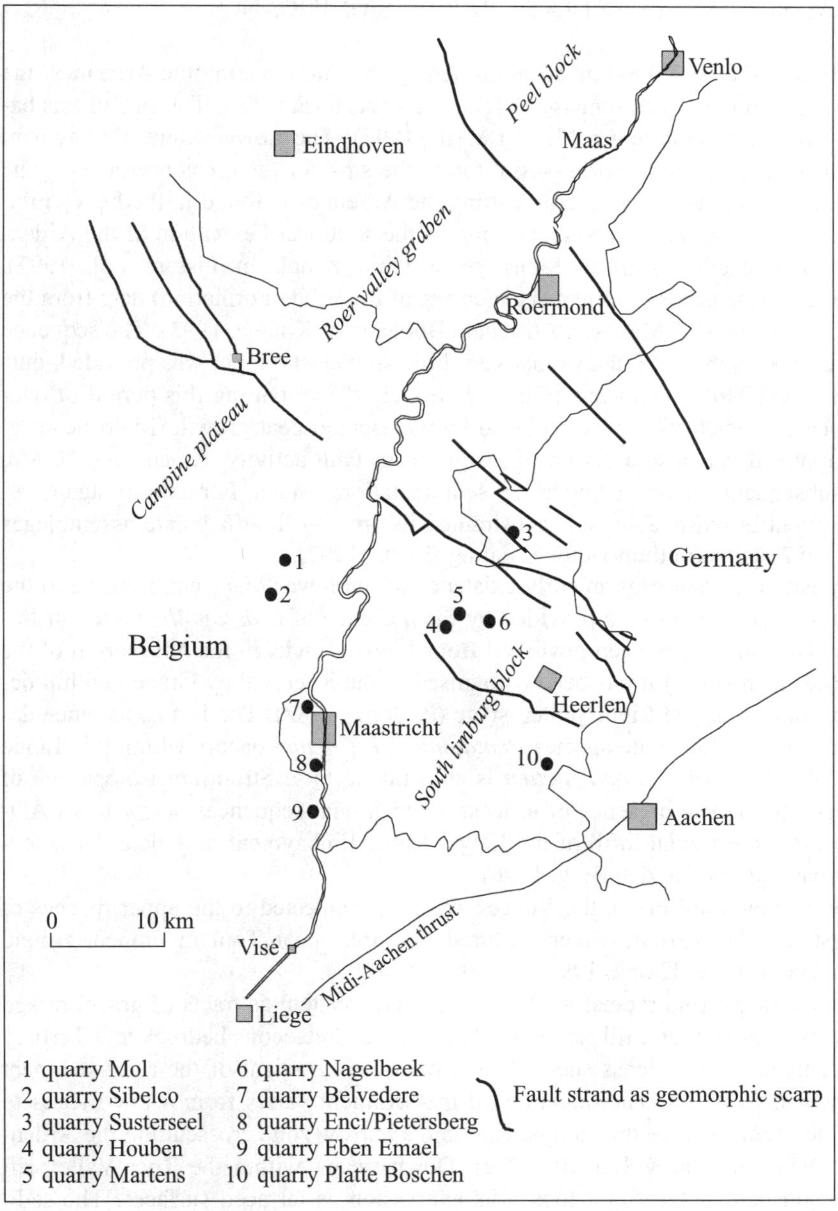

Figure 2. Locations of quarries.

the terrace sediments (palaeomagnetic-, pollen-, TL- and palaeo-sol data) are sparse but well distributed over a significant part of the flight. We will use these conventional chronstratigraphic data to tie the continental record to high-resolution, tuned oxygen-isotope records. We will also use the terrace-superposition to infer a correlation of the terraces which are not directly dated with the isotope record.

2 REGIONAL SETTING AND STRATIGRAPHIC SIGNIFICANCE

The River Maas drains a 33,000 km^2 wide rainfed catchment, covering the Ardennes (the main supply area) and the eastern marges of the Paris Basin (Fig. 1A). The size of this basin evolved through several stages (Pissart et al., 1997). The Lower Rhine Embayment (LRE), as part of the southern North Basin, forms the sink for the erosion products. The earliest evidence of a palaeo- Maas crosscutting the Ardennes is found in the heavy minerals of marine sands. A certain association marks the southward extention of the Ardennes Meuse into the catchment of the Meuse of Lorraine (Krook, In: Pissart et al., 1997). The correlative marine sands (Heksenberg member of the Breda Formation) date from the Early Middle Miocene (~15 Ma) (van Adrichem Boogaert & Kouwe, 1993). The sequence of continental infill in the LRE shows that very little siliciclastic input was provided, during the Early – and Middle Miocene (Gliese & Hager, 1978). During this period of over ten million years, siliciclastics were replaced by widespread peat growth (Main Seam or Hauptflozgruppe). It was also a period of insignificant fault activity. As late as ~10 Ma, significant subsequent fluvial siliciclastic sedimentation (Inden Formation) again resumed. This event is reasonable well constrained in time by dinoflagellate assemblages (Herngreen, 1987) and strontium-isotopes dating (Beets, 1992).

The first positive evidence of an early existence of the lower Maas contributing to the Inden Formation sedimentation is provided by the presence of *Classopollis* pollen in fluvial deposits. This pollen has been reworked from Jurassic rocks in the NE margin of the Paris Basin (Zagwijn, 1961) It has been recognised in the Roer Valley Graben within deposits that are ranked as the Linne pollen-stage (de Jong, 1982). The last appearance datum (LAD) of the dinoflagellate species *Pentadinium Lactictum* occurs within this Linne pollen zone. The LAD of *P. Lacticinctum* is constrained by a Strontium isotope age of 8.5 Ma (Beets, 1992). The presence of at least seven fluvial sequences below this LAD-level suggests that the fluvial infill of the Lower Rhine Embayment may be at least several million years older (Van den Berg, 1996).

At least during the Pleistocene the Vosges was also connected to the upper reaches of the Maas system. The Moselle River captured this area, near Toul in France, around 250 ka. (Zonneveld, 1949; Krook, 1993; Pissart et al., 1997).

In the study area we find remnants of the Maas river system as tracts of gravel mixed with sand, shaped into cut and fill terraces. They overlie Cretaceous bedrock and Tertiary sands and are mantled by a loess sheet. The downstream gradient of the river sediment surface is about 0.75 m/km. The thickness of the sediment varies from a few metres to over 25 m. These units can be traced upstream into a narrow zone crosscutting the Ardennes (Macar, 1954; Juvigné & Renard, 1992). Downstream, within the Roer Valley rift system, the main part of the flight looses its expression as terraced surfaces. The sediments here partly interfinger with Rhine sediments (van den Berg, 1994). After OIS 15 Maas sediments are spread side by side on the tilted Peel high within this rift (Figs 3 and 4). This downstream morphological change is important for the interpretation of crustal uplift being the main driving force for the uplift of the region forming the terrace record, rather than sea-level fall. Otherwise, closer to the shoreline, a similar terraced landscape would be expected.

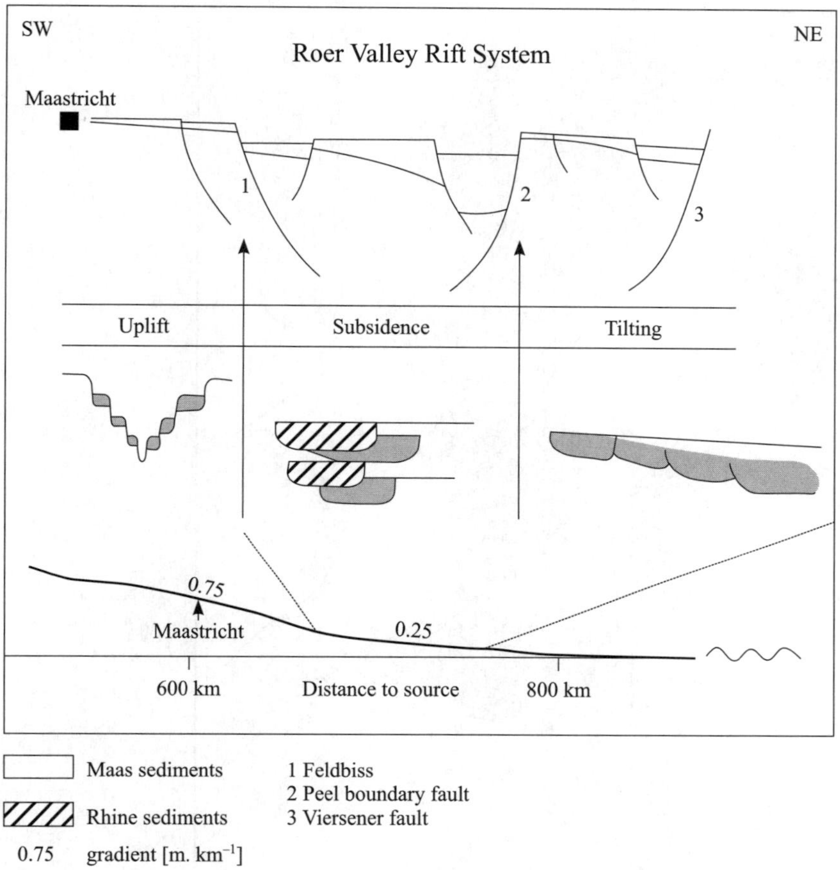

Figure 3. Cartoon of the Roer Valley Rift system. Various modes in tectonic style are reflected in the morpho-sedimentological expression of the lower Maas. The section is drawn parallel to the Maas (see also Fig. 1).

2.1 *Tectonic controls*

The main tectonic factor controlling the distribution of the terraces is the interplay of the compression dynamics of the Ardennes massif and the extensional dynamics of the Roer valley rift system. At a local scale the drainage network reflects a joint system with a characteristic pattern that fits with the above mentioned deformation directions (Van den Berg et al., 1994).

We distinguish three zones (Fig. 4):

1. A zone running SW-NE parallel to the frontal thrust of the Ardennes. This zone contains, in the Netherlands, a palaeo-valley called the 'East Maas'. Towards the SW, downstream of Liège (Belgium), the present Maas river still occupies this zone. This major drainage pathway can be followed northeastwards into Germany (Boenigk, 1978). The northern margin of this palaeo-valley may have formed in response to Ardennes-foreland compression by reactivation of the Hercynian Waubach anticline during the Neogene (van den Berg, 1994).

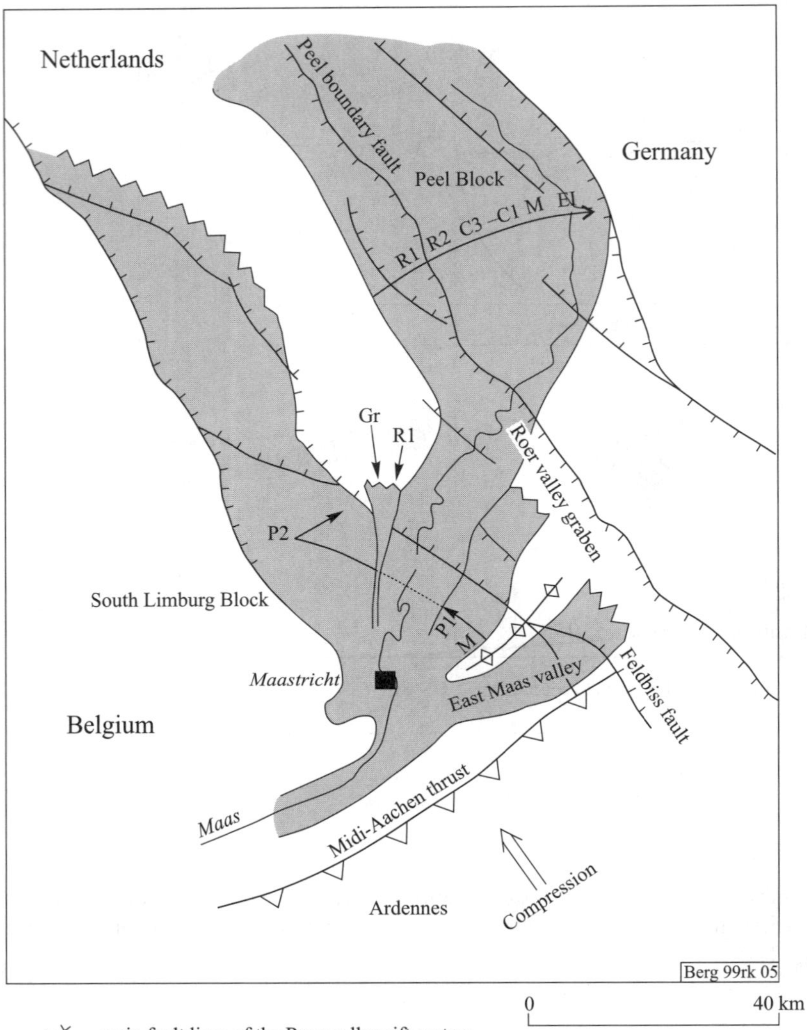

main fault lines of the Roer valley rift system

maas-terrace deposits Interfingering with Rhine sediments

P2 Pietersberg-2 level; for other abbr. see fig. 5

Waubach anticline

Figure 4. Main morpho-tectonic elements of the Lower Maas 'fan', as mentioned in the text. The re-activated Hercynian Waubach Anticline separates the older East Maas valley from the Pleistocene West Maas valley. Arrows indicate migration routes of the channel axis.

2. The West Maas valley. Zone 1 became abandoned in the Early Pleistocene (Kuyl, 1980). Younger sediments were dispersed in a fan-shape northwest of the Waubach anticline, forming the West Maas valley. The distribution of the related terrace remnants on its right bank shows that the West Maas river slipped off the north flank of the Waubach anticline and progressively migrated northwestwards throughout the Pleistocene. The

position of the Maas river during the Pietersberg-2 phase (and that of the Rhine in a slightly younger period), mark the onset of a major paleogeographic change in the region (van den Berg, 1994). During a period of c. 200 ka, the rivers Maas and Rhine migrated far to the west and returned again.

3. After phase 2, a 'gorge'-like valley developed within the West Maas area. This morphology characterises another phase in the uplift history. The 'fan-apex' migrated 70 km in a downstream direction. From three successive terraces stages within this phase only very small patches have been preserved in the course of this incision (Pietersberg-3, Gravenvoeren and Rothem-1; terrace levels P-3, Gr. and R-1, respectively, in Fig. 4). This restricted preservation may be indicative for a period of accelerated uplift.

3 GEOMORPHOLOGIC ANALYSIS

3.1 *Morphology*

The area studied (shown in Fig. 4) is roughly triangular with an apex near the Dutch-Belgium border where the river debouches from the Ardennes. It measures 35 km long and its base is 80 km wide and includes the adjacent graben. The Maastricht section covers a part of this measuring 30 × 35 km (M, in Fig. 1). This smaller area shows a general tilted surface from c. 250 m in the SE, to c. 40 m in the NW.

Periglacial erosion, has heavily affected the terraces, only very few individual terrace levels still form a long continuous tract. The majority of the terraces cap interfluves and plateaus as flat-topped remnants. They become progressively smaller with increasing age. The original terrace morphology is even more camouflaged by a cover of periglacial loess and associated colluvium.

3.2 *Mapping*

We used an extensive dataset of soil-, geomorphological- geological and topographic maps to identify the outlines of individual sediment spreads; in particular the combination of an elevation map with altitudinal information at 1:10,000 scale, and a regular net of boreholes (10 holes per km^2) reaching into the underlying fluvial sediments. Besides the shallow boreholes a less dense grid reaches the base of the fluvial sediments.

The upper and lower bounding surfaces of a typical terrace sediment unit are quite different in nature. The base is strongly undulating with relief up to several metres. The tops of the alluvium vary in general only one metre or so. The tops of the sediments therefore form a less variable- and much better constrained marker of a particular terrace fragment than the bases of the sediments.

Borehole-profiling showed that the loess-surfaces in general conform to the surfaces of the underlying river sediments. This loess blanket is almost devoid of palaeosol sequences and therefore could not be used to date the underlying alluvial terraces as has been done in other areas (e.g. Antoine, 1994; Porter et al., 1992).

The surface elevation map, issued by the Topographical Survey of the Netherlands, is based on third-order levelling data points arranged in a regular grid of 100 m × 100 m. The altitude information is given with 0.1 m accuracy. To find the outlines of the individual terrace remnants, we contoured the loess-surface with 0.5 m intervals. For altitudinal information of the fluvial sediments we relied on borehole data only. Borehole information also helped to discriminate buried river terrace scarps from other terrain irregularities. We scrutinised all interfluves for sub-horizontal parts that may reflect buried fluvial terraces. In this way we found a regular vertical spacing for the West Maas valley levels with steps of about 8 m. The separation of the East Maas terraces levels is more variable (between 4 and 9 m).

The top of the alluvium varies in general less than a metre in central parts of the flat-lying terrain of a particular level. Palaeo-channels within the tops of the terraces are under-represented. Probably many of them developed at a later stage into parts of the secondary drainage system; others may be degraded by being filled with late-stage overbank deposits (Hochflut-lehm) or with products of surface denudation due to periglacial levelling. Based on the lithological composition of bars of the last glacial river, we assume that only the sand-rich upper metre of these bars is prone to periglacial denudation. The underlying gravel most likely prevented substantially deeper areal surface erosion before a pavement develops as a lag-concentrate. The inaccuracy introduced by this method, in the determination of the average height of a particular terrace fragment, is minor compared with the 8 m steps between the various West Maas levels. This makes it unlikely that any terrace levels have been missed.

4 NUMBER OF TERRACE STEPS

The terrace staircase is composed of many terrace fragments that we have reconstructed into terrace levels (Fig. 5). To do this we constructed long profiles through altitudes of river terrace-tops. This (1) allowed a consistent downstream correlation between fragments and (2) provided us with the reconstruction of the total number of steps in the terrace-flight.

On the topographical map we selected flat terrain patches and determined the height of the underlying fluvial gravels. These values are plotted in Figure 5, showing altitude against distance. As a measure for the distance-axis we took the shortest distance to the apex of the fan (for this purpose here taken at the east bank of the present river near the bridge at Visé in Belgium). Single dots in Figure 5 represent small patches (less than 2 km^2) of terrace remnants. Here the highest top of the underlying fluvial sediments was chosen from all available drillhole data. Other buried surfaces can be followed over several kilometres. In those cases we took several height values in the downstream direction. These data are shown as a continuous line. These lines show the average terrace-gradient. It is clear from Figure 5 that the gradients of relatively long, continuous levels are sub-parallel to the gradient of the Holocene Maas (H in Fig. 5). We use this parallelism to attribute certain isolated remnants to a level. These correlations are indicated in the figure by the dashed lines. This procedure worked well for the West Maas river terraces as well as for the lower two levels of the East Maas valley (levels S-1 and S-2), because these remnants are distributed over a sufficient long distance.

In Figure 5 we restricted our investigations to the Dutch part of the Maas terraces because we only had access to a limited number of terrace-base data on the Belgian side of Maas fan. Terrace-base values are not directly compatible with our terrace-surface values. Moreover confusion exists in the literature about the correlation across the border as well as in the downstream correlation due to the different techniques used by various autors. To evaluate these aspects requires a separate paper (see Westaway, this volume).

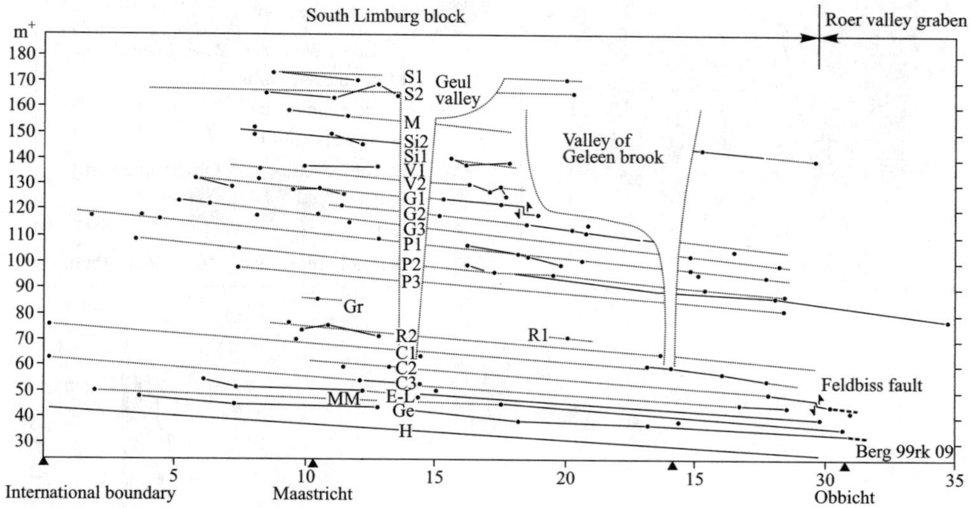

Terrace names

H	Holocene valley floor	C_{1-3}	Caberg	G_{1-3}	Geertruid	S_{1-2}	Simpelveld
Ge	Geistingen	R_{1-2}	Pothem	V_{1-2}	Valkenburg		
MM	Mechelen aan de Maas	Gr	Gravenvoeren	Si_{1-2}	Sibbe		
E-L	Eisden-Lankaar	P_{1-3}	Pietersberg	M	Margraten		

Figure 5. Downstream correlation of 'non eroded' terrace sediment surfaces provides the reconstructed gradients of the West Maas terraces and the lower two East Maas terraces (levels S1 and S2). Dots: highest altitude of Maas gravels of terrace remnants of less than a few square kilometres, drawn lines: longer continuous terraces, pecked lines: interpreted surface. (After: Van den Berg, 1989).

A number of complicating factors prevented us following the West Maas correlation procedure for the remaining higher levels in the East Maas valley. First, the remaining remnants of the East Maas river terrace flight are too sparse along this palaeovalley or their tops are so heavily eroded that no flat surface is preserved. Second, this poor preservation was complicated further by the possibility of large scale karst solution of the chalk underlying the fluvial deposits of the older levels higher up on the interfluves. Third, only fifteen kilometres of this palaeovalley is undisturbed by crosscutting faulting. Long distance correlation was therefore uncertain due to the presence of many crosscutting fault lines that coincide with sidevalleys. In the correlation of fragments across such valleys it is hardly possible to distinguish vertical displacement of a level from originally separated levels. We therefore restricted our survey to the non-faulted transect.

For the West Maas valley we found 21 terrace steps and for the East Maas valley we found 10. The latter number is derived from the correlation of three sections along nearby interfluves (Fig. 6). The 31 levels in the whole flight exceed earlier work by Breuren (1945) who suggested 13 levels in this area. Systematical mapping allowed Felder & Bosch (1989) to extend this number to 20. The difference is not only explained by the increase in knowledge over the years, but also by the method used. As discussed above, our number is based on the terrace tops, whereas the other studies relied on the bases of the sediments. To name the levels we followed as much as possible the names introduced by these earlier workers and extended an original name with a suffix 1, 2, or 3 when a level was further subdivided.

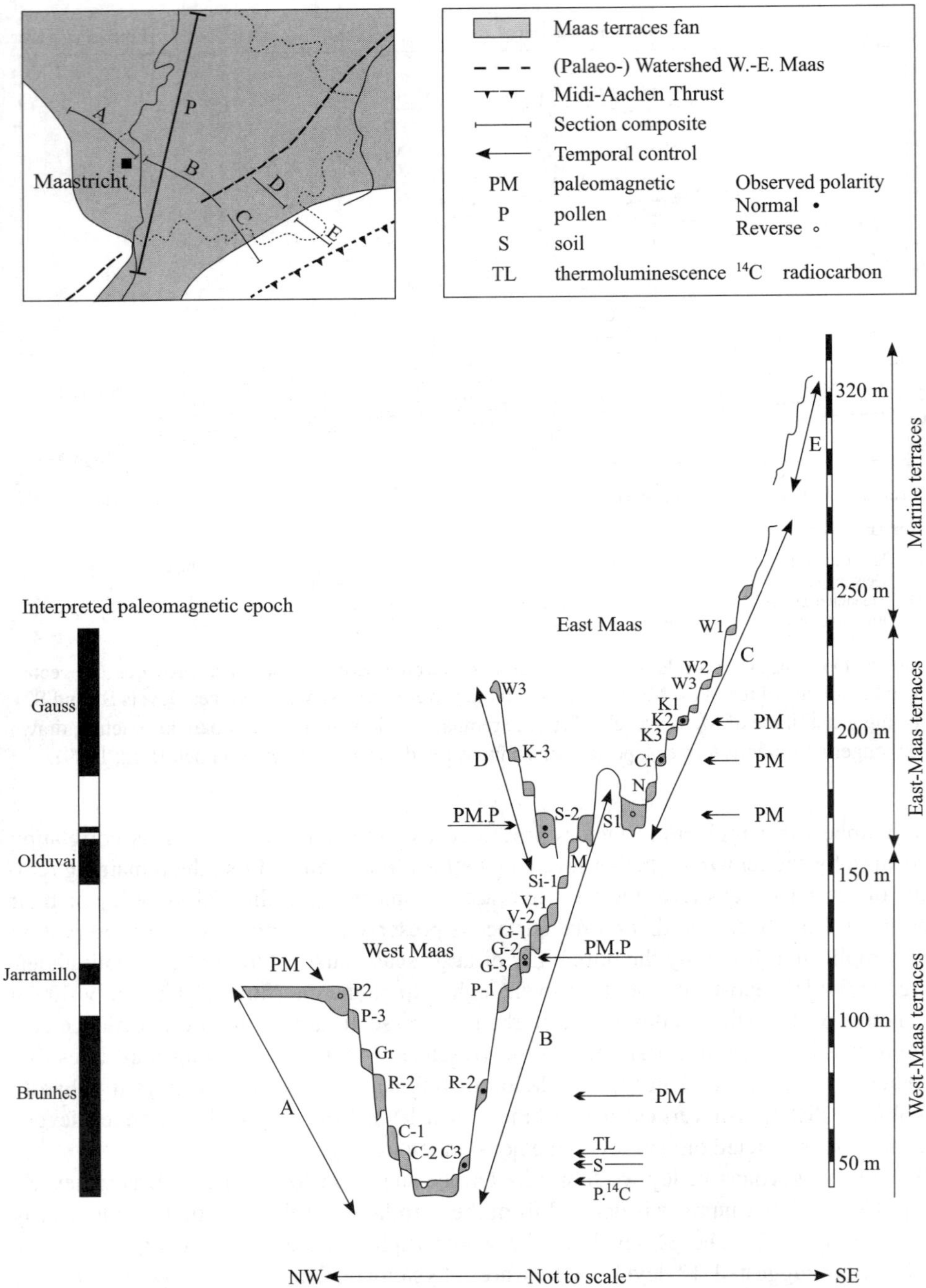

Figure 6. Schematic composite cross section of through the upper part of the Maas terraces fan (some levels have not been indicated due to the choice of the sections. Abbreviations: (e.g. p2) refer to the terrace names).

When all mapped terrace fragments from the sides of the 35 km long section of the lower West Maas valley are assigned to a level, a regularly spaced staircase is apparent. The plan view of the terrace surfaces has been published in Van den Berg, 1989. Felder & Bosch (1989) published a map showing the terrace bases.

5 CHARACTERISTICS OF THE SEDIMENTS.

Many temporary small-scale quarries in the various levels have existed over the years. At present only a few large-scale exposures can be studied. From older descriptions and current exposures, gravel and coarse sands dominate. Clay-beds are rare. This resembles the Scott-type system as defined by Miall (1978) (Ruegg, 1994) (e.g. Fig. 7). From two levels more sandy facies are known. (1): The Winterslag Sands, occupying the western margin of the Campine Plateau (= Pietersberg-2 level) in Belgium (Gullentops, 1974). (2): the Geertruid-2 terrace that contains a gravel rich facies on top of a sand rich facies, each about 12 m thick and with cold-climate indicators. We assume that the currently accessible exposures are representative of all the gravel-rich levels on the basis of the parallellism of the sediment-surface slopes and the information from the many boreholes that penetrated and recovered full terrace sequences. This makes it unlikely that important contrasts in sedimentation are hidden within the un-exposed sediment series.

The most conspicuous element in the internal architecture of the terrace sediments are stacked sedimentary units (parasequences). These are composed of several truncated braided-channel fills. Lag concentrates, characterised by an abundance of boulders, mark the parasequence boundaries (Fig. 7). These lags are directly overlain by a poorly sorted clast-rich package, which grades into a stacked series of laterally accreted bars. Towards the top of each unit, the distal parts of these bars frequently merge into sand wedges or, rarely, into clay bands. A boulder rich bed again overlies the finer top series. The stacking indicates net aggradation. The boulder-rich beds undulate, but are horizontal overall and they show a floodplain-wide persistency. The dominance of gravel lithofacies throughout the lower Maas valley indicates that a coarse bed-load-dominated river with a braided morphology deposited them. This morphology follows from: (1) the absence of fine-grained lateral accretion pointbar sediments, as would be found in single-channel systems. (2) The sheet-like geometry of the gravel/sand units, suggesting a shallow and wide channel plain. In some of the large exposures, low-angle gently dipping accretionary surfaces have been observed. Such an internal bar-architecture can be readily interpreted as braid-bars.

Several associated features point towards cold-climate conditions during the aggradation:

- Angular blocks of loose sands, obviously transported in a frozen state
- Syn-sedimentary ice-wedge casts and involutions (Krook, 1963; VandenBerghe, 1985; 1993). Such ice wedge casts have also been observed at the contact between the underlying bedrock and the fluvial gravel in the Obgrimbie quarry (Pietersberg-2 level).
- Metre-size boulders presumably transported by ice-flows;
- Remains of cold-climate fauna are occasionally preserved (Van Kolfschoten, 1985).

These features have been recognised either individually or in combination, within the units of the staircase from the Kosberg-levels downwards. Indications for sedimentation of river terraces in a permafrost environment are also reported from other NW European rivers. This includes the Rhine in Germany (Brunnacker et al., 1982) the Thames in England,

Figure 7. Part of the Geertruid-2 terrace sequence showing the braided stream alluvium (Scott-type) with typical planar erosion surfaces delineating (four) reactivation phases. The height of the gravelly sands approximately amount to 12m. The fluvial sequence is overlain by a loess mantle and underlain by Miocene marine sands. Quarry Houben in Spaubeek.

(Bryant, 1983); The Allier (Veldkamp, 1991) and Somme (Antoine, 1994) in France and the Ohre in the Czech Republic (Tyracek, 1983).

Within the Lower Maas terraces major, but step-by-step, shifts in the sediment composition occur. Such shifts are found in: gravel composition (Van Straaten, 1946); the heavy-mineral composition and in the bulk geochemical composition of the sand fraction (Zonneveld, 1949; Krook, 1993; Riezebos, 1971; Riezebos et al., 1978; Moura & Kroonenberg, 1990; Slomp, 1990). An important and very characteristic change in the gravel composition occurs around the Kosberg levels in the East Maas valley. Older terraces (the Waubach group) are quartz-rich, and dominated by (ultra-) stable minerals. Younger levels are progressively enriched in fresh rock components indicative of strong periglacial weathering in the drainage basin. The properties of the levels of the Kosberg-Crapoel group take an intermediate position between those of Miocene sediments and those of Pleistocene sediments. Their mineralogy is closely related to the Tertiary sediments (cf. Zonneveld, 1949), their average grain-size is coarser and their quartz/rest-group ratio is more like the Pleistocene terraces.

6 TERRACE CHRONOLOGY

The present interglacial conditions show an incised and shrunken floodplain relative to the floodplain formed during the previous period of net aggradation. The low rate of Holocene lateral channel migration shows that significant widening of the floodplain in this stretch of the system is not likely to occur during interglacials. Therefore the preservation potential of interglacial fluvial sediments in this part of the system is very limited. This is consistent with the absence of a positively identified fluvial deposit of the previous interglacial: the Eemian age (the equivalent of deep-sea stage 5e).

Within the Maastricht section one Late Glacial terrace is preserved: the Geistingen terrace (Ge in Fig. 5) (Paulissen, 1973; Van den Berg, 1989). Downstream of the Feldbiss fault, three Late Glacial levels have instead been mapped (van den Berg, 1996).

The older terraces are considered from their sediment characteristics to be aggradational, deposited during cool or cold climatic conditions. In particular the last net aggradation period is well constrained in time to the last glacial. This is based on the combined evidence of pollen, sedimentology, morphology and ^{14}C dates (see below). Net aggradation occurred between the onset of OIS 5-d through to the end of stage 2 (i.e., between about 110 ka and 16 ka) (Van den Berg, 1996, Chapter 5). So the stage 4 and 3 river phases are included into one prominent morpho-stratigraphic unit: the Mechelen aan de Maas terrace (MM in Fig. 5).

The last interglacial has been recognised in a palaeosol capping the next higher terrace (the Eisden- Lanklaar terrace). The next terrace above (Caberg-3 level) is also capped by interglacial deposits which date from the Belvédère interglacial (Kolfschoten et al., 1993; Vandenberghe, 1993). Thermoluminescence dating of associated burned artefacts revealed an age of 250 +/– 20 ka (Huxtable, 1985, 1993), suggesting time-equivalence with oxygen isotope stage 7c.

The morpho-stratigraphic record of the last three aggradational terraces shows that for the long-term fluvial-landscape evolution of the last 300 ka we may apply the assumption that aggradation matches cold-climatic conditions and erosion, and steady-state conditions occurs during interglacials. Subtle refinements are possible in the sequence of events leading to the end forms (Vandenberghe, 1993), but they do not alter the straightforward response of the fluvial system to the climatic evolution on the longer time scale. This conclusion forms the base for a further assessment of the chronology of our long record.

To support the age assignments of the older levels in the flight, we made use of the combined evidence provided by superposition, pollen, and palaeomagnetic determinations, in combination with arguments derived from lithological composition and palaeosols.

6.1 *Pollen*

Over the last fifty years several researchers have attempted to extract pollen from exposed clay layers in Maas terraces. These clays rarely contain pollen (A. Brouwer, pers. comm.). Even the fauna-rich interglacial deposits at the top of the Caberg-3 terrace (Belvédère interglacial) did not preserve any pollen. Two exceptions form palaeo-channel infills in the Early Pleistocene St. Geertruid-2 and Simpelveld-1 terraces. The local setting of these exceptions has not provided us with a clue as to why pollen was preserved. The gradual change from gravelly sands into the clays suggests a normal fining-upward sedimentary sequence. We therefore consider the pollen bearing strata to be significant for an age approximation of the underlying sedimentary sequence.

6.1.1 *St. Geertruid-2 terrace:*

In a brickyard quarry near Susterseel (at present just across the border with Germany), in the hanging wall of the Feldbiss fault, Maas sediments underlie Rhine sediments of the Weert heavy-mineral zone. Between these is an 8 m thick succession of clay and loess (Fig. 8). This unit contains at least four palaeosols as indicated by horizons with clay-illuviation (Bruins, 1981). The lowest palaeosol is a pollen bearing, humic clay band about 0.6-0.8 m thick. According to Zagwijn & de Jong (1983), this represents the Bavel 3-5 pollenzones, formed in the later part of the Bavel interglacial (OIS 31).

6.1.2 *Simpelveld-1 terrace:*

Gravel of the Simpelveld-1 terrace is overlain by about seven metres of a loam (in its lower part, sandy-loam) at Platte Boschen quarry near Simpelveld (PBS) (Fig. 9). Between the sandy-loam and the loam at about 172 m altitude (4 m below the surface), a 1.2-1.5 m thick horizon with pollen-bearing peaty clays is intercalated. The lithological succession in this quarry suggests to us that the organic layer is the top of a fining-upwards sequence starting with the underlying gravel. Sedimentary structures as well as the grain size of the upper loam point to a reworking of aeolian loess. Pollen assemblage from the organic layer represents a cool-temperate forest phase of a Lower Pleistocene interstadial or a cooler part of an interglacial (De Jong & Zagwijn, 1963; Kuyl, 1980, Fig 44). Zagwijn, (1974) attributed this assemblage to the Tiglian C-6-zone. The diagram indeed strongly resembles the Tiglian C-6 zone as published by Zagwijn (1963; enclosure 6) from one of the Tiglian type sections: the claypit Russel-Tiglia-Egypte at Tegelen (60 km north of our site). Differences do occur in conifer pollen: *Tsuga* is rare at Tegelen, but contignuously present in low values in the PBS diagram; *Picea* reaches values of > 40% in PBS, this in contast with Tegelen where it is much lower (between 10% and < 30%). The associated palaeomagnetic data (see below) suggests to us that the PBS diagram is older and probably part of the Tiglian B.

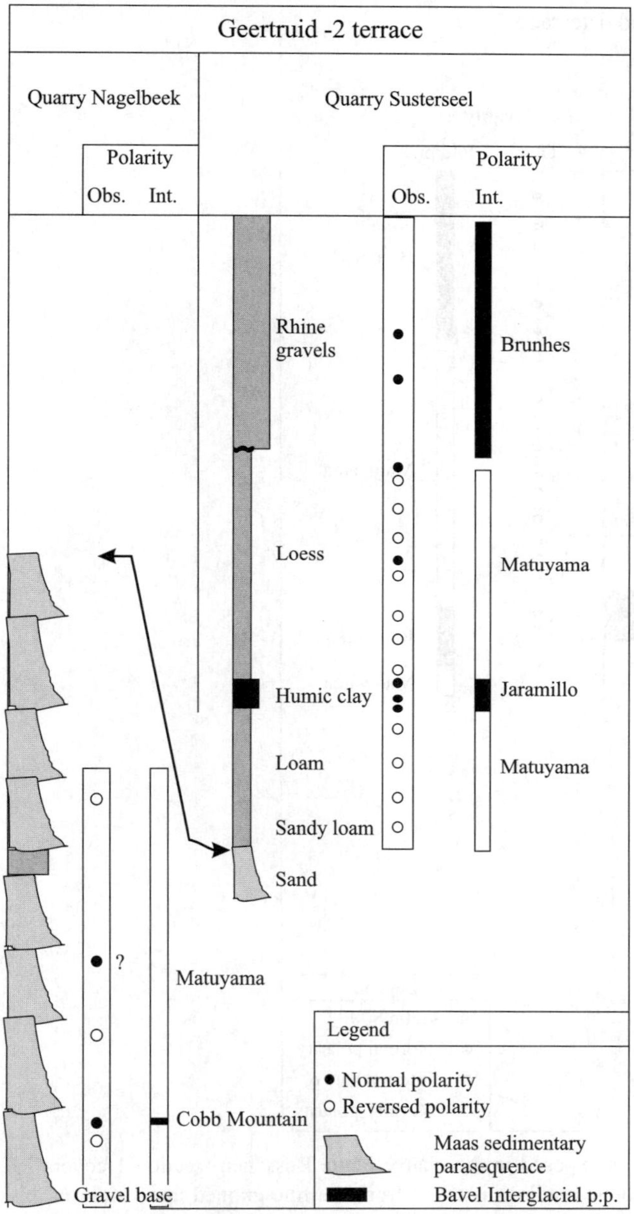

Figure 8. Several outcrops from the Geertruid-2 terrace have been sampled for palaeomagnetic dating. The quarries: Nagelbeek, Mertens, Houben, provided a stacked series of sands and gravels. Clay beds at the top of these sequences mainly provided reversals, except two levels in the lower part of quarry Nagelbeek, the lowest shows a well expressed normal demagnatisation. This is interpreted as a short normal-polarity event: from its position possibly the Cobb Mountain event. From the top of the third sequence 6 samples varied between normal and intermediate directions, we ignored this observation. In Susterseel the top sequence only was exposed, showing a fully developed fining-up sequence. This provided the upper part of the Matuyama, including the Jaramillo. The overlying Rhine gravels belong to the Weert heavy mineral zone deposited during the Brunhes. (Susterseel exposure length amount 15 m. Exposure Nagelbeek amount 14 m.).

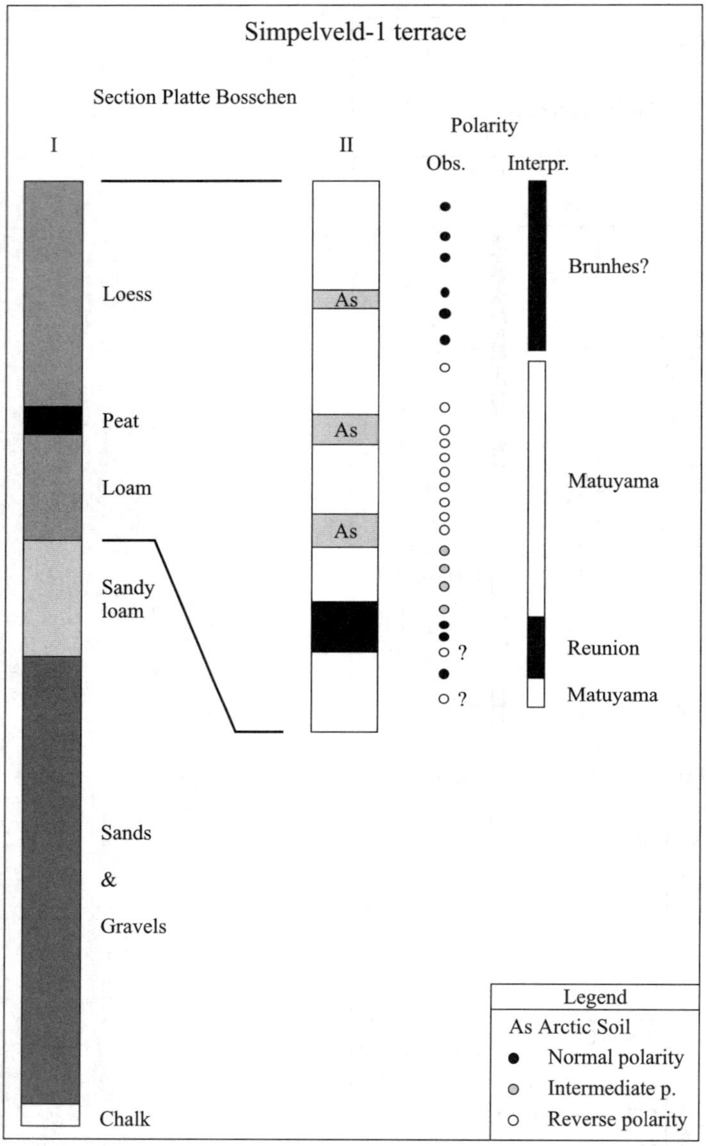

Figure 9. The Simpelveld-1 terrace is exposed in the quarry Platte Bosschen (section 1 covers 17 m). A long sequence of palaeomagnetic readings is taken from the fine-grained upper 4.85 m, including the peat bed with a Tiglian pollen association. (1) P. Bosch pers. comm.; (2) Bruins (1981).

6.2 *Palaeomagnetism*

The ages of reversals of the geomagnetic field are quite well known from astronomically-tuned oxygen isotope records (e.g. Shackleton et al., 1990; Hilgen, 1991); consequently, a geomagnetic polarity time scale can be established. For magnetostratigraphic dating the polarity pattern of a sequence of rocks is compared with this geomagnetic polarity time scale. It is therefore required that the sequence of rocks is deposited with a continuous rate of sedimentation, like deep-sea sediments. Fitting of obtained polarity patterns in fluviatile sediments to the standard geomagnetic polarity time scale is very difficult because: (1) the sequences cover a too brief a part of the geological time; (2) in fluviatile deposits hiatuses in sedimentation are, of course, inevitable and (3) only the lithologies with high clay content are suitable for palaeomagnetic purposes, and these occur only rarely, usually in very small lenses intercalated in the coarse deposits. However, the magnetic polarity of the natural remanent magnetisation (NRM) in the clay lenses can confirm age estimates determined by correlating the terraces with the cold stages in the ODP site 677 $\delta^{18}O$ record (Shackleton et al., 1990). For this purpose we have used two sets of data: first, we have reinterpreted the palaeomagnetic data by Bruins (1981), who studied several Maas terraces, and, secondly, an additional set of data, collected in April 1991, are analysed after sampling and measuring. All together 10 levels out of the 30 have been accessible.

6.2.1 *Data collection*

Fresh exposures of clay beds were sampled by pushing cylinders into the outcrop. After orientation, the cylinders were cut out of the outcrop. The remanence of the samples was measured with a cryogenic magnetometer at Utrecht-University. Bruins (1981) performed paleomagnetic measurements using progressive thermal (TH) or alternating field (AF) demagnetisations. Before the AF demagnetisations Bruins sometimes heated the samples to between 110° and 170° C, to remove the high water content present in the samples. Because the TH demagnetisations generally give better results than AF demagnetisations, we only performed TH demagnetisations, starting with 100°C and followed by steps of 50°C.

6.2.2 *Results*

About half of the paleomagnetic directions, from the 10 terrace levels, show intermediate directions. Intermediate directions of the geomagnetic field occur very rare in time and therefore intermediate paleomagnetic directions from samples taken randomly in rocks should also be observed exceptionally. Hence, it is statistically very unlikely that all intermediate directions represent the geomagnetic field. The intermediate directions that are close to Normal (N) or Reversed (R) (cf. Figs 10: STPI 116A, and PLB21A) were accepted as N or R, respectively. Other intermediate magnetic directions were considered unreliable.

N results are either secondary or primary. Discriminating between the two is difficult for two reasons: (1) in young rocks the primary N direction is very close to the present day field direction; (2) in sediments more commonly used for magnetostratigraphic studies, secondary directions are easily detected because they are generally bound to hematite and goethite. Because in fluviatile sediments hematite and goethite may also be detrital, these minerals can carry secondary as well as primary directions. These considerations mean that the observed N polarities can only be correlated with the geomagnetic time scale with great caution. The comparison between the expected and observed polarities is presented in stratigraphical order in Table 1.

We regard R magnetisations as more reliable. Only three terrace levels display those (Table 1): Pietersberg-2, Geertruid-2, and Simpelveld-1.

6.2.3 *Pietersberg-2 terrace*

Sediments of this level were exposed in three quarries: Eben-Emael, Mol and Sibelco, all on the Belgian side of the present Maas. Mol and Sibelco are only a few hundred metres

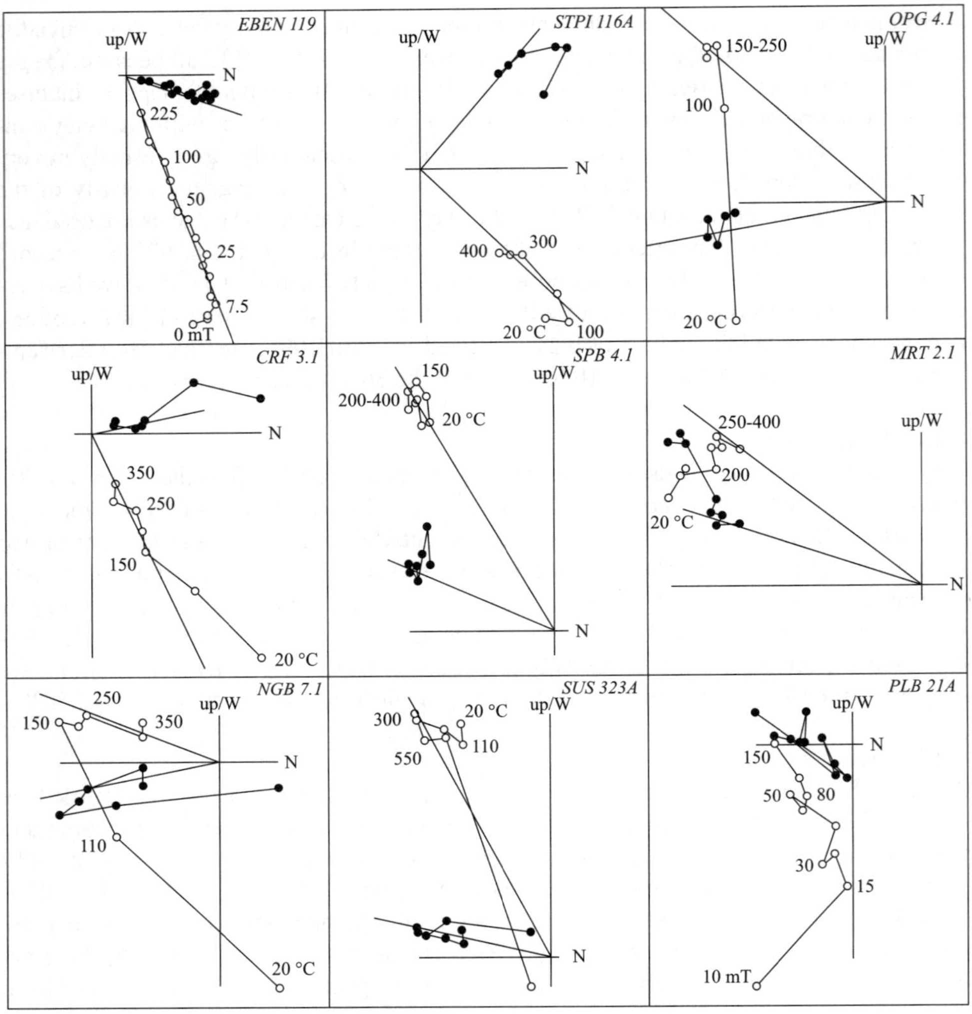

Figure 10. Examples of stepwise thermal and alternating field demagnetisation diagrams of representative specimens of various terrace levels. Numbers denote demagnetisation steps in degrees Celsius. Direction and intensity of magnetisation during demagnetisation were analysed from orthogonal projection plots (Zijderveld, 1967) and principle component analyses. Full dots indicate projections on the horizontal plane, circles indicate projections on a vertical plane. All original data both by Bruins, (1981) and this study are stored in the files of the Paleomagnetic Laboratory 'Fort Hoofddijk', Fac. of Earth Sc., Utrecht University, the Netherlands.
Abbreviations refer to the various quarries. Eben = quarry Eben Emael: Pietersberg-2 level. STPI = quarry ENCI: Pietersberg-3 level. OPG = quarry Sibelco near Obgrimbie: Pietersberg-2 level. CRF = quarry Curfs: Pietersberg-1. SPB = quarry Houben, Spaubeek: Geertruid-2 level. MRT = quarry Mertens: Geertruid-2 level. Ngb = Quarry Nagelbeek: Geertruid-2 level. SUS = quarry Susterseel: Geertruid-2 level. PLB = quarry Platte Bosschen: Simpelveld-1 level.

Table 1. Sample sites for palaeomagnetic demagnitisations of Maas terraces with observed polarity and interpreted chron. Italics denotes results from Bruins (1981). Interpreted isotope stages refer to the stage where terrace aggradation ended. Stage numbers: ODP sites 677 and 846.

site	terrace	observed polarity	interpr. chron	interpr. isotope stage
Amby	Rothem-2	N	Brunhes	14
ENCI	Pietersberg-3	N	Matuyama	20
Eben Emael	Pietersberg-2	N?	Matuyama	22
Sibelco	Pietersberg-2	N? + R	Matuyama	22
Kruisberg	Pietersberg-1	N	Jaramillo	24
Susterseel	top Geertruid-2	R-N-R	Matuyama + Jaramillo	31
Nagelbeek	base Geertruid-2	R (+N)	Matuyama (+ Cobb M.)	34
Mertens	base Geertruid-2	R	Matuyama	34
Spaubeek	base Geertruid-2	R	Matuyama	34
Rode Put	Simpelveld-2	N	Matuyama	78
Platte Bosschen	Simpelveld-1	R-N	Matuyama + Reunion	82
Reymerstok	Simpelveld-1	N	Matuyama	82
Scheiberger bos	Crapoel	N	Gauss	G4
Crapoel	Crapoel	N	Gauss	G4
Banholt	Kosberg-3	N	Gauss	G10
Landsrade	Kosberg-2	N	Gauss	G16

apart, whereas Eben lies about 18 km upstream from Sibelco. The section from quarry Sibelco (Fig. 2, Fig. 11) has N-R sequence with a clear reversal and with a questionable normal demagnetisation, while samples from Eben-Emaël section show weak N directions. All samples were taken from the middle to lower part of the sedimentary sequence.

6.2.4 *Geertruid-2 terrace*

Various localities gave access to this level (Fig. 2, Fig. 8, Table 1). The fine-grained top unit has been sampled in the Süsterseel quarry; the coarse grained main unit has been sampled in the Spaubeek (Houben), Martens and Nagelbeek quarries.

The Süsterseel sediments display a R-N-R sequence. The N part coincides with a layer in which the abundant pollen marks the Bavel interglacial (see above). The observed N polarity within the Bavel interglacial is in line with previous results by van Montfrans (1977) and Kasse (1988) in Rhine sediments. We interpret the R to N sequence as the lower Jaramillo boundary followed by an hiatus. The latter is here expressed by a major change in lithology. The subsequent loess layer bears the upper R polarity.

The basal, sand-rich, part of Geertruid-2 showed mainly R directions and a few N directions. These N directions could represent the Cobb Mountain event (Fig. 8).

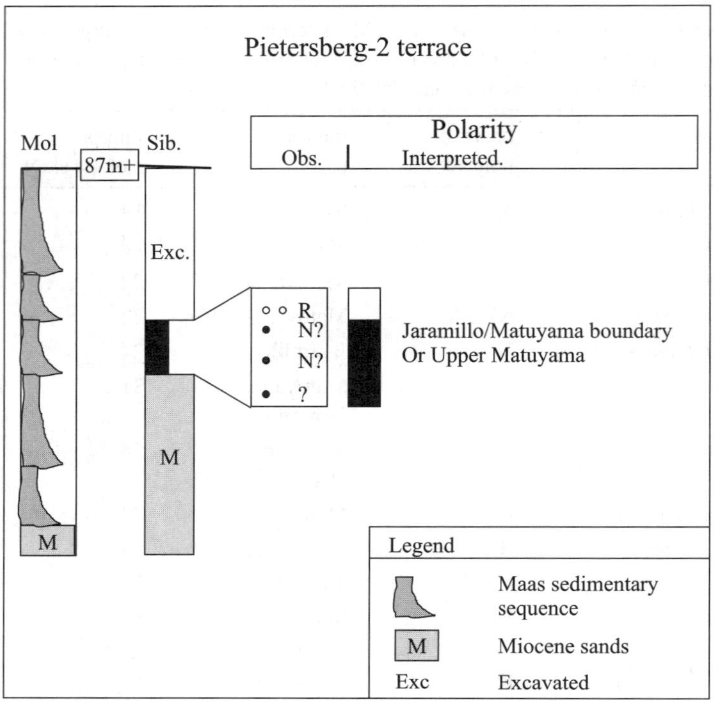

Figure 11. Two adjacent outcrops in the Campine Plateau were available from the Pietersberg-2 terrace. The quarry Mol showed 5 fining upward sedimentary sequences. In the old quarry Sibelco a 3.5 m thick channel fill at the base of the terrace sequence was left after excavation. We sampled from this clay unit our polarity sequence.

6.2.5 *Simpelveld-1 terrace*

The polarity data from the Platte Bosschen section (Fig. 9), come from the upper 4.85 m fine-grained series of this terrace. (Bruins, 1981). They show from bottom to top 0.8 m normal (N) polarity, followed by 0.75m with intermediate (I) values, in turn overlain by reversed R values. The N-I transition falls in the middle of the peat layer (0.66m-1.80m). This peat layer separates the underlying fluvial deposits from the overlying loess layers with arctic soils with a reversed polarity. If the peat layer represented the Late Tiglian C6 pollenzone, than the normal polarity should continue through the overlying arctic soil (e.g. the climate sequence of the upper part of the Olduvai in Fig. 12). There is no reversed interval within the Olduvai thus any such observations are due to a secondary overprint due to chemical reactions between iron and organics, (pers comm. C.G. Langereis). In our opinion the observed polarity in the PBS sequence fits better with the Reunion event, because this event is much shorter than the Olduvai and the next cold stage (= stage 78) is already reversed. This correlation is also compatible with the absence of any sign of a warm interval between the 'cold' gravels and the peat. as would be expected if the pollen assemblage represented Tiglian C 6 (Fig. 12).

In summary: the limited value of magnetostratigraphy in this type of sediments is caused by the difficulties in discriminating primary from secondary magnetizations. Based on the

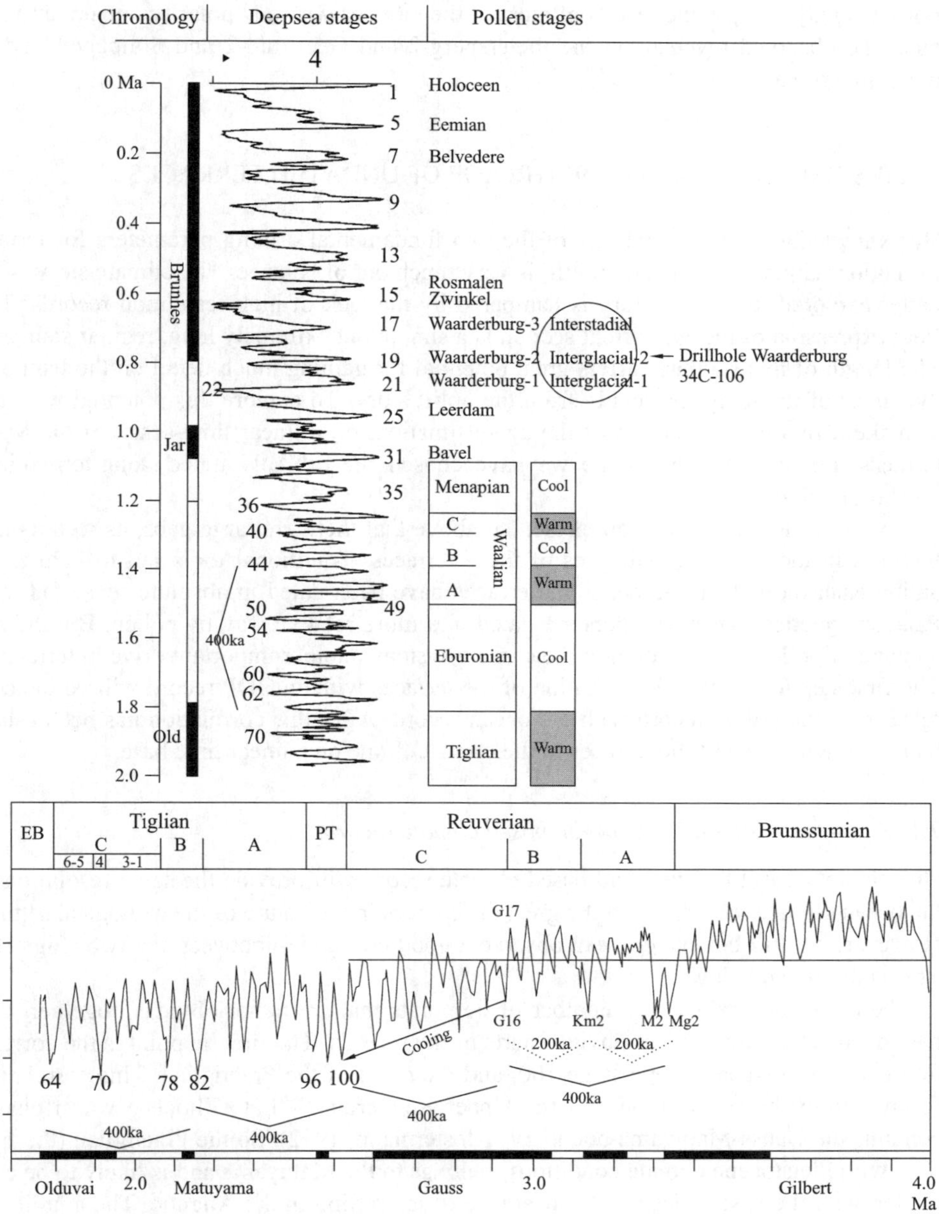

Figure 12. Proposed corrrelation between the pollen-based land climate record (modified Zagwijn, 1998) and the ocean based signal from ODP Site 677 (Shackleton et al., 1991). The upper tie point for this correlation is the position of the Brunhes/Matuyama boundary; the lower boundary is taken at the correlation between pollen stage Tiglian C4c and marine stage 70 within the Olduvai. For discussion see text. Note: the Belvedere interglacial is not defined by pollen but by faunal remains. Abbreviations: EB = Eburonian; PT = Praetiglian.

polarity analysis only, the results of most of the terraces with a N polarity remain ambiguous. The observed reversals in the Pietersberg-2 and Geertruid-2 and Simpelveld-1 terraces are critical.

7 FIRST APPROXIMATION OF THE AGE OF UNDATED TERRACES

Our knowledge on the variability of the two fundamental driving parameters for terrace formation: climate-change and uplift, is very much out of balance. The climate story is far better explored; the uplift story is hampered by the lack of high-resolution records. The field expression of the Maastricht section is a simple but extremely long, regular staircase. The length of this staircase offers great potential for gaining much detail on the temporal dynamics of terracing, especially from the uplift story. To explore this potential we want to make a first approximation of the age-estimation, on a linear time-scale, of the Maas terraces. For a linear time-scale we have chosen the orbitally tuned, long-term high-resolution ODP records.

It will be clear from the data presented above that there are large gaps, as well as differences in the dating possibilities of these terraces. Traditional tools fail to help us in dating each individual terrace, some terraces have been dated in absolute sense (TL and Palaeomagnetics) whereas others are dated in a more relative way by pollen. But the advantage of pollen is its relation to the climate system, another important drive in terracing. The first step to an overall correlation of the terraces with an ODP record will be the correlation of the pollen record with the ocean record. After this correlation has been established we will try to fit the terraces in the climate frame on a linear time base.

7.1 *Correlation of land- and ocean based climate records*

One cannot expect that the land-based climate record will provide the same resolution as the ocean climate records given the great differences in the nature of the geological setting. However, we still believe one can observe a good correlation between the two (Figs 12). This is discussed below.

The correlation relies on a number of tie points that are rather obvious, together with pattern-matching for the less obvious parts of the records. The first tie point is the correlation between oxygen isotope stage 100, and the onset of the Praetiglian. This correlation follows from the observation that the Upper Reuverian (= Late Pliocene warm phase) contains the Gauss-Matuyama boundary (Klostermann, 1992). So the Praetiglian (the first time with Pleistocene climate conditions) belongs to the Matuyama and is likely to be correlated with the first widespread appearance of ice-rafting in the Atlantic. The amplitudinal variations in the ODP 677 record suggest that such conditions lasted during the stages 100, 98 and 96. We group these stages together into the Praetiglian equivalent. The Praetiglian pollen zone is restricted to a cool assemblage so we correlate its upper boundary with the transition of stage 96 to 95 because younger stages show warmer conditions.

The Tiglian, forms an episode dominated by relative warmth, although some cooler intervals are included (see below). The sedimentary record shows that the Tiglian spans a long period of time. The lowest part of the following Eburonian, another long period but dominated by cool episodes, shows a normal polarity and falls within the Olduvai (van Montfrans, 1971). Within the Olduvai, the stages 68 and 62 show the highest $\delta^{18}O$ values.

We tentatively correlate the younger of the two with the onset of the Eburonian while this stage forms the onset of a 300 ka long period dominated by cooler conditions.

Zagwijn observed two important cooler intervals between the onset of the Tiglian and the onset of the Eburonian: the Tiglian B and the Tiglian-C4c. In the Eindhoven II borehole (located in a depocentre in the Roer valley graben) the major Tiglian pollen zones are represented (Zagwijn, 1963). The Tiglian-B falls in the middle of this sedimentary record whereas the Tiglian-C4c zone is recognised close to the overlying Eburonian. Bearing this in mind, we suggest a correlation between the Tiglian-B and the stages 78 though 82 (Fig. 12). Subsequently, we believe deep-sea stage 70 to be the most likely candidate to correlate with the cold Tiglian-C4c pollen stage. From this it follows that the pollen zones Tiglian C5-6 are represented by the stages 69-65. The record of ODP site 846 shows that these stages can be separated into a warmer older part and a relative cooler younger part. A similar expression is shown by the pollen sequence. This is important for our correlations because from this it follows that the cooler Tiglian C-6 zone falls within the upper Olduvai and is therefore older than stage 64.

The Bavel interglacial serves as the next tie point. This interglacial contains the lower Jaramillo (Van Montfrans, 1971; Bruins, 1981; Zagwijn & de Jong, 1983). This interglacial presumably forms the continental equivalent of oxygen isotope stage 31. This correlation is in contrast to Zagwijn, 1996 who suggested a correlation with oxygen isotope stage 25, but he did not consider the palaeomagnetic evidence at that time (pers. comm.).

The continental climate record of the Lower Pleistocene shows three major subdivisions. Longterm cool-warm-cool episodes separate the Tiglian from the Bavelian, respectively the Eburonian (= cool), the Waalian (= warm) and the Menapian (= cool) (Fig. 12). There are no additional constraints other than these long-term patterns in conjunction with the variations within the main pollen stages. We simply correlated on the bases of pattern matching, but relied heavily on the typical short-term variations known from the pollen record.

The last tie point is the Brunhes-Matyama boundary. This boundary has been identified within the Dutch stratigraphic record between the sediments of Cromerian interglacials I and -II (Zagwijn et al., 1971). Various high-resolution deep-sea cores give apparent different positions with respect to isotope stage for the Brunhes/Matuyama polarity boundary (for a review see: Langereis et al., 1997). This position shifts between the stages 20 and 19. The most reliable estimates place the boundary within stage 19.2.

Glacial advances during the Brunhes, or their correlative outwash, did not reach far enough south to serve as an additional tie to the staircase.

We used these tie points to correlate the sequence of pollen stages with the records of ODP site 677 and ODP site 846. These have been tuned to the frequencies of the orbital parametres (Shackleton et al., 1990; Shackleton, 1995). The correlation therefore provides us with a quantitative linear time frame for the qualitative pollen-based time scale.

Moreover we identified, by visual inspection, an overprint of a 400 ka cycle on the 41 ka obliquity signal. The latter dominates the Lower Pleistocene (Ruddimann, 1989; Mudelsee & Stattegger, 1997). This observation emerged from comparing the palaeotemperature curve for northwest Europe, as drawn by Zagwijn (1998), with oxygen isotope records (Fig. 12). The following sequence of Early Pleistocene relative deep cooling events: the Praetiglian, the Tiglian-B, the onset of the Eburonian and the Waalian- B. are separated by 400 ka. These cycles probably correspond to the supercycles as identified by Kukla & Cilek, (1996.) i.e. respectively their cycles MC-10 and MC-9, and the boundary between MC-6/7. Remarkable shifts in the oxygen isotope values apparently occur with a 400 ka cycle, even before the Praetiglien e.g. the shift from G17 to G16 forms the onset of significant cooler Pliocene conditions during the Gauss, culminating in the Praetiglian. Remarkably high values are reached in the earlier warm Pliocene during stages KM2 and the cluster M2/MG2; two 200 ka cycles. This observation is important because these 400 ka (or sometimes 200 ka) cycles apparently force some groups of 41 ka cycles to cooler conditions than others, which is expressed by vegetation and probably also by river discharge variations on the continent. We will use this additional aspect in our correlations.

7.2 *The Maastricht section*

Three portions of the Maastricht section are relatively well-constrained in time within the period spanning from slightly older than the Cobb-Mountain subchron (1.19 Ma) to the present:

- The aggradation period of the Geertruid-2 terrace ended before the interglacial of oxygen isotope stage 31. We believe we have identified the Cobb Mountain subchron in its lower parasequences. Therefore the whole aggradation period for this terrace spans the interval from stage 36 to stage 31.
- The sediment suite of the Pieterberg-2 terrace is composed of several fining upward sequences: (5, in quarry Mol, Fig. 11). Similar parasequences, which together compose a full glacial terrace, could be correlated with stadial-interstadial cycles within the last glacial cycle (van den Berg, 1996). This suggests for the Pietersberg-2 terrace that its formation lasted several stadial-interstadial cycles. Moreover it covers a very large area (the Campine plateau), much larger than any of the other terraces. Such characteristics together, suggest that its formation must have taken a substantial period of time. The available polarity data can be interpreted in various ways: Matuyama + top Jaramillo; or Brunhes with a reversal excursion. We prefer the Jaramillo/Matuyama boundary with the option that the top half of the sedimentary sequence belongs to the upper Matuyama. Out of the available cold stages within the Upper Matuyama, the duration of stage 22 is about three times longer than any other oxygen isotope stage. Oxygen isotope stage 22 therefore seems the most likely candidate. Stage 21 finished this aggradation phase. Stage 21 correlates with the first Cromerian interglacial (= Interglacial-I according to Zagwijn et al., 1971; Zagwijn, 1996), see below.
- The Caberg-3 correlates with OIS 8 while it is capped by the Belvedere interglacial (equated with OIS 7c).
- Eisden-Lanklaar and Mechelen aan de Maas terraces correlate respectively with stages: 6 and 5b through 2. (see above).

The latter rhythm matches the 100 ka frequency, which holds until stage 17, (Mudelsee & Stattegger, 1997) so for our correlation we simply count back: Caberg-2 = stage 10; Caberg-1 = stage 12; Rothem-2 = stage 14; Rothem-1 = stage16.

The OIS 15 interglacial takes an intermediate position between the younger and the older stages. It is double peaked, with a marked cool episode in between. Whether it is likely that a terrace remained from this cool interval, forms an important question for the correlation of terraces with ODP records. To tackle this question we incorporate the sedimentary record preserved within the adjacent Roer Valley graben (RVG), because the correlative sediments of the terraces of the Maastricht section are found here in a stacked position and separated by interglacials or temperate phases (Fig. 3). In total, four Middle Pleistocene temperate phases are recognised in the RVG, of which three occur in the Waardenburg borehole (39C-106). This borehole serves as a reference section (Zagwijn et al., 1971), because it contains the B/M polarity boundary in the lower part of this section. From base to the top, these three are here referred to as Waardenburg-1, Waardenburg-2 and Waardenburg-3. Waardenburg-3 has an interstadial character (Zagwijn et al., 1971). Waardenburg-1 shows a reversed polarity, the younger ones have a normal polarity. Thus the B/M polarity boundary is observed in between the Waardenburg-1 and -2 interglacials. This boundary allows us, by upward counting, to correlate the interglacials with the OIS record. We correlate the three interglacials respectively with the OIS stages 21-(B/M at the base of) 19-17. The Waardenburg-3 interglacial is overlain by Maas deposits of a glacial-interglacial cycle. The latter is known as the Rosmalen/Zwinkel Interglacial. It is found over large areas in this graben. (Zagwijn, 1996). The stratigraphic position of this interglacial suggests a correlation with OIS 15.

Previously a sand unit of subdued thickness (~2-3 m) was not recognised between the clay layers containing respectively the Rosmalen and Zwinkel interglacials. The pollen records of this sand unit show a cooler episode. This combined sedimentary- and pollen sequence strongly resembles the double-peaked expression of stage 15. A circumstantial confirmation of our above established correlation. The sediments of this cooler unit between the Rosmalen and Zwinkel interglacial are much finer and their thickness is subdued with respect to thickness of full glacial terraces (~10 m). We use these observations as an argument for not assigning a separate terrace within stage 15.

We now return to the Maastricht section where a correlation of the levels Gravenvoeren and Pietersberg-3 remains to be argued because above we have already established the correlation of Pietersberg-2 with OIS 22. The main problem for the remaining period is the change in the frequency mode of climate-fluctuation over the time-span between the stages 22 and 17. ('the so called: mid-Pleistocene climate revolution' of Mudelsee & Stattegger, 1997; Raymo, Oppo & Curry, 1997). This implies that the simple rhythm applied in the correlation above does not hold any more. Three stages: 18-2, 18-4 and 20 represent cold episodes. The amplitude of the stage 19 suggests that this represent a full interglacial (above correlated with the Waardenburg-2 (= Westerhoven) interglacial) and therefore an important erosion phase. We tentatively correlate this interglacial with the erosion phase between the Gravenvoeren and Pietersberg-3 terraces. This results in the correlation of Gravenvoeren with stage 18 and Pietersberg-3 with stage 20. We observed a normal polarity associated with the Pietersberg-3 terrace, suggesting the terrace was formed during the Brunhes, but delayed NRM acquisition (Van Hoof & Langereis, 1991) may have caused the observed normal polarity of Pietersberg-3.

7.3 *Older West-Maas terraces*

The next spike that may serve as a tie-point of the Maastricht section to the ODP records is the combination of a normal magnetisation with the pollen record of the Simpelveld-1 terrace. This pollen assemblage has been assigned to either a cool interglacial or an interstadial within the Tiglian. We argued above that the Simpelveld-1 terrace corresponds in our opinion with the OIS 82-80.

ODP 677 shows 22 cool episodes between stage 80 and the Cobb-Mountain event within stage 35. Such a mis-match between 7 terraces and 22 candidate cool periods may have various reasons: either we missed many terraces or not all candidate stages are equally qualified. We addressed this problem under the following assumptions:

- The staircase does not show any obvious hiatuses like oversized steps between two levels. This observation makes the low number of terraces seem reasonable.
- The modelling approach by Veldkamp & Van den Berg (1993) showed, under the assumption of a constant uplift rate for this period, 6 terraces are likely to be preserved in a Maas-type fluvial system. Below we will show that the uplift-rate assumption was too simplistic, but it served as a starting point. In this modelling exercise, the non-terrace periods cover a much longer time span than during the Middle Pleistocene. These results suggest that the solution may be found in a differentiation of potential cool episodes.

Above we mentioned the observation of an overprint of the 400 ka Milankovitch cycle on the (Early Pleistocene) 41ka cycles. The interference of the various frequencies had forced the (groups of) stages (82, 78), (70), (64, 62, 60), (54, 52 50), (44, 42, 40) into series of 41 ka cycles marked by higher $\delta^{18}O$ values (Fig. 12).

Stage 82 matches the, above established, prerequisites of a cold episode followed by temperate conditions in concert with normal magnetisation during the Tiglian (Réunion 2 event) so we took that stage as the older tie-point for the correlation of the Simpelveld-1 terrace with the ODP site 677 record. The other younger terraces have been divided tentatively accordingly to their morphostratigraphic position over the listed stages, taking into account the amplitudinal variations of the warm cycles, as well as the trend in the amplitude. The latter is based on the assumption that the lower $\delta^{18}O$ value of a particular cycle, the greater the change, hence such a cycle is considered to represent an erosion phase. Cycles with intermediate $\delta^{18}O$ values in contrast represent interstadials which may form parasequence boundaries within a terrace gravel sequence (van den Berg, 1996).

8 THE EAST MAAS VALLEY TERRACES

We distinguish 10 terrain steps on the East-Maas valley flanks underlain by fluvial deposits. This number is based primarily on the morphology of the interfluve between the Geul and Gulp brooks (section C and D in Fig. 6).

From old to young, our series comprises the following levels: Waubach-1 (= W-1) to 3, Kosberg-1 to 3, Crapoel, Noorbeek, and Simpelveld-1 and -2. The vertical distance between the various levels varies between 15 m and 4 m (Table 2). The altitude differences between East Maas terraces appear to occur with steps of around 4-5 m, or their multiples. This suggests little variability in the ultimate combination of factors controlling terrace formation.

8.1 *Age model*

Up to now, age models for the terraces older than the Simpelveld levels are very loose, due to the lack of a conventional biostratigraphic time constraint. The widest likely age-range is between the 13.8 Ma Middle Miocene sea-level high stand and the 2.5 Ma Northern Hemisphere ice event (equal to OIS 100-96).

The Neurath Sands record the 13.8 Ma transgression into the Lower Rhine embayment (Zagwijn & Hager, 1987; Herngreen, 1987). We correlate the coastline of this youngest

Table 2. Inferred age-altitude relation for the Maastricht terrace flight. The ages of the end of cold stages (= end of aggradation) are taken. Bold names mark terraces with age control.

Inferred age (Ma)	Altitude (m)	Terrace name
0.003	38	Holocene floodplain
0.011	45	Geistingen
0.014	49	Mechelen a/d Maas
0.13	52	Eisden Lanklaar
0.245	57	Caberg-3
0.33	63	Caberg-2
0.42	70	Caberg-1
0.51	76	Rothem-2
0.62	81.5	Rothem-1
0.715	86	Gravenvoeren
0.78	97.5	Pietersberg-3
0.87	104.5	Pietersberg-2
0.955	112.5	Pietersberg-1
1.03	119.5	Geertruid-3
1.09	124	Geertruid-2
1.28	130	Geertruid-1
1.5	136	Valkenburg-2
1.57	139	Valkenburg-1
1.69	146	Sibbe-2
1.77	150	Sibbe-1
1.87	159	Margraten
2.06	167	Simpelveld-2
2.14	175	Simpelveld-1
2.44	184	Noorbeek
2.69	190	Krapoel
2.81	198	Kosberg-3
2.94	202	Kosberg-2
3.04	206	Kosberg-1
3.14	215	Waubach-3
3.3	220	Waubach-2
3.98	235	Waubach-1
13.8	250	youngest marine cliff

transgression that intruded far enough southward with a low fossil cliff at 250 m altitude in the East Maas valley (e.g. section C in Fig. 6). This cliff is marked by terrain step mantled with well-rounded beach pebbles. This correlation is thus based on both a palaeogeographical (the strike and position of the coastline) and a stratigraphical argument (transition of marine facies to a fluvial facies).

After the Waubach terraces, the Kosberg terrace deposits are the first to contain large boulders transported by ice-flows. This feature suggested to Felder & Bosch, (1989) a correlation with the Praetiglian (in Fig. 12: P-T at 2.5 Ma) because the Praetiglian is the first period for which pollen records indicate ice-age conditions in the Netherlands. However, this correlation of the Kosberg levels is, in our view, problematic as it leaves too little time available with cold episodes for the Noorbeek- and Crapoel terraces (Figs 12, 13). Alternatively a correlation of the Crapoel terrace with stage 82 and consequently Noorbeek with 78 and Simpelveld-1 with stage 70 would allow sufficient time. But as argued at length before, the field data from Simpelveld-1 terrace are incompatible with such a correlation. Sediment characteristics of the Kosberg terraces (high quartz content; heavy minerals marked by a stable association; palaeosols with the characteristics of a red-yellow podsolic soil) might suggest a Pliocene age (Kuyl, 1980). Therefore we believe that the climostratigraphic position of the Kosberg terraces has to be placed back into the Pliocene. The general valley-side morphology does not indicate an important lapse in geomorphic processes but rather a continuing succession in the evolution of the landscape. However, if one adds any feasible age model to the morphology one will find a strong reduction in incision rate back in time. It is evident that uplift rates for the period between 13.8 Ma and 2.5 Ma were lower than those after 2.5 Ma, if simply measured by dividing observed differences in altitude over interpreted difference in age.

We took ODP 677 as a reference for the Pleistocene terraces, but for the Pliocene we use ODP site 846 (leg 138, Pacific ocean) because this was the first, astronomically calibrated, high-resolution OSDP record covering the Pliocene (Shackleton et al., 1993). This record shows many cooling events that preceded the 2.5 Ma event, similar to other deep-sea records from the Atlantic Ocean e.g. Fig. 15, suggesting global significance of this variability in the climate. The signal is quite different from the details of the pollen based land records where this high-frequency signal has not yet been recognised (van den Berg, 1996). However, the general pattern of the pollen record (Zagwijn, 1960) is in fair agreement with ODP site 864. The long-term trend (see Fig. 16) shows a general warm part of the Pliocene (the Brunssumian p.p) between 4.5 Ma and 3.4 Ma followed by a cooler part between 3.4 Ma and 2.5 Ma (the Reuverian). This cooler episode can be divided into two 400 ka periods with a break after stage G17 at 2.95 Ma. The first half of the Reuverian (3.4 Ma-2.95 Ma) can be interpreted as 2 cycles of 200 ka, separated by a cooling at 3.15 Ma (KM2). The oldest of the two may correspond to the Reuverian A; the younger, showing two subdivisions like the pollen record, corresponds to the Reuverian B. The decline from stage G17 to the 2.5 Ma glacial event corresponds to Reuverian C in the pollen zoning. This subdivision can also be recognized in other ocean records (Fig. 14). Some records show a gradual build up in intensity e.g. $\delta^{18}O$ in ODP site 864, with occasionally deep cooling events. Similar to this is the pattern of variations in abundance in the sum of six nannofossil species of genus *Discoaster* in ODP Site 606 (Backman & Pestiaux, 1987) (Fig. 14). Other records suggest a sudden intensification similar to the pollen record e.g. DSDP Site 552A (Shackleton & Hall, 1984). The latter record displays similar frequency in climatic cycles as the others, only the amplitude differs. The strong covariance in the various climate proxies of ocean records for the period between 2.5 Ma and 2.9 Ma strongly suggests that climatic variability before and after 2.5 Ma is basically alike in its frequencies of change. This observation may provide more time for the cold climate Kosberg terraces, because it means that ages in the range of 3.1-2.5 Ma are feasible.

All this circumstantial evidence together favours a simple solution that assumes that the East Maas terraces represent Late Pliocene climatic events rather than similar events from

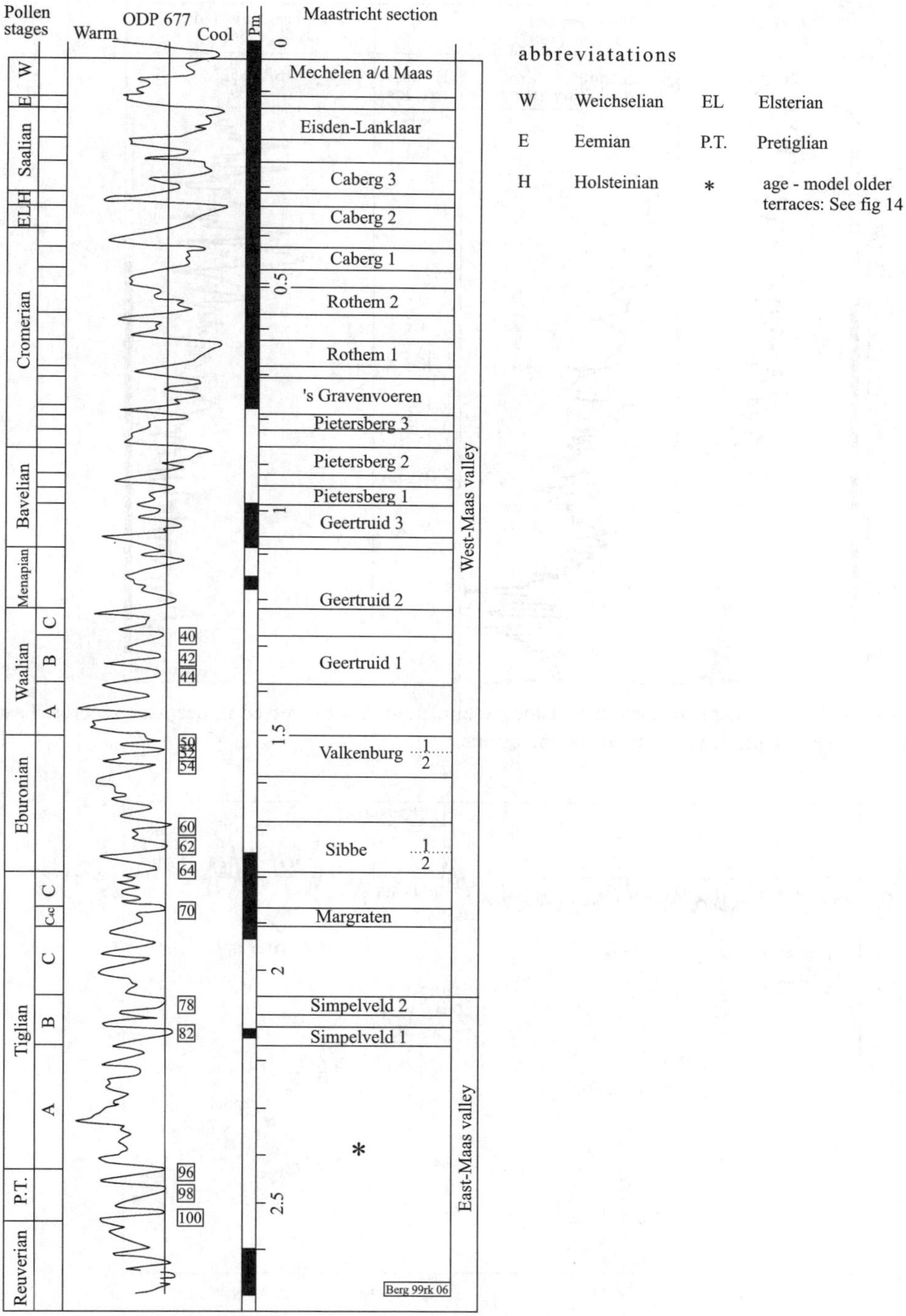

Figure 13. Stratigraphic chart showing the correlation between the Northwestern European pollen stages (Zagwijn, 1985), the ODP 677 ^{18}O ocean record (Shackleton et al., 1991) and the Pleistocene terraces of the Maas.

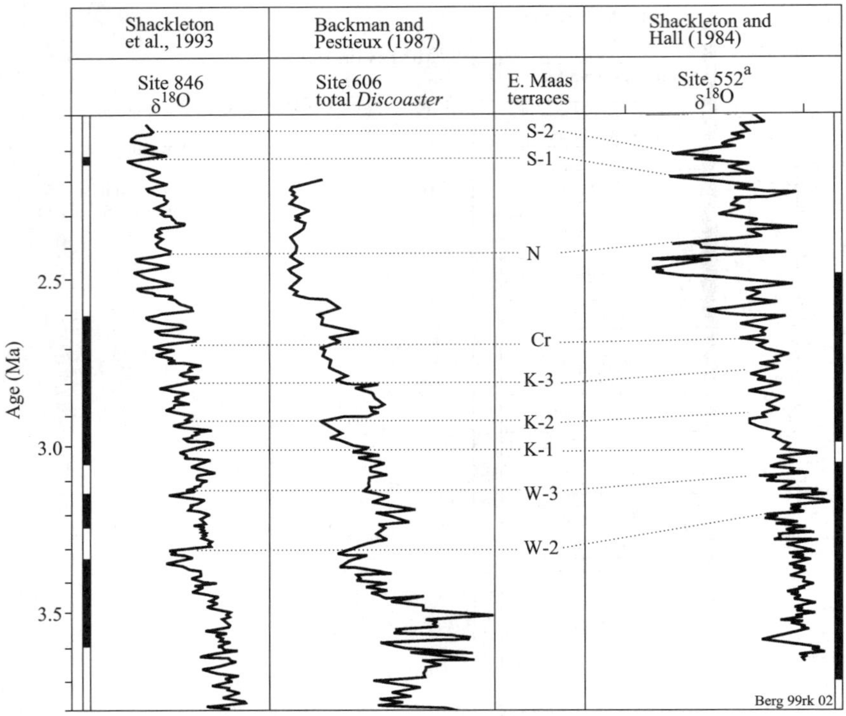

Figure 14. Late Pliocene to Earliest Pleistocene climate trends expressed in deep-sea records. East Maas terraces are plotted at marked cooling events.

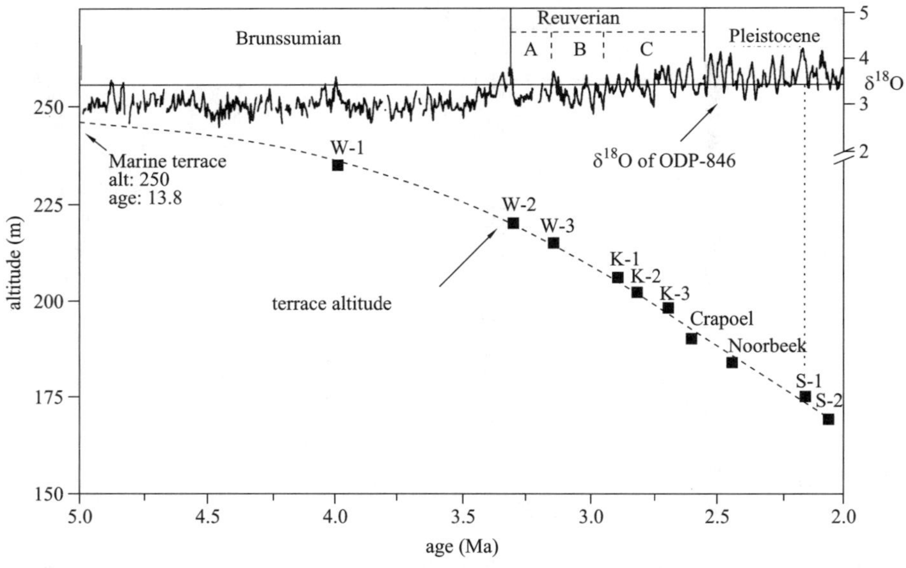

Figure 15. Altitude-age plot for the East Maas terraces. As a measure for the age assessment we took the age of a cool-warm shift in ODP 846 record (Shackleton et al., 1993) under the considerations mentioned in the text. The fitted function includes the youngest marine terrace. This graph is considered as a proxy of the regional uplift.

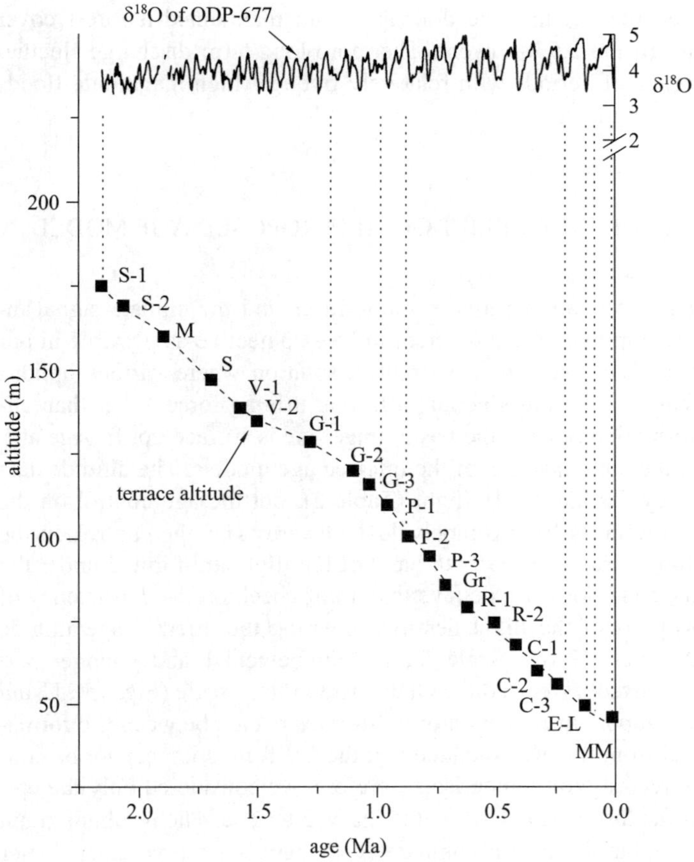

Figure 16. Altitude-age plot of the youngest 21 terrace levels. Dashed vertical lines mark the correlations with oxygen isotope stages as discussed in text.

the Late Miocene or the Early Pliocene, as has been proposed earlier by van den Berg (1994).

We took as our working hypotheses:

- Periods with clear evidence of surface-water temperature fluctuations are either evidenced by ice-rafted sediments in the North Atlantic Ocean or by events with strong fluctuations in the *Discoaster* spp. record (Backman & Pestiaux, 1987).
- The normal polarity found in the sediments of the Crapoel level does not reflect a present day field direction.
- We accept a correlation of the Simpelveld-1 level with oxygen isotope stages 82/81 on the basis of combined circumstantial evidence of palaeomagnetism and pollen.

In Figure 15 we worked back in time from the altitude/time position of the Simpelveld-1 level and plotted subsequently the height of the various older East Maas terrace-top levels against cool to warm shifts at the end of palaeoclimatic deterioration events with clear evidence of global cooling.

It is evident from pollen records that the drainage basin maintained a forest cover throughout the period prior to the 2.5 Ma event. Therefore long-term discharge fluctuations related to the occurrence of periods with relatively frequent high-magnitude floods may be inferred as a cause of terracing

9 CONSEQUENCES FOR SURFACE UPLIFT OF THE PROPOSED AGE MODEL

The above discussion on the correlation between the terraces and the climate signal inevitable hinges on many assumptions. We have tried to be as objective as possible in our correlation but in most cases we opted for the simplest solution where various options could be considered. As stated in the introduction, a further driving force, other than climate change, for terrace formation beyond the basin hinge line is surface uplift. Age/altitude graphs give insight in the consequences of the inferred age models. The altitude data are of a comparable accuracy for the whole flight (Table 2), but the age control on the Maas is in fact restricted to terraces from Simpelveld-1 onward. The age control on the older levels is extremely loose. However, as both parts of the flight are formed under the same geomorphic boundary conditions, we believe that a fair check on the consistency of the age models of the two parts of the flight lies in comparing the inferred age/altitude graphs. We took the ODP site 677 time-scale for the Simpelveld-1 and younger part (Fig. 16); and for East Maas terraces we used the ODP site 846 time-scale (Fig. 15). Using Figure 16 it appears that an important reduction in uplift rates occurs between the formation of V-2 through G-2 and from this we conclude that the uplift function cannot be constant throughout the whole record. For comparison therefore, we considered only the episode between formation of the last marine terrace and the V-2 terrace. The resultant graph through both parts of the flights displays a comparable fit between the younger, better constrained, part and the older, poorly constrained part (Fig. 17). This observation tends to support our assumption that both the Pliocene and the Early Pleistocene parts of the record have been shaped under the same set of geomorphic constrains. Moreover it suggests that the applied correlation models are at least consistent.

The empirically fitted curve through the various terrace-time/altitude points is interpreted as an approximation of the Late Miocene through Pliocene to Early Pleistocene regional surface uplift. The information obtained from Figure 17 depends on the scaling of the altitude and age axes. A strong exaggeration of both parameters displays an initial acceleration in uplift slightly after 10 Ma. However, using less exaggeration reveals a second and probably much more important, similar feature at 4 Ma. These estimates also depend on the inferred age of the oldest terraces. Giving them a Miocene age, such an acceleration occurs later, around 3 Ma. (pers. comm. R Westaway, 2000). The main problem with the introduction of a possible Miocene age for these terraces is that we end up with a major gap in time with many important cool spells but no sign in the morphology or in the lithology of the terraces that might hint towards such a hiatus. Neither is there any indication of changes (increase) in uplift rates which might have led to poor terrace preservation potential and hence lead to such a gap in the record. Therefore we prefer the simple model of gradual change in the uplift rate. The inferred 10 Ma event roughly matches the onset of the fluvial infill (Inden Formation) in the adjacent Lower Rhine Embayment. This also may point to an increase in regional tectonic dynamics.

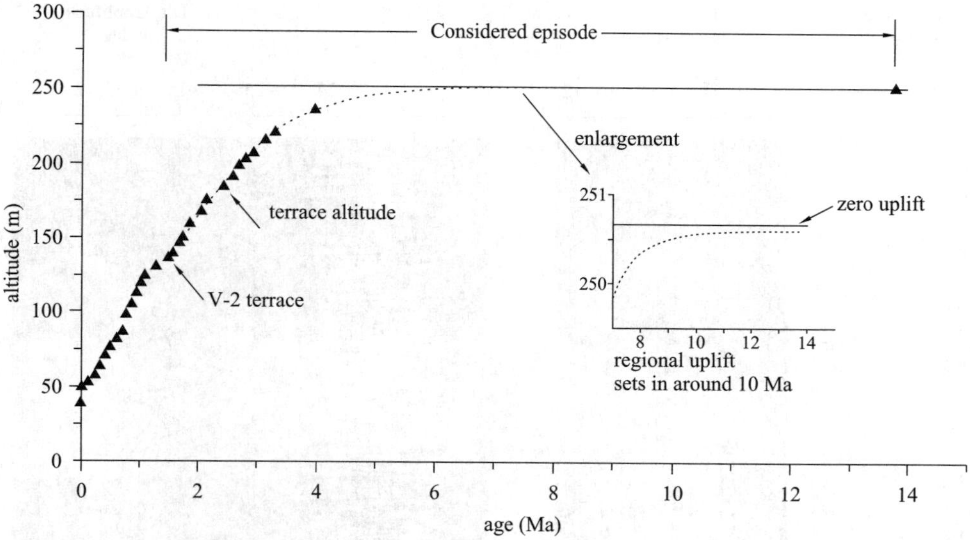

Figure 17. A combined East – and West Maas terrace record plotted in the discussed age/altitude relationship. The fitted function through the terraces older than V 2 is considered as a proxy for the regional uplift. It shows that uplift sets in around 10 Ma and reaches a maximum rate of change around 3 Ma.

The best age-constraints on the Maastricht section lie within the record formed over the last 1.2 Ma, a period mainly dominated by the very regular 100 Ka climate cycles. Therefore this section of the derived uplift graph (Fig. 16) may be the most reliable and hence most informative in terms of variations in the long-term uplift trend. Similar 'short term' variations can also be observed in the older parts of the graphs. In order to better express them, we plotted the relative variations in uplift as a separate graph (Fig. 18). The accelerations and decelerations that follow from this form of presentation seem to be controlled by a mechanism with a wavelength of 1.2 Ma. This wavelength strongly mirrors the long-term variations in tilt amplitude. We observe the highest uplift rates at tilt amplitude minima. The evaluation of this observation will be a subject of an other paper as it requires to be tested using other high-resolution tectonic records. Peaks in uplift rates show a linear increase with time reaching a maximum value of 11 cm per ky, these are superimposed on a basal rate of almost 3 cm per ky. The youngest uplift peak is probably responsible for an important change in the overall valley form of the Maas; the shift from the formation of so called 'plateau- terraces' or Main Terraces towards the formation of valley-side-terraces or Middle Terraces (Fig. 19); a highly characteristic feature of the whole of the ARS. Below we will discuss this in the wider context of the ARS. This strong acceleration in uplift sets in at the transition from G-2 to G-3 and culminates during the deposition of the 'Gravenvoeren' level. This level is one of the most poorly preserved levels from the whole flight. These two observations are consistent with each other as higher uplift rates produce unfavourable conditions for the preservation of terraces (Veldkamp, 1992).

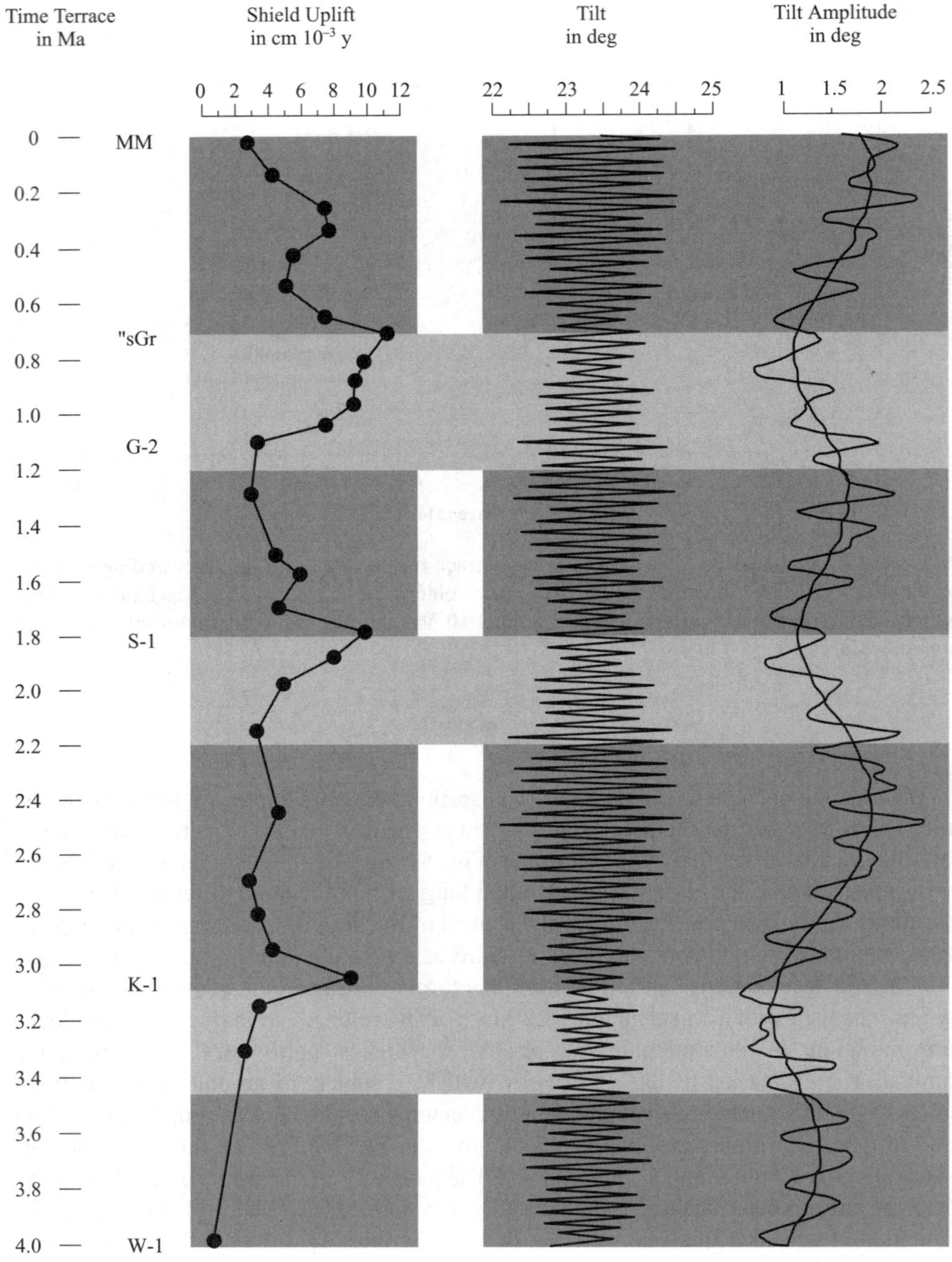

Figure 18. Uplift of the land surface based on the age-altitude position of the terraces of the Maastricht section (the black dots). This pattern matches the filtered record of the variations in tilt amplitude of the Earth axis. Under the heading 'terrace' abbreviations of some terrace names are given for reference purposes. (MM = Mechelen aan de Maas; 'sGr = Gravenvoeren; G-2 = Geertruid-2; S-1 = Sibbe-1; K-1 = Kosberg-1; W-1 = Waubach-1).

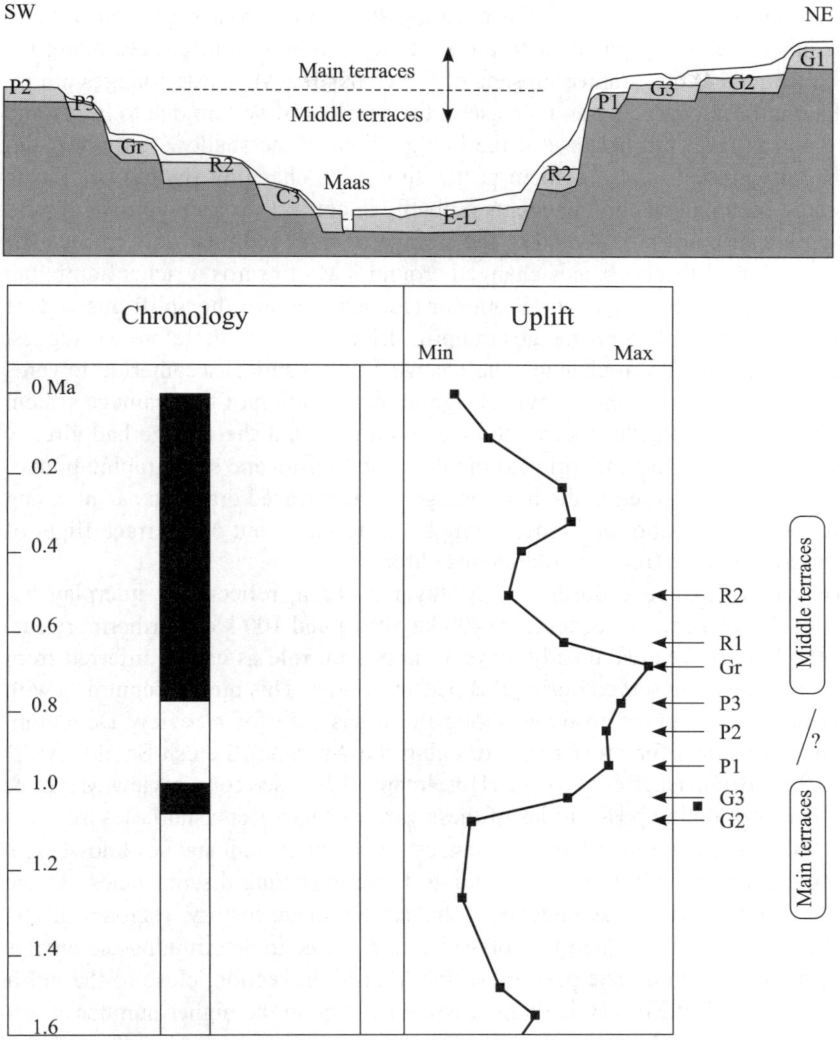

Figure 19. The characteristic geomorphic feature of the separation in the landscape between the group of Main Terraces and the group of Middle terraces can not be pinpointed to the level of the individual 100 ka terraces and it matches only *roughly* a major uplift pulse.

10 GENERAL DISCUSSION

A palaeo-Maas system, draining Northern France and the Ardennes peneplain into the Lower Rhine Embayment (LRE), could develop and evolve at least since the onset of the Late Miocene. The Middle Miocene sediments in the LRE suggest relative low sediment supply rate into this shallow sea-bight, associated with minor tectonic activity. A separate Maas river supply has not yet been recognized for this period. The Miocene drainage basin probably was that of a low gradient, non-terraced system supplying mainly quartz-rich sandy sediments. These general characteristics changed in the Latest Miocene through to the Pliocene. River terraces came into existence together with a coarse grained but still quartz-rich sediment supply (equivalent to the so called Kiesel-oolith terraces along the Mosel, Middle-Rhine and the Trainee mosane of the Upper Maas). These changes are interpreted as having been caused by an increase in the gradient of system due to hinterland uplift rather than sea-level fall because of the low gradient of the shallow seafloor. Contemporaneous delta growth is also evident at this time. The changing rhythm of climate change with more pronounced cooling events probably in concert with an increase in rainfall due to the plateau uplift contributed to the storage of river sediment in terraces. The sediment composition of these terraces changed around 3 Ma towards a richer admixture of physically weathered rock fragments. From our reconstruction of the uplift this change in composition cannot be linked to changes in uplift-driven incision, therefore we suggest this is probably a climate-driven change. The change from chemical weathering towards more physical weathering is the most obvious sign in the evolution of the drainage system of the gradually approaching Pleistocene climate conditions, but the change had already begun at least a half a million years prior to the Pliocene/Pleistocene stratigraphic boundary. This boundary is not marked in the morphology of the terraced drainage basin. Along the river Maas near Maastricht, the highest lying seven terraces, out of a terrace flight of 31 levels are inferred to date from pre-Matuyama chrons.

The terrace staircases shaped during the Matuyama chron, reflects the interplay between climate cycles of various frequencies (400 ka, 40 ka and 100 ka). Furthermore surface uplift in the Maastricht area already plays an important role as can be inferred from the 13 levels that became preserved during this period of time. This number contrasts with the 5 terraces preserved further upstream along the Maas (see for a review Demoulin, 1995) and similar numbers for other rivers draining the Ardenne-Rhenish Shield (ARS) (Brunnacker, 1978; Brunnacker et al., 1982; Hoselmann, 1996; see for a review Meyer & Stetts, 1998). The poor dating possibilities of these various Pleistocene staircases makes it virtual impossible to correlate over longer distances given our current state of knowledge. Only careful field mapping will bring a solution to these intriguing discrepancies. As we assume that the whole of the ARS underwent the same climate history, regional differences in uplift regime must have therefore played a crucial role in determining the precise nature of the preserved record. The position of the Maastricht section, close to the uplift center of the ARS (theWest Eifel/Hohe Fenn area), may explain the higher number of terraces in this area during the Matuyama part of the staircase. At greater distance (> 270 km) from the ARS uplift center only one or two terraces have been formed during the Late Matuyama e.g. the upper Weser river (Rohde, 1989) and the river Somme (Antoine, 1994), flowing respectively east and west of the ARS uplift center. Longer records with an early start near the Plio-Pleistocene boundary are known from the Thames (Bridgland, 1994; Maddy, 1997), the Ohre in the Bohemian Massif (Tyracek, 1983), the Allier in the Massif Central (Veldkamp, 1991, 1992) and the ancient Danube in the Alpine foreland (R. Becker-Haumann, pers. comm.). The geographical distribution of these classes of uplift records (which range from long to very short) appear to relate to the paleaogeography at the onset of the Oligocene as reconstructed by Ziegler (1982). Longer records appear to be associated with the continental areas whereas the shorter records occur in areas which emerge much later. This timeslice is important because it forms the onset of a new phase in the tectonic history of NW Europe (Zijerveld et al., 1992) and the contours of Late Neogene main structural elements and drainage systems emerge (Grimberieux et al., 1995; Meyer & Stets, 1998). This palaeogeography is probably the consequence of mantle driven processes and during the Plio-Pleistocene we observe an enhancement of this old topogragraphic pattern. The original highs register earlier the enhancement than the former lows.

During the Brunhes chron every 100 ka cold cycle produces a terrace in the lower Maas valley, a feature not only to be observed in the Maas and other NW European upland rivers but also suggested by the morphological expression with available age control of rivers from other structural parts of the European continent e.g. the Danube (Gabris, 1993), rivers in Central Italy (Coltorti, 1993) and in Spain (Garzon & Alonso, 1993).

This last main episode in the evolution of the Maas river valley coincides with the transition from the formation of the group of Main Terraces (wide and flat plateau-valley terraces) of Matuyama age to the group of the so called Middle Terraces (terraces along the steep valley sides of the deeply incised present day rivers) (Fig. 19). A very characteristic feature of the overall morphology of river valleys in the ARS. This feature cannot be observed in the vertical spacing of the longitudinal profiles of Maas terraces, but although clear in the landscape it is difficult to pinpoint which terrace forms the boundary, this highly depends on the position in the length of the river valley. Along the west side of the Maas valley, the Pietersberg 2 terrace is marked by a sharp edge against the channel valley, whereas on the east side of the valley the Pietersberg-1 terrace is in the same morphological position. On the bases of this criterion they both would belong to the last of the Main Terraces. In our chronology these levels were formed during the Upper Matuyama, an age 100-200 ka earlier than the age of terraces in a similar position along the Middle Rhine. There the transition from Main Terrace to Middle Terrace is placed in the earliest Brunhes (Meyer & Stets, 1998), this shows that great care has to be taken in cases where morphological marker are used for high resolution analyses and long distance correlations. Meyer and Stets attribute the formation of this morphological marker to a sudden increase in uplift rate of the ARS. Our data suggest a similar conclusion, but in the case of the Maastricht section this signal occurred earlier. This forms an additional argument for the above mentioned expansion in the geographical differentiation in the uplift pattern of the ARS; earlier near the uplift centres and later further away.

11 CONCLUSIONS

1. Fluvial sediments in the Lower Rhine Embayment, dated to the Late Miocene (~10 Ma), show the first evidence of a river Maas draining northern France and the Ardenne peneplain. Marine coastal sediments indicate an even earlier existence around 15 Ma.
2. The Maas river terrace staircase of the Maastricht section developed over the last 4 Ma. The number of steps in the staircase is probably one of the highest for any river in the world. Conventional dating techniques, including: pollen statigraphy, palaeomagnetics, and thermoluminescence resolved ages for six levels. Interpretations have been made for the remaining levels by correlation with high-resolution ODP oxygen isotope records.
3. Terraces formed younger than the Brunhes/Matuyama boundary match the 100 ka climate cycles. The formation of older terraces shows the importance of a differentiation in the 40 ka cycles by an overprint of a 400 ka on the 40 ka cycles.
4. Altitude differences between river terraces forms an important tool for high resolution analyses of crustal motion.

5. The amount of incision indicates ~200 m of surface uplift. Uplift accelerated during the Middle Pliocene. The rate of uplift fluctuated between 2 and 11 cm/ka. Uplift-rate maxima increase progressively with decreasing age. These fluctuations show a cyclic pattern with a wavelength of 1.2 Ma matching the filtered record of the variations in tilt amplitude of the Earth axis.
6. The youngest uplift pulse dates from 1.0 Ma. This pulse caused the prominent geomorphic knickpoint in the overall valley shape between the Main terraces and the Middle terraces.
7. The difference in age of the onset of the youngest uplift pulse between the middle Rhine area and the Maastricht area suggests that the Ardenne/Rhenish Shield 'plateau uplift' is driven by a source which is radiating from a center.
8. Pleistocene terrace sediments have been formed during cold stage river bed aggradation and each aggradation unit was incised during the following temperate stage probably due to retarded hillslope sediment supply. These changes in the rate of sediment transport by the river do not stop at the basin hinge line and therefore will be expressed in two different ways: in an uplifting landscape as terraces, in sediment accumulating subsiding basins, terrace surfaces may correspond to the sedimentary sequence boundaries. This implies that the formation of sequence boundaries can also be a climate-driven change.

ACKNOWLEDGEMENTS

Many people contributed to the realisation of this paper in one way or another. Werner Felder and Peter Bosch are thanked for discussions on the original field data. Cor Langereis assisted both with collecting and interpretation of the palaeomagnetic measurements. Salomon Kroonenberg, Sierd Cloetingh, Tom Veldkamp and Jan Jaap van Dijke encouraged us to write this manuscript. The manuscript benefited greatly from a very thorough review by Rob Westaway and David Bridgland. Darrel Maddy is thanked for his patience waiting for it.

REFERENCES

Antoine, P. 1994. The Somme valley terrace system (northern France); a model of river response to Quaternary climatic variation since 800.000 bp. *Terra Nova*, 6: 453-464.

Backman, J. & Pestiaux P. 1987. Pliocene Discoaster variations, deep sea drilling project site 606: chronology and paleoenvironmental implications. In: Ruddiman, W.F. et al. *Init. Repts. DSDP 94*, 903-910.

Beets, C.J. 1992. *Calibration of Late Cenozoic marine strontium isotope variations and its chronostratigraphic and geochemical applications*. PhD Thesis, Free University Amsterdam, the Netherlands.

Boenigk, W. 1978. Die Flussgeschichtliche Entwicklung der Niederrheinischen Bucht im Jungtertiär und Altquartair. *Eiszeitalter und Gegenwart*, 28: 1-9.

Breuren, J.W.R. 1945. Het terrassenlandschap van Zuid-Limburg. *Meded. Geol. Sticht., Serie C-VI-1.*

Bridgland, D.R. 1994. *Quaternary of the Thames*. Geol. Conserv. *Review Series* No.7 Chapman and Hall London.

Bruins, H.J. 1981. *Paleomagnetische datering van de Maasterrassen en Vroeg-Pleistocene Lössafzettingen in Zuid-Limburg*. Int. Rep. Landb. Hogeschool, Wageningen.

Brunnacker, K. 1978. Neuere Ergebnisse über das Quartär am Mittel-und Niederrhein. *Fortschr. Geol. Rheinld. U. Westf.*, 28: 111-122.

Brunnacker, K., Loscher, M., Tillmans, W. & Urban, B. 1982. Correlation of the Quaternary terrace sequence in the Lower Rhine valley and the northern Alpine foothills of central Europe. *Quaternary Research*, 18: 152-173.

Bryant, I.D. 1983. Facies sequences associated with some braided river deposits of late Pleistocene age from Southern Britain. *Spec. publs. Int. Ass. sediment.*, 6: 267-275.

Demoulin, A. ed. 1995. *L'Ardenne Essai de geographie physique, Hommage au Prof. A. Pissart* Dep. De Geographie physique et Quaternaire, Univ. de Liege.

De Jong, J. 1982. Pollenanalytisch onderzoek aan de boringen: Linne (58D/313), Schinveld (60D/1025) en Nederweert (58A/89), alsmede een herinterpretatie van de Koningsbosch (60B/27) – intern rapp. Paleobotanie 919- Rijks geol. Dienst, Haarlem.

Coltorti, M. 1993. River evolution in Umbria-Marchi Apenines during the Pleistocene- Abst. Eur. *Union of Geologists VII Congress, Strasbourg.*

Felder, W.M. & Bosch, P.B. 1989. Geologisch Kaart van Zuid-Limburg en omgeving, schaal 1:50.000. Afzettingen van de Maas. *Rijks Geol. Dienst*, Heerlen.

Gabris, Gy. 1993. Pleistocene evolution of the Danube in the Carpathian Basin Abst. Eur. *Union of Geologists VII Congress, Strasbourg.*

Garzon, G. & Alonso, A. 1993. Recent river evolution of a medium sinuosity gravel bed river in central Spain Abst. Eur. *Union of Geolgists VII Congress, Strasbourg.*

Gibbard, P.L. 1988. The history of the great north-west European rivers during the past three million years. *Phil. Trans. Ro. Soc. London*, ser. B., 318: 559-602.

Grimberieux, J., Laurant, A. & Ozer, P. 1995. Les rivieres s'installent in: *L'Ardenne Essai de geographie physique, Hommage au Prof. A. Pissart Dep. de Geographie physique et Quaternaire*, A. Demoulin, edit. Univ. de Liege.

Gliese, J. & Hager, H. 1978. On browncoal resources in the Lower Rhine Embayment (West Germany). *Geol. en Mijnbouw*, 57 (4): 517-527.

Gullentops, F. 1974. Buitengewone vergadering 1974 van de Vereniging voor Geologie en de Societe Geologique de Belgique, excursiegids.

Herngreen, G.W.F. 1987. Correlation between Miocene beds of the SE Netherlands and Italy based on dinoflagellate biozonation. *Meded. Werkgroep voor Tertiaire en Kwartaire Geologie*, 24 (1-2): 31-40.

Hilgen, F.J. 1991. Extension of the astronomically calibrated (polarity) time scale to the Miocene/Pliocene boundary. *Earth Plan. Sci. Lett.*, 107: 349-368.

Hoselmann, C. 1996. Der Haupterrassen-komplex am unteren Mittelrhein. *Z dt.geol. Ges*, 147: 481-497.

Hoffmann, R. 1996. Die quartäre Tektonik des sűdwestlichen Schiefergebirges begrűndet mit der Höhenlage der jungere Hauptterrasse der Mosel und ihrer Nebenflűsse. *Bonner geowiss. Schriften*, 19, Wiehl.

Huxtable, J. 1993. Further thermoluminescence dates for burnt flits from the Palaeolithic site Maastricht- Belvédère and a finalised age for the Unit IV Middle Palaeolithic sites. *Meded. Rijks Geol. Dienst*, 47: 41-44.

Huxtable, J. & Aitken, J. 1985. Thermoluminescence dating results for the palaeolithic site Maastricht-Belvedere. *Meded. Rijks Geol. Dienst*, 39-1: 41-44.

Juvigné, E. & Renard, F. 1992. Les terraces de la Meuse de liege a Maastricht. *Ann. de la Soc. Geol. de Belgique*, T. 115 (fasc.-1): 167-186.

Kasse, C. 1988. *Early Pleistocene tidal and fluvial environments in the southern Netherlands and northern Belgium.* PhD Thesis, Free Univ., Amsterdam, the Netherlands.

Klostermann, J. 1992. Das Quartar der Niederrheinischen Bucht. *Geol. Landesamt Nordrh. Westf.*

Kukla, G. & Cilek, V. 1996. *Plio-Pleistocene megacycles: record of climate and tectonics Palaeo-3*, 120: 171-194.

Krook, L. 1961. *Verslag van het onderzoek naar de klimaatscondities tijdens de afzetting van de Maasterrassen*. M-Sc thesis, Fysische Geografie, University of Utrecht, the Netherlands.

Krook, L. 1993. Heavy minerals in the Belvédère deposits. *Meded. Rijks Geol. Dienst*, 47: 25-31.

Langereis, C.G., Dekkers, M.J., de Lange, G.J., Paterne, M. & van Santvoort, P.J.M. 1997. Magnetostratigraphy and astronomical calibration of the last 1.1Myr from an eastern Mediterranean piston core and dating of short events in the Brunhes. *Geophys J. Int.*, 129: 75-94.

Macar, P. 1954. *Les terraces fluviales et la Haute Belgique au Quaternaire*. Prodrome d'une description géologique de la Belgique.

Maddy, D. 1997. Uplift-driven valley incision and river terrace formation in southern England *Journ. Quatern. Sc.*, 12(6): 539-545.

Meyer, W. & Stets, J. 1998. Junge Tektonik im Rheinischen Schiefergebirge und ihre Quantifizierung. *Z. dt,geol.Ges.*, 149/3: 359-379.

Miall, A.D. 1978. Lithofacies types and vertical profile models in braided river deposits: a summary. In: Miall, A.D. (ed.) Fluvial sedimentology. *Can. Soc. Petr.Geol. Mem.*, 5: 597-604.

Moura, M.L. & Kroonenberg, S.B. 1990. Geochemistry of Quaternary fluvial and eolian sediments in the southeastern Netherlands. *Geol. en Mijnbouw*, 69: 359-374.

Mudelsee, M. & Statteger, K. 1997. Exploring the structure of the mid-Pleistocene revolution with advanced methods of time-series analysis. *Geol.Rundsch.*, (86): 499-511.

Paulissen E. 1973. De morfologie en de Kwartairstratigrafie van de Maas vallei in Belgisch Limburg. Verh. Kon. Acad. Wetensch. Lett. en Schone Kunsten, *Klassieke Wetensch.*, 35 (127): 1-266.

Porter, S.C., Zhisheng, A. & Hongbo, Z. 1992. Cyclic Quaternary alluviation and terracing in a nonglaciated drainage basin on the north flank of the Qinling Shan, central China. *Quaternary Research*, 38: 157-169.

Raymo, M.E., Oppo, D.W. & Curry, W. 1997. the mid-pleistocene climate transition: A deep sea carbon isotopic perspective. *Paleoceanography*, 12(4): 546-559.

Pissart A, Harmand, D. & Krook, L. 1997. L'Evolution de la Meuse de Toul a Maastricht depuis le Miocene: Correlations chronologiques et traces des captures de la Meuse Lorraine d'apres les mineraux dense. *Geographie physique et Quaternaire*, 51(3): 267-284.

Riezebosch, P.A. 1971. A contribution to the sedimentary petrological description of the Maas deposits in southern Limburg (the Netherlands). *Geol. en Mijnbouw*, 50: 505-514.

Riezebosch, P.A., Bisdom, E.B.A. & Boersma, O. 1978. Composite grains in Maas (Meuse) sediments: a survey and a discussion of their opaque components. *Geol. en Mijnbouw*, 57: 417-431.

Rohde, P. 1989. Elf Pleistozäne Sand-kies-terrassen der Weser: Erläuterung eines Gliederungsschema für das obere Weser Tal Eisz. *Alter u. Gegenw.*, 39: 42-56.

Ruddiman, W.F., Raymo, M.E., Martinson, D.G., Clement, B.M. & Backman, J. 1989. Pleistocene evolution: Northern hemisphere ice sheets and North Atlantic Ocean. *Paleoceanography*, 4: 353-412.

Ruegg, G.H.J. 1994. Alluvial architecture of the Quaternary Rhine-Meuse river system in the Netherlands. *Geol. en Mijnbouw*, 72: 321-330.

Ruyters, H.M.J., Bosch, P.B., Kisters, P.J.M. & Felder, W.M. 1995. Geologische Kaart van Zuid-Limburg en omgeving, schaal 1:50.000. Paleozoïcum. *Rijks Geol. Dienst*, Heerlen.

Shackleton, N.J. & Hall, M.A. 1984. Oxygen and carbon isotope stratigraphy of deep sea drilling project hole 552A: Plio-Pleistocene glacial history. In: Roberts, D.G., Schnitker, D. et al., Init. Repts, *Deep-Sea Drilling Program,* 81: 599-609.

Shackleton, N.J., Berger, A. & Peltier, W.R. 1990. An alternative astronomical calibration of the Lower Pleistocene time scale based on ODP site 677. *Trans. Roy. Soc. Edinburgh Earth., Sci.*, 81: 251-261.

Shackleton, N.J., Crowhurst, S., Hagelberg, T., Pisias, N. & Schneider, D.N. 1993. A new Late Neogene timescale: application to ODP leg 138 sites. *Proc. ODP Sci. Results.*, 138: 73-101.

Slomp. C. 1990. *Geochemie en mineralogie van tertiare en kwartaire sedimenten in Zuid Limburg: een methode voor de bepaling van achtergrondgehalten aan zware metalen*. Msc Thesis, Landbouw Univ. Wageningen sect. Bodemkunde en Geologie/Rijks Geol. Dienst, Haarlem.

Tyraçek, J. 1983. River terraces: important paleoclimatic indicator. In: Billards, O., & Shotton, P.W. (eds) *Quaternary Glaciations in the Northern Hemisphere*. ICGP Project- 73/1/24 Rept. 9 Unesco ICGP, Paris: 34-41.

Van Adrichem Boogaert, H.A. & Kouwe, W. 1993. Stratigraphic nomenclature of the Netherlands, revision and update by RGD and NOGEPA. Meded Rijks Geol. Dienst nr 50.

Van den Berg, M.W. 1989. Toelichting op kaartblad 59-62, Geomorfologische kaart Nederland, 1:50.000 – DLO-Staring Centrum, Wageningen/Rijks Geol. Dienst, Haarlem.

Van den Berg, M.W. 1994. Neo-tectonics of the Roer Valley Rift System. Style and rate of crustal deformation inferred from syn-tectonic sedimentation. In: Symposion on the Roermond Earthquacke T. Van Eck (ed.). *Geol. en Mijnbouw*, 73: 143-156.

Van den Berg, M.W. 1996. *Fluvial sequences of the Maas: a 10 Ma record of neotectonics and climate change at various time-scales*. Phd Thesis, University of Wageningen, the Netherlands.

Van Hoof, A.A.M. & Langereis, C.G. 1991. *Reversal records in marine marls and delayed acquisition of remanent magnetization Nature*, 351: 223-225.

Van Kolfschoten, T. 1985. the Middle Pleistocene (Saalien) and Late Pleistocene (Weichselian) mammal faunas from Maastricht-Belvédère Southern Limburg. In: Van kolfschoten, T. & W. Roebroeks (eds); *Maastricht-Belvédère: stratigraphy, palaeoenvironment and archaeology of the Middle and Late Pleistocene deposits*. Meded. Rijks Geol. Dienst 39(1): 45-74.

Van Kolfschoten et al. 1993. Archaeology of the Middle and Late Pleistocene deposits. *Meded. Rijks Geol. Dienst* 47: 51-60.

Van Kolfschoten, Th & Gibbard, P.L. 1998. The dawn of the Quaternary: an introduction. *Meded. Ned. Inst.Toeg. Geowet.TNO*, Nr 60: 13-17.

Van Montfrans, H.M. 1971. *Palaeomagnetic dating in the North Sea Basin*. Thesis Amsterdam.

Vandenberghe, J. 1985. Solution slots or ice-wedge casts? In: Van kolfschoten, T. & W. Roebroeks (eds); Maastricht-Belvédère: stratigraphy, palaeoenvironment and archaeology of the Middle and Late Pleistocene deposits. *Meded. Rijks Geol. Dienst*, 39(1): 45-74.

Vandenberghe, J. 1993. River terrace devellopment and its relation to climate: the Saalian Caberg terrace of the Maas river near Maastricht (the Netherlands). *Meded. Rijks Geol. Dienst*, 47: 19-24.

Van Straaten, L.M.J.U. 1946. Grindonderzoek in Zuid Limburg. In: *Meded. Geol. Stichting serie C-VI-2*.

Veldkamp, A. 1991. *Quaternary river terrace formation in the Allier basin, France. A reconstruction based on sand bulk geochemistry and 3-D modelling*. PhD-Thesis, Agricultural University, Wageningen, the Netherlands.

Veldkamp, A. 1992. A 3-D model of Quaternary terrace development, simulations of terrace stratigraphy and valley asymmetry, a case study in the Allier terraces (Limagne, France). *Earth Surface Processes and Landforms*, 17: 487-500.

Veldkamp, A. & van den Berg M.W. 1993. Three-dimensional modelling of Quaternary fluvial dynamics in a climo-tectonic dependent system. A case study of the Maas record (Maastricht, the Netherlands). *Global and Planetary Change* 8: 203-218.

Verhoef, P. 1966. *Geomorphological and pedological investigations in the Redange-sur-Attert area* (grand Duche of Luxemburg,Thesis, publ. No. 8 Fysisch Geografisch Lab. University of Amsterdam).

Zagwijn, W.H. 1963. Pollen-analytic investigations in the Tiglian of the Netherlands. *Meded Geol. Sticht. Nw Serie,* no. 16: 49-71.

Zagwijn, W.H. 1960. Aspects of the Pliocene and Pleistocene vegetation in the Netherlands. *Meded. Geol. Sticht.*, C-III-1- no. 5: 78 p.

Zagwijn, W.H. 1961. Pollenanalytisch onderzoek van twee kleilagen in het Maasdal bij Roermond. *Intern. Rapp. Paleobot.* 268 Rijks Geol. Dienst, Haarlem.

Zagwijn, W.H. 1974. Pollenanalytisch onderzoek van een humeuze kleilaag in de groeve bij Susterseel en een opmerking n.a.v. rapp. 369. *Intern. Rapp. Paleobot.* 719, Paleobotanie, Rijks Geol. Dienst, Haarlem.

Zagwijn, W.H. 1985. An outline of the Quaternary stratigraphy of the Netherlands. *Geol. en Mijnbouw*, 64 (1): 17-24.

Zagwijn, W.H. 1989 The Netherlands during the Tertiary and the Quaternary. A case history of coastal lowland evolution. *Geol. en Mijnbouw*, 68: 107-120.

Zagwijn, W.H. 1996. The Cromerian Complex Stage of the Netherlands and correlation with other areas in Europe In: The early Middle Pleistocene in Europe, Turner, C. (ed.) *Proceedings of the SEQS Cromer symposium Norwich (UK) September 1990*, Balkema/Rotterdam/Brookfield: 145-172.

Zagwijn, W.H. 1998. Borders and boundaries: a century of stratigraphical research in the Tegelen-Reuver area of Limburg (the Netherlands). *Meded Ned. Inst.Toeg. Geowet. TNO*, Nr 60: 19-34.

Zagwijn W.H. & de Jong, J. 1963. Pollenanalytisch onderzoek van materiaal uit een groeve te Platte Bossen. *Intern. Rapp. Palaeobot* 369. Rijks Geol. Dienst, Haarlem.

Zagwijn, W.H., van Montfrans, H.M. & Zandstra, J.G. 1971. Subdivision of the 'Cromerian' in the Netherlands; Pollen-analysis, Palaeomagnetism and Sedimentary Petrology. *Geol. Mijnbouw*, 50: 41-58.

Zagwijn, W.H. & de Jong, J. 1983. Die Interglaziale von Bavel und Leerdam und ihre Stratgraphische Stellung in die Niederländische Früh-Pleistozän. *Meded Rijks Geol. Dienst*, 37(3): 155-169.

Zagwijn, W.H. & Hager, H. 1987. Correlations of continental and marine Neogene deposits in the south-eastern Netherlands and the Lower Rhine District. *Meded. werkgroep voor Tert. en Kwartaire geologie*, 24(1-2): 59-78.

Zijerveld, L., Stephenson, R., Cloetingh, S., Duin, E. & van den Berg, M.W. 1992. Subsidence analyses and modelling of the Roer Valley Graben (SE Nethelands) *Tectonophysics*, 208: 159-171.

Zijderveld, J.D.A. 1967. Demagnetization of rocks: Analysis of results. In: *Methods in Palaeomagnetism*, Collinson, D.W., Creer, K.M. & Runkon, S.K. (ed.) Elsevier, Amsterdam: 254-286.

Zonneveld, J.I.S. 1947 Zand-petrologische onderzoekingen in de terrassen van Zuid-Limburg. *Meded. Geol. Stichting*, nieuwe serie no. 3: 103-123.

Zonneveld, J.I.S. 1947 Het Kwartair van het Peelgebied en de naaste omgeving. *Meded. Geol. Stichting Ser. C-VI-3.*

4. Flow in the lower continental crust as a mechanism for the Quaternary uplift of the Rhenish Massif, north-west Europe

ROB WESTAWAY
16 Neville Square, Durham, UK

1 INTRODUCTION

Quaternary uplift of many coastal regions is revealed by sequences of marine terraces. In many localities this uplift relates to tectonic deformation caused by plate motions. However, many people have suggested that continental interiors that are distant from plate boundaries have also uplifted in the recent geological past. Some of the evidence to support this view is of course ambiguous, enabling for instance Molnar & England (1990) to propose that it can be explained instead as a Quaternary increase in relief – caused by increasing rates of erosion – with no change in average altitude. This idea has been criticised, for instance by Summerfield & Kirkbride (1992). However, a further difficulty has been the absence of any known mechanism which can cause widespread uplift of continental interiors or other localities which are distant from plate boundaries.

Eyles (1996) has noted the systematic Quaternary uplift, by hundreds of metres, of land areas bordering the North Atlantic Ocean, including Britain and the north-western part of mainland Europe. The present study will examine in greater detail the Quaternary uplift history of one of these upland regions: the Rhenish Massif (Rheinisches Schiefergebirge), centred in western Germany and including the Ardennes upland in Belgium (Fig. 1). Most outcrop in this study region comprises Palaeozoic rocks, metamorphosed during the Hercynian orogeny. The land surface, which rises to ~800 m, is dissected by major gorges, notably of the Rhine (Rhein/Rijn/Rhin), Maas (Meuse) and Mosel (Moselle) rivers (Fig. 1), which are incised by up to ~400 m. This systematic gorge incision, revealed by sequences of river terraces, has long been regarded as evidence for Quaternary uplift of this land surface (e.g. Illies et al., 1979; Gibbard, 1988).

Recent studies by van den Berg (1996) and Maddy (1997) have attempted for the first time to quantify Quaternary uplift rates in northwestern Europe and southern England using differences in altitude of river terraces. Maddy's (1997) study of the Thames terrace sequence in southern England indeed indicated uplift rates of ~0.07 mm a^{-1} for the past million years, similar to the range of values deduced here for the Rhenish Massif (see below). However, Kiden & Tornqvist (1998) criticised these conclusions, first, because Maddy (1997) did not justify equating gorge incision to uplift of the surrounding land surface. Second, they noted that Maddy's (1997) Quaternary uplift rate estimate was at least an order of magnitude greater than any feasible uplift rate during the Neogene. Kiden & Tornqvist (1998) considered the most likely explanation of the observed terrace sequences to be that climate change had caused erosion rates of interfluves to increase during the Quaternary, and the associated isostatic response has required gorge incision

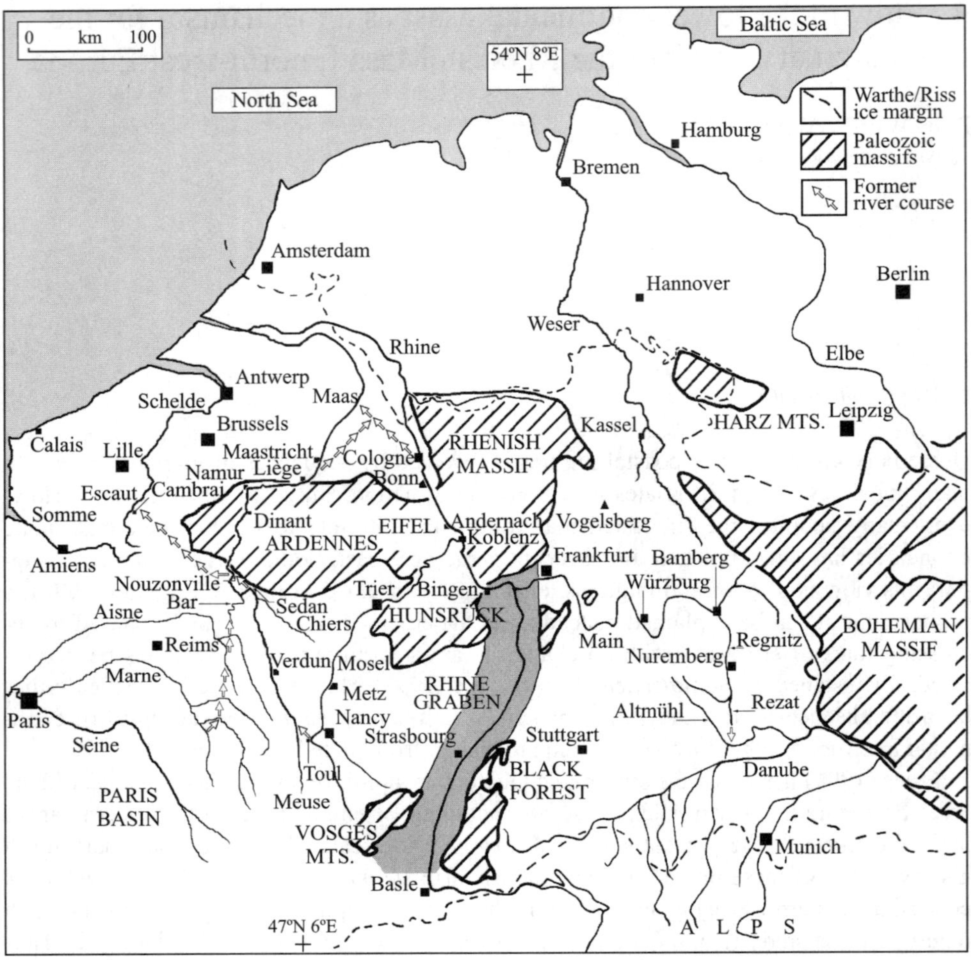

Figure 1. Location map, showing present-day and former river courses in the study region. Adapted from Brunnacker et al. (1982b). Sources of information for river channel adjustments are discussed in the text.

with no increase in the regional average altitude of the land surface. Kiden & Tornqvist (1998) thus re-stated the familiar Molnar & England (1990) point of view in the 'chicken and egg' dispute versus Summerfield & Kirkbride (1992) as to whether relief which develops due to lateral and/or temporal variations in rates of erosion may be mistaken for systematic uplift of a land surface. Maddy (1998) indeed conceded that the isostatic response to erosion of interfluves may be significant in southern England, such that although the Earth's surface may be experiencing net uplift, its rate may be substantially less than local rates of gorge incision.

Most recently, Whipple et al. (1999) have suggested that an increase in erosion rates may well *reduce* the relief in a region, not increase it as Molnar & England (1990) thought, because their calculations indicate that, in the resulting the response of a landscape, interfluves erode faster than gorges incise. Their results imply that systematic uplift of land surfaces during the Quaternary *cannot* be mistaken for the isostatic response to increasing rates of erosion, and thus does indeed require a separate cause. However, these results were obtained from analysis of regions of extreme relief and high drainage density, and their relevance to the present study region is thus unclear.

In the light of these difficulties, the present study has set out to be careful. First, it establishes that although in general rates of gorge incision can bear no relation to uplift rates of the surrounding land surface (e.g. Gilchrist et al., 1994; Mitchell & Westaway, 1999), such a relationship is well-justified, at least as a first approximation, in parts of the present study region. Second, it establishes that erosion rates of interfluves in this study region are negligible. The isostatic response to such erosion has thus had a negligible effect on this region's Quaternary uplift history, which evidently requires a separate cause to explain the ~100 m or more of gorge incision which has occurred in the last million years. Third, it proposes and tests an alternative explanation for this uplift: that it is the isostatic response to atectonic crustal thickening caused by flow of lower crust induced by transient loading effects caused by global sea-level fluctuations and the growth and decay of ice sheets in adjacent regions. Maddy (1998) noted that the precise timing of the latest Pliocene order-of-magnitude increase in the uplift rate of the land surface adjoining the River Maas around Maastricht should provide a critical constraint on models for the cause of this region's uplift. However, before this uplift can be modelled, it is necessary to investigate the conflicting river terrace chronologies proposed, for instance, by van den Berg (1996) and Pissart et al. (1997). This requires scrutiny of the available evidence.

The aims of this study are thus as follows. First, it will summarise observational evidence of sequences of Quaternary river terraces in the study region. Second, it will establish a chronology for these terraces, linked to the cold stages of the oxygen isotope time scale, and thus deduce the uplift histories at key localities. Third, it will develop quantitative theory which can account for the observed uplift rates and their timing. Fourth, it will show that the related increase in crustal thickness is sufficient to have caused the region's Quaternary volcanism. Finally, it will discuss the global implications of this uplift process, which can be expected to also occur elsewhere. These include, first, the possibility that it may account for large-scale changes in global relief which have affected the global climate. Previous studies (e.g. Ruddiman & Raymo, 1988) have argued that Quaternary climate change has resulted from surface uplift with an independent cause. It is suggested here, instead, that the interrelationship between relief and climate change is more complex, with each coupled to the other. An existing régime involving glacial to interglacial ice- and water-loading transitions will cause crustal thickness changes involving surface uplift, which will modify the global relief and hence the climate, which will affect the magnitude of loading transitions in future, which will determine how the relief will subsequently develop.

2 BACKGROUND INFORMATION

From its source in north-eastern France (Fig. 1), the Maas/Meuse flows for more than 200 km across the Paris Basin before entering a gorge through the Ardennes upland. Between Liège and Maastricht, it exits the Ardennes and enters the North European coastal plain (Fig. 1). Near Basle, almost 300 km from its present source in the Alps, the Rhine enters the Rhine Graben, an Oligocene normal fault zone reactivated in the Quaternary (e.g.

Illies & Baumann, 1982) (Fig. 1). After flowing axially along this graben for more than 300 km, the Rhine enters a gorge through the Rhenish Massif at Bingen. Near Bonn, it flows out of this gorge and onto the North European plain in the Lower Rhine embayment.

Sediments in this study region have been assigned to many stages of the local stratigraphy. In the absence of direct age control evidence, this stratigraphy is largely defined using pollen assemblages. Criteria for assigning particular sediments to particular pollen stages have been discussed by de Jong (1988), Zagwijn (1996), and many other studies, and are not repeated here.

Until the Early Pleistocene, the lower Maas and Rhine adjoined the lower reaches of the 'Baltic River', which flowed westward through what is now the Baltic Sea and drained much of eastern Germany, Poland, and southern Scandinavia. This course no longer exists due to relative elevation changes, including subsidence of the Baltic Sea floor to below sea level. The last unequivocal evidence of fluvial transport of Scandinavian material to the Netherlands is in sediments of Waalian age (i.e., > 1.2 Ma) (e.g. Bijlsma, 1981). Material of Scandinavian provenance is found in overlying sediments as young as Early Cromerian (i.e., ~0.8 Ma), but is thought to have been transported south by an ice sheet of Menapian (i.e., ~1.2 Ma) age, then reworked by northern European rivers (e.g. Gibbard, 1988; Tebbens et al., 1995). The most likely timing of this glaciation was in oxygen isotope stage 36, the first very severe Quaternary glaciation (e.g. Shackleton et al., 1990), which occurred during the Menapian.

The courses of the lower Maas and Rhine have changed repeatedly during the Quaternary (e.g. Zagwijn, 1974, 1989), being influenced by differential subsidence caused in part by vertical slip on active normal faults (Fig. 2). Ice sheets have reached parts of the lower Rhine (Figs 1 and 2) in some glaciations. However, the localities investigated in detail in this study have not been glaciated: hence the preservation of intact river terrace sequences.

Many of these terraces contain sedimentary features indicative of a cold climate (e.g. Brunnacker, 1978; van den Berg, 1996): rivers in the study region are indeed not aggrading at present. It is thought that the main factor leading to terrace formation is the greater rate of erosion under cold conditions (caused by periglacial processes such as frost shattering and by the reduction in vegetation whose presence would stabilise hillslopes) combined with a reduction in flow rates due to reduced rainfall (e.g. Gibbard, 1988; van den Berg, 1996). The question whether terrace aggradation has typically been associated with cold-climate-induced braided channel development, and incision with meandering regimes, has been widely debated (e.g. Ruegg, 1994, 1995; Törnqvist, 1995). The issue is complicated by the fact that the lower Rhine developed braided channels during part of the Holocene (Törnqvist, 1994), raising the possibility that similar complexity may have also occurred farther upstream along the Maas or Rhine.

Near Maastricht in the south-eastern Netherlands, directly north of the Rhenish Massif (Figs 2 and 3), a sequence of 31 terraces of the River Maas has been identified by van den Berg (1996) (Figs 4 and 5c). This detailed study has measured surface altitudes with precision, and established age control: the sequence begins in the late Middle Miocene, when the palaeocoastline lay along the northern flank of the Rhenish Massif (e.g. Steininger & Papp, 1979). In contrast, along the Rhine no more than 16 terraces have been directly identified (see below). Sediments in the subsurface do reveal additional alternations between warm and cold conditions, which have not left recognisable surface evidence of terrace formation. However, even so, fewer cycles have been identified than for the Maas.

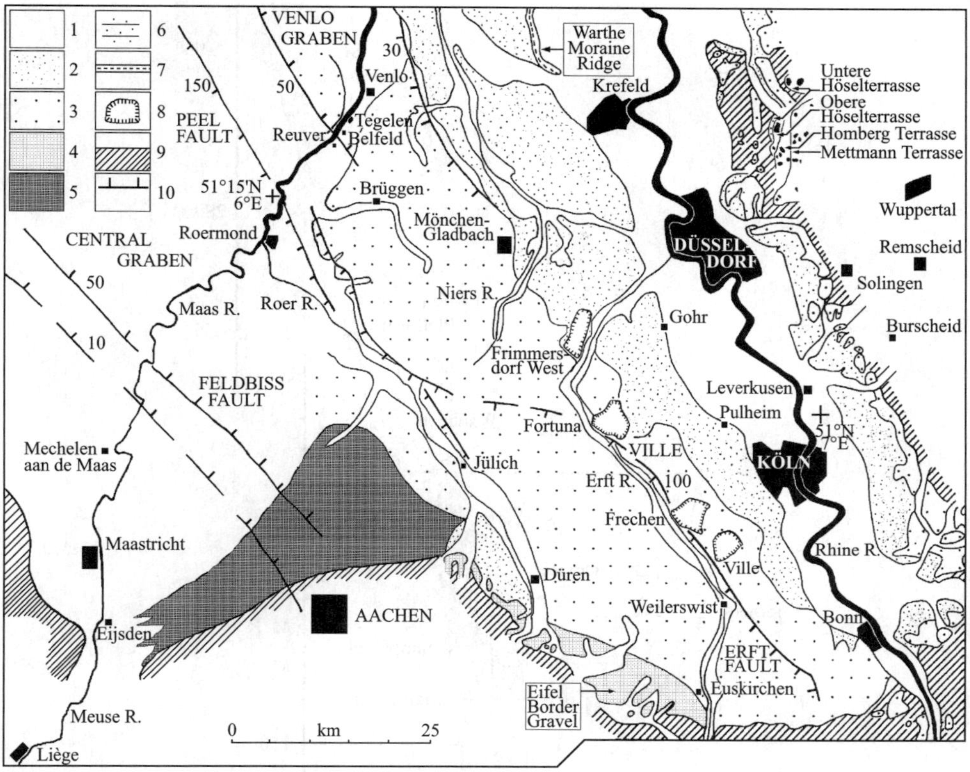

Figure 2. More detailed map of the study region. Key denotes: 1, Rhine Low Terrace (Niederterrasse/NT); 2, Rhine Middle Terraces (Mittelterrassen/MT); 3, Rhine Main Terraces (Hauptterrassen/HT); 4, Eifel border gravel; 5, Early Pleistocene alluvium of the East Maas; 6, gravels of the oldest Rhine terraces (the lines bounding the outcrop areas are for identification purposes and do not indicate original terrace boundaries); 7, Terminal moraine ridge from the Warthe (latest Saalian; oxygen isotope stage 6) glaciation; 8, Opencast lignite mines; 9, Bedrock outcrop; 10, Normal faults, labelled with vertical offset of base Pleistocene, in metres. Information is from Zagwijn (1974), Brunnacker et al. (1982a, b) and van den Berg (1996).

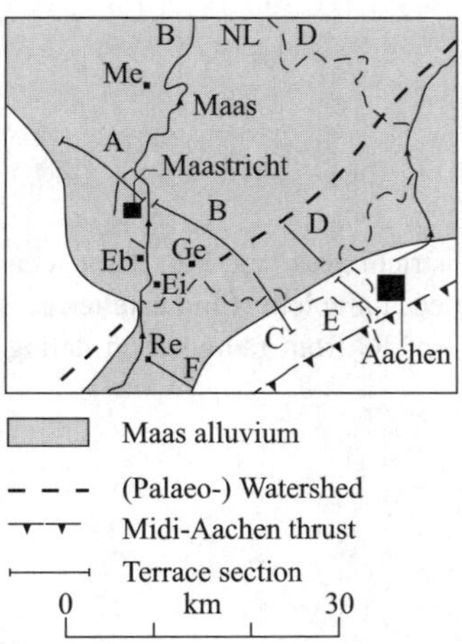

Figure 3. Map of the Maastricht area showing locations of Maas river terrace profiles A to E in Figure 4, profile F which yielded the additional data in Figure 5c, frontiers between Belgium, Germany and the Netherlands, and other details. Me is Mechelen aan de Maas; Ge is St. Geertruid; Eb is Eben-Emaël; Ei is Eijsden; Re is St. Remy. Redrawn from van den Berg (1996) with additional information from Juvigné & Renard (1992).

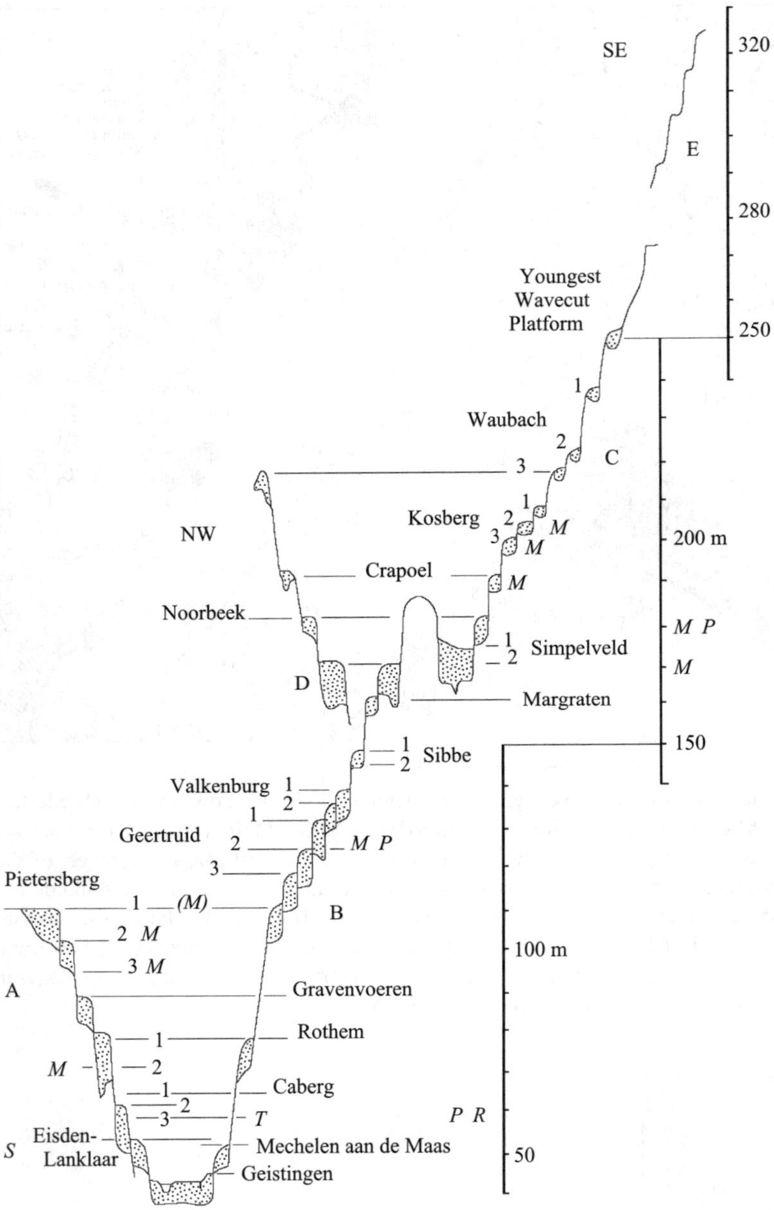

Figure 4. Transverse profiles of Maas river terraces from the Maastricht area (see Fig. 3 for location), adapted from van den Berg (1996). Terrace alluvium is stippled. Italic letters indicate terraces with dating evidence: M, from magnetostratigraphy; P, from pollen; R, from radiocarbon dating; and S, from a soil profile.

2.1 *Quaternary volcanism*

In addition to its uplift, Quaternary volcanism has also occurred in the western part of the Rhenish Massif, the Eifel upland (Fig. 1): the only part of north-western Europe with volcanism documented on this time scale. The nearest adjoining locality with Quaternary volcanism is the Massif Central upland of southern France (e.g. Wilson & Downes, 1991), another region where river terraces provide evidence of more than 100 m of Quaternary uplift (e.g. Veldkamp, 1992). From K-Ar dating, volcanism began in the latest Pliocene in the Western Eifel, where the land surface is typically at ~600 m, and around three quarters of a million years ago in the Eastern Eifel (e.g. Schminke & Meertes, 1979) where the land surface, at ~500 m, adjoins the Rhine near Andernach (Fig. 2). Radiocarbon dating indicates that this volcanism has continued into the Holocene (e.g. Windheuser et al., 1982).

The petrology and geochemistry of the young Eifel volcanics have been investigated by many people (e.g. Illies et al., 1979; Wilson & Downes, 1991). They are alkali basalts, commonly basanites, strongly enriched in potassium and other incompatible elements (K_2O content typically > 3%), and bearing phenocrysts of olivine and titanaugite. They have been interpreted as the result of extraction of metasomatic fluids from the asthenosphere following a small degree of partial melting. However, they also contain mafic and ultramafic xenoliths, interpreted as fragments of a layer of underplating at the base of the crust. Seismic refraction studies (e.g. Mooney & Prodehl, 1978) reveal a layer starting at 28 km depth, with a seismic velocity (v_P) of 7.3 km s^{-1}: an appropriate value for mafic underplating.

Duncan et al. (1972) was the first of many studies to suggest that the Eifel volcanism may be caused by an upwelling mantle plume. They thought that it is the youngest part of a westward younging progression of volcanism across Europe, requiring that the Eurasian plate has moved east relative to this plume by ~700 km since the Oligocene. However, this apparent age progression is refuted by K-Ar dating (Illies et al., 1979). If the plume is instead assumed stationary beneath the plate, it is unclear why it should suddenly cause volcanism in the Quaternary, when the previous significant volcanism in western Germany was in the Miocene (e.g. Schminke & Meertes, 1979).

Other studies (e.g. Illies et al., 1979) have suggested that this volcanism is extension-related, caused by a plate reorganisation which created a component of extension across a zone linking the North Sea, the Rhine Graben (Fig. 1), and the central Mediterranean. Illies et al. (1979) suggested that although the Rhenish Massif is not cut by major active normal faults, it may overlie a zone of incipient extension at depth 'like a front flap of a lederhosen protecting the fly of the trousers'. However, although the present phase of extension in southern Italy did indeed begin about three quarters of a million years ago, it is well-explained in local terms (e.g. Westaway, 1993b) and does not require a throughgoing plate boundary connected to the North Sea. Furthermore, attempts to model global plate kinematics treating western Europe (west of the North Sea and Rhine Graben) as a separate plate from the rest of Eurasia (e.g. Albarello et al., 1995) do not significantly improve the global fit, and require east-west shortening – not extension – across the Rhine Graben and North Sea.

Windheuser et al. (1982) suggested instead that the similarity in timing between Quaternary volcanism in the Eifel, glaciations, and the Quaternary uplift of the Rhenish Massif, requires these processes to be causally related, and suggested two possible mechanisms.

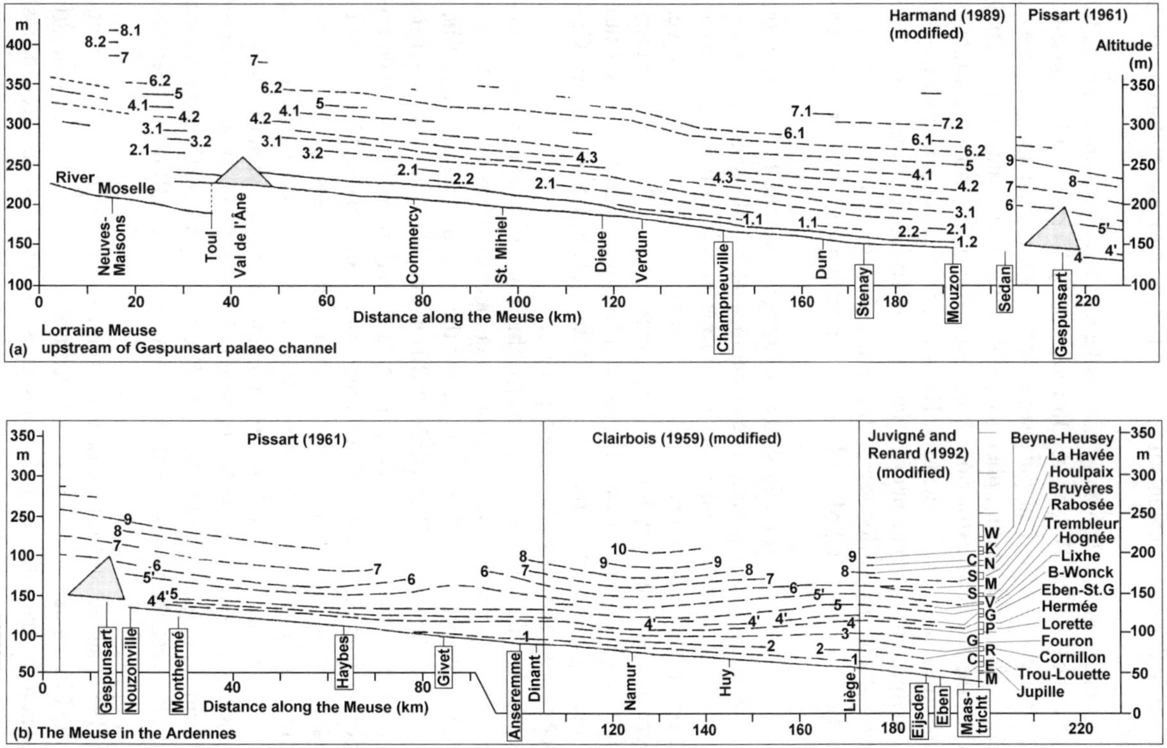

Figure 5a, b. Longitudinal profiles of Maas/Meuse river terraces: (a) in Lorraine (eastern France); and (b) in the Ardennes (France and Belgium), adapted from parts of Figure 2 of Pissart et al. (1997). Figures are labelled to indicate original sources of information, most of which was reinterpreted by Pissart et al. (1997), plus names and numbers of individual terraces. Shading indicates the extent of alluvium deposited in the Val de l'Âne and Gespunsart palaeo channels after they ceased to carry throughgoing drainage. The original Pissart et al. (1997) Figure had no horizontal scale. An approximate horizontal scale has now been provided. Part (b) is also labelled to compare terrace altitudes in Belgium with those at Maastricht (Fig. 4) and to indicate the correlations proposed by Pissart et al. (1997). Some of these do not appear correct given the mismatches in altitude (see main text, also Table 3, for further discussion).

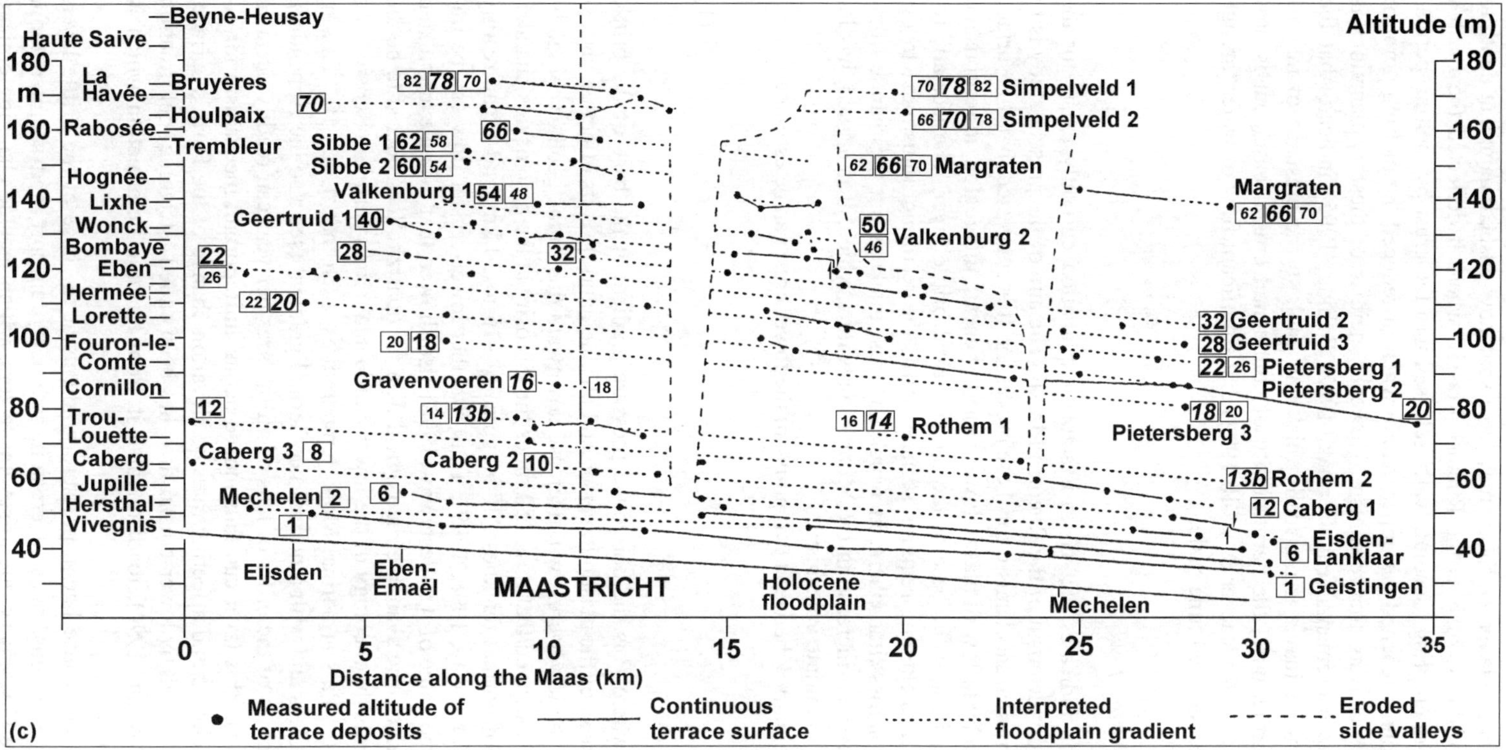

Figure 5c. (c) Maas terraces in the Maastricht area, adapted from van den Berg (1996, Fig. 4.1). Note the offset of the Caberg 2 terrace across one of the strands of the Feldbiss fault zone north of Mechelen aan de Maas. Column to left of figure shows the terrace sequence documented by Juvigné & Renard (1992) at St. Remy (Fig. 3). Terraces are labelled with the oxygen isotope stage numbers in which they formed. Numbers in normal type are from van den Berg (1996); numbers in italics are revisions investigated in this study. In cases where these differ, the larger size type indicates the stage which is preferred as a result of the modelling in this study. Note that the age order proposed for the Bruyères, La Havée, and Houlpaix terraces in (c) was rearranged for no stated reason by Pissart et al. (1997) in (b).

First, loading of Scandinavia by ice sheets may have somehow pushed magma at depth to beneath adjoining localities, such as the Rhenish Massif, where it deflected the Earth's surface upward before erupting. Second, water released into the crust by melting permafrost may have somehow reached the upper mantle where it reduced the melting point of basaltic magma, thus causing the volcanism. However, neither of these explanations is tenable: heat flow measurements (e.g. Cermak, 1982; Balling, 1995) indicate that the lithosphere beneath Scandinavia is unusually thick and cool, so the presence of magma there is unlikely. Furthermore, the permeability of continental crustal basement is now known to be so low (e.g. Rojstaczer & Wolf, 1994) that significant flow of water through it cannot occur on the required time scale.

2.2 *Role of lower-crustal flow*

The above discussion indicates that no satisfactory explanation currently exists for the Quaternary uplift and volcanism of the Rhenish Massif. The aim of the present study is to propose and test a physical mechanism for these processes. It is suggested that this surface uplift is being caused by inflow of plastic lower crust to beneath this region, in response to a lateral pressure gradient at the base of the upper crustal brittle layer. The isostatic response to this crustal thickening requires surface uplift; the resulting adjustment to the geothermal gradient across lithosphere, in which the crust thickens but the mantle lithosphere does not, causes an increase in the Moho temperature which, it is argued, has locally been sufficient to initiate volcanism.

The mean flow velocity V between two points in the lower crust can be estimated as

$$V = \frac{PW^2}{12\eta_e} \quad (1)$$

(Westaway, 1998), where P is the horizontal pressure gradient at the base of the brittle layer, W is the thickness of the plastic lower-crustal channel, and η_e is its effective viscosity. The viscosity of the lower crust will decrease downward due to the geothermal gradient. Westaway (1998) showed that η_e is ~50 times the viscosity at the base of this channel. In continental crust of normal thickness, W is typically ~15 to 20 km, and P can readily reach ~1 kPa m^{-1} (Westaway, 1998, 1999). Thus, whether lower-crustal flow can be important depends on the value of η_e. Westaway (1998) also showed that, when extrapolated to geological strain rates, the available rheological data for quartzite (considered to be the material which controls the rheology of the lower continental crust) permits viscosities no greater than ~10^{16}-10^{17} Pa s at normal Moho temperatures, such that η_e is ~10^{18} Pa s in crust of normal thickness and with a normal geothermal gradient (see also the Appendix). From (1), flow velocities of the order of tens of millimetres per year can readily occur in the lower crust. Flow in this layer can thus indeed act as an important mechanism for altering the crustal thickness and affecting rates of elevation change of the Earth's surface. Crustal deformation caused by lower-crustal flow has been called 'atectonic' deformation (e.g. Kaufman & Royden, 1994), to distinguish it from tectonic deformation which is caused by plate motions.

Several processes can cause lateral pressure gradients at the base of the brittle layer. For instance, in some regions, rates of erosion, sediment transport, and sedimentation, have increased in the Quaternary due to several factors, which include changing climate and destabilisation and subsequent reworking of young, unlithified sediments as a result of the drawdown in sea level during glacials (e.g. Westaway, 1996). The resulting increase in pressure at the base of the brittle upper crust beneath the depocentre (caused by sedimentation) and its reduction beneath the eroding sediment source can cause lower crust to flow to beneath this sediment source, thus adding to the surface relief and allowing this process to be self-sustaining (e.g. Mitchell & Westaway, 1999). However, tests discussed below suggest that this process has not been significant in the present study region: local rates of sediment transport are too low.

A second process is therefore considered: the net effect of pressure variations at the base of the brittle layer caused by changes in surface loading. The main features of this process are illustrated in Figure 6 for loading caused by the changing sea-level between glacials and interglacials: they will result in net flow from beneath the continental shelf (where the lower crust is initially slightly cooler due to the crust being slightly thinner) to beneath land areas (where the lower crust is initially slightly warmer due to the crust being slightly thicker), thus enhancing whatever initial relief existed in a region at the start of glaciation. This process is analysed algebraically in the Appendix, leading to an equation for predicting the surface uplift rate in the region where the crust is initially warmer as a function of time since the start of cyclic loading. At any given time, this uplift rate depends, first, on the temperature difference, near the base of the crust, between the cooler crust which has flowed in from outside and the crust originally present. Second, it also depends on the ratio of dimensions of the uplifting region to those of the adjacent cooler region which has supplied the lower crust. The source of the lower crust which is interpreted as having sustained the surface uplift of the present study region, is unclear. Thus, neither its dimensions nor its temperature contrast are known. However, the dependence of uplift rate on both these factors can be combined by defining a single parameter ΔT_e, the effective temperature difference, which can be varied to match predicted uplift rates against observations. Resolution of the resulting ΔT_e values into regional dimensions and actual temperature differences is beyond the scope of this study.

An additional loading effect is also possible, due to the cyclic growth and decay of ice sheets. Algebraically, this process is equivalent to Figure 6, the main differences being, first, that the loading is during glacials, not interglacials, and, second, loading by ice sheets can be more than an order of magnitude greater: a ~3 km thick ice sheet will load the underlying crust by ~30 MPa, compared with ~1 MPa for a ~100 m increase in sea-level. Given the analysis in the Appendix, cyclic ice loading will cause net outflow of crust from beneath the loaded region to beneath its surroundings, provided the crust in the loaded region is cooler. Thus, provided a locality has crust which is initially thicker or has higher heat flow than in an adjoining region which is periodically covered by ice, this locality will experience net inflow of lower crust causing uplift of the Earth's surface.

2.3 *Magnitude of surface processes in the study region*

Table 1 lists discharge and sediment transport data for the Rhine, Maas, and other major rivers which drain into the southern North Sea. Their total sediment load is roughly 0.002 $km^3\ yr^{-1}$, requiring a spatial average denudation rate across their drainage basin area of ~0.005 mm yr^{-1}. The total volume of Quaternary sediment in and around the southern North Sea is ~5000 km^3 (e.g. Cameron et al., 1992), which happens to equal the presentday

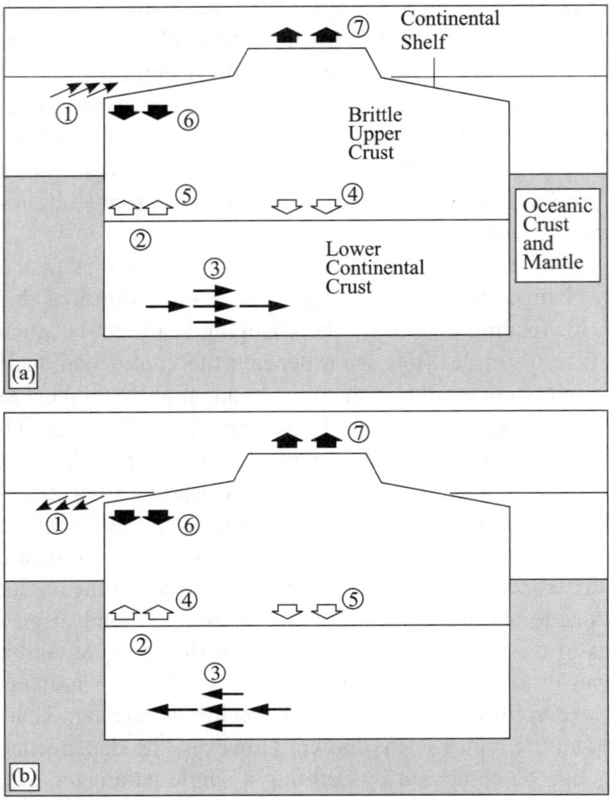

Figure 6. Summary of model. (a) During interglacial marine highstands, the rise in sea level (1) causes an increase in the pressure at the base of the brittle layer beneath the offshore shelf (2). This causes a landward pressure gradient, which drives relatively cool lower crust from beneath the shelf to beneath the land (3). This inflow of relatively cool lower crust perturbs the geothermal gradient beneath the land, causing the base of the brittle layer to advect downward in accordance with Equation (A11) (4). The effect of earlier flow of warm lower crust to beneath the offshore shelf causes a corresponding perturbation to the offshore geothermal gradient, which causes the base of the brittle layer to advect upward in accordance with Equation (A11) (5). Taking account of the need to balance overall crustal volume within a closed system, the overall isostatic response to these processes requires the sea floor of the shelf to adjust downward, causing net thinning of the offshore crust and deepening of the shelf sea (6), and thickening of the crust beneath the land, causing the land surface to uplift (7). (b) During glacial marine lowstands, the fall in sea level (1) causes an decrease in the pressure at the base of the brittle layer beneath the offshore shelf (2). This causes a seaward pressure gradient, which drives relatively warm lower crust from beneath the land to beneath the shelf (3). This inflow of relatively warm lower crust perturbs the geothermal gradient beneath the shelf, causing the base of the brittle layer to advect upward in accordance with Equation (A11) (4). The effect of earlier flow of cool lower crust to beneath the land causes a corresponding perturbation to the onshore geothermal gradient, which causes the base of the brittle layer to advect downward in accordance with Equation (A11) (5). Taking account of the need to balance overall crustal volume within a closed system, the overall isostatic response to these processes requires the sea floor of the shelf to adjust downward, causing net thinning of the offshore crust and deepening of the shelf sea (6), and thickening of the crust beneath the land, causing the land surface to uplift (7).

Table 1. Sediment transport by major rivers in north-west Europe.

River	Area km^2	Load 10^9 kg yr^{-1}	Yield 10^3 kg km^{-2} yr^{-1}	Denudation rate mm a^{-1}
Schelde	22,000	1.0	45	0.023
Maas	29,000	0.70	24	0.012
Rhine	170,000	0.72	4	0.002
Weser	38,000	0.33	8	0.004
Elbe	130,000	0.84	6	0.003
Total	389,000	3.59	9.23	0.005

Data are from Milliman & Meade (1983) and Milliman & Syvitski (1992). Yield is load divided by area. Spatial average denudation rates are estimated as yield divided by a notional sediment density of 2000 kg m^{-3}.

annual sediment load multiplied by the duration of the Quaternary. However, during glacials, when a Scandinavian ice sheet has covered much of northern Europe, major rivers have flowed westward around the southern ice margin into the southern North Sea (e.g. Zagwijn, 1986a). A substantial part of the ~5000 km^3 of Quaternary sediment in southern North Sea, plus the ~15,000 km^3 in the central North Sea (e.g. Gatliff et al., 1994), has thus probably originated from east of the modern drainage basin, either from these transient rivers or from the Early Pleistocene Baltic River mentioned earlier. An additional complication affecting the overall erosion versus sedimentation budget for this study region arises through the widespread deposition of loess during glacials. Some of the Quaternary sediment in the North Sea was presumably deposited subaerially as loess at times of lowered sea-level, in addition to the fluvial input. Other deposits of loess also cover much of the landscape in the study region. Some of this material may be subsequently reworked and contribute to the present-day sediment loads of the region's rivers. Taking these factors into account, it is difficult to say whether the present-day sediment loads transported by the Rhine and Maas are less than, or greater than, their time-averaged loads throughout the Quaternary. However, it is unlikely that their upper bound is more than, say, double their present-day value.

A substantial proportion of the total Quaternary sediment load of rivers in the study region can be presumed to result from valley incision: including both incision of bedrock and reworking of older terrace deposits. Denudation rates of interfluves are thus less than the spatial average of ~0.005 mm yr^{-1}, perhaps by an order of magnitude, especially in localities where erosion-resistant Palaeozoic rocks crop out at the Earth's surface. It is not even clear that these regions experience net erosion: in the absence of detailed measurements it is unclear to what extent the deposition of loess during glacials offsets the erosion during interglacials. It is nonetheless evident that the surface uplift rates, which reach ~0.2 mm yr^{-1} in some upland parts of the study region (see below), are orders of magnitude greater than any feasible net local erosion rate. One is thus readily justified in adopting in the present study region, the assumptions that (a) rates of erosion from upland localities within interfluves are negligible, and (b) the isostatic response to such erosion is not significant. These assumptions will be followed throughout this study.

3 CHRONOLOGIES OF RIVER TERRACES

3.1 *Plio-Quaternary stratigraphy of north-west Europe*

Global variations in the $^{18}O/^{16}O$ isotope ratio in microfossils in marine sediments provide the primary modern definition of the Pliocene and Quaternary time scale: the oxygen isotope time scale (e.g. Shackleton et al., 1990), in which odd- and even-numbered stages mark interglacials and glacials. The magnetostratigraphic time scale, based on polarity reversals of the Earth's magnetic field, is calibrated relative to the oxygen isotope stages (e.g. Shackleton et al., 1990). The series of local stage definitions of geological time for north-western Europe is, as already mentioned, based on pollen assemblages from interglacials and interbedded glacial sediments (e.g. Zagwijn, 1974, 1986a, 1986b, 1989, 1992, 1996).

The global definition of the Plio-Quaternary boundary is presently based on a marine stratotype in southern Italy. It occurs shortly after the Olduvai normal magnetic polarity subchron, and coincides with several marine biostratigraphic changes including the first local appearance of the cold water mollusc *Arctica islandica* (Aguirre & Pasini, 1985). The age of this event is currently estimated as ~1.81 Ma by tuning the sedimentary record at the stratotype to the orbital forcing stratigraphy (e.g. Hilgen, 1991a, 1991b). In and around north-western Europe, a more dramatic change in the sedimentary record occurs at ~2.5 Ma, when evidence of ice rafting in marine sediments indicates the start of lowland glaciation in the northern hemisphere (Shackleton et al., 1984). This change occurs just after the end of the Gauss normal-polarity chron, marks the start of the Pretiglian stage of the local stratigraphy (in which pollen evidence from the Netherlands reveals the first major cooling), and has been adopted by Dutch and British investigators as the local definition of the Plio-Quaternary boundary (e.g. Zagwijn, 1992; Cameron et al., 1992; Gatliff et al., 1994).

Within the Pleistocene, the main uncertainty concerns the duration of the Pretiglian stage. Although some studies (e.g. Brunnacker, 1986; Gatliff et al., 1994) have accepted that this ends around the time of the Réunion normal-polarity subchrons at ~2.1-2.0 Ma, others (e.g. Grube, 1986; van den Berg, 1996) have regarded the Pretiglian as much briefer, ending several hundred thousand years earlier. A consistent interpretation of the available pollen and magnetostratigraphic evidence from the Rhine terraces requires the longer duration of the Pretiglian (see below, also Brunnacker, 1986), but this raises some difficulties over constraining the chronology of the Maas terrace sequence, for which the recent work by van den Berg (1996) has assumed the shorter duration of the Pretiglian (see below).

3.2 *Quaternary evolution of the Maas river system*

Many studies have investigated the evolution of the Maas/Meuse river system and the development of its terrace sequence (e.g. Davis, 1895; Macar, 1945; Clairbois, 1959; Pissart, 1961, 1975; de Heinzelin, 1964; Zonneveld, 1975; Felder et al., 1989; Harmand, 1989, 1992; Juvigné & Renard, 1992; Ruegg, 1994, 1995; van den Berg, 1996; Pissart et al., 1997). De Heinzelin (1964) first proposed that this river adopted its present overall form, linking the Lorraine region of eastern France and the south-eastern Netherlands via the Ardennes, during the Late Miocene. However, Pissart et al. (1997) argued, using provenence evidence of heavy minerals in some of the oldest terraces, that this present overall form developed in the Middle Miocene. Previously, the upper Meuse flowed west of the Ardennes to the palaeo coastline near Cambrai, in what is now the drainage basin of the Schelde/Escaut (e.g. Pissart, 1975) (Fig. 1).

Figure 1 shows some of the principal changes which have occurred to the geometry of this river system during the Quaternary. First, in the vicinity of Maastricht, the Maas originally occupied the East Maas valley which led northeastward towards the Rhine. The complex interrelationships between sediments of the lower Rhine and lower Maas have been discussed by many people (e.g. Zonneveld, 1975; Zagwijn, 1989; Ruegg, 1994; Tebbens et al., 1995; van den Berg, 1996) and are not repeated here. The main effect of this complexity on the present study is to make the Maastricht area the northernmost locality with an intact terrace sequence which spans the Latest Pliocene and Quaternary (Fig. 5). The Simpelveld 2 terrace of van den Berg (1996) (Fig. 4) is the youngest which indicates drainage through the East Maas palaeochannel. Zagwijn (1974) first dated this drainage adjustment to the Late Tiglian stage of geological time (see below).

Second, as Davis (1895) first noticed, the Meuse south of the Ardennes has also experienced major changes (Fig. 1). In the Early Pleistocene, it occupied the Gespunsart palaeo-channel between Sedan and Nouzonville (e.g. Pissart et al., 1997). At this time, the headwaters of the Marne and Aisne rivers (which now drain west into the Seine) flowed north into the Bar which joined the Meuse at Nouzonville (e.g. Pissart et al., 1997). According to Pissart et al. (1997), the loss of the original headwaters of the Bar and the capture of the Meuse by the lower Bar (causing the abandonment of the Gespunsart palaeochannel) both occurred between the times of formation of Meuse terraces 5 and 4' of Pissart (1961). Pissart et al. (1997) correlated these with the Geertruid 1 and 2 terraces of the Maastricht area (Fig. 5b), and dated them to ~1.0 and ~0.85 Ma. In contrast, van den Berg (1996) dated the formation of the Geertruid 1 and 2 terraces at ~1.3 and ~1.1 Ma. However, careful comparison of the terrace profiles from Pissart et al. (1997) (Fig. 5b) and van den Berg (1996) (Fig. 5c) reveals that Middle Meuse terraces 5 and 4' in fact match van den Berg's (1996) Geertruid 3 and Pietersberg 1 terraces (Table 3). My revised chronology (see below) dates these to ~1.0 and ~0.9 Ma, indicating that the major changes to the Meuse indeed occurred around 0.95 Ma. Van Balen et al (2000) have instead correlated Middle Meuse terraces 5 and 4' with the Geertruid 2 and 3 terraces in the Maastricht area. However, although they did not quote an age for the Geertruid 2 terrace, they dated the Geertruid 3 terrace to ~0.85 Ma (oxygen isotope stage 22). Thus, in their scheme, these major diversions also occurred around ~0.9 Ma.

The final important adjustment, also first noted by Davis (1895), was the capture near Toul of the original Meuse headwaters, which drained the Vosges mountains, and their diversion into the Moselle causing the abandonment of the Val de l'Âne palaeochannel. This event is now well-dated to within oxygen isotope stage 6. The best lines of evidence are, first, that distinctive pebbles derived from the Vosges (for instance, of granite) are not found in the Maas terraces around Maastricht younger than the Caberg 3 from oxygen isotope stage 8 (van den Berg, 1996; see also below). Second, along the Moselle upstream of Toul, there are three terraces known locally as Fx1, Fx2, and Fx3, which are regarded as of Saalian age (Taous, 1994). The older two, Fx1 and Fx2, can be traced through the Val de l'Âne into the modern Meuse drainage basin, but Fx3 continues along the Moselle downstream of the Toul capture angle (Harmand et al., 1995). Terrace Fx3 is evidently latest Saalian, having aggraded after the peak of the glaciation during oxygen isotope stage 6; the Val de l'Âne was thus abandoned before this time. Third, along the Moselle between Toul and Trier, the terrace equivalent to Fx3 contains clasts of Vosges granite, unlike the older terraces (e.g. Negendank, 1983; Pissart et al., 1997).

3.3 *Terraces of the Maas around Maastricht*

Relationships between the 31 documented terraces of the Maas around Maastricht are shown in Figure 4. Altitude and interpreted age data are compiled in Table 2 and illustrated in Figure 7. Van den Berg (1996) devised a chronology (summarised in Fig. 7) for these terraces based on evidence from the Maastricht area of the Netherlands. In contrast, Pissart et al. (1997) has proposed a different chronology (also summarised in Fig. 7) for terraces of the Meuse in France and Belgium. Many individual terraces can be traced continuously across the Belgium-Netherlands border ~10 km south of Maastricht (Fig. 3), although their interpreted ages differ from one country to the other. Van den Berg (1996) did not attempt to reconcile his interpretation with previous schemes for France and Belgium, such as Pissart (1975) whose chronology is essentially the same as that of Pissart et al. (1997). Nor did Pissart et al. (1997) mention the new terrace evidence reported by van den Berg (1996).

It is clear that much of the Pissart et al. (1997) chronology is invalidated by van den Berg's (1996) new evidence (notably, his new magnetostratigraphic evidence and geomorphological evidence of additional terraces which need to be fitted into the sequence). However, some pieces of information appear to permit multiple interpretations. For instance, the Tiglian pollen assemblage in sediments of the Simpelveld terraces with normal magnetic polarity led van den Berg (1996) to deduce a match to the Réunion subchron but led Pissart et al. (1997) to deduce a match to the Olduvai. As both schemes have fitted neighbouring terraces subject to this constraint, the possibility exists that one of them contains a systematic error of up to several hundred thousand years for much of the Early Pleistocene. The rest of this section discusses the available dating evidence from the Maastricht area, with the aim of establishing which parts of van den Berg's (1996) chronology are well-established and which can be adjusted while remaining consistent with the available evidence. Attempts will later be made to model the uplift history of this area, using van den Berg's (1996) chronology as well as considering the revision which is tentatively suggested here.

3.3.1 *Dating evidence*

Pollen evidence, radiocarbon dating, sedimentology, and geomorphological evidence indicates that the Mechelen aan de Maas terrace developed during the last glacial cycle: aggradation began in oxygen isotope stage 5d (~110 ka) and ended around the end of stage 2 at ~13-14 ka (e.g. Kasse et al., 1995; van den Berg, 1996). The Geistingen terrace, a few metres lower, thus aggraded in the earliest Holocene (~10 ka) (van den Berg, 1996). The Eisden-Lanklaar terrace is capped by a palaeosol which is correlated with the Eemian interglacial (oxygen isotope stage 5e) (van den Berg, 1996). A slightly older age of 130 ka is thus adopted for the end of aggradation of this terrace. The Caberg 3 terrace is also overlain by interglacial deposits, from the Belvédère interglacial (van den Berg, 1996), the local equivalent of the Treene interglacial in oxygen isotope stage 7. Thermoluminescence dating of associated burned artefacts has revealed an age of 250 ± 20 ka (Huxtable, 1993).

These three river terraces thus all formed during cold stages. Although no direct evidence of aggradation preceding interglacials exists for the older terraces, many show sedimentological evidence of cold conditions, such as ice-wedge casts, metre-sized boulders (presumed to have been transported by ice floes), angular blocks of loose sand (presumed to have been transported in a frozen state), and evidence of cold-climate faunas (van den Berg, 1996). These older terraces thus also formed during cold periods, with aggradation

Table 2. Maas river terraces at Maastricht.

Name	H (m)	S	t (ka)	δH (m)	U (m)
Geistingen	46	1	10	0	4
Mechelen aan de Maas	49	2	14	0	7
Eisden-Lanklaar	52	6	130	0	10
Caberg 3	57	8	245	0	15
Caberg 2	63	10	330	0	21
Caberg 1	70	12	420	0	28
Rothem 2	76	13b	470	0	34
Rothem 1	81.5	14	510	0	39.5
Gravenvoeren	86	16	620	0	44
Pietersberg 3	97.5	18	700	0	55.5
Pietersberg 2	104.5	20	780	0	62.5
Pietersberg 1	112.5	22	870	0	70.5
Geertruid 3	119.5	28	1030	0	77.5
Geertruid 2	124	32	1090	0	82
Geertruid 1	130	40	1280	0	88
Valkenburg 2	136	46	1420	0	94
Valkenburg 1	139	48	1460	0	97
Sibbe 2	146	54	1570	0	104
Sibbe 1	150	58	1650	0	108
Margraten	159	62	1740	0	117
Simpelveld 2	167	66	1820	0	125
Simpelveld 1	175	68	1870	0	133
Noorbeek	184	78	2060	0.7	142.7
Crapoel	190	82	2140	1.0	149
Kosberg 3	198	-	2690	2.9	158.9
Kosberg 2	202	-	2810	3.3	163.3
Kosberg 1	206	-	2940	3.8	167.8
Waubach 3	215	-	4500	9.3	182.3
Waubach 2	220	-	6000	14.5	192.5
Waubach 1	235	-	10,000	28.6	221.6
Marine wavecut platform	250	-	13,800	42	250

This chronology is devised in the present study to illustrate, by comparison with van den Berg's (1996) chronology, the points at which uncertainties lie, as an aid to modelling. Terrace nomenclature and altitudes (H) are from van den Berg (1996). Van den Berg's (1996) count of 31 terraces includes the modern floodplain as well as the 30 terraces listed here. Terrace ages (t) are interpreted in this study. See Figure 7 for comparisons between this chronology and the alternative schemes by van den Berg (1996) and Pissart et al. (1997). δH is the correction for conversion between incision and uplift, calculated using Equation (2) assuming that the coastline prograded at a uniform rate between 13800 ka and 1870 ka during which time the Maas maintained a constant gradient. The resulting estimated value of δH takes values of up to 42 m, consistent with 60 km of progradation at the 0.7 m km^{-1} present-day gradient of the Maas at Maastricht, or 120 km of progradation at a gradient of 0.35 m km^{-1}. Amounts of uplift (U) are estimated as H + $\delta H - H_o$ for each terrace, where H_o is a local reference level, also of 42 m, which approximates the Holocene floodplain level of 38 m.

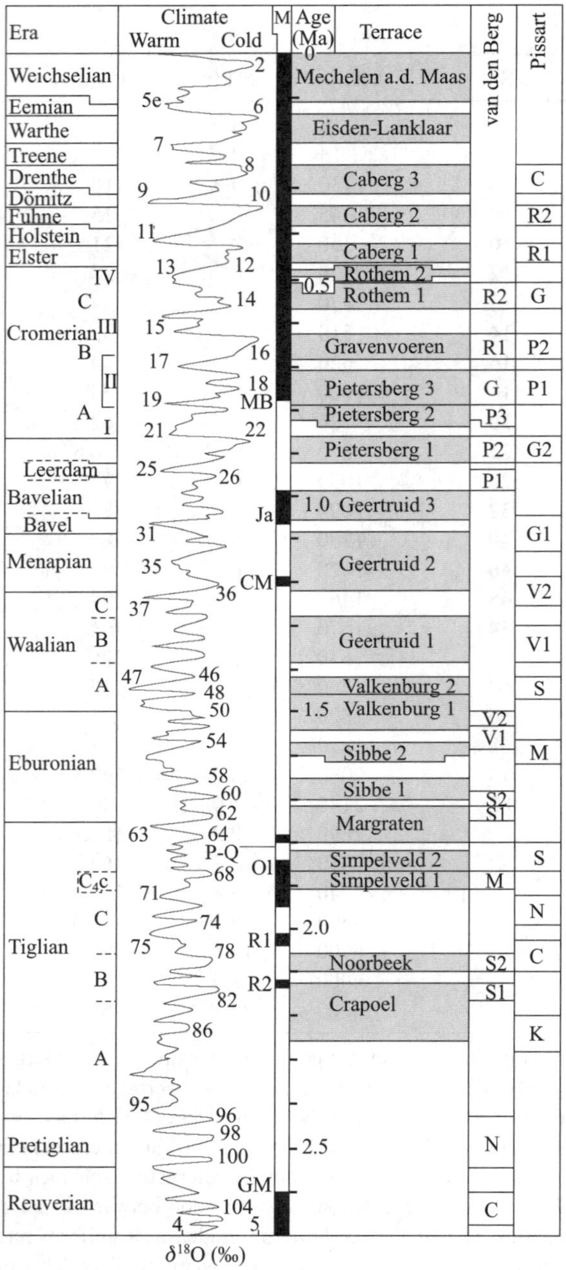

Figure 7. Interpreted chronology of Maas river terraces in relation to regional stratigraphy, oxygen isotope stages, and the chronology of magnetic polarity reversals. Intervals of time over which terrace gravels are interpreted as aggrading are stippled. The alternative terrace chronologies, from van den Berg (1996) and Pissart et al. (1997), are shown for comparison. Please note that this Figure shows the age assignments by Pissart et al. (1997), whereas Table 3 shows the age assignments which would follow by using their proposed correlations between the Maastricht and Liège areas, then re-correlating the terraces by altitude between these two areas and taking into account the extra terraces newly-discovered by van den Berg (1996). The apparent ages of some terraces attributed to Pissart et al. (1997) differ between Figure 7 and Table 3, apparently because some Maastricht terraces were wrongly correlated by Pissart et al. (1997). Preferred correlations of Cromerian interglacials and glacials follow discussion in this study. Timings of stages of the Tiglian era are taken from van den Berg (1996). Note that these differ from the values used in Figure 13. Timings of polarity reversals of the Earth's magnetic field summarise orbital-forcing stratigraphic results of Shackleton et al. (1990) and Hilgen (1991a), and other data from the review by Würm (1997). P-Q marks the conventional Pliocene-Quaternary boundary, from Hilgen (1991a,b).

ceasing shortly before interglacials. Each terrace can therefore be identified using the youngest oxygen isotope cold stage which it spans.

Some age bounds can be placed on the older terraces using magnetostratigraphic and pollen evidence. The Rothem 2 terrace contains sediment with normal magnetic polarity, and thus evidently formed during the Brunhes chron (van den Berg, 1996).

The Pietersberg 3 terrace yielded specimens with normal polarities, whereas the underlying Pietersberg 2 terrace yielded both reverse- and normal-polarity specimens (van den Berg, 1996). Van den Berg (1996) assigned these terraces to oxygen isotope stages 20 and 22, apparently in order to extrapolate the sequence of one terrace per glacial cycle. However, this places the formation of both terraces within the Matuyama chron, which requires the normal magnetic polarities in both to be explained as results of overprinting by the Earth's magnetic field with its modern polarity. Another possibility is thus to assign these terraces to oxygen isotope stages 18 and 20. If so, Pietersberg 3 falls within the Brunhes chron, and its normal magnetic polarity may thus reflect primary magnetisation and not overprinting.

Van den Berg (1996) reported that the Pietersberg 1 terrace has also yielded specimens of normal magnetic polarity. Neither his chronology nor mq proposed revision can account for this, which presumably also reflects overprinting.

However, if the Pietersberg 3 terrace is matched to oxygen isotope stage 18, as I propose, there is now one terrace too many in the Maas sequence for each to match one cycle of Middle Pleistocene glaciation. One possible explanation for this would use the complexity of oxygen isotope stage 15, which is known to consist of two or three interstadials between ~620 and ~580 ka (e.g. Shackleton et al., 1990; Bassinot et al., 1994), which produced separate marine terraces in some localities (e.g. Pirazzoli et al., 1991). In principle, the Rothem 1 terrace could have formed during the intervening interstadial(s), leaving a subsequent sequence of one terrace per glacial cycle. However, oxygen isotope stage 15 corresponds to the Cromerian III interglacial identified biostratigraphically, which in the Netherlands is known as the Rosmalen interglacial (e.g. Turner, 1996). At the stratotype at Rosmalen, the sediments from these interstadials are separated by layers, up to ~1.5 m thick, of sand and gravel from cold stadials (see Zagwijn, 1996, p. 153). However, the gravel forming the Rothem 1 terrace in the Maastricht area is much thicker than this (see Fig. 4, also van den Berg, 1996).

Another possibility, which I prefer, reflects the complexity of the subsequent interglacial stage 13. In addition to other climate fluctuations revealed by oxygen isotope evidence, it appears that the main interglacial at ~520 ka was followed by a cold stage, then a second warm stage at ~470 ka, which seems to have been more significant given recent high-resolution records (e.g. Bassinot et al., 1994) rather than older ones (e.g. Shackleton et al., 1990; Fig. 7). If the Rothem 1 terrace formed during stage 14, then Rothem 2 could have formed during the cold spell between ~520 and ~470 ka (which I tentatively name stage 13b). The Caberg 1 terrace could then have formed during stage 12 (the Elster glaciation), as was originally proposed by van den Berg (1996).

Knowledge of the age of the Pietersberg terraces is quite important for modelling the overall uplift history, because these marked an approximate doubling in the uplift rate compared with during deposition of the preceding Geertruid terraces (Figs 8 and 9). As is discussed in more detail later, it is much easier to account, via physically plausible models, for the later timing of this increase in the uplift rate indicated by my revised terrace ages than the earlier increase indicated by van den Berg's (1996) ages.

The Geertruid 2 terrace contains a reverse – normal – reverse magnetic polarity sequence which van den Berg (1996) interpreted as spanning the Cobb Mountain subchron. An overlying palaeosol contains pollen characteristic of the Bavel interglacial (van den Berg, 1996), which marked oxygen isotope stage 31. Aggradation of this terrace thus ended at the end of stage 32, around 1090 ka.

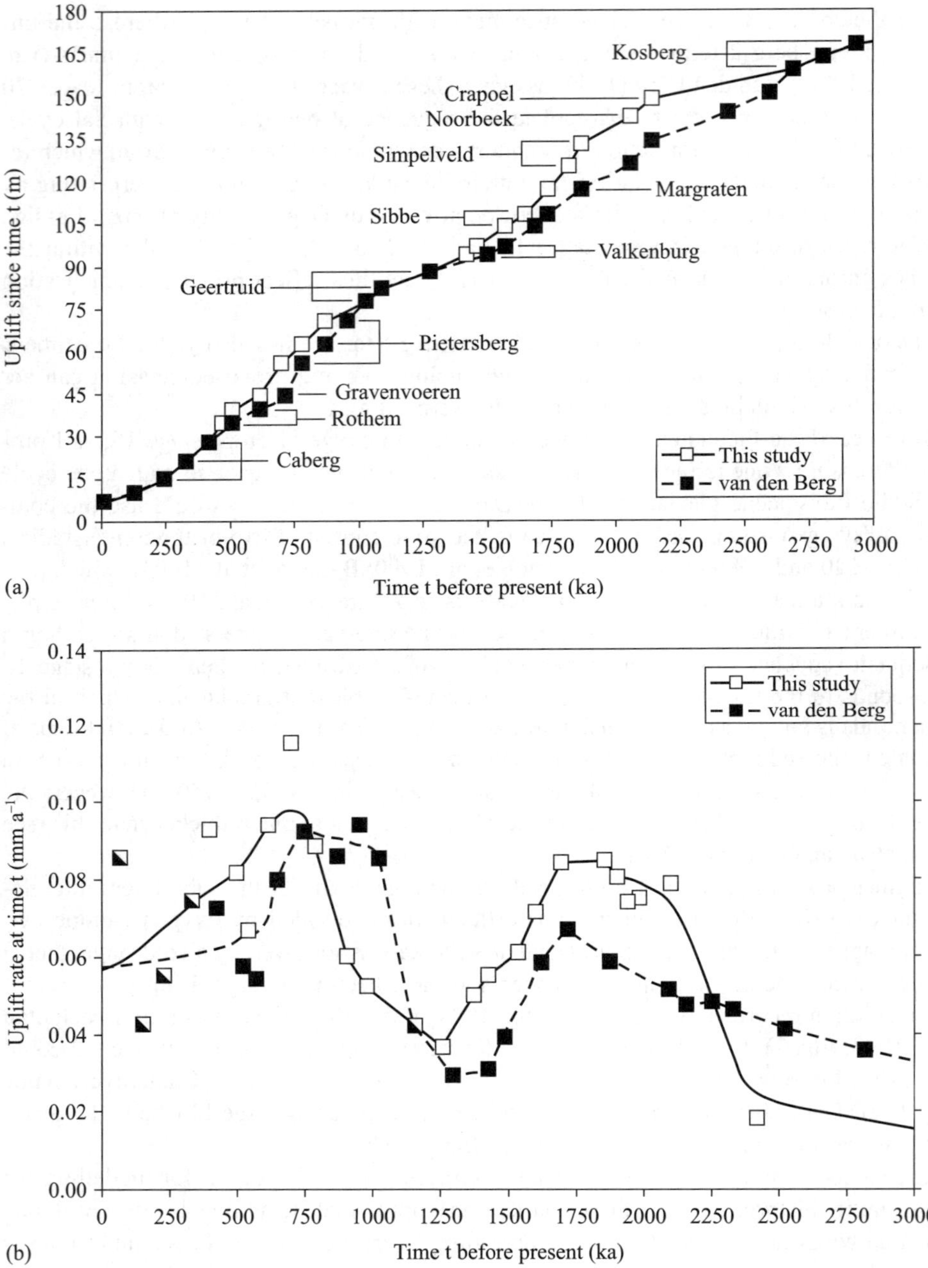

Figure 8. Comparison of estimates of uplift since deposition of each of the Maas terraces at Maastricht using the chronology of van den Berg (1996) (Fig. 7) and incorporating the revisions tentatively suggested in this study (Table 2). Estimates of uplift assume that the Maas channel lengthened by 60 km with a palaeo gradient of 0.7 m km^{-1} (or 120 km with a palaeo gradient of 0.35 m km^{-1}) at a uniform rate between 13.8 Ma and oxygen isotope stage 68 at 1.87 Ma, the Tiglian C_4c, which corresponds to the age of the Margraten terrace in van den Berg's (1996) chronology and the Simpelveld I terrace in the revision. (a) uplift; and (b) uplift rates. See text for discussion.

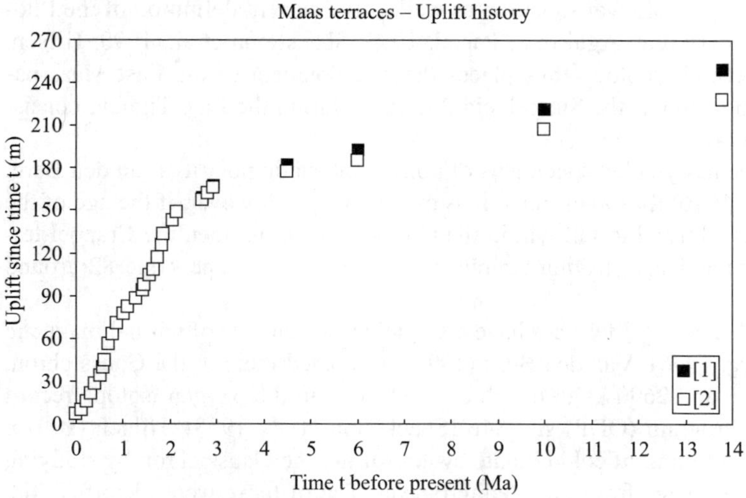

Figure 9. Uplift histories of the Maastricht area, derived from terrace altitudes. Data series 1 shows the data from Table 2 for which, in the absence of channel lengthening, incision is equated to uplift of the adjoining land surface. For this data series, channel length to the palaeocoastline is assumed to have increased by 60 km, with the river gradient maintained at 0.7 m km^{-1}, at a uniform rate between 13.8 Ma and 1.87 Ma. Data series 2 shows the alternative uplift history estimated for the same chronology of channel lengthening but with the river gradient maintained at 0.35 m km^{-1}. The two chronologies differ negligibly since ~3 Ma, since when points for data series 1 are omitted from this figure, and have been identical since 1.87 Ma due to the absence of systematic channel lengthening.

The Valkenburg terraces underlie loess with reverse magnetic polarity (van Monfrans, 1971). They thus formed during the Matuyama chron.

The Simpelveld 2 and Simpelveld 1 terraces yielded specimens with normal magnetic polarity (van den Berg, 1996). The Simpelveld 1 terrace is also overlain by peat which has yielded pollen of Late Tiglian age, and has thus previously been assigned to Late Tiglian time (e.g. Zagwijn, 1986a). Van den Berg (1996) thus correlated these two terraces with the Réunion 1 and 2 normal-polarity events during oxygen isotope stages 82 and 78. However, these oxygen isotope stages fall within the Early to Middle Tiglian, not the Late Tiglian (Fig. 7). It is thus evident that adoption of van den Berg's (1996) chronology requires a relaxation of Zagwijn's (1986a) age constraint. This can arguably be justified on the grounds that the pollen data available to Zagwijn were not particularly clear, and could possibly indicate the Early or Middle Tiglian, not just the Late Tiglian (M. van den Berg., pers. comm., 1999).

Another possibility, which I consider in addition to van den Berg's (1996) scheme, instead correlates both Simpelveld terraces with the Olduvai subchron during the Late Tiglian. Such a match was indeed suggested by Pissart et al. (1997). If so, the Simpelveld 1 terrace aggraded during oxygen isotope stage 68 which ended around 1870 ka. Van den Berg (1996) correlated stage 68 with the Tiglian C_4c, a distinctive cold stage known as the Beerse glacial (e.g. Kasse, 1998). The Simpelveld 2 terrace probably formed during the subsequent sequence of rather moderate climate fluctuations to stage 66 at ~1820 ka (Fig. 7), or possibly during time up to the rather more important cold stage 64 at ~1780 ka which marks the end of the Olduvai subchron as well as the current definition of the Plio-cene-Pleistocene boundary (e.g. Aguirre & Pasini, 1985; Shackleton et al., 1990; Hilgen, 1991a). This alternative chronology thus places the abandonment of the East Maas pa-laeochannel, after deposition of the Simpelveld 2 terrace, during the Late Tiglian; consis-tent with Zagwijn (1974).

The Crapoel terrace has yielded specimens of normal magnetic polarity (van den Berg, 1996). Van den Berg (1996) thus interpreted it as of Gauss age. However, if the age of the Simpelveld terraces is adjusted to fall within the Olduvai subchron, then the Crapoel ter-race could have Réunion 1 age, having aggraded up to oxygen isotope stage 82 around 2140 ka.

The Kosberg 2 and Kosberg 3 terraces have also yielded specimens of normal magnetic polarity (van den Berg, 1996). Van den Berg (1996) thus dated them to the Gauss chron, assigning ages of 2810 and 2690 ka to match cold intervals in the oxygen isotope record from Ocean Drilling Program (ODP) site 846 (Shackleton et al., 1993). Hilgen (1991a, 1991b) investigated the timing of cold climate cycles during the Gauss chron by studying sapropels in marine sediments from the Mediterranean. Ten of these were identified, the oldest having an age of ~3100 ka, any of which could correspond to the three Kosberg ter-races. The oldest, the Kosberg I terrace, could thus have an age of ~3100 ka, rather than the younger age of 2940 ka proposed by van den Berg (1996) to match another cold inter-val in the ODP site 846 record.

These Kosberg terraces evidently pre-date the ~2500 ka start of significant lowland glaciation of the northern hemisphere (e.g. Shackleton et al., 1984). Given their age, their periglacial character (including evidence of ice-rafted deposition of large boulders) ap-pears surprising. However, such ice-rafted material does not necessarily imply periglacial conditions at low levels at this time, as it may have been derived from valley glaciers or a localised high-level ice sheet in the Vosges mountains, which in the Late Pliocene were linked to the Maastricht area by the Val de l'Âne palaeo channel (Fig. 1). The youngest sapropels in the Mediterranean sea occur at times of rapid melting of ice sheets, when the influx of low-density fresh water from rivers draining Europe causes the vertical circula-tion of the Mediterranean Sea to stagnate (e.g. Cita et al., 1977; Thunell et al., 1977; Vergnaud-Grazzini et al., 1977; Aksu et al., 1987). This presence of sapropels from ~3100 ka onwards thus indicates that localised glaciation in upland parts of Europe started around 3100 ka.

Like the Kosberg terraces, van den Berg (1996) assigned ages to the Waubach terraces by matching successive cold intervals in oxygen isotope records from ODP site 846 (Shackleton et al., 1993) and other sites, working backwards in time using no independent age control: thus assigning ages of 3140, 3300, and 3980 ka. It is evident that this method will result in the *youngest* feasible ages for these terraces. The cooling at 3140 ka led to the formation of the earliest Mediterranean sapropel (Hilgen et al., 1991b), which could instead mark the Kosberg 1 terrace (see above). Earlier substantial cooling events evident at ODP site 846 occur at ~4.8 Ma and ~5.7 Ma (Shackleton et al., 1993). It is thus possible that the three Waubach terraces formed at ~5.7, ~4.8, and ~4.0 Ma. It is even possible that one or more of them is older than ~6 Ma, and thus pre-dates the current limit of orbital-forcing stratigraphic age control. Van den Berg (1996, Fig. 7.5, p. 130) indeed considered one such solution, with the Waubach I, II, and III terraces tentatively assigned ages of ~9.6, ~5.7, and ~4.1 Ma. In such a solution, there is an abrupt, order-of-magnitude in-crease in the uplift rate around 3 Ma (Fig. 9). In contrast, van den Berg's (1996) younger ages for the Waubach terraces would indicate a more gradual increase in the uplift rate mostly during the interval between ~4.0 and ~3.5 Ma.

The youngest marine terrace above the Maas river terrace sequence was correlated by van den Berg (1996) with marine deposits of Middle Miocene age elsewhere in the Netherlands, and assigned an age of 13.8 Ma.

The three Waubach terraces reveal no evidence of glacial conditions (van den Berg, 1996). Their mineralogy is also different from the younger terraces, being dominated by stable minerals such as quartz. Possible reasons for this difference include differences in provenance caused by the changing geometry of the Meuse river system (Fig. 1), and post-depositional weathering of the less stable minerals (e.g. Pissart et al., 1997). Alternatively, during the relatively mild climate in the Pliocene, vegetation may have facilitated low rates of weathering and sediment transport, causing the development on the land surface of a regolith of stable minerals in which the less stable mineral grains decomposed during the substantial time lags between erosion from bedrock and deposition in river terraces (e.g. van den Berg, 1996). Enhanced dissection of the landscape under later cold conditions is expected to first remove the regolith, then erode fresh material (e.g. Gibbard, 1988). The Noorbeek and younger terraces are indeed composed mainly of larger rock clasts, indicative of more rapid weathering under periglacial conditions. The Kosberg and Crapoel terraces have transitional characteristics, including an increase in grain size and a reduction in the proportion of quartz (data from van den Berg, 1996). For the three Waubach terraces, I will consider nominal ages in the early and late Late Miocene, and Early Pliocene, respectively (Fig. 9), as well as the possible age alternatives proposed by van den Berg (1996).

To make the rest of this revised chronology – tentatively devised to accommodate an Olduvai age for the Simpelveld terraces rather than a Réunion age – internally consistent, the terraces which are not directly dated are assumed to have aggraded during significant cold stages of the oxygen isotope time scale which lie between the control points provided by the dated terraces. The Noorbeek terrace is thus adjusted to stage 78 at 2060 ka, the Margraten to stage 62 at 1740 ka, the Sibbe terraces to stages 58 and 54 at 1650 and 1570 ka, and the Valkenburg terraces to stages 48 and 46 at 1460 and 1420 ka. Van den Berg's (1996) dating of the Geertruid terraces is not adjusted, including the Geertruid 3 terrace at oxygen isotope stage 28 at 1030 ka. However, rather than accepting his interpretation that the Pietersberg 1 terrace formed during stage 26 up to 955 ka, I assign it to stage 22 at 870 ka in keeping with my younger age assignments for other terraces. In this revised interpretation the major interglacial of oxygen isotope stage 25, known in the Netherlands as the Leerdam interglacial (e.g. Zagwijn, 1986a), thus separates the Geertruid and Pietersberg terraces; no aggradation occurred around this time, which marked an increase in the uplift rate in this revised chronology (Fig. 8).

3.3.2 *Comparison of this terrace chronology with other schemes*

This revised chronology differs substantially from other schemes, notably those by van den Berg (1996) (Fig. 8) and Pissart et al. (1997) (Fig 7). Van den Berg (1996) showed for the first time that several previously-recognised 'terraces' are resolvable into groups of two or more terraces, each from distinct glacial stages. This led him to suggest that the earliest identified terraces were older than was previously thought. Pissart et al. (1997) did not cite this work, and indeed treated several of van den Berg's (1996) terrace groups as single terraces. As Figure 7 shows, the van den Berg (1996) and Pissart et al. (1997) schemes agree for the last three glacial cycles, but diverge earlier in time, with van den Berg's (1996) solution systematically older. For instance, the Crapoel terrace formed by 2.0 Ma according to Pissart et al. (1997), making it Tiglian, but before 2.6 Ma according to van den Berg (1996), making it Reuverian.

The suggestion by Pissart et al. (1997) that the Pietersberg 2 terrace formed during oxygen isotope stage 16 violates the magnetostratigraphic evidence of a reverse polarity, which requires it to be older by at least two glacial cycles. The deduction by Pissart et al. (1997) that the Geertruid 2 terrace formed during oxygen isotope stage 22 also violates the evidence of a R→N→R polarity sequence, which fits the Cobb Mountain subchron around stage 36. However, the Pissart et al. (1997) scheme is consistent with the older magnetostratiographic evidence, notably the normal polarity for the Simpelveld terraces, which in their scheme fall within the Olduvai subchron rather that the Réunion as van den Berg (1996) thought.

Van Balen et al. (2000) have proposed another alternative chronology for the Maas terrace sequence around Maastricht: by tracing the terraces northward, correlating them in the subsurface with Rhine terraces, and applying an existing chronology for the Rhine terrace sequence. Of course, chronologies for the Rhine terrace sequence are themselves subject to considerable uncertainty (see below). The van Balen et al. (2000) Maas terrace chronology indeed differs quite dramatically from van den Berg's (1996) scheme, but is closer to my solution. For instance, the Geertruid 3 and Pietersberg 1 terraces which formed around the time of the most recent increase in uplift rates, are respectively assigned to oxygen isotope stages 28 and 26 by van den Berg (1996), 22 and 18 by van Balen et al (2000), and 28 (or possibly 26) and 22 in this study. My model solutions (discussed below) could be adjusted to fit the typically younger terrace ages proposed by van Balen et al (2000), keeping the same principle of the model but choosing different parameter values, should their chronology be confirmed by other work in future.

3.3.3 *Calculation of uplift from valley incision*

The available terrace altitude data, which indicate amounts of valley incision, need to be converted to measurements of uplift of the adjacent land surface. In general, such conversion is difficult (e.g. Gilchrist et al., 1994). However, if a river maintains a uniform gradient, erosion rates of interfluves are negligible and their spatial extent is much greater than of the river gorges (such that the isostatic response to gorge incision is negligible), and the coastline does not prograde, then the difference in level between successive river terraces at a given locality will equal the surface uplift of the interfluves. If, instead, the coastline progrades, but a river maintains a uniform gradient to each successive coastal position, then the difference in level between successive river terraces at a given locality will *underestimate* the surface uplift of the interfluves. In these circumstances, the amount δU by which surface uplift of interfluves exceeds valley incision can be estimated as

$$\delta U = \delta L \tan(\alpha), \qquad (2)$$

where δL is the distance by which the coastline has prograded and α is the river gradient. If the river gradient is non-uniform, then the correction is more complex.

According to Zagwijn (1974, 1989) the coastline of the Netherlands during major interglacials has remained near its present position since Late Tiglian time. To estimate uplift of interfluves, I assume that coastlines during times of terrace aggradation have likewise not prograded since oxygen isotope stage 68, and had prograded beforehand at a uniform rate since the Middle Miocene. I also assume that the Maas has maintained a uniform gradient, tan(α), in the vicinity of Maastricht. It is evident that these assumptions only crudely reflect the more complex history of the region (e.g. Zagwijn, 1974, 1989). For instance, the present gradient of the Maas is ~0.7 m km^{-1} at Maastricht, but decreases both upstream and downstream to ~0.4 m km^{-1} or less within a few tens of kilometres (e.g. van den Berg, 1996). Thought thus needs to be given to the effect on the conversion of incision to uplift of different choices of values for tan(α). Nonetheless, this rough and ready method for calculating uplift from incision is clearly better in this case than making no correction.

Figure 9 shows the resulting uplift history of the land surface around Maastricht. A similar graph was obtained by van den Berg (1996), although he did not convert gorge incision rates into uplift rates of interfluves. It is evident that the uplift rate in this locality increased by an order of magnitude in the Late Pliocene from ~0.007 to 0.07 mm a^{-1}. Since this change, this rate has fluctuated, with maxima around 1800 and 750 ka (see below).

3.4 *Effects of changing base level*

It is important to consider the potential significance of changes in base level between different times of terrace formation on the estimation of uplift relative to a fixed reference frame. Levels of river terraces which form near the coastline, within a few metres of sea level, of course directly reflect changes in sea level (e.g. Törnqvist et al., 1998). However, Törnqvist (1998) has also suggested that even though it is hundreds of kilometres inland, the river terrace deposition and incision investigated in the present study also reflects fluctuations in global sea-level, rather than climatic influence on hydrology and sediment transport as had previously been thought.

The evidence from the most recent glacial cycle reveals two river terraces: from before ~14 ka (after the peak of the Weichselian glaciation) and around 10 ka (during the Younger Dryas cold stage). However, as the sea level had risen by at least 30 m between these times (e.g. Fairbanks, 1989), the possibility exists that the other river terraces studied also formed with the sea at different global levels. Törnqvist (1998) has suggested that the observed geometry of the low terraces, which are ~4 m apart in the vicinity of Cologne with the younger terrace the lower, but converge near the modern North Sea coastline, is as expected for terrace formation controlled by global sea level. On the contrary, the fact that a ~30+ m difference in global sea-level has accompanied formation of river terrace levels, hundreds of kilometres inland, only a few metres apart, suggests to me that these levels are virtually insensitive to global sea level. Törnqvist's (1998) reasoning seemed anyway geared to the need to explain the lower level of the younger Low Terrace relative to the older one in the Lower Rhine embayment around Cologne. Farther upstream, for instance in the Middle Rhine gorge around Remagen, the older Low Terrace is instead buried several metres beneath the younger one (e.g. Quitzow & Thome, 1975). Furthermore, at the times when both these terraces formed the contemporaneous palaeo-coastline was hundreds of kilometres seaward of the modern coastline (e.g. Gibbard, 1995; Bridgland & D'Olier, 1995). As Törnqvist (1998) himself reported that over this reach the two terraces are indistinguishable, direct control by sea level of their levels at localities which are farther inland appears difficult to accept.

It is concluded that differences in level between the two Low Terraces which formed close in time are of no particular significance: they evidently have nothing directly to do with sea-level (see above), nor do they relate to surface uplift (which will be only a fraction of a metre over 4 ka), but presumably instead reflect local effects such as the relative ease of eroding the alluvium along the Lower Rhine versus the relative difficulty in eroding the metamorphic rocks along the Middle Rhine gorge. It thus also follows that altitudes of terraces which formed at any given inland locality during previous glacial cycles are also insensitive to global sea level. The unresolved question as to whether global sea levels may have been systematically higher during, for instance, the Early Pleistocene compared with during recent glacial cycles thus has no bearing oin this study. It is concluded instead that from one glacial cycle to the next, the terrace altitudes reflect the development of successive similar river profiles through an uplifting landscape, and, subject to the assumptions and approximations already noted, do indeed make it possible to estimate amounts of uplift of this land surface.

3.5 *Terrace sequences upstream of Maastricht*

Figure 5a and b show the profile of Meuse terraces upstream of Maastricht. Although altitudes of many terraces are well-constrained, except at Maastricht there is virtually no dating evidence. Dating the terraces along the long reaches of the Meuse studied by Clairbois (1959), Pissart (1961) and Harmand (1989) thus depends on matching these terraces to this dated sequence (Table 3).

The main difficulty in achieving such a match is due to the uncertainty over the sequence of terraces between Liège and Maastricht. Juvigné & Renard (1992) studied these in detail, but many of their results were disputed and reinterpreted by Pissart et al. (1997). A major justification given by Pissart et al. (1997) for this reinterpretation was that Juvigné & Renard (1992) found more terraces than anyone had previously identified along this reach. However, as Figure 5b shows, due to discrepancies in altitude several of the correlations with Maastricht terraces proposed by Pissart et al. (1997) are not sustainable. Table 3 thus proposes an alternative correlation scheme based on matching altitudes between the Maastricht terraces identified by van den Berg (1996) and the terraces upstream to Liège identified by Juvigné & Renard (1992) and reinterpreted by Pissart et al. (1997).

In this revised correlation scheme, the major changes to the Meuse south of the Ardennes (the loss of the left bank tributaries and the abandonment of the Gespunsart palaeo channel, between the times of formation of terraces 5 and 4') occurred around 0.95 Ma, roughly when the uplift rate began to increase towards its second maximum in the Maastricht area (Figs. 8 and 9). This diversion lengthened the course of the Meuse between Sedan and Nouzonville by ~20 km, from ~20 to ~40 km, reducing the river gradient along this reach. Figures 5a, b shows that the terraces older than 5, which pre-date this diversion, remain roughly equispaced throughout the Meuse, whereas 4' and the younger terraces converge upstream and merge near the reach with reduced gradient beside the Gespunsart palaeo-channel.

I suggest that the roughly uniform incision before ~1 Ma implies that at this time the Meuse developed successive similar states of equilibrium with sea-level, and the the land surface in the Ardennes and farther upstream was uplifting at a roughly uniform rate. I also suggest that after 1 Ma, the reduced gradient of the Meuse upstream of Nouzonville has reduced its ability to incise its bed, to such an extent that overall equilibrium has never been re-established. At present, the Meuse upstream of Nouzonville has a gradient of barely 0.1 m km^{-1}, compared with a typical gradient of ~0.4 m km^{-1} through the Ardennes. It thus follows that, for the last million years, the reach of the Meuse between Maastricht and Nouzonville has thus been responding in part to (a) regional uplift of the land surface

Table 3. Meuse river terrace sequence.

van den Berg S	van den Berg t (ka)	Alternative S	Alternative t (ka)	Harmand (1989); Lorraine. Number	Clairbois (1959); Pissart (1961); Ardennes. Number	Juvigné & Renard (1992); below Liège. Name	Maastricht Name: This study	Maastricht Name: Pissart et al. (1997)
1	10 *	1	10 *			Vivegnis	Geistingen	
2	14 *	2	14 *			Hersthal	Mechelen/Maas	
6	130 *	6	130 *	1.2	1	Jupille	Eisden-Lanklaar	Caberg
8	245 *	8	245 *	1.1 (8–10)	2	Caberg	Caberg 3	
10	330 *	10	330 *			Trou-Louette	Caberg 2	Rothem 2
12	420 *	12	420 *				Caberg 1	
14	510	13b	470 *			Cornillon	Rothem 2	Rothem 1
16	620	14	530 *				Rothem 1	
18	715	16	620 *		3	Fouron-le-Comte	Gravenvoeren	Gravenvoeren
20	780	18	700 *	2.2 (18–22)		Lorette	Pietersberg 3	Pietersberg 2
22	870	20	780 *		4	Hermée	Pietersberg 2	Pietersberg 1
26	955	22	870 *		4’	Eben (- St. Geertruid)	Pietersberg 1	Geertruid 2
28	1030 *	28	1030 *	2.1 (28–32)	5	Bombaye	Geertruid 3	Geertruid 1
32	1090 *	32	1090 *			Wonck	Geertruid 2	
40	1280 *	40	1280 *	3.2		Lixhe	Geertruid 1	Valkenburg 2
50	1500 *	46	1420	3.1	5’	Hognée	Valkenburg 2	Valkenburg 1
54	1570 *	48	1460				Valkenburg 1	
60	1690 *	54	1570	4.3	6	Trembleur	Sibbe 2	Sibbe
62	1740 *	58	1650	4.2		Rabosée	Sibbe 1	Margraten
68	1870	62	1740	4.1	7	Houlpaix$^+$	Margraten	Noorbeek$^+$
78	2060	66	1820	5	8	La Havée$^+$ Bruyères$^+$	Simpelveld 2	Crapoel$^+$ Simpelveld$^+$
82	2140	68	1870			Haute-Saive	Simpelveld 1	

Table 3. (Continued).

van den Berg		Alternative		Harmand (1989); Lorraine.	Clairbois (1959); Pissart (1961); Ardennes.	Juvigné & Renard (1992); below Liège.	Maastricht Name	
S	t (ka)	S	t (ka)	Number	Number	Name	This study	Pissart et al. (1997)
96	2440 *	78	2060	6.2	9	Beyne-Heusey	Noorbeek	Kosberg
102	2600 *	82	2140	6.1	10		Crapoel	Brunssum=Waubach
-	2690 *	-	2690 *	7.2			Kosberg 3	
-	2810	-	2810 ?				Kosberg 2	
-	2940	-	3140 ?	7.1			Kosberg 1	
-	3140	-	4500 ?	8.2			Waubach 3	
-	3300	-	6000 ?				Waubach 2	
-	3980	-	10,000 ?	8.1			Waubach 1	

Correlation scheme between the results of this study, the Meuse terrace chronologies for Lorraine (from Harmand, 1989; as modified by Pissart et al., 1997), the Ardennes (Clairbois, 1959; Pissart, 1961), and the reach downstream of Liège (Juvigné & Renard, 1992) (their original version, not the reinterpretation by Pissart et al., 1997). It also shows the correlations between terraces around Liège and around Maastricht from this study, from van den Berg (1996), and from Pissart et al. (1997). For the three Liège terraces labelled $^{+}$, Pissart et al. (1997) for some reason rearranged their age order to be La Havée, Houlpaix, Bruyères, in order of decreasing age: hence the out-of-sequence correlations with the Maastricht area. This table indicates, for instance, that terrace 4.1 in Lorraine correlates with terrace 7 in the Ardennes, and the Houlpaix terrace below Liège, and matches the Margraten terrace in the Maastricht area in my terrace scheme; on the other hand Pissart et al. (1997) thought the Houlpaix terrace around Liège equivalent to the Noorbeek terrace around Maastricht. Symbols (*) after the interpreted ages of terraces in the columns for this study and van den Berg (1996) indicate which ages are preferred, given the modelling results. The proposed correlation of the "Eben - St. Geertruid" terrace of Juvigné and Renard (1992) with the Pietersberg 1 terrace rather than a member of the Geertruid terrace group warrants comment. Fig. #5b shows the Eben terrace of the left bank of the Meuse in the vicinity of Eben-Emaël in Belgium. Comparison with Fig. #5c indicates that this is the same as the Pietersberg 1 terrace. The St. Geertruid terrace of Juvigné and Renard (1992) is on the right bank near St. Geertruid in the Netherlands, ~7 km east of Eben-Emaël and ~6 km east of the river. This appears to be the same as the Geertruid 3 terrace of van den Berg (1996), which is locally ~8 m higher than the Pietersberg 1 or Eben terrace. Others (e.g., Zonneveld, 1975) have also previously recognised the Eben terrace as part of the Pietersberg group. Elsewhere, the Bombaye and Wonck terraces were regarded as distinct by Juvigné and Renard (1992) but were treated as a single terrace by Pissart et al. (1997); they may well be the local equivalents of the Geertruid 2 and 3 terraces of the Maastricht area.

and in part to (b) the absence of significant incision around Nouzonville caused by the low river gradient. It is presumed that, moving upstream from Maastricht to Nouzonville, the effect of (a) on incision decreases and the effect of (b) increases.

Along the upper Meuse (Fig. 5a), terraces 8.1, 8.2, 7.1, and 7.2 are analogous to the Waubach and Kosberg groups of the Maastricht area (Table 3). The ~50 m of incision spanning terraces 8.1 to 6.1 thus occurred during ~10 to ~2.5 Ma at a time-averaged rate of ~0.007 mm a^{-1}. By analogy with the Maastricht area, it is presumed that this incision reflected contemporaneous uplift of the land surface relative to sea-level at a rate of this order. The ~100 m of incision between terraces 6.1 and 2.1 occurred during ~2 to ~1 Ma at a time-averaged rate of ~0.07 mm a^{-1}, and is presumed to reflect accelerated uplift of the land surface, as at Maastricht. The ~30-40 m of incision which has occurred in the last million years, since terrace 2.1 formed, indicates a time-averaged rate no greater than ~0.04 mm a^{-1}. This local incision is assumed to be limited by the disequilibrium caused by the lack of incision around Nouzonville, and may well substantially underestimate the uplift rate of the land surface.

3.6 *Chronology of Rhine river terraces*

The terraces of the lower and middle Rhine were studied in detail some years ago (e.g. Kandler, 1970; Bibus & Semmel, 1977; Brunnacker, 1978; Illies et al., 1979; Brunnacker et al., 1982a,b; Windheuser et al., 1982). As a result, terrace chronologies have been developed (e.g. Brunnacker et al., 1982b; Brunnacker, 1986), using magnetostratigraphic and palynological evidence. However, apart from the youngest terraces, no attempt at correlation with oxygen isotope stages was made. More recent investigations of this terrace sequence include Hoselmann (1994, 1996), Meyer & Stets (1998), and Semmel (1999).

The documented Rhine terraces (Fig. 10) have long been known to span time since the Late Pliocene. As for the Maas, the oldest terraces of the lower Rhine consist of quartz gravels, known locally as 'kieseloolite'. Most studies have accepted that the relative absence of less stable minerals has resulted from the relatively stable conditions before the Quaternary when erosion rates within the surrounding drainage basin were low (e.g. Illies et al., 1979; Brunnacker, 1986; Semmel, 1991). However, Tebbens et al. (1995) have suggested from studies of rare heavy mineral grains that the headwaters of the Rhine propagated southward from the Vosges/Black Forest region into the foreland (molasse) basin of the Alps around the time of the Reuverian – Pretiglian boundary. The resulting erosion of this basin and re-deposition of its material may have affected the lithology of the Rhine terraces.

Only 16 Rhine terraces have been identified, compared with 31 for the Maas. This appears to be due to the greater deposition, which results in terrace deposits overlapping (Fig. 11) rather than cropping out at distinct altitudes (Fig. 4) and thus makes terrace surfaces more difficult to identify. This greater deposition may be due to the lower gradient of ~0.2 m km^{-1} for the Rhine, compared with ~0.4 to ~0.7 m km^{-1} for the Maas.

3.6.1 *Terrace nomenclature*

The standard nomenclature for the lower Rhine (e.g. Brunnacker, 1978; Brunnacker et al., 1982a, b) begins with a Low Terrace or Niederterrasse (NT), which overlies clay from the Eemian interglacial and thus aggraded during the last glacial cycle. In some localities, two or even three separate levels are evident within this, but all relate to different parts of this cycle (e.g. Brunnacker, 1986). Törnqvist (1998) and others have established that the

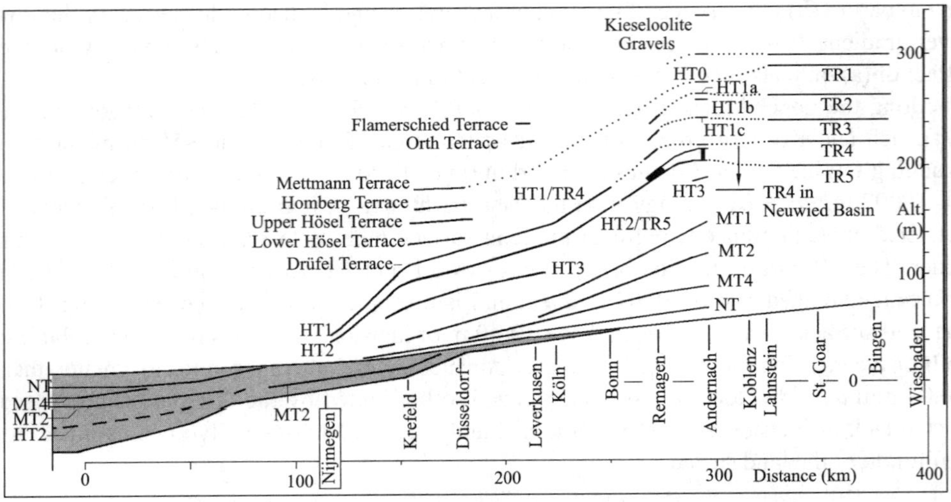

Figure 10. Longitudinal profiles of river terraces along the Middle and Lower Rhine. Information is from Quitzow & Thome (1975), Bibus & Semmel (1977), Illies et al. (1979), Brunnacker et al. (1982a, b) and Semmel (1991).

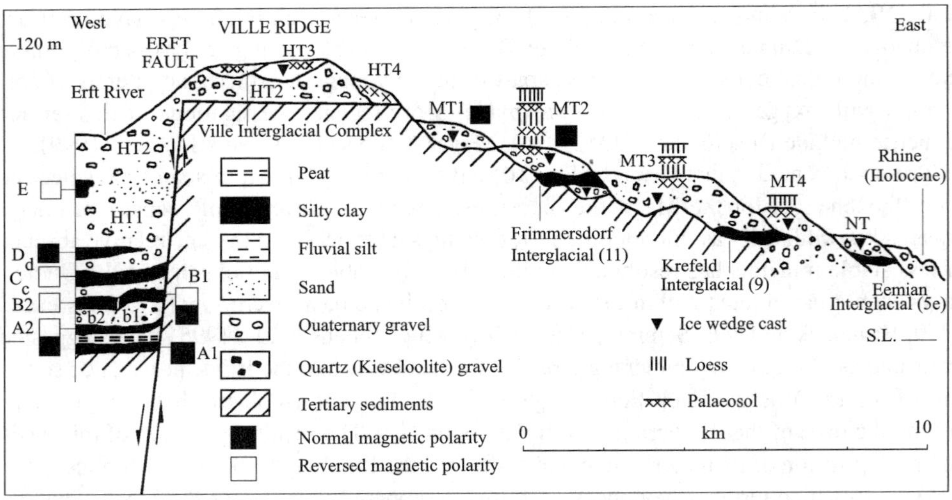

Figure 11. Composite cross-section across the Ville ridge between the Erft and Rhine rivers in the vicinity of Cologne, showing the river terraces and other Quaternary deposits which are discussed in the text. Redrawn from Figure 3 of Urban (1982). This cross-section was originally prepared by Brunnacker (1978).

clearest levels, here designated as the Older and Younger Low Terraces, formed before ~14 ka (after the peak of the Weichselian glaciation) and around 10 ka (during the Younger Dryas cold stage). These two terraces are ~4 m apart in the vicinity of Cologne but converge near the modern North Sea coastline.

Above these are four Middle Terraces or Mittelterrassen (MT1 to MT4) and, higher still, four Main Terraces or Hauptterrassen (HT1 to HT4). Fragmentary gravel outcrops indicating even older terraces are also evident, notably along the Bergisch Randhöhen (the abrupt flank of the Rhenish Massif along the right bank of the Rhine) north of Cologne, which have been interpreted in terms of four High Terraces or Hoheterrassen (e.g. Brunnacker et al., 1982a). Individual gravel terraces can be correlated between outcrop along the flanks of the Rhine valley and the subsurface of the Lower Rhine embayment. For instance, along the Ville ridge (Fig. 11), which has uplifted relative to its surroundings in the footwall of the Erft normal fault, Quaternary sediments have been revealed in the subsurface by opencast mining of the underlying Tertiary lignite. Successive terrace gravels are typically separated by fine-grained fossiliferous sediments from interglacials or interstadials, which have been dated using pollen and magnetostratigraphy.

Around Andernach (Table 4) along the Middle Rhine gorge through the Rhenish Massif, four separate levels within Main Terrace 1 have been identified (e.g. Zagwijn, 1989), suggesting that – like the terrace groups around Maastricht which are now resolved into individual terraces – it locally has an internal 'cut and fill' stratigraphy with successive levels younging downward. This contrasts with the evidence from the Lower Rhine embayment at Ville (Fig. 11) where the equivalent deposits form a 'layer cake' stratigraphy younging upwards.

Between Koblenz and Andernach (Fig. 1) the Rhine flows through a small sedimentary basin – the Neuwied Basin. The basin interior has subsided relative to its surroundings (e.g. Windheuser et al., 1982), revealed for instance by the lower level of terrace TR4 (e.g. Illies et al., 1979) (Fig. 10). This means that correlation of terraces between the adjoining reaches of the Rhine is not straightforward.

Upstream of Koblenz, a different terrace nomenclature has been devised (e.g. Bibus & Semmel, 1977; Illies et al., 1979). Although discrepancies have arisen over the interpretation of some terraces, several studies (e.g. Illies et al., 1979; Semmel, 1991) have tentatively correlated terraces TR4 and TR5 of the Middle Rhine with HT1 and HT2 of the Lower Rhine. Unlike downstream of Koblenz, where terraces converge downstream, farther upstream they remain parallel to each other at virtually uniform altitudes. Since deposition of terrace TR5, the Rhine has incised the landscape by up to 130 m (Fig. 12). The proposed correlation for these terraces (Figs 10, 13) indicates much faster incision during this interval than beforehand, which is presumed to reflect an increased rate of uplift of the adjoining land surface relative to sea level.

3.6.2 *The Middle Pleistocene and late Early Pleistocene*

A constraint on the chronology is provided by the presence of minerals from the Quaternary basalts of the East Eifel which, according to Brunnacker et al. (1982b), first appear in the Rhine terrace sequence in a thin layer of interglacial deposits between terraces HT2 and HT3. The oldest K-Ar date from this volcanism is 671 ± 30 ka (Schminke & Meertes, 1979). Terraces HT2 and HT3 thus formed during successive glacials around 0.7 Ma. At Ville, terrace HT2 is underlain by a layer of clay (clay E in Figs 11 and 13) which is dated to the Matuyama chron given its reverse magnetisation (e.g. Brunnacker et al., 1982b). The Ville 'interglacial complex', between terraces HT3 and HT4 (Fig. 11), indicates alternations in climate between three warm stages and two intervening cold stages. However, these sediments lack age-diagnostic pollen (e.g. Urban, 1982). Some studies (e.g. Urban, 1982; Brunnacker et al., 1982b) have suggested that they span three interglacials and two glacials. However, their thinness (Fig. 11) raises the possibility that they instead represent fluctuations in climate within a single glacial cycle (see below).

Table 4. Rhine river terraces.

Name	H (m)	S	t (ka)	U (m)
Bingen (H_o 75 m; H_r 75 m)				
TR5 (Y)	200	16	620	125
TR5 (O)	200	22	870	125
TR4 (Y)	225	26	955	150
TR4 (O)	225	62	1740	150
TR3	245	68	1870	170
TR2	265	78	2060	190
TR1	290	82	2140	215
Kieseloolite	300		3000	225
Andernach (H_o 67 m; H_r 60 m)				
NT	70	2	14	3
MT4	85	6	130	18
MT2	120	10	330	53
MT1	135	12	420	68
HT4	150	13b	470	83
HT3	190	14	530	123
HT2 / TR5 (Y)	205	16	620	138
HT2 / TR5 (O)	220	22	870	153
HT1d / TR4 (Y)	220	26	955	153
HT1d / TR4 (O)	220	62	1740	153
HT1c	245	68	1870	178
HT1b	260	74	1960	193
HT1a	265	78	2060	198
High terrace (HT0)	275	82	2140	208
Kieseloolite gravel (Y)	300		3000	233
Kieseloolite gravel (O)	330		5000	263
Remagen (H_o 65 m; H_r 50 m)				
Younger NT	66	1	10	1
Older NT	62	2	14	-
MT4	80	6	130	15
MT2	105	10	330	40
MT1	125	12	420	60
HT2 (Y) / TR5	180	16	620	115
HT2 (O) / TR5	200	22	870	135
HT1 (Y) / TR4	210	26	955	145
HT1 (O) / TR4	210	62	1740	145
High Terrace 2	225	68	1870	160
High Terrace 1	240	78	2060	175

Table 4. (Continued).

Name	H (m)	S	t (ka)	U (m)
Ville (H_o 40 m; H_r 36 m)				
NT	44	2	14	4
MT4	55	6	130	15
MT3	69	8	245	29
MT2	79	10	330	39
MT1	90	12	420	50
HT4	106	13b	470	66
HT3	113	14	530	73
Leverkusen (H_o 45 m; H_r 35 m)				
Younger NT	44	1	10	-
Older NT	48	2	14	3
MT4	55	6	130	10
MT2	59	10	330	14
MT1	74	12	420	29
HT3	113	14	530	68
HT2 (Y)	126	16	620	81
HT2 (O)	126	22	870	81
HT1 (Y)	141	26	955	96
HT1 (O)	141	62	1740	96
Orth	220	-	2690	175
Flamerscheid	240	-	2810	195
Düsseldorf (H_o 33 m; H_r 30 m)				
Younger NT	32	1	10	-
Older NT	36	2	14	3
MT4	41	6	130	8
MT1	54	12	420	21
HT3	87	14	530	54
HT2 (Y)	105	16	620	72
HT2 (O)	105	22	870	72
HT1 / Drüfel (Y)	123	26	955	90
HT1 / Drüfel (O)	123	62	1740	90
Lower Hösel	131	68	1870	98
Upper Hösel	140	74	1960	107
Homberg	157	78	2060	124
Mettmann	188	82	2140	155

Terrace nomenclature and altitudes (H) are from Quitzow & Thome (1975), Bibus & Semmel (1977), Illies et al. (1979), Brunnacker et al. (1982a,b), Windheuser et al. (1982), Zagwijn (1989), Semmel (1991), and Törnqvist (1998b). Terrace ages (t) are interpreted in this study. (Y) or (O) in the name column indicates a young or old age bound for a particular terrace. The age of 3 Ma for the kieseloolite gravel at Andernach, from Illies et al. (1979), is taken here as an upper bound. Amounts of uplift (U) are estimated as $H - H_o$ for each terrace, where H_o is a local reference level that in each locality is close to the river level (H_r).

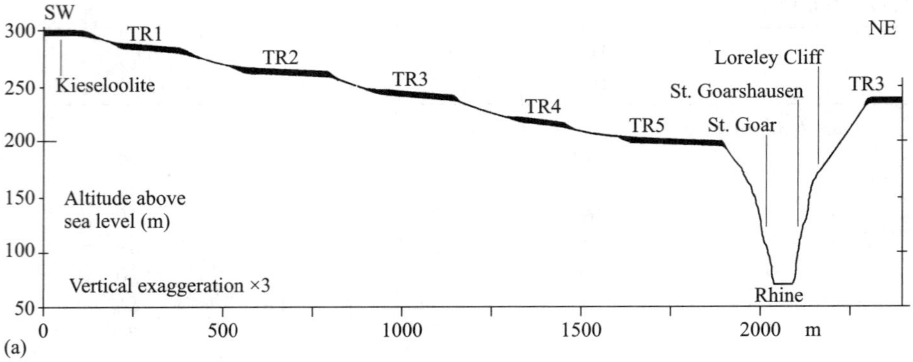

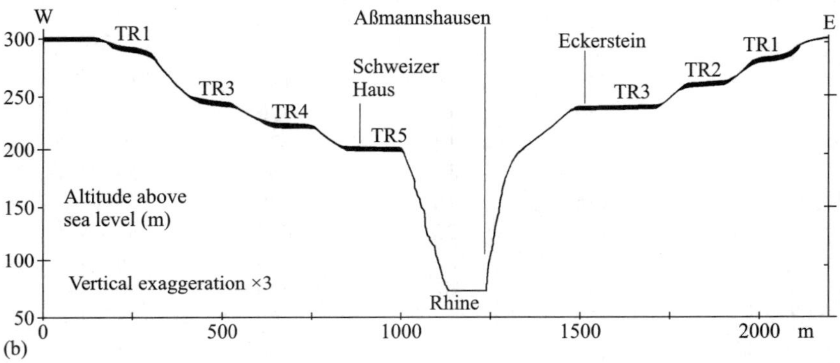

Figure 12. Profiles of Middle Rhine terraces upstream of Koblenz, at St. Goar (a) and Aßmannshausen (~4 km north of Bingen; b), localities which are 22 km apart. Adapted from parts of Figure 2 of Bibus & Semmel (1977). Note the transition from a broad floodplain to the present much narrower cliff-lined gorge around the time of depositon of terrace TR5, which is interpreted here as a consequence of faster incision caused by an increased rate of uplift of the adjacent land surface.

Terrace MT4 at Ville is overlain in turn by a palaeosol and a layer of loess (e.g. Brunnacker et al., 1982b). These have been interpreted as marking the Eemian interglacial and Weichselian glacial, dating this terrace to oxygen isotope stage 6. Terrace MT3 is overlain alternately by two palaeosols and two loess layers, and thus formed during stage 8, whereas MT2 is overlain alternately by three such layers and can thus be tied to stage 10.

The earlier Rhine terraces are more difficult to assign to particular glacial cycles. Matters are not helped by the inconsistencies between different interpretation schemes: for instance Brunnacker (1986) stated that terrace MT1 formed during the Elster glaciation (oxygen isotope stage 12), whereas Zagwijn (1986a) asserted that it formed during Cromerian glacial C. In Zagwijn's schemes, even as recent as Zagwijn (1996), the Elster glaciation has been placed in oxygen isotope stage 10, not 12, and stage 12 is Cromerian glacial C. However, others (e.g. Turner, 1996) have found this impossible to accept, and regard Cromerian glacials B and C as marking oxygen isotope stages 16 and 14, with the preceding interglacials III and IV (the Rosmalen and Noordbergum interglacials) marking stages 15 and 13 (as in Figs 7 and 13). The earlier Cromerian is also difficult to tie to the

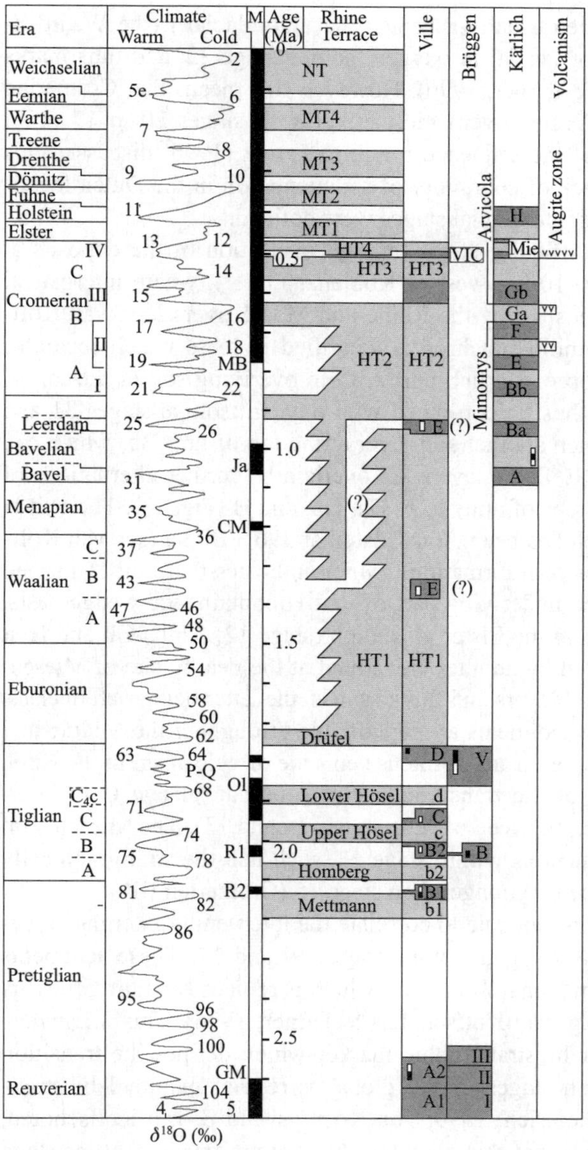

Figure 13. Interpreted chronology of Rhine river terraces in relation to regional stratigraphy, oxygen isotope stages, magnetic polarity reversals, and Quaternary volcanism of the Eifel. Intervals of time over which terrace gravels are interpreted as aggrading are stippled. Intervals of time with interglacial sediments are given darker shading. Evidence from subsurface sites at Ville and Brüggen, and from Kärlich and Miesenheim (Mie) in the Neuwied basin, is also shown. VIC at Ville marks the Ville Interglacial Complex. Two possible stratigraphic positions for clay E at Ville are shown, with their implications for the age of the transition from terrace HT1 to terrace HT2. B at Brüggen marks the Belfeld clay. Character of Kärlich beds A to H, timings of early Eifel volcanism from interbedded clasts, timing of later 'augite zone', and timing of *Mimomys-Arvicola* extinction event, use data from Zagwijn (1986a), von Koenigswald & van Kolfschoten (1996), and van Kolfschoten & Turner (1996). Rest of chronology is from the same sources as in Figure 7, except for the timing of the Tiglian stage which is discussed in the text.

oxygen isotope time scale. It has been usually accepted that interglacial I (the Waardenburg) immediately follows the major glacial in oxygen isotope stage 22, and thus marks stage 21 (e.g. Zagwijn, 1986a, 1996; Turner, 1996). However, this means that Cromerian glacial A and interglacial II (the Westerhoven) must either span stages 20 to 17 or fit somewhere within them with part of the sequence missing. Turner (1996) discussed this difficulty, which is caused by the lack of continuity of sedimentation in the Dutch localities where the Cromerian glacial and interglacial stages were defined.

One possible resolution of this difficulty arises through inspection of the deposits at Kärlich within the Neuwied basin (~10 km west of Koblenz; Fig. 1), where interglacial sediments interbed with gravels deposited by the Rhine and Mosel rivers (e.g. van Kolfschoten & Turner, 1996). The Jaramillo subchron is identified in loess which separates underlying interglacial soil at the top of Kärlich unit A from overlying soil at the top of unit Ba. These two soil layers can thus be correlated with oxygen isotope stages 31 and 25. Unit Ba is overlain by gravel, then another soil layer at the top of unit Bb, which can thus be assigned to stages 22 and 21. These layers are overlain by further alternations of loess and soil, the soils being at the tops of units E, F, Ga, Gb, and H (Fig. 13). The unit H soil is in turn overlain by tuff, which has been Ar-Ar dated at 396 ± 20 ka (see van Kolfschoten & Turner, 1996, for sources of information), which indicates that unit H formed during oxygen isotope stage 11. The underlying part of unit H contains ice wedge casts, reflecting the glacial conditions during the Elster glaciation (stage 12). Units Gb and H at Kärlich are stratigraphically separated by another unit found at the nearby site of Miesenheim. This shows an alternation of soil layers and fluvial sands, the latter being rich in clasts of augite. This 'augite zone', whose sediments are identifiable throughout the Middle and Lower Rhine, preserves material erupted from an intense episode of volcanism in the Eifel, and is well constrained as starting at the transition from Cromerian glacial C to interglacial IV (e.g. Zagwijn, 1986a) (i.e., between oxygen stages 14 and 13). The Miesenheim sediments thus reflect climate fluctuations within stage 13, such that the stratigraphically older soil in bed Gb at Kärlich can be no younger than stage 15 (Cromerian III).

Given these constraints, it seems reasonable to correlate the intervening interglacial deposits of Kärlich units E and F with oxygen isotope stages 19 and 17. These sediments contain fossils of the extinct micromammal *Mimomys*, which is replaced in unit Gb by its more modern analogue *Arvicola* (e.g. van Kolfschoten & Turner, 1996). This *Mimomys-Arvicola* extinction event is a clear biostratigraphic marker which defines the transition between the Biharian and Toringian stages of the global terrestrial mammal biostratigraphic time scale (e.g. Fejfar & Heinrich, 1990; von Koenigswald & van Kolfschoten, 1996). The above discussion suggests that this boundary lies within oxygen isotope stage 16, which marked a major glaciation (e.g. Shackleton et al., 1990) that can be expected to have caused extinctions. Cromerian III sites invariably contain *Arvicola*, whereas Cromerian II sites contain *Mimomys* (e.g, von Koenigswald & van Kolfschoten, 1996). This age assignment for the Kärlich sequence (Fig. 13) implies that Cromerian II sites may span all, or fragments of, stages 19 through 17. As stage 18 involved a weak glaciation (e.g. Shackleton et al., 1990), the apparent grouping of stages 19 to 17 as an apparent single interglacial – the Cromerian II – can be explained.

Reviews by Turner (1996) and Zagwijn (1996) have drawn attention to the early Middle Pleistocene sedimentary sequences of eastern Poland and adjoining parts of the former Soviet Union. These regions provide pollen and other evidence of the Ferdinandow/Shklov era, which consists of two interglacials with distinct temperate floras separated by a cool era with boreal forest. Mammals from this weak 'glacial' stage include *Mimomys*, demonstrating its early Middle Pleistocene age (Turner, 1996). It thus seems reasonable to assign the lower and upper Ferdynandow interglacials to oxygen isotope stages 19 and 17 and the intervening weak glacial to stage 18. Furthermore, pollen diagrams for this Ferdinandow era (see e.g. Zagwijn, 1996; Rzechowski, 1996) can be compared with Dutch Cromerian II sites. For instance, Zagwijn's (1996) summary of Cromerian II pollen data from Westerhoven and Waardenburg I match well the lower Ferdynandow interglacial (stage 19), whereas the stratigraphically younger Cromerian II Waardenburg II site matches the upper Ferdynandow (stage 17). It is difficult to accept Zagwijn's (1996) alternative suggestion that the Ferdynandow era spans the Cromerian II and III (i.e., oxygen isotope stages 17 to 15), because the intervening stage 16 or glacial B involved a major glaciation.

Having established as a working hypothesis these relationships (Fig. 13) between litho- and bio-stratigraphic markers, oxygen isotope stages, and conventional palynologically-defined stages of the Cromerian era, it is now possible to assign ages to the remaining early Middle Pleistocene and late Early Pleistocene Rhine terraces.

The uppermost part of terrace HT2, as defined by Brunnacker (1978), when extrapolated into the Netherlands coincides with distinctive sediments (the Weert zone, the uppermost part of the Sterksel formation) which from their stratigraphic position are dated to Cromerian glacial B (e.g. Zagwijn, 1986a). Deposition of HT2 thus ended during oxygen isotope stage 16.

Terrace HT3, as defined by Brunnacker (1978), when extrapolated into the Netherlands correlates with other distinctive sediments (the Rosmalen zone, the lowest part of the Veghel formation) which began to be deposited during the Cromerian III (e.g. Zagwijn, 1986a).

Terrace HT4, as defined by Brunnacker (1978), contains a higher proportion of clasts of material derived from the Eifel volcanism than the older terraces (e.g. Brunnacker, 1978). It thus post-dates the start of oxygen isotope stage 13. As it pre-dates the Elster glaciation when terrace MT1 was deposited, it evidently formed during the cold stage within oxygen isotope stage 13, already discussed (stage 13b), which ended around 470 ka. By implication, this scheme predicts that the Ville Interglacial Complex, between terraces HT3 and HT4 (VIC in Fig. 13), reflects climate fluctuations around the early peak (stage 13c; ~520 ka) of oxygen isotope stage 13; it correlates with the early part of the Miesenheim bed in the Neuwied basin (Fig. 13).

The key event which remains unresolved in age is the transition between the end of deposition of HT1 and the start of deposition of HT2. This transition is marked by the reverse-magnetised clay E at Ville (Fig. 13). Urban (1982) found the pollen in this layer difficult to interpret, and concluded that 'baased on the stratigraphic position and biostratigraphy' it most likely has Waalian B age. The main interglacial during this interval is in oxygen isotope stage 43 around 1350 ka (Fig. 13). However, as Urban (1982) regarded the Ville Interglacial Complex as spanning the early Cromerian and terraces HT1 and HT2 as Menapian, the possibility of a substantially younger age for clay E would appear to have not been considered. The youngest feasible age for clay E would appear to be oxygen isotope stage 25 (950 ka), the last major pre-Cromerian interglacial, which lies within the Matuyama chron (Fig. 13).

3.6.3 *The latest Pliocene and early Early Pleistocene*

The older part of the Rhine terrace sequence is even more problematical. Matters are not helped by the fact that previous interpretation schemes, by Brunnacker et al. (1982b) and

Brunnacker (1986), differ from each other, with no reasons given and neither being tied to the oxygen isotope time scale. At Ville, a sequence has been identified comprising clays A1 and A2, then terrace gravel b1, clay B1, gravel b2, clay B2, gravel c, clay C, gravel d, and clay D, the latter being overlain by the oldest recognised Main Terrace (HT1) (Fig. 11). Clays A1 and A2 are dated as Reuverian from their pollen. A1 is normally-magnetised, but A2 indicates a transition from normal to reverse polarity which has been interpreted as the Gauss-Matuyama (e.g. Brunnacker et al., 1982b; Brunnacker, 1986). At Brüggen, three clay beds (I, II, and III) have been interpreted as of equivalent age, although the youngest (clay III) spans the Reuverian-Pretiglian boundary (e.g. Urban, 1982).

At Brüggen, a younger clay layer (clay V) contains pollen indicative of the latest Tiglian (C5 to C6) and exhibits a transition from reverse to normal magnetic polarity (e.g. Brunnacker et al., 1982b), which is assumed here to be the Matuyama-Olduvai. The temperate conditions indicated at this time, known locally as the Van Eyck interglacial, are thus interpreted here as marking oxygen isotope stage 63 (Fig. 13). Clay D at Ville, which is equivalent to the normally-magnetised upper part of clay V at Brüggen (e.g. Brunnacker et al., 1982b), is evidently of the same age. I thus interpret the overlying HT1 gravel as having begun to aggrade during stage 62.

Clay B1 at Ville contains a Pretiglian pollen assemblage and a normal to reverse polarity transition (e.g. Brunnacker et al., 1982b). The overlying clay B2 contains pollen indicative of Tiglian A, and is reverse magnetised (e.g. Brunnacker et al., 1982b). It has been interpreted as slightly younger than the Belfeld clay near Brüggen (the type locality for Tiglian A), which is reverse-magnetised (e.g. Brunnacker et al., 1982b). I interpret these two polarity transitions as Matuyama – Réunion 2 and Matuyama – Réunion 1. Gravel b1, which underlies clay B1 and is equivalent to the Mettmann terrace, is thus assumed to immediately pre-date the Réunion 1, and thus aggraded during oxygen isotope stage 82. Gravel b2, which lies between clays B1 and B2 and is equivalent to the Homberg terrace, aggraded during stage 78. Gravels c and d, which aggraded between the Réunion 1and Olduvai subchrons, are associated here with oxygen isotope stages 74 and 68.

Gravel b1 at Ville, equivalent to the Mettmann terrace farther north (e.g. Brunnacker et al., 1982b), is the youngest quartz-dominated gravel or kieseloolite terrace: other minerals predominate in the younger terraces. In Fig. 13 this terrace is matched to oxygen isotope stage 82.

Gravel d is another key stratigraphic marker, as according to Brunnacker et al. (1982b) it marks the first evidence of influx of distinctive sediments indicating an expansion of the Rhine drainage basin. It includes chert from the Alps and lydite from the Bohemian Massif, indicating that the uppermost Rhine (upstream of the Rhine Graben) and Main rivers had attained their present geometries by this time (Fig. 1) (see also Boenigk, 1982). According to Brunnacker et al. (1982b), before this time the upper Main flowed southward from Bamberg along the course of the Regnitz and Rezat rivers to the Altmühl tributary of the Danube (Fig. 1). In the lower Rhine, gravel d also contains flint from the Paris basin, indicating that when it was deposited, the East Maas still formed a tributary of the Rhine. In Figure 13 this gravel is matched to oxygen isotope stage 68, equivalent to the Simpelveld 1 terrace of the Maas in the revised chronology in Figure 7, which indeed pre-dates the final abandonment of the East Maas channel (Figs 2 and 4).

3.6.4 *Remaining uncertainties*

If Urban's (1982) Waalian B interpretation of clay E at Ville is accepted, then terrace HT1 formed over stages 62 to 44 and terrace HT2 over stages 42 to 16. Localities are known where both terraces contain sand layers, fossil soils, and weathering fabrics (e.g. Brunnacker et al., 1982b), suggesting that they did each aggrade over multiple glacial to interglacial cycles. The lack of evidence of distinct terraces from these intervals of time also suggests that the contemporaneous incision rate, and thus the uplift rate, within the Lower Rhine Embayment was lower than at later times. If this age constraint on clay E is excluded, its youngest feasible age instead adjusts to that of the Leerdam interglacial (oxygen isotope stage 25). It thus follows that terrace HT1 may have continued aggrading as late as stage 26, and HT2 may have not begun to aggrade until stage 22. Both alternative schemes are summarised in Figure 13, the latter being preferred in other figures and tables.

The resulting overall chronology in Figure 13 differs significantly from some other schemes, a notable point of difference being the inference that the Ville Interglacial Complex represents just part of one interglacial, rather than a sequence of three interglacials and two glacials as others (e.g. Brunnacker et al., 1982b; Urban, 1982; Brunnacker, 1986) have previously thought. These studies deduced that the Ville Interglacial Complex matches Cromerian interglacials I to III and/or beds Ba to E at Kärlich. A match to the Cromerian I to III would place terrace HT3 no later than oxygen isotope stage 22; a match to the Kärlich sequence given its position in Figure 13 would place terrace HT3 no later than stage 26. Both these alternatives would thus place terraces HT2 and HT3 so much earlier in time compared with my preferred solution that there would be no possibility of reconciling their adjusted ages with the Dutch stratigraphy for the lower Rhine. The previously proposed correlation between the Ville Interglacial Complex and Kärlich beds Ba to E seems to have been based on the similarity of soil layers in these two localities, which were thought to yield a unique match (e.g. Brunnacker et al., 1982b).

3.6.5 *Summary of uplift history*

Unlike for the Maas, no attempt is made to quantify differences between gorge incision and surface uplift. This is, first, because the lower present-day gradient of the Rhine means that any such correction is smaller in magnitude (Equation (2)). Second, there is geomorphological evidence that along the Middle Rhine the gradient was even lower in the earliest Pleistocene (e.g. Illies et al., 1979; Semmel, 1991) (Fig. 12). In this case, the uncertainties in estimating any correction are probably much greater than the correction itself. Third, the Rhine has not experienced any diversion analogous to that which created the Gespunsart palaeo channel on the Meuse, nor any such dramatic change in gradient (although its gradient does decrease upstream from ~0.3 m km^{-1} through the Rhenish Massif to ~0.2 m km^{-1} in the northern part of the Rhine Graben).

Table 4 compiles data on the Quaternary uplift of the Rhine terraces at the representative localities of Bingen, Andernach, Remagen, Ville, Leverkusen, and Düsseldorf, assuming the chronology in Figure 13. At some of these localities, it is *a priori* unclear whether the uppermost levels of terraces HT1 and HT2 formed at the start of the development of these terraces, which aggraded to successively lower levels during later glacials (forming a 'cut and fill' stratigraphy); or whether each aggraded to successively higher levels during later glacials (forming a 'layer cake' stratigraphy), such that the upper terrace surfaces have the youngest ages.

The uplift rate of the land surface adjoining the Rhine increases upstream: from Table 4, the time-averaged rate over the past half-million years has increased from ~0.10 mm a^{-1} at Düsseldorf to ~0.23 mm a^{-1} at Andernach, reflecting the upstream divergence of the river terraces in Figure 10. Similar uplift rates are evident given the differences in altitude of the older terraces from around 2 Ma (Fig. 14). However, across the time in between, while terraces HT1 and HT2 were forming, the uplift rate in each locality was much less. Although uplift rates differ from place to place, the overall pattern (Fig. 14) is similar to that for the Maas (Fig. 9).

The altitudes of the Rhine terraces across the southern flank of the Rhenish Massif over the ~50 km distance between Bingen and Koblenz are nearly constant (Figs 10 and 12). This evidence of apparent zero river palaeo-gradients has been highlighted before (e.g. Bibus & Semmel, 1977; Illies et al., 1979; Semmel, 1991). One possible explanation is that since these terraces formed, the uplift rate has locally decreased southward, cancelling an initial northward river gradient. If this initial gradient was, say ~0.1 m km^{-1}, then from (2) the Bingen area has uplifted since these terraces formed by only ~5 m less than Koblenz and Andernach. Thus, although the Andernach area appears to have the highest uplift rate in the study region, localities farther south have not uplifted much less. This point

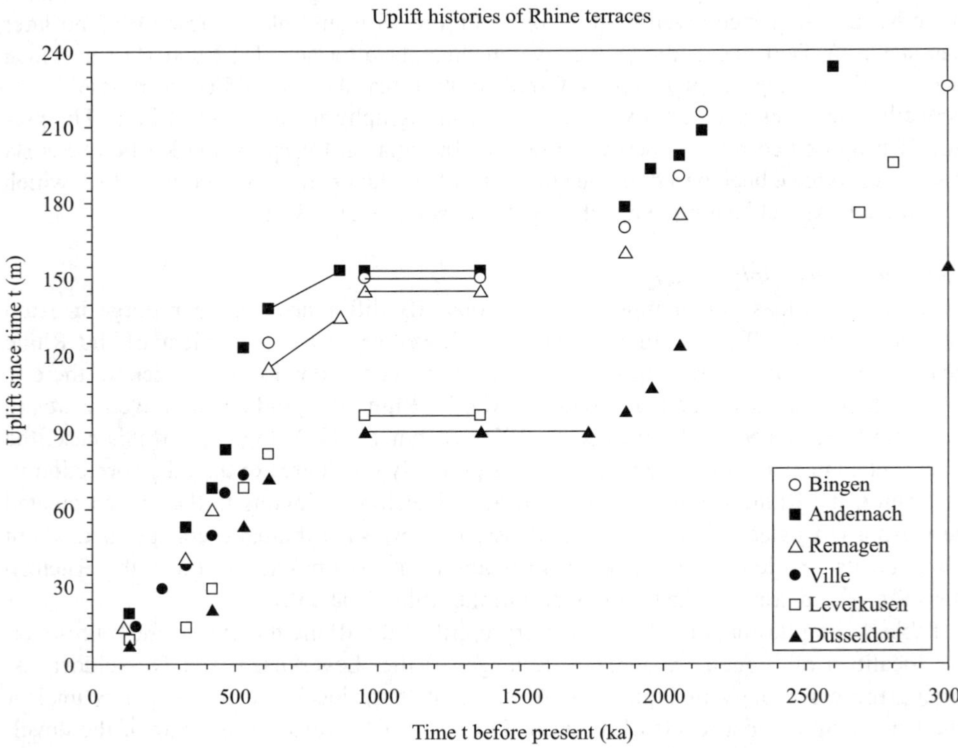

Figure 14. Chronologies of uplift for the clearest terraces of the Rhine at the localities listed in Table 4. Symbols for terrace HT2 are joined to indicated the estimated full range of time for deposition of this terrace. Symbols for terrace HT1 are joined to indicate the range of possible ages for the end of aggradation, except at Düsseldorf where the estimated full age span is depicted. The age plotted for the Kieseloolite gravel at Andernach is the youngest feasible for this deposit.

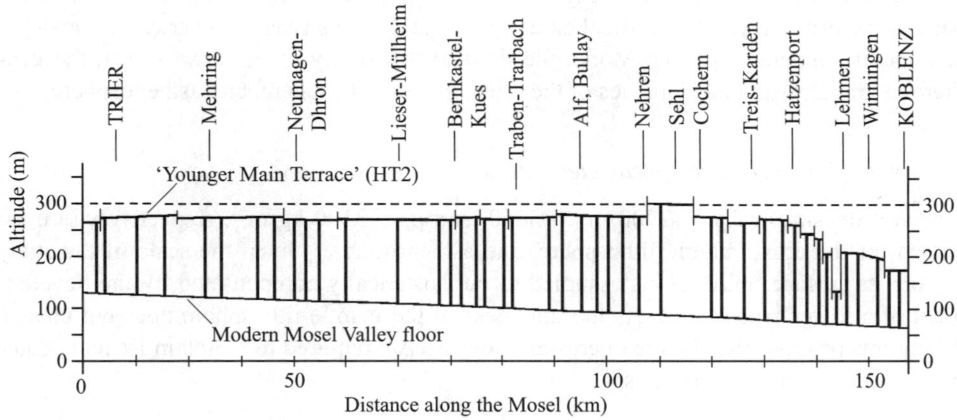

Figure 15. Longitudinal profile of the Mosel in the Rhenish Massif between Trier and Koblenz, adapted from part of Figure 8 of Meyer & Stets (1998). Meyer & Stets (1998) interpreted the lateral variations in incision as the result of components of vertical slip on active faults between observation points, but more gradual variations are also possible. See text for discussion.

has also been noted before: for instance, Semmel (1991) proposed that a large region including much of the drainage basin of the River Main (Fig. 1) has experienced a very similar history of Quaternary uplift to the land surface adjoining the Rhine gorge through the Rhenish Massif.

For comparison, Figure 15 shows the profile of the equivalent of terrace HT2 along the reach of the Mosel within the Rhenish Massif, between Trier and its confluence with the Rhine at Koblenz (Fig. 1). As on the Rhine, these terrace gravels are assumed here to just pre-date the Middle Pleistocene increase in uplift rates, and to thus have been deposited during oxygen isotope stage 22. The maximum incision since this time is ~210 m just upstream of Cochem, adjacent to the southern limit of the East Eifel Quaternary volcanism, which requires a *time-averaged* uplift rate of the adjoining land surface of ~0.25 mm a^{-1} since ~870 ka. The uplift decreases downstream of Cochem, into the Neuwied Basin. It also decreases more gradually upstream of Cochem, such that overall between Trier and Cochem terrace HT2 is subhorizontal, although with a slight net *upward* tilt in the downstream direction. Figure 15 thus indicates much more incision durng the Middle and Late Pleistocene than anywhere along the Middle Rhine, and currently represents the maximum known on this time scale on any river in north-western Europe.

4 MODELLING OF UPLIFT HISTORIES

The theory in the Appendix enables the surface uplift which results from net inflow of lower continental crust to beneath a region, to be modelled. This theory assumes that flow in the lower crust is horizontal laminar flow between parallel horizontal boundaries. The modelling is two-dimensional in the vertical plane, and thus does not incorporate any components of horizontal convergence or divergence in the flow. These limitations mean that the local complexities in lower-crustal flow which can be anticipated adjacent to migrating coastlines or perimeters of ice sheets are not considered.

The modelling requires knowledge of several parameters, including the depth of the base of the brittle layer, the vertical extent of the underlying plastic lower crust, the depth and initial temperature at the Moho, the thermal diffusivity of the lower crust, the geothermal gradient, and the densities of the crust, mantle lithosphere, and asthenosphere.

4.1 *Estimation of required parameter values*

Standard densities of ρ_c = 2700, ρ_m = 3300, and ρ_a = 3100 kg m^{-3}, respectively, are assumed for the crust, mantle lithosphere, and asthenosphere. Given the scale of the study region, its surface uplift ΔU is assumed to be isostatically compensated by the development of a lower-crustal root. As the thickness of the mantle lithosphere does not change during this process, the change in crustal thickness ΔH required to maintain isostatic equilibrium will approximately satisfy

$$\Delta H = \Delta U \frac{\rho_a}{\rho_a - \rho_c} \quad . \tag{3}$$

More complex expressions for isostasy in such circumstances have been developed, taking account for instance of the effect of thermal expansion due to the temperature increase around the Moho as the crust thickens (e.g. Westaway, 1995), but differ minimally from this simple form.

The thermal diffusivity, κ, of any material depends on its density ρ, specific heat capacity c, and thermal conductivity k, as

$$\kappa \equiv \frac{k}{c\rho} \quad . \tag{4}$$

Many studies have used a value of 1.0 m^2 s^{-1} for κ for mantle lithosphere. As continental crust has a ~20% lower density, from (4) a κ value ~20% higher is reasonable: 1.2 m^2 s^{-1} is thus adopted.

Seismic refraction profiles across the Rhenish Massif have long been interpreted with 28 km of continental crust with P-wave velocity v_p up to ~6.7 km s^{-1}, underlain by a layer of mafic underplating (v_p ~7.3 km s^{-1}) and then mantle lithosphere with v_p ~8.0 km s^{-1} (e.g. Mooney & Prodehl, 1978). Several studies have adopted a 4 km thickness for this underplated layer (e.g. Illies et al., 1979; Mengel et al., 1991; Wilson & Downes, 1991), indicating a present-day Moho depth of 32 km. Theilen & Meissner (1979) suggested instead that the seismic refraction data fit 25 km thick continental crust with v_p up to ~6.2 km s^{-1} and with the Moho at 29 km. Other studies (summarised by Cermak & Bodri, 1995) have obtained similar results. However, it is difficult to accept that such thin crust, in isostatic equilibrium, could maintain a land surface at almost 1 km altitude.

Many studies (e.g. Hurtig & Oelsner, 1977; Cermak & Hurtig, 1979; Morgan, 1982; Cermak & Zahradnik, 1982; Cermak & Bodri, 1995; Cermak, 1995) have compiled surface heat flow data for the study region. In the Eastern Eifel, this is measured as 73 mW m^{-2} (e.g. Morgan, 1982) to ~80 mW m^{-2} (e.g. Cermak, 1995). Assuming a thermal conductivity for the crust of 2.5 to 3.0 W m^{-1} °C^{-1}, a near-surface geothermal gradient of 24 to 29°C km^{-1} is indicated. Shallow geothermal gradients of ~30°C km^{-1} are also typical in the Netherlands (e.g. Visser, 1982). However, the modelling in the present study requires the geothermal gradient in the lower crust, which is less than near the surface due to radiogenic heat production in the upper crust. Cermak & Bodri (1995) estimated the heat flow through the Moho in the study region as ~50 mW m^{-2}. Using their value for the thermal conductivity in the crust of 2.5 W m^{-1} °C^{-1}, this indicates a geothermal gradient in the lower crust of ~20°C km^{-1}.

The ~300 m of recent surface uplift in the Rhenish Massif (e.g. Illies et al., 1979) requires ~2 km of crustal thickening to be isostatically compensated (Equation 3). The mafic underplated layer at the base of the crust is known, from samples of it forming xenoliths in Quaternary basalts, to be ancient and not the result of the Quaternary magmatism (e.g. Okrusch et al., 1979; Wilson & Downes, 1991). Its degree of metamorphism requires palaeo-temperatures of ~800°C and pressures of ~1.2 GPa (e.g. Okrusch et al., 1979) at some time since the underplating occurred, indicating a palaeo-crustal thickness of ~40 km. Studies of Sr, Nd, and Pb isotope ratios suggest that this underplating occurred at ~450 Ma (Rudnick & Goldstein, 1990). This layer was thus already present in the Late Pliocene, when the estimated thickness of continental crust was ~26 km with the Moho at ~30 km.

Allowing for the tapering of the geothermal gradient from ~25°C km^{-1} at the surface to ~20°C km^{-1} in the lower crust, the Moho temperature can be estimated as ~650-700°C. Values of this order have long been considered reasonable for much of north-west Europe (e.g. Hurtig & Oelsner, 1977; Cermak & Bodri, 1995). Extrapolating the 20°C km^{-1} geothermal gradient across the mantle lithosphere, the asthenosphere (the convecting part of the mantle, at ~1400°C) is predicted at a depth of ~70 km, again similar to estimates from previous studies (e.g. Cermak, 1982). Cermak & Bodri (1995) estimated a slightly lower lithosphere thickness of ~65 km from heat flow modelling. Studies of the dispersion of seismic surface waves suggest even thinner lithosphere, ~50-60 km thick (e.g. Blundell et al., 1992, p. 67; Cermak & Bodri, 1995). Furthermore, as the geothermal gradient is known, the depth of the base of the brittle layer, z_b, at which the temperature is ~350°C, can be estimated.

Modelling of the uplift response also requires knowledge of the timing of the start of each phase of water and/or ice loading related to intensified glaciations, each of which may in principle contribute to ΔT_e for each of the study localities. Six potential contributions are considered. First, is the cyclic water loading to the continental shelf caused by global sea-level fluctuations linked to the moderate cyclic glaciations of Antarctica which, from interpretation of oxygen isotope studies, are believed to have begun in the Miocene (e.g. Molnar & England, 1990). The start timing of these fluctuations is not well resolved (see below), and is assigned a nominal value of 18 Ma. For reasons connected with limitations resulting from the approximations used in the modelling (see below, also Appendix), and from the relatively sparse available data, this contribution of Miocene and Early Pliocene glaciations cannot be well modelled at present. It is included in the modelling simply to roughly match the slow rates of surface uplift before the Late Pliocene.

Van den Berg's (1996) preferred chronology for the Waubach terraces, discussed earlier, indicates an increase in the uplift rate from less than 0.01 mm a^{-1} to more than 0.02 mm a^{-1} between 4.0 and 3.5 Ma (Fig. 16c). To model this change would require an additional forcing contribution starting no later than 3.8 Ma. However, I am not aware of any major event involving an increase in surface loading starting at that time, which could cause an increase in uplift rate with this timing. The Dutch pollen stratigraphy indicates a mid-Pliocene cooling, at the boundary between the Brunsummian and Reuverian A stages. Zagwijn (1986b) correlated this boundary with the Zanclean – Piacenzian of the Mediterranean (the conventional boundary between Early and Late Pliocene), which is very near the Gilbert-Gauss geomagnetic reversal that is dated to 3.58 Ma from orbital forcing (e.g. Hilgen et al., 1991b). However, other studies (e.g. Ruddiman et al., 1987) identify a later mid-Pleistocene global cooling event at ~3.3 Ma, within the Mammoth subchron. Van den Berg (1996) suggested that this may mark the Dutch Brunsummian – Reuverian A boundary, instead. Nonetheless, both these timings for the start of a possible enhanced forcing contribution, ~3.6 and ~3.3 Ma, are too late to explain the jump in uplift rate suggested within the Waubach terraces given van den Berg's (1996) preferred chronology. However, if van den Berg's (1996) alternative chronology for the Waubach terraces (or a similar scheme, such as that in Table 2) is adopted, the uplift rate remains low (no greater than ~0.01 mm a^{-1}) until ~3 Ma, and there is thus no need to consider potential causes of forcing at earlier times.

The second event considered is thus the start of significant upland glaciations in Europe, already mentioned, which is dated at ~3.1 Ma by the start of cyclic deposition of sapropels in Mediterranean marine sediments (e.g. Hilgen,, 1991a, 1991b) and which appears to coincide with the start of the Reuverian B pollen stage of the Netherlands (e.g. Zagwijn, 1986b). The third event is the start of lowland glaciation of the northern hemisphere at ~2.5 Ma, revealed by the evidence of ice-rafting in Atlantic Ocean sediments (e.g. Shackleton et al., 1984). From this time onward, the variations in $^{18}O/^{16}O$ ratios indicate drawdown of global sea-level by ~60 m during glacials, at least double the amplitude of the variations occurring beforehand. The fourth event is the apparent intensification of northern hemisphere glaciations around 1.8 Ma, revealed for instance by the palaeobotanic evidence of a cold climate during the Tiglian C_4c (oxygen isotope stage 68; ~1.87 Ma) and the cooling at the conventionally-defined Pliocene-Pleistocene boundary (oxygen isotope stage 64; ~1.78 Ma). The fifth event is the start of intensified glaciations around 1.2 Ma (oxygen isotope stage 36), since when the sea level has been drawn down by up to ~120 m as larger ice sheets have developed during glacial stages (e.g. Shackleton et al., 1990). The final event is the abrupt buildup in mean ice volume, time-averaged throughout each glacial cycle, which occurred around 0.9 Ma (e.g. Mudelsee & Schulz, 1997) during the transition to the glaciation in oxygen isotope stage 22. Many oxygen isotope records indeed reveal that stage 22 was the first glaciation comparable in intensity to the largest in the Middle and Late Pleistocene.

All the parameter values needed can thus be independently constrained, except for the contributions to ΔT_e from each of the changes summarised above (designated as ΔT_{e1} to ΔT_{e6}). As is discussed in the Appendix, these quantify, for each of the stages of the uplift response, the difference in temperature at the depth z_i where lower-crustal flow is concentrated, between the uplifting study region and the source region of its added lower crust, scaled by the ratios of areas of these regions. However, the overall dimensions of the onshore uplift are yet to be determined, and the source region of the lower crust is also not clear (the North Sea?; the Baltic?) and so its dimensions also cannot be estimated (although there is some further discussion in the Appendix). ΔT_{e1} to ΔT_{e6} are thus treated as free parameters, whose values are adjusted to match the observed uplift histories. Conversion of these values to actual temperature differences and length scales of lower-crustal flow is beyond the scope of this study.

4.2 *Modelling the Maastricht terrace sequence*

Two key reasons exist for selecting Maastricht as the locality for modelling uplift of Maas river terraces. First, due to the work of van den Berg (1996) it has the best documented terrace record. Second, being the farthest point downstream with an intact terrace sequence, it is the locality farthest from, and thus least likely to be affected by, the disequilibrium caused by the decrease in the river gradient around Nouzonville (Fig. 1). However, it is necessary to take into account in the modelling the uncertainty caused by the range of possible age interpretations between the chronology of van den Berg (1996) and the possible revision to it tentatively suggested in this study (Fig. 8).

Using van den Berg's (1996) chronology, including his young age estimates for the Waubach terraces, the uplift rate was less than ~0.01 mm a^{-1} before ~4 Ma (Figs 8a and 16c). It then rose abruptly to ~0.02 mm a^{-1} by ~3.5 Ma, then increased gradually to a peak of ~0.06 mm a^{-1} around 2.0 Ma. If, instead, the Waubach terraces are older, then the uplift rate remained very low until later, ~3.0 Ma. It then increased abruptly through ~0.03 mm a^{-1} during the ~3.0-2.5 Ma time of formation of the Kosberg terraces (Fig. 8b), then continued to increase gradually, peaking as before at more than 0.06 mm a^{-1} around 2.0 Ma. It then decreased more abruptly, falling below ~0.03 mm a^{-1} by ~1.3 Ma (Fig. 8b). It then increased abruptly to more than ~0.09 mm a^{-1} around the time of the end of formation of the Geertruid 3 terrace and the start of deposition of the Pietersberg 1 terrace, which was shortly after 1.0 Ma using this chronology. Subsequently, it has gradually decreased to ~0.06 mm a^{-1} (Fig. 8b).

Using my revised chronology, the apparent small increase in uplift rate during the time of formation of the Kosberg terraces may be disregarded, given the possibility that these terraces may have formed over a longer range of cold climate cycles within the Gauss chron than was assumed by van den Berg (1996) (for instance, if the oldest Kosberg terrace matches the oldest Mediterranean sapropel, its age may be ~3.1 Ma, not ~2.9 Ma as was deduced by van den Berg, 1996). The uplift rate instead remained below ~0.01 mm a^{-1} until after ~2.5 Ma, then increased abruptly after deposition of the Crapoel terrace, reaching ~0.08 mm a^{-1} by ~2.0 Ma (Fig. 8b). The uplift rate then decreased gradually to ~0.04 mm a^{-1} by ~1.2 Ma (Fig. 8b). It then increased more abruptly, after deposition of the Pietersberg I terrace around 0.9 Ma, to values of the order of ~0.1 mm a^{-1}, before gradually decreasing to ~0.06 mm a^{-1} at the present day (Fig. 8b). Compared with the results of using van den Berg's (1996) chronology, the increase in uplift rate around ~1 Ma occurs later and appears more gradual, with the uplift rate reaching ~0.09 mm a^{-1} by ~0.8 Ma instead of shortly after ~1.0 Ma (Fig. 8b). This is due to the younger ages assigned for the Pietersberg terraces.

For both terrace chronologies, the Late Miocene and Early Pliocene uplift is modelled as the tail of the response to lower-crustal flow beginning at a nominal time of 18 Ma with the values of ΔT_{e1} listed in the captions to Figs 16 and 17. These parts of the models are not well resolved for several reasons. First, similar predicted uplift rate variations would result from later start times and smaller values of ΔT_e, or earlier start times and larger values of ΔT_e. There is thus strong trade-off between the start time and ΔT_e, which is not resolved by the available data. In principle, it could possibly be resolved if the older Miocene fossil shoreline levels in profile E of Figure 3 could be dated; but that is beyond the scope of the present study. Second, in order to simplify the calculations the modelling uses an approximation, whereby the requirement that the temperature at the Earth's surface is constrained, is not enforced (see Appendix). As a result, the method used will only crudely calculate the response at time scales of many millions of years after a given régime of lower-crustal flow has begun. Third, the mismatches between observed and predicted amounts of uplift since deposition of the Waubach terraces are also not significant. For these terraces, the calculations of 'observed' uplift are sensitive to the parameters used for the channel-lengthening correction for the Maas (Fig. 9) and to the age of each terrace, neither of which is well-resolved. In particular, adoption of a palaeo-gradient of less than 0.7 m km^{-1} for the Maas (and thus, to maintain internal consistency, a proportionately greater channel lengthening and a younger age bound for the end of this lengthening) would reduce the estimate of 'observed' uplift since deposition of each Waubach terrace (Fig. 9), thus giving a better match to the predictions.

4.2.1 *The match to van den Berg's (1996) chronology*

Figure 16 shows a fit between observed and predicted uplift histories for the Maas terrace sequence at Maastricht, the observations being dated using van den Berg's (1996) chronology with his older alternative set of ages for the Waubach terraces, and the prediction being calculated using the parameters listed in the Figure caption. In this solution, the increase in uplift rate after 3.0 Ma is well matched to the observations as the response to incipient upland glaciation at 3.1 Ma. The subsequent gradual increase in uplift rates is fairly well matched as the superposition of responses to this early upland glaciation plus the incipient lowland glaciation and associated sea-level fluctuations starting at 2.5 Ma. There is no signal in the observations which can be modelled as a consequence of intensified glaciation around 1.8 Ma: the peak in the observational uplift rate record after ~2 Ma is about 0.3 million years too early for that.

An attempt has been made to match the observed increase in uplift rate around 1 Ma as a consequence of intensified glaciation starting at 1.2 Ma. However, the resulting fit is not good, as the abrupt rise in the observational uplift rate record peaks at ~0.09 mm a^{-1} about 0.2 million years too early. To illustrate that this conclusion is not sensitive to details of calculation of uplift rate, Figures 8 and 16 present the data in different ways. In Figure 8, the earliest data point with a ~0.09 mm a^{-1} uplift rate occurs at 1.0225 Ma and is based on the difference in age and altitude of the Geertruid 2 and Pietersberg 1 terraces. In Figure 16, the corresponding data point occurs at 0.9925 Ma and is derived from the age and altitude of the Geertruid 3 and Pietersberg 1 terraces. In both figures, the slightly later data point indicating a similar uplift rate is at 0.95 Ma and uses the age and altitude of the Geertruid 3 and Pietersberg 2 terraces. Given the timing constraint on oxygen isotope stage 36 and the observational constraint on crustal thickness, the only way to significantly improve this match, by shifting the predicted increase in uplift rate to an earlier time, would require reducing the distance $z_i - z_b$ between the base of the brittle layer and the depth where lower-crustal flow is concentrated. However, this would not make physical sense, as it would imply an upward *decrease* in the geothermal gradient through the crust. It would also degrade the match between observed and predicted uplift rates in other parts of the model. Finally, there is no observational evidence for an uplift response to the intensification in glaciation around 0.9 Ma.

4.2.2 *The match to the revised chronology from this study*

Figure 17a and Figure 17b show a match between observed and predicted uplift histories for the Maastricht area, with the observations expressed using the modified chronology derived in this study. Figure 17d and Figure 17e show the corresponding variations in uplift rate. Like the earlier solutions, there is no evidence of any contribution to forcing the observed uplift starting around 1.8 Ma. The main differences compared with Figure 16a and Figure 16b are, first, that the younger ages adopted for terraces around the conventional Pliocene-Pleistocene boundary mean that the start of accelerated uplift is delayed relative

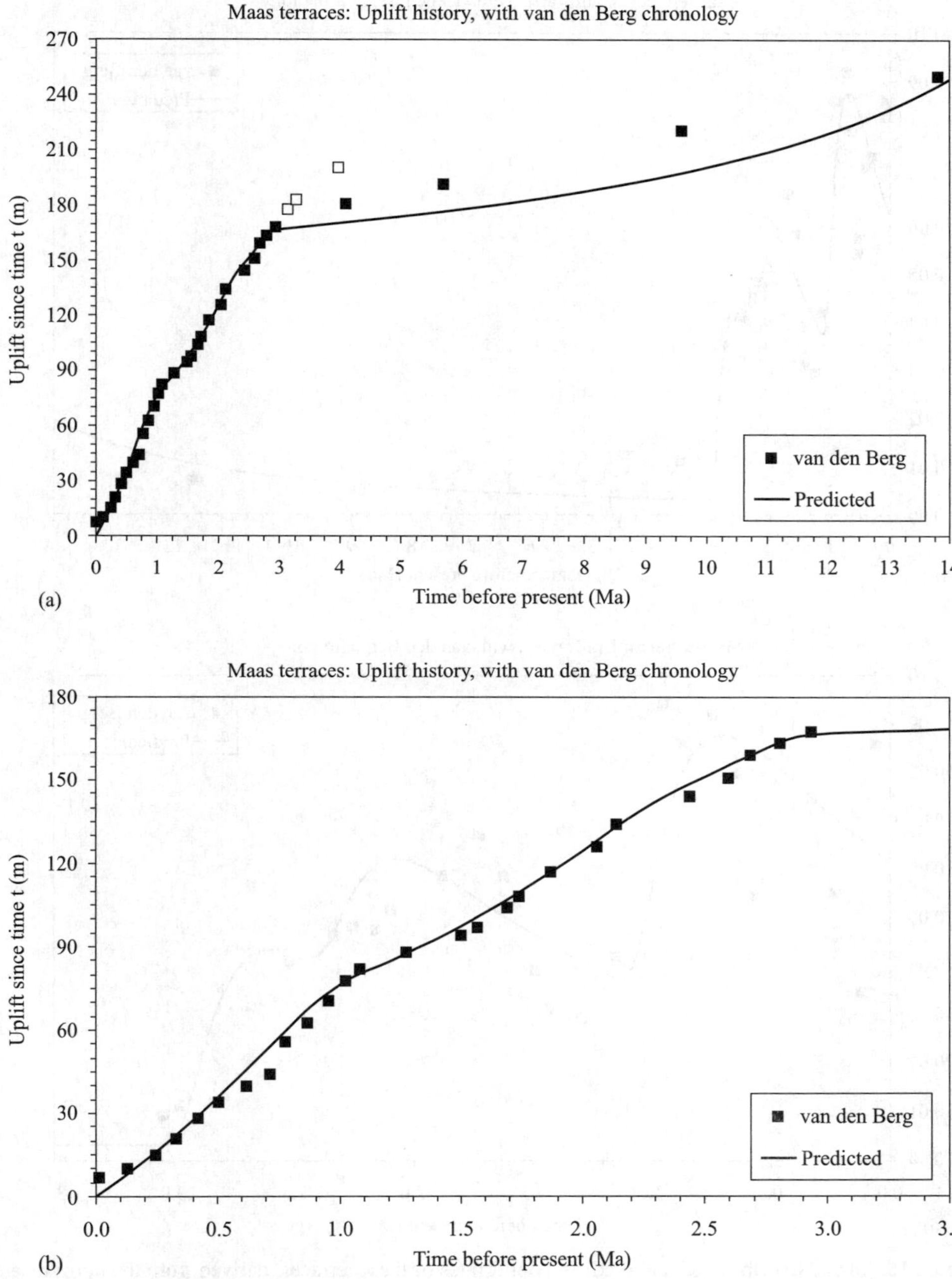

Figure 16. (a) and (b) Observed and predicted uplift history of Maas river terraces at Maastricht since 14 Ma (a) and 3.5 Ma (b), assuming van den Berg's (1996) chronology. Observed terrace ages and amounts of uplift are from Figure 8.

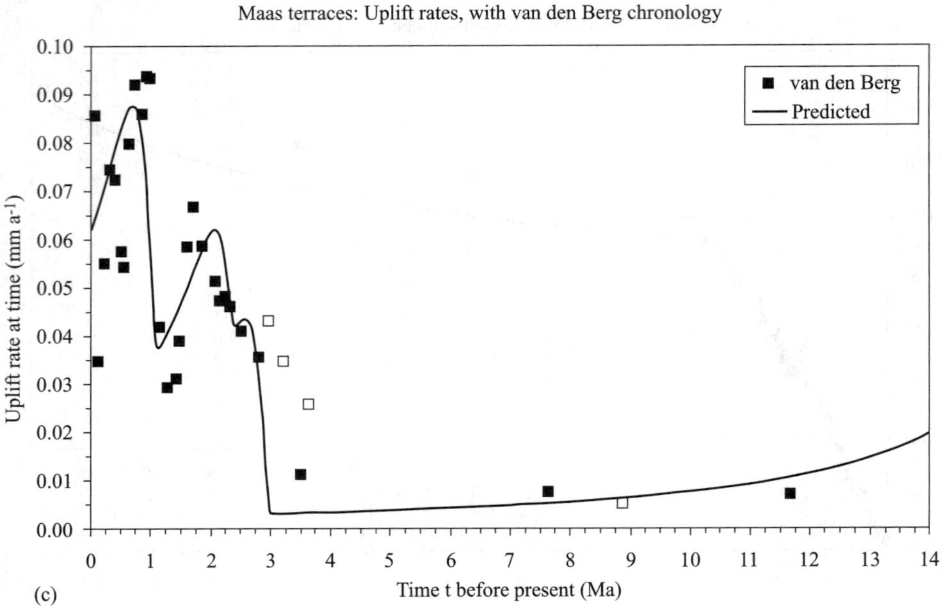

(c)

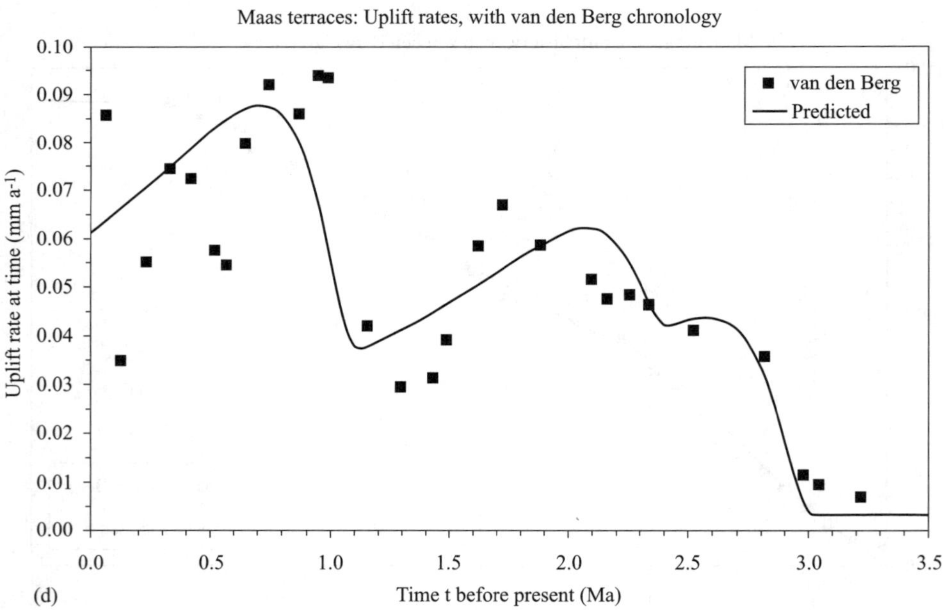

(d)

Figure 16. (c) and (d) Observed and predicted uplift rates of these terraces, derived from the uplift data in (a) and (b), time-averaged over intervals of time between two to four terraces spanning, typically, ~200 ka. The predicted uplift rate curve is calculated using Equation (A25) with κ 1.2×10^{-6} m^2 s^{-1}, u 20°C km^{-1}, z_b 15 km, and z_i 26 km, superposing solutions with ΔT_{e1} –7.5°C (starting at 18 Ma), ΔT_{e2} –2.8°C (3.1 Ma), ΔT_{e3} –2.0°C (2.5 Ma), ΔT_{e4} 0°C (1.8 Ma), ΔT_{e5} –4.0°C (1.2 Ma), and ΔT_{e6} 0°C (0.9 Ma). The uplift curves in (a) and (b) are obtained by numerically integrating this uplift rate curve. Open symbols in (a) and (c) indicate the younger ages for the Waubach terraces in van den Berg's (1996) preferred chronology. No attempt is made to fit these data here. See text for discussion.

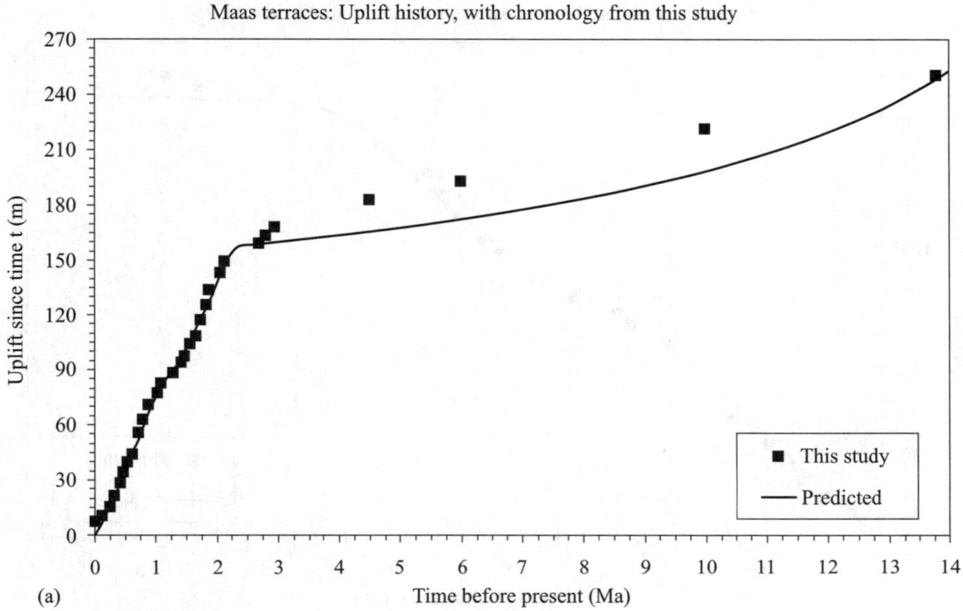

Figure 17a. See Figure 17g for caption.

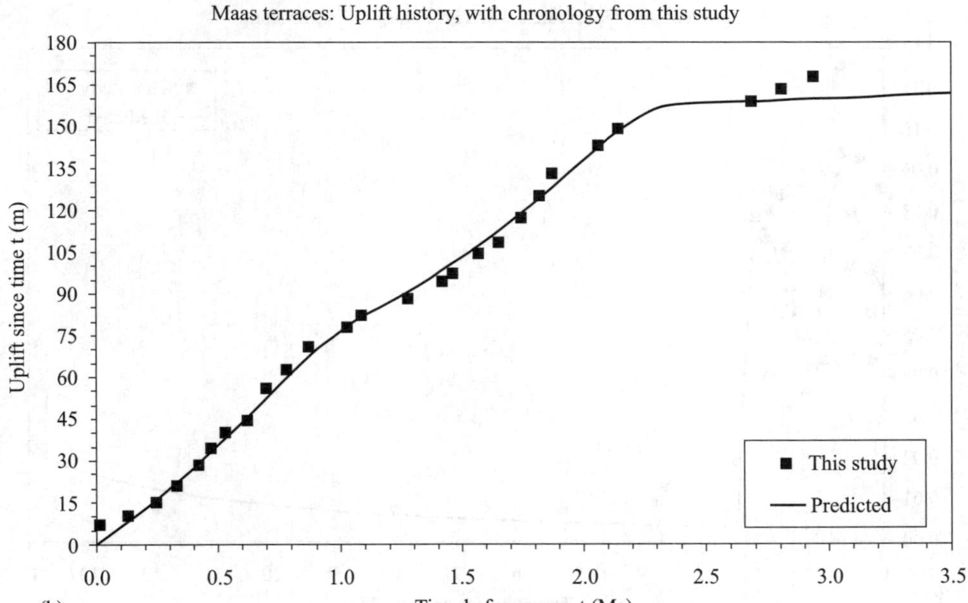

Figure 17b. See Figure 17g for caption.

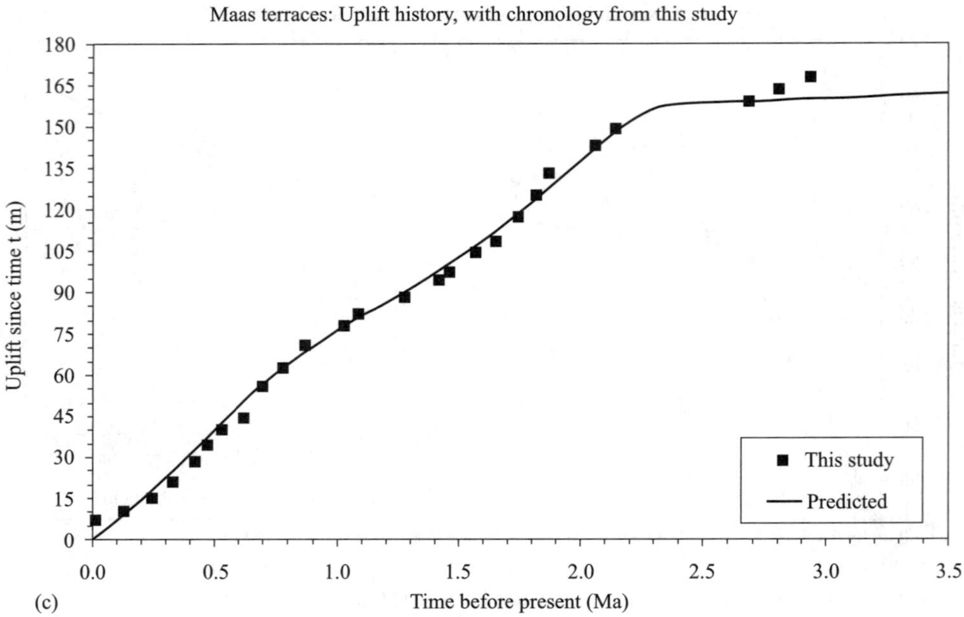

Figure 17c. See Figure 17g for caption.

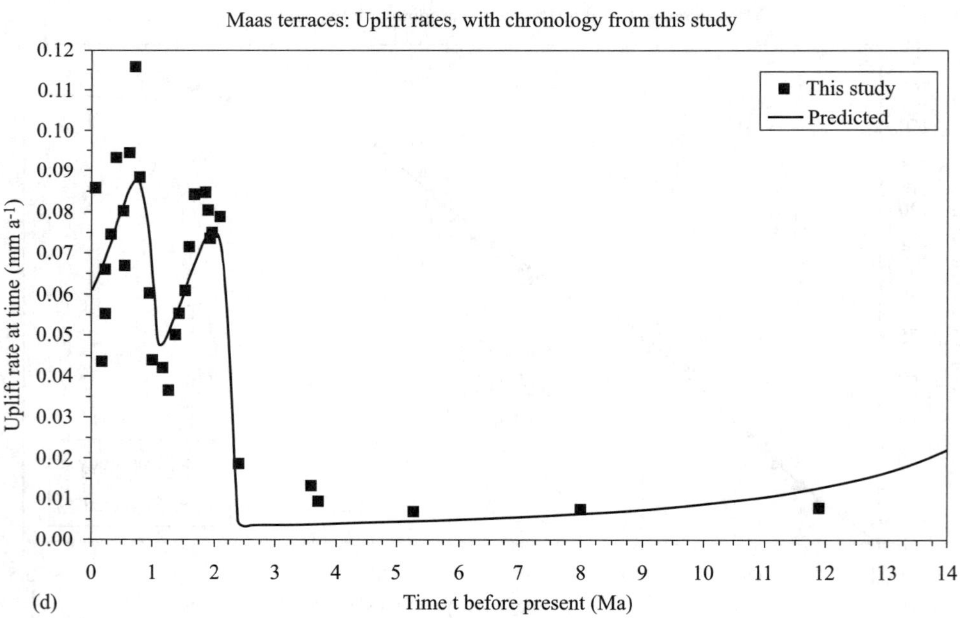

Figure 17d. See Figure 17g for caption.

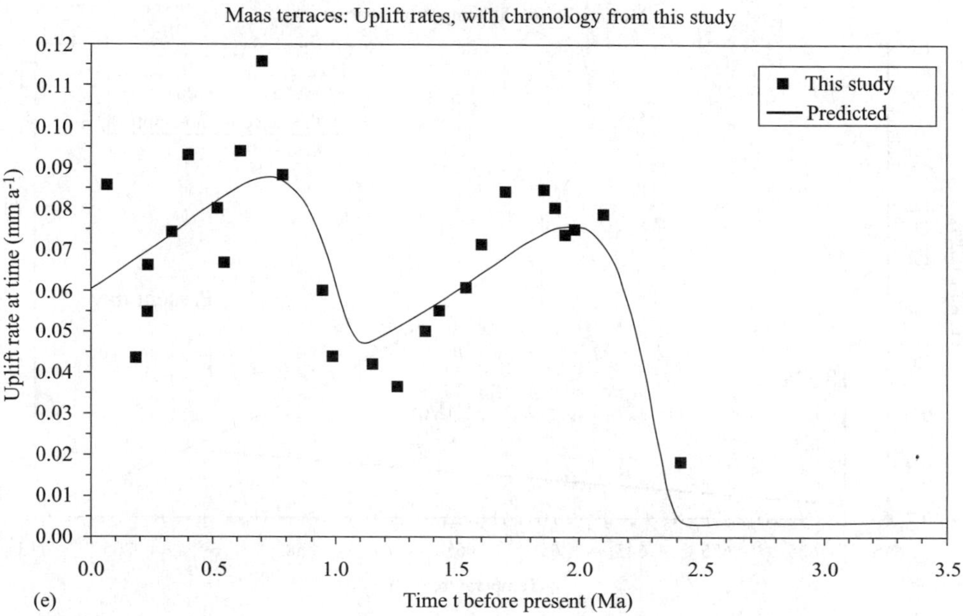

Figure 17e. See Figure 17g for caption.

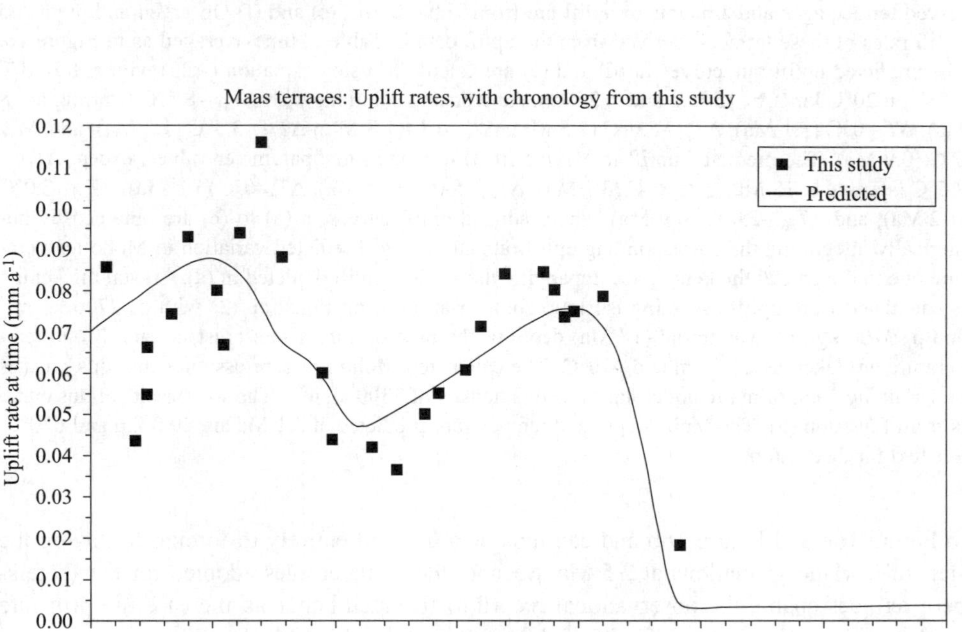

Figure 17f. See Figure 17g for caption.

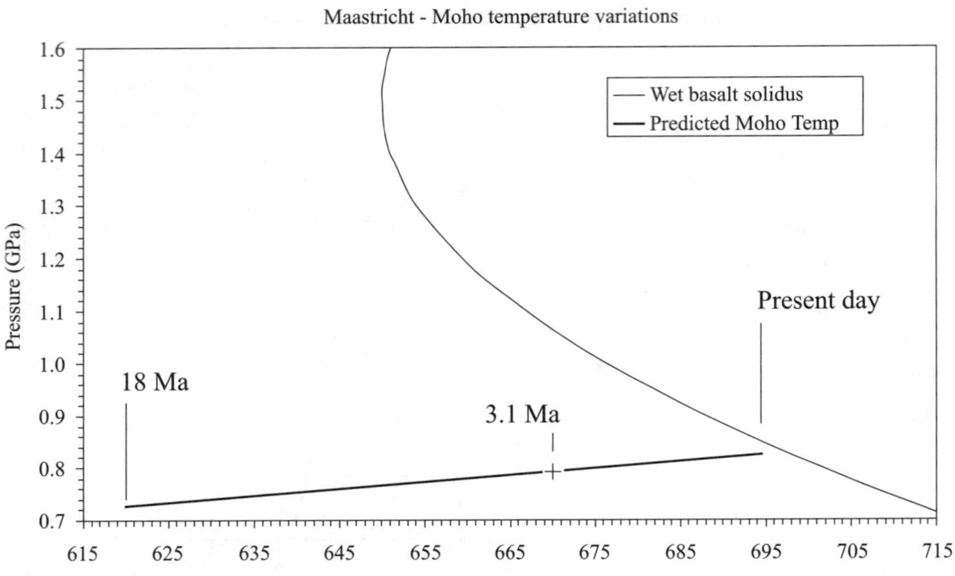

(g)

Figure 17g. (a), (b) and (c) Observed and predicted uplift histories of Maas river terraces at Maastricht since 14 Ma (a) and 3.5 Ma (b) and (c), assuming the revised chronology from this study. Observed terrace ages and amounts of uplift are from Table 2. (d), (e) and (f) Observed and predicted uplift rates of these terraces, derived from the uplift data in Table 2, time-averaged as in Figure 16. The predicted uplift rate curves in (d) and (e) are calculated using Equation (A25) with κ 1.2×10^{-6} m^2 s^{-1}, u 20°C km^{-1}, z_b 15 km, and z_i 26 km, superposing solutions with ΔT_{e1} –8.5°C (starting at 18 Ma), ΔT_{e2} 0°C (3.1 Ma), ΔT_{e3} –5.0°C (2.5 Ma), ΔT_{e4} 0°C (1.8 Ma), ΔT_{e5} –3.5°C (1.2 Ma), and ΔT_{e6} 0°C (0.9 Ma). The predicted uplift rate curve in (f) uses the same parameter values, except ΔT_{e1} –8.5°C (starting at 18 Ma), ΔT_{e2} 0°C (3.1 Ma), ΔT_{e3} –5.0°C (2.5 Ma), ΔT_{e4} 0°C (1.8 Ma), ΔT_{e5} –2.0°C (1.2 Ma), and ΔT_{e6} –2.0°C (0.9 Ma). The predicted uplift curves in (a) to (c) are obtained by numerically integrating the corresponding uplift rate curves. (g) Predicted variation in Moho temperature due to the crustal thickening accompanying the surface uplift depicted in (a). Crustal thickening is calculated from uplift assuming isostatic compensation using Equation (3) with ρ_c 2700 kg m^{-3} and ρ_a 3100 kg m^{-3}. The initial (18 Ma) depth of the base of continental crust, z_w, and Moho temperature are taken as 22.5 km and 620°C. The calculated Moho pressure assumes that this crust is underlain by 4 km of mafic underplating with a density of 3300 kg m^{-3}. The wet basalt solidus curve is from Equation (5). The Moho depth and temperature predicted at 3.1 Ma are 29.0 km and 670°C. See text for discussion.

to Figure 16a and Figure 16b and can thus be attributed entirely to forcing following the start of lowland glaciations at 2.5 Ma. Second, the younger ages adopted for the Pietersberg terraces enable the observational record to fit much better an increase in uplift rate predicted as a consequence of enhanced forcing starting at 1.2 Ma.

Figure 17c shows an alternative solution, in which the increase in uplift rate at the end of the Early Pleistocene is attributed as a consequence of forcing starting at 1.2 Ma, followed by an additional contribution starting at 0.9 Ma. This revised solution fits better the observational record from shortly after the start of the increase in uplift rate, although some later parts of the Middle Pleistocene record are fitted marginally less well (compare Fig 17f and Fig 17c with Fig 17e and Fig 17b).

4.3 *Modelling the Rhine terrace sequence at Andernach*

Andernach was identified earlier as the locality along the Middle Rhine gorge with the highest uplift rate. Local terrace altitudes (Table 4) are displayed in Figures 14 and 18a. As was discussed earlier, these altitudes are assumed to directly reflect the uplift of the surrounding land surface, with no correction for channel lengthening. Figure 18a also shows a solution for the predicted uplift history, calculated using the parameter values listed in its caption, for which the only forcing contributions start at 3.1 Ma and at 0.9 Ma. Its predicted uplift rate rises to ~0.07 mm a^{-1} by ~2.5 Ma, then decreases gradually to ~0.03 mm a^{-1} by ~0.8 Ma. It then increases to a peak value of ~0.23 mm a^{-1} at ~0.35 Ma before decreasing to ~0.20 mm a^{-1} at the present day. The sequence of terraces dated to the Latest Pliocene, up to about the time of the conventional Pliocene-Pleistocene boundary, fall within the early peak in uplift rate. The start of the second peak coincides with the aggradation of terrace HT2, and the preceding time span with a low uplift rate marked the aggradation of HT1.

Figure 18b shows an alternative solution, in which the uplift forcing involves contributions starting at 3.1 Ma, 2.5 Ma, 1.2 Ma, and 0.9 Ma, not just at 3.1 Ma and 0.9 Ma. The small contribution starting at 1.2 Ma causes a small increase in uplift rate starting shortly after 1.1 Ma, ahead of the much more rapid increase around 0.8 Ma (Fig. 18c). Although such complexity is not demonstrable by direct evidence, with an uplift history of this form the incision expected during the major interglacial of oxygen isotope stage 25 (0.95 Ma) – given the contemporaneous climate – can account for the distinction in the terrace record between the end of aggradation of HT1 and the start of aggradation of HT2. There is, likewise, no direct evidence to support the fluctuations in uplift rate predicted in the latest Pliocene by this more complex forcing history (Fig. 18c). However, they coincide with the estimated ages of the terraces which are recognisable from this era. Relatively rapid uplift of the land surface at these times provides a reason why these terraces are relatively readily distinguished. In contrast, the much lower uplift rate predicted during the later stages of aggradation of terrace HT1 can explain why it is much less clear: deposits from many successive glacial cycles can be expected at much more similar levels, making them more difficult to resolve.

4.4 *Summary of results*

The available direct evidence is clearly insufficient to uniquely constrain the terrace chronology and uplift history for either the Maas or Rhine. However, if both localities are compared, several points of similarity emerge. First, the closest resemblance between solutions for the latest Pliocene arises when one compares the Andernach solution in Figure 18b and Figure 18c with the Maastricht solution in Figure 16. Both involve forcing starting at 3.1 Ma followed by a less important contribution starting at 2.5 Ma. This resemblance implies that van den Berg's (1996) Maas terrace chronology for the latest Pliocene is more likely to be correct than the alternative tentatively proposed in this study.

However, problems remain with trying to reconcile van den Berg's (1996) chronology for the Maas (Fig. 7) with the chronology for the Rhine (Fig. 13). Notably, Van den Berg's (1996) Maas chronology assigns the peat overlying the Simpelveld I terrace to stage 81, but places it from pollen evidence in the Tiglian (Fig. 7). On the contrary, the

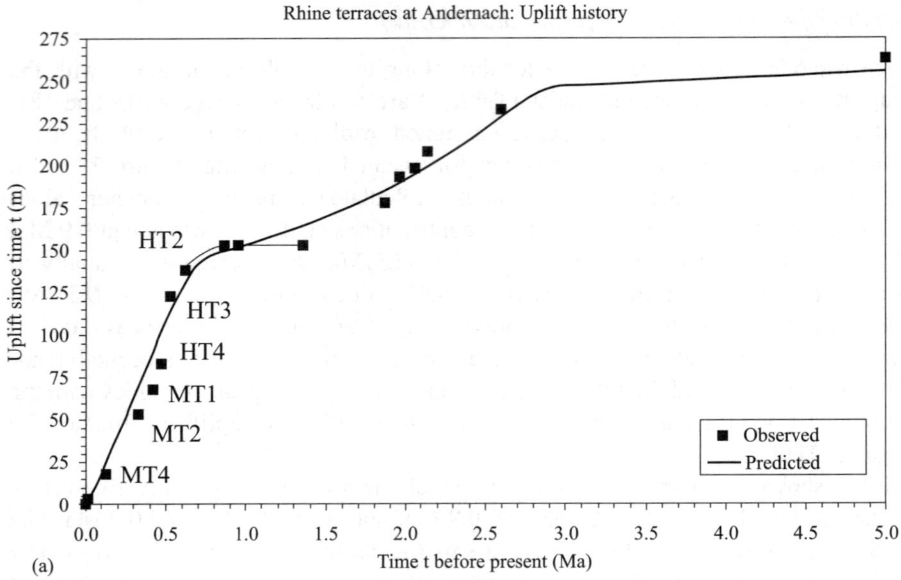

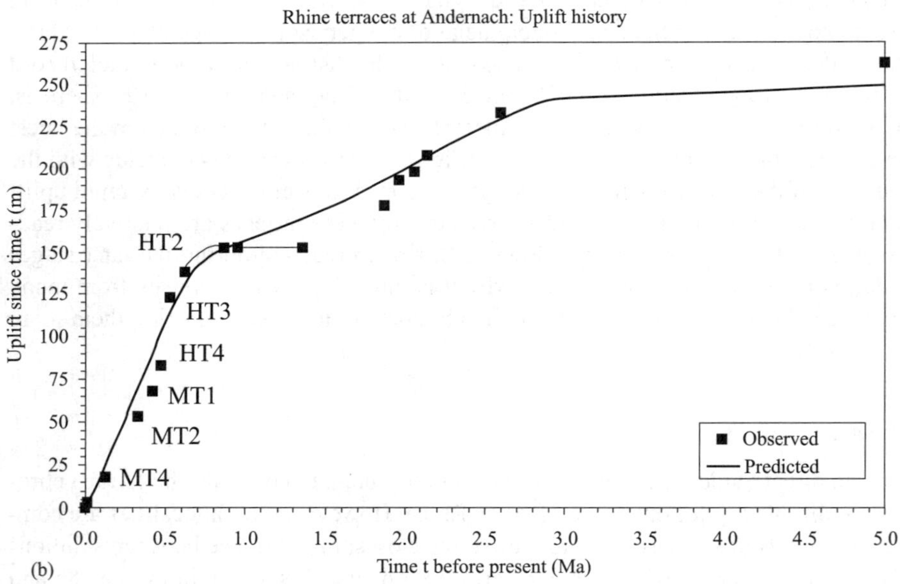

Figure 18a, b. (a) Observed and predicted uplift history of Rhine river terraces at Andernach since 5 Ma. Observed terrace ages and amounts of uplift are from Table 4. The symbols for terraces HT1 and HT2 indicate the estimated uncertainty in the young age bound for HT1 and full age span for aggradation of HT2. The predicted uplift curve is calculated by numerical integration of Equation (A25) with κ 1.2×10^{-6} m^2 s^{-1}, u 20°C km^{-1}, z_b 15 km, and z_i 26 km, superposing solutions with ΔT_{e1} –9.0°C (starting at 18 Ma), ΔT_{e2} –4.5°C (3.1 Ma), ΔT_{e3} 0°C (2.5 Ma), ΔT_{e4} 0°C (1.8 Ma), ΔT_{e5} 0°C (1.2 Ma), and ΔT_{e6} –14.0°C (0.9 Ma). The uplift data are too sparse for presentation of observed uplift rates as in Figure 17(d to f) (b) An alternative predicted uplift curve, calculated using the same procedure and parameter values as in (a), except ΔT_{e1} –9.0°C (starting at 18 Ma), ΔT_{e2} –2.8°C (3.1 Ma), ΔT_{e3} –1.5°C (2.5 Ma), ΔT_{e4} 0°C (1.8 Ma), ΔT_{e5} –1.0°C (1.2 Ma), and ΔT_{e6} –13.0°C (0.9 Ma).

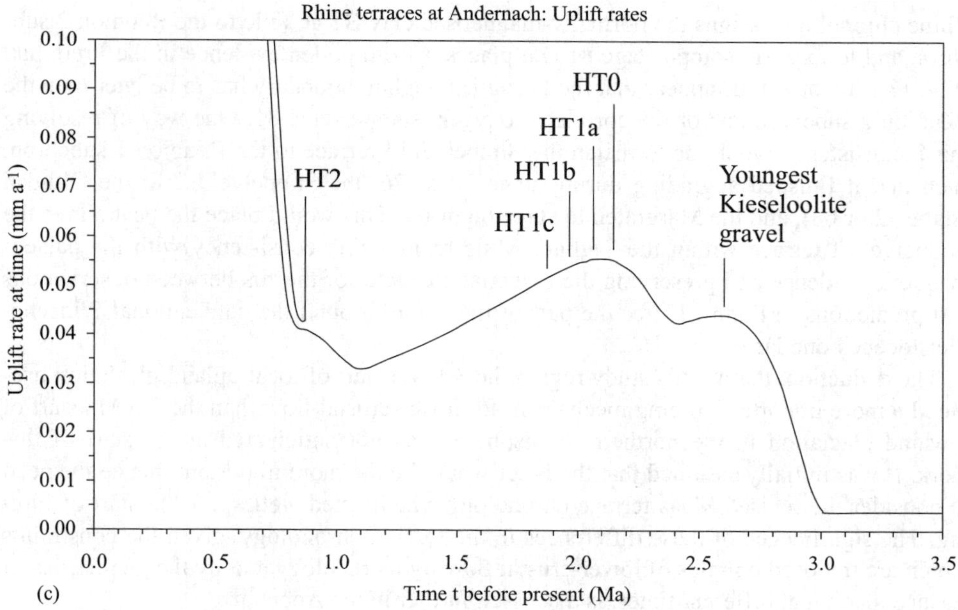

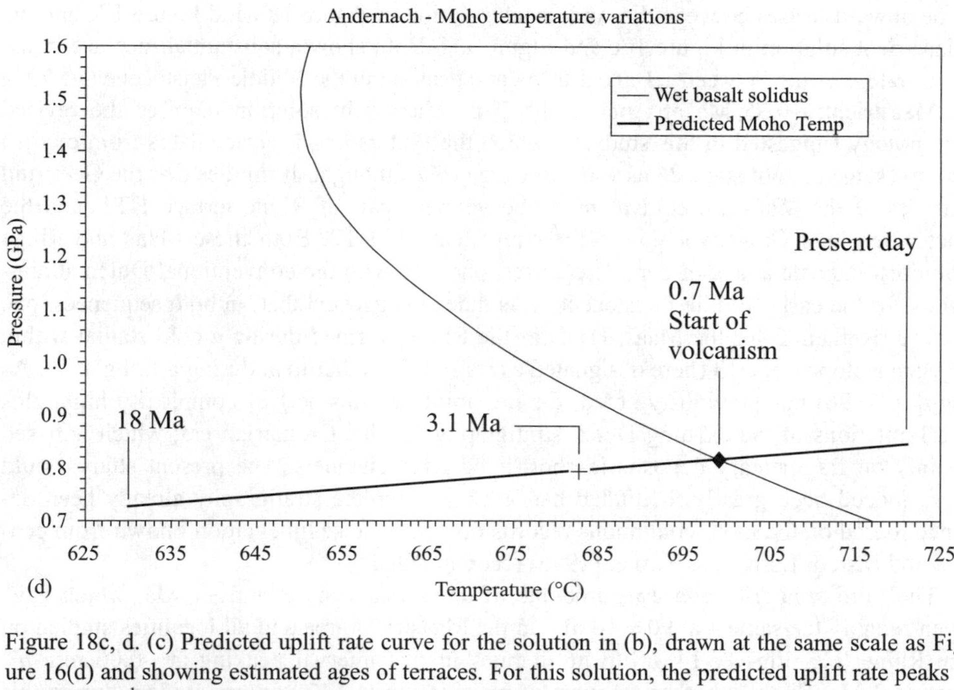

Figure 18c, d. (c) Predicted uplift rate curve for the solution in (b), drawn at the same scale as Figure 16(d) and showing estimated ages of terraces. For this solution, the predicted uplift rate peaks at ~0.23 mm a^{-1} around 0.35 Ma. (d) Predicted variation in Moho temperature due to the crustal thickening accompanying the uplift depicted in (a). The calculation procedure is the same as for Figure 17g, except the initial (18 Ma) Moho temperature is taken as 630°C. The Moho depth and temperature predicted at 3.1 Ma are 29.1 km and 683°C. See text for discussion.

Rhine chronology assigns the normally-magnetised clay B1 at Ville to the Réunion 2 subchron and to oxygen isotope stage 81 and places it from pollen evidence in the Pretiglian (Fig. 13). It would thus appear that the Pretiglian-Tiglian boundary has to be later than the Réunion 2 subchron and/or the correlated oxygen isotope stage 81. One way of resolving this inconsistency would be to match the Simpelveld I terrace to the Réunion 1 subchron, such that it finished aggrading during stage 78 or 76, the Simpelveld 2 to the Olduvai (stage 72 or 68), and the Margraten to stage 66 or 64. This would place the peat above the Simpelveld I terrace within the Tiglian, while maintaining consistency with the palaeomagnetic evidence and preserving the essential character of the fits between observations and predictions in Figure 16 for the part of the record around the conventional Pliocene-Pleistocene boundary.

The deduction, that in this study region the 3.1 Ma start of local upland glaciation provided a more important forcing mechanism for lower-crustal flow than the 2.5 Ma start of lowland glaciation in the northern hemisphere, was not anticipated at the start of this work. It was initially assumed that the latter would be the more important; hence the need to consider the revised Maas terrace chronology which fitted well a 2.5 Ma start of forcing. The significance of these differences in the forcing chronology, given the constraints which are imposed on rates of lower-crustal flow by its rheology and by the geometries of surface loading at different times, is discussed further in the Appendix.

Second, the closest resemblance between uplift chronologies for the late Early Pleistocene onwards arises between the Andernach solution in Figure 18b and Figure 18c and the Maastricht solution in Figure 17c and Figure 17f. Both show a substantial increase in the uplift rate, starting between 1.1 and 1.0 Ma and peaking in the Middle Pleistocene (~0.5 Ma at Maastricht; ~0.35 Ma at Andernach). This Maastricht solution requires the revised chronology suggested in this study, in which the Pietersberg I terrace dates from oxygen isotope stage 22, not stage 26 as van den Berg (1996) thought. It implies that the Geertruid terraces of the Maas are equivalent to the younger part of Rhine terrace HT1, and the Pietersberg and Gravenvoeren are the equivalents of HT2. Both these Maas and Rhine solutions indicate a lack of complete correspondence with the conventional Dutch stratigraphy for the early Middle Pleistocene. It is indeed suggested that, in both sequences, one terrace (Rothem 2 for the Maas; HT4 for the Rhine) formed during a cold stadial within oxygen isotope stage 13 (here designated as stage 13b) rather than during a full glacial. As Turner (1996) has previously noted, the possibility of this sort of complexity highlights the limitations of the existing Dutch stratigraphy for this Cromerian era, which is based mainly on fragmentary deposits in short-lived river channels. The present study would have indeed been greatly facilitated had a more complete stratigraphy already been defined, based on the more continuous records of lacustrine sedimentation known from central and eastern Europe, as Turner (1996) recommended.

The third similarity concerns amounts of uplift between ~3 and ~1 Ma, which have been roughly the same – at 90 ± 10 m – in the Maastricht area and all localities studied on the Rhine (e.g. Figs 8, 14), with uplift rates in this interval peaking at ~0.06 mm a^{-1} around 2 Ma. Slightly higher amounts of incision, up to ~120 m, are evident during this time span on much of the Lorraine Meuse (Fig. 5), which covers terraces 7.1 to 2.1 given the correlation scheme in Table 3. On the contrary, the clearest difference affecting the later parts of the terrace sequences has been that the *subsequent* uplift has varied dramatically from place to place (Figs 4 and 14).

Maddy (1997) estimated a typical uplift rate of central southern England of ~0.07 mm a^{-1} from terraces of the River Thames. Beside the Upper Thames west of Oxford, the Combe terrace (from oxygen isotope stage 22) has been incised by 60 m (Maddy, 1997), which Maddy (1997) took as indicating the local uplift rate (60 m/~870 ka or ~0.07 mm a^{-1}). Maddy (1997) also noted that Middle Pleistocene marine terraces at Boxgrove, Sussex, on the English Channel coast, reveal a similar uplift rate, and suggested that regional uplift at this uniform time-averaged rate has been widespread.

The land surface adjoining the River Somme in northern France (Fig. 1) has a similar uplift rate. Near Amiens, the highest well-defined Somme river terrace, the Grâce terrace, is 52 m above its latest Pleistocene counterpart (Antoine, 1993). Its fluvial gravels and overlying finer-grained deposits are reverse-magnetised and, according to Antoine (1993), also aggraded during stage 22 (although Bates (1994) suggested from amino acid dating that these gravels may be one glacial cycle younger). The resulting subsequent time-averaged incision rate is thus 52 m/~870 ka or ~0.06 mm a^{-1}. The Allier, a major tributary of the Loire in the French Massif Central, also has a long-term terrace sequence (e.g. Veldkamp, 1992). At Randan near Vichy (~46°N, ~3.4°E), ~70 m of incision post-dates terrace fV_a, which is isotopically dated to ~0.8 Ma (Veldkamp & Jongmans, 1990; Veldkamp, 1992). This terrace also probably formed during stage 22, an incision rate of ~0.08 mm a^{-1} being indicated.

For comparison, in the present study region the uplift since deposition of the equivalent terrace has ranged from 153 m at Andernach (at an average rate of ~0.18 mm a^{-1}) to 72 m at Düsseldorf (~0.08 mm a^{-1}) (Table 4), and has been 70.5 m at Maastricht (also ~0.08 mm a^{-1}) (Table 2). The highest estimated instantaneous uplift rate in this study has been ~0.23 mm a^{-1} at Andernach around 0.35 Ma (Fig. 18c); the highest rate at Maastricht having been ~0.09 mm a^{-1} (Fig 17e and Fig 17f). On this time scale, uplift rates and total amounts of uplift have thus varied between localities by a factor of about three. The upper limit to rates of atectonic uplift during the Middle and Late Pleistocene in other parts of north-western Europe thus remains a significant unanswered question.

The Massif Central provides another interesting comparison with the present study region. It is argued that the Quaternary crustal thickening of the Rhenish Massif has been sustained by net outflow of lower crust from beneath adjoining regions to the north which have been cyclically loaded by ice sheets. The Massif Central has repeatedly been heavily glaciated (e.g. Veyret, 1986), whereas its surroundings have not, and yet it is itself uplifting relative to its surroundings. At first sight, these two examples appear to contradict each other. However, the theory described in the Appendix indicates that the sense of net lower-crustal flow between two adjoining regions depends on which is initially hotter at lower-crustal depths, not which is loaded. In north-western Europe, the net lower-crustal flow is thus towards the Rhenish Massif rather than towards Scandinavia, because the lower crust is much cooler beneath Scandinavia than beneath the Rhenish Massif (e.g. Cermak & Bodri, 1995). Likewise, the Massif Central has much higher heat flow than its surroundings: ~110 mW m^{-2} against ~80 mW m^{-2} (e.g. Vasseur, 1982). Although this is in part due to the high radiogenic heat production in granites at shallow depths, it is also in part due to higher heat flow from the mantle (Lucazeau et al., 1984). Lucazeau et al. (1984) indeed estimated that the mantle heat flow beneath the Massif Central is up to ~60 mW m^{-2}, against ~40 mW m^{-2} in its surroundings, with the Moho temperature up to ~800°C against ~650°C. The Massif Central is evidently another potential case study locality which may demonstrate atectonic Quaternary uplift in response to lower-crustal flow induced by cyclic surface loading, and thus warrants more detailed investigation in future.

5 ATECTONIC INTRAPLATE MAGMATISM

I now investigate whether it is feasible to regard the Quaternary volcanism of the Eifel as a consequence of the crustal thickening which accompanies the observed surface uplift to maintain isostatic equilibrium.

I assume, following McKenzie (1985, 1989), that the asthenosphere is constantly experiencing small-volume-fraction partial melting, and incompatible elements are concentrated into the resulting metasomatic melt which percolates upward into the mantle lithosphere. Due to its low concentration, this melt will remain at each level in thermal equilibrium with its surroundings, and will thus freeze at the depth where the temperature and pressure match the melt's solidus. Frozen metasomatic melt will thus accumulate over time at a particular depth within the mantle lithosphere. I further assume that a subsequent temperature rise can re-melt this frozen metasomatic melt, causing it to rise and freeze again at a shallower depth. However, if the conditions at the Moho reach the melt solidus, then the melt will escape into the crust and rise to the Earth's surface, causing volcanism.

To test this idea, one needs to know the conditions which mark the solidus for the frozen metasomatic melt. I will assume that the appropriate solidus is that for 'wet' basalt, i.e., for melt of basaltic composition in the presence of excess water. The experimentally-determined form of this solidus has been published many times (e.g. by Lambert & Wyllie, 1972). At low pressures, the solidus temperature decreases with increasing pressure, reflecting the melting-point reduction caused by mixing with water; at higher pressures, it increases with pressure, due to the increased stability of hydrous minerals such as amphibole (e.g. Sood, 1981, p. 116; Gill, 1981, pp. 127-128). The minimum solidus temperature is 650°C at a pressure of 1.5 GPa (e.g. Lambert & Wyllie, 1972).

This solidus curve has a quite complex empirical shape, but over the range of pressure, p, from 0.5 to 2.0 GPa it can be well approximated as a parabola of form

$$T = A\,(p - p_o)^2 + B \tag{5}$$

where A, B, and p_o are constants, with A = 105°C GPa^{-2}, B = 650°C, and p_o = 1.5 GPa (Figs 17c, 18b). Below 1.2 GPa, the mineral assemblage produced at the solidus is a mixture of amphibole, plagioclase, clinopyroxene, olivine, and water (e.g. Lambert & Wyllie, 1972). On depressurisation, the amphibole will decompose, releasing its water of crystallisation, which along with the original excess water will escape, leaving a rock of basaltic composition.

For each study locality, the initial depth of the base of the continental crust, z_w, is calculated from the initial z_b and z_i values using Equation (A01). The pressure at this depth can then be calculated, and when added to the increase in pressure caused by the assumed 4 km of mafic underplating gives the initial Moho pressure. Equation (3) is then used to estimate the crustal thickening which accompanies the observed surface uplift. The resulting increase in Moho pressure can then be calculated.

Assumed values for the initial Moho temperature are listed in the captions to Figures 17g and 18d. The increase in Moho temperature due to crustal thickening is calculated, approximately, assuming that a geothermal gradient of 20°C km^{-1} is maintained across the added crustal thickness.

Figure 17g shows the resulting predicted variation in Moho temperature and pressure at Maastricht. After rising from its assumed initial value of 620°C in the Middle Miocene, the Moho temperature reached ~670°C by the latest Pliocene, but locally remains below the solidus, consistent with the absence of Quaternary volcanism in this locality. However, if the present rates of surface uplift and crustal thickening are maintained, local volcanism can be expected several hundred thousand years in the future.

Figure 18d shows the equivalent calculation for Andernach. In this locality, the start of volcanism can be estimated as ~0.7 Ma from the isotopic dating evidence mentioned earlier (Schminke & Meertes, 1979) and from the presence of clasts of Eifel basalt in fluvial gravels of Kärlich bed F which are correlated with oxygen isotope stage 18 in Figure 13. The assumed initial Moho temperature of 630°C in the Middle Miocene was chosen to make its value exceed 680°C by the latest Pliocene, then intersect the solidus at 0.7 Ma, the estimated start of Quaternary volcanism in the Eastern Eifel. The latest Pliocene start of volcanism in the Western Eifel can also be explained, assuming the crust is locally ~700 m thicker than in the East Eifel, such that the Moho conditions at the start of Quaternary uplift were adjacent to the solidus. Assuming the inflow of lower crust has thickened the crust equally in the eastern and western Eifel, the ~100 m typical difference in surface altitude reflects a ~700 m difference in crustal thickness both at present and at the start of the Pleistocene.

6 DISCUSSION

The uplift chronologies in the study region are well-matched by model predictions (Figs 16, 17, and 18) which require a minimum of free parameters. The crustal thickening accompanying the observed uplift since the Early Pleistocene is too fast to be explicable as a result of pressure variations caused by sea-level variations (see Appendix), but it is within the range attributable to the larger pressure variations caused by cyclic ice sheet loading in adjacent regions. The order-of-magnitude smaller uplift rate occurring before the latest Pliocene (Fig. 9) may instead have resulted from pressure variations caused by sea-level fluctuations due to cyclic variations in the volume of the Antarctic ice sheet, there having been no significant northern hemisphere ice sheets at this time.

The presence of a chain of upland regions across central Europe, such as the Rhenish Massif, the Harz Mountains, the Bohemian Massif, and the Carpathians, which bound the area to the north which has been repeatedly glaciated (Fig. 1), thus appears not a coincidence. Given the results of this study, it is possible that all these uplands have been produced by the same process: crustal thickening caused mainly by southward lower crustal flow, driven by the loading effects of Scandinavian glaciations, given the greater local heat flow than beneath the adjoining ice-loaded region (e.g. Hurtig & Oelsner, 1977; Cermak & Hurtig, 1979; Cermak & Zahradnik, 1982; Cermak, 1995). Balling (1995) estimated that the Moho temperature and heat flow are as low as ~500°C and ~20 mW m^{-2} beneath much of Scandinavia, indicating a lower-crustal geothermal gradient of only ~8°C km^{-1}. Eyles (1996) has identified other examples of systematic Quaternary uplift in localities on the peripheries of major transient northern hemisphere ice sheets.

Although Quaternary volcanism has so far affected only a small part of western Europe, the calculations suggest that the crustal thickening induced in other localities has brought them much closer to the condition for volcanism (as in Fig. 17c) than they were at the start of the Quaternary. This raises the possibility that with continued crustal thickening, more localities will experience equivalent volcanism over the next few hundred thousand years. As several European countries are currently investigating sites intended for the burial of high-level radioactive waste, which will need to be isolated over much longer time scales, this possibility of future atectonic volcanism thus represents a hazard which warrants systematic investigation. The calculations whose results are depicted in Figures 17g and 18d indicate that the faster the surface uplift is in a region, the greater the crustal thickening and thus the greater the Moho temperature rise. Other factors (such as initial Moho temperature) being equal, it follows that the faster the surface uplift is, the greater the likelihood that the Moho temperature will cross the wet basalt solidus, and thus the greater the likelihood of atectonic magmatism in a region. Studies of rates of atectonic uplift can thus bear upon hazard analyses, by indicating possibilities of atectonic magmatism in the future.

A further point concerns the relationship between the lower crustal flow causing the uplift of the Rhenish Massif and the active normal faulting in surrounding localities, including the Rhine Graben (Fig. 1) and the Lower Rhine Embayment (Fig. 2). In both these localities, normal faults began to slip at significant rates around the start of the Quaternary, having not been significantly active for millions of years beforehand (e.g. Zagwijn, 1974, 1989; Illies & Baumann, 1982). Westaway (1998, 1999) has shown that lower-crustal flow can become vigorous enough to significantly reorient the local stress tensor. Future work will investigate whether it is feasible that the flow beneath the present study region has adjusted the local stress tensor sufficiently to permit shear failure on these normal faults.

My proposed mechanism implies that this faulting is confined to the crust and so is unaccompanied by corresponding deformation of the mantle lithosphere. If so, one would not expect the active normal faulting in the Lower Rhine embayment (Fig. 2) to be accompanied by thinning of the underlying mantle lithosphere. Thus, whether or not this region is underlain by thinned mantle lithosphere provides a potential test of this mechanism. Many measurements have been made of the lithosphere thickness variations beneath the study region, using travel times of seismic waves from earthquakes, seismic refraction profiling, surface wave dispersion, and heat flow modelling (see, e.g. Blundell et al., 1992, for a compilation). Although some interpretations show a localised region of thin lithosphere beneath the Lower Rhine Embayment, others show either a much broader zone of thinned lithosphere, whereas the heat flow data (e.g. Cermak & Bodri, 1995) shows no correlation between this structure and lithosphere thinning. The inconsistency between these different methods of estimating the lithosphere thickness means that this point cannot be resolved at present.

The process which has been proposed as responsible for the Quaternary uplift of the study region will affect other localities also. Many studies have deduced that the mean altitude of continental interiors has increased systematically during Late Cenozoic and Quaternary time (e.g. De Sitter, 1952; King, 1955; Flint, 1957; Geyl, 1960; Holmes, 1965; Damon, 1971; McKee & McKee, 1972; Bond, 1978; Lucchitta, 1979; Partridge & Maud, 1987; Veldkamp, 1996; Murray-Wallace et al., 1996). Many of these localities, for instance Australia (Geyl, 1960; Murray-Wallace et al., 1996) and southern Africa (King, 1955; Partridge & Maud, 1987), are distant from any plate boundaries, making it impossible for plate motions to cause the observed uplift. Despite speculation in most of the references cited above, no clear mechanism for this uplift has emerged. Lithgow-Bertelloni & Silver (1998) suggested that the anomalously high present-day altitude of the land surface in southern Africa is being dynamically supported by an unusually large mantle plume. However, the up to ~900 m of surface uplift in this region since the Late Pliocene (Partridge & Maud, 1987) requires this plume to have become more vigorous on this time scale which – as in the equivalent argument over the Eifel volcanism – is not explained.

The question of whether large-scale regional uplift has occurred within the continents has become important, through attempts to explain global Quaternary climate change (e.g. Ruddiman & Raymo, 1988; Molnar & England, 1990; Raymo & Ruddiman, 1992). Two clear climatic factors require explanation. First, what caused the start of lowland glaciation in the northern hemisphere around 2.5 Ma? And, second, what caused its intensification after 1.2 Ma, when from oxygen isotope studies the amplitude of global sea-level variations roughly doubled? Some studies (e.g. Ruddiman & Raymo, 1988) have appealed to regional uplift of continental interiors, which they regard as an independent cause. There are several mechanisms by which surface uplift will facilitate glaciations: for instance, the air will be colder over higher ground, inhibiting the melting of snow; and the greater relief along edges of continents will increase orographic precipitation. However, the global distribution of plates is essentially the same now as it was in the Late Pliocene. Given the absence of any clear reason for systematic uplift of continental interiors, Molnar & England (1990) proposed that their mean altitude has remained constant but the relief has increased as a result of the isostatic response to faster valley incision.

Others have observed that volcanism has not occurred at uniform rates in Neogene and Quaternary time (e.g. McBirney et al., 1974; Kennett & Thunell, 1975). It has instead been concentrated in Middle Miocene and Quaternary times, with a less clearly-defined Pliocene episode. These concentrations, which have not previously been explained, may have resulted from episodes of atectonic thickening of the continental crust caused by glacial sea-level fluctuations.

Instances of magmatism correlated with apparently atectonic uplift are quite numerous. For instance, the uplift of the Colorado Plateau in the south-western United States has involved crustal thickening unaccompanied by significant local shortening, and can be presumed to have resulted from inward flow of lower crust from beneath its surroundings (e.g. McQuarrie & Chase, 2000). A pulse of magmatism occurred in the Colorado Plateau in the Middle Miocene (~15 Ma) (e.g. Krieger et al., 1971), while this region was uplifting (e.g. McKee & McKee, 1972; Lucchitta, 1979). In the Middle East, widespread magmatism occurred within the northern Arabian platform in the Middle Miocene (also ~15 Ma). At this time this region had just changed from being a prolonged shallow-marine depocentre along a seaway linking the eastern Mediterranean Sea and Persian Gulf to a land bridge linking Africa and Asia (e.g. Steininger & Rögl, 1984; Karig & Kozlu, 1990; Westaway & Arger, 1996). Magmatism resumed abruptly in the latest Pliocene in this region, whose land surface now stands typically ~500 – 1000 m above sea level (e.g. Westaway & Arger, 1996). Arger et al. (2000) recently applied the methods described in this study to modelling this instance of surface uplift correlated with Miocene and Quaternary magmatism. The French Massif Central experienced basaltic volcanism in the Miocene starting at ~20 Ma, then renewed volcanism in the Quaternary (e.g. Lucazeau et al., 1984; Wilson & Downes, 1991). Since the Early Miocene the land surface is estimated to have uplifted by several hundred metres (Lucazeau et al., 1984). River terraces indicate incision by ~110 m since the latest Pliocene and ~80 m since oxygen isotope stage 22 (Veldkamp, 1992), suggesting the amounts of uplift on these time scales. In eastern Europe, adjacent to the Carpathian mountains, an abrupt and widespread episode of magmatism occurred in Slovakia and western Ukraine at ~13 Ma (e.g. Mikhaylova et al., 1974; Steininger et al., 1976). This event marked a change in this region from being submerged by a shallow shelf sea, open to the global marine environment, to a land area adjoining the landlocked Paratethys Sea (e.g. Steininger & Papp, 1979). Contemporaneous brackish marine deposits of the Paratethys are now found at altitudes above 500 m across much of eastern Europe. The Arctic island group of Svalbard (Spitsbergen) has experienced many hundreds of metres of Late Cenozoic and Quaternary uplift, along with episodes of magmatism (e.g. Vågnes & Amundsen, 1993). In contrast, the outer part of the adjoining continental shelf appears to have subsided by hundreds of metres, making it now too deep to support a grounded ice sheet during glacial maxima (e.g. Flower, 1997), in contrast with conditions during the Early Pleistocene. In the south-eastern part of South Australia, a well-dated sequence of raised beaches and fossil dune ridges indicates that the coastline has prograded by up to ~100 km since oxygen isotope stage 21 (~0.85 Ma), as the land surface has uplifted by between 60 and 110 m (e.g. Huntley et al., 1993, 1994; Murray-Wallace et al., 1996). This sequence is terminated to landward by a fossil marine cliffline (e.g. Huntley et al., 1993), which suggests that the local uplift rate increased from lower values to the subsequent values of ~0.07 to ~0.13 mm a^{-1} shortly before stage 21: thus resembling the timing and rates of uplift at the localities investigated in the present study. The localities with the highest uplift rates in this part of Australia have experienced Quaternary basaltic volcanism (the Mount Gambier/Mount Schank volcanic field) (e.g. Sheard, 1990; Murray-Wallace et al., 1996).

Across much of central Germany, including the Vogelsberg volcano and the Northern Hessian Depression near Kassel (Fig. 1), Miocene basaltic volcanism began at ~20 Ma and peaked around 14 Ma (e.g. Wedepohl, 1985; Wilson & Downes, 1991). According to Wilson & Downes (1991) these localities are underlain by crust with a much thicker underplated layer than the Eifel: it has ~7 km thickness with the Moho at ~33 km. According to Cermak & Bodri (1995) the temperature is ~680°C at 30 km depth beneath Vogelsberg, making it ~740°C at the estimated ~33 km deep Moho. With a Moho temperature of 740°C and a Moho pressure of ~0.86 GPa (estimated from the thickness of crust and underplating) the uppermost mantle in this locality currently lies well above the wet basalt solidus in Figures 17(g) and 18(d). It is thus possible that its Miocene phase of atectonic crustal thickening brought the Moho above the wet basalt solidus, thus accounting for this Miocene volcanism and, by removing all the volatile material from the mantle lithosphere, its lack of Quaternary volcanism. On the contrary, more recent seismic refraction interpretations have yielded different results: a Moho at 30 ± 1 km depth, with little or no evidence of underplating in any locality (Blundell et al., 1992, p. 49). Another interpretation (summarised by Cermak & Bodri, 1995) places the base of the crust at 28.5 km depth beneath both the Vogelsberg area and the Northern Hessian Depression, with no underplating at Vogelsberg but 1.5 km of it in the Northern Hessian Depression. It thus remains unclear whether this Miocene volcanism can be explained as a result of atectonic crustal thickening: it will require separate, more detailed, investigation in future.

The gradual global cooling during Middle Miocene time, evident from oxygen isotope studies (e.g. Molnar & England, 1990), has been interpreted as marking the growth of the East Antarctic ice sheet to roughly its present size. Field evidence (e.g. Sugden et al., 1995; Bruno et al., 1997) now confirms that it had indeed reached roughly its present scale by the Late Miocene. The very slow uplift rate occurring in the present study region from Middle Miocene onward (Fig. 9) may thus represent the isostatic response to small-scale fluctuations in the global sea-level caused by cyclic fluctuations in Antarctic ice volume at this time. More detailed modelling of this and the other examples of Miocene uplift and atectonic magmatism cited earlier may thus help to constrain the start of the global sea-level fluctuations which have led to the build-up of the Antarctic ice sheet more tightly than other available methods. This would not only benefit studies of climate change; by explaining many instances of magmatism in terms of a single global process it would eliminate the need to devise separate explanations in each case, such as local plate motions or mantle plumes.

This study thus raises the possibility of a hitherto unknown global coupling mechanism linking crustal deformation, magmatism, and climate change. It suggests that low-amplitude fluctuations in sea-level in Miocene time caused net flow of crust to beneath the land, which slightly increased the mean altitude of the Earth's surface. This increase in altitude added to whatever global cooling was occurring independently, facilitating growth of ice sheets and the associated increase in amplitude of sea-level variations. The lower-crustal flow induced by these loading variations added to the increase in surface relief, which led to further global cooling, which further amplified the sea-level variations, which further affected the relief. The resulting positive feedback loop suggests that this process and its effects will intensify until they are countered by the increasing effective viscosity of the lower crust in areas of atectonic crustal thinning.

7 CONCLUSIONS

The uplift of land surfaces adjoining the Rhine and Maas rivers, revealed by Quaternary river terraces, has been modelled using theory for the isostatic response of the crust to the effects of adjacent loads from ice sheets and glacial-to-interglacial sea-level fluctuations, in the presence of a low-viscosity layer in the lower continental crust. The observed fluctuations in uplift rates, which reach upper bounds of ~0.09 mm a^{-1} at Maastricht on the Maas and ~0.23 mm a^{-1} at Andernach on the Rhine, are well-matched to predictions which assume major step changes to the intensification of Quaternary glaciation at 3.1 and 0.9 Ma, plus smaller-magnitude changes at 2.5 and 1.2 Ma. The resulting fits between observations and predictions suggest that the observed surface uplift is being caused by atectonic crustal thickening due to inward lower-crustal flow to beneath the study region. The resulting increase in Moho temperature can account for the Quaternary volcanism of the Eifel region beside the Rhine.

The same uplift process may well be occurring in other regions, also. Its anticipated consequences are a systematic increase in global surface relief, which may help explain the global cooling which has occurred during late Neogene time and since the start of the Quaternary.

ACKNOWLEDGEMENTS

I thank Darrel Maddy for many stimulating discussions and for originally drawing my attention to the Maas river terrace dataset, and Tom Veldkamp and Meindert van den Berg for thoughtful and constructive reviews. I am also grateful to the following for helpful discussions and correspondence: Jürgen Herget, Christian Hoselmann, Otto Kandler, Salomon Kroonenberg, Frank Sirocko, Chris Stemerdink, and Jef Vandenberghe.

APPENDIX: THEORY FOR QUATERNARY UPLIFT RATES

Horizontal flow in the plastic lower continental crust can be regarded as occurring between planar horizontal boundaries at depths z_b, the base of the upper-crustal brittle layer, and z_w, the base of the lower-crustal channel. It is assumed that in continental crust with mafic underplating, depth z_w marks the top of the underplated layer, with the Moho at depth z_m at its base. The horizontal velocity v_x is assumed to satisfy boundary conditions that $v_x(z_b) = 0$ and $v_x(z_w) = 0$, indicating no horizontal motion, relative to its external surroundings, at the boundaries of the lower-crustal channel. As a result of the geothermal gradient, the viscosity of the lower continental crust decreases downward. Together with the boundary conditions, this causes the flow velocity to be strongly peaked just above the base of the lower crust (Fig. A1a). Westaway (1998) considered a range of possible flow profiles, depending on different assumptions about lower-crustal rheology. He assumed a crust with z_b = 15 km and $z_m = z_w$ = 35 km, with a temperature of 350°C at $z = z_b$. In his preferred solutions, the lower-crustal flow peaked at z ~33 km, roughly 90% down through the lower crust. The general relation between z_b, z_i, and z_w is thus roughly expressed as

$$z_w = (10\, z_i - z_b)/9\ . \tag{A01}$$

The predicted effective viscosity of the lower crust, η_e, defined as the viscosity of an isoviscous layer which would transport the same flux of material in response to a given horizontal pressure gradient as a real lower-crustal layer with a vertical variation in viscosity (Equation 1), varies from ~4 × 10^{19} Pa s for a geothermal gradient of 10°C km^{-1} (giving a Moho temperature of 450°C) to ~3 × 10^{18} Pa s for 15°C km^{-1} and 550°C, and ~3 × 10^{17} Pa s for 20°C km^{-1}and 650°C (Westaway, 1998).

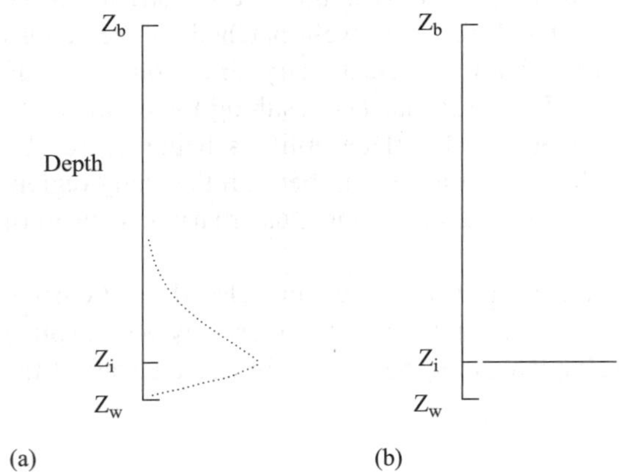

Figure A1. (a) Sketch of the general form of the vertical profile of horizontal flow velocity in the lower continental crust, from Westaway (1998). This velocity $v_x(z)$ is constrained to satisfy boundary conditions that $v_x(z_b) = 0$ and $v_x(z_w) = 0$, where z_b and z_w are the depths of the base of the brittle layer and of the base of the plastic lower-crustal channel. If the crust has experienced mafic underplating, as in the present study region, then z_w will be less than the Moho depth z_m. The velocity peaks at a depth z_i which is 90% of the distance down from z_b to z_w. (b) Approximation to the profile in (a) used here to calculate heat flow perturbations. The horizontal velocity is regarded as a unit spike located at depth z_i See text for discussion.

At times of high sea level during interglacials, the increase in water loading of the offshore continental shelf will exert an excess pressure, which will act to drive lower crust to beneath any adjoining land. Conversely, during times of low sea level during glacials, the decrease in water loading of the offshore continental shelf will give rise to a reduction in pressure, which will act to drive lower crust from beneath any adjoining land. If, before this process begins, the onshore and offshore crust are in isostatic equilibrium with each other, then the requirement that the offshore crust is slightly thinner (to maintain the Earth's surface below sea level) means that the Moho temperature is somewhat lower. Any crust that is drawn from beneath the offshore shelf will thus on average be somewhat cooler than the crust which originates beneath the land. Conversely, any crust that is drawn from beneath the land will on average be somewhat warmer than the crust which originates offshore. The result of this cyclic flow of lower crustal material to and from beneath the land and beneath the offshore continental shelf is thus a net cooling of the onshore crust and a net warming of the offshore crust.

This analysis will assume that the isostatic response of the brittle layer involves vertical deflection, with a hinge line at the transition from uplift to subsidence at the position of the coastline midway between interglacials and glacials. Calculations are made for the case where the horizontal scale of the associated lower-crustal flow is small enough that changes to crustal thickness are supported by the rigidity of the mantle lithosphere and not isostatically compensated. A correction can later be applied to deal with larger-scale flows, by which a lower-crustal root develops along with the surface uplift. It is further assumed that the flexural rigidity of the brittle layer is low enough to allow this warping to occur given the scale of variations in the vertical deflection. Localities are indeed known with warping of the continental crust with the scale of a few kilometres (e.g. Westaway, 1993a).

Consider the addition of material to beneath the onshore region. The profile of horizontal velocity to beneath the land is strongly peaked about a depth z_i, ~90% down through the lower crust between z_b and z_m (see above and Fig A1a). Suppose this injection of material at depth z_i causes local cooling by an amount ΔT. Once such material is injected, its heat content will cause a negative temperature perturbation at adjacent depths. This perturbation will gradually diffuse away from depth z_i, eventually affecting the temperature at the initial depth of the base of the brittle layer. To calculate the magnitude of this perturbation in the vicinity of the base of the brittle layer, ~10 to 20 km shallower than z_i, one may approximate the form of this temperature perturbation as that of a spike of appropriate amplitude located at depth z_i (Fig. A1b). This approximation is reasonable, because the rate at which the temperature perturbation diffuses away from the depth at which lower crust is injected is much greater than the rate at which material is added to the lower crust. One may thus write that

$$\delta T(z,0) = \Delta T \, \delta\,(z_i - z)\,, \qquad \text{(A02)}$$

where $\delta(z)$ is the Dirac delta function, a unit spike located at $z = 0$.

A.1 *Temperature perturbation of constant magnitude*

This heat flow problem requires solution of the one-dimensional diffusion equation, $\kappa\, \partial^2 T/\partial z^2 = \partial T/\partial t$. The simplest solution involves a temperature perturbation of constant magnitude at depth z_i starting at time $t = 0$, in accordance with Equation (A02), with

$\delta T(z,0) = 0$ for $z \neq z_i$. It is convenient to define a new vertical co-ordinate z', measured upward from a zero at $z = z_i$, such that z' equals $z_i - z$. The layer at which the temperature is known lies at $z' = 0$, and can thus be regarded as bounding a halfspace for $0 \leq z' \leq \infty$.

The solution can now be obtained using the standard Green's function method as

$$\delta T(z,t) = \frac{1}{2\sqrt{(\pi\kappa t)}} \int_0^\infty \delta T(\zeta,0)[\exp(-(z'-\zeta)^2/(4\kappa t)) - \exp(-(z'+\zeta)^2/(4\kappa t))]d\zeta$$
$$+ \frac{1}{2\sqrt{(\pi\kappa)}} \int_0^t \delta T(z'=0,\tau) \frac{\exp(-z'^2/(4\kappa t))}{(t-\tau)^{3/2}} d\tau \tag{A03}$$

(Carslaw & Jaeger, 1959, p. 357), where ζ and τ are dummy variables for depth and time. This solution is valid provided the time scale covered is sufficiently short for the calculated temperature perturbation at the Earth's surface (at $z = 0$ or $z' = z_i$) to be negligible. In the real Earth, the surface temperature is constrained to a fixed value of ~0°C. In the model, no such constraint is imposed.

With $\delta T(\zeta,0)$ satisfying Equation (A02), the integration over position on the right hand side of Equation (A03) is zero as the integrals of its constituent terms cancel. The integration over time gives:

$$\delta T(z,t) = \Delta T[1 - \mathrm{erf}(z'/(2(\kappa t)^{1/2}))] \;, \tag{A04}$$

where erf(x) is the Gaussian error function, defined as

$$\mathrm{erf}(x) \equiv \frac{2}{\sqrt{\pi}} \int_0^x \exp(-w^2)dw \;, \tag{A05}$$

such that erf(−∞) = −1, erf(0) = 0, and erf(∞) = 1. The complementary error function, erfc(x), is itself defined for $x \geq 0$ as

$$\mathrm{erfc}(x) \equiv 1 - \mathrm{erf}(x) \;, \tag{A06}$$

such that erfc(0) = 1 and erfc(∞) = 0.

Restoring the original co-ordinate system, the solution thus becomes

$$\delta T(z,t) = \Delta T\, \mathrm{erfc}[(z_i - z)/(2\,(\kappa t)^{1/2})] \;. \tag{A07}$$

Note that this requires that $\delta T(z_i,t) = \Delta T$, indicating that for as long as lower crust is injected, the temperature perturbation at the depth of injection remains constant.

With $\Delta T < 0$, the relatively cool material which is injected will cool the lower crust beneath the land, causing the base of the brittle layer to advect downward. To determine its rate of advection I calculate

$$\frac{\partial T(z,t)}{\partial t} = \frac{\Delta T(z_i - z)}{2\sqrt{(\pi\kappa t^3)}} \exp(-(z_i - z)^2/(4\kappa t)) \;, \tag{A08}$$

and

$$\frac{\partial T(z,t)}{\partial t} = \frac{\Delta T}{\sqrt{(\pi\kappa t)}} \exp(-(z_i - z)^2/(4\kappa t)) + \frac{\partial T_o(z)}{\partial z} \quad . \tag{A09}$$

If u is the initial geothermal gradient in the vicinity of the initial depth of the base of the brittle layer (i.e., $\partial T_o(z_b)/\partial z = u$), then the rate of change of depth z at which a temperature of T_b is encountered, v_a, can be calculated as $(-\partial T(z_b,t)/\partial t)/(\partial T(z_b,t)/\partial z)$ or

$$v_a = \frac{-(z_i - z_b)/(2t)}{2t(\sqrt{(\pi\kappa t)}u \exp((z_i - z_b)^2/(4\kappa t))/\Delta T + 1)} \tag{A10}$$

which approximates to

$$v_a = \frac{-(z_i - z_b)\Delta T \exp(-(z_i - z_b)^2/(4\kappa t))}{2\sqrt{(\pi\kappa t^3)}u} \tag{A11}$$

The corresponding net heating of the offshore part of the continental shelf will lead to upward advection of the base of the brittle layer relative to the sea floor. Isostatic equilibrium requires that the pressure at the base of the brittle layer is the same beneath the offshore shelf compared with beneath the land. If it occurs in isolation, the downward advection of the onshore base of the brittle layer will cause an increase in the pressure at this point. As a result, the pressure at the offshore base of the brittle layer is also required to increase, despite its upward advection. The only way this new state of equilibrium can be possible is for the overall isostatic adjustment to involve an increase in water depth, and hence in water pressure, offshore. This, in turn, means that the adjustment involves either (a) net thinning of the offshore crust and net thickening of the onshore crust or (b) a corresponding change in the relative crustal thicknesses in these two regions plus a vertical deflection of the Moho. Whether this isostatic response involves any deflection of the mantle lithosphere will depend on the spatial scale of the lower crustal flow (see below).

A.2 *Rates of elevation change following a temperature perturbation of constant magnitude*

Consider a small isolated island of area A_1 surrounded by an area A_o of offshore continental shelf. At some initial time, the pressure at the base of the brittle layer beneath the island is p_{bio}:

$$p_{bio} = \rho_c \, g \, z_{bio} \tag{A12}$$

where ρ_c is the density of continental crust, g is the acceleration due to gravity, and z_{bio} is the depth of the base of the brittle layer below the Earth's surface beneath the island. The pressure at the base of the brittle layer beneath the offshore shelf which is covered by water of initial depth z_{wo} is thus

$$p_{boo} = \rho_c \, g \, z_{boo} + \rho_w \, g \, z_{wo} \tag{A13}$$

ρ_w being the density of seawater, and z_{boo} being the initial depth of the offshore base of the brittle layer below the sea floor.

For the island and the offshore shelf to be in isostatic equilibrium with each other at this initial time requires that there be no lower-crustal flow between them. There is thus no horizontal pressure gradient, or

$$p_{bio} = p_{boo} \tag{A14}$$

Suppose that at some later time the pressure at the base of the brittle layer is p_{bi1} beneath the land and p_{bo1} beneath the offshore shelf. Equations (A12) and (A13) thus adjust to

$$p_{bi1} = \rho_c\, g\, z_{bio} + \rho_c\, g\, \Delta h_{ia} \tag{A15}$$

and

$$p_{bo1} = \rho_c\, g\, z_{boo} + \rho_w\, g\, z_{w1} + \rho_c\, g\, \Delta h_{oa} \tag{A16}$$

where z_{w1} is the new water depth, Δh_{ia} is the gain in thickness of the brittle layer due to the temperature perturbation caused by flow of cool lower crust to beneath the land, and Δh_{oa} is the gain in thickness of the brittle layer beneath the offshore shelf due to the temperature perturbation caused by flow of lower crust to beneath its area (Δh_{oa} will of course be negative, as flow of warm lower crust to beneath the offshore shelf will reduce the thickness of the brittle layer, in accordance with Equation (A11) with $\Delta T > 0$).

The value of Δh_{ia} which develops over time Δt can be estimated as $v_a\, \Delta t$ where v_a is given by Equation (A11)). It is proportional to the temperature contrast ΔT at the depth of injection between the land and the offshore shelf. As already noted, different parts of the continental shelf will have different crustal thicknesses and Moho temperatures, and thus different values of ΔT. To devise an approximate solution, I assume that the value of ΔT applicable for both the land and the offshore shelf corresponds to the difference in crustal thickness for crust which is in isostatic equilibrium at sea level at times of interglacial marine highstands and glacial marine lowstands. Subject to this assumption, ΔT for the land equals $-\Delta T$ for the offshore shelf, and hence:

$$\Delta h_{ia} = -\Delta h_{oa}\,. \tag{A17}$$

This simplification is equivalent to assuming that the crust which flows to beneath the land comes predominantly from localities which are just submerged during marine lowstands. This is reasonable, because crust from the outer shelf will be cooler and thus more viscous, and will flow more slowly in response to a given pressure gradient, as well as having farther to travel (Fig. A2).

Assuming that the adjustment to sea level persists for long enough for the resulting transient lower-crustal flow to reach a new state of isostatic equilibrium, then one may equate the new pressures at the base of the brittle layer beneath the land and the offshore shelf, or:

$$p_{bi1} = p_{bo1}\,. \tag{A18}$$

Given the finite (~5000 yr) durations of maxima and minima of sea level, this condition can only be satisfied if the lower-crust has a low viscosity. I will later test whether the ranges of viscosity deduced by Westaway (1998) allow this condition to be satisfied.

Finally, conservation of crustal volume requires that

$$A_i\,(\Delta u_i + \Delta h_{im}) + A_o\,(\Delta h_{om} - \Delta h_w) = 0 \tag{A19}$$

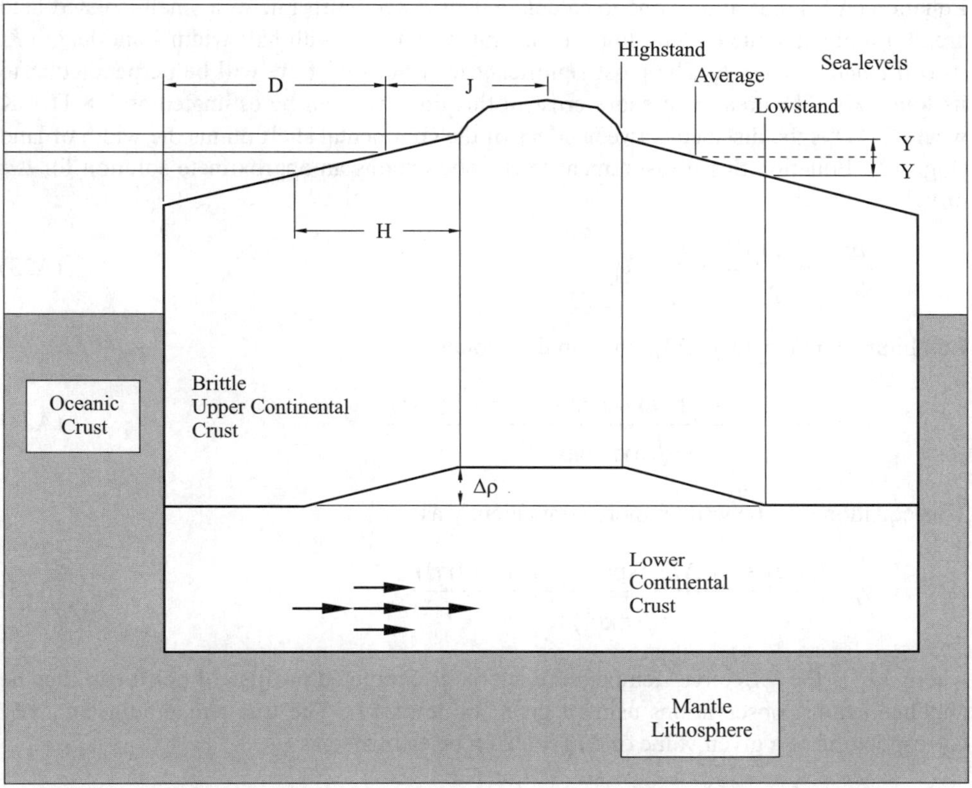

Figure A2. Sketch, not to scale, of the physical model developed here, showing the definitions of the length parameters D, H, and J. See text for discussion.

where A_i and A_o are the areas of the land area and of the adjoining continental shelf, Δh_{im} and Δh_{om} are the increase in Moho depth beneath the land and the offshore shelf, Δu_i is the change in altitude of the onshore land surface, and $\Delta h_w = z_{w1} - z_{wo}$ is the change in water depth.

Provided both the land and the adjoining shelf are narrow compared with the ~100 km wavelength for flexure of the mantle, Δh_{im} and Δh_{om} can be approximated as zero. Subject to this approximation, one may substitute (A12) and (A13) into (A14), then substitute (A15) and (A16) into (A18), then subtract (A14) minus (A18), to obtain:

$$\rho_c \, g \, \Delta h_{ia} = \rho_w \, g \, \Delta h_w + \rho_c \, g \, \Delta h_{oa} \quad . \tag{A20}$$

which simplifies, using (A17) to

$$\Delta h_{ia} = \frac{\rho_w \Delta h_w}{2\rho_c} \quad , \tag{A21}$$

or, using (A19),

$$\Delta u_i = \frac{2\rho_c \Delta h_{ia} A_o}{\rho_w A_i} \quad . \tag{A22}$$

Equation (A22) thus allows one to calculate the surface uplift rate of a small isolated land area for a given value of Δh_{ia}. For a rectangular land area with half width J and length K, A_i will equal $2 \times J \times K$. The most significant lower-crustal flow will be perpendicular to its long axis. The area of offshore crust in this direction can be estimated as 2 × D × K where 2 × D is the distance between edges of the continental shelf minus the width of land (Fig. A2). Equating this measurement to A_o, one obtains an approximate solution for Δu_i or v_u

$$\frac{\Delta u_i}{\Delta h_{ia}} = \frac{2\rho_c D}{\rho_w J} = \frac{v_u}{v_a} \; . \qquad \text{(A23)}$$

Combining (A11) with (A23), one can thus obtain

$$v_u = \frac{-\rho_c (z_b - z_i)\, D \Delta T \exp(-(z - z_i)^2 / (4\kappa t))}{\sqrt{(\pi \kappa t^3)} u \rho_w J} \; . \qquad \text{(A24)}$$

This equation may be written more conveniently as

$$v_u = \frac{-(z_b - z_i) \Delta T_e \exp(-(z - z_i)^2 / (4\kappa t))}{2\sqrt{(\pi \kappa t^3)} u} \; . \qquad \text{(A25)}$$

where ΔT_e is the 'effective' temperature anomaly. Predicted profiles of uplift can thus be matched against observations using a given value of ΔT_e. The true temperature anomaly corresponding to a given value of ΔT_e can then be estimated as

$$\Delta T = \frac{J \rho_w \Delta T_e}{2 D \rho_c} \; . \qquad \text{(A26)}$$

With ρ_c = 2700 kg m^{-3} and ρ_w = 1000 kg m^{-3}, ΔT and ΔT_e are numerically equal if J is roughly five times larger than D (Fig. A2).

A.3 *Test whether the lower crust is mobile enough for this method to be valid*

Equation (A18) is only valid provided pressure at the base of the brittle layer can re-equilibrate within the time-scale of marine highstands or lowstands. To assess the feasibility of this assumption, I use Equation (1). Per unit distance perpendicular to the flow, the horizontal flux of material in response to a given pressure gradient P persisting for time t is $V\,W\,t$, whereas the flux required to sustain a crustal thickening rate v_t beneath a land area with half-width J (Fig. A2) is $2\ v_t\ J\ t$. However, during each glacial cycle, the value of P will vary continuously. This variation can thus be approximated as a series of time steps each of duration t. Thus, if

$$V\,W \geq 2\ v_t\,J, \qquad \text{(A27)}$$

flow of lower crust is likely to be fast enough for pressure at the base of the brittle layer to re-equilibrate from an initial pressure gradient P on the time-scale of each time step, such that Equation (A18) is satisfied by the end of the time step. If the loading history and instantaneous crustal thickening rate were both known for each time step, the feasibility of this process could be assessed for each time step separately. However, as neither is known, a simpler test is needed. One approach would be to assume, on average, that (A27) is satisfied throughout each loading cycle, with V the time-averaged mean flow velocity. To test this I define a dimensionless ratio Ω where

$$\Omega \equiv \frac{VW}{2v_t J} = \frac{PW^3}{24\eta_e v_t J} \tag{A28}$$

where P is the mean pressure gradient induced by the loading. If $\Omega \geq 1$, then Equation (A18) is justified. If instead $\Omega < 1$, then the conditions will only be able to partially re-equilibrate on the time scales of marine high- or low-stands. The proposed process will still be expected to occur to some extent, but because Equation (A18) will not be satisfied the quantitative method presented in this study will not be formally applicable.

Crustal thickening rates are estimated in the main text as roughly 7 times the surface uplift rate. Peak uplift rates are ~0.09 mm a^{-1} at Maastricht and ~0.23 mm a^{-1} at Andernach on the Rhine, making peak values of v_t ~0.6 and ~1.6 mm a^{-1}. At Andernach, the peak rates of surface uplift and crustal thickening occurred when the Moho temperature reached ~700°C (Fig. 18b). With a geothermal gradient of ~20°C km^{-1} (see main text), the temperature at the base of the plastic lower crust was ~620°C. Table A1 summarises viscosity estimates for the lower continental crust from Westaway (1998), which constrain η_e to between ~10^{17} Pa s and ~6×10^{18} Pa s for this set of conditions. W, the thickness of the plastic lower-crustal channel, is ~15 km, and J, the half-width of the Rhenish Massif, is ~100 km (Fig. 1).

A rise in sea level by ~100 m will load the continental shelf by ~1 MPa. Taking the typical horizontal distance H (Fig. A2) between coastlines in the present study region (Fig. 1) at high- and lowstands as 100 km, the maximum of P can be estimated as ~1 MPa/~100 km or ~10 Pa m^{-1} making the mean value ~5 Pa m^{-1}. From Table A1, this would make the proposed model feasible provided the lower crust in the study region has $\eta_e <$ ~2×10^{17} Pa s.

Table A1. Estimates of lower-crustal viscosity.

G (kJ mol^{-1})	η_L (Pa s)	T_m (°C)	η_m (Pa s)	F	η_e (Pa s)	P (Pa m^{-1})
100	1.67×10^{10}	620	1.18×10^{16}	12	1.41×10^{17}	5
150	6.67×10^{7}	620	3.95×10^{16}	30	1.19×10^{18}	44
200	2.00×10^{5}	620	9.94×10^{16}	60	5.97×10^{18}	219

G is the assumed molar activation energy for plastic deformation of lower crust with a quartzite rheology. η_L is a temperature-independent material property of quartzite, defined in Equation (B11) of Westaway (1998). T_m is the temperature adopted in this study for the deepest crust beneath the Rhenish Massif (see main text). η_m is the estimated viscosity of crust at this temperature, calculated from G, η_L and T_m using Equation (B11) of Westaway (1998). F is the factor, estimated by Westaway (1998), by which η_e, the effective viscosity of the lower crust, in general exceeds η_m. Its value depends only on G. The resulting value of η_e is then calculated as $F \times \eta_m$. P is the pressure gradient in the lower crust which is required for $\Omega = 1$, estimated using Equation (A28) with $W = 15$ km, $J =$ 100 km, and $v_t = 1.6$ mm a^{-1}, the upper bound for the Rhenish Massif (see text).

However, larger transient pressure gradients could also be induced by transient loading of adjacent areas by ice sheets. Boulton et al. (1985) examined models for the tapering geometry of the Weichselian Scandinavian ice sheet, which can provide rough estimates of the induced pressure gradient. In one such model, with the ice assumed to have been in contact with rigid bedrock, this ice sheet was estimated to have reached a maximum thickness of ~3500 m and to have tapered from 500 m thickness to zero in ~40 km distance near its southern margin. Taking the density of ice as ~900 kg m^{-3}, this geometry would have induced maximum pressure gradients in the lower crust of ~110 Pa m^{-1} and thus mean values of ~55 Pa m^{-1}. A second ice sheet model incorporated lateral variations in boundary conditions at the base of the ice, due to the presence of deformable sediments in some localities, notably beneath the Baltic Sea. The resulting rapid ice movement in the 'Baltic Ice Stream' would restrict the maximum ice sheet thickness to ~2000 m and cause tapering near its southern margin from 500 m thickness to zero in ~150 km distance, making the maximum pressure gradient at the base of the brittle upper crust ~30 Pa m^{-1} and the mean value ~15 Pa m^{-1}. From Table A1, pressure gradients of ~15 to ~55 Pa m^{-1} would make the proposed process feasible for η_e up to ~3 × 10^{17} Pa s to ~2 × 10^{18} Pa s.

Less is known about the tapering geometries of earlier ice sheets. However, some which were rather larger than during the Weichselian, notably in oxygen isotope stages 12 and 6 (the Elster and Warthe glaciations), may have tapered more steeply. Even less is known of the scale of Early Pleistocene or Late Pliocene glaciations of northern and eastern Europe. Mitchell & Westaway (1999) have recently reviewed this topic, which is therefore not debated in detail here. It was established long ago that the outflow of meltwater from the earliest major lowland glaciation of eastern Europe caused the Akchagyl highstand of the Caspian Sea (e.g. Moskvitin, 1961). This event was defined in the local stratigraphy as Late Pliocene, and for many years was thought to have occurred no later than ~1.8 Ma. Mitchell & Westaway (1999) suggested instead that it marked oxygen isotope stage 36 around 1.2 Ma. However, whatever direct evidence may once have existed as to the scale of glaciations of Scandinavia at this time has been obliterated by later glaciations. The indirect evidence, mentioned in the main text, provided by the former existence of the 'Baltic River' and its disruption during and after a major glaciation which probably marked stage 36, suggests that the first very large Scandinavian ice sheet developed at this time. However, whether the presence of ice-rafted material in Atlantic Ocean sediments after ~2.5 Ma (e.g. Shackleton et al., 1984) resulted from smaller ice sheets in Scandinavia or elsewhere is not currently known. Likewise, although the Mediterranean sapropel evidence mentioned in the main text implies that the Alps and/or other mountain ranges bordering the Mediterranean Sea were glaciated during glacial cycles after 3.1 Ma, and the Kosberg terraces of the Maas preserve evidence of analogous glaciations of the Vosges, the extent of contemporaneous glaciation of Scandinavia is unknown and one thus cannot even begin to speculate about the overall geometry of flow in the lower crust at this time. All that can usefully be said at this stage is that models which require vigorous forcing of lower-crustal flow, with large values of ΔT_e, starting at either of these times (such as that in Fig. 17a, b, d, e) are less readily reconciled with the likely small scale of contemporaneous glaciation than models which require smaller values of ΔT_e. The apparent high values of ΔT_e which appear in the models in Figures 16 to 18 for the Miocene components of forcing are not significant. As has been mentioned in the main text, smaller values could have been used along with later start times. Alternatively, the Miocene and Early to Middle Pliocene uplift could have been modelled as superpositions of multiple components of forcing with very small values of ΔT_e and different start times (see below).

This discussion thus indicates that the lower-crustal flow regime required for the Middle and Late Pleistocene is feasible, given the known scale of glaciations and the resulting pressure gradients inferred from this ice loading. Whether the flow regime required for the Late Pliocene Early Pleistocene is feasible depends on the scale of contemporaneous glaciations.

A related topic concerns the subsidence of the Baltic floor. It is evident from the geometry of the 'Baltic River' that the present Baltic Sea did not exist at the start of the Quaternary. As much of this sea is now ~200 m deep, the level of the rock surface in this area has thus been lowered by ~200 m, at least, on this time scale. Subglacial erosion has previously been assumed to have been the cause of this lowering (e.g. White, 1972; Gibbard, 1988), although this idea has been criticised by Sugden (1976). However, erosion on this scale will have been isostatically compensated. In conventional Airy-type isostatic calculations ~200 m of erosion-induced subsidence would require ~7 × 200 m or ~1.4 km of erosion, which is two orders of magnitude more than in some adjacent land areas (e.g. Niini, 1968). More recent calculations, by Mitchell & Westaway (1999), indicate that the lower-crustal flow response to rapid erosion will anyway involve inflow of lower crust and local *uplift* – not subsidence – of the Earth's surface.

A.4 *Temperature perturbation comprising a series of steps*

As the magnitude of the temperature perturbation at depth z_i will depend on the amplitude of sea level fluctuations between glacials and interglacials, it can be expected to vary gradually over time. Rather than ΔT being constant, a more realistic situation involves approximating it as a series of steps. I thus adopt the notation that $\Delta T = \Delta T_1$ for $0 \leq t < t_1$, $\Delta T = \Delta T_2$ for $t_1 \leq t < t_2$, and so on.

Equation (A07) thus applies for $0 \leq t < t_1$, with $\Delta T = \Delta T_1$. At time t_1, the temperature perturbation is thus

$$\delta T(z, t_1) = \Delta T_1 \ \mathrm{erfc}[(z_i - z)/(2\ (\kappa t_1)^{1/2})] \ . \qquad \text{(A29)}$$

At the start of time step 2, the boundary conditions are thus: Equation (A29) for $z \neq z_i$, and

$$\delta T(z, t_1) = \Delta T_2 \ \delta(z_i - z) \ , \qquad \text{(A30)}$$

for $z = z_i$.

The solution for $t > t_1$ can now be obtained using Equation (A03) as

$$\delta T(z,t) = \frac{1}{2\sqrt{(\pi \kappa t)}} \int_0^{\infty} \delta T(\zeta, t_1)[\exp(-(z' - \zeta)^2 /(4\kappa t)) - \exp(-(z' + \zeta)^2 /(4\kappa t))] d\zeta$$
$$+ \frac{1}{2\sqrt{(\pi \kappa)}} \int_{t_1}^{t} \delta T(z' = 0, \tau) \frac{\exp(-z'^2 /(4\kappa t))}{(t - \tau)^{3/2}} d\tau \ . \qquad \text{(A31)}$$

This complicated integration may be rewritten as

$$\delta T(z, t) = I_1 + I_2 \ . \qquad \text{(A32)}$$

Using Equations (A29) and (A30), one may write

$$I_1 = \frac{\Delta T_1}{2\sqrt{(\pi \kappa t)}} \int_0^{\infty} \operatorname{erfc}[\zeta /(2\sqrt{(\kappa t_1)})] \times[\exp(-(z' - \zeta)^2 /(4\kappa t)) - \exp(-(z' + \zeta)^2 /(4\kappa t))]\, d\zeta \ , \tag{A33}$$

and

$$I_2 = \frac{\Delta T_2}{2\sqrt{(\pi \kappa)}} \int_{t_1}^{t} \frac{\exp(-z'^2 /(4\kappa t)}{(t - \tau)^{3/2}} d\tau \ .$$

I_2 is straightforward to evaluate. Restoring the original co-ordinate system, it becomes:

$$I_2 = \Delta T_2 \ \operatorname{erfc}[(z_i - z)/(2\ (\kappa\ (t - t_1))^{1/2})] \ , \tag{A35}$$

and thus has the same form as Equation (A07) with a different amplitude and time lag. The other analytic integral, I_1, is intractable: general analytic solutions to integrals with its form are not given in texts such as Erdelyi (1954), Abramowitz & Stegun (1965), and Gradshteyn & Ryzhik (1994). However, I_1 can instead be evaluated numerically; it turns out to be negligible compared with the other terms for the spatial and time scales relevant to this study. This means that, to a good approximation, the uplift response to forcing by multiple cyclic surface loads starting at different times can be calculated by summing the individual responses with appropriate time lags. This is the procedure used in the modelling described in the main text.

REFERENCES

Abramowitz, M. & Stegun, I.A. 1965. *Handbook of Mathematical Functions*, Dover, New York.

Aguirre, E & Pasini, G. 1985. The Pliocene-Pleistocene boundary. *Episodes*, 8: 116-120.

Aksu, A.E., Piper, D.J.W. & Konuk, T. 1987. Late Quaternary tectonic and sedimentary history of outer Izmir and Candarli bays, western Turkey. *Mar. Geol.*, 76: 89-104.

Albarello, D., Mantovani, E., Babbucci, D. & Tamburelli, C. 1995. Africa – Eurasia kinematics: main constraints and uncertainties, *Tectonophysics*, 243: 25-36.

Antoine, P. 1994. The Somme valley terrace system (northern France); a model of river response to Quaternary climatic variations since 800,000 BP. *Terra Nova*, 6: 453-464.

Arger, J., Mitchell, J. & Westaway, R. 2000. Neogene and Quaternary volcanism of south-eastern Turkey. In: *Tectonics and Magmatism of Turkey and the Surrounding Area*, Bozkurt, E., Winchester, J.A. & Piper, J.D.A. (eds) Geological Society of London Special Publication 173: 459-487.

van Balen, R.T., Houtgast, R.F., van der Wateren, F.M., Vandenberghe, J., & Bogaart, P.W. 2000. *Sediment budget and tectonic evolution of the Maas catchment in the Ardennesd and the Roer Valley rift system*, Global and Planetary Change, submitted.

Balling, N. 1995. Heat flow and thermal structure of the lithosphere across the Baltic Shield and northern Tornquist Zone, *Tectonophysics*, 244: 13-50.

Bassinot, F.C., Labeyrie, L.D., Vincent, E., Quidelleur, X., Shackleton, N.J. & Lancelot, Y. 1994. The astronomical theory of climate and the age of the Brunhes-Matuyama magnetic reversal, *Earth Planet. Sci. Lett.*, 126: 91-108.

Bates, M.R. 1993. Quaternary aminostratigraphy in northwestern France. *Quat. Sci. Rev.*, 12: 793-809.

van den Berg, M.W. 1996. *Fluvial sequences of the Maas: a 10 Ma record of neotectonics and climate change at various time-scales*. Ph.D. Thesis, University of Wageningen, the Netherlands.

Bibus, E. & Semmel, A. 1977. Über die Auswirkung quartärer Tektonik auf die altpleistozänen Mittelrhein Terrassen, *Catena*, 4: 385-408.

Blundell, D., Freeman, R. & Mueller, S. 1992. *A Continent Revealed: The European Geotraverse*, Cambridge University Press.

Bond, G. 1978. Evidence for Late Tertiary uplift of Africa relative to North America, South America, Australia and Europe, *J. Geol.*, 86: 47-65.

Boenigk, W. 1982. Der Einfluß des Rheingraben-Systems auf die Flußgeschichte des Rheins, *Z. geomorph., N.F., Suppl.* 42, 167-175 (in German with English summary).

Boulton, G.S., Smith, G.D., Jones, A.S. & Newsome, J. 1985. Glacial geology and glaciology of the last mid-latitude ice sheets, *J. Geol. Soc. London*, 142: 447-474.

Bridgland, D.R. & D'Olier, B. 1995. The Pleistocene evolution of the Thames and Rhine drainage systems in the southern North Sea Basin. In: Preece, R.C. (ed.), *Island Britain: A Quaternary Perspective*. Geol. Soc. London, Spec. Publ. 96: 27-45.

Brunnacker, K. 1978. Gliederung und Stratigraphie der Quartär-Terrassen am Niederrhein, *Kölner Geographische Arbeiten*, 36: 37-58.

Brunnacker, K. 1986. Quaternary stratigraphy in the Lower Rhine area and the northern Alpine foothills, *Quat. Sci. Rev.*, 5: 373-379.

Brunnacker, K., Farrokh, F. & Sidiropoulos, D. 1982a. Die altquatären Terrassen östlich der Niederrheinischen Bucht, *Z. Geomorph., N.F., Suppl.* 42, 215-226 (in German with English summary).

Brunnacker, K., Löscher, M., Tillmanns, W. & Urban, B. 1982b. Correlation of the Quaternary terrace sequence in the Lower Rhine valley and northern Alpine foothills of Central Europe, *Quat. Res.*, 18: 152-173.

Bruno, L.A., Baur, H., Graf, T., Schlüchter, C., Signer, P. & Wieler, R. 1997. Dating of Sirius Group tillites in the Antarctic Dry Valleys with cosmogenic ^{3}He and ^{21}Ne, *Earth and Planetary Science Letters*, 147: 37-54.

Cameron, T.D.J., et al. 1992. *The geology of the southern North Sea*. United Kingdom Offshore Regional Report, British Geological Survey, Edinburgh.

Carslaw, H.S., & Jaeger, J.C. 1959. *Conduction of Heat in Solids*, Second Edition, Oxford University Press.

Cermak, V. 1982. *Regional pattern of the lithosphere thickness in Europe*. In: *Geothermics and Geothermal Energy*, Cermak V. & Haenel, R. (eds) pp. 1-10, Schweitzerbart, Stuttgart, Germany.

Cermak, V. 1995. A geothermal model of the central segment of the European Geotraverse. *Tectonophysics*, 244: 51-55.

Cermak, V. & Bodri, L. 1995. Three-dimensional deep temperature modelling along the European geotraverse. *Tectonophysics*, 244: 1-11.

Cermak, V. & Hurtig, E. 1979. The preliminary heat flow map of Europe and some of its tectonic and geophysical implications, *Pure and Applied Geophysics*, 117: 93-103.

Cermak, V. & Zahradnik, J. 1982. *Two-dimensional correlation of heat flow and crustal thickness in Europe*. In: *Geothermics and Geothermal Energy*, Cermak V. & Haenel, R. (eds) pp. 17-25, Schweitzerbart, Stuttgart, Germany.

Cita, M.B., Vergnaud-Grazzini, C., Robert, D., Chamley, H., Ciaranfi, N. & D'Onofrio, S. 1977. Paleoclimatic record of a long deep-sea core from the Eastern Mediterranean. *Quat. Res.*, 8: 205-235.

Clairbois, A.M. 1959. L'évolution de la Meuse entre Liège et Anseremme au cours de la Quaternaire. *Annales de la Société géologique de Belgique*, 82: 213-234.

Damon, P.E. 1971. The relationship between Late Cenozoic volcanism and tectonism and orogenic-epeiorogenic periodicity. In: *The Late Cenozoic Glacial Ages*, Turekian, K.E. (ed.) pp. 15-35, Yale University Press, New Haven, Connecticut.

Davis, W.M. 1895. La Seine, la Meuse et la Moselle, *Annales de Géographie*, 5: 25-49.
de Heinzelin, J. 1964. Le réseau hydrographique de la région gallo-belge au Néogène. Essai de reconstitution. *Bulletin de la Société belge de Géologie*, 72: 137-148.
De Jong, J. 1988. Climatic variability during the past three million years, as indicated by vegetational evolution in northwest Europe and with emphasis on data from the Netherlands, *Phil. Trans. R. Soc. London, Ser. B.*, 318: 603-617.
Duncan, R.A., Petersen, N. & Hargraves, R.B. 1972. Mantle plumes, movement of the European plate, and polar wandering, *Nature*, 239: 82-86.
Erdelyi, A. 1954. *Tables of Integral Transforms, volume II*, McGraw-Hill, London.
Eyles, N. 1996. Passive margin uplift around the North Atlantic region and its role in northern hemisphere late Cenozoic glaciation, *Geology*, 24: 103-106.
Fairbanks, R.G. 1989. A 17,000-year glacio-eustatic sea level record: influence of glacial melting rates on the Younger Dryas event and deep-ocean circulation, *Nature*, 342: 637-642.
Felder, W.M., Bosch, P.W. & Bisschops, J.W. 1989. Geological map of deposits of the Maas in southern Limburg, 1:50000 scale, and accompanying explanatory text. National Geological Service, Haarlem, the Netherlands (in Dutch).
Flint, R.F. 1957. *Glacial and Pleistocene Geology*, John Wiley & Sons, London.
Flower, B.P. 1997. Overconsolidated section on the Yermak Plateau, Arctic Ocean: Ice sheet grounding prior to ca. 660 ka? *Geology*, 25: 147-150.
Gatliff, R.W. and 14 others 1994. *The geology of the central North Sea*. United Kingdom Offshore Regional Report, British Geological Survey, Edinburgh.
Geyl, W.F. 1960. Geophysical speculations on the origin of stepped erosion surfaces, *Journal of Geology*, 68: 154-176.
Gibbard, P.L. 1988. The history of the great north-west European rivers during the past three million years. *Phil. Trans. R. Soc. London, Ser. B.*, 318: 559-602.
Gibbard, P.L. 1995. The formation of the Strait of Dover. In: Preece, R.C. (ed.), *Island Britain: A Quaternary Perspective*, Geol. Soc. London, Spec. Publ. 96: 15-26.
Gilchrist, A.R., Summerfield, M.A. & Cockburn, H.A.P. 1991. Landscape dissection, isostatic uplift, and the morphologic development of orogens, *Geology*, 22: 963-966.
Gill, J.B. 1981. Orogenic Andesites and Plate Tectonics, Springer-Verlag, Berlin, Germany.
Gradshteyn, I.S. & Ryzhik, I.M. 1994. *Tables of Integrals, Series, and Products* (5th edition) (English translation, Jeffrey A. (ed.)), Academic Press, Boston, Massachusetts.
Grube, F. and 6 others 1986. Glaciations in north west Germany, *Quat. Sci. Rev.*, 5: 347-358.
Harmand, D. 1989. *La Meuse lorraine. Contribution à l'étude des alluvions anciennes de la Meuse entre Pagny-sur-Meuse et Mouzon (Ardennes). Tentative d'une reconstitution paléogéographique et dynamique actuelle du bassin*. Ph.D. thesis, University of Nancy II, France.
Harmand, D. 1992. *Histoire de la vallée de la Meuse lorraine*. Nancy University Press, Nancy, France.
Harmand, D., Kartit, A., Ochietti, S. & Weisrock, A. 1995. L'âge de la capture: corrélations entre les formations fluviatiles saaliennes de la Haute Moselle et de la Meuse, *Revue géographique de l'Est*, 3-4: 269-290.
Hilgen, F.J. 1991a. Astronomical calibration of Gauss to Matuyama sapropels in the Mediterranean and implications for the Geomagnetic Polarity Time Scale, *Earth Planet. Sci. Lett.*, 104, 226-244.
Hilgen, F.J. 1991b. Extension of the astronomically calibrated (polarity) time scale to the Miocene/Pliocene boundary. *Earth Planet. Sci. Lett.*, 107: 349-368.
Holmes, A. 1965. *Principles of Physical Geology*, 2nd ed., Nelson, London.
Hoffmann, R. 1996. Die quartäre Tektonik des südwestlichen Schiefergebirges begründet mit der Höhenlage der jüngeren Hauptterrasse der Mosel und ihre Nebenflüsse. *Bonner geowissenschäftliche Schriften*, 19: 1-156.
Hoselmann, C. 1994. Stratigraphie des Hauptterrassenbereichs am unteren Mittelrhein. *Sonderveröffentlichungen der Geologischen Institut der Universität zu Köln*, 96: 1-235.

Hoselmann, C. 1996. The main terrace complex in the lower Middle Rhine valley, *Zeitschrift der Deutschen Geologischen Gesellschaft*, 147: 481-497 (in German with English summary).

Huntley, D.J., Hutton, J.T. & Prescott, J.R. 1993. The stranded beach-dune sequence of south-east South Australia: A test of thermoluminescence dating, 0-800 ka. *Quat. Sci. Rev.*, 12: 1-20.

Huntley, D.J., Hutton, J.T. & Prescott, J.R. 1994. Further thermoluminescence dates from the dune sequence in the south-east of South Australia. *Quat. Sci. Rev.*, 13: 201-207.

Hurtig, E. & Oelsner, C. 1977. Heat flow, temperature distribution and geothermal models in Europe: Some tectonic implications, *Tectonophysics*, 41: 147-156.

Huxtable, J. 1993. Further thermoluminescence dates for burnt flints from the Palaeolithic site Maastricht- Belvédère and a finalised age for the Unit IV Middle Palaeolithic sites. *Meded. Rijks. Geol. Dienst*, 39-1: 41-44.

Illies, J.H. & Baumann, H. 1982. Crustal dynamics and morphodynamics of the Western European Rift System, *Z. Geomorph., N.F., Suppl*. 42: 135-165.

Illies, J.H., Prodehl, C., Schminke, H.-U. & Semmel, A. 1979. The Quaternary uplift of the Rhenish Shield in Germany, *Tectonophysics*, 61: 197-225.

Juvigné, E. & Renard, F. 1992. Les terrasses de la Meuse de Liège à Maastricht. *Annales de la Sovietégéologique de Belgique*, 115: 167-186.

Kandler, O. 1970. Untersuchungen zur quartären Entwicklung des Rheintales zwischen Mainz/ Wiesbaden und Bingen/Rüdesheim. *Mainzer Geographische Studien*, 3: 1-92.

van Kalfschoten, T. & Turner, E. 1996. Early Middle Pleistocene faunas from Kärlich and Miesenheim I and their biostratigraphical implications. In: *The early Middle Pleistocene in Europe*, C. Turner (ed.): 227-253. Balkema, Rotterdam.

Karig, D.E. & Kozlu, H. 1990. Late Paleogene – Neogene evolution of the triple junction region near Maras, south-central Turkey, *J. Geol. Soc. London*, 147: 1023-1034.

Kasse, K., Vandenberghe, J. & Bohncke, S. 1995. Climatic change and fluvial dynamics of the Maas during the late Weichselian and early Holocene, *Paläoklimaforschung*, 14: 123-150.

Kasse, K. 1998. Early Pleistocene fluvial and estuarine archives in the southern Netherlands and northern Belgium: Impact of climatic and environmental change (abstract). Presented at the *September 1998 Quaternary Research Association conference on 'River Basin Sediment Systems'*, Cheltenham, England.

Kaufman, P.S. & Royden, L.H. 1994. Lower crustal flow in an extensional setting: Constraints from the Halloran Hills region, eastern Mojave Desert, California, *Journal of Geophysical Research*, 99: 15723-15739.

Kennett, J.P. & Thunell, R.C. 1975. Global increase in Quaternary explosive volcanism, *Science*, 187: 497-503.

Kiden, P. & Törnqvist, T.E. 1998. Can river terrace flights be used to quantify Quaternary tectonic uplift rates? *J. Quat. Sci.*, 13: 573-574.

King, L.C. 1955. Pediplanation and isostacy: An example from South Africa, *Quarterly Journal of the Geological Society of London*, 111: 353-359.

von Koenigswald, W. & van Kolfschoten, T. 1996. The Mimomys-Arvicola boundary and the enamel thickness quotient (SDQ) of Arvicola as stratigraphic markers in the Middle Pleistocene. In: *The early Middle Pleistocene in Europe*, C. Turner (ed.): 211-226. Balkema, Rotterdam.

Krieger, M.H. Creasey, S.C. & Marvin, R.F. 1971. Ages of some Tertiary andesitic and latitic volcanic rocks in the Prescott-Jerome area, north-central Arizona, *US Geological Survey Professional Paper 750-B*: B157-B160.

Lambert, I.B. & Wyllie, P.J. 1972. Melting of gabbro (quartz eclogite) with excess water to 35 kilobars, with geological applications, *Journal of Geology*, 80: 693-708.

Lithgow-Bertelloni, C. & Silver, P.G. 1998. Dynamic topography, plate driving forces, and the African superswell, *Nature*, 395: 269-272.

Lucazeau, F., Vasseur, G. & Bayer, R. 1984. Interpretation of heat flow data in the French Massif Central, *Tectonophysics*, 103: 99-119.

Lucchitta, I., 1979. Late Cenozoic uplift of the southwestern Colorado Plateau and adjacent lower Colorado River region, *Tectonophysics*, 61: 63-95.

Macar, P. 1945. L'étrange capture de la Meuse par la Bar. *Annales de la Societé géologique de Belgique*, 68: 198-213.

Maddy, D. 1997. Uplift-driven valley incision and river terrace formation in southern England. *J. Quat. Sci.*, 12: 539-545.

Maddy, D. 1998. Reply: Can river terrace flights be used to quantify Quaternary tectonic uplift rates? *J. Quat. Sci.*, 13: 574-575.

McBirney, A.R., Sutter, J.F., Naslund, H.R., Sutton, K.G. & White, C.M. 1974. Episodic volcanism in the central Oregon Cascade Range, *Geology*, 2: 585-589.

McKee E.D. & McKee, E.H. 1972. Pliocene uplift of the Grand Canyon region – Time of drainage adjustment, *Geological Society of America Bulletin*, 83: 1923-1932.

McKenzie, D. 1985. The extraction of magma from the crust and mantle, *Earth and Planetary Science Letters*, 74: 81-91.

McKenzie, D. 1989. Some remarks on the movement of small melt fractions in the mantle, *Earth and Planetary Science Letters*, 95: 53-72.

McQuarrie, N. & Chase, C. 2000. Raising the Colorado Plateau. *Geology*, 28: 91-94.

Mengel, K., Sachs, P.M., Stosch, H.G., Wörner, G. & Loock, G. 1991. Crustal xenoliths from Cenozoic volcanic fields of West Germany: Implications for structure and composition of the continental crust, *Tectonophysics*, 195: 271-289.

Meyer, W. & Stets, J. 1998. Upper Pleistocene to Recent tectonics in the Rhenish Massif and its quantitative analysis, *Zeitschrift der Deutschen Geologischen Gesellschaft*, 149: 359-379 (in German with English summary).

Mikhaylova, N.P., Glevasskaya, A.M. & Tsykora, V.N. 1974. Geomagnetic polarity epochs in the Neogene, *Doklady of the Academy of Sciences of the USSR, Earth Sciences, English Translation*, 214, 112-114 (Russian original: *Doklady Akademii Nauk SSSR*, 214: 1145-1148).

Milliman, J.D. & Meade, R.H. 1983. Worldwide delivery of river sediment to the oceans, *J. Geol.,* 91: 1-21.

Milliman, J.D. & Syvitski, J.P.M. 1992. Geomorphic/tectonic control of sediment discharge to the ocean: The importance of small mountainous rivers, *J. Geol.,* 100: 525-544.

Mitchell, J., & Westaway, R. 1999. Chronology of Neogene and Quaternary uplift and magmatism in the Caucasus: Constraints from K-Ar dating of volcanism in Armenia. *Tectonophysics*, 304: 157-186.

Molnar, P. & England, P. 1990. Late Cenozoic uplift of mountain ranges and global climate change: Chicken or egg? *Nature*, 346: 29-34.

van Monfrans, H.M. 1971. Paleomagnetic dating in the North Sea basin, Earth Planet. *Sci. Lett.*, 11: 226-236.

Mooney, W.D. & Prodehl, C. 1978. Crustal structure of the Rhenish Massif and adjacent areas: A reinterpretation of existing seismic-refraction data, *Journal of Geophysics*, 44: 573-601.

Morgan, P. 1982. Heat flow in rift zones. In: Continental and Oceanic Rifts, edited by G. Palmason, *Geodynamics Series*, vol. 8, pp. 107-122, American Geophysical Union, Washington D.C., USA.

Moskvitin, A.I. 1961. The flood-plain terraces of the Volga and early Caspian transgressions as related to glaciations, *Doklady Acad. Sci. USSR, Earth Sci., Engl. Transl.* 136, 76-78 [Russian original: *Doklady Akad. Nauk SSSR* 136, 689].

Mudelsee, M. & Schulz, M. 1997. The Mid-Pleistocene climate transition: onset of 100 ka cycle lags ice volume build-up by 280 ka, *Earth Planet. Sci. Lett.*, 151: 117-123.

Murray-Wallace, C.V., Belperio, A.P., Cann, J.H., Huntley, D.J. & Prescott, J.R. 1996. Late Quaternary uplift history, Mount Gambier region, South Australia. *Z. Geomorph., N.F., Suppl.* 106: 41-56.

Negendank, J. 1983. Trier und Umgebung. *Sammlung geologischer Führer*, 60: 195 pp.

Niini, H. 1968. A study of rock fracturing in valleys of Precambrian bedrock, *Fennia*, 97: 1-60.

Okrusch, M., Schröder, B. & Schnütgen, A. 1979. Granulite-facies metabasite ejecta in the Laacher See area, Eifel, West Germany, *Lithos*, 12: 251-270.

Partridge, T.C. & Maud, R.R. 1987. Geomorphic evolution of southern Africa since the Mesozoic, *South African Journal of Geology*, 90: 179-208.

Pirazzoli, P.A., Radtke, U., Hantoro, W.S., Jouannic, C., Hoang, C.T., Causse, C. & Borel Best, M. 1991. Quaternary raised coral-reef terraces on Sumba Island, Indonesia. *Science*, 252: 1834-1836.

Pissart, A. 1961. Les terrasses de la Meuse et de la Semois. La capture de la Meuse lorraine par la Meuse de Dinant, *Annales de la Société géologique de Belgique*, 84: M1-M108.

Pissart, A. 1975. La Meuse en France et en Belgique. Formation du bassin hydrographique. Les terraces et leurs enseignements. In: P. Macar (ed.), L'évolution quaternaire des bassins fluviaux de la mer du nord méridionale. *Centenaire de la Société géologique de Belgique*, Liège, pp. 105-131 (in French with English abstract).

Pissart, A., Harmand, D. & Krook, L. 1997. L'évolution de la Meuse de Toul à Maastricht depuis le Miocène: Corrélations chronologiques et traces des captures de la Meuse Lorraine d'après les minéraux denses, *Géographie physique et Quaternaire*, 51: 267-284.

Quitzow, H.W. & Thome, K.N. 1975. Les terrasses du Rhin de Sinzig et Cologne. Pléistocène ancien de la cuvette de l'Erft. In: P. Macar (ed.), L'évolution quaternaire des bassins fluviaux de la mer du nord méridionale. *Centenaire de la Société géologique de Belgique, Liège*, pp. 308-318.

Raymo, M.E. & Ruddiman, W.F. 1992. Tectonic forcing of Late Cenozoic climate, *Nature*, 359: 117-122.

Rojstaczer, S. & Wolf, S. 1994. Hydrologic changes associated with the earthquake in the San Lorenzo and Pescadero drainage basins. In: *The Loma Prieta, California, Earthquake of October 17, 1989 – Hydrologic Disturbances*, Rojstaczer, S. (ed.), US Geological Survey Professional Paper 1551-E: E51-E64.

Ruddiman, W.F., Backman, J., Baldauf, J., Hooper, P., Keigwin, L., Miller, K., Raymo, M. & Thomas, E. 1987. Leg 94 paleoenvironmental results. In: W.F. Ruddiman, R.B. Kidd, E. Thomas et al. (eds), *Initial Reports of the Deep-Sea Drilling Program*, 94: 1207-1215.

Ruddiman, W.F. & Raymo, M.E. 1988. Northern hemisphere climate régimes during the past 3 Ma: Possible tectonic connections, *Philos. Trans. R. Soc. London, Ser. B*, 318: 411-430.

Ruddiman, W.F., Raymo, M. & McIntyre, A. 1986. Matuyama 41,000 year cycles: North Atlantic Ocean and northern hemisphere ice sheets, *Earth and Planetary Science Letters*, 80: 117-129.

Rudnick, R.L. & Goldstein, S.L. 1990. The Pb isotopic compositions of lower crustal xenoliths and the evolution of lower crustal Pb, *Earth and Planetary Science Letters*, 98: 192-207.

Ruegg, G.H.J. 1994. Alluvial architecture of the Quaternary Rhine-Meuse river system in the Netherlands, *Geologie en Mijnbouw*, 72: 321-330.

Ruegg, G.H.J. 1995. Discussion: Alluvial architecture of the Quaternary Rhine-Meuse river system in the Netherlands. Reply by the author, *Geologie en Mijnbouw*, 74: 187-190.

Schminke, H.-U. & Meertes, H. 1979. Pliocene and Quaternary volcanic phases in the Eifel volcanic fields, *Naturwissenschaften*, 66: 614-615.

Semmel, A. 1991. Neotectonics and geomorphology in the Rhenish massif and the Hessian basin, *Tectonophysics*, 195: 291-297.

Semmel, A. 1999. Landeschaftsentwicklung am Oberen Mittelrhein. In: Hoppe A. & Steininger F.F. (eds), Exkursionen zu Geotopen in Hessen und Rheinland-Pfalz sowie zu naturwissenschaftlichen Beobachtungspunkten Johann Wolfgang von Goethes in Böhmen. *Schriftenreihe der Deutschen Geologischen Gesellschaft*, 8: 127-149.

Shackleton, N.J., Berger, A. & Peltier, W.R. 1990. An alternative astronomical calibration of the lower Pleistocene timescale based on ODP site 677, *Trans. R. Soc. Edinburgh, Earth Sci.*, 81: 251-261.

Shackleton, N.J., Crowhurst, S., Hagelberg, T., Pisias, N.G. & Schneider, D.A. 1993. A new Late Neogene timescale: Application to ODP leg 138 sites. In: Pisias, N.G., Mayer, L.A., Janecek, T.R., Palmer-Julson, A. & van Andel T.H. (eds) *Proc. Ocean Drilling Program, Sci. Res.*, 138: 73-101.

Shackleton, N.J. and 16 others 1984. Oxygen isotope calibration of the onset of ice-rafting and history of glaciation in the North-east Atlantic, *Nature*, 307: 620-623.
Sheard, M.J. 1990. A guide to the Quaternary volcanoes in the lower south-east of South Australia. *Mines and Energy Review, South Australia*, 157: 40-50.
de Sitter, L.U. 1952. Pliocene uplift of Tertiary mountain chains, *Am. J. Sci.*, 250: 297-307.
Sood, M. 1981. *Modern Igneous Petrology*, John Wiley, New York.
Steininger, F.F. & Papp, A. 1979. Current biostratigraphic and radiometric correlations of the Late Miocene Central Paratethys stages (Sarmatian s. str., Pannonian s. str., and Pontian) and Mediterranean stages (Tortonian and Messinian) and the Messinian Event in the Paratethys, *Newsl. Stratigr.*, 8: 100-110.
Steininger, F.F. & Rögl, F. 1984. Paleogeography and palinspastic reconstruction of the Neogene of the Mediterranean and Paratethys. In: Dixon, J.E. & Robertson, A.H.F. (eds) The Geological Evolution of the eastern Mediterranean, *Geological Society of London Special Publication*, 17: 659-668.
Steininger, F., Rögl, F., & Martini, E. 1976. Current Oligocene/Miocene biostratigraphic concept of the central Paratethys (Middle Europe), *Newsletters of Stratigraphy*, 4: 174-202.
Sugden, D.E. 1976. A case against deep erosion of slields by ice sheets, *Geology*, 4: 580-582.
Sugden, D.E., Marchant, D.R., Potter Jr, N., Souchez, R.A., Denton, G.H., Swisher III, C.C. & Tison, J.-L. 1995. Preservation of Miocene glacier ice in East Antarctica, *Nature*, 376: 412-414.
Summerfield, M.A. & Hulton, N.J. 1994. Natural controls of fluvial denudation rates in major world drainage basins, *J. Geophys. Res.*, 99: 13,871-13,883.
Summerfield, M.A. & Kirkbride, M.P. 1992. Climate and landscape response, *Nature*, 355: 306.
Taous, A. 1994. *Le système alluvial de la moyenne terrasse de la Moselle en Lorraine méridionale (approche sédimentaire et pétrographique)*. Ph.D. thesis, University of Nancy II.
Tebbens, L.A., Kroonenberg S.B. & van den Berg, M.W. 1995. Compositional variations of detrital garnets in Quaternary Rhine, Meuse and Baltic River sediments in the Netherlands, *Geologie en Mijnbouw*, 74: 213-224.
Theilen, F. & Meissner, R. 1979. A comparison of crustal and upper mantle features in Fennoscandia and the Rhenish Shield, two areas of recent uplift, *Tectonophysics*, 61: 227-242.
Thunell, R.C., Williams, D.F. & Kennett, J.P. 1977. Late Quaternary paleoclimatology, stratigraphy and sapropel history in eastern Mediterranean deep-sea sediments, *Mar. Micropaleontol.*, 2: 271-388.
Törnqvist, T.E. 1994. Middle and late Holocene avulsion history of the River Rhine (Rhine-Meuse delta, Netherlands). *Geology*, 22: 711-714.
Törnqvist, T.E. 1995. Discussion: Alluvial architecture of the Quaternary Rhine-Meuse river system in the Netherlands, by G.H.J. Ruegg, *Geologie en Mijnbouw*, 74: 183-186.
Törnqvist, T.E. 1998. Longitudinal profile evolution of the Rhine-Meuse system during the last deglaciation: interplay of climate change and glacio-eustasy. *Terra Nova*, 10: 11-15.
Törnqvist, T.E., van Ree, M.H.M., van't Veer, R. & van Geel, B. 1998. Improving methodology for high-resolution reconstruction of sea-level rise and neotectonics by paleoecological analysis and AMS ^{14}C dating of basal peats. *Quat. Res.*, 49: 72-85.
Turner, C. 1996. A brief survey of the early Middle Pleistocene in Europe. In: *The early Middle Pleistocene in Europe*, C. Turner (ed.), pp. 295-317. Balkema, Rotterdam.
Urban, B. 1982. Quaternary paleobotany of the Lower Rhine basin, *Z. Geomorph., N.F., Suppl.* 42: 201-213.
Vågnes, E. & Amundsen, H.E.F. 1993. Late Cenozoic uplift and volcanism on Spitsbergen: Caused by mantle convection? *Geology*, 21: 251-254.
Vasseur, G. 1982. Synthése des résultats du flux géothermique en France. *Ann. Geophys.*, 38: 189-201.
Veldkamp, A. 1992. A 3-D model of Quaternary terrace development, simulations of terrace stratigraphy and valley asymmetry: A case study for the Allier terraces (Limagne, France). *Earth Surface Processes and Landforms*, 17: 487-500.

Veldkamp, A. 1996. Late Cenozoic landform development in East Africa: The role of near base level planation within the dynamic etchplanation concept. *Z. Geomorph., N.F., Suppl.* 106: 25-40.

Veldkamp, A. & Jongmans, A.G. 1990. Trachytic pumice weathering, Massif Central, France: geochemistry and micromorphology. *Chemical Geology*, 84: 145-147.

Vergnaud-Grazzini, C., Ryan, W.B.F. & Cita, M.B. 1977. Stable isotopic fractionation, climate change and episodic stagnation in the eastern Mediterranean during the late Quaternary. *Mar. Micropaleontol.*, 2: 353-370.

Veyret, Y. 1986. Quaternary glaciations in the French Massif Central. *Quat. Sci. Rev.*, 5: 395-396.

Visser, W.A. 1982. Geothermal investigation in the deep sedimentary basin of the Netherlands. In: *Geothermics and Geothermal Energy*, Cermak V. & Haenel, R. (eds) pp. 249-256, Schweitzerbart, Stuttgart, Germany.

Westaway, R. 1993a. Neogene evolution of the Denizli region of western Turkey, *J. Struct. Geol.*, 15: 37-53.

Westaway, R. 1993b. Quaternary uplift of southern Italy, *J. Geophys. Res.*, 98: 21741-21772.

Westaway, R. 1995. Crustal volume balance during the India-Eurasia collision and altitude of the Tibetan plateau: A working hypothesis, *Journal of Geophysical Research*, 100: 15173-15194.

Westaway, R. 1996, Quaternary elevation change of the Gulf of Corinth in central Greece. *Philosophical Transactions of the Royal Society of London, Series A*, 354: 1125-1164.

Westaway, R. 1998. Dependence of active normal fault dips on lower-crustal flow regimes, *Journal of the Geological Society of London*, 155: 233-253.

Westaway, R. 1999. The mechanical feasibility of low-angle normal faulting, *Tectonophysics*, still in press.

Westaway, R. & Arger, J. 1996. The Gölba debasin, southeastern Turkey: A complex discontinuity in a major strike-slip fault zone, *J. Geol. Soc. London*, 153: 729-743.

Whipple, K.X., Kirby, E. & Brocklehurst, S.H. 1999. Geomorphic limits to climate-induced increases in topographic relief. *Nature*, 401: 39-43.

White, W.A. 1972. Deep erosion by continental ice sheets. *Geol. Soc. Am. Bull.*, 83: 1037-1056.

Wilson, M. & Downes, H. 1991. Tertiary-Quaternary extension-related alkaline magmatism in western and central Europe, *Journal of Petrology*, 32: 811-849.

Windheuser, H., Meyer, W. & Brunnacker, K. 1982. Verbreitung, Zeitstellung und Ursachen des quartärer Vulkanismus in der Osteifel, *Z. Geomorph., N.F., Suppl.* 42: 177-194 (in German with English summary).

Würm, H.-U. 1997. A link between geomagnetic reversals and events and glaciations, *Earth Planet. Sci. Lett.*, 147: 55-67.

Zagwijn, W.H. 1974. The palaeogeographic evolution of the Netherlands during the Quaternary, *Geologie en Mijnbouw*, 53: 369-385.

Zagwijn, W.H. 1986a. The Pleistocene of the Netherlands with special reference to glaciation and terrace formation, *Quat. Sci. Rev.*, 5: 341-345.

Zagwijn, W.H. 1986b. Plio-Pleistocene climatic change: Evidence from pollen assemblages. *Mem. Soc. Geol. Italy*, 31: 145-152.

Zagwijn, W.H. 1989. The Netherlands during the Tertiary and the Quaternary: A case history of coastal lowland evolution, *Geologie en Mijnbouw*, 68: 107-120.

Zagwijn, W.H. 1992. The beginning of the Ice Age in Europe and its major subdivisions, *Quat. Sci. Rev.*, 11: 583-591.

Zagwijn, W.H. 1996. The Cromerian Complex stage of the Netherlands and correlation with other areas in Europe. In: *The early Middle Pleistocene in Europe*, by C. Turner (ed.), pp. 145-172. Balkema, Rotterdam.

Zonneveld, J.I.S. 1975. The terraces of the Maas (and the Rhine) downstream of Maastricht. In: P. Macar (ed.), L'évolution quaternaire des bassins fluviaux de la mer du nord méridionale. *Centenaire de la Société géologique de Belgique*, Liège, pp. 133-157.

Part 3: Climate forcing: records of Pleistocene and Holocene river behaviour

5. Early Pleistocene fluvial and estuarine records of climate change in the southern Netherlands and northern Belgium

C. KASSE & S. BOHNCKE
Department of Quaternary Geology and Geomorphology, Faculty of Earth Sciences, Vrije Universiteit, the Netherlands Centre for Geo-ecological Research (ICG), Amsterdam, the Netherlands

1 INTRODUCTION

Current understanding of the Early Pleistocene stratigraphy and climate of the Netherlands and Northwest Europe is essentially based on vegetation changes, largely reconstructed from pollen assemblages, that have been related to changes in climate. In the Netherlands, the Plio/Pleistocene boundary is traditionally placed at c. 2.3 million years BP (Zagwijn, 1989). Suc et al. (1997) proposed to lower the Plio/Pleistocene boundary to the Gauss-Matuyama magnetic reversal at c. 2.6 million years BP. Previously, within the Tiglian Stage, three warm (Tiglian A, Tiglian C3 and Tiglian C5) and two cold substages (Tiglian B, Tiglian C4c) have been recognized in the southeastern Netherlands (Venlo area) (Zagwijn, 1957, 1960, 1963; Westerhoff et al., 1998). According to De Jong (1988) and Zagwijn (1989) the Early Pleistocene glacials were not periods of extreme cold and the climatic oscillations lasted longer and had a lower amplitude than those of the Late Pleistocene. The deep-sea record, however, shows frequent and high-amplitude climatic oscillations (Shackleton et al., 1995) throughout this period and Funnell (1996) correlated the Tiglian Stage of the Netherlands with Oxygen Isotope Stages 63 to 91.

In northern Belgium (Beerse area),within the Early Pleistocene Kempen Formation, a cold phase (Beerse Glacial) coincided with a low sea level and was characterized by fluvial and aeolian sedimentary environments and periglacial structures (frost cracks and cryoturbations) (Dricot, 1961; Paepe & Vanhoorne, 1970; Kasse, 1988, 1993). This cold episode was correlated by Kasse (1988, 1993) with the Tiglian C4c of the type localities of the Tegelen Formation in the Venlo area (Zagwijn, 1963b). Later, Gibbard et al. (1995) described periglacial phenomena possibly of Tiglian C4c age at Öbel in Germany, east of the Venlo area. According to Funnell (1996) the Tiglian C4c corresponds with Oxygen Isotope Stage 68.

The interglacial periods are characterized by thermophilous trees and successive interglacials have previously been differentiated on the basis of their distinctive pollen assemblages (Zagwijn, 1960). The colder climatic intervals are characterized by similar pollen assemblages of pine, birch and herbs and are therefore not suited for stratigraphic reconstructions and correlations. The Early Pleistocene succession of warm and cold intervals has been reconstructed from discontinuous fluvial backswamp and channel-fill deposits at many different sites which are embedded within extensive, mostly pollen-free, sandy deposits.

This article documents a new glacial cycle in the Tiglian and thus demonstrates greater complexity in the terrestrial Early Pleistocene stratigraphy of northwestern Europe. In contrast to the continuous sequences of the deep ocean or lake sediment records, the extent to which the numerous climatic cycles of the Quaternary have been registered in fluvial and estuarine successions in the region is not well known. The presence of erosional hiatuses in the terrestrial Early Pleistocene record is shown and the mechanisms responsible for the incompleteness of the record are discussed. The location of the area at the southern rim of the North Sea Basin makes it a very sensitive region where the effects of external forcing (climate, tectonics) on the fluvial and estuarine environment could be expected to be registered.

The sites that we have studied are located close to the towns of Beerse in northern Belgium and Venlo in the southern Netherlands (Fig. 1). At Beerse the Early Pleistocene succession is exposed in clay pits on the west-east orientated Campine Microcuesta; thick estuarine clay beds were deposited here at the landward margin of the southern North Sea Basin. Due to Middle and Late Pleistocene uplift and preferential erosion, the more resistant clay beds form the present-day higher relief at c. 30 m above sea level (Kasse, 1988).

In the neighbourhood of Venlo the Early Pleistocene succession is exposed in many clay pits on the 'High Terrace' of the Rhine and the type sections of the Tiglian are located nearby at Tegelen (Zagwijn, 1960; Westerhoff & Cleveringa, 1996). During the Early Pleistocene and early Middle Pleistocene predominantly fluvial sequences of the Rhine and the Meuse were deposited. Later, uplift of the area resulted in erosion by the Meuse and terraces were formed at c. 45 m above present sea level.

Figure 1. Location map of the sites investigated at Beerse Ossenweg and Venlo Laumans.

2 EARLY PLEISTOCENE LITHOSTRATIGRAPHY AND PALYNOLOGY

2.1 *Beerse*

The clay pits near Beerse in northern Belgium have been studied for several decades (De Ploey, 1961; Dricot, 1961; Paepe & Vanhoorne, 1970, 1976; Kasse, 1988, 1990, 1993, 1996). The Early Pleistocene succession consists of three members: a basal clay unit (Rijkevorsel Member), an intermediate sand unit (Beerse Member) and an upper clay unit (Turnhout Member) (Figs 2, 3 and 4). These three members have been correlated by pollen analysis and magnetostratigraphy with the Tiglian substages C3, C4 and C5 respectively (Kasse, 1988, 1996). The Rijkevorsel Member is a bluish grey non-calcareous clay, sandy and silty at the base, with a clear fining upward tendency towards the top (Fig. 3a: unit 1). It has been interpreted as an estuarine tidal flat and salt marsh deposit (Dricot, 1961; Kasse, 1988). The Beerse Member overlies the lower clay without evidence for erosion. The member consists of fine sand with four distinct soil or peat horizons (Fig. 3a: unit 2). Frost cracks and small ice-wedge casts are common in the sand beds, while the lower three soil and peaty horizons are often strongly involuted (Fig. 5a) (Kasse, 1993). The Beerse Member shows a drying-upwards trend and the Beerse Member has been interpreted in terms of shallow braided rivers giving way upwards to aeolian sand sheets (Fig. 4). The contact with the overlying Turnhout Member has been subject to channel erosion (Fig. 3a: unit 3/5). Over short distances the Beerse Member can be completely preserved (Fig. 3a east side) or completely eroded (Fig. 3a west side). The Turnhout Member consists of a bluish to greenish grey non-calcareous clay with a clear fining upward trend. Heterolithic clay-sand deposits change upwards into stiff, crumbly clay, locally

CHRONOSTRATIGRAPHY			LITHOSTRATIGRAPHY		PROVENANCE	CLIMATE
M.PL	Cromerian		Sterksel Formation		R + M	complex
EARLY PLEISTOCENE	Bavelian		Kedichem Formation	Bavel Member	R + M	complex
	Menapian			Gilze Member	S + M	cold
	Waalian				S (+ R)	complex
	Eburonian				S	cold
	Tiglian	C5	Tegelen Formation	Turnhout Member	R (+ S)	complex
		C4		Beerse Member	S	cold
		C3		Rijkevorsel Member	R (+ S)	warm
		B		Hiatus		cold
		A				warm
	Praetiglian		Kiesel-oölite Formation			cold
PLIOCENE				Merksplas Sand		warm

Figure 2. Chrono- and lithostratigraphy of the Early Pleistocene deposits in the southern Netherlands and northern Belgium. R = Rhine, M = Meuse, S = Scheldt.

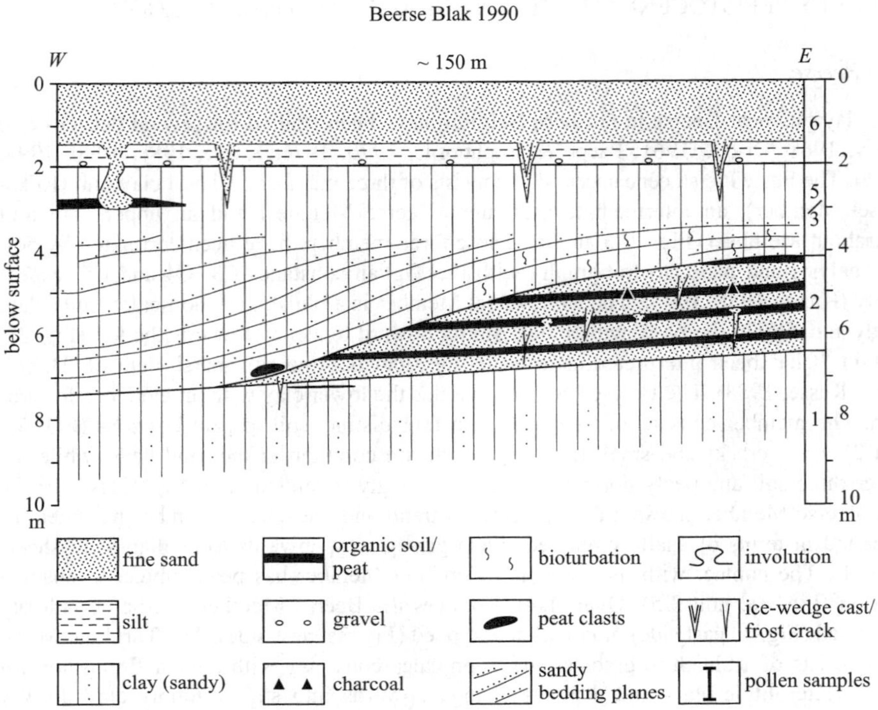

Figure 3a. Generalized section of the exposures at Beerse Blak in 1990. At Beerse Blak the glacial Beerse Member (unit 2) has been severely eroded by the tidal channels of the interglacial Turnhout Member (units 3/5). 1 = Rijkevorsel Member; 2 = Beerse Member; 3/5 = Turnhout Member; 6 = Twente Formation.

with an intercalated peat bed in the upper part. Heterolithic, lenticular and lateral accretion bedding in the deeper channels and bioturbation features are common (Fig. 3a) and the Turnhout Member has been interpreted as a tidal channel, flat and brackish marsh sequence (Dricot, 1961; Kasse, 1988).

The Early Pleistocene sequence is overlain by a Late Pleistocene (Weichselian) unit of aeolian silt and sand (Fig. 3a: unit 6) with large-scale periglacial features. At the transition from the Early to the Late Pleistocene units a gravel lag deposit is present that represents an erosional hiatus, comprising the Middle and part of the Late Pleistocene.

Since 1996 a new clay pit has been worked (Fig. 3b and 4) close to the previously studied Beerse Blak and Merksplas Strafinrichting pits (Kasse, 1988). In general, in comparison to the old pits, only minor differences in the stratigraphic record were observed:

- The Rijkevorsel Member contains two fining upward sequences of silty clay to clay locally with a thin organic bed (c. 10 cm) near the top of the upper fining-up sequence (Fig. 3b: unit 1).
- The lowermost organic soil of the Beerse Member almost directly overlies the Rijkevorsel Member (Figs 3b and 4: unit 2) and a thin intervening sand bed is present only locally between the two.

Beerse Ossenweg 1996

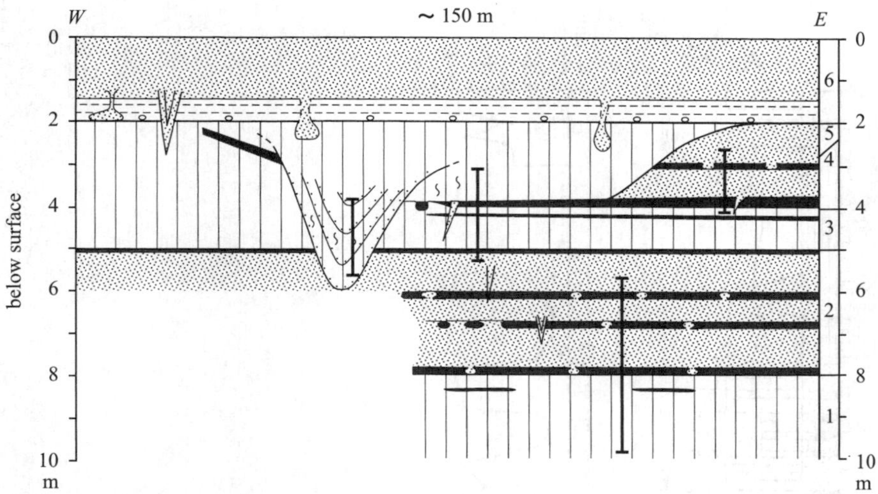

Figure 3b. Generalized section of the exposures at Beerse Ossenweg in 1996. At Beerse Ossenweg the glacial Beerse Member is fully preserved. Within the Turnhout Member a younger glacial sand unit (4) is locally preserved at the east side of the section. 1 = Rijkevorsel Member; 2 = Beerse Member; 3 = base of Turnhout Member; 4 = sand unit intercalated in Turnhout Member; 5 = upper part of Turnhout Member; 6 = Twente Formation.

- The Beerse Member is almost completely preserved, the upper organic bed, that directly overlies a strongly bleached eluvial horizon of a podsol soil, being present over the whole exposure. This means that tidal channel erosion, preceding the deposition of the Turnhout Member, did not occur here.

The Turnhout Member in three of the four pit faces consists of one clay unit without clear fining-upward sequences or lateral accretion bedding characteristic of channel migration (Fig. 3b: units 3 and 5). This indicates that, in comparison to the other pits, low-energy tidal environments prevailed. One of the pit faces showed significant differences to the succession detailed above. The eastern pit face showed a 2 m thick sand layer (Fig. 3b and 4: unit 4) that split the Turnhout Member into two clay units (Fig. 3b: units 3 and 5). The base of sand unit 4 conformably overlies unit 3 of the Turnhout Member. The sand is identical in grain size to the underlying Beerse Member (Fig. 4: unit 2 and 4). It is fine-grained and dominated by slightly deformed horizontal bedding of shallow fluvial or aeolian origin. At the base of the sand unit several (frost) cracks have been found that penetrate the underlying clay of unit 3 (Fig. 5b). The cracks are up to 70 cm deep and 5 cm wide and filled with pale yellow sand of unit 4. Within unit 4 a weakly developed podsol soil was found that was deformed by involutions of c. 30 cm amplitude (Fig. 5c). Sand unit 4 of the Turnhout Member has been preserved only locally (Fig. 3b). Clay unit 5 overlies unit 4 with a clear erosional boundary formed by channel erosion prior to the deposition of unit 5. Surprisingly however, in those cases where unit 4 has been eroded completely the erosional contact between clay units 5 and 3 can hardly be distinguished (Fig. 5b).

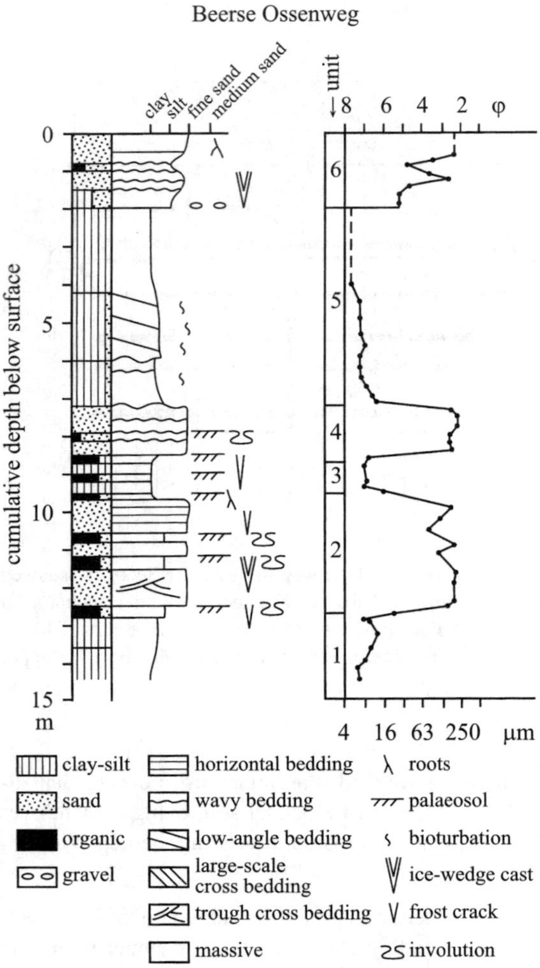

Figure 4. Sedimentological log of exposure Beerse Ossenweg. Note the similarity in grain size of the interglacial units 1, 3, 5 and the glacial units 2 and 4.

Forty samples were taken from the sequence for pollen analysis (Fig. 6) and the pollen sum was c. 200 grains per sample. Five major zones have been distinguished that show a strong correlation with lithofacies.

Zone 1 coincides with the Rijkevorsel Member and contains high amounts of thermophilous trees, especially *Quercus*, *Corylus* and *Alnus*, that point to interglacial conditions. The thermophilous trees show a decrease and the upland herbs a simultaneous increase to the top of the zone. First *Quercus* and *Ulmus* disappear, later followed by *Corylus* and finally by *Alnus*, which can be interpreted as an effect of gradual cooling. The Chenopodiaceae are important in the lower part of zone 1. This family is associated with tidal litter zones or salt marsh environments (Dricot, 1961).

Figure 5a. Periglacial phenomena of the Tiglian C4c glacial phase (unit 2) in Beerse Ossenweg. Small ice-wedge cast in the sands and soils of the Beerse Member (unit 2) indicating permafrost conditions. Trowel for scale is 25 cm.

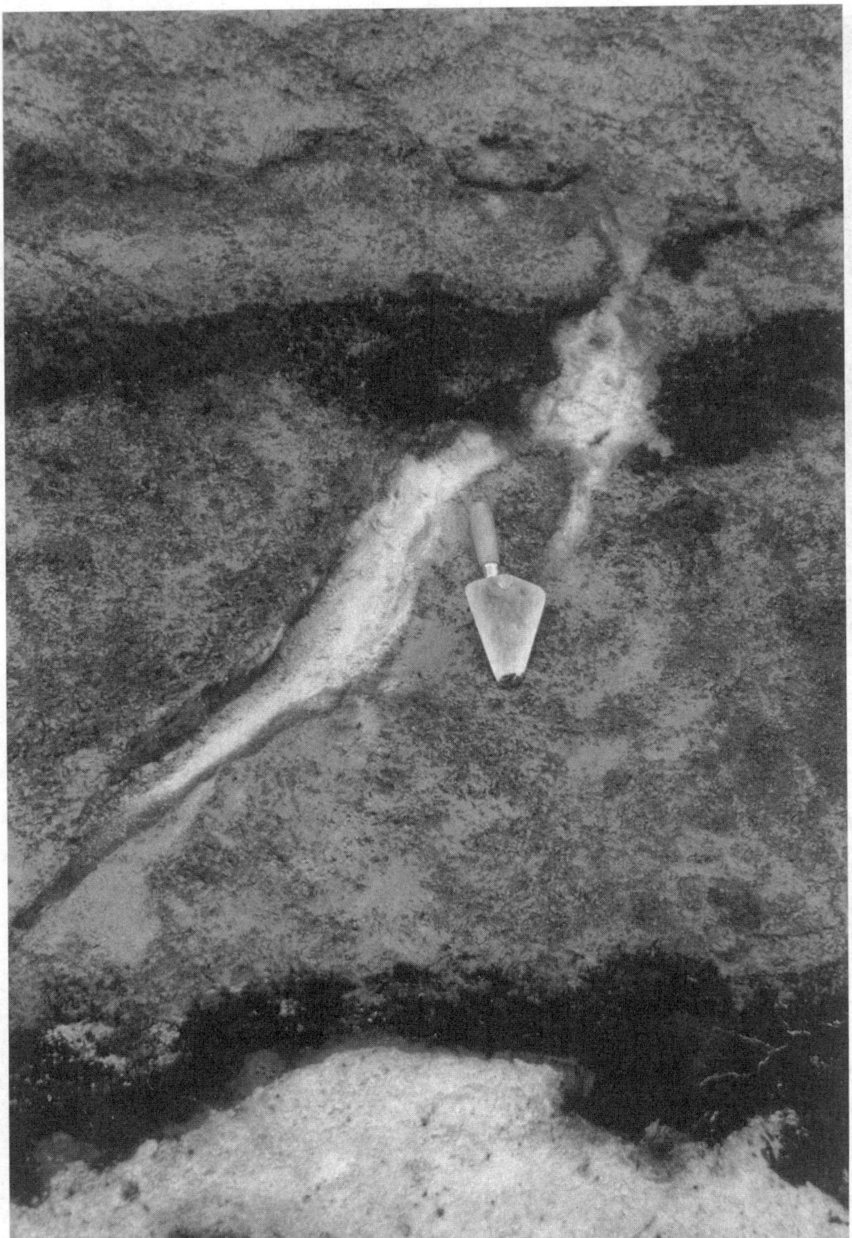

Figure 5b. Periglacial phenomena of the Tiglian C5b glacial phase (unit 4) in Beerse Ossenweg. Frost crack associated with glacial sand unit 4 (see Fig. 3b middle part). The frost crack has penetrated the interglacial clay unit 3 and the sand of unit 4 has been eroded completely here before the deposition of interglacial clay unit 5. Note that the erosional boundary left and right of the crack is hardly visible. At the base of the photo the white sand of the Beerse Member is exposed. Trowel for scale is 25 cm.

Figure 5c. Periglacial phenomena of the Tiglian C5b glacial phase (unit 4) in Beerse Ossenweg. Cryoturbations of the weakly developed podsolic soil in unit 4. Trowel for scale is 25 cm.

Zone 2 coincides with the Beerse Member, except for the uppermost organic layer of the Beerse Member, which is part of zone 3. The pollen diagram is characterized by a near absence of thermophilons trees. *Betula, Pinus* and *Salix* are the dominant three species but *Pinus* values are below 20% which probably indicates long-distance wind transport (Lotter et al., 1992). The vegetation is dominated by herbs (Gramineae and Cyperaceae). This pollen assemblage indicates glacial conditions with tundra vegetation. The presence of *Artemisia* indicates disturbed and bare soil conditions. The Chenopodiaceae, indicative of near-shore conditions and a high sea level in the previous zone, are almost absent in this zone, suggesting coastal retreat and sea-level fall.

Zone 3 represents the uppermost organic layer of the Beerse Member and the basal clay layer of the Turnhout Member (Fig. 3b: unit 3). The vegetation is strongly dominated by thermophilous trees, especially *Alnus*, indicating interglacial conditions. The highest *Alnus* values coincide with peaty beds at the top of the Beerse Member and intercalated in the clay unit, suggesting that *Alnus* was part of the peat-forming vegetation. The dominance of *Alnus* in the peaty, podsolized soil on top of the Beerse Member shows that the peat layer and the podsol were formed during interglacial times on a sandy surface of the previous Beerse Glacial. The high values of Chenopodiaceae in the clay layer indicate high sea level and near-shore conditions. Two samples from unit 3 were analyzed on the presence of the water-fern *Azolla tegeliensis*, a guide fossil for the Tiglian, but no megasporangia were found.

Zone 4 coincides with the basal organic unit of the newly discovered sand unit (Fig. 3b: unit 4). The sand itself and the intercalated podzolic soil are devoid of pollen. The vegetation

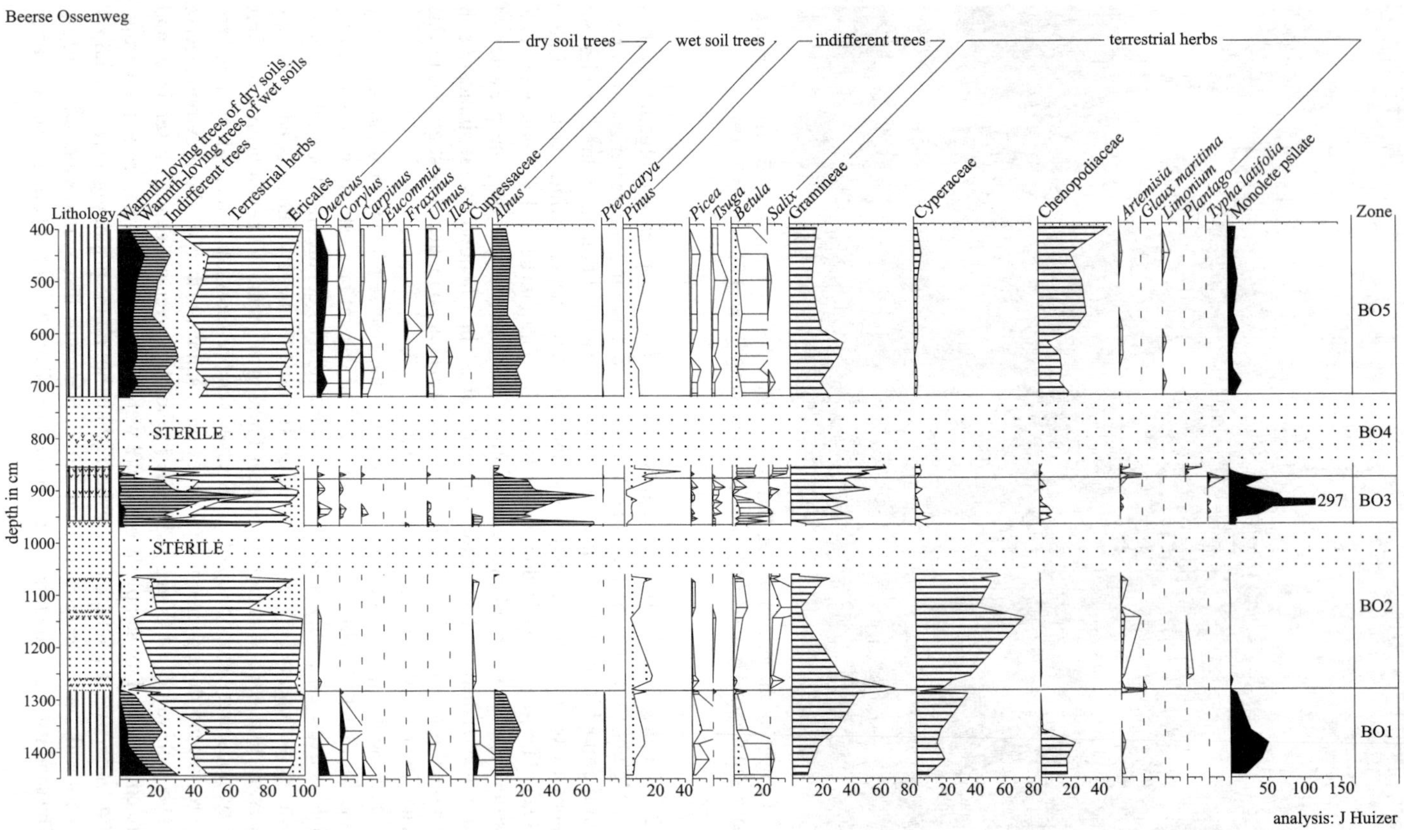

Figure 6. Pollen diagram of Beerse Ossenweg. The clay units are characterized by thermophilous trees and Chenopodiaceae indicating interglacial high sea level conditions; the sand units are characterized by herbs indicating glacial conditions.

is dominated by herbs, especially Gramineae. *Pinus*, *Betula* and *Salix* are the dominant trees. Pinus values up to 35% suggest local presence. *Artemisia* shows a maximum again. Like zone 2, the tundra or taiga vegetation of this zone points to glacial conditions. One megasporangium of *Azolla tegeliensis* was found at the base of unit 4.

Zone 5 coincides with the upper clay bed of the Turnhout Member (Fig. 3b: unit 5). The pollen composition strongly resembles zone 1. Thermophilous trees like *Quercus*, *Corylus*, *Carpinus*, *Fraxinus*, *Ulmus* and *Alnus* are important. In contrast to zone 1, *Tsuga* is a continuous element in the vegetation. The pollen assemblage indicates a warm temperate interglacial climate and the high presence of Chenopodiaceae and *Limonium* points to near-shore tidal flat, litter zone or salt marsh environments during a high sea level stand. Three out of five samples revealed the presence (up to 30 specimens) of megasporangia of *Azolla tegeliensis*.

2.2 *Venlo*

The clay pits in the Venlo area have been studied for more than a century (Zagwijn, 1998). The type sections of the Late Pliocene Reuverian and Early Pleistocene Tiglian stages have been defined here (Zagwijn, 1960). In the Venlo Laumans clay pit the Tegelen, Kedichem and Sterksel Formations of Early and Middle Pleistocene age are exposed (Figs 2, 7 and 8) and have previously been described by Van Straaten (1956) and Westerhoff & Cleveringa (1996).

The base of the pit shows a bluish grey clayey silt fining-upwards into a non-calcareous clay with a clear prismatic structure probably related to desiccation during soil formation (Fig. 7: unit 1, Tegelen Formation). Overlying this a massive clay with large white siderite nodules is present (Fig. 7: unit 1 upper part). Unit 1 can be interpreted as the low-energy upper part of a fluvial sequence, perhaps the uppermost fill of an abandoned channel or a back swamp deposit. The top of the clay in horizontal sections showed a large-scale polygonal network of wedges, of uncertain depths, filled with fine sand probably derived from unit 2 (Fig. 9a). The boundary between clay unit 1 and sand unit 2 is gradual (coarsening-upward sequence), suggesting continuous sedimentation.

Unit 2 (Kedichem Formation) is present only very locally in the pit because strong channel erosion preceded the deposition of coarse-grained unit 3 (Sterksel Formation) (Fig. 7). Unit 2 predominantly consists of fine-grained sand (Fig. 8). The low-angle trough cross bedding indicates shallow fluvial deposition and a current direction to the northwest. Some loamy beds indicate local low-energy conditions on a flood plain. Unit 2 is divided into two subunits by a humic layer. The absence of sedimentary structures and the presence of root traces indicate soil formation. In the lower part of unit 2, underlying the soil, periglacial phenomena are frequently found (Figs 7, 8 and 9). Involutions with an amplitude of 50 cm are very well developed with clear flat-bottomed bases (Fig. 9b) and thermal contraction cracks are common. In one case a poorly developed ice-wedge cast was found. In the upper part of unit 2, overlying the soil, periglacial phenomena are less frequent (Fig. 7). The soil has been penetrated by widely spaced, deformed wedge structures or involutions, some 70 cm deep, filled with white sand (Fig. 9c).

Unit 3 (Sterksel Formation) consists of coarse-grained gravelly sands (Figs 7 and 8) of the so-called High Terrace in this area. The base shows channel forms and lag deposits and large-scale planar cross bedding is dominant indicating straight-crested bars migrating in a braided river. Current direction is towards the northwest. Locally, fine-grained sands

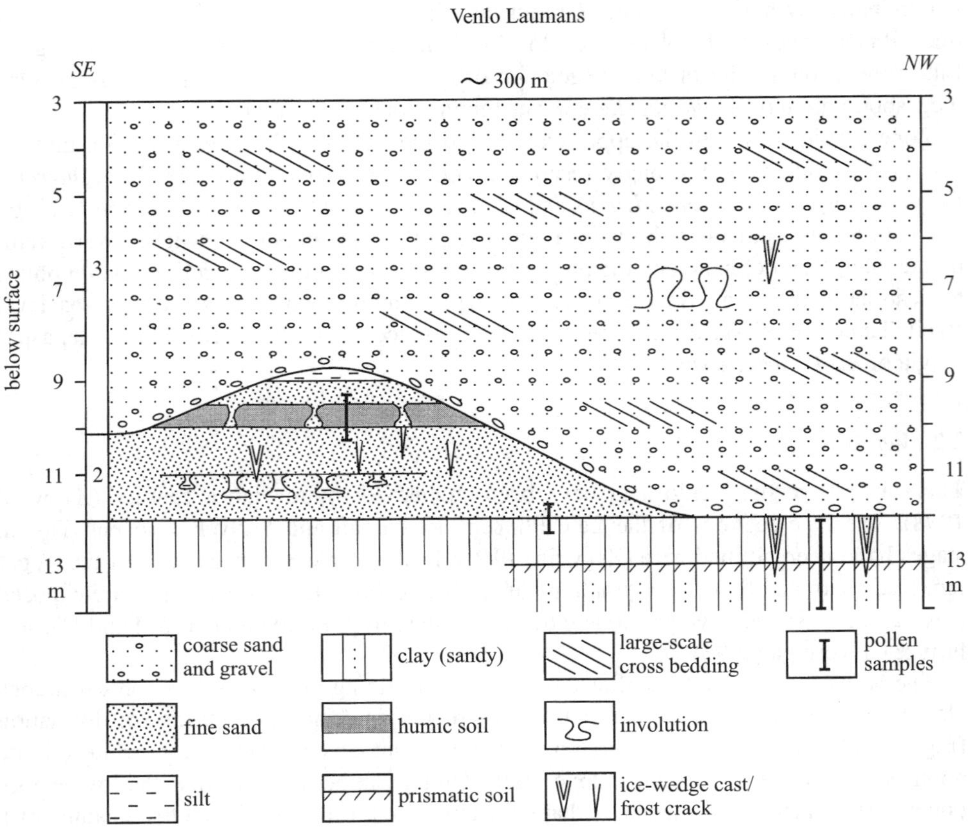

Figure 7. Generalized section of exposure Venlo Laumans. The Kedichem Formation (unit 2) is only locally preserved due to intense erosion prior to the deposition of the Sterksel Formation (unit 3). 1 = Tegelen Formation.

are present with major involutions up to one metre thick. An isolated ice-wedge cast was observed higher in the sequence.

Samples were taken for pollen analysis from units 1 and 2, but the lower part of clay unit 1 and the soil of unit 2 appeared to be poor in pollen. Pollen was recovered from the upper part of clay unit 1 and the coarsening-upward transition to unit 2 (Figs 7 and 10). Three pollen zones have been distinguished:

- Zone 1 is characterized by the presence of thermophilous trees of dry and wet soils, especially *Quercus*, *Corylus* and *Alnus*, but the values do not exceed 20%. Gramineae and Cyperaceae dominate the herb vegetation. *Typha latifolia* is present only in this zone.
- Zone 2 is dominated by *Pinus* and thermophilous trees have decreased.
- In zone 3 the values of the thermophilous trees are low. *Quercus, Ulmus*, *Pterocarya* and *Carya* have disappeared while *Corylus* and *Alnus* are still present. The Ericales show a strong increase at the start of this zone. *Artemisia, Plantago* and *Saxifraga* are present indicating an open vegetation cover.

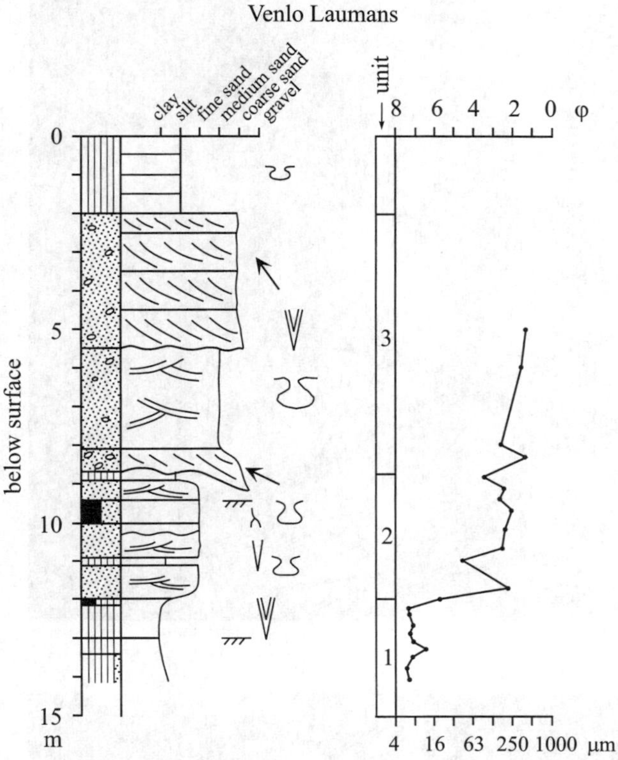

Figure 8. Sedimentological log of exposure Venlo Laumans. The clay of the Tegelen Formation (unit 1) is gradually overlain by fine sands of the Kedichem Formation (unit 2) with many periglacial features. See Figure 4 for legend.

3 DISCUSSION

3.1 *Early Pleistocene stratigraphy*

Previously, within the Tiglian complex, three warm (Tiglian A, Tiglian C3 and Tiglian C5) and two cold intervals (Tiglian B, Tiglian C4c) have been distinguished (Zagwijn, 1960, 1963) (Fig. 11). The Rijkevorsel and Turnhout Members in northern Belgium have been attributed to the Tiglian C3 and C5 respectively because of the interglacial character of the vegetation, the absence of *Fagus* (common in the Tiglian A in the Netherlands (Zagwijn, 1963b; Westerhoff et al., 1998)) and the presence of *Azolla tegeliensis* (Greguss & Vanhoorne, 1961; Kasse, 1988, 1993). The Beerse Member, intercalated between the Rijkevorsel and Turnhout Members, was associated with the Tiglian C4(c) glacial phase, because of its glacial pollen assemblage and periglacial structures (Kasse, 1993).

Our new results show, above the Beerse Glacial (Fig. 6: zone 2) attributed to the Tiglian C4c, a further episode of cold (Fig. 6: zone 4) that is characterized also by a low sea level, fluvial and aeolian sand deposition, cryogenic structures and cold vegetation with mainly *Pinus*, *Betula*, *Salix* and herbs. The sandy sediments of this latter cold phase are

a)

b)

Figure 9a and 9b. Periglacial phenomena of the Eburonian glacial phase (unit 2) in Venlo Laumans. Trowel for scale is 25 cm. (a) Horizontal section with polygonal wedge pattern in the top of clay unit 1 filled with sand of unit 2 indicating permafrost conditions at the start of the Eburonian. (b) Large-scale cryoturbations in the lower part of unit 2 attributed to the Eburonian glacial.

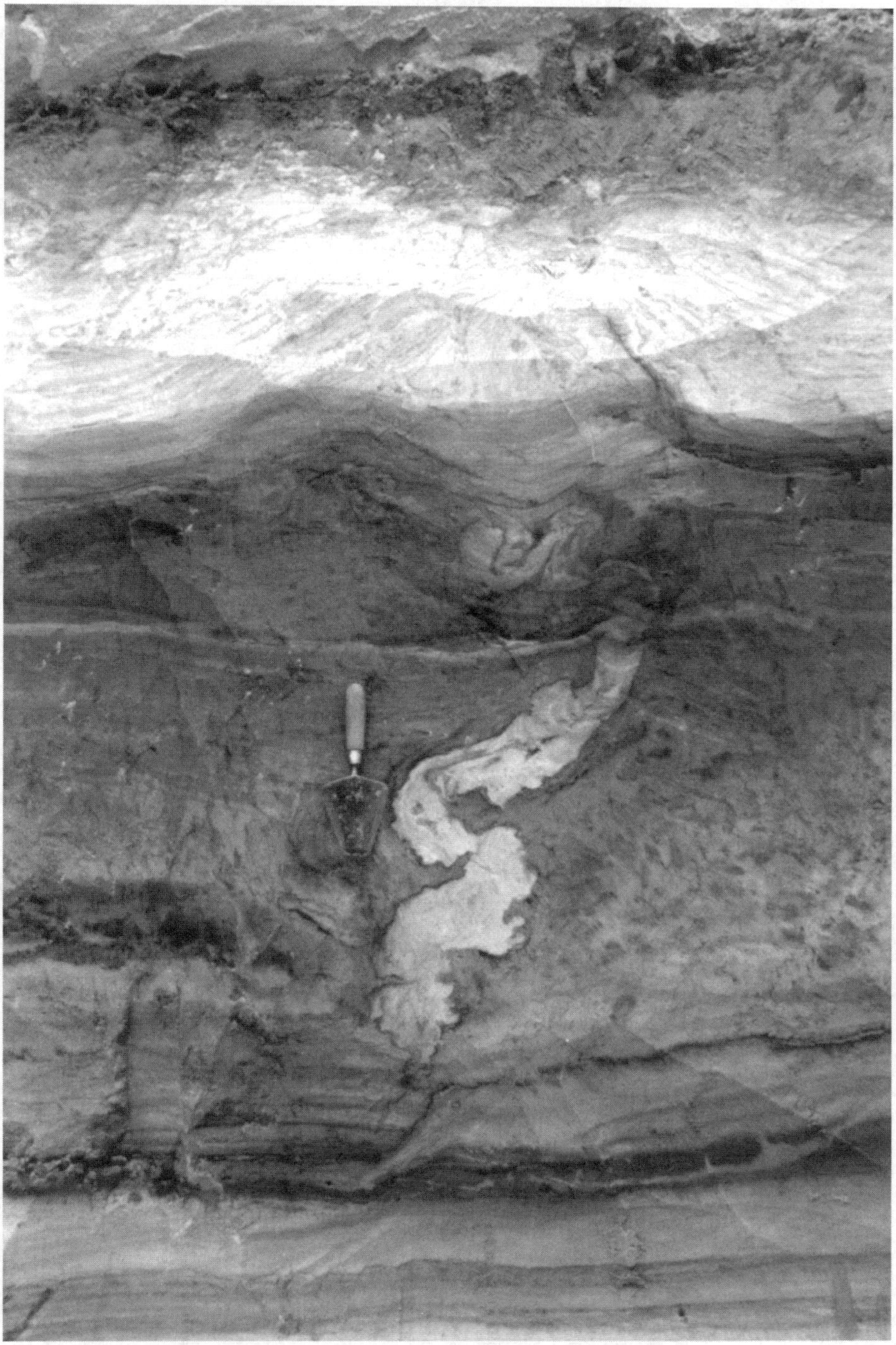

c)

Figure 9c. Periglacial phenomena of the Menapian glacial phase (unit 2) in Venlo Laumans. Deformed involutions of Menapian age with white sand disturbing the Waalian soil. Trowel for scale is 25 cm.

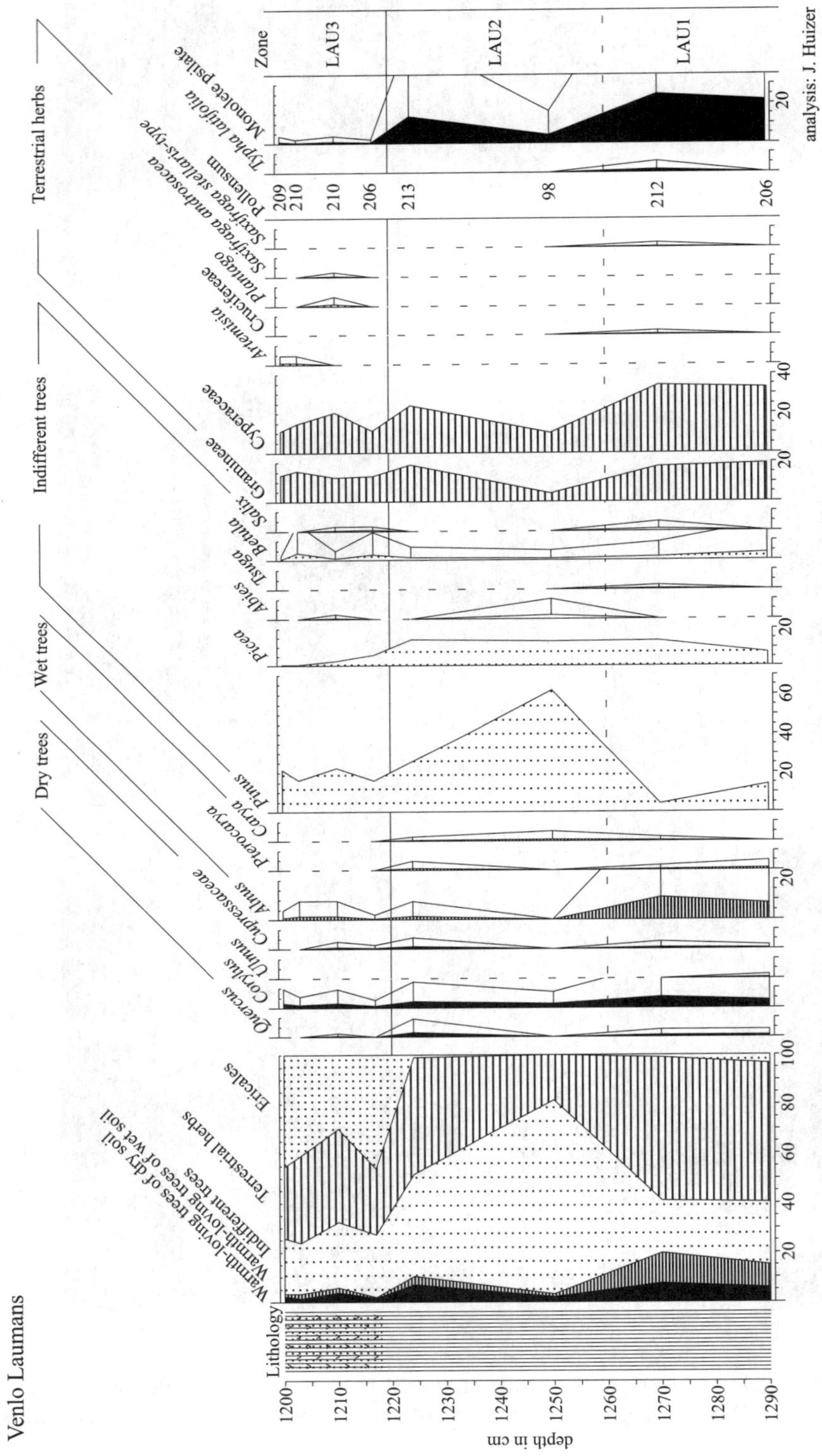

Figure 10. Pollen diagram of the transition from clay (unit 1) to sand (unit 2) in Venlo Laumans. The decrease in thermophilous trees and increase in herbs reflect the climatic cooling at the Tiglian – Eburonian boundary.

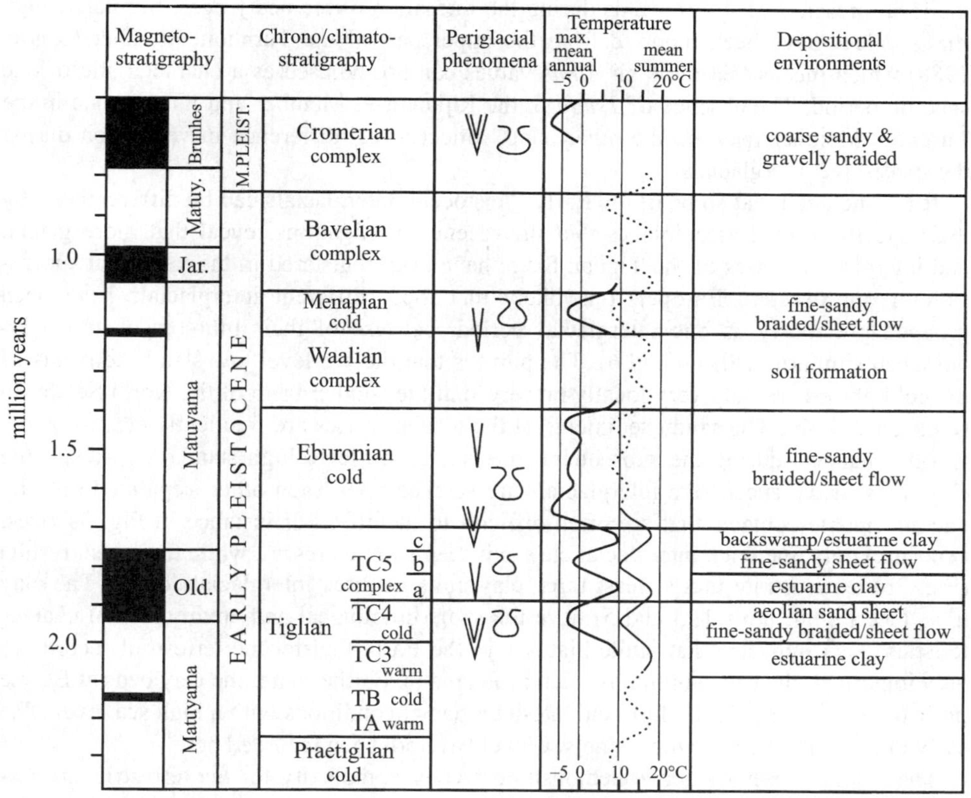

Figure 11. Synthesis of Early Pleistocene climatic evolution in the region. The glacial – interglacial temperature cycles reveal both high frequency and amplitude. Changes in depositional environment are often related to climate change. Magnetostratigraphy according to Cande & Kent (1995); chrono/climatostratigraphy according to Zagwijn (1989), Kasse (1988, 1993) and this study; periglacial phenomena according to this study; maximum mean annual temperature according to this study; mean summer temperature according to Zagwijn (1963, 1989) (dotted line) and this study (continuous line); depositional environments according to this study.

intercalated in the clayey Turnhout Member of Tiglian C5 age. This means that the Tiglian C5, previously defined as one undivided warm temperate phase, is a climatic complex of at least two interglacials (TC5a and TC5c, Fig. 6: zones 3 and 5) and one glacial. For the time being this new glacial phase (Fig. 6: zone 4) will be referred to as the intra-Tiglian C5 glacial (TC5b) (Fig. 11).

The results show that both the Tiglian glacials (Beerse Glacial or Tiglian C4c and new intra-Tiglian C5 glacial) are dominated by *Pinus*, *Betula*, *Salix*, Gramineae, Cyperaceae, Ericaceae, *Artemisia* and *Plantago* indicating a tundra or taiga vegetation (Fig. 6: zones 2 and 4). The Tiglian interglacials (Tiglian C3 and C5a and C5c) are characterized by deciduous trees (*Quercus*, *Corylus*, *Carpinus*, *Fraxinus*, *Ulmus* and *Alnus*) of warm temperate climates (Fig. 6: zones 1, 3, and 5). The pollen assemblages of the warm intervals resemble each other strongly. The major difference between the interglacial pollen assemblages is the higher *Alnus* content in pollen zone 3 (Fig. 6), though this seems to be due to the local presence of *Alnus* stands during the formation of the peaty beds. Similarly, high *Alnus* values have been reported from the upper part of the Turnhout Member (Kasse, 1988) which means that the high *Alnus* values can not be used as a characteristic of one specific period. The absence of *Tsuga* in the Rijkevorsel Member and its presence in the Turnhout Member may, on the other hand, reflect a real difference in vegetation during the successive interglacials.

It is concluded that some of the Early Pleistocene interglacials can be differentiated by their specific pollen assemblages, but the recent investigations reveal that more glacial and interglacial phases of the Tiglian Stage have been registered in the terrestrial succession of northwestern Europe. It is likely that these different interglacials have been grouped previously as one interglacial period, because of their lithological and palynological similarity (Figs 4 and 6). The point is that the sea-level low-stand sediments of the cold phases are only very locally preserved at the southern rim of the North Sea basin (Figs. 3a and 3b). The sandy sediments of the glacial phases are frequently eroded by estuarine channels during the subsequent interglacial sea-level high stand. As a result the clay units of the successive interglacials are stacked upon each other, separated only by unconformable contacts that are often difficult to identify. For instance in Fig. 3a (west exposure) only one thick interglacial clay unit seems to be present, while the recent results (Fig. 3b) demonstrate that at least three clay units of three interglacials occur. The clay units (see Fig. 4: units 1, 3 and 5) have the same lithological and environmental characteristics. This indicates that while hiatuses in the Early Pleistocene terrestrial record are very important, their recognition is often problematic. Furthermore, the clay beds at Beerse probably represent only the full- and late-interglacial conditions with a high sea level. The early interglacial periods with rising sea level have not been recorded here.

The results presented above demonstrate greater complexity for the terrestrial successions of the Tiglian Stage. In accordance with the deep-sea record, the duration of the glacial-interglacial cycles was short, although it is important to point out that individual glacials and interglacials are difficult to distinguish as they lack diagnostic characteristics.

3.2 *Temperature reconstructions*

Palaeotemperatures, especially the mean annual air temperature and mean temperature of the coldest month (winter temperature in this paper) can be estimated from the various periglacial structures (Figs 5 and 9) (cf. Maarleveld, 1976; Vandenberghe & Kasse, 1989; Vandenberghe & Pissart, 1993). In combination with the mean temperature of the warmest month (summer temperature in this paper), derived from the palaeobotanical record (cf. Zagwijn, 1963), the amplitude of the climatic oscillations on land can be reconstructed (Fig. 11). Frost cracks are normally associated with a mean annual air temperature lower than –1°C and ice-wedge casts indicate a mean annual air temperature lower than –6°C and a mean temperature of the coldest month lower than –20°C.

The Beerse Member of Tiglian C4c age at Beerse Ossenweg (Figs 3 and 4: unit 2) contains abundant frost cracks but ice-wedge casts are rare (Fig. 5a) and therefore a mean annual air temperature between –1 to –6°C is postulated (Fig. 11) (Kasse, 1993). The deposits of the new glacial phase in the Tiglian C5 (TC5b) (Figs 3 and 4: unit 4) include frost cracks and small cryoturbations (Fig. 5b and 5c) so a mean annual air temperature lower than –1°C is reconstructed. The tundra to taiga vegetation of the TC4c and TC5b glacials shows that the mean summer temperature was around or below 10°C. The combination of the mean annual air temperature (–1 to –6°C) and the mean summer temperature (10°C) indicates a mean winter temperature lower than –12°C and possibly lower than –22°C.

The clay units in Beerse Ossenweg show interglacial pollen assemblages (Fig. 6). According to Zagwijn (1963b), the presence of *Eucommia* indicates a mean summer temperature of c. 18°C. The presence of *Ilex*, a frost susceptible species, has been related to mean winter temperatures higher than –1°C (Iversen, 1944). Combining these summer and winter temperatures results in an estimate of the mean annual air temperature of around 8.5°C, which is comparable to the present day mean annual temperature of the Netherlands and northern Belgium.

The three pollen zones in Venlo Laumans suggest a climatic cooling at the end of an interglacial (Fig. 10). The thermophilous trees of zone 1 indicate the deciduous forest vegetation of an interglacial period. *Typha latifolia* points to a mean July temperature higher than 13°C. In zone 2 the thermophilous elements have decreased and are replaced by a pine forest vegetation. Finally, a more open vegetation established dominated by Ericales, *Artemisia* and other herbs (zone 3) indicating a drop of the summer temperature to c. 10°C or lower. The clay layer is part of the Tegelen Formation and its upper part is attributed in nearby pits to the Tiglian – Eburonian transition (Zagwijn, 1963b: Russel-Tiglia-Egypte pit). Therefore, the interglacial to glacial transition at Laumans is correlated with the late Tiglian – Eburonian boundary (Westerhoff & Cleveringa, 1996).

Within the Laumans pit a prominent soil has been found in unit 2 (Kedichem Formation) (Figs 7 and 8). A similar soil has been described previously from nearby pits and attributed on palynological grounds to the warm temperate Waalian interglacial (Zagwijn, 1960, p. 49 and 57). This means that the sand underlying and overlying the assumed Waalian soil (Figs 7 and 8: unit 2) can be attributed to the Eburonian and Menapian cold stages respectively. Periglacial phenomena in the Kedichem Formation have been recorded previously by Van Straaten (1956), Kortenbout van der Sluijs (1956) and Westerhoff & Cleveringa (1996). The large-scale wedge structures and polygons that formed in the top of clay unit 1 in Laumans (Fig. 9a) indicate very cold conditions, with mean annual air temperatures below –6°C and a mean winter temperature below –20°C at the beginning of the Eburonian glacial phase (Fig. 11). Above this level, but still below the Waalian soil, frost cracks and involutions are common (Fig. 9b), but only one indistinct ice-wedge cast has been found, which suggests somewhat higher mean annual air temperatures (between –1 and –6°C) than at the beginning of the Eburonian glacial.

The assumed Waalian soil has been disturbed locally by large-scale deformed wedges or involutions (Fig. 9c) which penetrate the soil and are filled with white sand from above. This means that their formation is younger than the Waalian and they probably date from the Menapian. A mean annual air temperature lower than –1°C has been postulated for such solitary forms (Vandenberghe, 1988).

The upper coarse-grained unit 3 at Laumans can be correlated with the Sterksel Formation of Cromerian age (Zagwijn, 1960). Locally within the sequence large-scale involutions and one ice-wedge cast have been seen which indicates that the mean annual air temperature was below –6°C during part of the Middle Pleistocene Cromerian Complex (Fig. 11).

Temperature reconstructions of the Early Pleistocene thus show large-amplitude climatic oscillations. Glacials phases had a mean annual air temperature below –1°C and occasionally below –6°C. The interglacials had temperature values comparable to the present day. The amplitude of these Early Pleistocene climatic oscillations was therefore comparable with those of the Late Pleistocene (Vandenberghe, 1992).

3.3 *Climate change and depositional environments*

The results show that there is a clear relationship between climate and depositional environment in these successions. In Beerse Ossenweg, the interglacials were characterized by clay deposition (Figs 3b and 4: units 1, 3 and 5). The sedimentary structures (heterolithic bedding, bioturbation) and the pollen assemblages (Chenopodiaceae, *Limonium*) indicate tidal channel, flat and salt marsh environments and a high sea level. During these high sea-level stands the area in northern Belgium was situated at the southernmost rim of marine deposition along the North Sea Basin. During the glacial phases shallow fluvial and aeolian sand sheet environments were dominant in the Beerse area (Fig. 3b and 4: units 2 and 4). The periglacial phenomena and pollen assemblages indicate cold conditions. The strong correlation between lithological changes and vegetational changes indicates that the alternation of clay and sand deposition was climatically-controlled and sea level related. During the glacial phases a fall in sea level occurred, the estuarine environment regressed and shallow fluvial systems extended their courses to the north. An associated change in sedimentary provenance also took place (Kasse, 1990). During subsequent interglacial phases, transgressions occurred and estuarine environments were established. Marked erosion by laterally migrating tidal channels took place and the sand unit of the previous glacial sea-level low stand was almost completely removed (Fig 3a: unit 2; Fig. 3b: unit 4). As a consequence, stacking of the interglacial estuarine deposits took place, separated by major, but almost indiscernible hiatuses. The preservation of the high-stand deposits is surprising as one would expect erosion during the low stands. It is suggested that, despite the sea level fall, the gradients of the rivers flowing into the shallow North Sea Basin did not change and, in addition, sediment supply increased leading to deposition in stead of erosion.

Farther upstream in the southeastern Netherlands (Venlo Laumans) clay deposition predominantly occurred during the late Tiglian interglacial in a low-energy, probably meandering fluvial system (Figs 7 and 8: unit 1). During the Eburonian and Menapian glacial phases (as reconstructed from the periglacial structures), sand deposition dominated by shallow braided systems (Figs 7 and 8: unit 2). The lithological change from clay to sand (coarsening upward) coincides with a change in vegetation from deciduous forest to tundra (Fig. 10). Thus climatic change is reflected by the vegetation, by periglacial phenomena and by a change in depositional environment. The climate-related change in fluvial style probably resulted from changes in discharge regime and sediment supply as shown by Kasse et al. (1995) for Late Pleistocene rivers. The climatic cooling at the Tiglian – Eburonian transition created a more open vegetation cover and hence increased sediment supply from hillslopes to river channels. The deep seasonal frost or permafrost of the Eburonian glacial period would have created a snow-melt related discharge regime. Both the increased sediment supply and more peaked discharge are responsible for the change in fluvial style and the relative lack of erosion of interglacial deposits, despite the lower base level.

4 CONCLUSIONS

1. The Early Pleistocene terrestrial record of the southern Netherlands and northern Belgium does not preserve a continuous record and hiatuses are common. In northern Belgium, periglacial sandy deposits formed during periods of low sea level were intensively eroded by tidal channeling during the subsequent interglacial high sea level

periods. This resulted in the stacking of interglacial units, separated by hardly distinguishable erosional contacts. In the more upstream fluviatile environments of the southeastern Netherlands, deposition during the interglacials was restricted to back swamp areas and abandoned channels of meandering rivers.

2. The Early Pleistocene record contains more preserved glacial and interglacial phases than previously recognised. In the Beerse area, a new glacial phase has been found in the late Tiglian, younger than the Beerse Glacial of Tiglian C4c age. This new glacial phase, separating two interglacials, has provisionally been assigned the term Tiglian C5b. This means that at least three interglacials and two glacials occurred in the Beerse area during the Tiglian C substage.
3. Early Pleistocene interglacials often have no or few diagnostic features and often cannot be distinguished from each other. The pollen assemblage, grain size and provenance of the successive interglacial deposits resemble each other. The same holds true for the glacials.
4. Changes in depositional environments are controlled by climate change and related sea-level fluctuations. In northern Belgium, predominantly clay deposition occurred in tidal channels, mud flats and marshes (Rijkevorsel Member and two clay units of the Turnhout Member) during three interglacial periods with high sea levels (Tiglian C3, C5a and C5c). Sand deposition occurred by shallow braided systems and aeolian sand sheets (Beerse Member and sand unit in Turnhout Member) during two glacial periods with low sea level (Tiglian C4c and C5b). The alternation of clay and sand units in the Beerse area is climate and sea-level controlled as indicated by vegetation reconstruction and periglacial phenomena. In the Venlo area, the change from low-energy fluvial clay deposition to shallow braided sand deposition coincides with a change from Tiglian interglacial to Eburonian glacial conditions as shown by the vegetation change and periglacial structures.
5. The Early Pleistocene climatic oscillations have a large amplitude, comparable with those of the Late Pleistocene. The deciduous vegetation of the interglacials indicates that the mean summer temperature was around 18°C, the mean winter temperature above –1°C and the mean annual temperature circa 8.5°C. The tundra to taiga vegetation of the glacials shows that the mean summer temperature was around 10°C. The periglacial phenomena (large-scale cryoturbations, abundant frost cracks and some ice-wedge casts) point to deep seasonal frost or permafrost. They show that the mean annual temperature during the glacials was at least lower than –1°C and sometimes lower than –6°C. The mean winter temperature was occasionally lower than –20°C. The start of the Eburonian glacial was extremely cold with continuous permafrost as indicated by large-scale wedge structures and polygons.

ACKNOWLEDGMENTS

The authors thank J. Huizer for the pollen counts, Wim Hoek for preparing the pollen diagrams, H. Sion for the figures and the sediment laboratory for the grain-size analysis. The manuscript benefitted from the careful reviews by Phil Gibbard and Jan Zalasiewicz.

REFERENCES

Cande, S.C. & Kent, D.V. 1995. Revised calibration of the geomagnetic polarity timescale for the late Cretaceous and Cenozoic. *Journal of Geophysical Research* 100, no. B4: 6093-6095.

De Jong, J. 1988. Climatic variability during the past three million years, as indicated by vegetational evolution in northwest Europe and with emphasis on data from the Netherlands. *Philosophical Transactions Royal Society London,* B 318: 603-617.

De Ploey, J. 1961. Morfologie en Kwartair-stratigrafie van de Antwerpse Noorderkempen. *Acta geographica Lovaniensia* 1.

Dricot, E.M. 1961. Microstratigraphie des Argiles de Campine. *Bulletin Société belge Géology, Paléontology, Hydrology,* 70: 113-141.

Funnell, B.M. 1996. Plio-Pleistocene palaeogeography of the southern North Sea Basin (3.75-0.60 Ma). *Quaternary Science Reviews,* 15: 391-405.

Gibbard, P.L., Krook, L. & Vandenberghe, J. 1995. Early Pleistocene depositional environments and stratigraphy at Öbel (Brüggen), Nordrhein-Westfalen, Germany. *Mededelingen Rijks Geologische Dienst* 52: 83-96.

Greguss, P. & Vanhoorne, R. 1961. Etude paléobotanique des argiles de la Campine à Saint-Léonard (Belgique). *Bulletin Institut royal des Sciences Naturelles de Belgique*, Tome 37, no. 33: 1-33.

Iversen, J. 1944. *Viscum, Hedera* and *Ilex* as climate indicators. *Geologiska Föreningens Förhandlingar* Bd. 66, H. 3: 463-483.

Kasse, C. 1988. *Early-Pleistocene tidal and fluviatile environments in the southern Netherlands and northern Belgium*. PhD Thesis, Vrije Universiteit Amsterdam.

Kasse, C. 1990. Lithostratigraphy and provenance of the Early-Pleistocene deposits in the southern Netherlands and northern Belgium. *Geologie en Mijnbouw,* 69: 327-340.

Kasse, C. 1993. Periglacial environments and climatic development during the Early Pleistocene Tiglian stage (Beerse Glacial) in northern Belgium. *Geologie en Mijnbouw,* 72: 107-123.

Kasse, C. 1996. Paleomagnetic dating and effects of Weichselian periglacial processes on the magnetization of early Pleistocene deposits (southern Netherlands, northern Belgium). *Geologie en Mijnbouw,* 75: 19-31.

Kasse, K., Vandenberghe, J. & Bohncke, S. 1995. Climate change and fluvial dynamics of the Maas during the late Weichselian and early Holocene. In: Frenzel, B. (ed.) *European river activity and climatic change during the Lateglacial and early Holocene.* ESF Project 'European Palaeoclimate and Man', Special Issue 9, Paläoklimaforschung – Palaeoclimate Research 14: 123-150.

Kortenbout van der Sluijs, G. 1956. The cryoturbations in the Tegelen region. *Geologie en Mijnbouw* 18: 421-422.

Lotter, A.F., Eicher, U., Siegenthaler, U. & Birks, H.J.B. 1992. Lateglacial climatic oscillations as recorded in Swiss lake sediments. *Journal of Quaternary Science,* 7: 187-204.

Maarleveld, G.C. 1976. Periglacial phenomena and the mean annual temperature during the last glacial time in the Netherlands. *Biuletyn Peryglacjalny,* 26: 57-78.

Paepe, R. & Vanhoorne, R. 1970. Stratigraphical position of periglacial phenomena in the Campine clay of Belgium, based on palaeobotanical analysis and palaeomagnetic dating. *Bulletin Société belge Géology, Paléontology, Hydrology,* 79: 201-211.

Paepe, R. & Vanhoorne, R. 1976. The Quaternary of Belgium in its relationship to the stratigraphical legend of the geological map. *Toelichtende Verhandelingen Geologische kaart en Mijnkaart van België*, no. 18.

Shackleton, N.J., Hall, M.A. & Pate, D.1995. Pliocene stable isotope stratigraphy of Site 846. *Proceedings of the Ocean Drilling Program Scientific Results,* 138: 337-355.

Suc, J.P., Bertini, A., Leroy, A.G. & Suballyova, D. 1997. Towards the lowering of the Plio/Pleistocene boundary to the Gauss-Matuyama reversal. *Quaternary International,* 40: 37-42.

Vandenberghe, J. 1988. Cryoturbations. In: Clark, M.J. (ed.) *Advances in Periglacial Geomorphology*, Chicester, John Wiley, 179-198.

Vandenberghe, J. 1992. Geomorphology and climate of the cool oxygen isotope stage 3 in comparison with the cold stages 2 and 4 in the Netherlands. *Zeitschrift für Geomorphologie Neue Folge, Supplement-Band* 86: 65-75.

Vandenberghe, J. & Kasse, C. 1989. Periglacial environments during the Early-Pleistocene in the southern Netherlands and northern Belgium. *Palaeogeography, Palaeoclimatology, Palaeoecology,* 72: 133-139.

Vandenberghe, J. & Pissart, A. 1993. Permafrost changes in Europe during the last glacial. *Permafrost and Periglacial Processes,* 4: 121-135.

Van Straaten, L.M.J.U. 1956. Structural features of the 'Papzand' Formation at Tegelen (Netherlands). *Geologie en Mijnbouw,* 18: 416-420.

Westerhoff, W. & Cleveringa, P. 1996. Upper-Pliocene and Lower-Pleistocene deposits in the Dutch-German border area near Tegelen and Reuver. Excursion guide INQUA-SEQS Symposium '*The Dawn of the Quaternary*', 16-21 June 1996, Kerkrade, the Netherlands, Geological Survey of the Netherlands, 1-13.

Westerhoff, W.E., Cleveringa, P., Meijer, T., Van Kolfschoten, T. & Zagwijn, W.H. 1998. The Lower Pleistocene fluvial (clay) deposits in the Maalbeek pit near Tegelen, the Netherlands. *Mededelingen Nederlands Instituut voor Toegepaste Geowetenschappen TNO* 60: 35-70.

Zagwijn, W.H. 1957. Vegetation, climate and time-correlations in the Early Pleistocene of Europe. *Geologie en Mijnbouw,* 19: 233-244.

Zagwijn, W.H. 1960. Aspects of the Pliocene and Early Pleistocene vegetation in the Netherlands. *Mededelingen van de Geologische Stichting Serie C*-III-1-No. 5.

Zagwijn, W.H. 1963a. Pleistocene stratigraphy in the Netherlands, based on changes in vegetation and climate. *Verhandelingen Koninklijk Nederlands geologisch mijnbouwkundig genootschap, Geologische Serie* 21-2: 173-196.

Zagwijn, W.H. 1963b. Pollen-analytic investigations in the Tiglian of the Netherlands. *Mededelingen Geologische Stichting*, N.S. 16: 49-69.

Zagwijn, W.H. 1989. The Netherlands during the Tertiary and the Quaternary: A case history of coastal lowland evolution. *Geologie en Mijnbouw,* 68: 107-120.

Zagwijn, W.H. 1998. Borders and boundaries: a century of stratigraphical research in the Tegelen – Reuver area of Limburg (The Netherlands). *Mededelingen Nederlands Instituut voor Toegepaste Geowetenschappen TNO* 60: 19-34.

6. Fluvial responses to external forcing: examples from the French Massif Central, the Texas Coastal Plain (USA), the Sahara of Tunisia, and the Lower Mississippi Valley (USA)

MICHAEL D. BLUM
Department of Geosciences, University of Nebraska, Lincoln, Nebraska, USA
ERIC C. STRAFFIN
Department of Geosciences, Edinboro University of Pennsylvania, Edinboro, Pennsylvania, USA

1 INTRODUCTION

Fluvial deposits provide a readily studied record of landscape evolution, and fluvial response to external forcing has long been a topic of interest in the Quaternary science community. At a most fundamental level, it is widely recognized that alluvial channel size, shape, and the spatial distribution of depositional landforms reflect discharge regimes, stream power, and sediment loads. Moreover, alluvial channels respond to changes in these fundamental controls through a series of morphological and sedimentary adjustments, and by storing or removing sediments from valley axes (Schumm, 1965; Knox, 1983; 1996a; Schumm & Brackenridge, 1987; Starkel, 1991; Bull, 1991; Vandenberge, 1995; Blum & Törnqvist, 2000). Climate change, glaciation, tectonism, and relative base-level change are the primary external forcing mechanisms that can alter discharge regimes, stream power, and/or sediment loads.

Within the context of this basic process framework a number of factors make interpretation of fluvial response to external forcing less than straightforward (Knox, 1983; Bull, 1991; Schumm, 1991; Blum & Törnqvist, 2000). Schumm (1991), for example, uses fluvial examples to illustrate a number of issues that can impact interpretations of Earth history. These include (a) spatial and temporal scale, (b) convergence of responses given different external forcing mechanisms, (c) divergence of response following similar external inputs, (d) differential sensitivity to external inputs, and (e) complexity of response due to internal system dynamics. Issues such as these suggest that fluvial responses to external forcing may be geographically circumscribed, non-deterministic due to antecedent conditions, and non-linear due to internal complex response mechanisms (Fig. 1).

This paper discusses four case studies of fluvial landscape evolution in medium to large systems from different climatic-tectonic settings. Our goal is to illustrate a variety of fluvial responses and the dominant signatures present in these different fluvial systems. However, in doing so two caveats must be kept in mind. First, this paper only draws on case studies we have participated in, which are not representative of the entire range of possibilities (for a similar approach in mostly upland non-alluvial rivers, see Bull, 1991). Second, many of the details, local and otherwise, that might be considered central to each case study have been left out of the discussion so as to distill some general themes (see the primary references for these details).

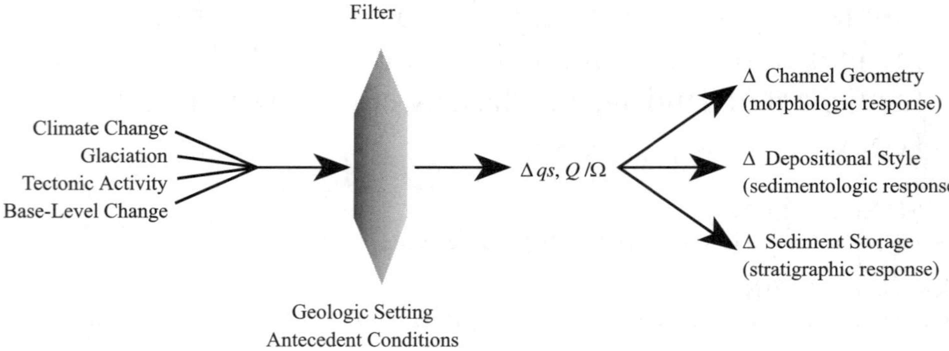

Figure 1. General model for fluvial responses to external forcing mechanisms. Forcing mechanisms, shown as higher-order independent controls on the left, are passed through a filter that defines how they are actually translated into changes in the lower-order independent controls of discharge (Q)/ stream power (Ω) and sediment supply (qs). Changes in these variables actually determine fluvial morphologic, sedimentologic, and stratigraphic responses.

2 LOIRE RIVER, MASSIF CENTRAL OF FRANCE

The upper Loire River, Massif Central of France (Fig. 2a), provides a record of landscape evolution within a midlatitude, mostly unglaciated continental interior. Since the study area is far-removed from the effects of sea-level change, and in a relatively quiescent tectonic setting, the dominant controls on changes in fluvial systems through time should be climate change and its effects on discharge regimes and sediment loads. The following is distilled from the work of Straffin et al. (1999), Colls (1999), and Straffin (2000).

2.1 *General Setting*

The Loire is one of the largest rivers in Western Europe, with a total drainage area of ~120,000 sq. km. The Loire and its major tributary, the Allier, head in the Massif Central at > 1500 m elevation, then flow north until joining at Never. After emerging from the Massif, the Loire continues northwest across the Paris Basin, then cuts across the Amorican Massif before discharging into the Atlantic. Within the Massif Central, the Loire is a 'wandering gravel bed river' (*sensu* Church, 1983), with gravelly point bars and thin flood plain deposits of fine sand and mud. Farther downstream, especially below the Allier confluence, the Loire becomes increasingly sandy, sinuosity decreases, width-to-depth ratio increases, and the channel becomes braided.

The upper Loire drainage includes elements of Oceanic, Mediterranean, and Continental climates, each with characteristic temperature and precipitation regimes, storm types and storm tracks, and corresponding hydrological regimes. Historic period climate fluctuations are linked to changes in circulation modes and storm tracks, as controlled in part by the North Atlantic Oscillation (NAO; see Rogers, 1997). Significant floods on the Loire are rainfall-generated, and follow the passage of mid-latitude cyclones during the late Spring or early Fall (Gustard et al., 1989). Peak discharges during the historic period within the upper Loire study area exceed 1900 m^3s^{-1}, with typical bankfull stages of 4-5 m or so above the low water channel (Straffin, 2000).

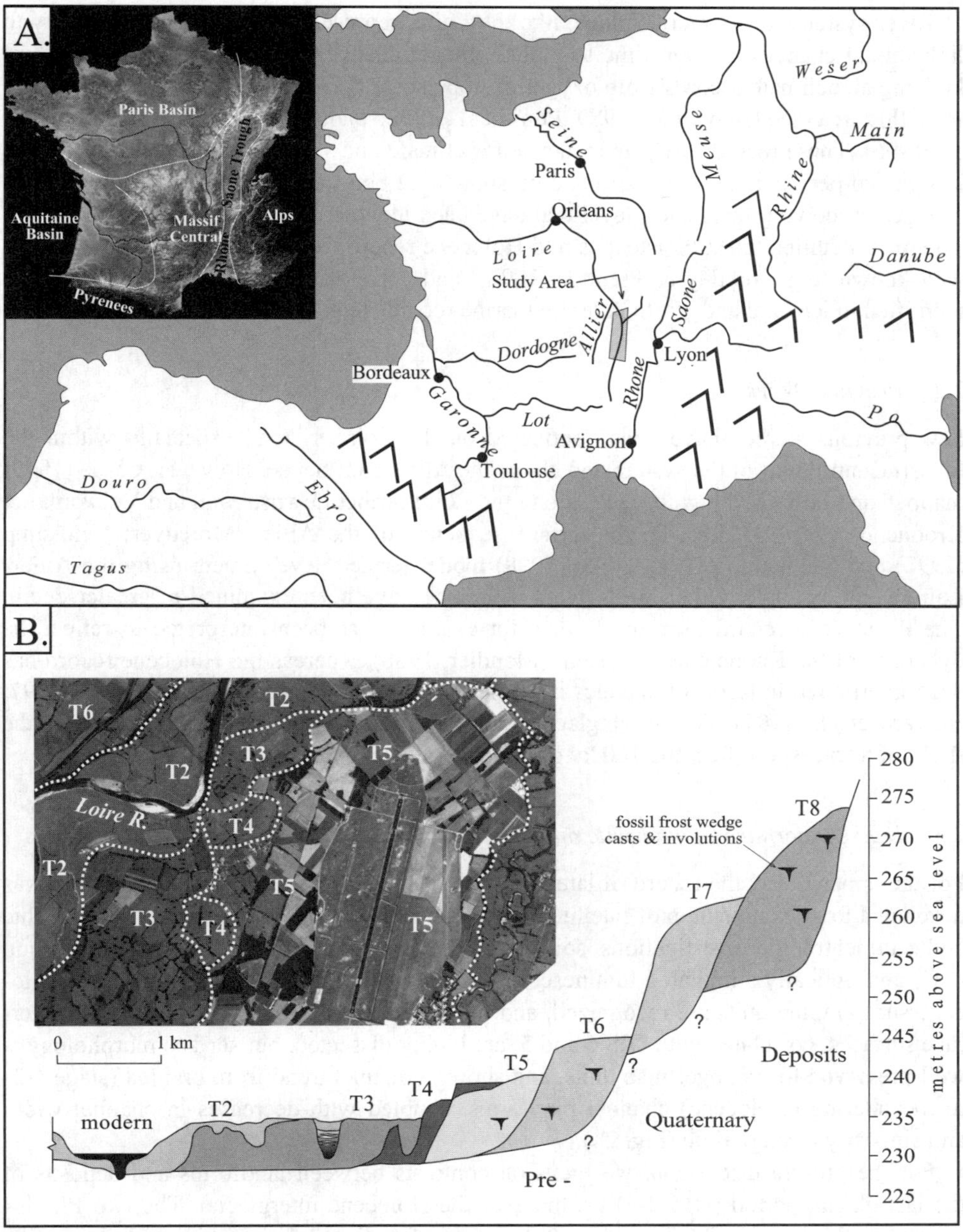

Figure 2. (a) Map showing location of Loire River and the study area discussed in the text. (b) Schematic cross-section of the Loire Valley in the Massif Central, illustrating common geomorphic, stratigraphic, and geochronological relations (modified from Straffin et al., 1999). Inset is an air photo from a small stretch of the Loire, upstream from the town of Digoin, illustrating contrasts in surface morphology.

River systems in southern France, like the Loire, represent an excellent opportunity to link fluvial changes through time to global climate change, since some of Europe's best known paleoclimate records from oxygen isotope stage (hereafter OIS) 6 to present come from this area (see Guiot et al., 1989; Pons et al., 1992; Thouveny et al., 1994; Williams et al., 1996). These records provide evidence for climate and vegetation changes through the last glacial period, and show a strong correspondence between regional climate and global changes in ice volume and sea level that have been identified from marine isotope curves. The more detailed latest Pleistocene and Holocene record from this part of Europe is also well-known (e.g. Huntley & Prentice, 1993), and correlates with the record of changes identified in ice core and North Atlantic marine records (e.g. Stuiver et al., 1995).

2.2 *Previous Work*

Few previous studies have been conducted on the Loire system, especially within the Massif Central part of the system and above the Allier confluence. However, Larue (1979) mapped and described high terraces along the Loire farther downstream, and Veldkamp & Kroonenberg (1993) describe the terrace sequence of the Allier. Moreover, Veldkamp (1992) and Veldkamp & Van Dijke (1998) model terrace development using the Allier-Loire as an example. Other large rivers in France have been examined in greater detail. The Pleistocene record from the Rhone, for example, has been interpreted to reflect the dynamics of the Rhone glacier system (Mandier, 1988), whereas the Holocene record has been interpreted in terms of changes in climate and vegetation (e.g. Bravard, 1992; 1997; Bravard et al., 1997). For the unglaciated Somme River, Antoine (1994) interprets the flight of terraces to reflect the 100 ky glacial-interglacial cycles of the last 800 ky.

2.3 *Late Quaternary Geomorphic and Stratigraphic Record*

For the upper Loire, the record of late Pleistocene and Holocene landscape evolution was developed from examination of satellite imagery and air photos, field mapping, stratigraphic and sedimentologic investigations, soil-geomorphic relations, and a series of radiocarbon (^{14}C) and optically-stimulated luminescence (OSL) ages. Seven late Pleistocene to Holocene stratigraphic units are recognized, and represent the last 140 ky (Fig. 2b). Older terrace surfaces, correlated with OIS 6 and 5, are highly dissected, but surface morphology is well-preserved for younger map units, and shows a distinct trend from braided (stage 4-2) to meandering (Holocene) channel planforms, coupled with decreases in channel width and sinuosity through time (Fig. 2b).

For the present discussion, we focus on contrasts between landforms and deposits of the last glacial period (OIS 4-2) vs. those of the Holocene interglacial. The two Pleistocene glacial-age units (T6-T5 in Fig. 2b) consist of distinct terraces with underlying fills at elevations of 12 and 17 m above the modern Loire channel, respectively. Each fill ranges from 6-10 m in thickness, and consists of channel facies dominated by multiple sets of trough cross-stratified sands and gravels. Intrastratal frostwedge casts and/or involuted bedding are common, indicating periglacial conditions during the time of deposition. Most conspicuous by their absence, fine-grained flood plain facies are not present on Pleistocene terrace surfaces, and surface soils have developed in upper channel bar facies (Fig. 3a). OSL ages of c. 40-50 ka and 18-23 ka indicate these 2 units were deposited during OIS 3-2 (Colls, 1999).

Figure 3. (a) Photograph of typical exposure in late Pleistocene deposits of the Loire, illustrating thick accumulation of sand and gravel, but a lack of flood plain facies. (b) Photograph of typical exposure in late Holocene deposits of the Loire, illustrating by thick flood plain facies (OF) resting on point bar gravel (PB), and punctuated by weakly-developed soils (P).

Early-middle Holocene units are inset into terraces and underlying fills from the OIS 2 glacial period, and characterized by high-sinuosity meandering planforms (Fig. 2b) with coarse gravelly lower point bar facies 2-4 m in thickness, overlain by chute and upper point-bar sands, and thin veneers of sandy overbank facies. Late Holocene units generally have lower sinuosity meandering planforms, with thinner channel/point bar gravels (< 3 m), but increasingly thicker flood plain sand and mud. Over the last 2000 years, fluvial activity included the scouring of low Holocene terraces and deposition of terrace veneers (as in Brackenridge, 1984) over weakly-developed paleosols (Fig. 3b), channel straightening and infilling of abandoned channels, and development of a mixed meandering to braided pattern. Multiple weakly-developed paleosols on terrace surfaces reflect episodic deposition of organics and fine sand/silt by large overbank floods, followed by periods of stability and soil development.

2.4 *Relations to External Controls*

In the Loire Valley the OIS 4-2 glacial period was characterized by 2-3 episodes of lateral migration, aggradation, and storage of gravel and coarse sand by braided streams, with each episode followed by valley incision and flood plain abandonment. Chronological controls are not sufficient to permit correlation between late Pleistocene units and specific climatic conditions. However, Straffin et al. (1999) suggest that major periods of aggradation represent high sediment concentrations along valley axes under periglacial climatic conditions, with discharge regimes dominated by moderate magnitude floods that were produced by snowmelt runoff. The conspicuous absence of fine-grained overbank facies on late Pleistocene units would seem to indicate a general paucity of deep overbank floods during time periods of net deposition and sediment storage. Intervening periods of valley incision would represent (a) diminished sediment supply due to increased slope stability or exhaustion of slope materials, (b) increased flood magnitudes and/or more frequent floods with increased competence and capacity, or (c) some combination of the above.

Valley incision accompanied the Pleistocene-Holocene transition, but little additional incision has occurred since that time. Transformations from braided to single-channel meandering planforms also accompanied the Pleistocene-Holocene transition, and throughout the Holocene coarse gravels have been concentrated along the valley axis by the lateral reworking of older deposits, with corresponding preferential downstream winnowing of finer gravel and sand. At a more detailed level, moderate magnitude floods associated with the moist oceanic climate of the early to middle Holocene produced high-sinuosity meandering channels, thin overbank deposits, and the accumulation of organic materials in cutoff channels. During the late Holocene (especially the last 2000 yrs), infrequent high-magnitude floods associated with an increasingly Mediterranean-style climate, and increasing human modification of the landscape, produced frequent episodes of channel straightening, an increasing frequency of overbank deposition, and episodic vertical accretion of fine sand and mud over paleosols.

3 COLORADO RIVER, GULF OF MEXICO COASTAL PLAIN, USA

The Gulf of Mexico coastal plain in Texas (USA) consists of coalescing Quaternary alluvial plains constructed by a series of medium to large rivers. In contrast to the erosional upland setting for the upper Loire, this is a slowly subsiding continental margin, and stra-

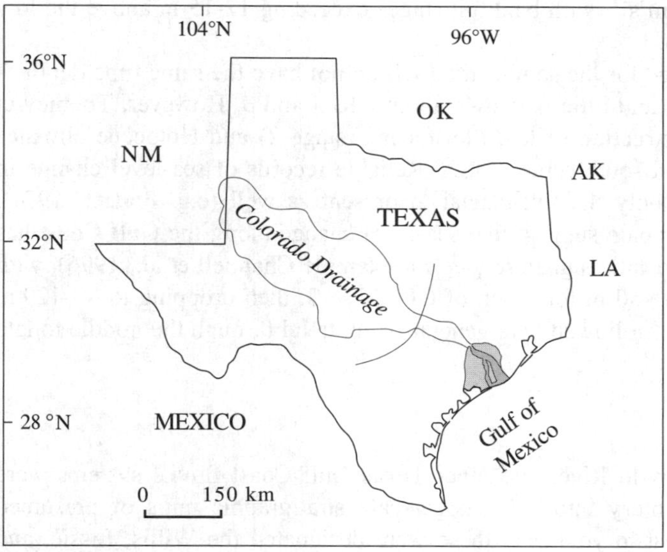

Figure 4. Map of Colorado River drainage, Texas. Stippled pattern indicates Pleistocene alluvial plain, whereas gray pattern indicates location of valley fill from the last glacial period.

tigraphic records illustrate fluvial response to interactions between climate change in lower midlatitude continental interior source regions and glacio-eustatic sea-level change in the Gulf of Mexico. The example discussed here is from the Colorado River (Fig. 4), and draws on material published by Blum (1993), Blum & Valastro (1994), Blum et al. (1994; 1995), and Blum & Price (1998).

3.1 *General Setting*

The Colorado River drains an area of ~110,000 km^2, with major tributaries originating on the Southern High Plains and Edwards Plateau of west and central Texas. Farther downstream, the Colorado trunk stream flows across the Inner Coastal Plain within a bedrock valley that contains an extensive flight of degradational terraces, then emerges onto a constructional Quaternary alluvial plain. The lower Colorado River is best described as a coarse-grained meandering stream dominated by large, high-relief (5-8 m) gravelly to coarse sandy point bars that are cross-cut by chute channels and chute bar complexes that form during flood stage (see McGowen & Garner, 1970). Modern flood plains typically consist of thick accumulations (3-5 m or more) of overbank sand and mud.

Climate of the Colorado drainage ranges from continental-semiarid on the Southern High Plains and Edwards Plateau to subtropical-subhumid along the coast (Larkin & Bomar, 1983). Precipitation maxima occur in late spring and early fall, and minima during the winter and summer months. Most major floods reflect the passage of mid-latitude cyclonic storms during the late Spring or early Fall, or the deep inland penetration of tropical cyclones during the late summer and early Fall. Exceptional flood events have also occurred during winter months in El Nino years (Blum, 1992). Largely due to proximity of the tropical moisture source, the largest floods in the historic period, prior to dam construction, exceed 8-10,000 m^3s^{-1}, with bankfull stages exceeding 12-15 m above the low water channel (Blum, 1992).

Records of climate change for the southcentral US do not have the same time depth as many records from Europe, and little is known about OIS 4 and 3. However, Toomey et al. (1993) provide a reconstruction of late Pleistocene (stage 2) and Holocene climates for the Edwards Plateau and south-central USA. Reliable records of sea-level change in the Gulf of Mexico cover only the full-glacial to present as well (e.g. Frazier, 1974). However, a variety of other data suggest that sea-level change along the Gulf Coast has closely followed a glacio-eustatic signature (see for example Chappell et al., 1996), with oscillations between –20 to –80 m for much of OIS 5c to 3, then dropping to ~ –120 m during the OIS 2 full-glacial, followed by a general rising trend through the middle to late Holocene.

3.2 *Previous Work*

Alluvial plains of the Colorado River and other Texas Gulf Coast fluvial systems were subdivided early in the century into three large-scale stratigraphic units of presumed Pleistocene age. From oldest to youngest, these were designated the Willis, Lissie, and Beaumont Formations (see Bernard & LeBlanc, 1965; Morton & Price, 1987; DuBar et al., 1991; Blum & Price, 1998 for reviews), with post-Beaumont deposits mostly undifferentiated. Early interpretations for alluvial plain surfaces were developed when the Pleistocene was divided into 4 glacial period sea-level lowstands separated by interglacial highstands. Following the model of Fisk (1944) for the lower Mississippi Valley, valley entrenchment and coastal plain sediment bypass was inferred for glacial periods, and depositional units were interpreted to represent alluvial-deltaic plains constructed during transgressions and interglacial highstands (Bernard & LeBlanc, 1965). The Beaumont Formation was assigned to the 'Sangamon' interglacial (now OIS 5), whereas post-Beaumont valleys were presumed to represent entrenchment during the 'Wisconsin' glacial (now OIS 4-2) and filling in response to Holocene sea-level rise and highstand.

Terraces within the bedrock valley on the inner coastal plain have also been the focus of investigations for almost 100 years. Early mapping differentiated a number of surfaces within the bedrock valley of the inner coastal plain on the basis of landscape position, and suggested correlations with the large-scale alluvial plain surfaces of the outer coastal plain (see DuBar et al., 1991; Blum & Valastro, 1994 for reviews). Prior to those reported on here, the most recent studies were by Baker & Penteado-Orellana (1977; 1978). They developed a complex and still widely cited model (see for example Knighton, 1998) for late Pleistocene and Holocene fluvial response to climatic change. Unfortunately, their work was mostly based on remote sensing and lacked critical stratigraphic and geochronologic control. For example, their late Pleistocene surfaces and deposits are now known to be Holocene in age, and their entire Holocene sequence is now known to represent the flood plain and channel belt of the last 200-1000 years (Blum & Valastro, 1994).

3.3 *Late Quaternary Geomorphic and Stratigraphic Record*

Throughout the upper Colorado drainage, as well as through the Inner Coastal Plain, the geomorphic and stratigraphic record of the Colorado River is typical of many stable to isostatically-uplifting continental interiors, and consists of a flight of terraces of Pliocene(?) through Holocene age that document episodes of sediment storage superimposed on a long-term trend of valley incision (Blum et al., 1994; Blum & Valastro, 1994). Farther downstream on the aggradational alluvial plain, the Beaumont Formation is now interpreted to consist of a series of cross-cutting valley fills that reflect the 100 ky glacial-interglacial cycles of coupled climate change and glacio-eustasy from OIS 9 to OIS 5 (Blum & Price, 1998).

The record from the last glacial period to present (OIS 4-1) is contained within the post-Beaumont valley, and can be subdivided into five to six stratigraphic units on the basis of geomorphic and stratigraphic relations (Fig. 5a and b). Chronological control is provided by a number of ^{14}C and thermoluminescence (TL) ages. Within the degradational bedrock valley, the OIS 4-2 glacial period is represented by 2-3 terraces with underlying fills (commonly referred to as the 'Deweyville' terraces), and terrace surfaces have well-preserved large-wavelength, high-amplitude relict meanders (Fig. 6a). By contrast, the latest Pleistocene through Holocene consists of 3 paleosol- and unconformity-bounded components within a larger complex valley fill that underlies an extensive terrace/flood plain surface. Holocene surfaces have a few preserved channel forms, and are instead dominated by cross-cutting ridge-and-swale floodplain accretionary topography (Fig. 6b).

As was the case for the Loire, glacial-period units are sedimentologically-distinct from units of Holocene age, and consist of 10-15 m of horizontally-bedded and trough cross-stratified gravel and sand, reflecting deposition by point and channel bars in a moderate sinuosity, coarse-grained meandering stream. Fines are concentrated in channel fills, and flood plain fine sand and mud is rare, with paleosols typically developed in upper point bar sand and gravel (Fig. 7a) or laterally equivalent channel fill mud. Since relict meanders and point bar morphologies are well-preserved, and paleosols are typically well-developed and intact, this lack of overbank fines is not simply a matter of non-preservation due to later erosion, but instead suggests that deep overbank flooding and vertical accretion of flood plains was not a significant process. Glacial-age units were deposited by channels graded to shorelines that were lower in elevation and farther basinward on the present-day continental shelf or at the shelf edge. In the lower 90 km of the present-day coastal plain glacial-age units are onlapped and buried by Holocene strata (Figs 5c, 7b).

Upstream from the limits of onlap, Holocene units have similar channel facies, but are most notable for thick accumulations of horizontally-bedded to undulatory fine sand and mud that represent vertical accretion in flood plain settings (Fig. 7c). A dominant theme within the Holocene, especially the last 3-5000 years, are millennial-scale periods of flood plain abandonment and soil development, followed by millennial-scale episodes of deep overbank flooding and burial of paleosols by vertical accretion facies (Fig. 7d). Hence Holocene stratigraphic units consist of a thick succession of point bar/chute bar sand overlain by flood plain facies, but also laterally equivalent veneers of fine sand and mud that rest unconformably on paleosols. Downstream from the onlap point, channel gradient decreases, and the Holocene record consists of a series of paleosol-bounded units within the overall aggradational part of the valley fill. Individual units consist of avulsion-dominated channel belts and splays encased in flood basin facies consisting of alternating paleosols and laminated mud (Aslan & Blum, 1999).

3.4 *Relations to External Controls*

The OIS 4-2 glacial period in the Colorado valley was characterized by 2-3 episodes of lateral migration, aggradation, and storage of gravel and coarse sand by meandering

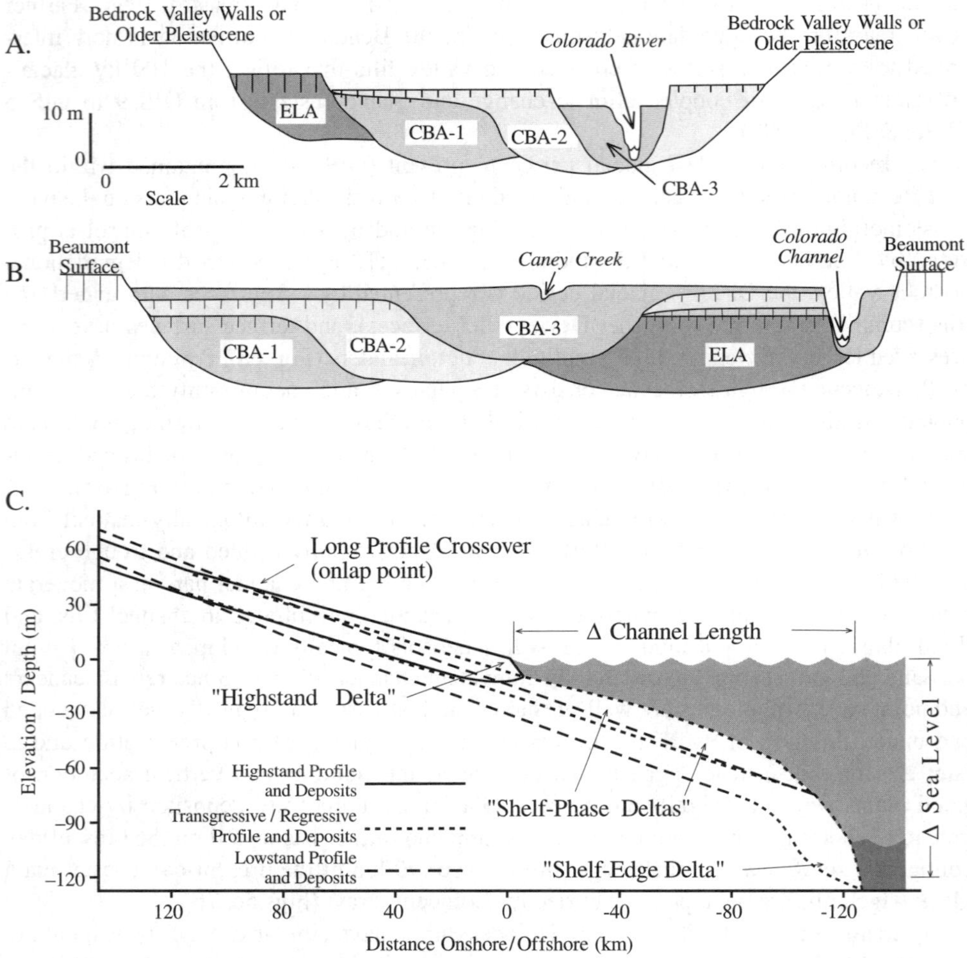

Figure 5. Late Pleistocene and Holocene stratigraphic framework for the Colorado River (last 25,000 yrs only). (a) Schematic cross-section of the bedrock-confined lower Colorado valley, illustrating common geomorphic and stratigraphic relations. (b) Schematic cross-section of the post-Beaumont valley of the lower Colorado River, downstream from the onlap point. ELA = Eagle Lake Alloformation (last full-glacial maximum), CBA-1 = Columbus Bend Alloformation Member 1 (latest Pleistocene to middle Holocene), CBA-2 = Member 2 (late Holocene), CBA-3 = Member 3 (last 1000 yrs). Nomenclature from Blum & Valastro (1994). (c) Schematic long profiles of stratigraphic units within post-Beaumont valley fill on the coastal plain, projected to temporally equivalent sea-level positions. Exercise shows probable thickness and geometric relations between different components of a valley fill sequence influenced by sea-level change.

streams with large meander wavelengths and amplitudes, and each episode was followed by valley incision and flood plain abandonment. The last major period of valley incision occurred during the latest Pleistocene, with little additional incision since that time. The absence of fine-grained flood plain facies on late Pleistocene units indicates a general paucity of deep overbank floods during time periods of net deposition and sediment stor-

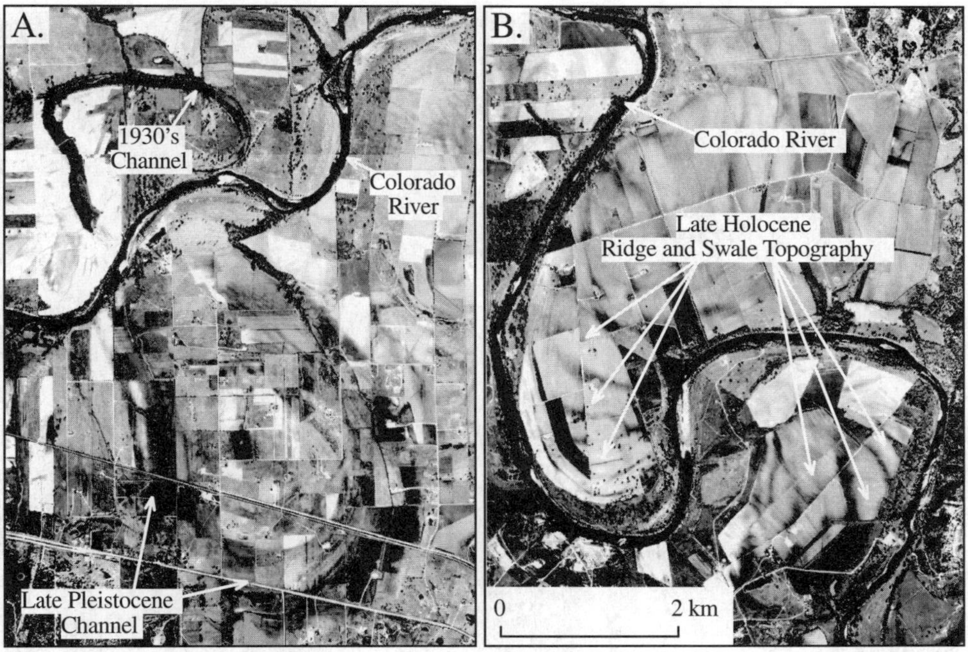

Figure 6. Air photos of the Colorado River, near La Grange, Texas. Scale is the same for both photos. (a) Late Pleistocene terrace with well-preserved abandoned channel morphology. Contrast the size of the abandoned late Pleistocene channel with a channel abandoned during the 1930's, and with the modern channel of Colorado River. (b) Late Holocene flood plain, illustrating clear ridge and swale morphology. Adapted from Blum & Valastro (1994).

age. This contrasts markedly with Holocene units, where thick accumulations of flood plain sand and mud, puncuated by multiple weakly-developed paleosols, indicate frequent periods of deep overbank flooding followed by periods of flood plain abandonment and soil development, then renewed overbank flooding and burial of paleosols.

As noted above, little is known about climate and sea-level change during OIS 4 and 3 for the south-central USA. However, for the late Pleistocene full glacial (OIS 2) through Holocene, Blum et al. (1994) and Blum & Valastro (1994) show strong correspondence between independently-derived records of climatic change (see Toomey et al., 1993), alluvial records of the upper Colorado drainage, and alluvial records from the lower Colorado River. Most importantly, the timing of changes in fluvial process regimes throughout the Colorado system, as represented by the succession of stratigraphic units, unconformities, and paleosols, corresponds with independently-identified periods of climatic and environmental change. Fundamental contrasts in morphology and depositional style between late Pleistocene and Holocene stratigraphic units also correspond to changes that accompanied the transition from glacial to interglacial conditions. Glacial period precipitation most likely resulted from mid-latitude cyclones and much of the bedrock landscape was covered by deep soils, whereas the Holocene witnessed the increased importance of high-intensity convectional storms and tropical cyclones, as well as widespread soil degradation. As a result, rates of surface runoff would have been at a minimum through the glacial

Figure 7. (a) Typical exposure in late Pleistocene deposits of the lower Colorado River, illustrating a complete lack of flood plain facies. (b) Exposure in late Pleistocene strata below the onlap point, illustrating paleosol (lowermost P) developed in point bar sand and gravel, again with no flood plain facies, then buried by Holocene flood plain facies and a second paleosol (uppermost P). (c) Typical exposure in late Holocene deposits of the lower Colorado River, illustrating thick flood plain sand. (d) Multiple paleosols (denoted by P's) developed in Holocene flood plain facies. Adapted from Blum & Valastro (1994).

period, and reached a maximum during the late Holocene when the present bedrock landscape was exposed. In contrast to the glacial period, then, Holocene fluvial systems have flashy discharge regimes, with floods that frequently exceed bankfull channels and deposit thick successions of flood plain facies.

In the bedrock-confined valley, fluvial landscape evolution during the last glacial cycle was independent of sea-level control. However, the downstream continuity of stratigraphic units and corresponding downstream changes in stratigraphic architecture (onlap of Pleistocene surfaces) indicate that sea-level changes played a role in shaping the record in far downstream reaches. Post-Beaumont valleys initially formed in response to sea-level lowering after OIS 5, when the channel incised and extended across the newly subaerial shelf. During the OIS 4-2 glacial period, the channel flowed through a laterally-confined valley that was extended to shorelines in mid-shelf or farther basinward positions (see Anderson et al., 1996), but went through 2-3 episodes of lateral migration with sediment storage, followed by renewed valley incision with terrace formation. With post-glacial sea-level rise, the lower 90 km of the present-day river system switched to a fully aggradational mode, with progressive onlap of glacial-age profiles by Holocene deposits.

4 WADIS OF THE SAHARAN MARGINS OF TUNISIA

Wadis of the northern Sahara in Tunisia discharge into a series of endorheic lacustrine basins developed in the Atlas foreland. Detailed study of wadis from the Chott Rharsa basin (Fig. 8) provides an example of the responses of desert fluvial systems to climatically-controlled fluctuations in lake levels within an active tectonic setting (Blum et al., 1998).

4.1 *General Setting*

Chott Rharsa is one of several dry lake basins within the Chotts Trough, a topographic depression that parallels the Atlas Mountains, and originated in the Miocene as a thrust-loaded foreland (Sahagian, 1993). The modern Chott Rharsa playa covers 620 km^2, with a contributing drainage area of about 22,000 km^2. The present playa floor is –10 to –25 m relative to modern sea level (RSL), and would remain an isolated basin until water surface elevations exceed –5 m RSL, at which time it would be connected to a series of playas in the Chotts Trough farther west in Algeria. Water surface elevations exceeding +25 m RSL would result in linkage with Chotts Djerid and el Fejej to the south and east, whereas water surface elevations exceeding +37 m RSL would result in spillover of the entire system to the Mediterranean Sea.

Major wadis of Chott Rharsa Basin head in the Atlas Ranges to the north and east, then flow in and around smaller-scale structural elements. Wadis are ephemeral, with broad and shallow channels that grade downstream into terminal fans, constructed where the channel discharges as sheet flow onto a low gradient playa floor (see Fig. 8c). Gypsiferous sand and mud flats (hereafter mud flats; also referred to in the literature as 'sabkhas'), interfinger up inactive wadi axes, fringe and grade into terminal fans, and extend basinward into the playa. Mud flats accrete by sheet flood and eolian processes when capillary water is at, or close to, the surface, and deflate when the water table falls. They in turn grade into gypsum crusts that demarcate the seasonally flooded playa.

Present climate of the Chotts region is arid, with the semiarid Mediterranean climate of the Atlas to the north, and the hyperarid Sahara farther south (Griffiths, 1972; Griffiths & Soliman, 1972). Annual precipitation for Chott Rharsa Basin ranges from 150-200 mm, with 250-300 mm in contributing areas to the north and east. Most precipitation occurs in the winter half-year when evaporation rates are low, but high rates of potential evaporation during the summer result in annual values that exceed precipitation by an order of magnitude. Past climates of the Sahara are reasonably well-known from a variety of lake, pollen, and other studies (see Hooghiemstra et al., 1992; Yan & Petit-Maire, 1994). In general, relatively moist semiarid phases in the Sahara correlate with interglacial/interstadial warm periods (i.e. OIS 5e, 3, and 1) of the northern hemisphere mid-latitudes, whereas arid to hyperarid phases correlate to glacial or stadial cool periods (i.e. OIS 6, 5c-4, and 2).

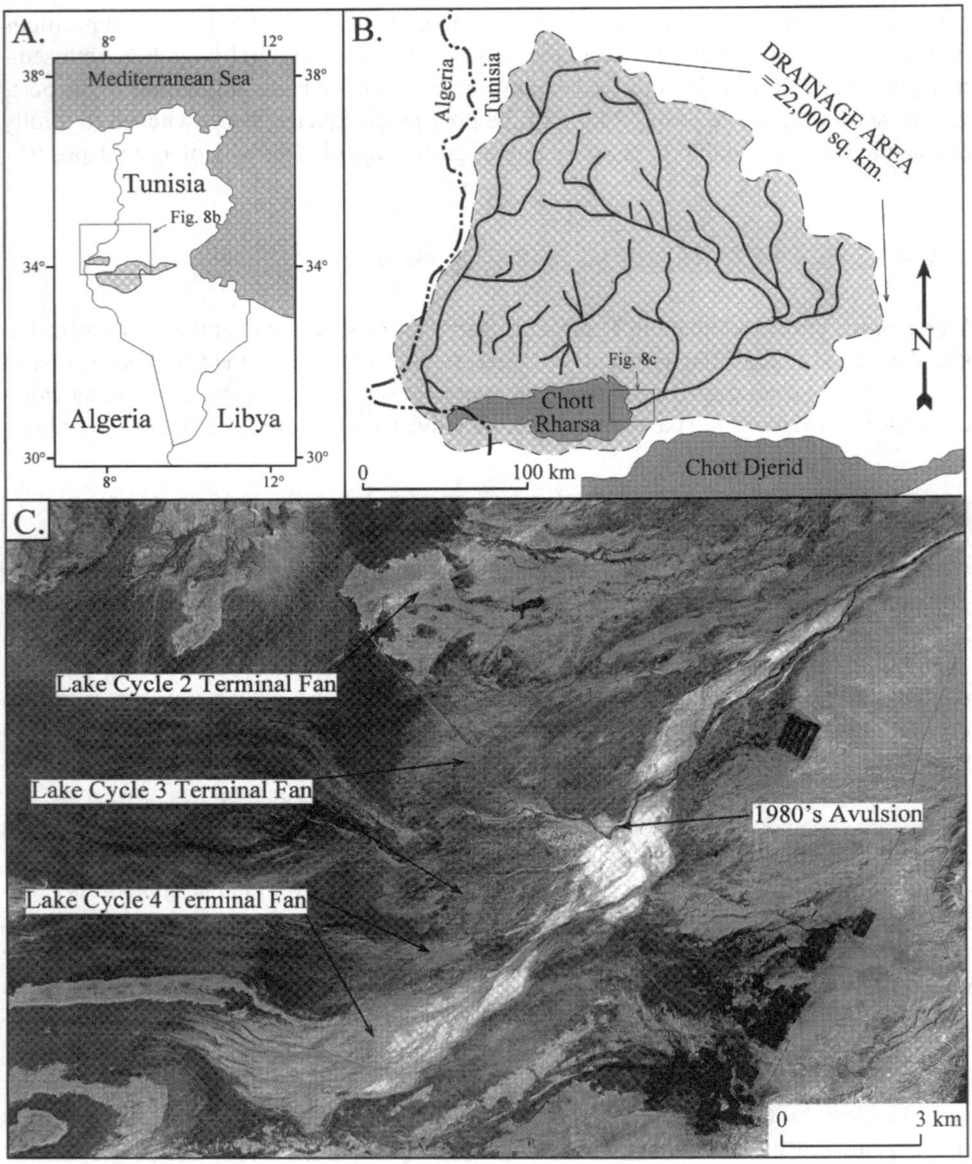

Figure 8. (a) Map of Tunisia, illustrating location of Chott rharsa basin study area. (b) Map of Chott Rharsa drainage basin in southern Tunisia. (c) SPOT panchromatic satellite image illustrating terminal fan surfaces from Lake Cycles 2, 3, and 4 along the eastern margins of Chott Rharsa basin.

4.2 *Previous Work*

A comprehensive geomorphological study of the Tunisian lake basins was conducted by Coque (1962), who identified three regionally-significant gravel-veneered pediments and terraces ('glacis'), and two lacustrine cycles ('lagoonal phases') with water surface elevations estimated at +45 m relative to modern sea level. A number of later workers recognized shoreline features from two major lacustrine cycles (e.g. Richards & Vita-Finzi, 1982; Causse et al., 1989), whereas Page (1972) identified a third lacustrine terrace. Coque (1962) and Richards & Vita-Finzi (1982) noted the Pleistocene stratigraphic record had been affected by tectonic deformation.

Coque (1962) suggested the older widespread 'glacis' were early to middle Pleistocene and middle to late Pleistocene in age based on associations with Achuelian and Mousterian artifacts, whereas the younger 'glacis' were presumed to be late Pleistocene and Holocene in age. Accordingly, his 'first lagoonal phase' was assigned a Plio-Pleistocene age and the 'second lagoonal phase' was presumed to be late Pleistocene in age. A number of workers (e.g. Page, 1972; Richards & Vita-Finzi, 1982) published radiocarbon ages of ca. 35-25 ka from mollusk shells associated with Coque's (1962) 'second lagoonal phase', and from shells associated with a 'younger regressive terrace'. More recently, Causse et al. (1989; also Fontes & Gasse, 1991) report U-series ages on mollusk shells from multiple settings within the Chotts that suggest high lake levels at ca. 150 ka and ca. 90 ka. Fontes & Gasse (1991) infer a major water level rise in Chott Djerid during the early Holocene based on radiocarbon ages from mollusk shells.

4.3 *Late Quaternary Geomorphic and Stratigraphic Record*

Blum et al. (1998) reexamined the geomorphic and stratigraphic record of Chott Rharsa using satellite imagery coupled with field investigations of geomorphic and stratigraphic relations, and a chronological framework provided by TL dating. Detailed field mapping showed the Chott Rharsa Basin preserves evidence for four distinct lacustrine cycles and wadi-lacustrine depositional sequences. Each underlies a mappable terrace, and is preserved as a series of valley fills that rest unconformably on Neogene strata (Figs 8c, 9). Terraces from the oldest Lake Cycle 1 occur as isolated remnants at the highest elevations within paleovalley axes, whereas terraces from Lake Cycles 2 and 3 are more widespread and continuous, and occur inset and at successively lower elevations. Deposits of Lake Cycle 4 occur within modern wadi axes.

Although considerable variability exists, a general transgressive-regressive pattern characterizes valley fill successions, regardless of the specific cycle. Many exposures begin at the base with medium to coarse wadi sand, with some gravel, most commonly 1-3 m in thickness. Wadi facies are commonly truncated upwards by a succession of fine sand of distal wadi/terminal fan origin that grades into mud flat facies, and/or open water lacustrine mud, with thicknesses ranging from 1-5 m. Mud flat or lacustrine facies are, in turn, commonly overlain by distal wadi/terminal fan sand, up to 2 m in thickness, that coarsen upwards to proximal wadi sand and gravel, also up to 2 m in thickness (Fig. 10). Each cycle shows a great deal of internal complexity, suggesting that smaller-scale fluctuations in lake level were superimposed on the overall large-scale transgressive-regressive patterns of lake level change. True beach landforms or deposits are rare, but strandlines rich in mollusk shells *(Cardium [Cerastoderma] glaucum)* occur where maximum highstand water levels from one lake cycle lapped on to terraces from an older cycle.

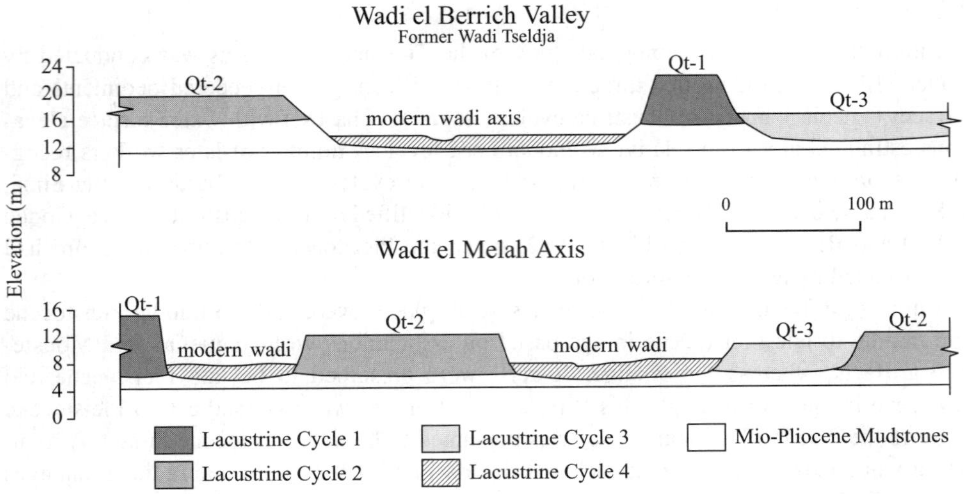

Figure 9. Schematic cross-sections of wadi axes in Chott Rharsa basin, illustrating common geomorphic and stratigraphic relations (adapted from Blum et al., 1998).

The four wadi-lacustrine cycles differ in the thickness and maximum up-valley limits of lacustrine and mud flat facies. In Lake Cycle 1, these facies occur at the highest elevations, are substantially thicker, and extend farther updip into each paleovalley axis, whereas similar facies from Lake Cycles 2 and 3 occur at progressively lower elevations, are progressively less thick, and extend upvalley progressively shorter distances. These observations indicate, in turn, that Lake Cycle 1 was the most laterally extensive, and perhaps the deepest, with successively younger cycles less extensive and perhaps not as deep. Open-water lacustrine facies from Lake Cycle 4 are not exposed on the north and east sides of the basin, but the relative elevation of mud flat facies within any paleovalley axis is several meters less than similar facies from Lake Cycle 3, which suggests Lake Cycle 4 was the smallest of the large-scale transgressive-regressive cycles.

Blum et al. (1998) also provide additional evidence for syndepositional tectonism in the basin. This includes: (a) wadi sediments from Lake Cycle 1 that drape up and over an anticlinal ridge, (b) warping of strandlines from Lake Cycles 2 and 3, and (c) mud flat deposits from Lake Cycle 3 with small intrastratal thrusts and folds, most likely from liquefaction associated with paleoearthquakes. As a consequence of tectonic activity, there have been shifts in the location of sediment input and principal fluvial depocenters through time due to lateral stepping of wadi courses around active anticlines, and a succession of wadi capture events (Fig. 11). Moreover, wadi-terminal fan surfaces from Lake Cycle 3 are incised when located over the active anticline axis, but undergoing burial by modern terminal fan and mud flat facies where located over the active synclinal axis.

Given the different interpretations provided by radiocarbon vs. U-series ages on lake cycles, a precise chronology for the Chotts has proven difficult to define. Blum et al. (1998) further complicate this issue with a limited number of thermoluminescence ages on regressive to lowstand terminal fan facies from Lake Cycles 1-3. Their data suggest a correlation between lake lowstands and glacial/stadial cool periods, and lake highstands with interglacial/interstadial warm periods over the last 200 ky (Fig. 12). Such an interpretation

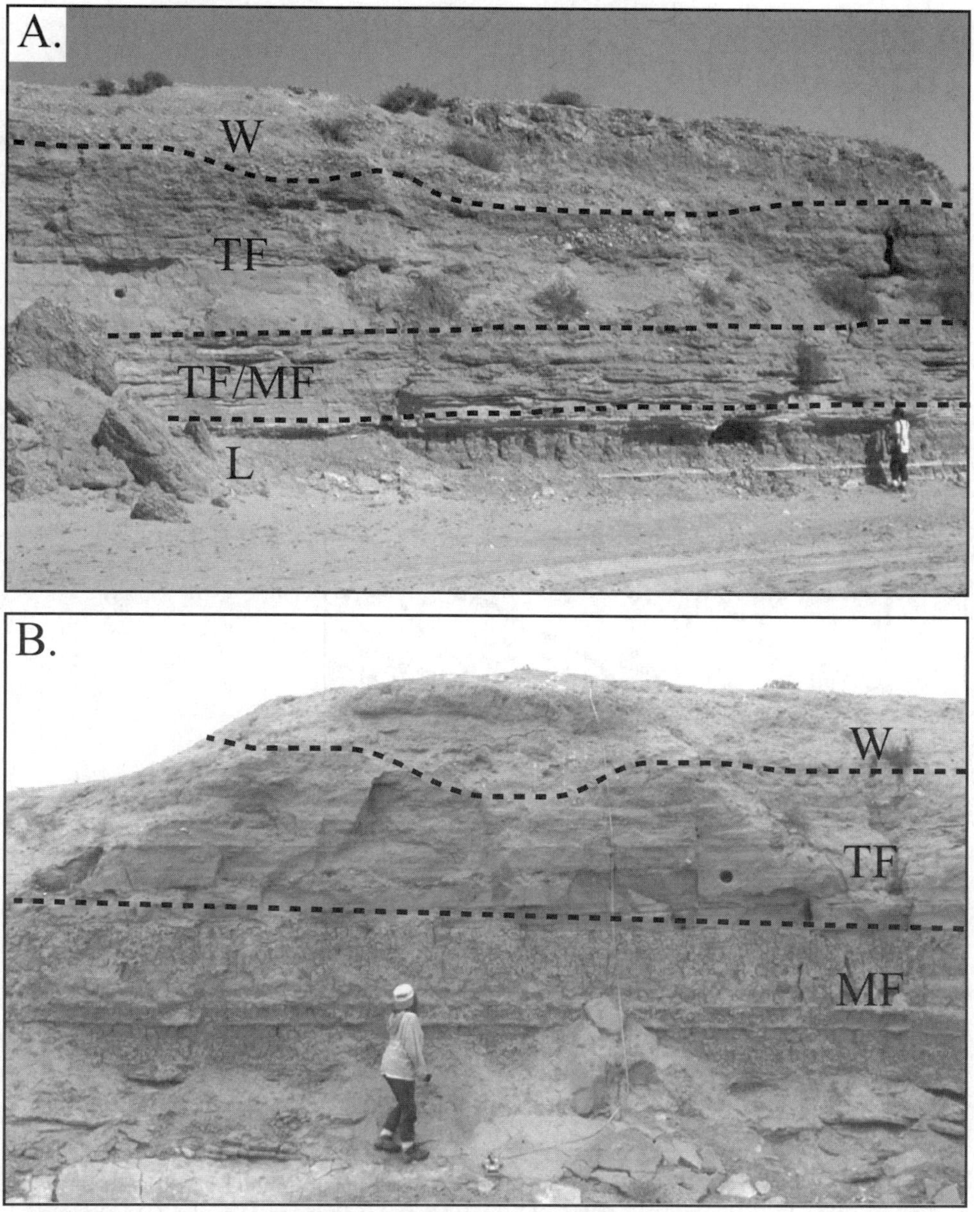

Figure 10. (a) Photograph of typical exposure in deposits from Lake Cycle 1. (b) Photograph of typical exposure in deposits from Lake Cycle 2. W = wadi sand and gravel, TF = terminal fan sand, and L = lacustrine muds and gypsum, and MF = mud flat facies.

certainly is at odds with both radiocarbon and U-series ages from these lake basins, but in agreement with other records from the Sahara (see Yan & Petit-Maire, 1994). Without additional data it is difficult to correlate Lake Cycle 4 with work done elsewhere, but it may record the sum of all relatively small-scale Sahara humid-arid subcycles since the

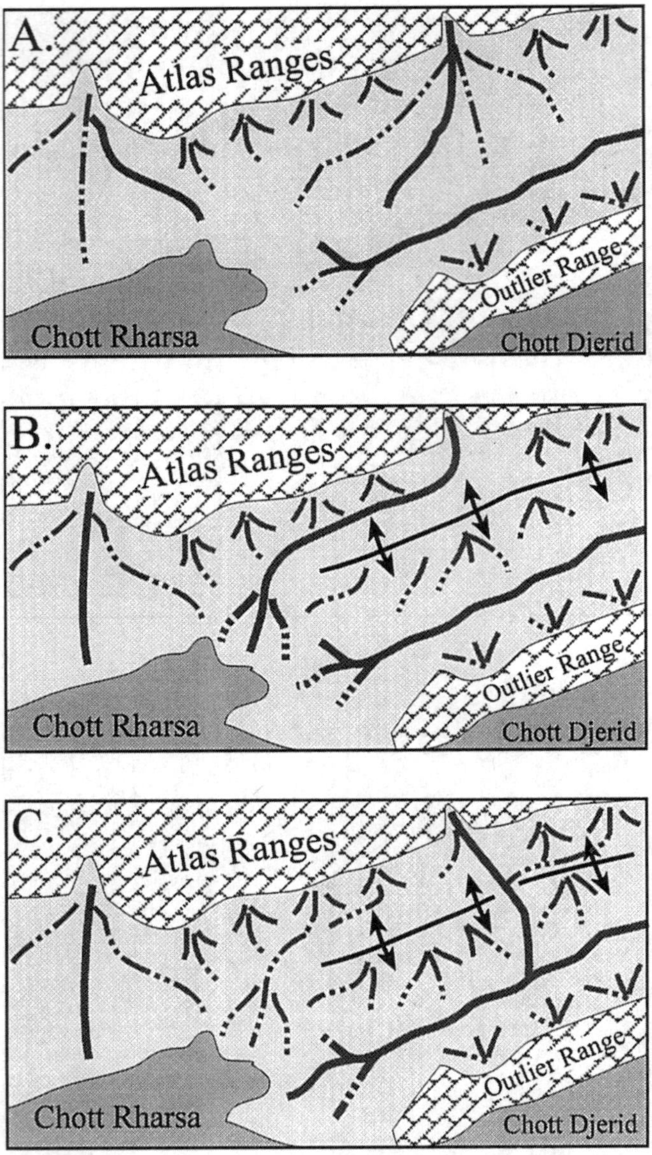

Figure 11. Model for tectonic disruption and diversion of wadi axes in the Chott Rharsa Basin due to syndepositional anticline growth followed by capture and rediversion of wadis. (a) Lake cycle 1, when the anticline had not yet started to move. (b) Lake Cycle 2, when movement forced lateral diversion of one of the main wadi axes. (c) Lake Cycle 3 to present, when stream capture rediverted one of the main wadis back to its original course.

OIS 2 full-glacial cool and dry period (e.g. Gasse & Fontes, 1992; Gasse & Van Campo, 1994). High frequency fluctuations such as these are difficult to define on the wadi-dominated aggradational NE side of Chott Rharsa basin, but have been documented on the highly deflationary southern margins (Swezey et al., 1999).

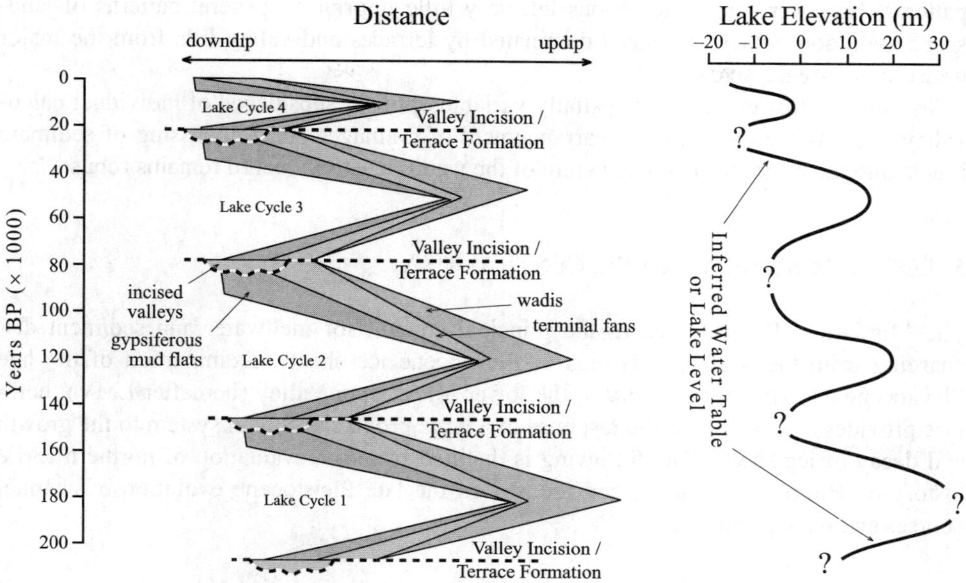

Figure 12. Summary stratigraphic and geochronologic model for transgressive-regressive wadi-lacustrine cycles of the Chott Rharsa basin (modified from Blum et al., 1998).

4.4 *Relation to External Control*

Geomorphic and stratigraphic data from Chott Rharsa Basin provide a record of four major wadi-lacustrine cycles. Individual cycles are preserved as a series of mappable terraces and incised valley fills, with stratigraphic successions that are interpreted to represent (a) lacustrine lowstand when wadi-terminal fan systems extend basinward to a deflated playa surface, (b) transgression and highstand, when wadi-terminal fan systems backstep up the valley axis, and valleys are filled with mud flat/lacustrine facies, and (c) regression and early lowstand, when wadi-terminal fan systems prograde across newly-exposed lacustrine plains.

All available chronological data place wadi-lacustrine strata from the Chott Rharsa basin within the last 2 glacial-interglacial periods. Lake Cycle 1 was the deepest and most areally-extensive, with Lake Cycles 2, 3, and 4 progressively smaller. Moreover, each successive intervening lowstand has deflated the basin to a deeper level. One possible interpretation would be that the cyclical changes in water balance over the last 2 glacial-interglacial periods are superimposed on a longer-term progressive dessication of the northern Sahara. More detailed correlations between lacustrine cycles and global to regional climate change are at present difficult. Although not without problems, the new TL ages reported by Blum et al. (1998) suggest that lacustrine highstands correlate with interglacial/interstadial warm periods (OIS 5e, 3, and 1), whereas major episodes of desiccation, basin-floor deflation, and valley incision correlate to glacial or stadial cool periods (OIS 6, 5c through 4, and 2). Regardless of precise age contsraints, it is clear that each wadi-lacustrine depositional sequence shows a great deal of internal complexity, suggesting that smaller-scale fluctuations in lake level were superimposed on the larger-scale patterns. However, such fluctuations left very little imprint on general patterns of landscape evolution, which is instead dominated by terraces and valley fills from the major transgressive-regressive cycles.

In spite of clear evidence of spatially-variable uplift of subsidence of individual paleovalley axes, as well as tectonic rearrangement of drainages and refocussing of sediment inputs into the basin, the cyclical nature of the wadi-lacustrine record remains robust.

5 LOWER MISSISSIPPI RIVER, CENTRAL USA

The Mississippi River served as the principal conduit for meltwater and sediment discharging from the southern margins of Pleistocene ice sheets. Examination of the late Pleistocene and Holocene history of the lower Mississippi Valley (hereafter LMV) therefore provides an example of the response of a large proglacial fluvial system to the growth and decay of ice sheets. The following is distilled from a reevaluation of northern LMV history by Blum et al. (2000), and focuses on the late Pleistocene evolution of channel courses and major landforms.

5.1 *General Setting*

The Mississippi River is a continental-scale fluvial system with the 4th largest drainage area in the world (Fig. 13a). The Mississippi heads in northern Minnesota, within the formerly glaciated Central Lowlands of the US, then flows south through a relatively narrow bedrock-confined valley for some 1000 km. Within this part of its course, the Mississippi is joined by the Missouri River, which drains much of the central and northern Rocky Mountains and US Great Plains. South of Cape Girardeau, Missouri, and especially below the confluence with the Ohio River, the Mississippi flows through a broad alluvial valley (the LMV) with an extensive suite of Late Quaternary fluvial and eolian landforms and deposits (Autin et al., 1991; Saucier, 1994). Farther downstream, the Mississippi picks up the Arkansas and Red Rivers, which drain the central and southern Rockies and Great Plains, before discharging into the Gulf of Mexico.

The northern half of the LMV is bounded on the west by the Ozark Plateau in Missouri and Arkansas, and to the east by the uplands of southern Illinois, Kentucky, and Tennessee (Fig. 13b). This part of the LMV can be subdivided into two distinct channel courses, the Western and Eastern Lowlands, which are separated by Crowleys Ridge, a north-south oriented linear topographic high. The Eastern Lowlands in turn contains two entry points, the Bell City-Oran and Thebes Gaps, with the latter a narrow bedrock gorge that contains the modern river. The Western Lowlands course, as well as the Eastern Lowlands course through Bell City-Oran Gap, consist of multiple Pleistocene braided stream surfaces (referred to as "valley trains") of the Mississippi River, that are cross-cut and/or reoccupied by Holocene meander belts of smaller streams. The Eastern Lowlands course downstream from Thebes Gap contains Pleistocene valley trains and Holocene meander belts of the combined Mississippi and Ohio Rivers (Autin et al., 1991; Saucier, 1994).

A variety of external controls might be called upon to explain the origin and genetic significance of Late Quaternary landforms and deposits in the LMV. First and foremost are the fundamental changes in discharge regimes and sediment flux that reflect the growth and decay of continental ice sheets, and the drainage of large proglacial lakes,

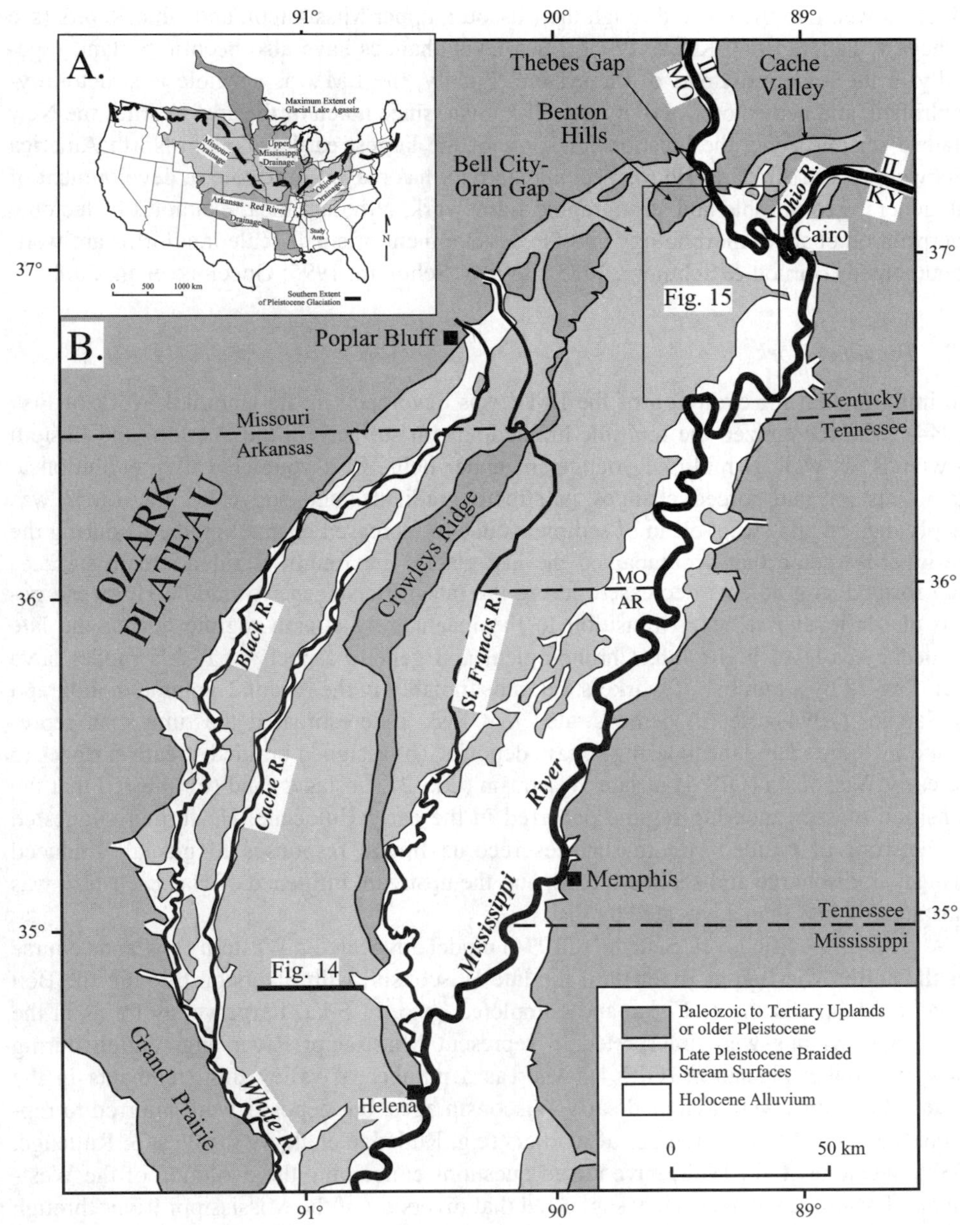

Figure 13. (a) Drainage basin of the Mississippi River relative to Pleistocene ice margins, and the location of study area in the Lower Mississippi Valley (LMV). Also shown is the approximate extent of Glacial Lake Agassiz, the largest of the late Pleistocene proglacial lakes. (b) Channel courses of the northern LMV (generalized from Saucier, 1994), showing prominent physiographic features, and the extent of Pleistocene valley trains (undifferentiated) vs. the Holocene meander belt. The Western Lowlands lie between the Ozark Plateau and Crowleys Ridge, whereas the Eastern Lowlands lies east of Crowleys Ridge and west of the uplands in southern Illinois, Kentucky, and Tennessee. Locations of study sites discussed in the text as shown.

which flowed into the LMV through the Missouri, upper Mississippi, and Ohio Rivers (see Kehew & Teller, 1994; Knox, 1996b). Sea-level changes have also been important, especially in the lowermost part of the system. Finally, the LMV as a whole is structurally-controlled, and neotectonic activity is well-known, since much of the area overlies the New Madrid seismic zone, the location for some of the largest earthquakes in North America (Schweig & van Arsdale, 1996). Tectonic activity has played little role in development of the general geomorphic and stratigraphic framework, although local examples of tectonic overprint on channel morphology and the development of small-scale landforms are well-documented (Burnett & Schumm, 1983; Boyd & Schumm, 1995; Guccione et al., 2000).

5.2 *Previous Work*

An initial model for evolution of the LMV was developed in the landmark work of Fisk (1944), who recognized the multiple braided stream surfaces in the Western and Eastern Lowlands, as well as multiple younger meander belts. Fisk coupled valley evolution to glacio-eustasy and forced changes in stream gradients, and suggested the LMV was deeply incised and swept clean of sediments due to increased channel gradients during the sea-level lowstand that accompanied the last glacial maximum. Braided stream surfaces then formed as gradients decreased during the latest Pleistocene to middle Holocene period of sea-level rise, with transition to the meandering stream regime during the late Holocene sea-level highstand. Chronological and genetic aspects of Fisk's model have been revised by a number of workers, but most notably in the recent 2-volume monograph by Saucier (1994). He (a) demonstrated that Fisk underestimated the time span represented by braided and meandering stream deposits, (b) assigned braided stream surfaces to the early Wisconsin (OIS 4) or late Wisconsin (OIS 2) substages, and (c) inferred that the transition to a meandering regime occurred in the early Holocene. He further suggested the network of braided stream surfaces records fluvial responses to glacially-induced changes in discharge and sediment load, and the upstream influence of glacio-eustasy was substantially less than advocated by Fisk.

At a more specific level, Saucier's (1994) model advocated a Western Lowlands course for the entire Mississippi River until the late Wisconsin, with diversion through the Bell City-Oran Gap starting c. 17 ka, and completed by c. 11.5 ka. Large valley trains in the Western Lowlands were interpreted to represent Mississippi River aggradation during early Wisconsin glaciation (OIS 4), whereas a number of valley train remnants in the Eastern Lowlands were assigned early Wisconsin ages, but deposits were inferred to represent the Ohio River. A number of workers (e.g. Rutledge et al., 1985; West & Rutledge, 1987; Guccione et al., 1988) have raised questions concerning the evolution of the Western and Eastern Lowlands. They suggested that diversion of the Mississippi River through Bell City-Oran Gap occurred prior to 60 ka, and the Western Lowlands was abandoned during the early to middle Wisconsin (OIS 4-3), then served as an overflow course during periods of late Wisconsin flooding (OIS 2).

5.3 *Late Quaternary Geomorphic and Stratigraphic Record*

LMV landforms and deposits are well-mapped but have been notoriously difficult to date for a variety of reasons. Hence Blum et al. (2000) examined relationships between loess units and fluvial deposits in the Western and Eastern Lowlands, so as to test the alternative models summarized above. Loess units are valuable as stratigraphic markers in the LMV because of the widespread nature of loess deposition on pre-existing landscapes, and an abundance of previous work on loess geochronology (see Rutledge et al., 1996). Blum et al. (2000) support and expand on the views presented by Rutledge et al. (1985) and West & Rutledge (1987).

For the Western Lowlands, stratigraphic relationships show that major braided stream surfaces are either (a) covered by the well-known 'Sangamon paleosol' that began developing during the OIS 5 interglacial, plus the early to middle Wisconsin Roxana Silt and late Wisconsin Peoria Loess units (Fig. 14), (b) covered by only thin veneers of late Wisconsin Peoria Loess, or (c) have no loess cover at all. Thinning and fining relationships for the early to middle Wisconsin and late Wisconsin loess units clearly indicate that silty materials were derived from deflation of valley trains in the Eastern Lowlands Bell City-Oran Gap course. Moreover, thick slackwater deposits in tributary valleys are late Wisconsin in age, and accumulated in valleys that were deeply incised during the early to middle Wisconsin. These data are interpreted to indicate that major valley trains of the Western Lowlands are either late Illinoian (OIS 6) or late Wisconsin (OIS 2) in age, and that valley trains of early to middle Wisconsin age (OIS 4 and 3) are not present, but rather the Western Lowlands was deeply incised during this time by tributary streams. Hence the Western Lowlands was the primary Mississippi course during OIS 6 deglaciation, but was completely abandoned during OIS 4 and 3 in favor of the Bell City-Oran Gap course, then episodically reoccupied as a secondary overflow course during OIS 2.

Even though the Mississippi flowed through the Eastern Lowlands during OIS 4 and 3, valley trains from this time period are not widely preserved, with the exception of a small remnant where the river broke through a gap in the southernmost margins of Crowleys Ridge and reentered the Western Lowlands. Instead, major valley trains in the Eastern

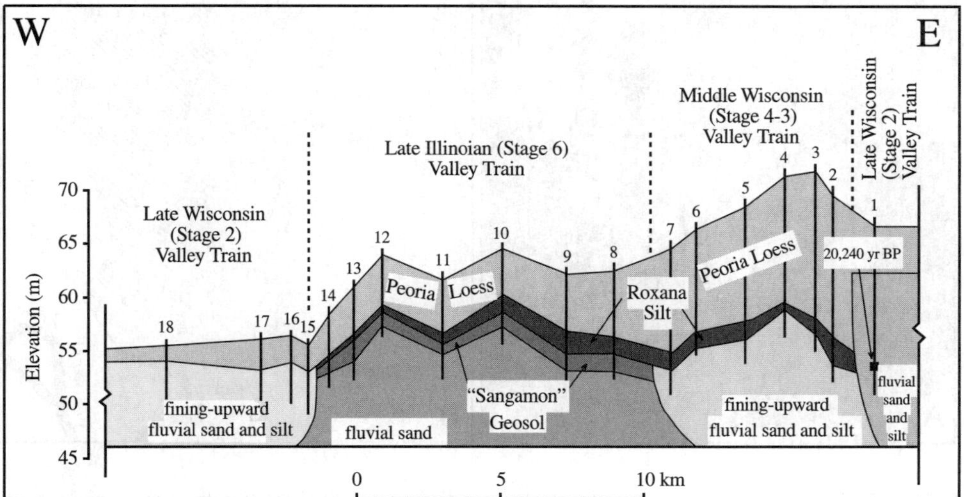

Figure 14. Cross-section illustrating stratigraphic relations between loess units and fluvial deposits in core transect (location in Fig. 13b). Data and interpretations provided by D. Wysocki, US Natural Resources Conservation Service (personal communication). Adapted from Blum et al. (2000).

Lowlands are OIS 2 in age, since they are covered by thin increments of late Wisconsin Peoria Loess resting directly on fluvial sand, or they are not loess covered at all (Fig. 15). A series of discontinuous terrace surfaces in the Bell City-Oran Gap course are now thought to represent dissected remnants of the primary OIS 2 full-glacial maximum and early late-glacial course of the Mississippi, or combined Mississippi and Ohio Rivers (Sikeston Ridge and correlative surfaces; see Figs 15, 16), whereas younger continuous braided stream surfaces of the Bell City-Oran Gap and Thebes Gap courses represent valley trains of late OIS 2 and earliest Holocene age (Pvl-2 and Pvl-1; Figs 15, 16). A variety of geomorphic data indicate that Thebes Gap was first breached during the latest part of OIS 2, while the main course remained in the Bell City-Oran Gap, then was permanently occupied in the earliest Holocene. Shortly thereafter, the Mississippi was transformed into the well-known meandering stream of the Holocene.

5.4 *Relations to External Controls*

Blum et al.'s (2000) model, summarized in Fig. 16, suggests (a) that major valley train surfaces of the LMV are either OIS 6 or OIS 2 in age, (b) there is no evidence for fluvial

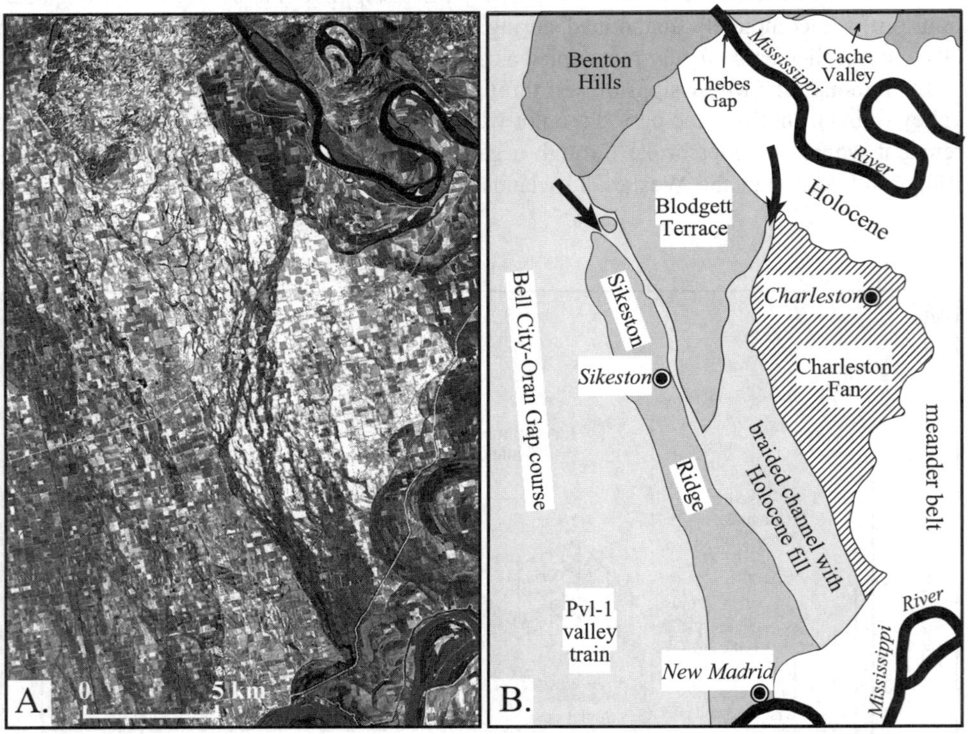

Figure 15. Satellite image and generalized map illustrating principal landforms of the Eastern Lowlands. Sikeston Ridge and Blodgett terrace are remnants of the full-glacial and early late-glacial valley train surface, Charleston Fan is a deposit that reflects the first breakthrough by the Mississippi River at Thebes Gap, and is late-glacial in age, whereas other valley trains and braid channels are latest Pleistocene to earliest Holocene age (compare with Fig. 16). Adapted from Blum et al. (2000).

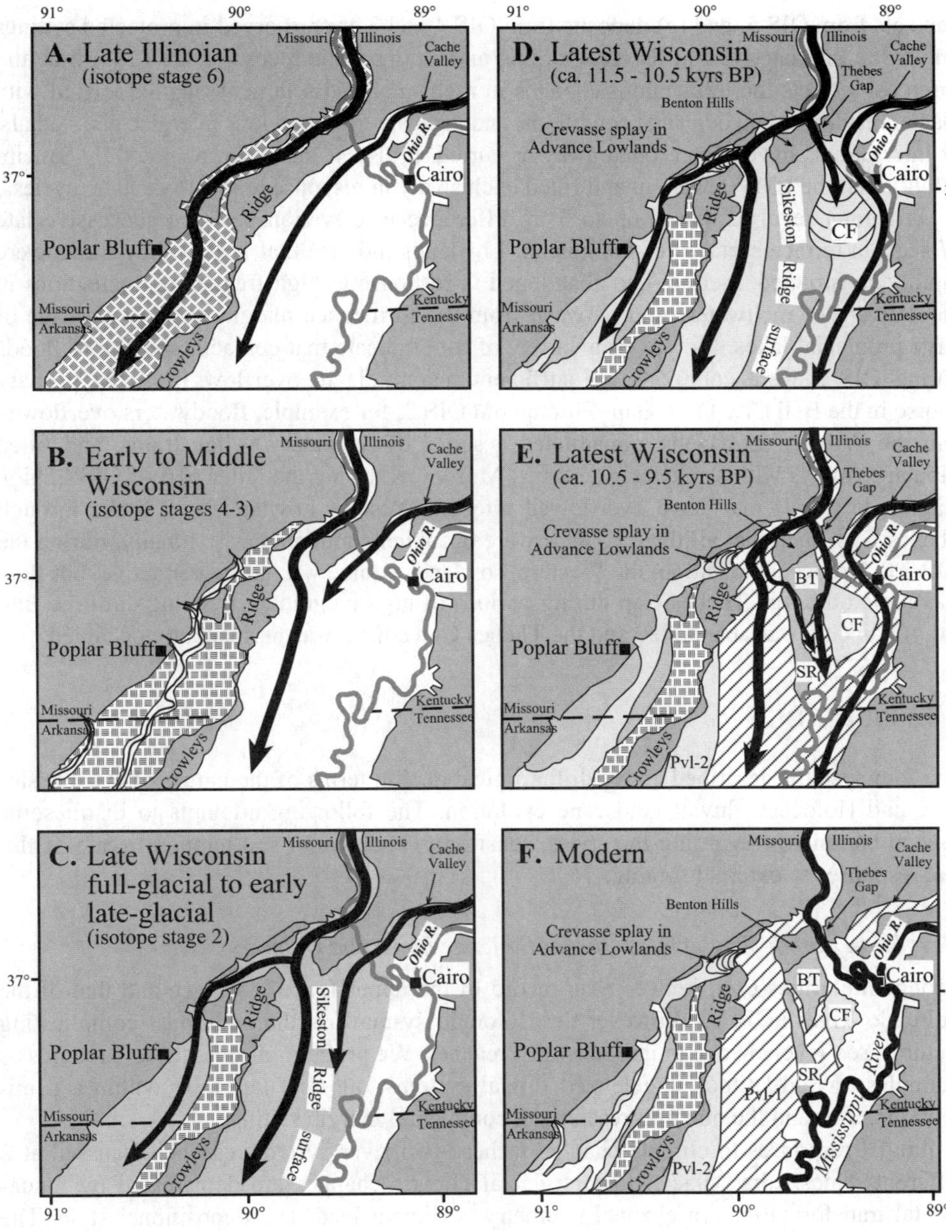

Figure 16. Model for evolution of the lower Mississippi River from Illinoian (OIS 6) deglaciation to present (from Blum et al., 2000). (a) The late Illinoian (OIS 6) glacial stage. (b) The early to middle Wisconsin (OIS 4-3) substage. (c) The late Wisconsin full-glacial to early late-glacial period (OIS 2). (d) Latest Wisconsin late-glacial period, perhaps ca. 11.5 to 10.5 ka. (e) Latest Wisconsin late-glacial period, after 10.5 ka. (f) The modern setting, with reinterpreted ages for key maps units. CR = Crowley's Ridge or northern unnamed extension into Missouri, Sikeston Ridge surface = full-glacial and early late glacial braided stream remnants (remnants indicated by SR and BT), Pvl-2 and Pvl-1 = late glacial to earliest Holocene braided stream surfaces as mapped by Saucier (1994a). Location of Ohio River is schematic only.

deposits from OIS 5, and (c) deposits from OIS 4 and 3 are preserved in protected settings only. One inference is that the geomorphic and stratigraphic record of the LMV first and foremost reflects the tremendous changes in meltwater and sediment flux associated with glacial maximum and deglacial conditions, and deposits from previous interglacials, stadials, or interstadials have been eroded away or buried. It also seems clear, as noted by Saucier (1994), that the LMV never cut and filled exclusively in response to glacio-eustatic cycles.

At a finer level of detail, up to 5 m difference in elevation between successive late Wisconsin terrace surfaces of the Eastern Lowlands indicate that large valley trains were rapidly constructed, incised, and abandoned in response to high-frequency fluctuations in the delivery of meltwaters and outwash from the former ice margin and the drainage of large proglacial lakes. Moreover, a variety of data indicate that deglacial meltwater floods during OIS 2 were commonly of sufficient magnitude to overflow from the primary course in the Bell City-Oran Gap. Throughout OIS 2, for example, floodwaters overflowed into the Western Lowlands, constructed a series of secondary valley trains, and filled tributary valleys with slackwater deposits. Moreover, during the latter part of OIS 2, high magnitude floods may have overflowed into the Western Lowlands and broke through Thebes Gap, such that all three courses were occupied simultaneously. Finally, during the OIS 2 to Holocene transition, the Western Lowlands course was no longer active, but flow was routed through Thebes Gap during periods of high magnitude flooding until the Bell City-Oran Gap was abandoned and the Thebes Gap course was permanently occupied.

6 DISCUSSION

The four systems described above differ a great deal in terms of the nature of late Pleistocene and Holocene fluvial landscape evolution. The following attempts to distill some general comments regarding the styles, sensitivity, and spatial and temporal scale of fluvial responses to external forcing.

6.1 *Mid-Latitude Unglaciated Fluvial Systems: Loire and Colorado Rivers*

Strong similarities exist between the record of the upper Loire in France and that of the Colorado River in Texas, however the Colorado system has the additional complicating factor of sea-level change in its lowermost reaches. We prefer to discuss these 2 records as examples of mid-latitude unglaciated fluvial systems, one that lies fully within a continental interior and the other representing a continental margin setting.

Fluvial responses to climate change in these two river systems can be discussed at 2 different scales. First, glacial to interglacial climate changes are demarcated by fundamental transformations in channel geometry, sediment load, and depositional style. The observation of channel metamorphosis at the end of the Pleistocene and through the Holocene is nothing new, and is widely recognized throughout Europe and elsewhere (see Starkel, 1991 for a review). We stress instead that glacial-period fluvial deposits are dominated by coarse-grained channel belts (braided channels in the Loire, large-wavelength meandering in the Colorado), and systematically lack fine sandy and muddy flood plain deposits, whereas Holocene units, especially those of late Holocene age, contain appreciable thicknesses of flood plain facies. Hence, deep overbank flooding was not a significant process during the glacial period, most floods were contained within bankfull channel perimeters, and fine sediments were mostly bypassed through the system. By contrast, deep overbank flooding has been increasingly important during the Holocene, and a significant volume of fine sediment is sequestered in flood plain settings.

Second, glacial vs. interglacial periods resulted in different amplitudes and frequencies of fluvial adjustment to climate change. For example, high-amplitude but low-frequency adjustments characterized the relatively long glacial period, with 2-3 extended periods of lateral migration and sediment storage puncuated by episodes of significant bedrock valley incision. By comparison, low-amplitude but high-frequency adjustments have been more typical of the relatively short Holocene, when there has been little net valley incision or net changes in sediment storage, and only minor adjustments in channel geometry, but relatively frequent changes in the magnitude and frequency of flood events and periods of overbank flooding. The record of these changes consists of the alternating periods of flood plain stabilization with soil development versus deep overbank flooding and burial of paleosols by flood plain facies. This type of high-frequency signature seems to be mostly lacking in landforms and deposits from the glacial period.

The glacial vs. interglacial dichotomy appears to be a pattern common to many large unglaciated midlatitude fluvial systems, even though recent studies of the climate record per se shows similar frequencies for climate cycles in the last glacial period vs. the Holocene (e.g. Bond et al., 1997). However, since 70-80% or more of any 100 ky glacial cycle is characterized by substantial ice volume, cooler land surface temperatures, and cooler-smaller ocean basins, glacial-period process regimes should be regarded as the norm, and the interglacial regime is relatively unique (see Porter, 1989). To take this a step further, we suggest (a) that large unglaciated midlatitude fluvial systems are essentially in long-term equilibrium with the glacial period environment, (b) that glacial period fluvial systems respond to major changes in climate, discharge regimes, and sediment loads through cycles of lateral migration and sediment storage, punctuated by episodes of valley incision, but (c) they may be relatively insensitive to higher-frequency changes. Short interglacials like the Holocene are, by comparison, periods of relatively lower-amplitude climate changes, but fluvial systems appear to exhibit a greatly increased sensitivity to subtle changes in climatically-controlled discharge regimes. This increased sensitivity produces frequent periods of disequilibrium that result in episodes of flood plain stability and soil development punctuated by high magnitude flooding and burial of paleosols.

The contrast between a continental interior setting for the Loire and a continental margin setting for the Colorado River illustrates the nature of interactions between climate and sea-level change. It is clear that sea-level change had no impact on the basic sequence of depositional, erosional, and pedogenic events in the lower Colorado valley, since these events can be traced from the degradational bedrock valley to the aggradational alluvial plain. However, in far downstream reaches, sea-level changes have exerted a strong control on geometric relationships between stratigraphic units, unconformities, and paleosols ('stratigraphic architecture') such that late Pleistocene through middle Holocene units, deposited when the shoreline was lower in elevation and farther basinward, are buried by those of late Holocene age, with depth of burial increasing downstream. Here again, the glacial period process regime, with river long profiles graded to shorelines on the mid-shelf or farther seaward, should be regarded as the norm and highstand conditions are relatively unique. Moreover, this style of interaction between upstream climatic controls on discharge regimes and sediment flux vs. downstream sea-level controls on stratigraphic architecture is common to many large river systems that discharge to subsiding continental margins (Blum & Törnqvist, 2000; also Törnqvist, 1998; Törnqvist et al., 2000).

6.2 *Subtropical Desert Fluvial Systems: Tunisian Wadis*

For the Tunisian wadis example, climatically-controlled changes in lake level play the fundamental role in landscape evolution. Several aspects of this particular case study deserve special mention.

First, the late Pleistocene record in Chott Rharsa basin most likely reflects major orbitally-forced changes in the Saharan water balance, and the dominant geomorphic and stratigraphic signature is the transgressive and regressive responses of wadi-terminal fan systems to high-amplitude/low frequency changes in lake level. Available chronological control is contradictory at best, however TL ages reported in Blum et al. (1998) suggest that major lacustrine highstands correlate with interglacial/interstadial warm periods, whereas major episodes of desiccation correlate to glacial or stadial cool periods. This interpretation contradicts previous workers, but is in general agreement with other data from the Sahara (e.g. Yan & Petit-Maire, 1994). Lower-amplitude/higher-frequency transgressive-regressive events are clearly superimposed on the larger cycles, but these are only recorded as minor facies changes within individual valley fills, and have not left a significant imprint on the overall pattern of landscape evolution.

Second, water table lowering and deflation of the basin floor has played a special role in formation of the geomorphic and stratigraphic record. In the Chott Rharsa basin, falling lake levels clearly do not promote incision of wadis, but rather progradation of wadi-terminal fan systems over the former lacustrine plain, since a coarsening-upward succession of terminal fan to wadi facies caps each depositional sequence. Nevertheless, the next cycle begins with wadis in a deeply incised position. We envision a situation where drying up of the lake basin does not signify the driest part of the climatic cycle, but instead the water table continues to fall until the basin floor begins to deflate, which lowers base level for the wadis, triggers incision and headward erosion, and in doing so partitions a new valley axis that will served as the locus of accumulation for the next lacustrine cycle. Similar water table- and deflation-controlled base levels may play an important role in many modern and ancient desert basins.

Finally, the active tectonic setting provides an opportunity to contrast the role of climate change vs. tectonism in formation of the geomorphic and stratigraphic record. It is clear that syndepositional tectonism has served to focus sediment input into the basin, change the locus of sediment input through time, partitioned the basin into topographic highs and subbasins, and will ultimately control the locations where the wadi-lacustrine record will be eroded away or preserved as part of the landscape and foreland basin fill. However, the cyclical nature of the wadi-lacustrine record, as measured by the sequence of transgressive and regressive events, is largely unaffected and remains robust throughout the basin. This further supports the conclusion that sedimentary rhythms reflect major climate changes.

6.3 *Continental-Scale Proglacial System: The Lower Mississippi River*

Landforms and deposits of the Lower Mississippi Valley testify to the tremendous scale, complexity, and dynamism of this continental-scale proglacial fluvial system, as do the changing interpretations through time. The revised model presented in Blum et al. (2000) is only the most recent installment.

In Fisk's (1944) model, the Mississippi responded deterministically to changes in sea level, and upstream controls on discharge and sediment load were not as significant. By contrast, Saucier (1994) deemphasized sea-level controls, suggested that aggradation and construction of major braided stream surfaces correlate to waning of stadial (OIS 4) or full glacial (OIS 2) conditions. Blum et al. (2000) argue instead that the LMV geomorphic and stratigraphic record represents relatively short periods of time (~10 ka.) at the end of successive 100 ky glacial cycles (OIS 6 and OIS 2) when there were tremendous fluctuations in discharge and sediment loads delivered from the ice margin. In their model, deposits from the last interglacial (OIS 5) or the early to middle Wisconsin (OIS 4 and 3) are not present at all or preserved in protected settings only. Holocene landforms will also most likely be removed during the next glacial cycle, and not preserved in the stratigraphic record over longer periods of time.

The above interpretation does not deny the importance of significant fluvial deposition during other periods of time, only that other periods of time are poorly preserved. This distinction between what actually happened and what is preserved and resolvable is an important one. For example, OIS 4 and 3 fluvial landforms and deposits are not widespread, but are present in selected settings that have been protected from subsequent erosion. The presence of isolated remnants clearly indicates (a) that fluvial landforms and deposits of this age were at one time more widespread and continuous, (b) that the LMV was not incised and swept clean of sediments due to sea-level fall in a manner envisioned by Fisk (1944) and other sea-level control advocates, and (c) these deposits were actually eroded and removed due to the tremendous meltwater and sediment discharges during deglaciation, a time period that actually correlates to rapid sea-level rise. Hence, the series of depositional and erosional events in the LMV largely reflect upstream controls on discharge and sediment flux, and were independent of sea-level change through most of the valley. However, the record is biased towards preservation of the erosional and depositional events associated with deglaciation at the end of successive 100 ky glacial cycles.

Sea-level influences did not extend as far upstream as many would envision, but the question remains as to how far did they extend? Using Saucier's (1994) maps, and Blum et al.'s (2000) reevaluation of age relations, late Pleistocene surfaces are onlapped and buried by Holocene strata some 250-350 km upstream from the average highstand shoreline, or 350-450 km upstream from the present river mouth. Following the criteria discussed in Blum & Törnqvist (2000), this would be defined as the maximum upstream limits of any significant sea-level influence.

7 CONCLUSION

The case studies described above illustrate clear contrasts in the spatial and temporal scale, sensitivity, complexity, and styles of fluvial responses to external forcing. We are in an exciting time when there is new enthusiasm for understanding the dynamics of Earth systems over time, and integration of a variety of data- and model-based approaches over a variety of spatial and temporal scales. However, as the variability in these case studies implies, we are far from the point where we can apply general models to reconstruct changes in past forcing mechanisms from the fluvial record with a great deal of confidence. Indeed, Blum & Törnqvist (2000) suggest that empirical research on fluvial response to external forcing lags behind (a) the global change community's understanding of the magnitude and frequency of climate and sea-level change as forcing mechanisms; (b) the sequence-stratigraphic community's desire to interpret ancient alluvial strata in terms of external forcing; and (c) the modeling community's ability to generate numerical and physical models of surface processes and their stratigraphic results.

It is therefore time for the Quaternary fluvial community to mount concerted efforts to catch up, and distinct challenges await at both ends of the temporal scale. On the one hand, it will be important to keep pace with research in high-frequency climate and sea-level change. Do fluvial systems of varying scale respond to, and preserve records of, Dansgaard-Oeschger and other high-frequency events that are so prominent in records from Greenland, the North Atlantic, and elsewhere? On the other hand, it is equally important to study longer-wavelength cycles of landscape evolution. For example, most detailed studies of fluvial response to climate change during the last 3-4 decades has focussed on the last 20-30 kyr, a relatively short period that encompasses only the tail end of a single 100 kyr glacial to interglacial cycle. Very little is actually known about fluvial responses to complete glacial-interglacial cycles, or how the first 70-80% of the last cycle preconditioned fluvial processes during the last 20-30%. To address such issues, we should avoid simplistic model-driven interpretations, and instead develop a comprehensive range of well-defined and well-dated empirical records of fluvial system changes from a variety of settings, from sources to sink, that can be compared with independently-derived records of changes in external forcing mechanisms themselves.

REFERENCES

Anderson, J.B., Abdulah, K., Sarzalejo, S., Siringin, F. & Thomas, M.A. 1996. Late Quaternary sedimentation and high-resolution sequence stratigraphy of the East Texas Shelf. In: De Batist, M. & Jacobs, P. (eds) *Geology of Siliciclastic Shelf Seas*. Special Publication 117, Geologic Society of London, 95-124.

Antoine, P. 1994. The Somme Valley terrace system (northern France): A model of river response to Quaternary climatic variations since 800,000 BP. *Terra-Nova,* 6: 453-464.

Aslan, A. & Blum, M.D. 1999. Contrasting styles of Holocene avulsion, Texas Gulf Coastal Plain. In: N.D. Smith & J.J. Rogers (eds) *Current Research in Fluvial Sedimentology: Proceedings of 6th International Conference on Fluvial Sedimentology*. Special Publication 28, International Association of Sedimentologists, 193-209.

Autin, W.J., Burns, S.F., Miller, B.J., Saucier, R.T. & Snead, J.J. 1991. Quaternary geology of the Lower Mississippi Valley. In: R.B. Morrison (ed.) *Quaternary Non-Glacial Geology of the Conterminous United States*, v. K-2. The Geology of North America: Geological Society of America, Boulder, Colorado, 547-582.

Baker, V.R. & Penteado-Orellana, M.M. 1977. Adjustment to Late Quaternary climate change by the Colorado River of Central Texas. *J. Geol.,* 85: 395-422.

Baker, V.R. & Penteado-Orellana, M.M. 1978. Fluvial sedimentation conditioned by Quaternary climate change in Central Texas. *J. Sed. Pet.*, 48: 433-461.

Bernard, H.A. & LeBlanc, R.J. 1965. Resume of the Quaternary Geology of the Northwestern Gulf of Mexico Province. In: Wright, H.E. & Frey, D.G. (eds) *The Quaternary of the United States*. Princeton University Press, 137-185.

Blum, M.D. 1992. *Modern Depositional Environments and Recent Alluvial History of the lower Colorado River, Gulf Coastal Plain of Texas*: Unpub. Ph.D. Dissertation, University of Texas at Austin.

Blum, M.D. 1993. Genesis and architecture of incised valley fill sequences: a Late Quaternary example from the Colorado River, Gulf Coastal Plain of Texas. In: Weimer, P. & Posamentier, H.W. (eds) *Siliciclastic Sequence Stratigraphy: Recent Developments and Applications*. Memoir 58, American Association of Petroleum Geologists, 259-283.

Blum, M.D. & Valastro, S.Jr. 1994. Late Quaternary sedimentation, Lower Colorado River, Gulf Coastal Plain of Texas. *Geo. Soc. Amer. Bull.*, 106: 1002-1016.

Blum, M.D. & Price, D.M. 1998. Quaternary alluvial plain construction in response to interacting glacio-eustatic and climatic controls, Texas Gulf Coastal Plain. In: Shanley, K.W. & McCabe, P.J. (eds) *Relative Role of Eustasy, Climate, and Tectonism in Continental Rocks*: Special Publication 59, Society for Sedimentary Geology, 31-48.

Blum, M.D. & Tornqvist, T.E. 2000. Fluvial response to climate and sea-level change: a review and look forward. *Sedimentology*. 47(suppl.): 2-48.

Blum, M.D., Toomey, R.S. III, & Valastro, S.Jr. 1994. Fluvial response to Late Quaternary climatic and environmental change, Edwards Plateau of Texas. *Paleogeogr. Paleoclimatol. Paleoecol.* 108: 1-21.

Blum, M.D., Morton, R.A. & Durbin, J.M. 1995. 'Deweyville' terraces and deposits of the Texas Gulf Coastal Plain: *Trans. Gulf Coast Assoc. Geo. Soc.*, 45: 53-60.

Blum, M.D., Kocurek, G., Deynoux, M., Swezey, C., Lancaster, N., Price, D., & Pion, J.-C. 1998. Quaternary wadi-lacustrine-aeolian depositional cycles and sequences, Chott Rharsa Basin, Tunisia. In: Alsharhan, A.S., Glennie, K., Whittle, G.L. & St.C. Kendall C.G. (eds) *Quaternary Deserts and Climatic Change*. Rotterdam: Balkema, 539-552.

Blum, M.D., Guccione, M.J., Wysocki, D. & Robnett, P.C. (in press). Late Pleistocene Evolution of the Mississppi Valley, Southern Missouri to Arkansas. *Geo. Soc. Amer. Bull.*

Bond, G., Showers, W., Cheseby, M., Lotti, R., Almasi, P., deMenocal, P., Priore, P., Cullen, H., Hajdas, I. & Bonani, G. 1997. A pervasive millennial-scale cycle in North Atlantic Holocene and glacial climates. *Science,* 278: 1257-1266.

Boyd, K.J.F. & Schumm, S.A. 1995.*Geomorphic evidence of deformation in the northern part of the New Madrid seismic zone.* Survey Professional Paper 1538-R, United States Geological Survey.

Brackenridge, G.R. 1984. Alluvial stratigraphy and radiocarbon dating along the Duck River, Tennessee. *Geo. Soc. Amer. Bull.*, 95: 9-25.

Bravard, J-P. 1992. Rapport No. 2: Les rythemes d'evolution morphologique des valles Francaises au tardiglaciaire et l'Holocene. *Bull. Assoc. Geogr. Franc.,* Paris, 207-226.

Bravard, J.-P. 1997. Tectonique et dynamique fluviale du Wurm a l'Holocene a la confluence Saone-Rhone (France).*Geographie Physique et Quaternaire*, 51: 315-326.

Bravard, J.-P., Bourrlely, A.V. & Franc, O. 1997. Paleodynamique du site fluvial de Lyon depuis le tardiglaciaire. In: Bravard, J.-P. & Prestreau, D. (eds) *Dynamique du Paysage – Entretiens de Geoarchaeologie*. Documents d'Archaeologie en Rhone-Alpes. CNRS – University of Lyon II, 177-201.

Bull, W.B. 1991. *Geomorphic responses to climate change*. Oxford University Press.

Burnett, A.W. & Schumm, S.A. 1983. Alluvial-river response to neotectonic deformation in Lousiana and Mississippi. *Science,* 272: 49-50.

Causse, C., Coque, R., Fontes, J.-C., Gasse, F., Gibert, E., Ben Ouezdou, H. & Zouari, K. 1989. Two high levels of continental waters in the southern Tunisian chotts at about 90 and 150 ka. *Geology,* 17: 922-925.

Chappell, J., Omura, A., Ezat, T., McCulloch, M., Pandolfi, J., Ota, Y. & Pillans, B. 1996. Reconciliation of late Quaternary sea levels derived from coral terraces at Huon Peninsula with deep sea oxygen isotope records. *Earth Planet. Sci. Lett.*, 141: 227-236.

Church, M. 1983. Patterns of instability in a wandering gravel bed channel. In: J.D. Collinson & J. Lewin (eds) *Modern and Ancient Fluvial Systems*. Special Publication 6, International Assocciaion of Sedimentologists, 169-180.

Colls, A.E. 1999. *Optical dating of fluvial sediments from the Loire Valley, France.* Unpublished M.Sc. thesis, School of Geography, University of Oxford.

Coque, R. 1962. *La Tunisie Presaharienne.* Paris: Armand Colin.

DuBar, J.R., Ewing, T.E., Lundelius, E.L., Otvos, E.G. & Winker, C.D. 1991. Quaternary geology of the Gulf of Mexico Coastal Plain. In: R.B. Morrison (ed.) *Quaternary Non-Glacial Geology of the Conterminous United States*, v. K-2. The Geology of North America: Geological Society of America, 583-610.

Fisk, H.N. 1944. *Geological Investigations of the Alluvial Valley of the Lower Mississippi River*: Mississippi River Commission, United States Army Corps of Engineers.

Fontes, J.C. & Gassse, F. 1991. PALHYDAF (Paleohydrology in Africa) program: objectives, methods, major results. *Paleogeogr. Paleoclimatol. Paleoecol.* 84: 191-215.

Frazier, D.E. 1974. *Depositional Episodes – Their Relationship to the Quaternary Stratigraphic Framework in the Northwestern Portion of the Gulf of Mexico Basin*: Geological Circular, Bureau of Economic Geology, University of Texas at Austin, 74-1.

Gasse, F. & Fontes, J. 1992. Climatic changes in northwest Africa during the last deglaciation (16-7 ka BP). In: E. Bard & W. Broecker (eds), *The Last Deglaciation: Absolute and Radiocarbon Chronologies*. Berlin: Springer-Verlag, 295-325.

Gasse, F. & Van Campo, E. 1994. Abrupt post-glacial climate events in West Africa and North Africa monsoon domains. *Earth Planet. Sci. Lett.,* 126: 435-456.

Griffiths, J.F. 1972. The Mediterranean zone. In: Griffiths, J.F. (ed.) *Climates of Africa.* World Survey of Climatology, 10. Amsterdam: Elsevier, 37-74.

Griffiths, J.F. & Soliman, K.H. 1972. The northern desert. In: Griffiths, J. F. (ed.) *Climates of Africa.* World Survey of Climatology, 10. Amsterdam: Elsevier, 75-132.

Guccione, M.J., Prior, W.L. & Rutledge, E.M. 1988. Crowley's Ridge, Arkansas. In: Hayward, O.T. (ed.) *South-Central Section of the Geological Society of America, Centennial Field Guide*. Geological Society of America, Boulder, Colorado, 225-230.

Guccione, M.J., Van Arsdale, R.B. & Hehr, L.H. (in press). Origin and age of the Manila high and associated Big Lake 'Sunklands', New Madrid seismic zone, north-eastern Arkansas. *Geo. Soc. Amer. Bull.*

Guiot, J. Pons, A., de Beaulieu, J.L. & Reille, M. 1989. A 140,000-year continental climate reconstruction from two European pollen records. *Nature,* 338: 309-313.

Gustard, A., Roald, L.A., Demuth, S., Lumadjeng, H.S. & Gross, R. 1989. *Flow Regimes from Experimental and Network Data.* Institute of Hydrology, Wallingford, UK.

Hooghiemstra, H., Stalling, H., Agwu, C. & Dupont, L. 1992. Vegetational and climatic changes at the northern fringe of the Sahara 250,000-5000 years BP: evidence from 4 marine pollen records located between Portugal and the Canary Islands. *Rev. Paleobot. Palynol.,* 74: 1-53.

Huntley, B. & Prentice, I.C. 1993. Holocene vegetation and climates of Europe. In: Wright Jr., H.E., Kutzbach, J.E., Webb III, T., Ruddiman, W.F., Street-Perrott, F.A. & Bartlein, P.J. (eds) *Global Climates since the Last Glacial Maximum*. University of Minnesota Press, Minneapolis, 136-168.

Kehew, A.E. & Teller, J.T. 1994. History of late-glacial runoff along the southwestern margin of the Laurentide Ice Sheet. *Quat. Sci. Rev.,* 13: 859-878.

Knighton, D. 1998. *Fluvial Forms and Processes*. Edward Arnold, London. 2nd Ed.

Knox, J.C. 1983. Responses of river systems to Holocene climates. In: Wright, H.E. & Porter, S.C. (eds) *Late Quaternary environments of the United States: Vol.2, The Holocene*. University of Minnesota Press, 26-41.

Knox, J.C. 1993. Large increases in flood magnitude in response to modest changes in climate. *Nature,* 361: 430-432.

Knox, J.C. 1996a. Fluvial systems since 20,000 yrs BP. In: Gregory, K.J., Starkel, L. & Baker, V.R. (eds) *Global Continental Palaeohydrology*. John Wiley and Sons, 87-108.

Knox, J.C. 1996b. Late Quaternary upper Mississippi River alluvial episodes and their significance to the lower Mississippi River system: *Eng. Geo*. 45: 271-274.

Larkin, T.J. & Bomar, G.W. 1983. *Climatic Atlas of Texas*. Texas Department of Water Resources.

Larue, J.P. 1979. *Les napes alluviales de la Loire et de ses affluents dans le Massif Central et dans le Sud de bassin Parisien: étude géomorphologique*. Thése géographie, Universite Clermont II.

Mandier, P. 1988. *Le Relief de la Moyenne Vallée du Rhone au Tertiaire et au Quaternaire*. Doc. du BGRM, No 151. Paris.

McGowen, J.H. & Garner, L.E. 1970. Physiographic features and stratification types of coarse-grained point bars, modern and ancient examples. *Sedimentology,* 14: 86-93.

Morton, R.A. & Price, W.A. 1987. Late Quaternary sea-level fluctuations and sedimentary phases of the Texas Coastal Plain and shelf. In: Nummedal, D. & Pilkey, O.H. (eds) *Sea-level Fluctuations and Coastal Evolution*. Special Publication 15, Society of Economic Paleontologists and Mineralogists, 181-198.

Page, W.D. 1972. *The Geological Setting of the Archaeological Site at Oued el Akarit and the Paleoclimatic significance of Gypsum Soils, Tunisia*. Unpublished MS Thesis, University of Colorado.

Pons, A., Guiot, J., de Beaulieu, J.L. & Reille, M. 1992. Recent contributions to the climatology of the last glacial-interglacial cycle based on French pollen sequences. *Quat. Sci. Rev.,* 11: 439-448.

Porter, S.C. 1989. Some geological implications of average Quaternary glacial conditions. *Quat. Res.*, 32: 245-262.

Richards, G.W. & Vita-Finzi, C. 1982. Marine deposits 35,000-25,000 years old in the Chott el Djerid, southern Tunisia. *Nature,* 295: 54-55.

Rogers, J.C. 1997. North Atlantic storm track variability and its association to the North Atlantic Oscillation and climate variability of Northern Europe. *J. Climate*, 10: 1635-1647.

Rutledge, E.M., West, L.T. & Oamkupt, M. 1985. Loess deposits on a Pleistocene-age terrace in eastern Arkansas. *J. Soil Sci. Soc. Amer.,* 49: 1231-1238.

Rutledge, E.M., Guccione, M.J., Markewich, H.W., Wysocki, D.A. & Ward, L.B. 1996. Loess stratigraphy of the Lower Mississippi Valley. *Eng. Geo.,* 45: 167-183.

Sahagian, D. 1993. Structural evolution of African basins: stratigraphic synthesis. *Basin Res.* 5: 41-54.

Saucier, R.T. 1994. *Geomorphology and Quaternary Geologic History of the Lower Mississippi Valley*: Waterways Experiment Station, US Army Corps of Engineers.

Schumm, S.A. 1965. Quaternary Paleohydrology. In: Wright, H.E. & Frey, D.G. (eds) *The Quaternary of the United States*. Princeton University Press, 783-794.

Schumm, S.A. 1991. *To Interpret the Earth: Ten Ways to be Wrong*. Cambridge University Press.

Schumm, S.A. & Brackenridge, G.R. 1987. River responses. In: Ruddiman, W.F. & Wright, H.E. (eds) *North America and Adjacent Oceans During the Last Deglaciation, Geological Society of America* v. K-3. The Geology of North America: Geological Society of America, Boulder, Colorado, 221-240.

Schweig, E.S. & Van Arsdale, R.B. 1996. Neotectonics of the upper Mississippi embayement. *Eng. Geo.,* 45: 185-204.

Starkel, L. 1991. Long-distance correlation of fluvial events in the temperate zone. In: Starkel, L., Gregory, K.J. & Thornes, J.B. (eds) *Temperate Palaeohydrology*. John Wiley and Sons, 473-495.

Starkel, L. 1983. The reflection of hydrologic changes in the fluvial environment of the temperate zone during the last 15,000 years. In: Gregory, K.J. (ed.) *Background to Palaeohydrology*. John Wiley and Sons, 213-235.

Straffin, E.C. 2000. *Fluvial Response to Climate Change and Human Activities, Burgundy Region of France*. Unpublished Ph.D. Dissertation, University of Nebraska-Lincoln.

Straffin, E.C., Blum, M.D., Colls, A. & Stokes, S. 1999. Alluvial stratigraphy of the Loire and Arroux Rivers, Burgundy, France. *Quaternaire,* 10: 271-282.

Stuiver, M., Grootes, P.M. & Braziunas, T.F. 1995. The GISP $\delta^{18}O$ climate record of the past 16,500 years and the role of the sun, ocean and volcanoes. *Quat. Res.,* 44: 341-354.

Swezey, C.S., Lancaster, N., Kocurek, G., Deynoux, M., Blum, M.D., Price, D.M. & Pion, J.-C. 1999. Response of aeolian systems to Holocene climatic and hydrologic changes on the northern margin of the Sahara: a high resolution record from the Chott Rharsa basin, Tunisia. *The Holocene,* 9: 141-147.

Thouveny, N and 10 others 1994. Climate variations in Europe over the past 140 kyr deduced from rock magnetism. *Nature,* 371: 503-506.
Toomey, R.S. III, Blum, M.D. & Valastro, S.Jr. 1993. Late Quaternary climates and environments of the Edwards Plateau, Texas. *Global Planet. Change*, 7: 299-320.
Törnqvist, T.E. 1998. Longitudinal profile evolution of the Rhine-Meuse system during the last deglaciation: interplay of climate change and glacio-eustasy? *Terra Nova*, 10: 11-15.
Törnqvist, T.E., Wallinga, J., Murray, A.S., De Wolf, H., Cleveringa, P. & De Gans, W. (in press). Response of the Rhine-Meuse system (west-central Netherlands) to the last Quaternary glacio-eustatic cycles: a first assessment. *Global Planet. Change*.
Vandenberghe, J. 1995. Timescales, climate and river development. *Quat. Sci. Rev.,* 14: 631-638.
Veldkamp, A. 1992. A 3-D model of terrace stratigraphy and valley asymmetry, a case study of the Allier terraces (Limagne, France). *Earth Surf. Proc. Landforms,* 17: 487-500.
Veldkamp, A. & Kroonenberg, S.B. 1993. Late Quaternary chronology of the Allier terrace sediments (Massif Central, France). *Geol. en Mijnbouw,* 72: 179-192.
Veldkamp, A. & Van Dijke, J.J. 1998. Modelling long-term erosion and sedimentation processes in fluvial systems: A case study for the Allier/Loire system. In: Benito, G., Baker, V.R. & Gregory, K.J. (eds) *Palaeohydrology and Environmental Change*. John Wiley, Chichester, 53-66.
West, L.T. & Rutledge, E.M. 1987. Silty deposits of a low Pleistocene-age terrace in eastern Arkansas. *J. Soil Sci. Soc. Amer.*, 49: 1231-1238.
Williams, T., Thouveny, N. & Creer, K.M. 1996. Paleoclimatic significance of the 300 ka mineral magnetic record from the sediments of Lac du Bouchet, France. *Quat. Sci. Rev.,* 15/2-3: 223-235.
Yan, Z. & Petit-Maire, N. 1994. The last 140 ka in the Afro-Asian arid/semi-arid transition zone. *Paleogeogr. Paleoclimatol. Paleoecol.,* 110: 217-233.

7. River terrace formation in synchrony with long-term climatic fluctuation: supporting mammalian evidence from southern Britain

D.R. BRIDGLAND
Department of Geography, University of Durham, UK

D.C. SCHREVE
Department of Geography, Royal Holloway University of London, Egham, Surrey, UK

1 INTRODUCTION

Modelling of river terrace systems in southern Britain, especially the Thames, suggests that climate-driven terrace formation has been in operation, against a background of progressive uplift (Bridgland, 1994, 1995a; Bridgland & Allen, 1996; Maddy, 1997a; Maddy & Bridgland, 2000). Correlation with the oceanic oxygen isotope sequence can be attempted using stratigraphical markers, biostratigraphy and absolute dating. The mammalian biostratigraphy of interglacial deposits within fluvial sequences has proved particularly effective in Britain, with distinctive assemblages enabling separation of individual oxygen isotope stages or even, in the case of the Lower Thames, substages (Schreve, 1997, 1998, in press).

In this paper the Lower Thames sequence is taken as an exemplar for the post-Anglian (post-Elsterian) Pleistocene, since there are extensive mammalian assemblages from interglacial deposits within four different terraces in the Thames valley within and downstream from London (Figs 1 and 2). Post-Anglian terrace formation in the Thames valley would seem to have occurred in complete synchrony with orbitally forced climatic fluctuation at the $\delta^{18}O$ stage level; i.e. at the 100 ka cycle glacial/interglacial level (Bridgland, 1994, 1995a, 2000). Thus the four interglacials represented within the Lower Thames sequence (Fig. 1) are those equivalent to OIS 11, 9, 7 and Substage 5e of the oceanic record.

Taking the evidence from the Lower Thames as a template with which to compare sequences in valleys elsewhere, it appears that other rivers in southern Britain may have generated terraces at somewhat different frequencies. In some of these cases mammalian evidence can assist in correlation with the Lower Thames exemplar and, thereby, with the global oceanic record.

2 THE LOWER THAMES SEQUENCE

As stated above, four post-Anglian interglacials are recognized within the Lower Thames terrace sequence, represented within the sedimentary sequences forming four different terraces and thought to correlate with OIS 11, 9, 7 and 5e (Fig. 2). This interpretation was originally proposed by Bridgland (1994; Bridgland & Harding, 1993) on the basis of lithostratigraphy, although from the outset the model drew support from the distribution within the terrace sequence of faunal remains and Palaeolithic artefacts, as well as from amino-acid geochronology (Bowen et al., 1989, 1995).

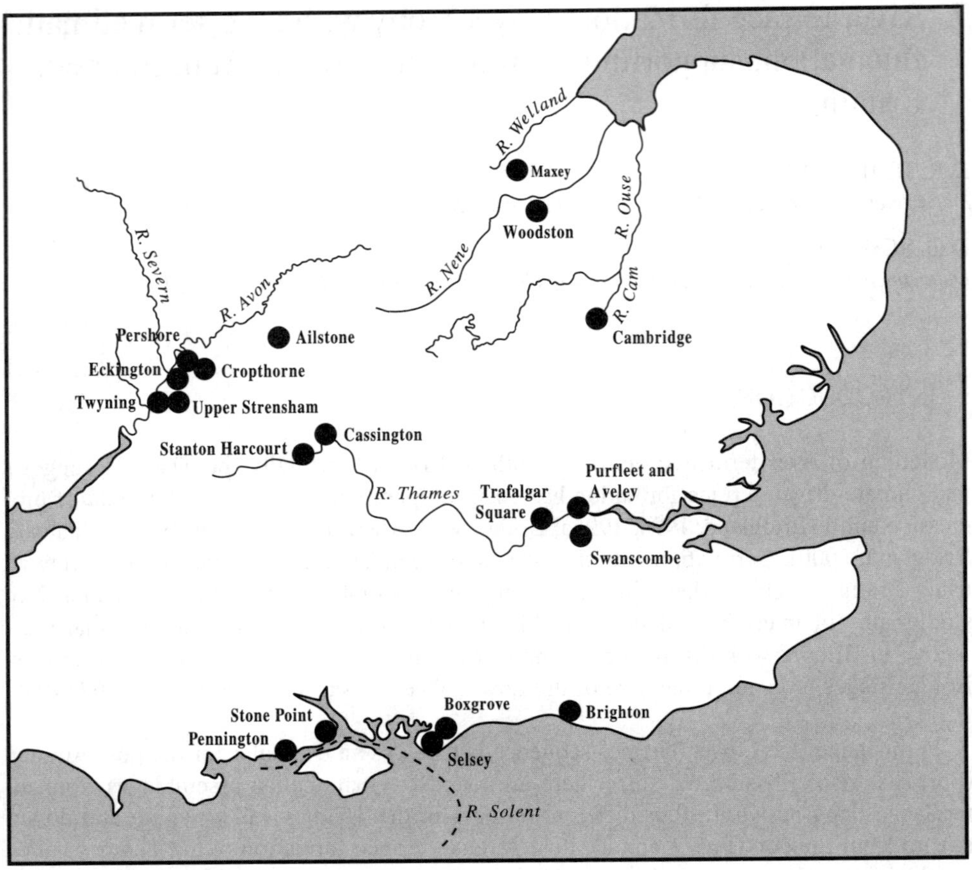

Figure 1. Location of sites mentioned in the text.

Detailed appraisal of the mammalian assemblages from the Lower Thames interglacial sites was subsequently undertaken by Schreve (1997). She found four distinctive mammalian signatures to match the four post-Anglian interglacials, providing both corroboration of the Bridgland model and a basis for interpreting sites over a wider area (Table 1; Fig. 2).

Sites that do not fall within the sort of framework provided by terrace sequences are much more difficult to put into stratigraphical context, despite the fact that they often have excellent, sometimes high-resolution records of a particular interglacial. This is particularly true of lake sites, especially those where the lake has been infilled in a single climatic cycle. Schreve (1997, in press) has been able to use mammalian biostratigraphy to correlate many such sites with the sequence of interglacials recognized in the Lower Thames. The mammalian story from the Lower Thames is thus of considerable value as a yardstick for correlation with assemblages elsewhere.

Essential details of the mammalian record from the Lower Thames are given in Table 1. The remainder of the paper will explore the extent to which comparable evidence can be obtained from other river systems in southern Britain.

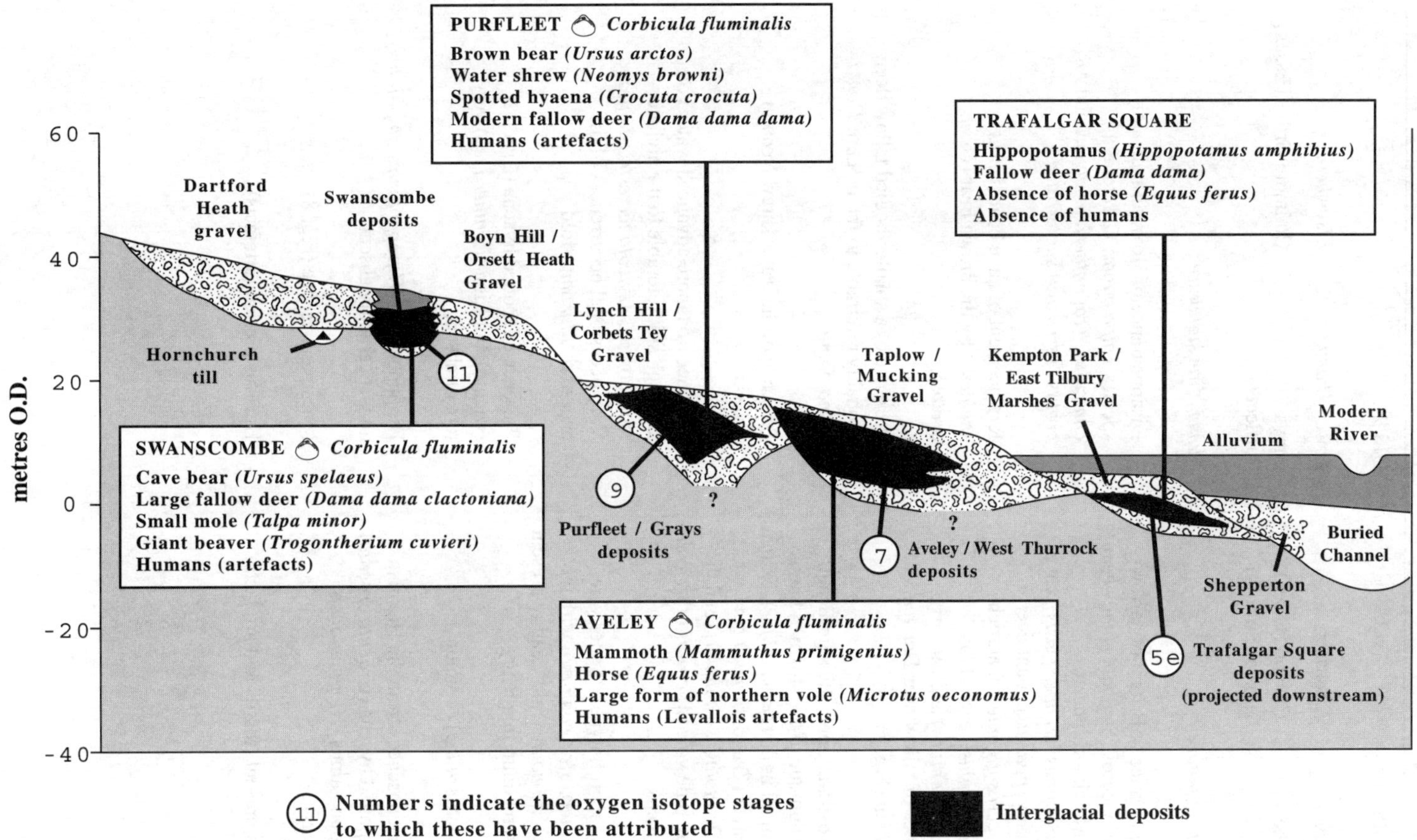

Figure 2. Idealized transverse section through the Lower Thames terrace sequence, showing the stratigraphical positions of interbedded interglacial sediments. Correlations with the oceanic oxygen isotope sequence are suggested. The most significant characteristics of the mammalian assemblages are indicated. Modified from Bridgland (1994).

Table 1. Summary of the mammalian biostratigraphy of interglacial deposits within the Lower Thames sequence (from Schreve, 1997). Important correlative sites elsewhere are also listed.

Hoxnian (OIS 11):

Lower Thames	*Elsewhere*
Swanscombe	Clacton (Essex) Hoxne (Suffolk)

Mammalian signature:

- First appearance in Britain of narrow-nosed rhinoceros (*Stephanorhinus hemitoechus*), Merck's rhinoceros (*Stephanorhinus kirchbergensis*), cave bear (*Ursus spelaeus*), giant deer (*Megaloceros giganteus*), aurochs (*Bos primigenius*), extinct 'ass' (*Equus hydruntinus*) and the large subspecies of fallow deer *Dama dama clactoniana* (Fig. 7)
- Other biostratigraphically significant elements unique to the Hoxnian fauna: extinct small mole (*Talpa minor*), extinct giant beaver (*Trogontherium cuvieri*), rabbit (*Oryctolagus cuniculus*) and the European pine vole (*Microtus* (*Terricola*) *subterraneus*)
- Primitive morphotype of water vole, *Arvicola terrestris cantiana*, with the ancestral '*Mimomys* fold' feature present at low to medium frequency
- Lion (*Panthera leo*) common
- Important absences from the Hoxnian fauna: spotted hyaena (*Crocuta crocuta*) and hippopotamus (*Hippopotamus amphibius*)
- Humans present (fossils and artefacts)

OIS 9:

Lower Thames	*Elsewhere*
Grays Purfleet	Cudmore Grove (Essex)

Mammalian signature:

- Disappearance of Hoxnian indicator species *Talpa minor, Oryctolagus cuniculus, Trogontherium cuvieri* and *Microtus (Terricola) subterraneus*
- Replacement of cave bear (*Ursus spelaeus*) by the brown bear (*Ursus arctos*)
- Presence of a smaller-bodied fallow deer, probably referrable to the modern *Dama dama dama*
- Presence of the water shrew *Neomys browni*
- Presence of a more advanced water vole, although still referable to the archaic morphotype, *Arvicola terrestris cantiana.* Very occasional occurrences of the ancestral '*Mimomys* fold'
- Presence of spotted hyaena (*Crocuta crocuta*), but Lion (*Panthera leo*) very rare or absent
- Hippopotamus (*Hippopotamus amphibius*) almost certainly absent
- Macaque presence (Fig. 8)
- Humans present (artefacts)

Table 1. (Continued).

OIS 7:		Ipswichian (Substage 5e):	
Lower Thames	*Elsewhere*	*Lower Thames*	*Elsewhere*
Ilford (Uphall)	Stanto4n Harcourt (Oxfordshire)	Trafalgar Square	Brentford (Middlesex)
Aveley	Marsworth (Bucks)	Peckham	Bobbitshole, Ipswich
Northfleet	Bielsbeck (Yorks)		Barrington (Cambridgeshire)
Crayford	Stoke Tunnel, Ipswich		

OIS 7 *Mammalian signature:*	Ipswichian *Mammalian signature:*
– Water voles significantly larger and with a more advanced dental morphology than witnessed previously	– Assemblages characterized by hippopotamus (*Hippopotamus amphibius*), straight-tusked elephant (*Palaeoloxodon antiquus*), fallow deer (*Dama dama*), narrow-nosed rhinoceros (*Stephanorhinus hemitoechus*) and bison (*Bison priscus*)
– Earlier part of the interglacial characterized by abundance of temperate woodland indicators, including co-occurrence of *Palaeoloxodon antiquus* and fallow deer (*Dama dama*)	– Disappearance from Britain of Merck's rhinoceros (*Stephanorhinus kirchbergensis*)
– Later temperate part of the interglacial characterized by co-abundance of woolly mammoth (*Mammuthus primigenius*) and horse (*Equus ferus*) and absence of *D. dama*	– Absence of horse (*Equus ferus*) and humans
– The OIS 7 *Mammuthus primigenius* is represented frequently (but not exclusively) by the 'Ilford type' (100% at some localities). The presence of this form (Fig. 9) is considered here to be of particular biostratigraphical significance in demonstrating an OIS 7 age	
– Merck's rhinoceros (*Stephanorhinus kirchbergensis*) present but always in much smaller numbers than *S. hemitoechus*	
– Presence of a large form of northern vole (*Microtus oeconomus*)	
– Hippopotamus (*Hippopotamus amphibius*) absent	
– Humans present (artefacts)	

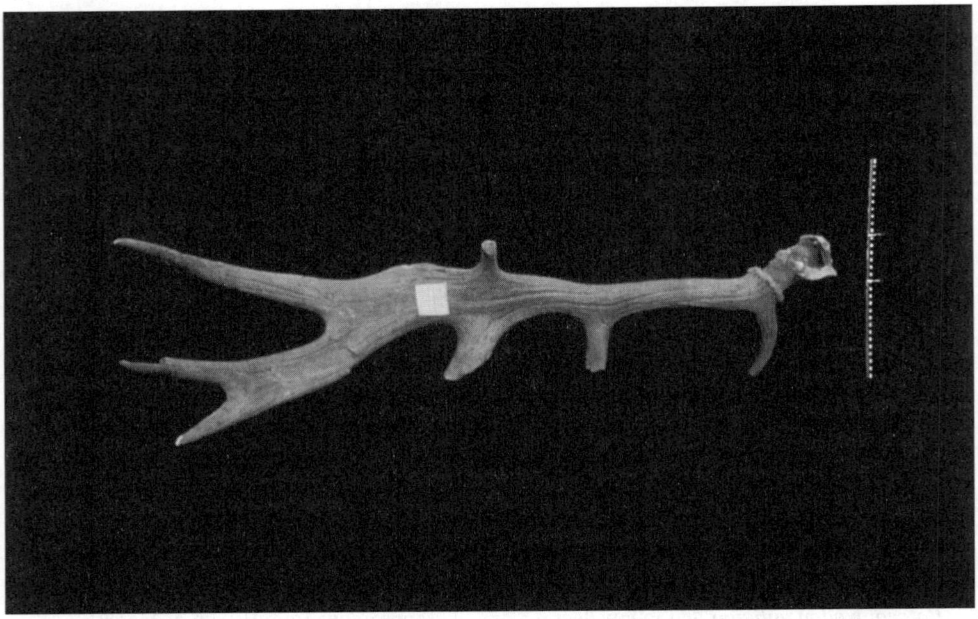

Figure 7. Antler of large fallow deer (Dama dama clactoniana) from Swanscombe.

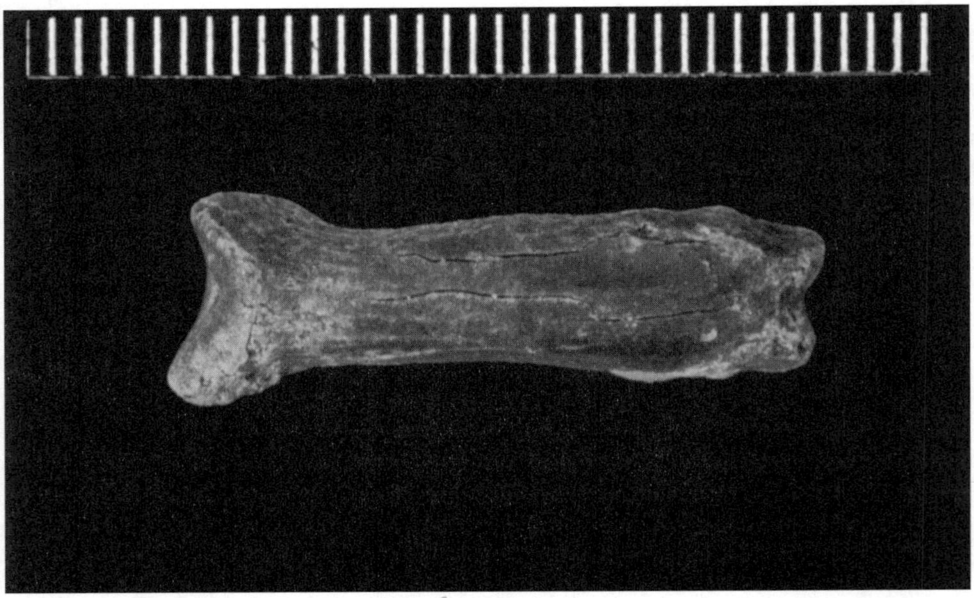

Figure 8. First phalanx of macaque (Macaca sylvanus) from Purfleet.

Figure 9. Tooth of 'Ilford type" (Mammuthus primigenius).

3 THE WORCESTERSHIRE/WARWICKSHIRE AVON

The limestone-rich gravels of the Worcestershire/Warwickshire Avon have yielded significant mammalian remains, preservation undoubtedly having been assisted by the calcareous nature of the deposits and of the local ground water.

Many of the mammaliferous sites are in the New Inn Member of Maddy et al. (1991), which forms Avon Terrace No. 3, as defined by Tomlinson (1925). Assemblages from sites on this terrace (Fig. 3) are generally dominated by *Hippopotamus amphibius*, an unequivocal biostratigraphical indicator in Britain of the Ipswichian Interglacial (OIS 5e). Remnants of Avon Terrace No. 3, exposed in a railway cutting at Eckington (Worcestershire), have yielded important remains of this species, in association with the richest molluscan fauna known from the Avon valley (Strickland, 1842, 1858; Keen & Bridgland, 1986; Bridgland et al., 1989). Hippopotamus is thus a useful marker fossil for the identification of deposits of Ipswichian age within the Avon sequence.

Limited mammalian assemblages have also been recovered from Terraces No. 4 and 5 of the Avon. Recent reappraisal of the Avon terrace stratigraphy (Bridgland et al., 1989; Maddy et al., 1991) has demonstrated unequivocally that the Terrace No. 4 deposits are not underlain by Terrace No. 3 sediments, as suggested by Tomlinson (1925), but instead the two terraces were formed by entirely separate aggradations (Fig. 3). The Cropthorne Member, which forms the higher No. 4 Terrace, is the older of the two. This is supported by important differences in the molluscan faunas from the two terraces. Basal Terrace

No. 4 deposits at Ailstone (Worcestershire) are characterized by the presence of *Corbicula fluminalis* and absence of *Belgrandia marginata*, whereas in Terrace No. 3 deposits at Eckington and the New Inn cutting at Cropthorne (Worcestershire), the reverse is true (Keen, 1990). The Ailstone fauna has accordingly been correlated with the post-Hoxnian, pre-Ipswichian warm episode represented by OIS 7, an attribution upheld by aminostratigraphy (Bridgland et al., 1989; Maddy et al., 1991). At Twyning (Gloucestershire), the cold-climate gravels of No 4 Terrace, which overlie these *Corbicula*-bearing deposits, have yielded a mammalian assemblage characteristic of cold, open environments, including *Mammuthus primigenius*, *Equus ferus* and *Coelodonta antiquitatis* (Whitehead, 1989). Although no species of biostratigraphical significance is present, an OIS 6 age has been inferred for this assemblage, based upon its stratigraphic position between the Ailstone OIS 7 deposits and the Ipswichian deposits of Terrace No. 3 (Schreve, 1997; Fig. 3).

Further evidence has come from Upper Strensham (Worcestershire), where an organic horizon at the base of a fluvial and colluvial sequence has yielded rich faunal and floral remains of interglacial character, including a disarticulated skeleton of *Mammuthus primigenius* and a single antler of *Cervus elaphus* (de Rouffignac et al., 1995). The Strensham site lies beneath a well-defined terrace surface at 43 m O.D., which occupies an intermediate position 7 m below the projected Terrace No. 5 surface and 7 m above the projected surface of Terrace No. 4 in this area of the valley. Furthermore, the Strensham deposits are separated from the sediments underlying Terrace No. 4 by a bedrock step, and thus form a morphostratigraphically separate sediment body, estimated to be broadly time-equivalent to the OIS 7 deposits at Ailstone (de Rouffignac et al., 1995). Although the assemblage is small, the mammalian evidence is consistent with a late OIS 7 age for the Strensham deposits, since *M. primigenius* is not known from any other interglacial in Britain (Schreve, 1997). The presence of un-reworked bones of *M. primigenius* in deposits containing evidence of fully temperate conditions must therefore be viewed as significant. This conclusion has also received support from amino-acid geochronology and from coleopteran biostratigraphy, in the latter case from the abundance of the dung beetle *Oxytelus gibbulus*, which has also been noted as the dominant taxon at other sites now widely regarded as being of OIS 7 age, such as Aveley, Marsworth (lower channel), Stanton Harcourt and Stoke Goldington (cf. Coope, in Shotton, 1983a).

The highest (and oldest) of the Avon terraces (No. 5) was considered by Tomlinson (1925) to have been deposited as outwash from the main glaciation of the English Midlands. On the basis of faunal evidence obtained from deposits underlying the gravels of Avon No. 5, it was suggested by Shotton (1983b) that a temperate-climate episode was represented within the early part of this aggradation. More recently, Maddy et al. (1991, 1995; Maddy, 1997b) have correlated the interglacial deposits of Terrace No. 5 with OIS 9 and the overlying gravel with OIS 8. This is based partly on aminostratigraphic comparison with the Bushley Green Terrace of the River Severn, widely believed to be the direct correlative of Avon No. 5 Terrace (Wills, 1938; Bridgland et al., 1989). A very small number of remains of *Equus ferus* and *Cervus elaphus* were collected from the infill of a channel in Avon Terrace No. 5 near Pershore (Whitehead, 1989). The mammalian remains themselves are too limited to be of any assistance in pinpointing the precise age of the deposits. However, the identification of a *Hippopotamus* fauna of Last Interglacial age in No. 3 terrace and of interglacial deposits attributed to OIS 7 underlying Terrace No. 4 at Ailstone (see above) lends support to an earlier age for the Pershore interglacial.

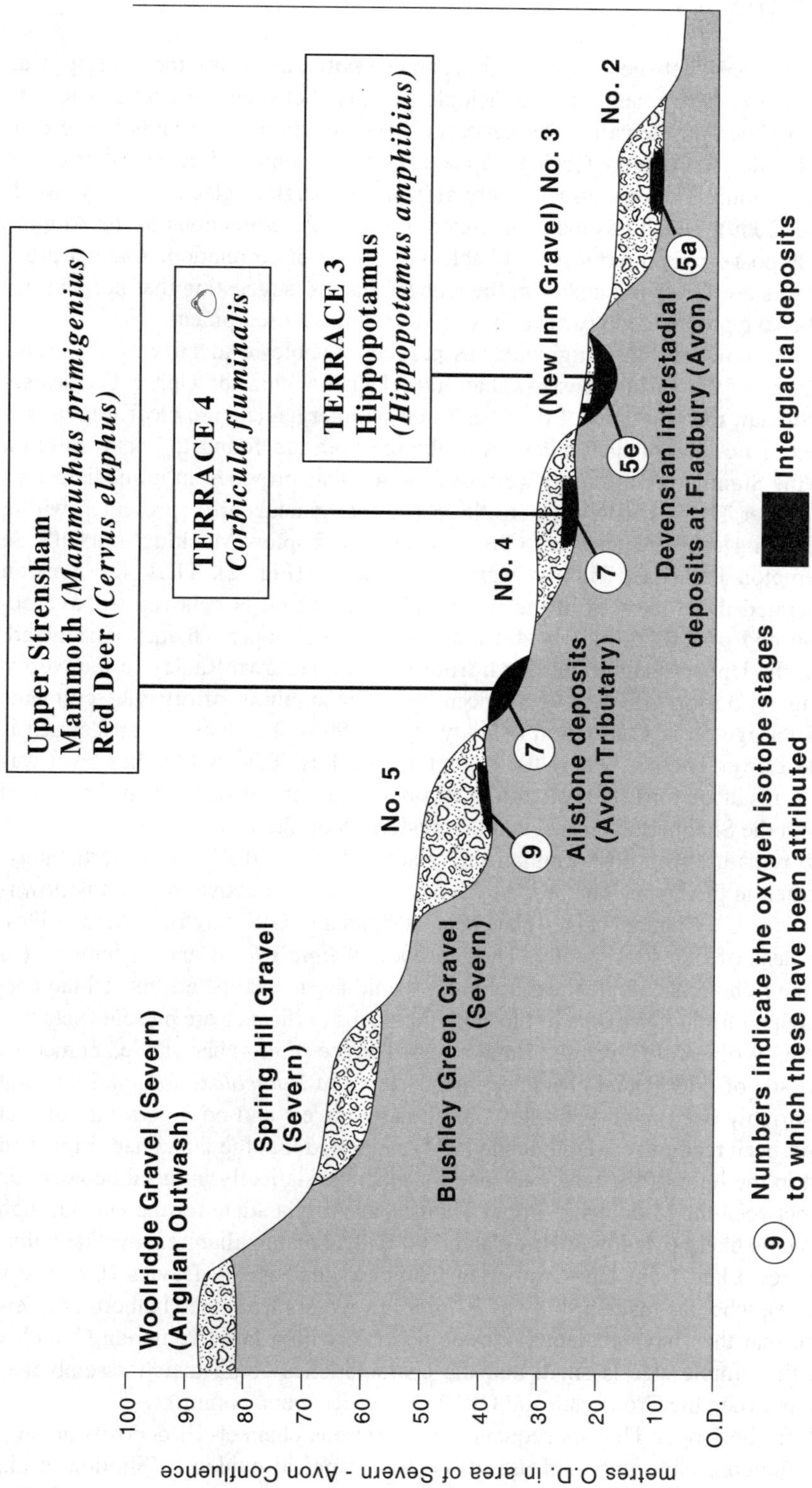

Figure 3. Idealized transverse section through the terraces of the Worcestershire-Warwickshire Avon, showing the stratigraphical positions of interglacial sediments. Correlations with the oceanic oxygen isotope sequence are suggested. The most significant characteristics of the mammalian assemblages are indicated. Data from Maddy et al. (1991); figure modified from Bridgland (in press).

4 THE UPPER THAMES

Correlation of terraces between the Upper Thames, upstream from the Goring Gap through the Chilterns escarpment, and the Middle Thames, between the Goring Gap and London, has long been problematic. The general absence of terrace remnants in the confined section of valley through the Goring Gap is part of the problem. Recently it has been realised that the Middle Thames terraces were affected by Anglian glacio-isostasy (Maddy & Bridgland, 2000), which has made upstream long-profile projections of the Anglian and immediately post-Anglian terraces unreliable as a means of correlation. Another problem is that terraces are fewer in number in the Upper Thames, suggesting that not all rejuvenation events have propagated upstream beyond the Chiltern escarpment.

Once again, mammalian biostratigraphy has proved valuable in identifying the stratigraphical positions of the various post-Anglian interglacials within the Upper Thames sequence. In particular, the positions of the OIS 7 and 5e interglacials are clearly indicated by their characteristic mammalian suites, even though both are found to occur within a single terrace, the Summertown-Radley Terrace (Fig. 4). This provides an immediate contrast with the Lower Thames valley, where these two interglacials are represented within the deposits forming separate terraces, OIS 7 within the Taplow/Mucking and OIS 5e within the Kempton Park/East Tilbury Marshes sediments (Fig. 2). Thus, the incision event that separated these terraces in the Lower Thames, which is believed to have occurred near the end of OIS 6, clearly did not occur in the Upper Thames (Bridgland, 1994). Instead, the Upper Thames cut down from the Summertown-Radley to the Northmoor level (Fig. 4) during OIS 5, as evidenced by the recognition of organic sediments attributed to Substage 5a at Cassington (Maddy et al., 1998). The Northmoor Terrace is the Middle and Lower Thames, where the Kempton/Park East Tilbury Marshes level was abandoned as a result of further downcutting (probably late in OIS 2), when the Buried Channel beneath the Shepperton Gravel and the modern floodplain was incised.

Mammalian remains from post-Anglian interglacials before OIS 7 are, unfortunately, poorly represented in the Upper Thames. At the next terrace level above the Summertown-Radley, the Wolvercote Channel (Fig. 4) has been attributed to OIS 9 by Bridgland (1994), purely on the basis of lithostratigraphy. The presence of *Equus ferus* and of humans (as witnessed by the rich archaeological assemblage) would argue against an Ipswichian correlation, although no further elements of biostratigraphical significance are present (Schreve, 1997). Above the Wolvercote level, the Hanborough Terrace (Fig. 4) has yielded numerous mammalian fossils of interglacial character, mostly teeth of *Palaeoloxodon antiquus*, but these have invariably been reworked into cold-climate gravels and no *in situ* interglacial deposit has ever been recognized. Bridgland (1994) suggested that the fauna had originated in OIS 11, but in the light of the new evidence for glacioisostatically-induced neotectonic displacement between the Middle and Upper Thames, an Anglian date for the Hanborough Terrace now seems likely (Maddy & Bridgland, 2000). The mammalian remains have thus probably been reworked from late Cromerian Complex deposits. P. Davies (University College, London), who has made a study of *P. antiquus* molars from the Hanborough Terrace, has found that they have characteristics compatible with a late Cromerian Complex age, although the sample size is small and the characteristics could match assemblages ranging in age between the Cromerian and OIS 7 (P. Davies, pers. comm.).

Higher still in the Upper Thames sequence, fossiliferous channel-fill deposits at Sugworth, near Abingdon, have provided useful biostratigraphical evidence (Shotton et al.,

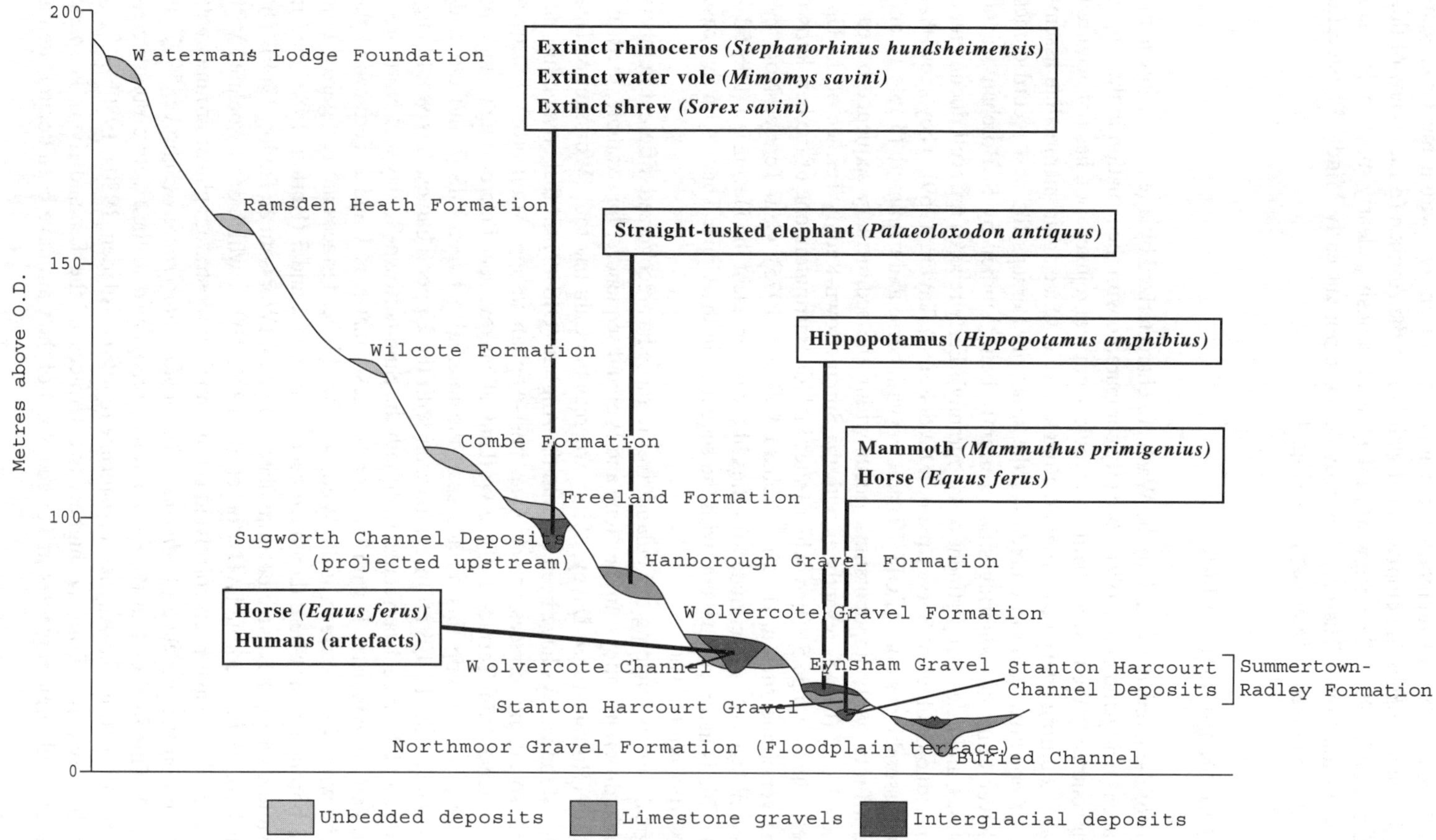

Figure 4. Idealized transverse section through the Upper Thames terrace sequence, showing the stratigrapical positions of interbedded interglacial sediments. Correlations with the oceanic oxygen isotope sequence are suggested. The most significant characteristics of the mammalian assemblages are indicated. Modified from Bridgland (1994).

1980). The mammals from Sugworth include *Stephanorhinus hundheimensis*, *Mimomys savini* and *Sorex savini*, all unimpeachable indicators of an age within the Cromerian Complex, *M. savini* suggesting emplacement well before the Anglian (for a review of this site, see Bridgland, 1994). The Sugworth interglacial deposits, clearly much older than any of those in the Lower Thames, thus provide an important early Middle Pleistocene marker within the Upper Thames sequence (Fig. 4).

5 THE FENLAND/WASH RIVERS

The valleys of the rivers draining into the Wash are characterized by a relatively low number of terraces (Bridgland et al., 1991), broadly comparable with the situation in the Upper Thames. In contrast to the Lower Thames, it would seem that rejuvenation has not occurred in these valleys during every post-Anglian climatic cycle. Once again mammalian assemblages are of potential value in elucidating the situation. Perhaps the best record comes from the River Cam, in Cambridgeshire, currently being studied by S. Boreham of the University of Cambridge (Department of Geography). Early records are well documented in the Cambridge Geological Survey memoir (Worssam & Taylor, 1969). They show that mammalian assemblages both of OIS 7 character, with horse and mammoth (Table 1), and of OIS 5e affinity, with hippopotamus, occur within the sedimentary sequence beneath Cam Terrace No. 2 (Fig. 5), exactly as with the Summertown-Radley Terrace of the Upper Thames. The Substage 5e (Ipswichian) sites include Barrington, one of the best known Pleistocene vertebrate sites in Britain (Gibbard & Stuart, 1975). Cam Terrace No 1 has yielded assemblages of cold-climate affinities likely to represent the Devensian (OIS 5d-2 inclusive). This is again comparable with the sequence in the Upper Thames, in this case from the Northmoor Gravel.

Geological Survey mapping of Fenland river terraces has recognized different numbers of terraces in different valleys. In the Nene and Welland sequences, for example, there are three terraces (Horton, 1989; Davey, 1991), whereas the Cam has four (Worssam & Taylor, 1969). All the valleys have been cut into the Anglian (OIS 12) till that covers this part of England, so all the terraces are *de facto* post-Anglian in age. Multi-proxy biostratigraphical evidence from Terrace 1 of the Welland, at Maxey, near Peterborough, suggests that the sediments forming this terrace span the interval between OIS 7 and the mid-Devensian (Davey et al., 1991). In comparison with the Upper Thames, this would suggest that Welland Terrace 1 correlates with both the Summertown-Radley and Northmoor terraces. Likewise, comparison with the Cam suggests that Nos 1 and 2 Terraces of the latter equate with No. 1 Terrace of the Welland. Two higher terraces are recognized both in the Peterborough area, where they have been numbered 2 and 3 (Horton, 1989), and in the Cam valley, where they have been numbered 3 and 4 (Worssam & Taylor, 1969). The Woodston Beds at Peterborough (Horton et al., 1991, 1992), which are associated with Nene Terrace No. 3, have been correlated on mammalian biostratigraphical grounds with OIS 11, based on the presence of *Microtus* (*Terricola*) *subterraneus* (Schreve, 1997; in press; Fig. 5). This implies that the three terraces recognized in the rivers around Peterborough span the entire post-Anglian Pleistocene (*contra* Castleden, 1980a, 1980b).

Mammalian evidence from the highest two terraces of the Fenland rivers is sparse elsewhere, athough important sites in Terrace No. 3 of the Cam have been documented at

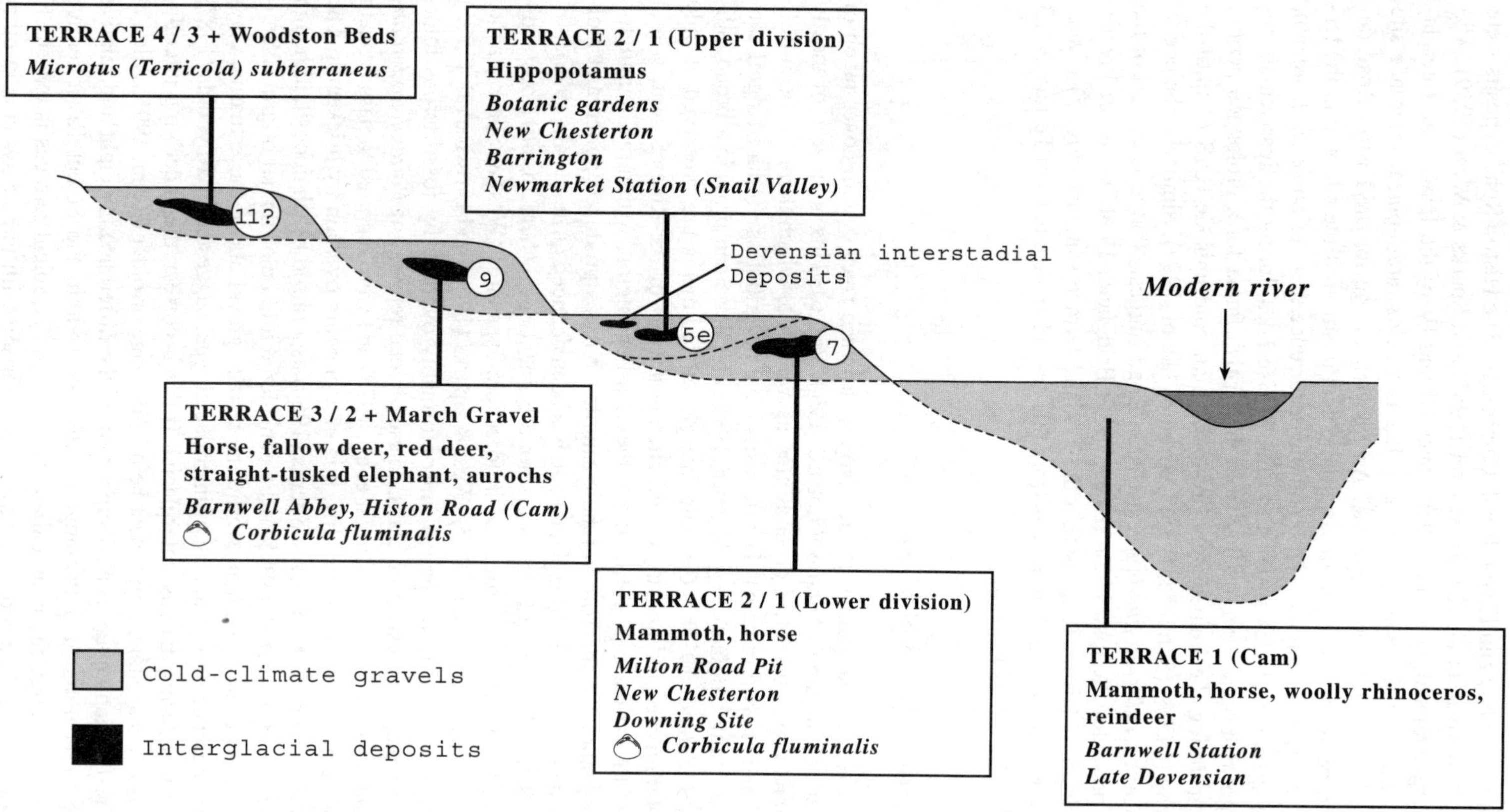

Figure 5. Idealized Fenland river terrace sequence (transverse section), showing the stratigrapical positions of interbedded interglacial sediments. Combines data from Horton (1989), Davey (1991) and Bridgland et al. (1991) from the Peterborough area, supplemented with that provided by Worssam & Taylor (1969) for the Cam. For explanation of terrace numbering in the different rivers, see text. Correlations with the oceanic oxygen isotope sequence are suggested. The most significant characteristics of the mammalian assemblages are indicated.

East Road, Barnwell Abbey and Histon Road, Cambridge. The Histon Road deposits were described by Hollingworth et al. (1950), Walker (1953) and Sparks & West (1959), who ascribed the interglacial sediments to pollen zones III and IV of the Ipswichian Interglacial. For many years, the site was widely regarded as a key stratigraphical reference site for the latter half of this temperate episode. An Ipswichian age for the Histon Road deposits now seems problematic, given the stratigraphical position of the site within the terrace sequence (Worssam & Taylor, 1969) and the occurrence there of horse (S. Boreham, pers. comm.), a species that was absent from the British Isles during the Ipswichian. Indeed, it is thought that horse was missing from the British fauna for a prolonged period, from some point during the immediately pre-Ipswichian cold episode (OIS 6) until the Middle Devensian (OIS 3) (Sutcliffe, 1995; Currant & Jacobi, 1997). Similarly, the occurrence of the bivalve *Corbicula fluminalis* at Histon Road suggests that at least part of the sequence there pre-dates the Ipswichian (Keen, 1990; Bridgland, 1994; Meijer & Preece, 1995). The position of Histon Road within the Fenland terrace staircase (Fig. 5) is suggestive of OIS 9, but the mammalian faunas from this and other sites in Cam Terrace No. 3 have yet to be fully evaluated.

6 THE ERSTWHILE RIVER SOLENT

Extensive gravel terraces laid down by this largely drowned river system cover much of coastal south-central England (Allen & Gibbard, 1994; Fig. 6). This is an area of mostly sandy bedrock and consequent acid groundwater, as a result of which there has been little preservation of mammalian remains. The deposits have nevertheless produced significant numbers of Palaeolithic artefacts (e.g. Wymer, 1999) and these can provide a broad indication of age (Bridgland, 1995b). Genuine interglacial sediments are preserved at only two localities within the predominantly cold-climate gravels of the River Solent, within the lower part of the staircase (Fig. 6). One of these, at Pennington, within the Pennington Gravel, has been studied only from borehole samples, but has produced pollen and molluscs from which an Ipswichian Interglacial age has been deduced (Allen et al., 1996).

The second Solent interglacial site is at Stone Point, Lepe, where fossiliferous sediments within the Lepe Gravel are revealed in foreshore exposures (West & Sparks, 1960; Brown et al., 1975; Green & Keen, 1987). West & Sparks (1960) and Brown et al. (1975) recorded the presence of a lower gravel, unconformably overlain by fossiliferous interbedded freshwater peats and estuarine clays. These yielded remains of *Palaeoloxodon antiquus* and *Dama dama* (Reid, 1893; Brown et al., 1975) and were reported to pass beneath the coarse upper gravel that forms the cliff, this in turn being overlain by brickearth. The upper gravel was interpreted by West & Sparks (1960) as a raised beach deposit, although this was disputed by Brown et al. (1975), who considered it to be of fluvial origin. Examination of the brickearth deposits that overlie the terrace gravel above the estuarine clays has revealed the presence of a palaeosol of interglacial type, suggested by Reynolds (1987) to have been formed during the Last Interglacial. If this is correct, the underlying interglacial fossiliferous muds at Stone Point must be of OIS 7 age or older (Keen, 1995; Allen et al., 1996). A pre-Ipswichian age is also supported by the difference in height and relative position of the Stone Point and Pennington sequences within the Solent River terrace staircase (Fig. 6). Although the mammalian record from the Solent terraces is extremely sparse, it is possible to gain a broad picture by comparison with terrace systems in other

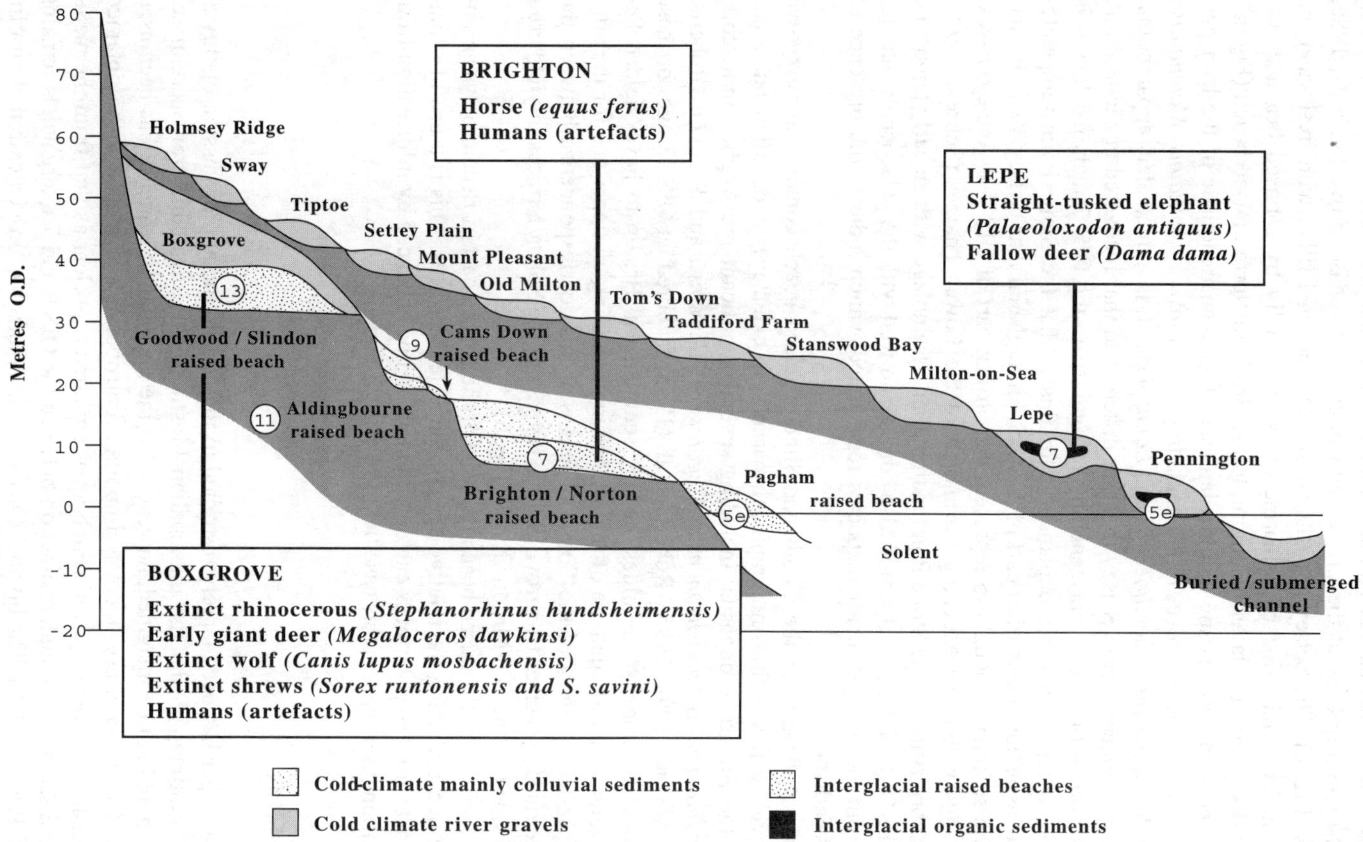

Figure 6. Idealized transverse section through the terraces of the Solent River, showing the stratigraphical positions of interbedded interglacial sediments. The raised beach deposits of the South Coast are shown for comparative purposes. Correlations with the oceanic oxygen isotope sequence are suggested. The most significant characteristics of the mammalian assemblage from Lepe (Stone Point) and the raised beaches at Boxgrove and Brighton are indicated. Modified from Bridgland (2000).

southern British rivers and with evidence from raised beach deposits on the south coast, where better mammalian evidence is preserved (Parfitt, 1998).

The elevation of the Stone Point deposits is closely comparable with other temperate-climate deposits at Selsey (West Sussex), which lie at the seaward end of a 'staircase' of raised beaches on the Sussex coastal plain, ranging in altitude from the inter-tidal zone at Selsey itself to 43.5 m O.D. at the Cromerian Complex site of Boxgrove (Roberts, 1986, 1990; below; Fig. 6). The Selsey deposits consist of a channel filled with freshwater organic muds and overlain first by estuarine clays and finally by a raised beach gravel (West & Sparks, 1960), the latter belonging to the Brighton/Norton raised beach (Fig. 6). The deposits contain a diagnostic mammalian assemblage, characteristic of the later part of the OIS 7 interglacial (Schreve, 1997), including *Palaeoloxodon antiquus*, *Mammuthus primigenius*, *Stephanorhinus hemitoechus* and *Equus ferus*. It is immediately apparent that the mammalian evidence from Selsey is rather different to that recovered at Stone Point, in that *Dama dama* (a species not generally recognized in the OIS 7 mammal fauna) is present at the latter. A possible explanation for this is that the Stone Point temperate-climate deposits relate to an earlier part of the OIS 7 interglacial than Selsey. *D. dama* has been found in sediments thought to represent the earliest part of OIS 7 at Aveley, Essex, one of the sites within the Mucking Formation of the Lower Thames (Schreve, 1997, Fig. 2). The correlation of the Stone Point mammalian assemblage with an early temperate episode in the OIS 7 interglacial would also correspond well with the palaeoecological information from the two mammalian species represented, which indicate the presence of deciduous woodland.

The littoral sediments of the Goodwood/Slindon raised beach complex at Boxgrove (Fig. 6) provide a further biostratigraphical marker. An extensive mammalian fauna has come from the Boxgrove deposits, including several significant pre-Anglian indicators, such as *Stephanorhinus hundsheimensis*, *Megaloceros dawkinsi*, and *S. savini* (Roberts, 1986; Roberts et al., 1994, 1995). Roberts et al. (1995) favoured an OIS 13 age for these deposits. They are amongst the oldest datable sediments in Britain to have yielded Palaeolithic artefacts and as such are of importance in suggesting an upper age for the artefact-bearing terrace gravels of the Solent. There is a close correspondence between the height of the raised beach at Boxgrove and the highest of the Solent terraces to yield artefacts, that formed by the Old Milton Gravel (Fig. 6).

Thus evidence from the raised beach deposits of the south coast of Britain plays an important role in establishing the relative ages of the Solent terraces. This is a useful example of how data from river terraces can complement and be enhanced by information from other depositional settings, in this case the littoral environment.

7 DISCUSSION

Thus far this paper has been largely restricted to the description and discussion of fluvial terrace lithostratigraphy and the mammalian biostratigraphy available from certain sequences. It should not be imagined, however, that the mammalian evidence has been used in isolation. On the contrary, there is important corroborative evidence to be obtained from molluscan assemblages, from which a pattern almost as clear as that from the mammals can be discerned, although distinction between the OIS 9 and 7 interglacials remains difficult (cf. Preece, 1995). The bivalve *Corbicula fluminalis* has proved extremely useful as a biostratigraphic indicator, since it occurs in deposits attributable to OIS 11, 9 and 7, but was apparently absent from Britain during OIS 5e (Keen, 1990; Bridgland, 1994; Preece, 1995). Its occurrence within the various terrace staircases discussed is illustrated (Figs 2-5).

Further support for the interpretation of the fluvial sequence in the Lower Thames, described above, has come from amino acid geochronology (Bowen et al., 1995). This latter method has provided the best dating evidence in the case of the Warwickshire-Worcestershire Avon (Maddy et al., 1991), from which the mammalian record is less complete than that from the Thames. The method can only be applied where molluscan fossils are preserved in quantity, which means it is largely restricted to areas with calcareous groundwater.

8 CONCLUSIONS

The mammalian biostratigraphical sequence from post-Anglian interglacials in southern Britain (Table 1) provides important evidence with which to correlate terrace sequences in different valleys, using the mammalian record from the Lower Thames as a template. In some cases it is advantageous to supplement data from fluvial sequences with information from other environments, as illustrated by the use of evidence from the raised beaches of the south coast in the interpretation of the Solent terraces. Attempts are currently underway to extend this avenue of research to rivers on the European continent, in an attempt to correlate terrace sequences and mammalian assemblages across NW Europe.

ACKNOWLEDGEMENT

The authors wish to acknowledge the Leverhulme Trust for funding their work on the project 'Middle Pleistocene mammalian biostratigraphy of NW European rivers'. DCS acknowledges the award of a Biotechnological and Biological Sciences Research Council (BBSRC) Research Studentship at University College London, 1993-6).

REFERENCES

Allen, L.G. & Gibbard, P.L. 1994. Pleistocene evolution of the Solent River of southern England. *Quaternary Science Reviews*, 12: 503-528.

Allen, L.G., Gibbard, P.L., Petit, M.E., Preece, R.C. & Robinson, J.E. 1996. Late Pleistocene deposits at Pennington Marshes, Lymington, Hampshire, Southern England. *Proceedings of the Geologists' Association*, 107: 39-50.

Bowen, D.Q., Hughes, S., Sykes, G.A. & Miller, G.H. 1989. Land-sea correlations in the Pleistocene based on isoleucine epimerisation in non-marine molluscs. *Nature,* 340: 49-50.

Bowen, D.Q., Sykes, G.A., Maddy, D., Bridgland, D.R. & Lewis, S.G. 1995. Aminostratigraphy and amino acid geochronology of English lowland valleys: the Lower Thames in context. In: Bridgland, D.R., Allen, P. & Haggart, B.A. (eds). *The Quaternary of the lower reaches of the Thames*. Quaternary Research Association Field Guide, Durham, 61-63.

Bridgland, D.R. 1994. *Quaternary of the Thames*. Geological Conservation Review Series 7, Chapman and Hall, London.

Bridgland, D.R. 1995a. The Quaternary sequence of the eastern Thames basin: problems of correlation. In: Bridgland, D.R., Allen, P. & Haggart, B.A. (eds). *The Quaternary of the lower reaches of the Thames*. Quaternary Research Association Field Guide, Durham, 35-52.

Bridgland, D.R. 1995b. Quaternary river terrace deposits as a framework for the Lower Palaeolithic record. In: Gamble, C. & Lawson, A. *The English Palaeolithic Reviewed.* Trust for Wessex Archaeology, Salisbury, 23-39.

Bridgland, D.R. 2000. River terrace systems in north-west Europe: an archive of environmental change, uplift and early human occupation. *Quaternary Science Reviews,* 19: 1293-1303.

Bridgland, D.R. & Allen, P. 1996. A revised model for terrace formation and its significance for the early Middle Pleistocene terrace aggradations of north-east Essex, England. In: C. Turner (ed.) *The Early Middle Pleistocene in Europe*, 121-34. Rotterdam, Balkema.

Bridgland, D.R., Keen, D.H & Davey, N.D.W. 1991. The Pleistocene sequence in the Peterborough district: possible correlation with the deep-sea oxygen isotope record. In: Lewis, S.G., Whiteman, C.A. & Bridgland, D.R. (eds) *Central East Anglia and the Fen basin.* Field Guide, Quaternary Research Association, London, 209-212.

Bridgland, D.R., Keen, D.H. & Maddy, D. 1989. The Avon terraces: Cropthorne, Ailstone and Eckington. In: Keen, D.H. (ed.) *West Midlands.* Field Guide, Quaternary Research Association, Cambridge, 51-67.

Bridgland, D.R. & Harding, P. 1993. Middle Pleistocene Thames terrace deposits at Globe Pit, Little Thurrock, and their contained Clactonian industry. *Proceedings of the Geologists' Association,* 104: 263-283

Brown, R.C., Gilbertson, D.D., Green, C.P. & Keen, D.H. 1975. Stratigraphy and environmental significance of Pleistocene deposits at Stone, Hampshire. *Proceedings of the Geologists' Association*, 86: 349-363.

Castleden, R. 1980a. Fluvioperiglacial pedimentation: a general theory of fluvial valley development in cool temperate lands, illustrated from western and central Europe. *Catena*, 7: 135-152.

Castleden, R. 1980b. The Second and Third Terraces of the River Nene. *Mercian Geologist*, 8: 29-46.

Currant, A.P. & Jacobi, R.M. 1997. Vertebrate Faunas of the British Late Pleistocene and the Chronology of Human Settlement. *Quaternary Newsletter*, 82: 1-8.

Davey, N.D.W. 1991. A review of the Pleistocene geology of the Peterborough district. In: Lewis, S.G., Whiteman, C.A. & Bridgland, D.R. (eds) *Central East Anglia and the Fen basin.* Field Guide, Quaternary Research Association, London, 150-162.

Davey, N.D.W., Bridgland, D.R. & Keen, D.H. 1991. Maxey gravel pit, near Peterborough (TF 1307). In: Lewis, S.G., Whiteman, C.A. & Bridgland, D.R. (eds) *Central East Anglia and the Fen basin.* Field Guide, Quaternary Research Association, London, 185-203.

Gibbard, P.L. & Stuart, A.J. 1975. Flora and vertebrate fauna of the Barrington beds. *Geological Magazine,* 112: 493-501.

Green, C.P. & Keen, D.H. 1987. Stratigraphy and palaeoenvironments of the Stone Point deposits: the 1975 investigation. In: Barber, K.E. (ed.) *Wessex and the Isle of Wight. Field Guide.* Cambridge: Quaternary Research Association, 17-22.

Hollingworth, S.E., Allison, J. & Godwin, H. 1950. Interglacial deposits from the Histon Road, Cambridge. *Quarterly Journal of the Geological Society of London*, 105: 495-509.

Horton, A. 1989. *Geology of the Peterborough District.* Memoir of the Geological Survey of Great Britain.

Horton, A., Keen, D.H. & Davey, N.D.W. 1991. Hicks No. 1 Brickyard, Fletton, Peterborough (TL 190956). In: Lewis, S.G., Whiteman, C.A. & Bridgland, D.R. (eds) *Central East Anglia and the Fen basin.* Field Guide, Quaternary Research Association, London, 163-171.

Horton, A., Keen, D.H., Field, M.H., Robinson, J.E., Coope, G.R., Currant, A.P., Graham, D.K., Green, C.P. & Phillips, L.M. 1992. The Hoxnian Interglacial deposits at Woodston, Peterborough. *Philosophical Transactions of the Royal Society of London*, B338: 131-164.

Keen, D.H. 1990. Significance of the record provided by Pleistocene fluvial deposits and their included molluscan faunas for palaeoenvironmental reconstruction and stratigraphy: case study from the English Midlands. *Palaeogeography, Palaeoclimatology, Palaeoecology*, 80: 25-34.

Keen, D.H. 1995. Raised beaches and sea-levels in the English Channel in the Middle and Late Pleistocene: problems of interpretation and implications for the isolation of the British Isles. In: Preece, R.C. (ed.) *Island Britain: a Quaternary perspective*. Geological Society Special Publication 96: 63-74.

Keen, D.H. & Bridgland, D.R. 1986. An interglacial fauna from Avon No. 3 Terrace at Eckington, Worcestershire. *Proceedings of the Geologists' Association*, 97: 303-307.

Maddy, D. 1997a. Uplift-driven valley incision and river terrace formation in southern England. *Journal of Quaternary Science*, 12: 539-545.

Maddy, D. 1997b. Midlands drainage development. In: Lewis, S.G. & Maddy, D. (eds) *The Quaternary of the South Midlands and the Welsh Marches. Field Guide*. London: Quaternary Research Association, 7-18.

Maddy, D., Green, C.P., Lewis, S.G. & Bowen, D.Q. 1995. Pleistocene geology of the Lower Severn Valley, U.K. *Quaternary Science Reviews*, 14: 209-222.

Maddy, D. & Bridgland, D.R. 2000. Accelerated uplift resulting from Anglian glacioisostatic rebound in the Middle Thames valley, UK: evidence from the terrace record. *Quaternary Science Reviews,* 19: 1581-1588.

Maddy, D., Keen, D.H., Bridgland, D.R. & Green, C.P. 1991. A revised model for the Pleistocene development of the River Avon, Warwickshire. *Journal of the Geological Society of London*, 148: 473-484.

Maddy, D., Lewis, S.G., Scaife, R.G., Bowen, D.Q., Coope, G.R., Green, C.P., Hardaker, T., Keen, D.H., Rees-Jones, J., Parfitt, S. & Scott, K. 1998. The Upper Pleistocene deposits at Cassington, near Oxford, England. *Journal of Quaternary Science*, 13: 205-231.

Marr, J.E. 1920. The Pleistocene deposits around Cambridge. *Quarterly Journal of the Geological Society of London*, 75: 204-244.

Marr, J.E. 1926. The Pleistocene deposits of the Lower Part of the Great Ouse Basin. *Quarterly Journal of the Geological Society of London,* 82: 101-143.

Meijer, T. & Preece, R.C. 1995. Malacological evidence relating to the insularity of the British Isles during the Quaternary. In: Preece, R.C. (ed.) *Island Britain: a Quaternary perspective*. Geological Society Special Publication 96: 89-110.

Parfitt, S.A. 1998. Pleistocene Vertebrate Faunas of the West Sussex Coastal Plain: their Stratigraphic and Palaeoenvironmental Significance. In: Murton, J.B., Whiteman, C.A., Bates, M.R., Bridgland, D.R., Long, A.J., Roberts, M.B. & Waller, M. (eds) *The Quaternary of Kent and Sussex. Field Guide*. London: Quaternary Research Association, 121-134.

Preece, R.C. 1995. Mollusca from interglacial sediments at three critical sites in the Lower Thames. In: Bridgland, D.R., Allen, P. & Haggart, B.A. (eds). *The Quaternary of the lower reaches of the Thames*. Quaternary Research Association Field Guide, Durham, 53-69.

Reid, C. 1893. A fossiliferous Pleistocene deposit at Stone on the Hampshire coast. *Quarterly Journal of the Geological Society of London*, 49: 325-329.

Reynolds, P.J. 1987. Lepe Cliff: the evidence for a pre-Devensian brickearth. In: Barber, K.E. (ed.) *Wessex and the Isle of Wight. Field Guide*. Cambridge: Quaternary Research Association, 21-22.

Roberts, M.B. 1986. Excavation of the Lower Palaeolithic Site at Amey's Eartham Pit, Boxgrove, West Sussex: A Preliminary Report. *Proceedings of the Prehistoric Society*, 52: 215-145.

Roberts, M.B. 1990. Amey's Eartham Pit, Boxgrove. In: Turner, C. (ed.) *SEQS. The Cromer Symposium Norwich 1990. Field Excursion Guidebook*. Cambridge, 62-76.

Roberts, M.B., Stringer, C.B. & Parfitt, S.A. 1994. A hominid tibia from Middle Pleistocene sediments at Boxgrove, UK. *Nature*, 369: 311-313.

Roberts, M.B., Gamble, C. & Bridgland, D.R. 1995. The earliest occupation of Europe: the British Isles. In: Roebroeks, W. & Kolfschoten, T. van, *The Earliest Occupation of Europe*. University of Leiden, 165-191.

Rouffignac, C. DE, Bowen, D.Q., Coope, G.R., Keen, D.H., Lister, A.M., Maddy, D., Robinson, J.E., Sykes, G.A. & Walker, M.J.C. 1995. Late Middle Pleistocene interglacial deposits at Upper Strensham, Worcestershire, England. *Journal of Quaternary Science*, 10: 15-31.

Schreve, D.C. 1997. *Mammalian biostratigraphy of the later Middle Pleistocene in Britain*. Unpublished PhD. thesis, University of London.

Schreve, D.C. 1998. Mammalian biostratigraphy of the later Middle Pleistocene in Britain. Thesis abstract. Quaternary Newsletter, 85: 59-60.

Schreve, D.C. in press. Differentiation of the British late Middle Pleistocene interglacials: the evidence from mammalian biostratigraphy. *Quaternary Science Reviews*.

Shotton, F.W. 1983a. Interglacials after the Hoxnian in Britain. In: Billard, A., Conchon, O. & Shotton, F.W. (eds) *Quaternary Glaciations in the northern hemisphere*. UNESCO International Geological Correlation Programme, Project 73/1/24, Report 9, Paris. 109-115. Reproduced in *Quaternary Newsletter*, 39: 20-25.

Shotton, F.W. 1983b. The Wolstonian stage of the British Pleistocene in and around its type area of the English Midlands. *Quaternary Science Reviews*, 2: 261-280.

Shotton, F.W., Goudie, A.S., Briggs, D.J. & Osmaston, H.A. 1980. Cromerian interglacial deposits at Sugworth near Oxford, England, and their relation to the Plateau Drift of the Cotswolds and the terrace sequence of the Upper and Middle Thames. *Philosophical Transactions of the Royal Society of London*, B289: 55-86.

Sparks, B.W. & West, R.G. 1959. The paleaoecology of the interglacial deposits at Histon Road, Cambridge. *Eiszeitalter und Gegenwart*, 10: 123-143.

Strickland, H.E. 1842. Memoir descriptive of a series of coloured sections of the cuttings on the Birmingham and Gloucester railway. *Transactions of the Geological Society of London*, Series 2(6): 545-555.

Strickland, H.E. 1858. Chapter VI. In: Sir William Jardine *Memoirs of Hugh Edwin Strickland*. London: J. van Voorst.

Sutcliffe, A.J. 1995. Insularity of the British Isles 250,000-30,000 years ago: the mammalian, including human, evidence. In: Preece, R.C. (ed.) *Island Britain: a Quaternary perspective*. Geological Society Special Publication 96: 127-140.

Tomlinson, M.E. 1925. The river terraces of the lower valley of the Warwickshire Avon. *Quarterly Journal of the Geological Society of London*, 81: 137-169.

Walker, D. 1953. The interglacial deposits at Histon Road, Cambridge. *Quarterly Journal of the Geological Society of London*, 108: 273-282.

West, R.G. & Sparks, B.W. 1960. Coastal interglacial deposits of the English Channel. *Philosophical Transactions of the Royal Society of London*, B 243: 95-133.

Whitehead, P.F. 1989. The development and sequence of deposition of the Avon valley river- terraces. In: Keen, D.H. (ed.) *West Midlands. Field Guide*. Cambridge: Quaternary Research Association, 37-41.

Wills, L.J. 1938. The Pleistocene development of the Severn from Bridgnorth to the sea. *Quarterly Journal of the Geological Society of London*, 94: 161-242.

Worssam, B.C. & Taylor, J.H. 1969. *Geology of the Country around Cambridge*. Memoir of the Geological Survey of Great Britain.

Wymer, J.J. 1999. *The Lower Palaeolithic occupation on Britain*. Wessex Archaeology and English Heritage, Salisbury. 2 volumes.

8. Lateglacial and Holocene environmental change inc by floodplain deposits of the Hessian Depression (Central Germany)

PETER HOUBEN, SABINE NOLTE, HOLGER RITTWEGER &
JÜRGEN WUNDERLICH
Institute of Physical Geography, J.W. Goethe – University, Frankfurt on Main, Germany

1 INTRODUCTION

Major advances in late Quaternary palaeoclimatology have been derived from ice cores and marine sediments (Anderson, 1997; Bond et al., 1997; Dansgaard et al., 1989). Previous research on changing fluvial systems showed the susceptibility of rivers to both major and subtle climatic perturbations in Western and Central Europe (e.g. Antoine, 1997; Buch, 1988; Huisink, 1998; Macklin & Lewin, 1993; Rose, 1995; Starkel, 1991; Vandenberghe, 1995).

Contrasting to trunk stream deposits the sediments of small to mid-scale rivers reflect more directly the processes of landscape change on a local to regional scale. The floodplain deposits in the Hessian Depression (Hessische Senke) contain much evidence for the environmental changes that have occurred since the Lateglacial period. The response of the fluvial system to varying climatic conditions during the Lateglacial/Holocene transition is reflected in deposits of different facies. The objective of this paper is to assess the interplay of climatic and local acting controls. These cause spatially divergent changes of the environment as shall be evidenced by fluvial sediments.

2 STUDY AREA

Investigations were carried out in two neighbouring small basins. Both the Amoeneburger Becken and the Wetterau are located north of Frankfurt am Main within the Hessian Depression (Fig. 1). The latter forms a zone of subsidence which represents the northern continuation of the Upper Rhine Graben. The Hessian Depression is bordered by the Rhenish Slate Mountains (Rheinisches Schiefergebirge) in the West and the basaltic Vogelsberg upland in the East. The basins are filled with Tertiary sediments which are covered with Pleistocene loess. Lying in the non-glaciated area between the Scandinavian inland ice sheet and the alpine glaciation the landscape in the study area formed in response to periglacial processes during Pleistocene glacial times.

The lowland areas of the Amoeneburger Becken and the Wetterau show mixed to suspended load types of single-thread rivers that drain a slightly undulated landscape. The investigations concentrated on the catchments of the Wetter and the Ohm rivers. Both rise from the nearby Vogelsberg.

The sites referred to are located in the central Wetter valley near Muenzenberg and in the upstream Wetter valley near Lich. In the Amoeneburger Becken the study areas are situated in the central part of the basin and at Niederwald in the Ohm valley (Fig. 1).

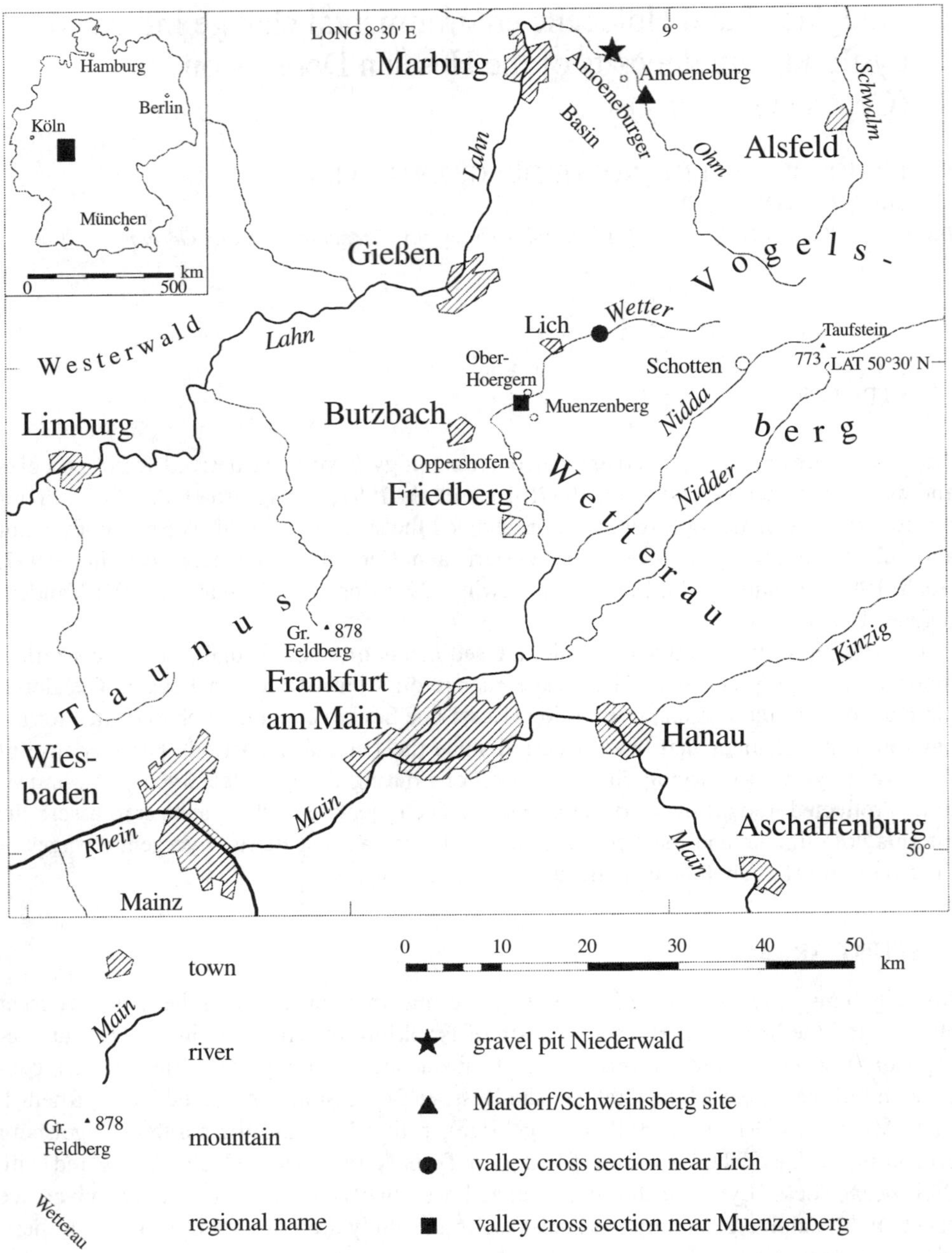

Figure 1. Location of the study area

The study sites of the Wetter valley show low gradient sections separated from each other by steeper reaches. The low gradient reaches are also characterised by floodplain widening. In the central Wetter valley, the record of fluvial sediments starts with sands and gravels, dated to the Younger Dryas, which are covered by fine grained overbank sediments. Older sediments could only be found in part of the valley near Oppershofen. The early to mid-Holocene 'Black Floodplain Soil' (BFS) with an extremely high clay content is preserved in parts of the floodplain. Late Holocene flood loam cap the alluvial sediment sequence across the whole valley (Nolte, 1999; Lang & Nolte, 1999).

The sedimentary record of the upstream Wetter reach near Lich was described by Houben (1997). The alluvial valley fill shows a considerable persistence of valley floor aggradation since the early Lateglacial. It starts with a well stratified series of Lateglacial overbank fines including palaeosol horizons, which are buried by tephritic sediments. These deposits contain tephra spread out by the Laacher See volcano eruption during the Allerød period. An overlying coarse grained deposit of Younger Dryas age grades into overbank fines that in turn are draped by the clayey formation of the BFS. The overbank fines which are conformably resting on the BFS correspond to the Holocene flood loam mentioned above.

The central part of the Amoeneburger Becken is characterised by small restricted catchments. Because the rivulets in these catchments caused no regular flooding the floodplain sediments are of special interest. Several small lakes developed from Lateglacial to mid-Holocene times. The lacustrine deposits are overlain by BFS, colluvial deposits and peat. The contained plant and animal remains document the local environmental history and the processes on the slopes within the restricted catchments (Rittweger, 1997).

In the Ohm valley more than 10 m of fluvial deposits, mostly sands and gravels, were accumulated during the Pleistocene and Holocene period. At the Niederwald site the upper part of the sedimentary sequence can be attributed to the Lateglacial and Holocene period. Gravels grade into fluvial sands and flood loam which were affected by soil formation. They are overlain by Laacher See Tephra (LST) of up to 0.5 m thickness. In some places the LST was removed by erosion, cutting downward to the Lateglacial gravels. Several channels developed in the reworked coarse material. Subsequently these channels were filled with silt, fine sand and organic deposits. These are covered by BFS. Overbank fines of up to 2 m constitute the uppermost layer of floodplain deposits.

3 METHODS

The results presented here were obtained by studying open sections as well as by closely spaced corings. In order to retrieve undisturbed samples several 5 cm diameter cores were also recovered, using a closed 1-m sampling rod with plastic liner. Sediment samples were subjected to grain size analysis, measurements of carbon, calcium carbonate, and metal contents (Fe, Mn, Al, Cu, Ni, Cr). The lithogenetic and lithostratigraphic approach is supported by applying site-specific additional methods such as pollen and macro fossil analysis, tephrochronological investigations and magnetic volume susceptibility measurements on the same sections. Organic matter was radiocarbon dated (conventional ^{14}C, AMS). The use of the biostratigraphic terminology refers to Litt & Stebich (1999) (Table 1), because this chronology currently provides the most recent high resolution record for the Lateglacial to Holocene period.

Table 1. Varve and biostratigraphy of the Lateglacial and early Holocene according to Litt & Stebich (1999).

	Bio-/Pollen zone		varve yr BP
Holocene	Boreal	V	
			10640
	Preboreal	IV	
			11590
Lateglacial	Younger Dryas	III	
			12680
	Alleroed	II	
			13350
	Older Dryas	Ic	13540
	Boelling	Ib	13670
	Oldest Dryas	Ia	13800
	Meiendorf		
			14450
	Pleniglacial		

4 RESULTS

4.1 *Pleniglacial to Allerød period*

In the central Wetter valley, pre-Allerød sediments were only encountered near Oppershofen as a relatively thin layer of grey silty sands beneath the Lateglacial sands and gravels. Organic matter for radiocarbon dating was not available, so the age remains speculative.

In the upper Wetter valley, near the town of Lich, Pleistocene deposits are represented by terrace remnants on valley slopes. Gravelly deposits overlain by sands provide the Pleniglacial base of the superimposed Lateglacial and Holocene valley fill (Fig. 2). The Lateglacial strata consist of two or more layers of fine-grained overbank deposits. Particularly the lowermost overbank fines can be related to abandoned channels. Pollen and macro remains (e.g. *Batrachium spec., Carex spec., Eleocharis palustris, Hippuris spec., Potamogeton spec., Alisma plantago-aquatica*) of these peaty and clayey fills suggest a shallowing channel. Radiocarbon measurements from the macro fossils of three channel fills bracket the time of shallowing at between 15745 to 15105 cal BP (HD 16877) and 14485 to 14135 cal BP (HD 16876) (late Pleniglacial to early Lateglacial, Fig. 2).

These fines also contain several weak fossil A horizons with a vertical extent of 1 to 4 cm. As the upper parts of these fines grade into each A horizon they are probably

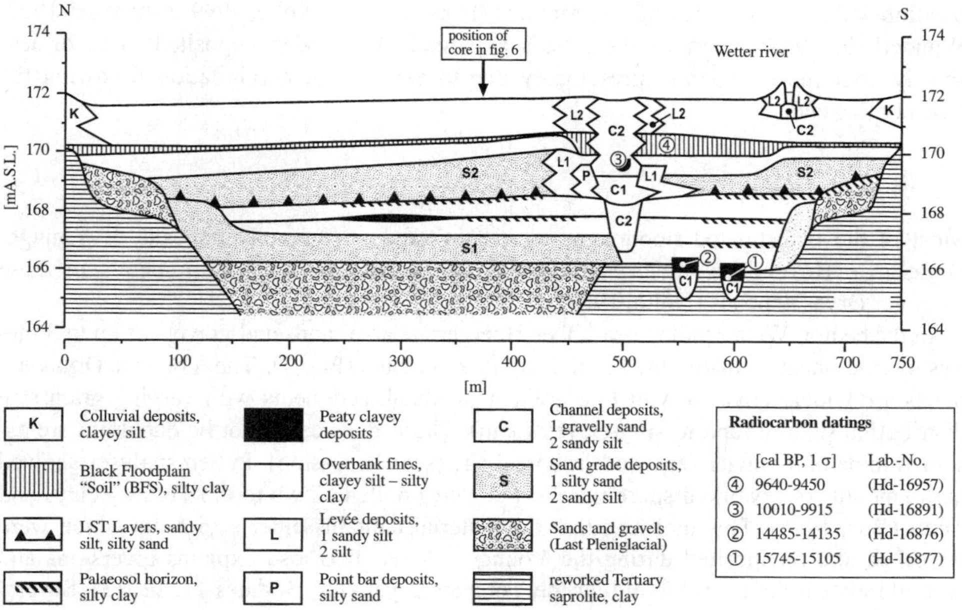

Figure 2. Schematic cross section of the upper Wetter valley near Lich. The alluvial stratigraphy depicted here summarises the main sedimentary features of two detailed cross sectional surveys.

autochthonous formations. Despite low pollen concentrations in these buried soils, the remaining pollen contents point to environmental conditions of a cold climate (Stobbe, pers. comm.). In the study site an uppermost well-developed soil formation of Allerød age was preserved at the floodplain margin (Fig. 2). This soil reveals a clayey and humic A horizon with a thickness of up to 30 cm. A sharp contact separates this soil from the overlying LST deposits (see below).

In the central part of the Amoeneburger Becken the lacustrine deposits point to an inwash with a gradual decrease in grain size. This was coeval with the development of peat and calcareous muds during the Bølling and Allerød periods. Analyses of pollen and molluscs from these deposits indicate an increase of forestation with birch (*Betula*) and pine (*Pinus*) during the first part of the Allerød period (Rittweger, 1997). Relatively high values of NAP, especially with taxa which are indicative of open grassland or heliophilous herbaceous communities (e.g. *Artemisia-, Botrychium-, Papaver rhoeas* – type and *Centaurea scabiosa*), show that the forests had not closed up to a large extent. Some thin layers of concentrated charred plant remains show that fire (caused by palaeolithic man?) might have been of special importance for the ecosystem (Rittweger, 1997). The second part of the Allerød is characterised by a further expansion of woodland with *Pinus* becoming dominant over *Betula* (Bos, 1998).

During this phase the eruption of the Laacher See volcano took place at about 12,900 cal BP, i.e. ca. 200 years before the onset of the cooling of the Younger Dryas at about 12,700 cal BP (Andres & Litt, 1999). In the study area the occurrence of LST serves as an isochronous marker horizon. Many floodplain sites exhibit different kinds of reworked and in situ LST bodies. Composition, grain size distribution, and thickness depend on the position within the depositional environment (Houben, 1997; Nolte, 1999; Rittweger, 1997; Wunderlich, 1998). Most of these beds apparently have been deposited more or less shortly after the tephra fall, presumably due to exceptional precipitation following the Laacher See event.

4.2 *Younger Dryas*

Most of the investigated floodplains consist of sandy to gravelly deposits of Younger Dryas age. Heavy mineral analyses proved the occurrence of LST minerals in these Younger Dryas deposits (Nolte, 1999).

In the central Wetter valley near Ober-Hoergern coarse sands and gravels of up to 3 metres overlie unconformably the basal Tertiary sediments (Fig. 3). The Younger Dryas deposits are widely covered with fine grained overbank sediments with varying grain size distribution (loamy sand to silty-clayey loam). These fines could not be dated but are assumed to be of Younger Dryas or Preboreal age (see Discussion). Pollen analyses showed high amounts of reworked sporomorphs and *Pinus* pollen, down to the Tertiary clays and sands (Bos, 1998). This indicates that the material contains older sediments which were reworked and redeposited during the Younger Dryas. This also explains inversions and several radiocarbon dates too old for the Younger Dryas (Fig. 3, dates 11, 12, 13). Several channels in the Younger Dryas deposits are filled with silts, organic silts (gyttja) and peaty deposits. The basal sequences of inorganic and organic silts were dated by radiocarbon determination and pollen analyses into the Younger Dryas (Fig. 3, 4). Sample 9 in

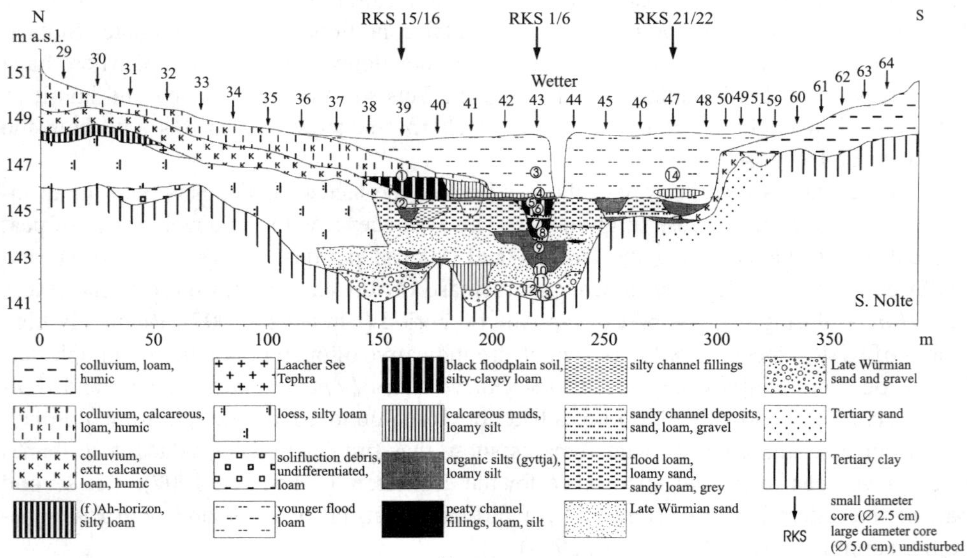

^{14}C-dates

	BP	cal BP, 1 Sigma	Lab.-No.		BP	cal BP, 1 Sigma	Lab.-No.		BP	cal BP, 1 Sigma	Lab.-No.
①	3602 ±67	3980-3780	HD-17826	⑥	7890 ±60	8940-8550	UtC 5415	⑪	11640 ±70	13700-13460	UtC 4407
②	5413 ±43	6282-6179	UtC 4905	⑦	9160 ±70	10280-10030	UtC 4403	⑫	11340 ±70	13350-13160	UtC 4408
③	830 ±80	790-670	UtC 5512	⑧	9470 ±70	10800-10370	UtC 4404	⑬	11360 ±70	13370-13180	UtC 4409
④	1991 ±48	1985-1875	UtC 5416	⑨	9523 ±60	10870-10420	UtC 4405	⑭	2040 ±60	2050-1900	UtC 4902
⑤	7715 ±66	8540-8410	HD-18458	⑩	10180 ±190	12270-11090	UtC 4406				

Figure 3. Valley cross section near Muenzenberg.

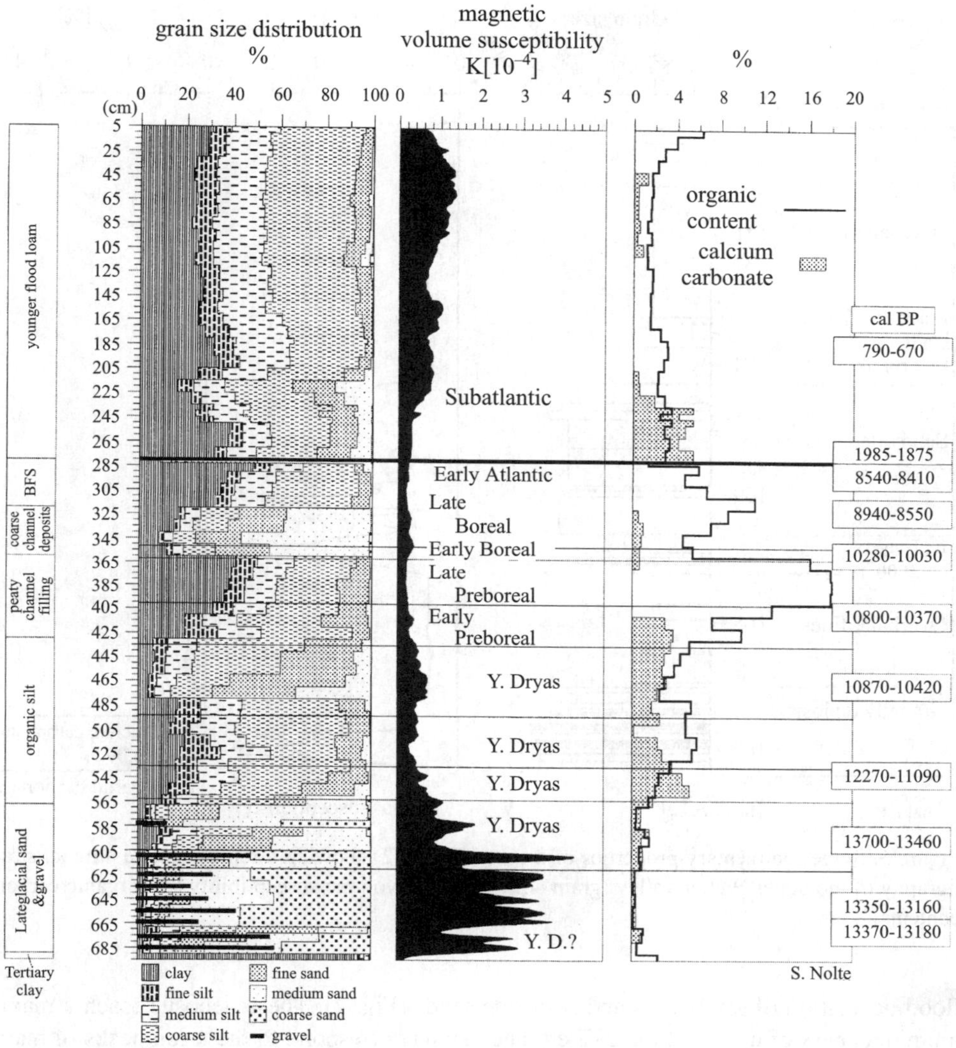

Figure 4. Core 1/6 from the valley cross section near Muenzenberg (sedimentology, magnetic volume susceptibility, and radiocarbon dates; grain size classes according to AG Boden (1994) standards).

Figure 3 shows a radiocarbon date too young for the Younger Dryas (10870-10420 cal BP, UtC 4405). This could possibly be explained by a phase of increased ^{14}C production toward the end of the Younger Dryas (Jöris & Weninger, 1997). According to pollen analyses from these fills the vegetation in the northern Wetterau was characterised by open pine forests with birch, dwarf birch, willow and juniper (Bos, 1998), showing that there was no considerable deforestation during the Younger Dryas. The overlying organic-rich sediments were deposited in the Preboreal (Bos, 1998, Fig. 5, radiocarbon dates in Fig. 3).

In the investigated upstream Wetter reach near Lich much of the sedimentation that took place on the Younger Dryas floodplain was brought about by the activity of sheet

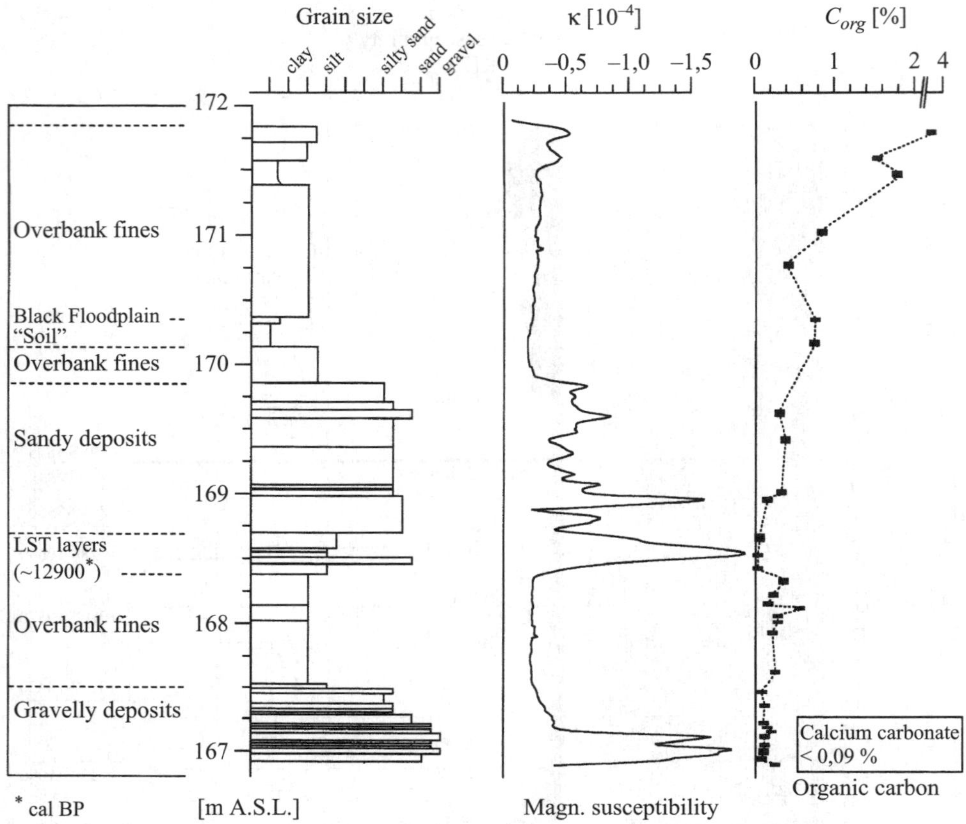

Figure 5. Some sedimentary properties of a core (cf. Fig. 2) showing a representative alluvial stratigraphy of the upper Wetter valley (grain size, magnetic volume susceptibility (κ), organic carbon contents).

flood deposition of sandy silts and silty fine sands (Fig. 2). These deposits reach a maximum thickness of up to 1.2 m. LST enriched strata correspond to the κ log peaks of magnetic susceptibility measurements (Fig. 5). The interstratified laminae consisting of LST suggest the deposition of the reworked light-weight tephra particles during waning floods. Differing from the mid-basin valley reaches as well as from the Niederwald site the Younger Dryas sands gradually change into silty clayey fines. Thus at the Lich site there is no indication of a sudden change in fluvial style during the Younger Dryas period.

In the Ohm valley at Niederwald channels developed in reworked Younger Dryas gravels which contain LST minerals. The fill of the investigated channel depicted in Figure 6 consists of silty mud with intercalations of fine sand and thin organic layers containing mosses and fragments of twigs. These sediments show diapir like structures which are attributed to cryoturbation. The disturbed mud and moss layers are unconformably overlain by peat deposits. The lower part of the peat has a layered structure mainly consisting of mosses (see below). It grades into a 30-40 cm thick peat containing large pieces of wood. Conventional and AMS radiocarbon dating of organic material from the highly disturbed sediments yielded ages which show that the deposition of the silty muds occurred during

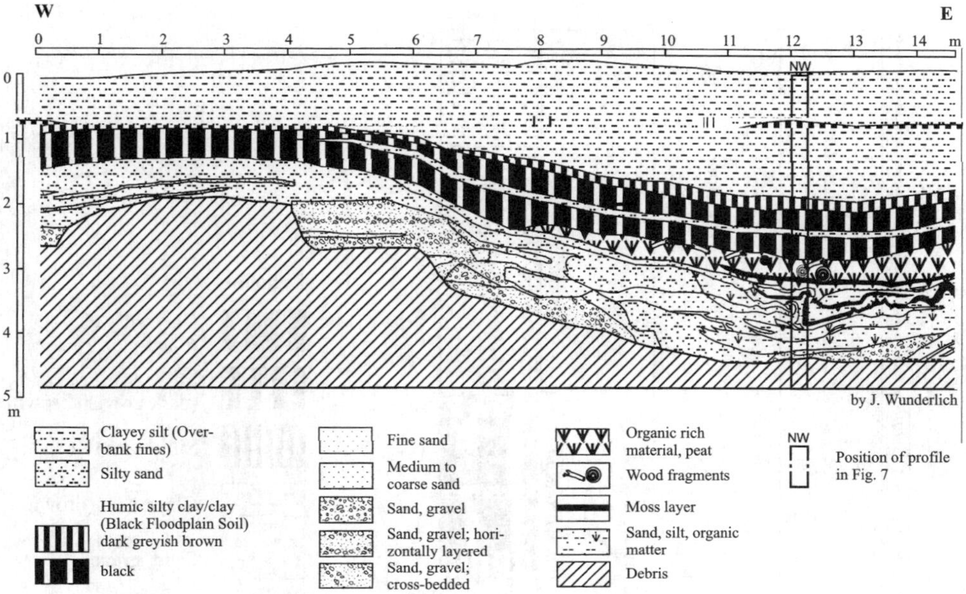

Figure 6. Cross section of the investigated palaeochannel at the Niederwald site in the Amoeneburger Becken.

the Younger Dryas period (Fig. 7). As it cannot be excluded that the dated material had been redeposited, all the dates are maximum ages. Thus the silty muds must have been deposited during the second half of the Younger Dryas period. A first analysis of the botanical macro remains shows that these late glacial channel deposits mainly have to be regarded as lacustrine sediments. Mosses like *Drepanocladus sendtneri* (dominant), *Cratoneuron filicinum* or *Philonotis tomentella* and aquatic plants such as *Myriophyllum spicatum*, *Hippuris vulgaris* or *Potamogeton pusillus* (s.l.) are indicating shallow and stagnant (or slowly flowing) water at least during the summer months (e.g. Rothmaler, 1986; Frahm & Frey, 1987 and Ellenberg et al., 1992). The water vegetation was accompanied by a sedge swamp with *Carex aquatilis* (frequent). In the surrounding area shrubs of *Betula nana* and willows were present. In addition there must have been heliophilous herb or pioneer communities which for example are indicated by taxa such as *Pastinaca sativa*, *Biscutella laevigata* and *Polygonum aviculare*. Similar results were presented by Bos (1998) from the Mardorf-Schweinsberg site in the SE of the Amoeneburger Becken.

4.3 *Early to mid Holocene period*

In the central Wetter valley the Preboreal channel fills mentioned above typically consist of a sequence of fine grained sediments starting with silts gradually changing into more peaty deposits (Fig. 4). The mean organic content in the peaty deposits is 9.6%, with a maximum of 24.7% (Fig. 5). This indicates the continuous input of minerogenic sediment. Since no coarse sediments were deposited and no hiatuses were found by pollen analysis, a decreased fluvial activity can be assumed for the duration of the Preboreal. In some floodplains of the Hessian Depression overbank fines were found between dated deposits

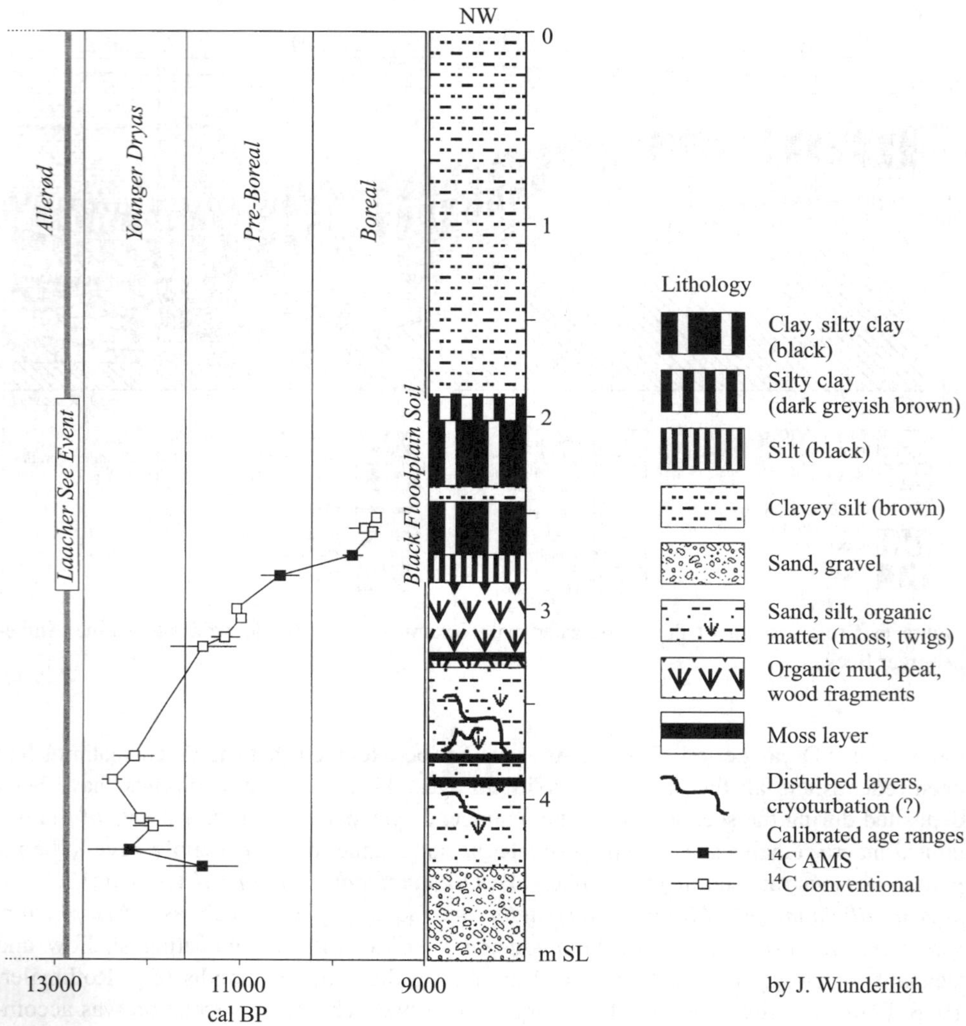

Figure 7. Lithological profile NW and results of radiocarbon datings from the central part of the palaeochannel at Niederwald (position of profile see Fig. 6).

of Younger Dryas and early Boreal age (e.g. Urz, 1995). Unfortunately it was not possible to date the fines in the Wetter valley due to the absence of organic material. Thus it is still questionable, if all these fines should be attributed to the terminal phase of the Younger Dryas or to Preboreal times (see Discussion).

The vegetation development during the early Holocene is primarily marked by a widespread expansion of pine forests. But open herbaceous communities and grassland had not disappeared completely which can be shown for different sites (Urz, 1995; Stobbe, 1996; Rittweger, 1997; Bos, 1998). As described from numerous sites in NW Europe the Younger Dryas/Preboreal transition is often indicated by a lithological change to more organic sediments (e.g. Bohncke & Vandenberghe, 1991). At the Niederwald site this transition occurs in a depth of about 3.30 m (Fig. 7). Here the disturbed sandy and silty sediments with moss layers are capped by an undisturbed laminated sequence consisting of thin peat and clay layers (Wunderlich, 1998). Between 3.20 m and 3.32 m the sediments are mainly formed by remains of the mosses *Aulacomnium palustre*, *Homalothecium nitens* and *Sphagnum* sp. and further by fruits and leaves of *Potamogeton pusillus* (s.l.), *Salix* cf. *repens* and *Carex rostrata*. They are overlain by a more humic mud rich in large wood remains of willow (*Salix* sp.), which are of Preboreal age (Wunderlich, 1998).

In the central part of the Amoeneburger Becken the climatic change is indicated by a change in sedimentation, too. A moss layer (*Drepanocladus* sp.) is overlain by a calcareous mud which developed as a consequence of a rising ground water table and a recurrence of the loess decalcification (Rittweger, 1997). The mollusc fauna of the Preboreal period indicates only a thin forest cover. Many species have to be regarded as relics of the Lateglacial (e.g. *Vertigo arctica*). Only for the following periods do land snails, such as *Acanthinula aculeata* or *Vitrea crystallina* point to an increase in forest cover (Rittweger, 1999).

In numerous Central European floodplains a dark brown to black clayey horizon is developed between Lateglacial/early Holocene and late Holocene deposits (Rittweger, 2000). In the study area this so-called Black Floodplain Soil (BFS) has a thickness of up to 1.7 m and is characterised by extremely high clay contents (mean 50%, max.: 84%). Despite its dark colour the BFS mostly shows very low contents of organic carbon ranging from 1 to 5 per cent. Pollen is generally poorly preserved. Macro remains are lacking. Radiocarbon age estimates yield age ranges of 9800 to 6300 cal BP (Nolte, 1999, see also Figs 2, 3) indicating a Boreal to Atlantic age of formation. In the central Wetter valley samples from the top of the BFS provided Subboreal ages (Fig. 4, Table 3).

In the upstream Wetter reach near Lich underlying overbank fines grade into the dark grey to black silty clay of the BFS (Fig. 2). The geomorphic geometry of the BFS is similar to that of older overbank fines mentioned above. The BFS dips to the floodplain's edges and achieves its maximum thickness adjacent to the contemporary channel belt. The accumulation of fines is suggested to to result from overbank flooding.

5 DISCUSSION

The late Pleistocene alluvial record of the Hessian Depression starts with gravelly deposits of the last Pleniglacial. A discernible series of Lateglacial sediments and soil formations was only found in the upper Wetter valley (Fig. 2). The sedimentation of Lateglacial deposits was preceded by an erosion of older sandy deposits. The preserved Lateglacial valley fill is represented by overbank fines, channel deposits and weak, presumable autochthonous A horizons. The sedimentation of fines reveals a major change in fluvial style at the end of the Pleniglacial. The system shifted from a braidplain to a fluvial environment consisting of one or very few stable channels. The development of vertically accreted fines and a single-thread course is assigned to a more continuous runoff reflecting a decrease in the magnitude of seasonal snowmelt events. According to the ^{14}C determinations of channel fills, this period already started in the very beginning of the Lateglacial (Meiendorf, Table 1). The dominant suspended load must have been supplied by the lower loess covered parts of the sub-catchment, which therefore have been affected by a phase of increased soil erosion. In addition, the pollen contents of the buried palaeosols suggest a cold climate during this phase of the Lateglacial (Stobbe pers. comm.). Accordingly, the formation of the A horizons containing pollen of plants growing under a cold climate may have been caused by an inhibited decomposition of organic litter due to the still cold climate.

In contrast to previous phases of the Lateglacial the Allerød period is seen to be a phase of an increasing expansion of woodland cover as it was shown for example shown by Rittweger (1997) for the Amoeneburger Becken. The locally occurring formation of a well-developed Allerød soil horizon also accounts for a more stable environment (Fig. 2, Bos, 1998). Overlying this soil distinct layers of LST reflect the more or less immediate inwash from the valley slopes after the cataclysmic fallout event.

The Younger Dryas period is generally regarded as a phase with a distinct colder climate (e.g. Andersen, 1997; Isarin, 1997; Isarin & Bohncke, 1999). Nearly all floodplains were affected by increased fluvial activity during the Younger Dryas. The reworking and deposition of coarse grained sediments clearly demonstrates a climate-driven change to a braided river system. On the other hand all known palynological data from the study areas show that in the lowland areas just a slight decrease of forestation took place (Bos, 1998; Rittweger, 1997). In addition, it could be proved that in the central part of the Amoeneburger Becken there even was no recurrence of solifluction on the slopes (Rittweger, 1997). Thus inferring the Younger Dryas climatic situation from the sedimentological records or pollen results may lead to opposite interpretations. These seemingly contrasting results arise from the fact that both approaches provide palaeoenvironmental information of different spatial scales although the samples were gained from the same sections. As the palaeobotanical record is mainly determined by climatic and edaphic conditions, the pollen results are primarily indicative of the specific regional vegetation (cf. Bos, 1998).

In the fluvial system climate is also recognised as being a major control. However, a greater complexity of extrinsic and intrinsic factors affects the fluvial system (e.g. Rhoads, 1992). Fluvial sedimentation is also determined by sediment availability due to erosional processes in the sediment source area, sediment transport processes, and the local inherited setting of the depositional environment.

Considering the reworking of material during the Younger Dryas, it is argued that fluvial sediments represent the fluvial activity of the entire upstream sub-catchment. As upland areas are generally affected by climatic conditions which progressively deteriorate with increasing elevation, the headwaters are more sensitive to changing meteorological conditions. Taking the relief situation into account climate changes do not affect the entire catchment area in the same way. The change to a braided river environment was probably caused by higher amounts of precipitation stored as snow in the upper catchment. This yielded high magnitude discharges during the snowmelt which were able to entrain and transport coarse materials. Additionally, vegetation cover ought to have played a minor role in the upland areas. Low gradients of downstream mid- to low-basin reaches are responsible for the sedimentation of the Younger Dryas coarse grained deposits. Hence the setting of the Wetter and Ohm catchments caused a depositional activity in downstream valley reaches, which was induced by a pronounced fluvial response of the upstream upland headwaters. At the same time the lowland reaches have been only slightly affected by the Younger Dryas cooling. Therefore the synopsis of palaeobotanical and sedimentological evidence reveals a spatially differentiated response of upland and lowland subcatchments to the Younger Dryas climate change.

In the central Wetter valley and the Amoeneburger Becken channel incision and redeposition of older deposits occurred in the Younger Dryas. Age estimates of basal channel fills show that the fluvial system has undergone a change within this period (Figs 3, 4, 7).

The grading from sandy deposits into fine grained overbank deposits in the upper Wetter valley points to the same change in sedimentation pattern. Inwash of fines into abandoned channels is ascribed to a more continuous base flow associated with overbank flooding processes. The decrease in stream power ought to be related to a decrease in snowmelt magnitude. Therefore the reflected change in fluvial style is suggested to be first of all a change in fluvial regime. These results correspond to similar conclusions on a climatic shift within the Younger Dryas of other authors (e.g. Kasse et al., 1995). From the fluvial sediments of the Hessian Depression no deductions regarding the amount of the change in total annual precipitation can be made. The recorded fluvial change results from a complex interplay of precipitation, soil moisture, evapotranspiration, vegetation cover etc. Thus, unlike Kasse et al. (1995) a change to an overall dryer climate cannot be inferred.

During the Preboreal pine forests in the Wetterau and Amoeneburger Becken expanded further and closed to a large extent (Bos, 1998; Rittweger, 1997; Stobbe, 1996). Soils were stabilised, surface runoff and the sediment load of the river decreased. This led to the development of thick peat deposits in shallowing abandoned channels.

Within most valley fills the BFS overlies Younger Dryas or Preboreal sediments (Figs 2, 3). Using pollen analyses and geomorphic criteria the BFS can be shown to reflects both slow sedimentation of fines and semi-terrestrial to hydromorphic soil formation processes. Since the ^{14}C-determinations agree with the pollen-analytical result, it is likely that the formation of the BFS started during the early Boreal. This accords with the results found by other authors, who give age ranges of the formation period of the BFS starting in the late Preboreal/early Boreal period and lasting until Atlantic or even Subboreal (e.g. Mäckel, 1969; Schirmer, 1983; Urz, 1995). In the small restricted catchments of the Amoeneburger Becken there is much evidence that the BFS was supplied with humic clay from the surrounding chernozems (Rittweger, 1997, 2000). In addition, the black clay material has to be regarded as a residue of the nearly totally decomposed organic matter. The decomposition is explained by one or more subsequent dry periods which led to a lowering of the ground-water table and lessivation even in former lacustrine deposits. These processes could be dated to the late Atlantic and (or) early Subboreal period (Rittweger, 2000).

The BFS can be interpreted as an indicator for a geomorphologically stable floodplain which was unique during the Holocene. The very high clay content can only be explained by its formation under a dense vegetation covering the floodplain. Therefore the name once given to this formation does not reflect its genesis in a appropriate way. Actually the BFS developed by a combination of pedogenesis and slow accretion of overbank fines. As a marker horizon, the BFS is of particular importance for the environmental history since it marks a last natural state of floodplain development in the study area. According to settlement history all overlying sediments may carry information regarding the anthropogenic impact on the landscape as for example shown for the Younger Neolithic period (late Atlantic) (Rittweger, 2000).

6 SUMMARY

The present paper aims to contribute to the understanding of Lateglacial and Holocene environmental changes. Detailed investigations of sedimentary and pedogenetic features developed in fluvial environments are conducted in selected areas within the Hessian

Depression. Changes in fluvial sedimentation of the Hessian depression generally correspond to known overall climatic changes. As far as can be deduced from preserved sediments, overbank fines demonstrate the occurrence of soil erosion during cool stages of the Lateglacial period. Pollen results and a prominent soil developed in overbank fines illustrate the climatic warming of the Allerød interval. Reworked Allerød LST deposits occur in nearly all studied floodplains. In particular regarding the Younger Dryas deposits it is shown that locally different environmental conditions developed according to the physiographic setting. Protected lowland basins exhibit a persistent vegetation cover throughout the cooling phase, whereas floodplains were subjected to the reworking of coarse grained sediments. This was probably caused by high-energy flows induced by increased snowmelt originating in the upstream headwater catchments. Proved by dated channel fills the sedimentation changed to fines during the later part of the Younger Dryas. Preboreal to mid-Holocene organic and fine grained sediments, especially the BFS, reflect stable environmental conditions due to an increased vegetation cover.

ACKNOWLEDGEMENTS

The research was carried out within Priority Programme 'Changes of the Bio-Geosphere during the past 15,000 years' under the guidance of W. Andres, Frankfurt am Main. We are grateful to the Deutsche Forschungsgemeinschaft for financial support. We also thank J. A.A. Bos and A. Stobbe for pollen analyses. Organic materials were radiocarbon dated in Heidelberg by Dr. B. Kromer (conventional ^{14}C) and in Utrecht by Dr. K. van der Borg (AMS).

All data from our research used in this paper are stored in the information system PANGAEA and can be retrieved from http://www.pangaea.de/Projects/Großio15k.

REFERENCES

Anderson, D.E. 1997. Younger Dryas research and its implications for understanding abrupt climatic change. *Progress Physical Geogr.*, 21: 230-249.

Andres, W., & Litt, T. (eds) (in press). The Last Termination in Central Europe. *Quaternary International.*

Antoine, P. 1997. Modifications des Systèmes fluviatiles à la transition pléniglaciaire-tardiglaciaire et à l'Holocène: l'exemple du bassin de la Somme (nord de la France). *Géographie physique et Quaternaire*, 51: 93-106.

Boden, A.G. 1994. *Bodenkundliche Kartieranleitung*. 4th ed.; Hannover.

Bohncke, S.J.P. & Vandenberghe, J. 1991. Palaeohydrological development in the southern Netherlands during the last 15000 years. In: Starkel, L., Gregory, K.J. & Thornes, J.B. (eds): *Temperate Palaeohydrology,* Chichester: 253-281.

Bond, G., Showers, W., Cheseby, M., Lotti, R., Almasi, P., deMenocal, P., Priore, P., Cullen, H., Hajdas, I. & Bonani, G. 1997. A Pervasive Millennial-Scale Cycle in North Atlantic Holocene and Glacial Climates. *Science*, Washington, 278: 1257-1266.

Bos, J.A.A. 1998. Aspects of the Lateglacial-Early Holocene vegetation development in western Europe. Palynological and palaeobotanical investigations in Brabant (the Netherlands) and Hessen (Germany). *LPP Contribution Series*, Utrecht, Den Haag, 10: 240 pp.

Buch, M.W. 1988. Zur Frage einer kausalen Verknüpfung fluvialer Prozesse und Klimaschwankungen im Spätpleistozän und Holozän – Versuch einer geomorphologischen Deutung von Befunden von Donau und Main. Zeitschr. *Geomorph. N.F., Suppl.*, Berlin, 70: 131-162.

Dansgaard, W., White, J.W.C. & Johnsen, S.J. 1989. The abrupt termination of the Younger Dryas climate event. *Nature*, London, 339: 532-533.

Ellenberg, H., Weber, H.E., Düll, R., Wirth, V., Werner, W. & Paulissen, D. 1992. Zeigerwerte von Pflanzen in Mitteleuropa. *Scripta Geobotanica*, XVIII: Göttingen.

Frahm, J.-P. & Frey, W. 1987. *Moosflora*. Stuttgart.

Houben, P. 1997. Lateglacial and Holocene fluvial sedimentation in a small upland catchment in Hesse (Germany). *Zeitschr. Geomorph. N.F.*, Berlin, 41: 461-478.

Huisink, M. 1998. *Changing river styles in response to climate change – examples from the Maas and Vecht during the Weichselian Pleni- and Lateglacial*. PhD thesis Vrije Universiteit Amsterdam, Amsterdam.

Isarin, R.F.B. 1997. Permafrost distribution and temperatures in Europe during the Younger Dryas. *Permafrost and Periglacial Processes*, Chichester, 8: 313-333.

Isarin, R.F.B. & Bohncke, S.J.P. 1999. Mean july temperatures during the Younger Dryas in Northwestern and Central Europe as inferred from climate indicator plant species. *Quat. Research*, Washington, 51: 158-173.

Jöris, O. & Weninger, B. 1997. Der absolutchronologische Rahmen spätpleistozäner Fundplätze im Rheinland – Probleme und Perspektiven – (mit besonderer Berücksichtigung der Laacher See-Eruption). *Protokoll Teilkolloquien Schwerpunktprogramm 'Wandel der Geo-Biosphäre während der letzten 15,000 Jahre'* 15.-17.12.1997 in Bonn: 15-16; Bonn (unpubl.).

Kasse, C., Vandenberghe, J. & Bohncke, S. 1995. Climatic change and fluvial dynamics of the Maas during the late Weichselian and early Holocene. In: Frenzel, B., Vandenberghe, J., Kasse, K., Bohncke, & Gläser, B. (eds): *European River Activity and Climatic Change During the Lateglacial and Early Holocene*. Stuttgart, Jena, New York, 123-150.

Lang, A. & Nolte, S. 1999. The chronology of Holocene alluvial sediments from the Wetterau, Germany, provided by optical and ^{14}C dating. *Holocene*, Sevenoaks, 9: 207-214.

Litt, T. & Stebich, M. (in press). Bio- and chronostratigraphy of the Lateglacial in the Eifel region. In: Andres, W. & Litt, T. (eds): Andres, W., & Litt, T. The Last Termination in Central Europe. *Quaternary International*; Amsterdam.

Macklin, M.G. & Lewin, J. 1993. Holocene river alluviation in Britain. *Zeitschr. Geomorph. N. F.*, Suppl. 88: 109-122.

Mäckel, R. 1969. Untersuchungen zur jungquartären Flußgeschichte der Lahn in der Gießener Talweitung. *Eiszeitalter u. Gegenwart*, 20: 138-174.

Nolte, S. 1999. *Auensedimente der Wetter als Indikatoren für die spätglaziale und holozäne fluviale Morphodynamik in der nördlichen Wetterau, Hessen*. PhD thesis Goethe-University; Frankfurt a. M. (unpubl.).

Rhoads, B.L. 1992. Fluvial geomorphology. *Progress in Physical Geogr.*, 16: 456-477.

Rittweger, H. 1997. Spätquartäre Sedimente im Amöneburger Becken – Archive der Umweltgeschichte einer mittelhessischen Altsiedellandschaft. *Materialien Vor- und Frühgeschichte Hessen*, Wiesbaden, 20: 242 pp.

Rittweger, H. 1999. Eine boreale-subboreale Molluskensukzession als Spiegel der Vegetationsgeschichte in der Ohmniederung bei Marburg/Lahn. *Frankfurter geowissenschaftliche Arbeiten*, Frankfurt a.M, D24: 197-220.

Rittweger, H. 2000. The 'Black Floodplain Soil': a lower Holocene marker horizon and indicator for an upper Atlantic to Subboreal dry period in Central Europe. *Catena Special Issue*.

Rose, J. 1995. Lateglacial and early Holocene river activity in lowland Britain. In: Frenzel, B. (ed.): *European river activity and climatic change during the lateglacial and early Holocene*. ESF project 'European palaeoclimate and man'; Special issue Paläoklimaforschung, Stuttgart, 14: 51-74.

Rothmaler, W. 1986. *Exkursionsflora für die Gebiete der DDR und BRD*. Berlin, 6th ed.

Schirmer, W. 1995. Valley bottoms in the late Quaternary – Der Talgrund im jüngeren Quartär. *Zeitschr. Geomorph., N.F.*, Suppl., Berlin 100: 27-51.

Starkel, L. 1991. Long-distance correlation of fluvial events in the temperate zone. In: Starkel, L., Gregory, K.J. & Thornes, J.B. (eds): *Temperate Palaeohydrology*, Chichester, John Wiley, 473-495.

Stobbe, A. 1996. Die holozäne Vegetationsgeschichte der nördlichen Wetterau – palökologische Untersuchungen unter besonderer Berücksichtigung anthropogener Einflüsse. *Dissertationes Botanicae*, Berlin, Stuttgart, 260: 216 pp.

Urz, R. 1995. *Jung-Quartär im Auenbereich der mittleren Lahn – Stratigraphische und paläontologische Untersuchungen zur Rekonstruktion vergangener Flußlandschaften*. PhD thesis, Fb Geowissenschaften, Univ. Marburg/L.

Vandenberghe, J. 1995. Postglacial river activity and climate: state of the art and future prospects. In: Frenzel, B. (ed.): *European river activity and climatic change during the lateglacial and early Holocene*. Special Issue ESF Project European Palaeoclimate and Man, Stuttgart, Jena, New York, 14: 1-9.

Wunderlich, J. 1998. *Palökologische Untersuchungen zur spätglazialen und holozänen Entwicklung im Bereich der Hessischen Senke – ein Beitrag zur internationalen Global Change-Forschung*. Unpublished Habil. thesis Marburg University.

9. Lateglacial and Holocene palaeohydrology of the lower Vychegda river, western Russia

ALEKSEY SIDORCHUK, ANDREY PANIN
Moscow State University, Moscow, Russia

OLGA BORISOVA
Institute of Geography, Academy of Sciences, Moscow, Russia

NIKOLAI KOVALYUKH
State Scientific Centre of Environmental Radiogeochemistry, Kyiv Radiocarbon Laboratory, Kyiv, Ukraine

1 INTRODUCTION

The palaeohydrology of the taiga zone of European Russia has been little studied. Fundamental reviews have been undertaken (Goretskiy, 1964), and several palaeohydrological reconstructions (Panin et al., 1992; Sidorchuk & Borisova, 2000) have been conducted for the rivers of the southern part of the East European Plain. However, there have been only a few attempts to reconstruct the development of the drainage network in the European taiga zone. Research on the evolution of the Northern Dvina, Mezen', and Pechora Rivers has been carried out (Kvasov, 1979; Potapenko, 1975), but this did not include palaeohydrological investigation of the ancient rivers. Nevertheless, this region contains a great deal of valuable information on past hydrological conditions (Sidorchuk et al., in press). Palaeochannels (often incised) are widespread in the small- and medium-sized river valleys in the region and their widths and meander lengths are commonly much larger than those of the recent river valleys. Fragments of palaeochannels of smaller sizes than those of the modern rivers also often occur. These morphological features reflect significant variability in humidity and water yield within this region in the past.

2 THE VYCHEGDA RIVER BASIN

The Vychegda River is the main right bank tributary of the Northern Dvina River. It is 1,130 km in length and the basin area is c. 121,000 km^2. The source of the river lies on the Timan Ridge between 200 to 300 m above sea level. The main part of the basin is a dissected plain with mean altitudes ranging from 140 to 160 m. The mean annual precipitation is 700 mm and of this 350 mm falls during the winter-spring period. January is the coldest month of the year (mean air temperature at Kotlas is –14°C), and July is the warmest with a mean air temperature of 17.2°C. The mean annual air temperature is 1.2°C. Mean annual water discharge at the mouth of the Vychegda River is c. 1160 m^3/s with a mean maximum value of 7500 m^3/s. The runoff depth is 300 mm, including 200 mm for the spring flood period. On average about 56% of the river flow passes during the flood. The flow distribution during the year is uneven as 61% of the flow occurs in spring (April to June), 30% in summer and in autumn (July to November), and only 9% in winter (December to March). Karst features control the water regime in the northeastern part of the basin.

The basin is densely forested (up to 98%), but its northeastern part contains swamps which occupy up to 5 to 10% of the basin area. Interfluve areas are covered by spruce forest together with some birch. Pine forests occupy sandy river terraces. In the southwestern part of the basin the spruce forests include fir and lime. Grasslands occur mainly on the lower levels of the floodplain and cover no more than 0.3% of the basin area.

3 FLUVIAL GEOMORPHOLOGY OF THE LOWER VYCHEGDA

The morphology of the lower Vychegda valley was investigated within a 120 km reach near the confluence with the Northern Dvina River. Within this reach valley bottom width increases from 8 to 10 up to 35 to 50 km at which point the valley of the Vychegda joins that of the Northern Dvina. Several terraces and a floodplain with complicated morphology have been identified in the area (Fig. 1).

A broad third terrace occupies up to 65% of the valley bottom. Its relative height is up to 15 to 25 m above low flow stage. The elevation of the third terrace decreases with distance down the lower Vychegda from about 80 m above mean sea level at 100 km from the mouth, to 65 m above mean sea level at the river mouth. The longitudinal slope of the terrace also decreases along this section from 0.2 to 0.12‰. Below the Vychegda River junction within the Northern Dvina valley, this terrace has a constant altitude above sea level of c. 65 m, such that its relative height is 35 m. The sediments which underlie the

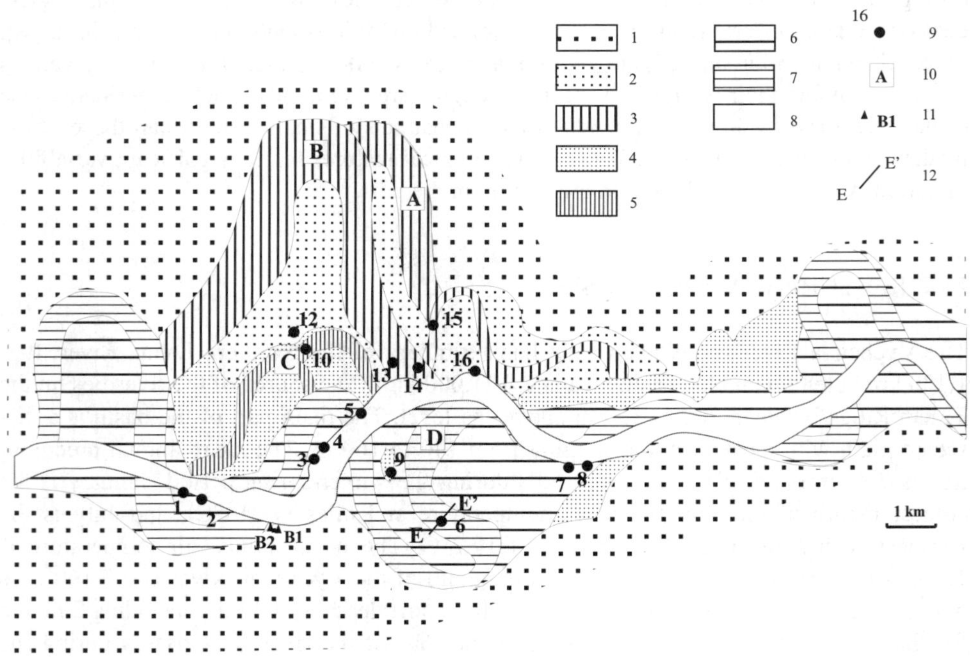

Figure 1. Morphology of the lower Vychegda valley: 1. the third terrace; 2. the second terrace; 3. palaeochannels within the second terrace; 4. the first terrace; 5. palaeochannels within the first terrace; 6. floodplain; 7. palaeochannels within the floodplain; 8. modern river channel; 9. numbers of sites with radiocarbon dates; 10. palaeochannel index; 11. location of sections Baika-1 and Baika-2; 12. location of the cross-section, shown on Figure 2.

third terrace are composed of fine- and medium-grained sand with horizontal and cross-bedded structures, with layers of loam and peat. The exposed section in the left bank near Baika village shows characteristic elements of this sedimentary sequence. There are nine main layers in the outcrop named Baika-1 (Table 1). Some of these layers were dated on the basis of their pollen spectrum and some (with larger quantities of organic matter) were dated by radiocarbon. It is important to point out that only non-calibrated radiocarbon dates are used in this work.

The surface of the third terrace is covered by (fluvial?) elongated, elliptical (on the plane and in profile) dunes. Individual dunes are 1 to 3.5 km long, 0.5 to 1 km wide and 3 to 7 m high. The dunes have an alternating pattern and their long axes have a general northwestern orientation.

Table 1. The sedimentary sequence beneath the third terrace of the Vychegda River near Baika village (16 km from the river mouth).

Depth	Sediment texture and structure	Sediment age based on radiocarbon dating or palynology
0.0-0.5 m	Red-brown clay loam and clay with wavy structure. Upper 10 to 15 cm transformed by podsolic soil formation.	
0.5-1.4 m	Yellow-brown fine sand with small wedges of red-brown loam.	
1.4-1.65 m	Red-brown clay loam and clay with abundant spots of iron oxidation.	
1.65-8.5 m	Yellow horizontal (in the upper part) and cross-bedded (in the lower part) well sorted fine sand. Sand wedges penetrate through the sediments. Rare lenses of fine gravel and small and medium size boulders (from 0.15 to 1.0 m in diameter) are in sharp contrast with the general fine texture of the sediments.	
8.50-9.1 m	Red-brown clay loam and clay. Thin (up to 1-1.5 mm) lenses of sand displaying a wavy structure in the loamy sediments.	Post Last Glacial Maximum time (palynology)
9.1-10.9 m	Yellow-brown horizontally bedded medium sand with rare layers of coarse sand (in the upper 30 cm of sediments) and sandy loam (mainly in the lower 80 cm of sediments).	The Last Glacial Maximum (palynology)
10.9-11.4 m	Dark grey to black clay loam. The lower 20 cm contain organic particles in increasing proportion, so that the lowermost 5 cm are represented by pure compressed peat.	^{14}C, 43,600 ± 2100 (Ki-6397)
11.4-15.1 m	Yellow-brown loamy fine sand with layers of grey sandy loam. Sand layers are horizontally and wavy bedded, their thickness being 15-30 cm. Layers and lenses of loam are 3-5 cm thick, with wavy boundaries. Narrow (35 cm at the top) sand wedges penetrate through the lower 60 cm of sediments.	
15.1-16.3 m	Grey soft loam with thin layers of grey sandy loam. The bottom 30 cm are filled with plant detritus. The layer continues below the water level.	^{14}C, 52,350 ± 3000 (Ki-6398)

The second terrace has a relative height of 12 to 14 m above low river stage and a total width of c. 2 to 4 km. Its altitude decreases along the lower Vychegda. It is 65 m above mean sea level 130 km from the mouth, and 55 m above mean sea level at the river mouth. The average longitudinal slope of the terrace in this part of the valley is about 0.08‰. An outcrop in the right bank of the river near Sol'vychegodsk revealed a sequence which comprised yellow horizontal and cross-bedded fine sands. A lens of grey sandy silt 1.5 m thick underlies the sand at the base of this outcrop. This silt layer continues under the water level. The organic matter extracted from the silt has a radiocarbon age of 34,200 ± 2900 (Ki-6410). This means that the sediments at the base of this exposure beneath the second terrace have the same Middle Valdai (Weichselian) age as those of the third terrace.

There are several steps of different age on the surface of the terrace, the younger features cutting across the older ones. The primary fluvial forms are quite clearly seen on the surface of these steps: there are palaeochannels, mid-channel and alternating bars, and former floodplains with levées and chutes. The curved bluff of the third terrace forms the right margin of the oldest palaeochannel A (Fig. 1) on the highest step of the second terrace. The palaeochannel is completely filled with very fine deposits and peat. The surface is now 6 to 7 m above the level of flooding. Its width is about 1200 m (Table 2). A chain of aeolian dunes traces the left margin of the oldest palaeochannel. This part of palaeochannel A was not dated directly. One of the secondary channels on the lower step of the terrace cut palaeochannel A, which indicates that the latter is older. This secondary channel was active more than 9200 years ago, because peat and plant remnants infilling deposits at sites 15 and 16 (Fig. 1) have yielded radiocarbon ages of 9255 ± 65 (Ki-6406) and 8950 ± 50 (MGU-1454) years BP.

The age of peat filling a similar large palaeochannel on the second terrace near the mouth of the Lokchim River and near the village of Gam (middle Vychegda valley) is 10,500 to 10,900 years BP (Potapenko, 1975; Arslanov et al., 1980). A section of palaeochannel A is also exposed near the village of Baika following erosion by the Vychegda River (Baika-2 section). There are eight main layers of sand, loam and peat in the outcrop and they are described in Table 3.

Table 2. Morphology of the palaeochannels in the lower Vychegda valley.

Surface	Relative altitude (m) above low stage	Channel pattern	Bankfull width (m)	Bankfull depth (m) at the riffle/ pool	Main meander wavelength (km)	Wavelength (km) of secondary sinuosity
Second terrace high step	13-14	Meandering with braids	1200	?/~16	~13	?
Second terrace low step	10-11	Meandering with braids	1300	9/?	13	8
First terrace	9-10	Meandering	600	6/?	7	-
Floodplain high steps	6-7	Meandering	800	6/12	9	-
Floodplain low steps	5-6	Sinuous with braids	1100	7/14	12	7

Table 3. The palaeochannel infilling at the Baika-2 section (16 km from the Vychegda River mouth). Radiocarbon dates are uncalibrated.

Depth, m	Sediment texture and structure	Sample depth (m)	Laboratory number	Radiocarbon age in years BP
0.0-0.7	Grey - brown soft loam with layers of grey sandy loam and very fine sand.			
0.7-1.3	Light brown peat with wood remnants.	0.70-0.74	Ki-7026	860 ± 60
	The lower 20 cm contain layers of gyttia.	0.98-1.02	Ki-7027	1120 ± 55
		1.26-1.30	Ki-7028	1670 ± 60
1.3-1.73	Dark grey sandy loam with wood remnants.	1.60-1.65	Ki-7029	2430 ± 50
	The lower part contains tree trunks.			
1.73-2.45	Dark-brown peat with wood remnants.	1.75-1.79	Ki-7030	3250 ± 65
	The lower 35 cm contain layers of gyttia	1.89-1.91	Ki-7031	3840 ± 55
	and thin layer of fine sand at the bottom.	2.09-2.13	Ki-7032	4505 ± 50
		2.31-2.35	Ki-7033	5820 ± 50
		2.41-2.45	Ki-7034	6155 ± 60
2.45-4.9	Grey sandy loam with wood remnants and	2.60-2.70	Ki-7526	8050 ± 70
	layers of fine brown sand at the lower part.	2.90-2.95	Ki-7527	8240 ± 60
	At a depth of 4.5 m pine cones were	4.5-4.55	Ki-7122	8775 ± 70
	found.		Ki-7024	8825 ± 60
			Ki-6403	8705 ± 65

A well-preserved meandering palaeochannel B with islands and alternating bars is found on the lowermost step of the second terrace near Sol'vychegodsk. Its mean width is 1.3 km (up to 1.5 km in broader sections). The wavelength of the main meander is about 12 to 14 km. Elongated aeolian dunes 7 to 10 m high on the older step of the second terrace mark the right margin of this palaeochannel. Coring at the riffles of the palaeochannel shows that the bankfull depth was about 8 to 10 m. The riffles are composed of fine sand. Plant remnants in the top layers of these alluvial sands related to the initial phase of palaeochannel filling, and have an age of 8655 ± 60 (Ki-6413) at site 12 (Fig. 1), 8630 ± 60 (Ki-6405) at site 13, and 8400 ± 70 years BP (Ki-6407) at site 14. Palaeochannel B was filled with very fine silty sand, sandy loam and peat, but the deposition was not sufficient to completely cover the primary fluvial relief. This very fine material covers well-sorted fine- and medium-grained alluvial sands. It is 1.5 to 2 m thick on the ancient bars, and about 4 to 5 m thick in the channels at the riffle sites. The surface of the second terrace within the palaeochannel is generally 3 to 4 m higher than the modern floodplain, but the lowermost sections are still susceptible to inundation by the river during flood events.

The first terrace is about 2 to 4 km wide and has a relative height above low flow level of 9 to 10 m. Several systems of meandering palaeochannels (C in Fig. 1) with natural floodplain levées and chutes are well preserved on its surface. Their mean widths are about 600 m and meander wavelengths are about 7 km. The longitudinal slope within these palaeochannels is about 0.07 to 0.08‰. The palaeochannels are filled with very fine silty sand, loam and peat. The layer covering the alluvial fine sand is about 3.5 m thick. Plant remains in the top layers of alluvial fine sand at site 10 (Fig. 1) have yielded a radiocarbon age of 8120 ± 50 years BP (Ki-6404).

The floodplain of the lower Vychegda River is 5 to 7 m high above the low water level. Its total width is about 8 km and the floodplain can be subdivided into four main steps of different age. Well preserved remnants of meandering palaeochannels (D in Fig. 1) with natural levees, chutes and oxbows exist on two higher steps. The palaeochannel is 800 m wide, meander wavelength is 9 km, and the bankfull depth on the riffles is 6 to 8 m. Water surface slope within the channel was 0.05 to 0.06‰. The palaeochannel is filled by sandy loam 1 to 2 m thick at the point bars and by sandy loam and silty fine sand up to 9 m thick at the former pools (Fig. 2). Some of the former pools remain in the modern relief as oxbow lakes up to 12 m deep (Table 2). Peat in the depressions between adjacent natural levées on the floodplain has an age of 4200 ± 50 (Ki-6401) at site 7 (Fig. 1), 4470 ± 60 (Ki-6402) at site 8, and 4670 ± 60 (Ki-6409) at site 5. Plant remnants and charcoal in the oxbow deposits (Fig. 2) are 3980 ± 60 (Ki-6395) and 1900 ± 50 (Ki-6390) years old respectively.

The modern meandering-braided channel of the Vychegda River forms the two lowermost steps of the floodplain. The river has a sinuous pattern with numerous braids within the main channel. The total width of the river is about 1100 m at the crossings, meander wavelength is about 12 km and the water surface slope is 0.07-0.08‰. Bankfull depth is about 7 m at the riffles and up to 12-14 m at the pools. Channel alluvium is composed of

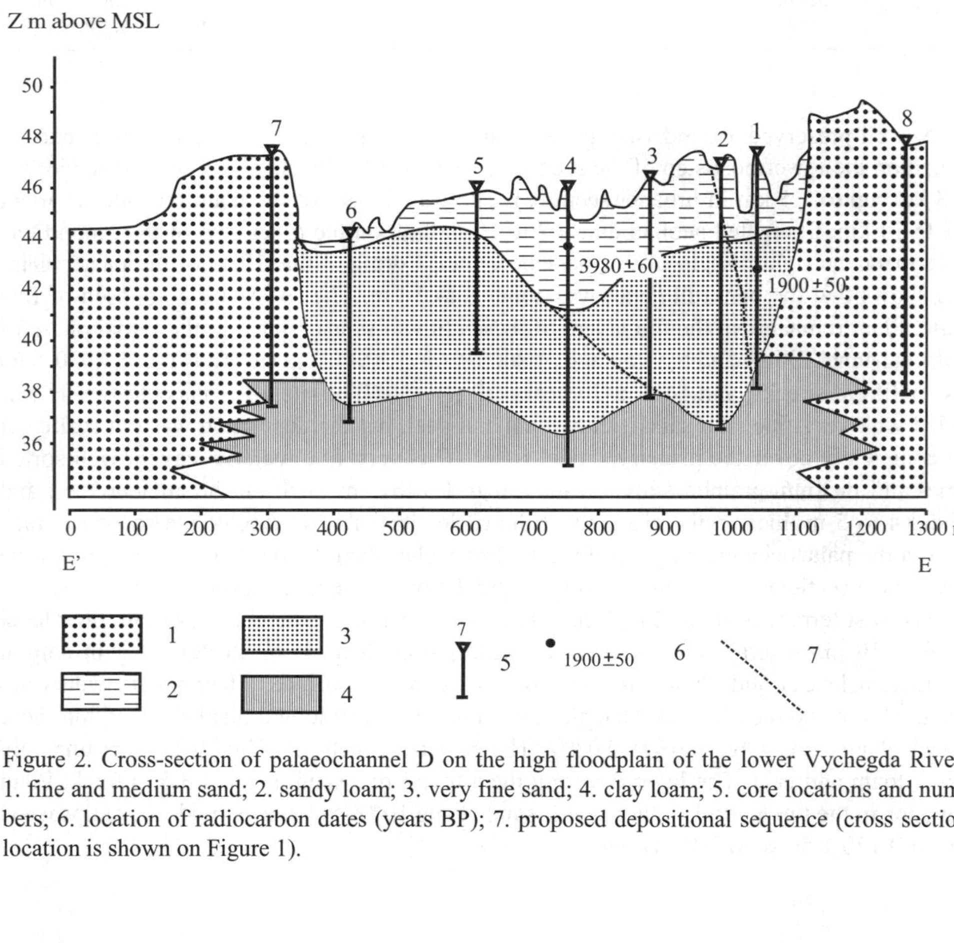

Figure 2. Cross-section of palaeochannel D on the high floodplain of the lower Vychegda River: 1. fine and medium sand; 2. sandy loam; 3. very fine sand; 4. clay loam; 5. core locations and numbers; 6. location of radiocarbon dates (years BP); 7. proposed depositional sequence (cross section location is shown on Figure 1).

fine and medium sand with the lenses of fine gravel. The floodplain is composed of fine and very fine sand with a loam cover 0.5 to 1 m thick. Peat in the depressions between adjacent natural levées on the floodplain has given radiocarbon ages of 1700 ± 70 (Ki-6391) at the site 1 (Fig. 1), 2040 ± 60 (Ki-6392) at the site 2, 2250 ± 60 (Ki-6393) at the site 3, and 2570 ± 60 (Ki-6394) at the site 4.

4 RADIOCARBON DATING

Radiocarbon dating of alluvial deposits involves specific difficulties, as the content of organic matter in them is usually very low, and plant macrofossils seldom occur. Most of the deposits filling in palaeochannels were dated in the Radiocarbon Laboratory of the State Scientific Centre of Environmental Radiogeochemistry (Ukraine). The organic matter used for dating was extracted from peat, wood and plant remnants. Dispersed organic material from the fossil soils was also used.

Radiocarbon dating is a complex process with many stages. It includes initial chemical treatment of the sample, obtaining lithium carbide and acetylene, converting it into benzol, purification, the addition of scintillators and assessment of its β-activity by sensitive low-background luminescent counters. To estimate the degree of isotope fractionation it is necessary to determine the ratio of stable carbon isotopes. The reliability of the results obtained depends fully on the correct execution of each operation.

The aim of the initial treatment of the samples is primarily to extract the reliable dating fraction of the carbon-containing matter and to clear out accidentally introduced carbon-containing substances, which distort the true age. Each sample is practically unique because of its environmental conditions. That is why it is necessary to find an optimum mode of treatment in each case. Universally accepted standard acid-alkaline treatment of carbon-containing substrata of vegetative origin often results in losses of useful material. This is brought about by dissolution of half-decomposed cellulose in the course of alkaline extraction of humic acids. The technique of selective extraction of introduced organic matter worked out in our laboratory makes this stage much safer and straightforward. The kernel of the method is in treatment of the samples of vegetative origin with 1-3% solution of hydrofluoric acid. In this case the covering silicate layer dissolves first. This layer is formed due to redeposition of the soil silicic acid and has a great sorbing capacity. When dissolving this layer, we transfer all absorbed humic and fulvic acids into a jelly-like state.

A new method of obtaining the lithium carbide by 'vacuum pyrolysis' considerably simplifies and accelerates two important types of radiocarbon dating application, such as dating of wood and of the soil organic matter (Skripkin & Kovaliukh, 1998). Prior to pyrolysis, samples of wood are crushed into 3 to 10 mm fragments, then after removing the roots of modern plants, they are treated with dilute hydrofluoric acid (1-3%). The main advantage of such a treatment over the standard acid-alkaline one is the maximum preservation of organic matter even from highly disintegrated wood. After washing and drying, the wood is placed into a reactor over melted lithium at 800°C in vacuum. Intense thermodestruction of wood takes place there, while the pyrolysis products are absorbed by lithium. In a certain temperature interval, only carbon and oxygen are absorbed while the hydrogen goes into a free state and can be removed by vacuum pump.

After complete destruction of the organic matter, the solid residue (charcoal) is also added to the reaction mixture. Thus, all carbon from the sample passes into the lithium carbide with minimum losses. The synthesis proceeds in one stage and takes only 20 to 30 minutes instead of 3 to 5 hours as in the traditional method.

Application of the 'vacuum pyrolysis' method for dating soil organic matter has even more advantages. In this case preliminary extraction of humic acids in the pure state is not required for soil samples with relatively high organic content (over 5%). The samples of soil are treated by hot water to remove fulvic acids. At the same time the roots of modern plants should be removed. After centrifuging, drying, and rough estimation of the content of organic matter, the sample should be mixed with the calculated quantity of manganese dioxide and placed into the reactor for synthesis of lithium carbide by vacuum pyrolysis. Heating the sample in a vacuum causes the thermodestruction of organic matter accompanied by absorption of the gaseous products of pyrolysis by lithium. After reaching 500°C, eduction of gaseous products of pyrolysis chiefly terminates, and the manganese dioxide begins to educe oxygen. Oxygen is educed gradually in the interval between 500 and 950°C. It oxidises the solid carbon resulting from pyrolysis in the entire sample of soil. Carbon monoxide and dioxide formed during the process are also absorbed by lithium. As a result of these reactions, practically all the carbon of the soil organic matter transforms into lithium carbide in one stage.

Implementation of direct synthesis by the vacuum pyrolysis method is inefficient for soil types with a low organic content. Nevertheless, the advantages of the new method make it possible to simplify the traditional method. One can use chemical reagents in high concentrations and boiling to extract pure humic acids. Using non-organic coagulants it is also possible to rapidly precipitate all the humic acids from an alkaline solution. In contrast to the traditional methodology, salts and amorphous silicon oxide, which fall out at this stage, do not contaminate lithium carbide with sulphur and phosphorus in the process of vacuum pyrolysis.

For the samples of loamy and sandy soils the following technique is used. The sample is macerated in hot water and filtered through a fine sieve to remove the roots of modern plants. Simultaneously, the major portion of fulvic acids passes into the hot water. After decanting or centrifuging, the sample is treated with 1% hydrochloric acid to destroy those soluble calcium and magnesium salts of humic acids that are not easily soluble. The mixture is separated again by decanting or centrifuging and then treated with hot sodium alkali (2 to 5%). In such conditions humic acids dissolve rapidly and with high yield. The solution of humic acids is separated from the mineral residue by centrifuging. After adding a coagulant (sodium sulphate), it is neutralised with hydrochloric or sulphuric acids to pH 4 to 6. Decanting or centrifuging separates the jelly containing amorphous silicon oxide and all kinds of humic acids. Later, if necessary, it is possible to fractionate the humic acids and to date them separately or selectively. Our experience of dating soil organics suggests that there is no need for such separation in the great majority of cases. In practically all cases, the predominant fraction of humic acids is dated.

After drying and determining the percentage of carbon in the obtained mixture of silicon oxide and humic acids, the calculated quantity of manganese dioxide is added, and the mixture is placed into the reactor for vacuum pyrolysis. These processes are identical to those for the soils rich in organics, described above. Silicon and other harmful impurities do not pass into carbide, which allows benzol to be obtained with high counting responses. The method of intensive extraction of humic acids in combination with that of vacuum pyrolysis makes it possible to minimise losses of carbon and to reduce the time needed in the chemical treatment of each sample.

A useful attribute of the manganese dioxide is its ability to link sulphur and phosphorus in the course of vacuum pyrolysis. An absence of minute amounts of phosphorus and sulphur compounds in acetylene is a necessary condition for further catalytic synthesis of benzol. In our laboratory we use a highly selective Vanadium catalyst. This Vanadium catalyst is very sensitive to harmful impurities but it allows purer and more stable benzol to be obtained from acetylene. The chemical installation for the purification and polymerisation of the acetylene comprises materials that do not absorb it. The interior volume of the vacuum line was minimised due to the optimum construction of the elements of the system and the use of a cryogenic pressure stabiliser. Therefore, the process of benzol synthesis is free of a 'memory effect'.

To completely standardise the counting responses of the obtained benzol we perform an additional chemical treatment. The calculated quantity of sulphuric acid is added to the freshly obtained benzol. In three hours the acid is removed with the help of a micropipette. For complete removal of sulphur-containing compounds dissolved in benzol, a small quantity metallic sodium foil is placed into the glass vessel for 48 to 72 hours. Then benzol is distilled off by the method of low-temperature sublimation. Such distillation allows avoiding drop transfer of harmful impurities as well as losses of benzol. This is especially important for dating samples with a low carbon content such as loamy fluvial and lacustrine sediments or micro-samples of wood and bone. After adding a complex of scintillators, the benzol is placed into Teflon vials of optimum size. Size optimisation makes it possible to minimise the influence of cosmic irradiation on accuracy and to achieve stability in determining the β-activity.

The radiocarbon content is measured by 'Quantulus-1220T'. To take into consideration all distorting factors when calculating a sample age, corrections based on the ratio of stable carbon isotopes are made. In our laboratory this ratio is determined in the fully purified benzol. A microscopic portion of benzol (1to 5 milligrams) is oxidised to carbon dioxide and tested with the use of a MI-2001 mass-spectrometer.

5 LANDSCAPE EVOLUTION IN THE LOWER VYCHEGDA VALLEY

To reconstruct the landscape and climatic conditions that existed at various stages in the evolution of the Vychegda valley, palynological studies of alluvium, peat, and lake sediments were carried out on three sections in the river terrace and floodplain sequence that were dated by radiocarbon methods. The use of palaeobotanical data for palaeoclimatic and palaeolandscape reconstruction implies that the flora of a particular region, or the composition of plant species growing there, is directly related to the natural environment as a whole and the climate in particular. The method used here of reconstructing vegetation and climate from the composition of fossil floras was developed by Grichuk (1969, 1985), based upon an earlier concept developed by Szafer (1946). Geographical analysis of the modern spatial distribution of all the plant taxa of certain fossil flora allows the location of the closest modern floristic analogue to the past vegetation at the site to be found. By identifying the region where all the species of plants grow at the present time it is possible to determine the closest modern landscape and climatic analogue to the past environment under consideration. Usually the conditions suitable for all the plants of a given fossil flora can be found within a comparatively small area. The present-day features of plant communities and main climatic indices of such a region-analogue should be close to if not identical to those that existed at the site in the past.

A series of 15 samples of fluvial deposits was subjected to detail palynological study. An attempt was made to achieve the highest possible taxonomic resolution: pollen identifications have been made to species or genus levels for arboreal plants and for certain groups of herbaceous plants. Where identifications were possible only to the family level, they were not included in the resulting lists of fossil floras. For three pairs of successive samples the floristic lists were combined to obtain more complete fossil flora. This was possible in the case of close resemblance between pollen spectra of successive samples. Landscape and climatic reconstructions using the palaeofloristic method were accomplished for eleven resulting fossil floras whose radiocarbon ages range from 52,000 to 1400 years BP (Table 4).

Table 4. Fossil floras of the lower Vychegda River deposits found in the outcrops Baika-1 and Baika-2 (according to palynological analysis).

1. Baika-1, lower peat, ~52,000 yr. BP	*Alisma plantago-aquatica; Alnus glutinosa; A. incana; Betula alba; B. humilis; Corylus avellana; Cystopteris sudetica; Ephedra distachya; Equisetum scirpoides; Hydrocharis morsus-ranae; Larix sp.; Lemna sp.; Lycopodium annotinum; L. selago; Menyanthes trifoliata; Nuphar luteum; Nymphaea alba; Osmunda cinnamomea; Picea abies; Pinus silvestris; Polemonium coeruleum; Polygonum bistorta; Sanguisorba officinalis; Selaginella selaginoides; Thalictrum angustifolium; Tilia cordata; Ulmus laevis; Valeriana officinalis;*
2. Baika-1, upper peat, ~44,000 yr. BP	*Betula alba; B. humilis; B. nana; Centaurium; Encalypta; Ephedra distachya; Larix sp.; Lycopodium tristachyum; Picea abies; Pinus silvestris; Polygonum bistorta; Selaginella selaginoides; Thalictrum flavum;*
3. Sol'vychegodsk, aleurite, ~34,000 yr. BP	*Alnaster fruticosus; Alnus incana; Betula alba; B. humilis; B. nana; Chenopodium album; Cystopteris sudetica; Ephedra distachya; Filipendula ulmaria; Lycopodium annotinum; Picea abies; Pinus sibirica; P. silvestris; Sanguisorba officinalis; Saxifraga sp.; Thalictrum minus;*
4. Baika-1, lower loam layer, estimated age: Last Glacial Maximum	*Alnaster fruticosus; A. incana; Betula alba; B. humilis; B. nana; Botrychium boreale; Chenopodium album; Cystopteris sudetica; Encalypta sp.; Lycopodium annotinum; L. appressum; L. pungens; L. selago; L. tristachyum; Picea abies; Pinus silvestris; Riccia sp.; Saxifraga sp.; Selaginella selaginoides;*
5. Baika-1, upper loam layer, estimated age: Post the Last Glacial Maximum	*Abies sp.; Alnaster fruticosus; Alnus incana; Atriplex hastata; Betula alba; B. humilis; B. nana; Chenopodium album; C. botrys; C. viride; Cystopteris sudetica; Eurotia ceratoides; Kochia prostrata; Lycopodium annotinum; L. clavatum; Picea abies; Pinus sibirica; P. silvestris; Pteridium aquilinum; Saxifraga sp.; Selaginella selaginoides;*

Table 4. (Continued).

6. Baika-2, loam, depth 435-445 cm, ~ 8800 yr. BP	*Alnaster fruticosus; Alnus glutinosa; A. incana; Betula alba; B. humilis; Cystopteris sudetica; Dryopteris phegopteris; Filipendula ulmaria; Lycopodium annotinum; L. clavatum; L. selago; L. tristachyum; Nuphar pumilum; Nymphaea candida; Picea abies; Pinus sibirica; P. silvestris; Polygonum amphybium; Selaginella selaginoides; Thalictrum simplex; Tilia cordata; Ulmus laevis; Valeriana officinalis;*
7. Baika-2, loam, depth 320-325 cm, ~ 8400 yr. BP	*Alisma plantago-aquatica; Alnus glutinosa; A. incana; Betula alba; B. humilis; Chenopodium album; Corylus avellana; Cystopteris sudetica; Lycopodium annotinum; L. clavatum; L. tristachyum; Lythrum salicarya; Menyanthes trifoliata; Picea abies; Pinus sibirica; P. silvestris; Sagittaria sagittifolia; Thalictrum angustifolium; Tilia cordata; Ulmus laevis;*
8. Baika-2, peat, depth 231-255 cm, ~ 6000 yr. BP	*Acer platanoides; Alisma plantago-aquatica; Alnus glutinosa; A. incana; Betula alba; B. humilis; Corylus avellana; Cystopteris sudetica; Lycopodium annotinum; L. clavatum; L. complanatum; L. selago; L. tristachyum; Lythrum salicarya; Nuphar luteum; N. pumilum; Nymphaea alba; Picea abies; Pinus silvestris; Quercus robur; Tilia cordata; Ulmus glabra; U. laevis; Valeriana officinalis; Viburnum opulus;*
9. Baika-2, peat, depth 209-211 cm, ~ 4500 yr. BP	*Alisma plantago-aquatica; Alnus glutinosa; A. incana; Betula alba; B. humilis; Corylus avellana; Cystopteris sudetica; Filipendula ulmaria; Lycopodium annotinum; L. clavatum; Lythrum salicarya; Menyanthes trifoliata; Nuphar luteum; Picea abies; Pinus silvestris; Quercus robur; Rhamnus frangula; Thalictrum angustifolium; Tilia cordata; Ulmus laevis; Viburnum opulus;*
10. Baika-2, peat, depth 175-191 cm, ~ 3500 yr. BP	*Acer platanoides; Alnus glutinosa; A. incana; Betula alba; B. humilis; Chamaepericlimenum suecicum; Corylus avellana; Lycopodium annotinum; L. clavatum; Lythrum salicarya; Menyanthes trifoliata; Picea abies; Pinus silvestris; Quercus robur; Rhamnus frangula; Tilia cordata; Ulmus laevis; Utricularia sp.; Viburnum opulus*
11. Baika-2, peat, depth 100-130 cm, ~ 1400 yr. BP	*Abies sibirica; Alnus glutinosa; Alnus incana; Athyrium crenatum; Betula alba; B. humilis; Corylus avellana; Calluna vulgaris; Filipendula ulmaria; Juniperus sp.; Lycopodium annotinum; L. clavatum; L. complanatum; Lythrum salicarya; Menyanthes trifoliata; Picea abies; Pinus sibirica; P. silvestris; Rhynospora alba; Ribes sp.; Tilia cordata; Ulmus laevis; Viburnum opulus;*

The earliest fossil flora (Number 1 in Table 4) derived from the palynological study of the peat layer at the base of the Baika-1 section (the third terrace of the Vychegda River) belongs to the relatively warm stage at the beginning of the Middle Valdai period. The region currently inhabited by the listed species was determined from map sources by superimposing their modern ranges. It lies in the Southern Ural Mountains, in the eastern part of the lower Ufa River basin. Cool mixed broad-leaved and dark coniferous forest (*Picea abies* with minor presence of *Ulmus laevis*, *Tilia cordata*, and *Corylus avellana*) and light coniferous mountain forest of Southern Ural type (*Larix sibirica, Pinus silvestris*) occupy the area at present. Tree birch in this region form both open forests and grow in the light coniferous forests. The area is characterised by a more continental and severe climate than that of the study area today. As climate is a key factor restricting the ranges of plants, it is assumed that the climatic indexes at the time when these species grew together in the lower Vychegda basin were identical to the present indexes of the region-analogue. This comparison indicates that the mean temperature of the coldest month (January) was 2°C lower than the present-day one, while the mean July temperature was close to the modern one at the site. Annual precipitation was about 750 mm.

Flora 2, which characterises the upper peat layer in the Baika-1 section with a radiocarbon age of 43,600 ± 2100 yrs BP, is similar to the first one, though the pollen assemblage reflects considerable changes in the composition of vegetation. It is likely that these changes occurred due to the cooling of climate, possibly accompanied by a decrease in precipitation. Birch forest became dominant in the area, while the dark coniferous taiga almost disappeared.

Fossil flora 3 obtained from the silt layer at the base of the sequence beneath the second terrace of the Vychegda River near Sol'vychegodsk (34,200 ± 2900 yr. BP), is very similar to the first one. At this time the dark coniferous forests had recovered their position due to an amelioration in climate. Such second order climatic oscillations are characteristic of the Middle Valdai period in the Russian Plain, and it is usually referred to as the 'mega-interstadial' to emphasise its complex structure. The region currently inhabited by the species of flora 3 lies at the headwaters of the Ufa River where dark conifer forest (*Picea abies, Pinus sibirica*), birch forest and light conifer (pine) forest with steppe elements in the herbaceous layer are found. The climate at this time interval was warmer than at the previous one but still slightly colder than in the first Middle Valdai warming characterised by flora 1. Mean January temperature was –17°C, that of July was about 17°C, and annual precipitation was 650 mm.

Pollen assemblage 4 of the lower loamy layer at a depth of 10 m in the Baika-1 section is separated from the previous one by a considerable period. As no organic material for radiocarbon dating was found in this layer, its age was estimated as being close to that of the maximum spread of the last (Late Valdai) glaciation in the area. At that stage an ice-dammed lake occupied part of the lower Vychegda valley. The pollen assemblage of this time interval is strongly dominated by spores of *Sphagnum*, Polypodiaceae and *Lycopodium*. It consists mainly of spores and pollen produced by local plant communities, which probably existed near the shoreline of the cold ice-dammed lake, while the tundra-like vegetation in the area fringing the ice-sheet margin was sparse and had very low pollen productivity. The flora includes such typical cryophytes as *Lycopodium pungens*, *Botrychium boreale*, and *Selaginella selaginoides* along with plants growing on barren (eroded) ground (*Riccia*, *Encalypta*). It is assumed that scarce pollen grains of spruce, pine and possibly also tree birch found in the sample were brought by wind from some distance, while the area at the time was treeless. A modern floristic analogue for this vegetation can be found in the Malozemelskaya tundra, near the Barents Sea coast. The region is characterised by very low summer temperatures (10-12°C), as well as by a very short frost-free period (c. 90 days). It is also situated within the limits of the permafrost zone.

Pollen assemblage 5 (loamy layer at 8.5 to 9.1 m depth in the Baika-1 section) reflects the spread of specific cryoxerotic vegetation of the Last Glacial Maximum – the so-called periglacial forest steppe (Grichuk, 1989). This pollen assemblage is characterised by the highest *Artemisia* pollen percentage and the greatest diversity of the Chenopodiaceae species of all the studied samples. Flora 5 includes species of light coniferous forest (*Pinus silvestris*), small-leaved deciduous (*Betula alba*), and dark coniferous forest (*Picea abies, Pinus sibirica, Abies sibirica*), as well as steppe and tundra species. An important characteristic of this assemblage is the spread of periglacial steppe communities including such xerophytes as *Ephedra distachya*, *Eurotia ceratoides*, *Kochia prostrata* and others, relatively indifferent with regard to temperature. At present these species grow in plant communities of dry steppe and semi-desert environments. *Ephedra*, however, occurs in the Pamir Mountains up to an altitude of c. 4000 m above sea level. Cryophytes (*Botrychium boreale*, *Selaginella selaginoides*) and plants, growing on barren (eroded) ground, are also typical of periglacial flora, and are commonly found growing on barren (eroded) ground. This type of flora has no direct contemporary analogue. The closest modern region-analogue is situated on the western slope of the Altai Mountains, east of the headwaters of the Irtysh River basin (Bukhtarma River). Within this small area, dark coniferous mountain forests (*Picea abies*, *Abies sibirica*) grow next to mountain meadow steppes and patches of wormwood-grass dry steppe. The area is characterised by a cold semi-arid and extremely continental climate. The mean January air temperature is –18°C and that of July is about 15°C. Positive temperatures occur on between 90 and 100 days per year and the mean annual precipitation varies between 500 and 600 mm, including 200 mm for the period from November until March. The region is situated near the boundary of the permafrost zone. This flora also has an additional centre of concentration in the Malozemelskaya tundra.

As the data published by Potapenko (1975) show, periglacial vegetation persisted in the Vychegda basin until the beginning of the Holocene. In the middle part of the Vychegda basin, three radiocarbon dates were obtained for a thin layer of peat in the sandy alluvial deposits: 10,900 ± 1300 (MGU-128), 10,560 ± 90 (MGU-90), and 10,460 ± 20 (St-3327). Pollen spectra from this layer are characterised by high concentrations of non-arboreal pollen (up to 60%), including about 20% of *Artemisia* pollen. The flora shows all the important features of the periglacial type, as it includes cold-resistant arboreal species, such as tree, shrub and dwarf birch, shrub alder (*Alnaster fruticosus*), typical cryophytes (*Selaginella selaginoides*, *Lycopodium pungens*) along with xerophytes (*Ephedra distachya*).

Fossil floras 6 to 11 characterise successive layers of the Baika-2 section (Fig. 3B). The alluvial deposits are represented by loam and sand with two layers of peat in the upper part of the sequence. These deposits accumulated during the Holocene. The pollen assemblages are dominated by spruce, pine and birch pollen throughout the whole sediment sequence, but substantial changes nevertheless took place in the composition of the vegetation. These changes are reflected by 'migration' of modern floristic analogue regions over the East European Plain beginning from the Middle and Southern Urals, as far west as the Sukhona River basin, and back to the site under study at the downstream end of the Vychegda River (Table 5, Fig. 3A).

A modern floristic analogue of flora 6 (~8800 yr. BP) is located at the headwaters of the Kolva River, at the boundary between dark coniferous mountain taiga forests (*Picea abies, Pinus sibirica*) and middle- and southern taiga pine forests (*Pinus silvestris*). The present-day climate of this region is much more continental and severe than that at the Vychegda site. Mean January temperature was 3.5°C lower and mean July temperature

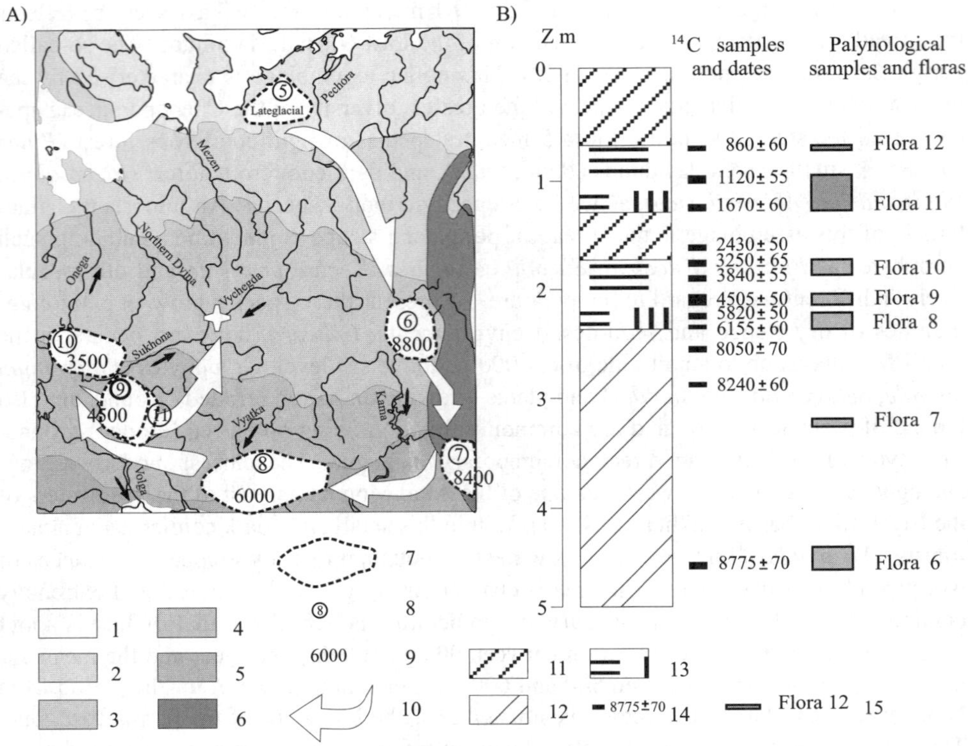

Figure 3. Evolution of the landscape in the lower Vychegda region during the Lateglacial to Holocene based on contemporary region-analogues (A), derived from the dated samples from the Baika-2 section with palaeofloristic analysis (B): 1. tundra and forest-tundra; 2. northern taiga; 3. middle taiga; 4. southern taiga; 5. broad-leaf forest; 6. mountain forest; 7. area of region-analogue; 8. number of region-analogue (the same as for fossil flora); 9. age of region-analogue; 10. sequence of region-analogues; 11. loam; 12. sandy loam; 13. peat; 14. radiocarbon sample location and date; 15. palynological sample location and fossil flora number as shown in Table 4.

0.5°C lower than at present (Table 5). The mean annual temperature was slightly below 0°C and annual precipitation exceeded the modern value by more than 150 mm.

A region-analogue for flora 7, with the age estimated by interpolation between the closest ^{14}C dates as 8400 yrs BP, lies near the headwaters of the Chusovaya River. In this area broad-leaved-dark coniferous mountain forests of the Ural type are found in the vicinity of dark coniferous mountain taiga forests (*Picea abies* with *Pinus sibirica*). The continentality of climate was still very high at this time, as the temperature of the coldest month was still 2.5°C lower than at present, while that of the warmest month was almost 1°C higher. Mean annual temperature was already above 0°C. The duration of the frost-free period was slightly longer than the present-day equivalent. Annual precipitation was slightly less than the present-day mean (630 mm).

A considerable shift to the west of the modern analogue for flora 8 (about 6000 yrs BP) corresponds to the rise in summer temperature, which exceeded the present-day mean by

Table 5. Positions of the paleofloristic analogues and contemporary hydro-climatic characteristics of the regions discussed in the text.

Fossil flora numbers as shown in Table 4	Region-analogue	Co-ordinates of the centre of each area		Region age, years BP	Recent air temperature (°C)		Recent runoff depth (mm)		Recent precipitation (mm)		Recent evapotranspiration (mm)	
		Northern latitude	Eastern longitude		January	July	Annual	Winter-spring	Annual	Winter-spring	Annual	Winter-spring
4-5	Malozemel'skaya tundra	67°	49°	Lateglacial	–17	11	420	270	600	300	180	30
6	Kolva River basin	61°	57°	8800	–17.5	16.5	380	260	850	410	470	150
7	Chusovaya River basin	57.5°	59°	8400	–17	18	190	140	630	310	440	170
8	Middle Vyatka River	57.5°	49°	6000	–14	19	200	160	640	360	440	200
9	Unzha River basin	58.5°	44°	4500	–13	18	250	180	720	360	470	180
10	Headwaters of Sukhona River	59.5°	39.5°	3500	–13	18	330	220	750	400	420	180
11	Unzha River basin	58.5°	44°	1400	–13	18	250	180	720	360	470	180
	Vychegda River basin	61.3°	47°	recent	–14	17	300	200	700	350	400	150

2°C, while the winter temperature reached the present level. This region, in the middle part of the Vyatka River basin, is covered at present by mixed broad-leaved and dark coniferous sub-Volga forests (*Picea abies* with *Quercus robur* and *Tilia cordata*). The frost-free period at that stage was almost one month longer than at present, though the annual amplitude of air temperature was still greater than now. Precipitation was approximately 640 mm.

A modern analogue for fossil flora 9 (4500 yr. BP) can be found in the northern part of the southern taiga dark coniferous forest in the Unzha River basin. The boundaries of the present-day ranges of such species as *Quercus robur* and *Thalictrum angustifolium* form its eastern boundary. Present-day climatic conditions of the region show that for the Vychegda River site, the mean January temperature exceeded the present value for the first time by about 1°C, while that of July decreased by 1°C compared to the previous stage. A substantial increase in precipitation is also suggested (by c. 80 mm per year). The continentality of climate reached its minimum value for the Holocene at this time and the period characterised by fossil floras 8 and 9 can be regarded as the climatic optimum of the Holocene.

A region-analogue for flora 10 (about 3500 yr. BP) is found in the Sukhona River basin, near the northern limit of the southern taiga dark coniferous forest, where large peat bogs are widespread. At that stage the mean January temperature was still about 14°C (1°C higher than the present), and that of July was about 18°C. Annual precipitation reached 750 mm – approximately 50 mm more than at present.

The modern analogue for fossil flora 11 is again found towards the east, into the Unzha River basin, where southern taiga dark coniferous forests are present along with middle- and southern taiga pine forests of the North European type. The climatic conditions of this region indicate that the climate of the lower Vychegda basin around 1400 yrs BP was still warmer and less continental than at present. The precipitation was about 720 mm per year.

The fossil flora of the upper peat layer in the Baika-2 section dated by ^{14}C as 860 ± 60 yr. BP (Ki-7026) does not include any taxa which are absent from the modern flora of the region, where the studied section is located. This suggests that the climatic conditions at that time were essentially similar to those of the present-day.

6 PALAEOHYDROLOGICAL RECONSTRUCTION

Close relationships between river morphology (channel pattern, width, depth, slope, meander wavelength) and grain size of alluvial sediments on the one hand and the main hydrological and hydraulic characteristics of the river flow (discharge, velocity) on the other hand represent the basis of morpho-palaeohydrology. Hydro-morphological (regime) relationships were first obtained by Fergusson for the Ganga River in 1883 (according to S. Leliavski, 1955). Later they were broadly used for investigations of both river channels and canals. After the works of Dury (1964, 1965) and Schumm (1965, 1968) hydro-morphological formulas have been widely used in palaeohydrological reconstructions (see, for example, Maizels, 1983; Williams, 1988; Starkel et al., 1996).

The evolution of the hydrological characteristics of the lower Vychegda was reconstructed on the basis of palaeochannel morphology (Table 2) and a relationship between mean annual discharge $\overline{Q}$ (m^3/s) and bankfull channel width W_b (m). This relationship was established for 185 sections of meandering rivers with broad floodplains on the Russian Plain in western and eastern Siberia (Fig. 4). The long-term observations of the Russian

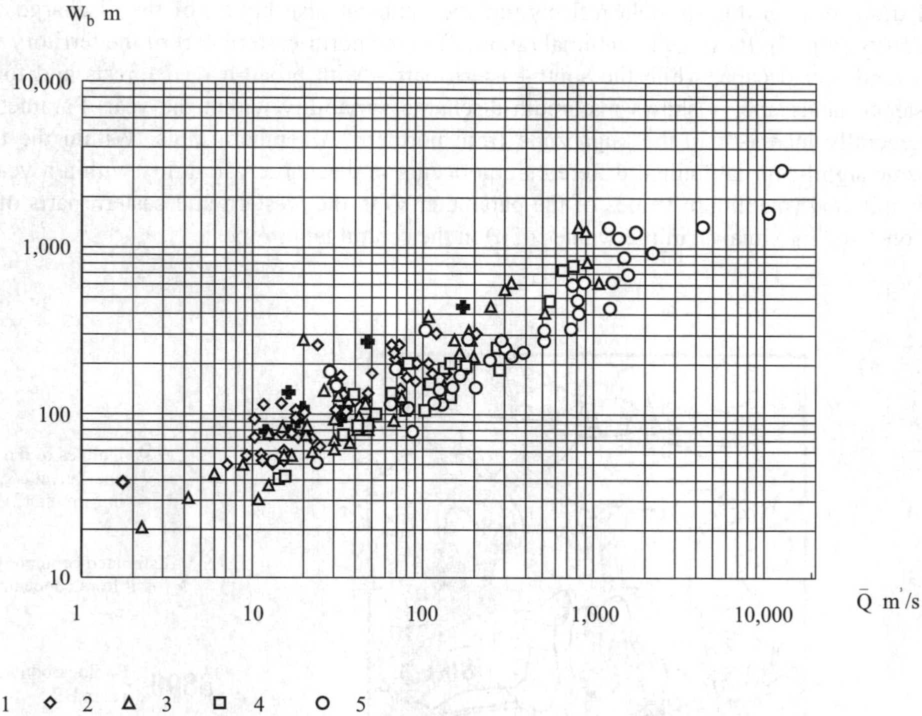

Figure 4. Correlation between channel bankfull width W_b and mean discharge Q for meandering rivers of the East European and West Siberian plains. Point symbols indicate the discharge variability within a year $y = 100\ Q/Q_{max}$. $1 - 0 \leq y < 5$; $2 - 5 \leq y < 10$; $3 - 10 \leq y < 15$; $4 - 15 \leq y < 20$; $5 - y \geq 20$.

Federal Hydrometeorological Survey were analysed. The average discharge values were calculated for the whole period of measurements (based mainly on data from the 1950s to 1980s). A width for each river site was estimated from width-stage relations for the initial stage of floodplain submersion. Regression analysis of these data leads to the formula

$$\overline{Q} = 0.012y^{0.73}\,W_b^{\,1.36} \qquad (1)$$

This formula allows calculation of the discharge with a multiple regression correlation coefficient of 0.9. The parameter y was used to decrease scatter in the relationship between $\overline{Q}$ and W_b. This parameter is related to the variability of river discharge during the year and can be calculated as $y = 100\ (\overline{Q}/Q_{max})$. Here Q_{max} is the average of annual maximum discharges for the period of measurement. It is evident from (1), that for the same mean discharge a river with a more variable hydrograph has a wider channel. The points on the graph corresponding to the rivers of the north-eastern Russian Plain have a general distribution similar to that of rivers of other regions.

The variability of discharge within the year, which is related to parameter y, changes significantly over the territory of the north-eastern Russian Plain and depends on the river basin area A (km^2):

$$y = aA^{0.125} \qquad (2)$$

Parameter *a* in this formula reflects the geographical distribution of the discharge variability (Fig. 5). Its value is minimal (about 2) in the north-eastern part of the territory with a tundra landscape, while the south-eastern part – with broad-leaved forests and forest-steppe landscapes – shows maximum discharge variability within the year. Parameter *a* generally increases in the taiga zone from northern to southern taiga. Within the taiga zone significant latitudinal differentiation occurs in discharge variability within a year. It is minimal (maximum values of the parameter *a*) in the western and eastern parts of the zone, and maximal (minimal values of *a*) in the central taiga zone.

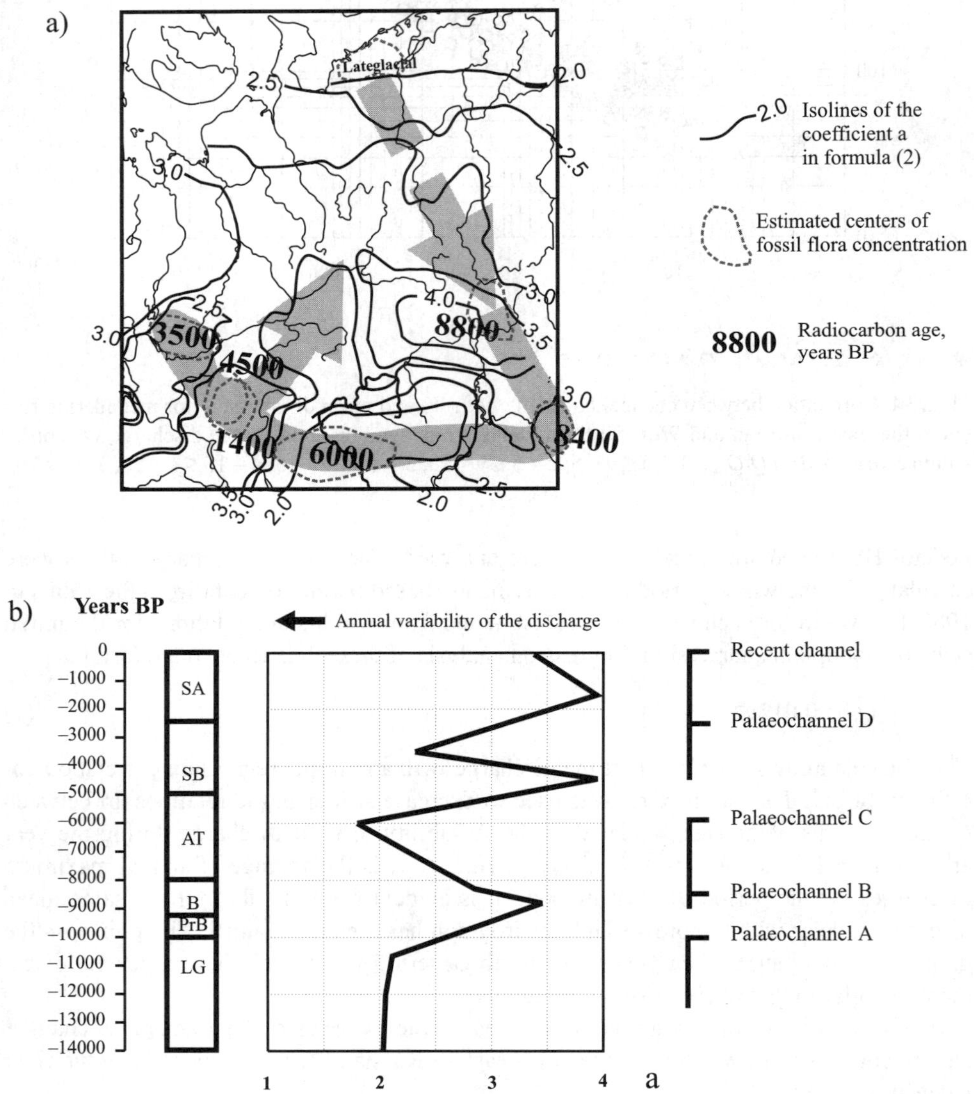

Figure 5. Annual discharge variability as characterised by coefficient *a* in Formula (2). (a) Spatial distribution in the north-eastern Russian Plain. (b) Change over time at the Vychegda study site.

The initial information for palaeohydrological calculations can be obtained from Tables 2 and 5. The region-analogue for the time of palaeochannel formation gives an estimate of parameter *a* in Formula (2), annual precipitation P_a, evapotranspiration E_a, and runoff depth X_a, and those for the winter-spring season P_{ws}, E_{ws}, and X_{ws} (Table 5). Mean annual discharge at the time of palaeochannel formation can be calculated directly from the palaeochannel width with the use of Formula (1), and parameter *y* estimated with Formula (2) for a given region – analogue. Mean maximum discharge is calculated from the mean discharge with the use of parameter y:

$$Q_{\max} = 100\overline{Q}/y^* \tag{3}$$

Annual runoff depth equals to

$$X_a = 31.536\overline{Q}/A \tag{4}$$

that for a winter-spring period – to

$$X_{ws} = \frac{X_{ws}^*}{X_a^*} X_a \tag{5}$$

A precipitation depth per annum is calculated from annual runoff depth with the use of the water budget equation:

$$P_a = X_a + E_a^* \tag{6}$$

or

$$P_a = \frac{P_a^*}{X_a^*} X_a \tag{6a}$$

For the winter-spring season it can be estimated from the water budget equation for this period:

$$P_{ws} = X_{ws} + E_{ws}^* \tag{7}$$

or

$$P_{ws} = \frac{P_{ws}^*}{X_{ws}^*} X_{ws} \tag{7a}$$

Formulae (6a) and (7a) are used if evapotranspiration in the region-analogue can not be accurately estimated. The characteristics taken from region-analogues (Table 5) are marked with an asterisk.

Reconstructions of the Lateglacial landscapes and palaeohydrology of the north-easten part of European Russia have been produced by several workers including Potapenko

(1975), Kvasov (1979) and Arslanov et al. (1980). The majority of researchers believe that a lake occupied the lower Northern Dvina and Vychegda River valleys during the period of damming of these valleys by the last ice sheet. Kvasov (1979) reconstructed the level of this lake up to 170 m above mean sea level so that surplus water presumably drained over the watershed into the Kama River valley.

The composition of the sediments comprising the third terrace in the lower Vychegda valley bears evidence of complicated processes of formation. Sediment structure and texture, radiocarbon dates and pollen analyses allow the deposits to be sub-divided into two main beds. The lower bed (layers 1-3 in Table 1 from the bottom) represents floodplain deposits of the Vychegda River which accumulated during Middle Valdai times. These deposits include layers of floodplain lake sediments (layers 1 and 3, and the lower silt layer near Sol'vychegodsk), and sediments of irregular flood flows (layer 2).

The upper bed is composed mainly of fine-grained horizontal and cross-bedded well sorted sand (layers 6, 8). These sands were transported by flows with velocities of approximately 0.3 to 0.5 m/s. The flow was apparently rather uniform over time, as the sediment grain size does not vary significantly. The inclination of cross beds indicates the general north-westward direction of the flow. The slope of the third terrace is in the same direction. The sand deposits include layers 5, 7, and 9 of red-brown loam with a wavy cryogenic structure. These layers could have been formed by sedimentation in a lake and by solifluction from adjacent slopes during periods of flow shifting to the opposite (right) bank of the valley. In general, the upper bed of the sediments beneath the third terrace represents the depositional stage of a broad delta formation in an ice-dammed lake.

The termination of the ice-dammed lake at the Northern Dvina and Vychegda valleys is correlated with the Bolling Interstadial (Kvasov, 1979). After recession of the Last Glacial ice sheet about 12,500 years ago the base level lowered considerably. River incision into deposits of the third terrace had begun, and the sequence of erosional terraces was formed. The most ancient surface of the second terrace on the lower reaches of the Vychegda corresponds to the initial stage of the valley evolution, which began after c. 12,500 years ago. The large size of palaeochannel A on this step indicates a high channel-forming discharge during that period. The presence of deep chutes and secondary channels on the former floodplain also evidence a high water yield during floods.

The most likely age of this ancient palaeochannel in the lower Vychegda valley is 12,500 to 10,000 years BP (the Lateglacial). Fossil floras 4 to 5 (Table 4) show that tundra and forest-tundra in north-eastern European Russia represent the closest through incomplete analogue for that period. The hydrological regime of rivers of that area was used as a contemporary hydrological analogue. The mean value of coefficient a in equation (2) is 2.02 for those rivers. The annual discharge variability was rather high during the Lateglacial (for the lower Vychegda mean y = 8.7), so that the mean maximum discharge of 10,300 m^3/s corresponds to the relatively low mean annual discharge of 900 m^3/s. The depth of the mean annual runoff was 235 mm. According to the same analogue, the depth of winter-spring runoff was 180 mm. In conditions of continuous permafrost, the runoff coefficients were very high, the mean annual value being about 0.7, and that for a flood period up to 0.9. The mean annual precipitation depth estimated with equations (6a) and (7a) is 335 mm, that of the winter-spring period is 200 mm, and that of the summer-autumn period is 135 mm.

On the basis of palaeobotanical data the Lateglacial period is usually characterised as an extremely dry period (Khotinskiy, 1977). The palaeohydrological information allows some amplification of this supposition. For the Vychegda River basin, annual precipitation was half the recent value. Such a deficit of humidity was probably related almost entirely to the summer and autumn period. The palaeobotanical data for the Lateglacial show the presence of typical xerophytes in the vegetation of this period. Their spread can be explained by a relatively warm and dry summer, though it was short (not longer than three months). The precipitation (mainly in the form of snow) during a long winter – spring period was not lower than in recent tundra. In combination with low permeability of the ground due to the permafrost, the thawing snow during the spring led to sharp high floods and to the formation of large wide river channels.

Palaeochannel B on the lower step of the second terrace was abandoned about 8400-8600 years ago. It was still active during early Boreal time between c. 8500 and 9000 years ago. According to palaeobotanical data, the region-analogue for that time is situated at the south-eastern part of the middle taiga. The territory was already free of permafrost, and the depth of seasonal freezing of the ground was close to the contemporary level. The hydrological analogues give a mean value of the coefficient in Formula (2) of $a = 3.48$. The annual distribution of the water flow during the early Boreal was also close to the present-day distribution (for lower Vychegda, mean $y = 15$). A large and wide palaeochannel under conditions of low discharge variability corresponds to a relatively high mean annual discharge of c. 1500 m^3/s. Mean maximum discharge was about 9900 m^3/s. The depth of mean annual runoff was 390 mm and the depth of spring runoff was 265 mm. In conditions of seasonal ground freezing the runoff coefficients were close to recent ones: they were about 0.45 for the whole year and up to 0.63 for a flood period. The air temperature and evapotranspiration at the lower Vychegda valley were close to those in the region-analogue. This gives a rather high estimate (860 mm) for the mean annual precipitation. Precipitation in the winter-spring period was twice that of the previous time interval (415 mm), and that of the summer and autumn period increased even more – up to 445 mm.

The hydrological regime of the lower Vychegda River changed significantly during the transition from the Lateglacial to the Holocene. The annual water yield increased by a factor of 1.7 and this was mainly due to a more humid low flow period. At the same time, maximum flood discharges decreased slightly due to more regular water flows in the basin and greater flood duration. Flood volume was rather stable during this transition, therefore the size and pattern of the river channel did not change. A meandering channel together with alluvial bars and islands shifted by 600 m down the valley and incised by about 3 m in approximately 2500 years.

Palaeochannel C on the first terrace of the lower Vychegda was abandoned about 8200 years ago. It was active during late Boreal time around 8200 to 8500 years ago. According to the palaeobotanical reconstruction the region – analogue is situated at the eastern part of the southern taiga. The depth of seasonal freezing of the ground was close to the contemporary level. The hydrological analogue gives a mean value of coefficient a in Formula (2) of 2.8. The annual distribution of flow during the late Boreal was more variable than at present, and humidity at that time was significantly lower. The palaeochannel on the first terrace is smaller than all other palaeochannels. Its width and curvature correspond to a mean annual discharge of about 440 m^3/s and the mean maximum discharge was about 3700 m^3/s. The depth of mean annual runoff was 115 mm and the depth of spring runoff was 90 mm. This corresponds to the minimum value of the winter-spring precipitation depth (260 mm) for the whole period studied. Mean annual precipitation was about 555 mm.

There are numerous palaeochannels of relatively small size on the first terrace of the Viled' River, a left bank tributary of the Vychegda (Fig. 6). The width and meander lengths of these palaeochannels correspond to the mean annual discharge of 29 m^3/s, which is about 1.6 times less, than the recent one. One of the latest palaeochannels of Viled' was abandoned about 7700 years ago (point 11 in Fig. 6). It means that the stage of low humidity belonged to the Atlantic period of the Holocene.

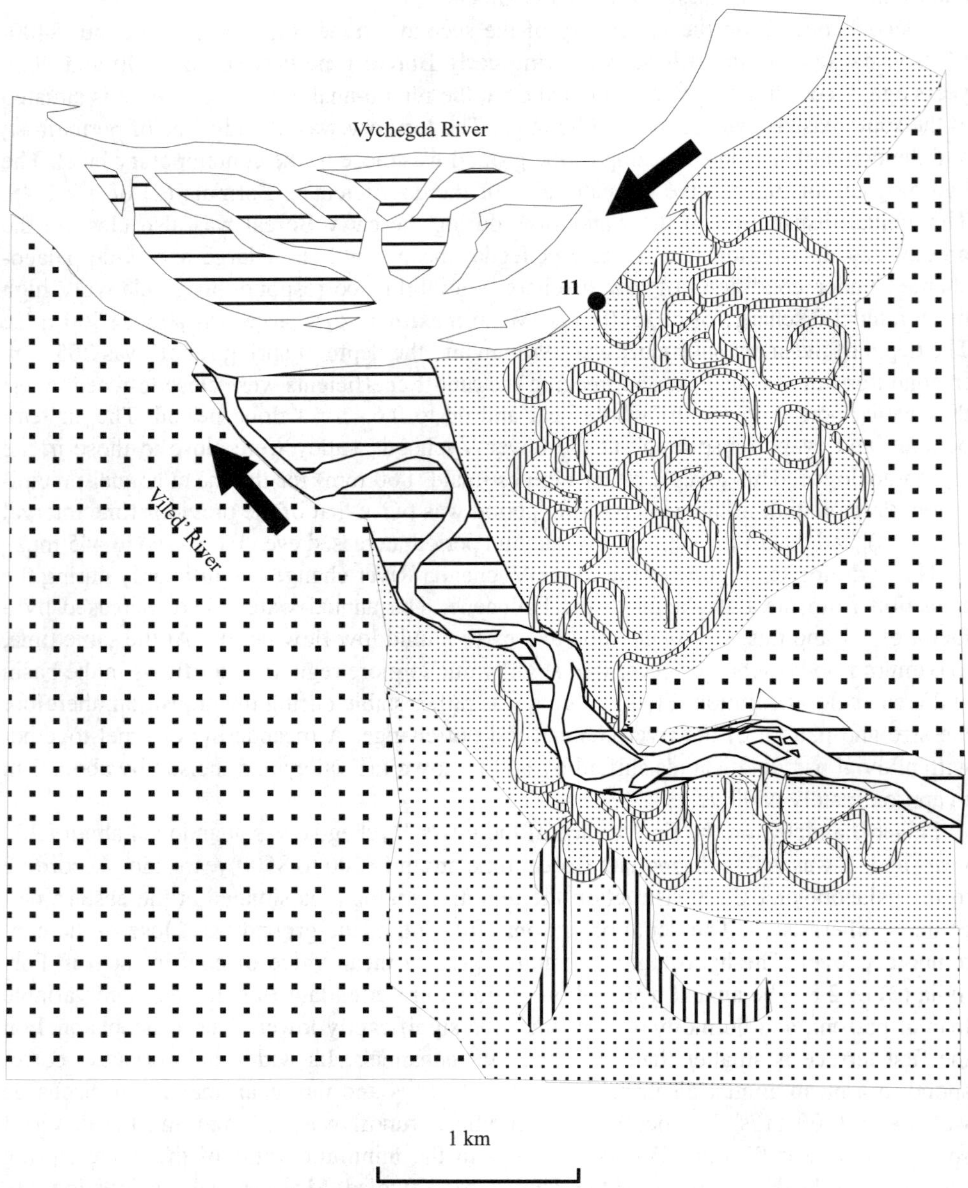

Figure 6. Morphology of the lower Viled' river valley. See Figure 1 for legend.

The region-analogues for this period, found by the palaeofloristic method, are also characterised by low precipitation. Palaeohydrological calculations for this period were conducted with the use of interpolated values of palaeochannel width (Table 6). Minimum water flow occurred at the lower Vychegda at the end of Atlantic, about 6000 years ago. The chain of aeolian dunes along the right border of the early Boreal palaeochannel of the second terrace presumably was formed during this relatively dry period. This was also a time of interruption in deposition on the floodplain at the Baika-2 section.

The most recent palaeochannel (D) in the lower Vychegda valley is well-preserved on the oldest steps of the floodplain. This channel evolved during the Sub-boreal period and was abandoned at the beginning of the Sub-atlantic, about 2500 years ago. The geometry of natural levées and chutes on the floodplain mark the main paths of a migrating meandering channel. The rate of channel migration down-valley is estimated to have been 1.6 m per annum near Durnitsino village for the period 4200-4500 years ago, according to radiocarbon dating of a sequence of natural levées and depressions between them (points 7 and 8 in Fig. 1). The migrating channel reworked deposits that form the first terrace and several remnants of this surface were preserved within the floodplain. The development of curved omega-shaped meanders indicates the low erosivity of the flow on the floodplain. It probably also indicates relatively low annual variability in water discharge.

As the location of the region-analogues 9 and 10 show, the Sub-boreal palaeochannels were formed during a time that experienced a significant decrease in climate continentality. The morphology of the last palaeochannel corresponds to these climatic conditions. Its width and curvature were formed by a mean annual discharge about 840 to 670 m^3/s, and the mean maximum discharge was about 5000 to 6700 m^3/s. The annual distribution of flow was characterised by low variability (mean $y = 16.4$) at the beginning of the period. It increased towards the end of the formation of the palaeochannel (Table 6). The depth of mean annual runoff was 175 to 220 mm and the depth of winter-spring runoff was 115 to 150 mm. Together with the evapotranspiration value for the region-analogue, this gives a mean annual precipitation depth equal to 600 to 690 mm. Precipitation in the winter-spring period was about 290 to 330 mm.

At the beginning of the Sub-atlantic period the morphological character of the river channel changed again as it developed a sinuous with braids form. The stream then abandoned the omega-shaped meanders of palaeochannel D, which are preserved as oxbows. This was connected with an increase in the maximum discharge at the end of Sub-boreal period and with the general humidity of the climate during the Sub-atlantic. Palaeodischarge calculations indicate that flow and precipitation reached their maximum about 1500 years ago, at the period of lowest intra-annual flow variability. By recent times both the calculated mean annual discharge and precipitation depth decreased to 1170 m^3/s and 700 mm respectively, and the mean maximum discharge increased to 7900 m^3/s due to a greater discharge variability within the year (Table 6).

7 DISCUSSION AND CONCLUSIONS

One of the main problems associated with palaeohydrological reconstruction is its accuracy. This depends on the appropriateness of the main palaeogeographical hypothesis, on the validity of the empirical formulas used, and on the errors in the input data.

Table 6. Runoff and precipitation at the lower Vychegda River basin in the Lateglacial and Holocene, reconstructed from palaeochannel (and recent channel) morphology and contemporary hydro-climatic characteristics of region-analogues (see Table 5).

region-analogue	^{14}C age (years BP)	Palaeo channel index	Palaeo channel width (m)	Coefficient a in Equation (2)	Discharge (m^3/s)		Runoff depth (mm)		Precipitation (mm)	
					Mean	Mean maximum	Annual	Winter-spring	Annual	Winter-spring
input					reconstruction					
Malozemel'skaya tundra	~12000	A	1200	2.02	900	10300	235	180	335	200
Basin of Kolva River	8800	B	1300	3.48	1490	9910	390	265	860	415
Basin of Chusovaya River	8400	C	600*	2.80	440	3670	115	90	555	260
Middle Vyatka River	6000	C	700*	1.80	400	5110	105	85	545	285
Basin of Unzha River	4500	D	800	3.99	840	4970	220	150	690	330
Headwaters of Sukhona River	3500	D	900*	2.32	670	6710	175	115	595	295
Basin of Unzha River	1400	recent	1100*	3.99	1270	7720	330	225	800	405
Basin of Vychegda River	recent	recent	1100	3.66	1170	7950	300	205	700	355

* – interpolated values.

The main palaeogeographical hypothesis formulated here as the principle of palaeogeographical analogue (Sidorchuk & Borisova, 2000) postulates an opportunity to transform the present-day spatial (geographical) relationships into localised temporal relationships. G. Kalinin (1966) demonstrated that this hypothesis is based on an ergodic theorem. Broad experience of interpolation and extrapolation of hydrological parameters in space and time suggested the possibility of using this theorem in palaeohydrology. Geographical influences on river flow bring about similar hydrological regimes of rivers in similar landscapes (Evstigneev, 1990). Geographical controls over river flow and their applications to palaeohydrology lead to the principle of palaeogeographical analogy:

1. Similar hydrological regimes were characteristic of palaeorivers in similar palaeolandscapes;
2. The hydrological regime of a palaeoriver within some palaeolandscape would be similar to that of a present-day river within the same type of landscape.

The second statement forms the basis of the method of palaeogeographical analogy. Palaeohydrological reconstructions are connected to the reconstructions of palaeolandscapes. The hydrological regime of modern rivers in a certain type of landscape can be used for estimations of palaeohydrological regime in the same type of landscape.

Successive application of the principle of palaeogeographical analogy employs indices selected to describe the spatial distribution of hydrological features. These include hydrological parameters, such as the maximum, annual and daily discharge, etc., statistical moments of their distributions, and the coefficients in empirical formulae. For example, maps of elements of the water budget are available for the entire globe. Maps of mean annual and mean maximum discharges and their temporal variability were compiled for the former Soviet Union. Special maps, including those portraying parameters of flood hydrographs or discharge distribution within the year, are available for some regions or can be compiled from existing information (e.g. Fig. 5). The principle of palaeogeographical analogy can be applied only to zonal bioclimatic landscapes and to river basins of representative size. For the East European Plain the representative basin has an area not less than 10,000 km^2 to overcome the influence of local factors, and not more than 75,000 to 100,000 km^2 to avoid inter-zonal effects.

The accuracy of space – temporal transformation depends on the ability to locate a potential region – analogue on the map, as much as on the accuracy of the age control. A centre of concentration of fossil flora can be determined with some degree of accuracy dependent both on the completeness of the fossil flora record and on topographical data for the geographical ranges of modern plant species. The area of a region-analogue is usually about 10,000 to 40,000 km^2. Variability of the estimated parameter within this area causes errors in spatially-dependent data. Difficulties may also be associated with the age control for palaeohydrological events. Erosive relief has to be dated with the use of samples taken from correlative sediments. Channel, bar and floodplain deposits of the period of palaeochannel development are the most reliable sources of samples for dating. Usually these deposits are composed of well-washed sand with organic matter contents that are too low even for AMS methods of radiocarbon dating. In most cases the samples for dating are obtained from the lowers layers of fine sediments, which already fill abandoned palaeochannels. The lag between the time of channel activity and the time of fine sediment deposition can make the age control for palaeohydrological events less certain.

Two empirical hydro-morphological Formulae, (1) and (2) were used here for palaeohydrological analysis. Formula (1) describes a well-known relationship between the mean annual discharge and channel width (Leopold et al., 1964). Usually hydro-morphological formulae are designed for morphological and hydrological parameters measured in a river channel for a certain water stage. This rule is broken in Formula (1), where the bankfull width is related to the mean annual discharge, although the latter generally corresponds to a lower water level and a smaller width. Such a hydro-morphological relationship was chosen due to the specific objectives of palaeohydrological reconstructions and the uncertainty in estimation of palaeochannel parameters. In the conditions of the Vychegda River, only the bankfull width of the palaeochannels can be reliably measured, while the mean and maximum discharges are the most important hydrological characteristics.

Usually hydro-morphological formulae with two variables allow calculation of a dependent value with some scatter. The correlation coefficient for the simple relationship $\overline{Q} \sim W^b$ for the same set of data, that was used to derive Formula (1), is 0.82 (Table 7). Additional parameters have to be used to reduce the scatter. The choice of additional parameters is also an empirical procedure. The proposed parameter y is well suited to the goals of palaeohydrological calculations. This parameter has a clear hydrological meaning, as a measure of the discharge variability within a year. It is clear that with an increase in variability, the channel width which corresponds to the same annual flow would also increase. This proposition is confirmed by the data on natural rivers used to derive Formula (1). Application of parameter y in hydro-morphological formulae increased the multiple correlation coefficient to 0.90. The mean relative error of discharge estimation with Formula (1) is ±18.6%.

Discharge variability within a year generally decreases with an increase in river basin area A. The parameter y also depends on basin area. This dependency is described by Formula (2). As Figure 5 shows, coefficient a in this formula has a distinctive spatial distribution. The map of coefficient a makes it possible to use the method of palaeogeographical analogue. Formula (2) describes the $y \sim A$ relationship only in the first approximation. Further analysis may also help to discover spatial differentiation of the exponent in Formula (2).

In some cases coefficient a varies significantly within a region-analogue (at the boundaries of hydrological regions). This variability also causes errors in palaeohydrological calculations. The relative error varies from one region-analogue to another from ±2 to ±10%. The accuracy of palaeochannel width measurements depends on the degree of preservation of an ancient fluvial form and on the extent to which the exact definition of the former river bank location can be identified. Width usually changes along the river channel and the bankfull width of meandering channels is greatest at the apexes of meanders

Table 7. Coefficients, exponents, and correlation statistics for various hydro-morphological relationships.

a	*b*	*c*	*R*	formula
0.028	1.55		0.82	$\overline{Q} = a\,W^b$
1.41	1.55		0.38	$\overline{Q} = a\,y^b$
0.012	1.36	0.73	0.90	$\overline{Q} = a\,W^b y^c$

and lowest at their crossovers (riffles). For palaeohydrological calculations the channel width should be measured at the crossovers, at the point of alteration of the curvature sign. Repeated measurements of the same width by several operators show that the relative error of measurements can be up to ±10% depending on palaeochannel preservation.

The maximum relative error of the combined application of Formulae (1) and (2) (including the errors of measurements) is ±20 to 40% (see the bars on Fig. 7). The relative errors in calculations of the hydrological parameters of modern rivers with the use of their modern analogues are mainly within ±10%, and up to 40%. Relative errors of palaeohydrological reconstructions are closer to the upper limit of this interval. The main way to decrease these errors is to increase the number of data points to derive equation (1) more accurately. The errors in the input values of palaeochannel width and parameter y vary randomly from site to site, so that it is impossible to decrease these errors in general.

The depth of runoff for each time interval, reconstructed from palaeochannel morphology, was compared with the depth of runoff for the modern river basins in the corresponding region-analogue (Fig. 7). The latter can be estimated with some level of accuracy, depending on the extent of the region-analogue and the variability of river flow within this area. The relative errors of this estimation are shown by open rectangles on Figure 7. This comparison shows good qualitative and quantitative correspondence between the two sets of data. Of course, these data are not fully independent: they are connected by parameter y, defined for the same region-analogues. But the influence of this parameter on the shape of the reconstructed curve is much less than the influence of palaeochannel width, which is totally independent from the location of the region-analogue. Such a similarity of the values of flow depth, obtained by significantly independent palaeohydrological and palaeofloristic methods, indicates the reliability of both of them.

The morphology of palaeochannels on the low terraces and the floodplain of the lower Vychegda River changes considerably through time during the Lateglacial and Holocene. Palaeochannels were larger than the modern ones in the Lateglacial to early Boreal period. They were significantly smaller than the modern ones during the late Boreal to Atlantic period, and increased again from the Sub-boreal until the present. The mean maximum discharge of the Vychegda changed in the same way. It reached its maximum during the Lateglacial, and was greater than the present value at the beginning of the Boreal period. It was half of the present-day maximum discharge during late Boreal to early Atlantic times, and since the end of Atlantic increased up to its modern values. These reconstructions seemingly contradict a general belief of the Lateglacial as a very dry period, and of the Atlantic as a relatively humid period.

The method of palaeogeographical analogy has been used to estimate not only the maximum flow, which is directly related to river channel size, but also other important parameters. The hydrological region-analogues were determined using the palaeofloristic method for several time periods. For each of these levels the ratio of mean annual and mean maximum discharges, the ratio of annual and winter-spring flow, and evapotranspiration were obtained from corresponding maps for the region-analogues. This provides an opportunity to calculate mean annual and mean maximum discharge, annual and seasonal depth of runoff, and annual and seasonal precipitation with Formulae (1)-(6) (Table 6).

Assessments of the flow distribution within a year (river regime) may provide an explanation for the first contradiction mentioned above. The Lateglacial period was characterised by very high differentiation of flow (marked seasonality). Precipitation during a long winter-spring period (mainly in the form of snow) was not less than in the present-day

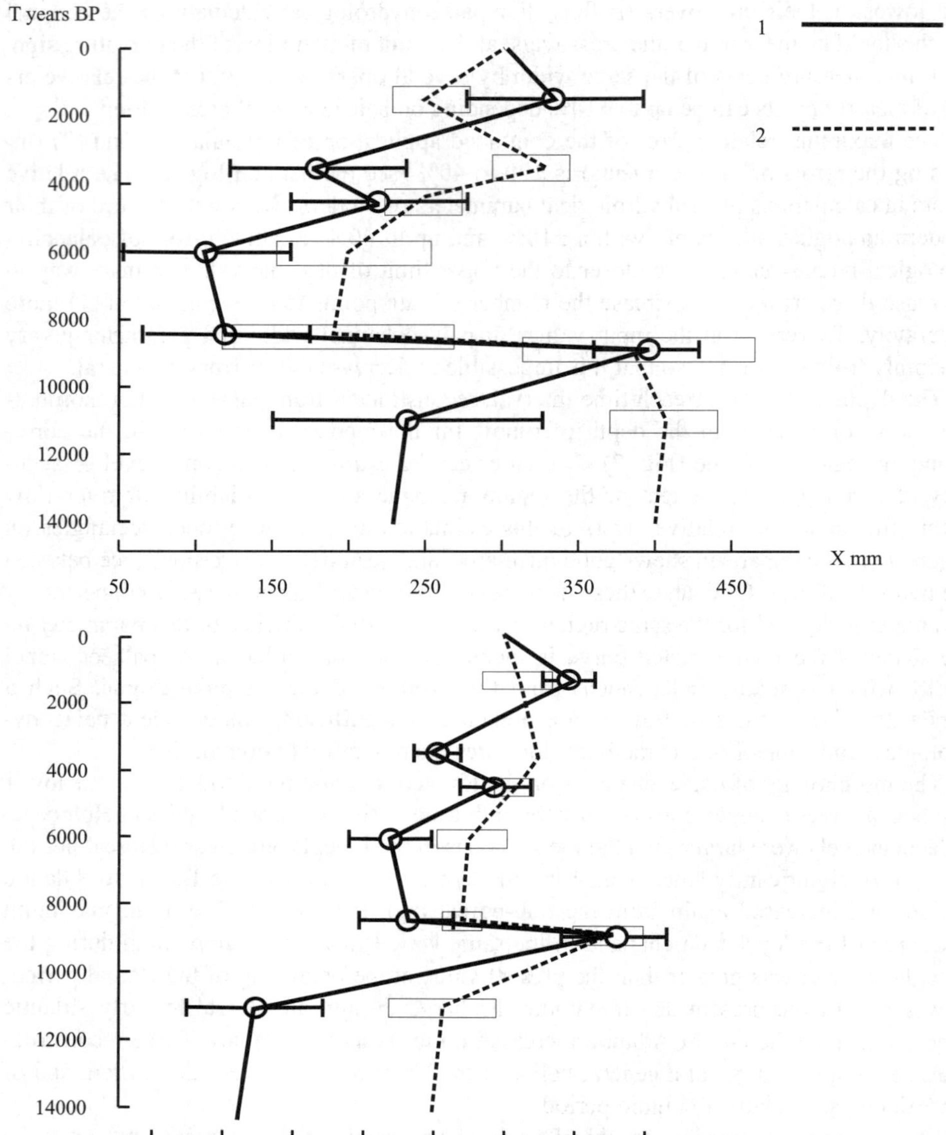

Figure 7. Comparison of annual runoff depth *X* and precipitation *P*, reconstructed with palaeochannel morphology (1) and obtained from floristic region – analogues (2). Bars shows the errors associated with the palaeohydrological calculations, boxes shows the variability of the characteristic within region-analogue.

tundra. Due to low ground permeability in the permafrost zone, the spring thaw led to sharp high floods and to the formation of large wide river channels. But the annual flow was less than the modern one. The apparent humidity deficit was related entirely to the summer-autumn season. In the Vychegda River basin summer precipitation was nearly half the present value. The large periglacial river channels were almost dry throughout the summer due to low precipitation inputs and negligible ground water flow. The same situation is common in modern periglacial landscapes, for example, on the Yamal Peninsula in arctic West Siberia.

The large palaeochannel of the early Boreal age was formed by much more uniform flows. The degradation of permafrost led to a decrease in the runoff rate and an increase in summer-autumn ground water inputs. In conditions of seasonal ground freezing, the runoff coefficients were close to the recent values. Estimates of the mean annual precipitation depth for the period are rather high (860 mm). Precipitation in the winter-spring season was double that of the previous period (415 mm) and the summer-autumn total increased even more – up to 445 mm. A large difference in precipitation regime between the Lateglacial and early Boreal time was derived from the reconstructed intra-annual flow distribution for these two periods. Palaeochannel magnitude is very similar during these time intervals.

The late Boreal and Atlantic periods were less humid. Minimum precipitation and runoff depth was calculated for the late Boreal to middle Atlantic. This temporal shift cannot be explained only by changes in flow variability. The main information about humidity during this period is derived from palaeochannel morphology. The calculated humidity minimum (Fig. 7, Table 6) is related to the smallest palaeochannels formed at that time. In the Sub-boreal and Sub-atlantic periods, the runoff depth and precipitation had a general tendency to increase. This trend was also calculated mainly from an increase in channel size, only local maximum and minimum on the curve being related to changes in flow variability.

The later direct palaeohydrological evidence for the lower Vychegda region contradicts recent climatic reconstructions for the Holocene of Europe. For example, Guiot et al. (1993) showed that the Atlantic period was the most humid in Europe, and humidity decreased during Sub-boreal and Sub-atlantic times. This apparent contradiction can be explained in terms of the metachronous principle (Markov, 1965). Khotinskiy (1977) showed that according to this principle the alteration of more and less humid periods in the territory of northern Eurasia in the Holocene varied significantly in different regions. In the western regions (East European plain) with a more oceanic climate, the Atlantic was more humid than the early Boreal and modern times, and the maximum of humidity took place during the Sub-boreal to Sub-atlantic. In the central regions of Siberia with a continental climate, the maximum of humidity was related to the Boreal period, and the minimum to Sub-boreal times. The climate of the lower Vychegda was more continental than at present during the Boreal and Atlantic times. The region – analogues for this period are situated at the eastern part of taiga zone near the Ural mountains. The change in humidity then followed with a continental climate scenario. During the Sub-boreal and Sub-atlantic, the climate of the lower Vychegda region was more oceanic than the present day, and the trend in humidity change follows the onset of an oceanic climatic scenario.

ACKNOWLEDGEMENTS

The work was funded by the Russian Foundation for Basic Research, grant 97-05-64708. The authors are very grateful to John Lewin, Paul Carling and Jamie Woodward for valuable suggestions which have been incorporated into the text.

REFERENCES

Arslanov, Kh.A., Lavrov, A.S., Lyadov, V.V., Nikiforova, L.D., Potapenko, L.M. & Tertychnaya, T.V. 1980. Radiocarbon geochronology and palaeogeography of Middle Valdai interval and last glacial sheet at the north-east of the Russian Plain. In: *Geochronologiya chetvertichnogo perioda*, Moskva, Nauka (in Russian).

Dury, G.H. 1964. Principles of underfit streams. *US Geological Survey Professional Paper* 452-A, Washington.

Dury, G.H. 1965. Theoretical implications of underfit streams. *US Geological Survey Professional Paper* 452-B, Washington.

Evstigneev, V.M. 1990. *River flow and hydrological calculations*. Izdatelstvo Moskovskogo Universiteta, Moskva, (in Russian).

Goretskiy, G.I. 1964. *Alluvium of the great pre-rivers of the Anthropogene in the Russian Plain*. Moskva, Nauka, (in Russian).

Grichuk, V.P. 1969. An attempt of reconstruction of certain climatic indexes of the Northern Hemisphere during the Atlantic stage of the Holocene. In: Neustadt, M.I. (ed.), *Golotsen*, Moskva, Nauka (in Russian).

Grichuk, V.P. 1985. Reconstructed climatic indexes by means of floristic data and an estimation of their accuracy. In: Velichko, A.A. & Gurtovaya, Ye.Ye. (eds), *Metody Rekonstruktsii Palaeoklimatov*, Moskva, Nauka (in Russian).

Grichuk, V.P. 1989. *The history of flora and vegetation of the Russian Plain in the Pleistocene*, Moskva, Nauka (in Russian).

Guiot, J., Harrison, S.P. & Prentice, I.C. 1993. Reconstruction of Holocene precipitation patterns in Europe using pollen and lake-level data. *Quaternary Research*: 40, N 1.

Kalinin, G. P. 1966. Space – temporal analysis and ergodicity of hydrological elements. *Vestnic Moskovskogo Universiteta*, ser. *Geografiya*, N 5 (in Russian).

Khotinskiy, N.A. 1977. *Holocene of the Northern Eurasia*. Moskva, Nauka (in Russian).

Kvasov, D.D. 1979. The Late Quaternary history of large lakes and inland seas of Eastern Europe. *Annales Academiae Scientarum Fennicae*, ser. A III, 127, Helsinki.

Leliavsky, S. 1955. *An introduction to fluvial hydraulics*. London.

Leopold, L.B., Wolman, M.G. & Miller, J.P. 1964. *Fluvial processes in geomorphology*. Freeman, San Francisco.

Maizels, J.K. 1983. Palaeovelocity and palaeodischarge determination for coarse gravel deposits. In: Gregory, K. (ed.), *Background to Palaeohydrology*, John Wiley and Sons, Chichester.

Markov, K. K. 1965. Space and time in Geography. *Priroda*: N 5 (in Russian).

Panin, A.V., Sidorchuk, A.Yu. & Chernov, A.V. 1992. Macromeanders on the rivers of the European USSR and problems of palaeohydrological reconstructions. *Vodnye Resursy*, N 4 (in Russian).

Potapenko, L.M. 1975. *Quaternary deposits and valley evolution of the lower Vychegda River*. PhD thesis, Moscow University (in Russian).

Schumm, S.A. 1965. Quaternary Palaeohydrology. In: Wright, H. & Frey, D. (eds), *The Quaternary of the United States*, Princeton University Press, Princeton.

Schumm, S.A. 1968. River adjustment to altered hydrologic regimen – Murrumbidgee River and palaeochannels, Australia. *US Geological Survey Professional Paper 598*, Washington.

Skripkin, V. & Kovaliukh, N. 1998. Recent developments in the procedures used at the SSCER Laboratory for the routine preparation of lithium carbide. *Radiocarbon*, Vol 40, No. 1.

Sidorchuk, A.Yu. & Borisova, O.K. 2000. Method of palaeogeographical analogues in palaeohydrological reconstructions. *Quaternary International*, 72: 95-106.

Sidorchuk, A., Borisova, O. & A. Panin. (in press). Fluvial Response To The Late Valdai/Holocene Environmental Change On The East European Plain. *Global and Planetary Changes*.

Starkel L., Kalicki T., Soja R. & Gebica P. (1996) Analysis of paleochannels in the valleys of the upper Vistula and the Wisloka. In: Starkel, L. (ed.), *Evolution of the Vistula River valley during the last 15000 years. Part. VI.* Wydawnictwo Continuo, Wroclaw.

Szafer, W. 1946. Flora pliocenska w Kroscienku nad Dunajcam. *Rozprawy Wydzialu Matematy-ceno-przyrodniczego*, Polska academia nauk, 72 (B. 1-2).

Williams, G.P. 1988. Paleofluvial estimates from dimensions of former channels and meanders. In: Baker, V. et al. (eds), *Flood Geomorphology*, John Wiley and Sons, Chichester, 321-334.

10. A 300-year history of flooding in an Andean mountain river system: the Rio Alizos, southern Bolivia

GLENN S. MAAS & MARK G. MACKLIN
Institute of Geography and Earth Sciences, University of Wales, Aberystwyth, UK

JEFF WARBURTON
Department of Geography, University of Durham, UK

JAMIE C. WOODWARD & EILIDTH MELDRUM
School of Geography, University of Leeds, UK

1 INTRODUCTION

Understanding the relative importance of natural and anthropogenic environmental controls on river behaviour in upland and mountain areas has been a central theme in many geomorphological investigations (e.g. Pope & van Andel, 1984; Macklin et al., 1992; Lewin et al., 1995; Maas et al., 1998; Merrett & Macklin, 1999). As development in these environments intensifies (Stone, 1992), many researchers have suggested the combined effects of natural- and human-induced stresses have increased the magnitude and frequency of geomorphic events in these fragile terrains (Gallart & Clotet-Perarnau, 1988; Molina Sempere et al., 1994; Woodward, 1995). This tends to be detrimental to both landscape stability and the livelihoods of those populations who inhabit these regions (Heine, 1987; Warburton et al., 1998). Mountain catchments in seasonally-dry environments are especially sensitive to changes in climate and land use. Steep relief, thin soils and highly irregular precipitation regimes often lead to flashy stream discharges capable of transferring large quantities of sediment from the headwaters to the trunk stream (Macklin et al., 1995; Poesen & Hooke, 1997). These environments are also prone to landslides, valley floor inundation by floods and large scale changes (vertical and horizontal) in the position of active channels (Selby, 1993; DeGraff, 1994; Prosser et al., 1994; Warburton et al., 1998). In the Mediterranean, for example, phases of river basin and channel metamorphosis have been linked to century and decadal scale changes in climatic parameters, such as those associated with the Little Ice Age climatic deterioration (Gutiérrez Elorza & Peña Monné, 1990; Maas et al., 1998) and to periods when human pressure on the landscape lead to significant increases in hillslope erosion and sediment transfer (Pope & van Andel, 1984; van Andel et al., 1990; Lewin et al., 1995).

Recent advances in dating techniques and methods of palaeoenvironmental reconstruction have significantly enhanced the resolution of late Holocene palaeoclimate histories in many regions of the world (Bradley, 1985; Bradley & Jones, 1995) and as a result, links between well-dated Holocene alluvial sequences and environmental change are becoming more widely established. Detailed analysis of instrumented hydro-climatological data spanning the last 100 years or so (Benito et al., 1996; Longfield & Macklin, 1999) in association with alluvial records from upland catchments (Rumsby & Macklin, 1994; Warburton et al., 1998; Merrett & Macklin, 1999) has enabled links between climate and river behaviour at annual, monthly and even event scales to be established. In many cases, observed trends in the magnitude and frequency of these events have been linked to large scale oceanic/atmospheric oscillations (e.g. El Niño/Southern Oscillation) and these appear to be particularly well correlated in certain parts of the world (e.g. Craig & Shimada, 1986; Wells, 1987, 1990). The increasing number of media reports has also highlighted the effects of recent catastrophic flood events in mountainous river basins throughout the world (Milne, 1986). The destructive forces and loss of life and valuable landscapes associated with these events underscores the need for further detailed assessment of the causal links between environmental change and river basin response in mountain environments (Lewin et al., 1995; Benito et al., 1998).

This paper reports a field-based study of a small river catchment which drains the Eastern Cordillera of the Andes in southern Bolivia. High resolution records of flood frequency and magnitude derived from the detailed geomorphological mapping, sedimentological analysis and lichen-based dating of coarse gravel deposits are evaluated in conjunction with local precipitation data and palaeoenvironmental data from the Andes. The main aim was to produce a record of river flooding during the last c. 300 years and to explore the influence of environmental factors such as climate change. The relationship between flood frequency and the Southern Oscillation is of particular interest as are the potential impacts of the Little Ice Age (Grove, 1988) upon river behaviour (e.g. Rumsby & Macklin, 1996; Maas et al., 1998). The degree to which human activity has affected geomorphic processes in small mountain catchments is equally important but has been little researched in the Andes. Moreover, previous investigations elsewhere have tended to concentrate on much larger catchments where the history of settlement is more fully documented (e.g. van Andel et al., 1990; Abbott & Valastro, 1995; Barker & Hunt, 1995). Although some general information is available for this region detailing changes in agrarian practices during the twentieth century (Preston et al., 1997), there are no data with spatial and temporal resolution of sufficient detail.

2 STUDY SITE

Our geomorphological fieldwork was centred on the Rio Alizos (21° 49' South, 64° 54' West), a small tributary of the Rio Camacho in the Tarija department of southern Bolivia (Fig. 1). The easterly flowing Rio Alizos drains the Eastern Cordillera and Altiplano (> 4000 masl) of the Andes where high relative relief within this region (> 2000 m) has resulted in river basins with steep channels (average gradient 0.03 m m^{-1}) and hillslopes. The lithological units within the basin range from Cambrian quartzites, sandstones, siltstones and conglomerates within the headwaters to Ordovician and Silurian siltstones and shales within the lower part of the catchment. The Alizos has a basin area of 51.3 km^2 and Holocene alluvial fills form a stepped sequence of terraces which expose unconsolidated and semiconsolidated pebbles, gravels, sands, silts and clay (Warburton et al., 1998). The modern river has several ephemeral headwater tributaries but beyond the front of the Altiplano mountain slope it is perennial with a braided channel pattern. Hillslope vegetation is sparse, but there are a variety of crops cultivated on the floodplain and river terraces.

Five trunk stream reaches were selected for investigation, spaced at approximately 2 km intervals downstream from the mountain front (Fig. 1 and Table 1). Each reach contains a number of coarse-gravel splays, berms and lobes (Wells & Harvey, 1987; Macklin et al., 1992) which appear to have been deposited during high magnitude flood events. These deposits form the basis of this investigation as they are suitable for both lichen-based dating and flood magnitude estimates using boulder size measurements. The modern channel

Table 1. Topographic characteristics of the five study reaches.

	Reach 1	Reach 2	Reach 3	Reach 4	Reach 5
Distance from mountain front (km)	0.0	2.2	4.0	6.3	7.9
Length of reach (m)	1600	750	900	700	600
Active channel slope (m m^{-1})	0.049	0.043	0.033	0.031	0.025
Local bedrock geology	Shales & siltstones		Shales, siltstones & sandstones		
Number of dated units in each reach	11	47	42	22	23

has incised into an alluvial fan at the mountain front, in reaches 1 and 2 (Fig. 2) and downstream in reaches 3, 4 and 5 (Fig. 3), it is confined between hillslopes and alluvial terraces. Much of the sediment in the Rio Alizos is derived from the headwaters in the mountain zone above reach 1, the tributaries in the middle and lower part of the catchment appear to be significant sources of sediment for the valley floor and channel system and Warburton et al. (1998) have suggested that the larger (> 0.8 km^2), steeper tributary catchments are dominant in terms of sediment delivery within the Alizos basin.

3 FIELD METHODS

3.1 *Geomorphological mapping*

Fieldwork was carried out during July and August between 1995 and 1998. Geomorphological maps of valley floor features within the five reaches were constructed by field survey using aerial photographs taken in 1983 as base maps. These maps were used to differentiate flood units, areas of recent sedimentation, palaeochannels and the course of the main active channel. Cross-section surveys at selected sites within the valley floor were also undertaken. For each alluvial unit identified within the valley floor the intermediate axes of the 20 largest clasts exposed at the surface were measured. These data were used to provided an estimate of flood magnitude, based on the assumption that the largest boulders transported relate to maximum flow competence during a particular flood event (Costa, 1983; Williams, 1983). Lichenometry was used to establish a chronology of flood events over the last 200 to 300 years and is discussed in detail below.

3.2 *Lichenometric dating*

Lichenometry has become an established method for dating rock surfaces in fluvial environments in the UK (Innes, 1983; Harvey et al., 1984; Macklin et al., 1992; Merrett & Macklin, 1999) and Mediterranean regions (e.g. Maas et al., 1998). Lichenometry is based upon the assumption that the largest lichen thalli are directly related to the date of substrate exposure. By establishing an empirical relationship between lichen thalli size and age by measuring the largest thalli growing on dated substrates (usually gravestones), this can then be related to those lichens present on the rock surfaces of interest, thus giving a minimum age estimate. A single crustose lichen species *Rhizocarpon geographicum* agg. was found growing on rock substrates within the study area and was used in this investigation.

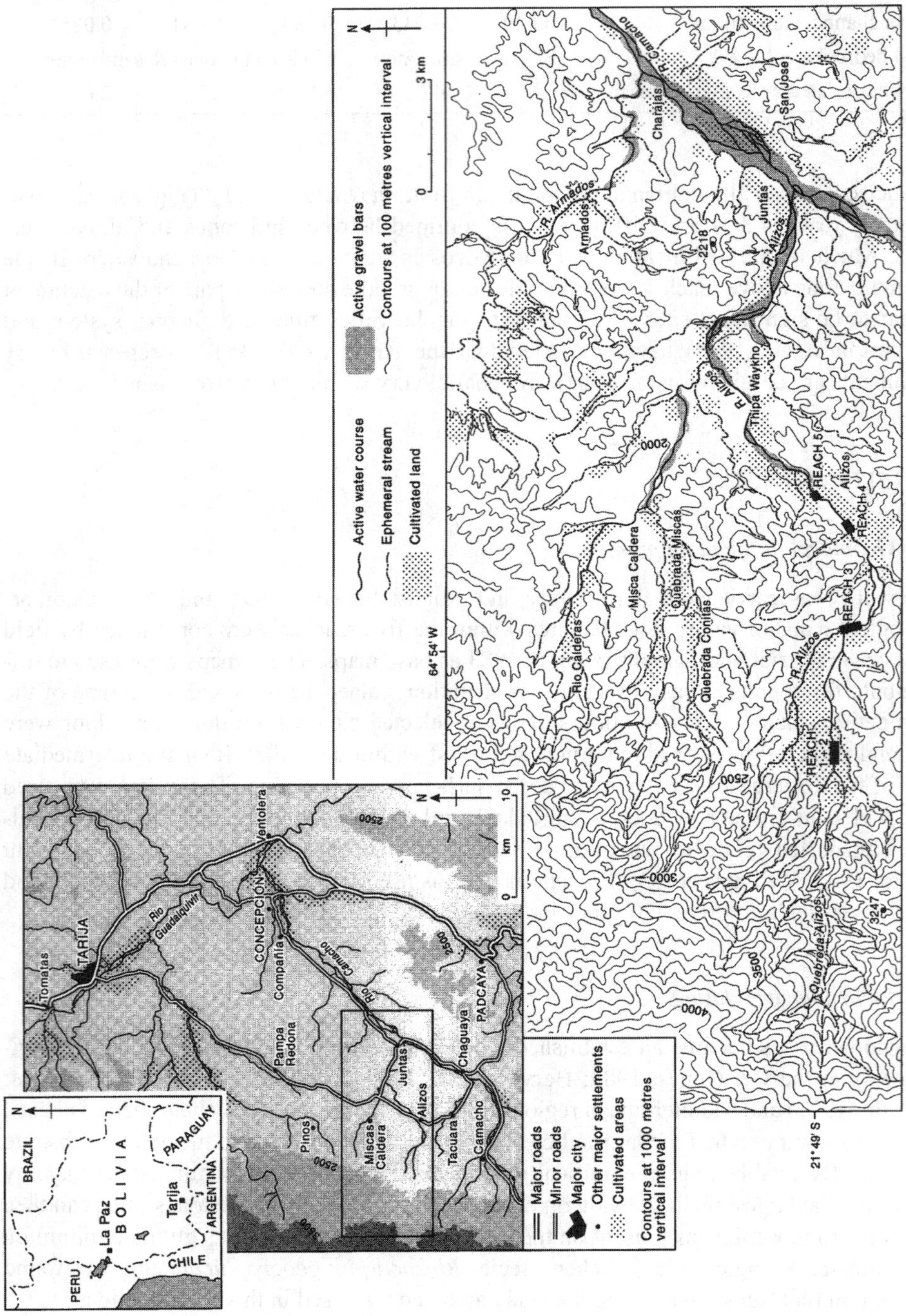

Figure 1. Location of the Rio Alizos catchment in southern Bolivia showing the position of the five study reaches.

Figure 2. View of reach 1 looking upstream from the mountain front towards the headwaters of the Rio Alizos. The modern channel is inset below the surface of an extensive trunk-channel alluvial fan.

Figure 3. View of reach 3 looking downstream. The modern valley floor is confined between the hillslopes and tributary alluvial fans. A recent phase of avulsion has switched the position of the modern channel from the right to the left-hand side of the valley floor.

Sixty one measurements on dated gravestones at locations within the Alizos and Tarija valleys and no more than 15 km from the study site, were used to construct the age–size relationship (Fig. 4). Each point on the age-size plot represents the mean of the five largest lichens on dated substrates. The longest thalli axes were measured to the nearest 0.1 mm and lichens showing obvious signs of senescence, coalition or irregular shape were excluded. The age-size plot covers the last 70 years or so and a linear trend line fitted to the data gives a lichen growth rate of 0.28 mm yr^{-1}. To extend the range of this relationship to date alluvial deposits older than 70 years we have assumed that the growth rate has remained constant over the last 300 years. However, the growth rate of lichens tends to decrease after an initial period of linear growth known as the 'great growth period' (Bradley, 1985), but the age at which this occurs depends upon many physiological and environmental factors specific to a particular lichen species (Easton, 1994). The duration of the initial period when the growth rate tends to be constant for *Rhizocarpon geographicum* agg. varies in different environments. For example, Benedict (1988) identified a linear growth rate during the first 100 years of the lichen species' life span in the Colorado Front Range, USA, whereas Merrett & Macklin (1999) working in the Yorkshire Dales of northern England, showed there was very little change in growth rate over the last 200

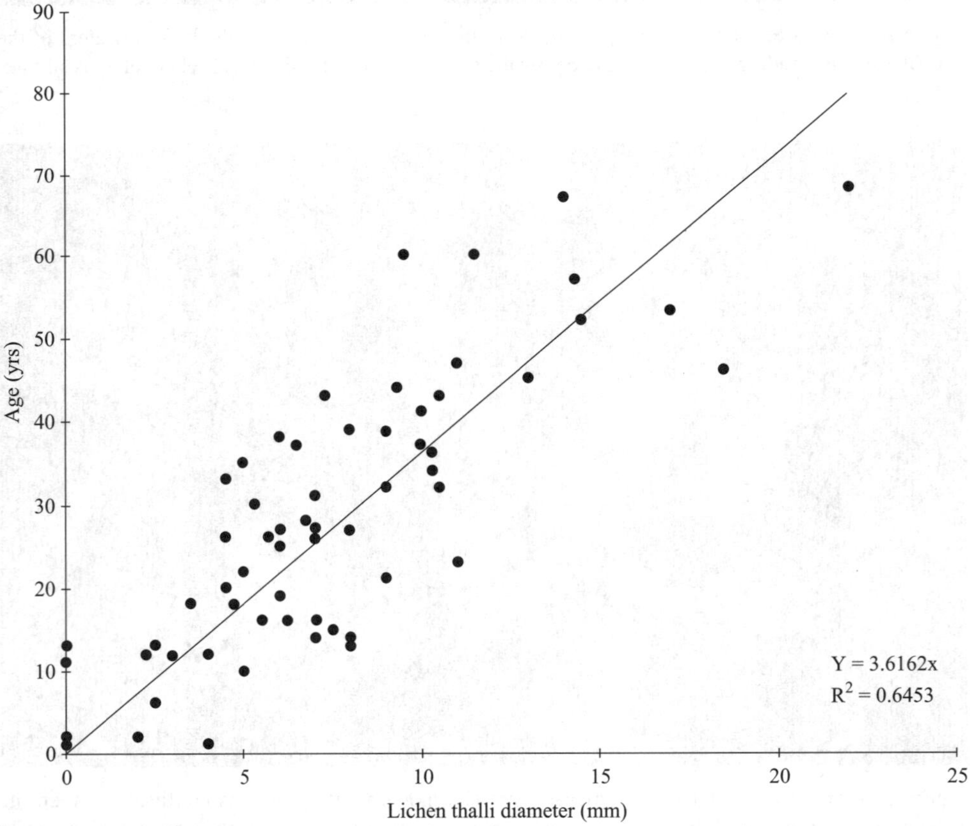

Figure 4. Age-size relationship for *Rhizocarpon geographicum* agg. based on the mean of the five largest lichens present on dated substrates.

years. The growth rate obtained from age-size data in the Tarija area over the last 70 years is taken to be linear in form and this relationship is assumed to remain constant over the last three centuries for the purpose of dating substrates older than c. 1930. It is important to bear in mind that lichen-based age estimates of alluvial deposits within the Rio Alizos will be underestimates if the growth rate of *Rhizocarpon geographicum* agg. decreases over time.

4 CLIMATE CHANGE

Instrumented and proxy climate data from southern Bolivia and the Andes are analysed in this paper to determine the nature and timing of climatic trends during the last c. 300 years and to explore any links between climate change and river response. This section commences with an appraisal of recent (last c. 100 years) climatic variations in southern Bolivia inferred from instrumented records in the Tarija area and then considers evidence for climate change associated with the Little Ice Age (c. 1450 to 1850) in the central Andes.

4.1 *Current climate and recent climatic trends in southern Bolivia*

Southern Bolivia is characterised by pronounced seasonal contrasts in precipitation with over 95% of the annual precipitation falling during the months of October through to April (Table 2). During the Austral summer, precipitation is generated by two dominant processes: 1) increased convective activity over the Altiplano associated with the 'Bolivian High' and 2) the formation of a low pressure cell which forms over central South America and draws in moisture from the South Atlantic (Aceituno & Montecinos, 1993). Annual precipitation totals within the Tarija region vary with altitude (Table 2) and tend to increase with elevation, though high variability suggests other factors such as aspect and barrier effects may be important. At Alizos mean annual precipitation for the period 1977 to 1993 was 850 mm with 825 mm falling in the wet season months (October to April). Precipitation events are generally characterised by extreme downpours generated during

Table 2. Precipitation records from the Tarija department, southern Bolivia.
(Data supplied by the Servico Nacional de Meteorologia e Hidrologia, Bolivia).

Station	Altitude (m)	Latitude Longitude	Period of record	Mean annual and mean wet season (Oct-Apr) Precipitation (mm)	Max/min annual precipitation (mm)
Rejara	3000	22°01'S, 64°59'W	1977-1994	1132/1129	2324/637
Alizos	2040	21°49'S, 64°52'W	1977-1993	850/825	1224/533
Colon Sud	2000	21°45'S, 64°39'W	1977-1993	367/360	551/100
Tarija Airport	1875	21°32'S, 64°43'W	1951-1994	604/590	918/310
Tarija University	1859	21°32'S, 64°44'W	1970-1994	606/591	845/349
Concepcion	1710	21°42'S, 64°37'W	1968-1993	596/586	1019/392
Canchas Mayu	1120	21°54'S, 64°54'W	1969-1993	927/896	1202/631

periods of convective instability. For example, rainfall records indicate that on average almost 25% of the total January precipitation falls within a single 24 hour period. Although high-intensity low-duration storm events are likely to be significant in the generation of runoff and channel flow (Selby, 1993), no continuous daily data are available from which analysis at the storm event scale can be made.

An analysis of climate time-series data from the Pacific coast of South America indicates large inter-annual variations in precipitation (Ropelewski & Halpert, 1987; Aceituno, 1988; Tapley Jnr. & Waylen, 1990; Stone, 1992). These variations have been attributed primarily to the quasi-periodic phases of the Southern Oscillation (e.g. Lockwood, 1984; Enfield & Cid, 1991; Allan et al., 1996). The Southern Oscillation refers to sea-surface pressure and temperature anomalies recorded at the eastern and western margins of the Pacific Ocean. The Southern Oscillation Index (SOI) is a measure of the magnitude of these anomalies, with a low (negative) index relating to El Niño phases and a high (positive) index correlating with anti-El Niño or La Niña phases. During strong El Niño Southern Oscillation (ENSO) events, drought commonly occurs on the Andean Altiplano in the Austral summer, while during anti-El Niño or La Niña phases precipitation increases (Tapley Jnr. & Waylen, 1990; Thompson et al., 1984). During strong El Niño events, climatic disturbances are generated by two main processes.

1. A westward shift of the convection zone, normally centred on Amazonia.
2. Blocking of polar frontal systems in a zone extending from southern Peru to southern Brazil, related to an enhancement of the subtropical jetstream (Kousky et al., 1984).

These conditions generally promote anomalously heavy rainfall in the northern littoral of Peru and anomalously low rainfall in Amazonia and the Bolivian Altiplano.

At present the detailed relationships between precipitation in the Tarija valley, southern Bolivia and the state of the Southern Oscillation are unknown. To analyse the associations between precipitation and the SOI in southern Bolivia, monthly precipitation from 7 stations within the Tarija area were categorised using the corresponding monthly SOI values obtained from the NOAA Climate Diagnostic Bulletin. The SOI represents the normalised monthly mean pressure differential at Darwin, Australia subtracted from that at Tahiti. For each rainfall station monthly precipitation totals during the Bolivian wet season (October to April) were grouped according to the sign and magnitude of the month's SOI value. Those months with an SOI greater than or equal to +1 represented one data set and those months with an SOI less than or equal to –1 represented a second set. Mean precipitation was calculated for those months with positive and negative SOI.

At 5 of the 7 stations, precipitation during months with a positive SOI was generally greater than during months with a negative SOI (Table 3). This relationship is most pronounced at Canchas Mayu where the difference in mean monthly precipitation was 21% and at Alizos where the value was 18%. Comparison of the two datasets from Tarija (Airport and University) does, however, illustrate some of the problems involved in interpreting climate data over different time periods. The shorter record (1970 to 1994) from Tarija University suggests a 3% increase in precipitation during those months characterised by a negative SOI compared with positive SOI months. The longer record (1951 to 1994) from Tarija Airport shows significantly more precipitation (11%) during phases with a positive SOI. Although the mean annual precipitation totals for each station are virtually the same (approximately 600 mm) one record suggests a 'positive SOI/wet' relationship, whereas the other shows a weak 'negative SOI/wet' relationship.

Table 3. Differences in mean precipitation totals between wet season months characterised by a strong negative (< –1.0) and a strong positive (> +1.0) SOI.

Station	Mean precipitation during *negative* SOI months (mm)	Mean precipitation during *positive* SOI months (mm)	Percentage differences in mean precipitation
Rejara	133	151	**14%**
Alizos	96	113	**18%**
Colon Sud	47	47	0%
Tarija Airport	80	89	**11%**
Tarija University	81	79	3%
Concepcion	79	85	**8%**
Canchas Mayu	116	140	**21%**

Values in bold indicate that on average greater amounts of precipitation fell during months with a positive SOI.

Table 4. Total precipitation during the 1982-83 wet season (October – April) expressed as a percentage of the average wet season precipitation.

Station	Percentage
Rejara	45%
Alizos	35%
Colon Sud	18%
Tarija Airport	64%
Tarija University	71%
Concepcion	108%
Canchas Mayu	72%

These analyses do, however, indicate significant differences in precipitation during the two opposite phases of the Southern Oscillation. Indeed, during the major El Niño event of 1982/83, six out of the seven stations showed a marked decrease in total wet season precipitation (Table 4). The greatest deficit was recorded at Colon Sud, where between October 1982 and April 1983 precipitation amounted to only 18% of the average for October to April. For the same period at Alizos, the value increases to 35%, whereas at Conception, precipitation during the 1982/83 El Niño event was actually greater than the long term average (108%) (Table 4).

Although a quasi-periodic cycle of 1 to 5 years is evident in the Southern Oscillation (Wells, 1987), a longer term trend also exists which has been dominant in the latter half of the present century. The period 1930 to 1977 had a mean SOI value of +0.02 which fell to –0.53 between 1978 and 1998 (Fig. 5). This latter period includes one of the strongest ENSO events of this century (1982 to 1983) and two ENSO events during the 1990s (1990 to 1995 and 1997 to 1998). Several causes for the reduction in the SOI have been postulated, including increased volcanic activity leading to a greater amount of sulphate

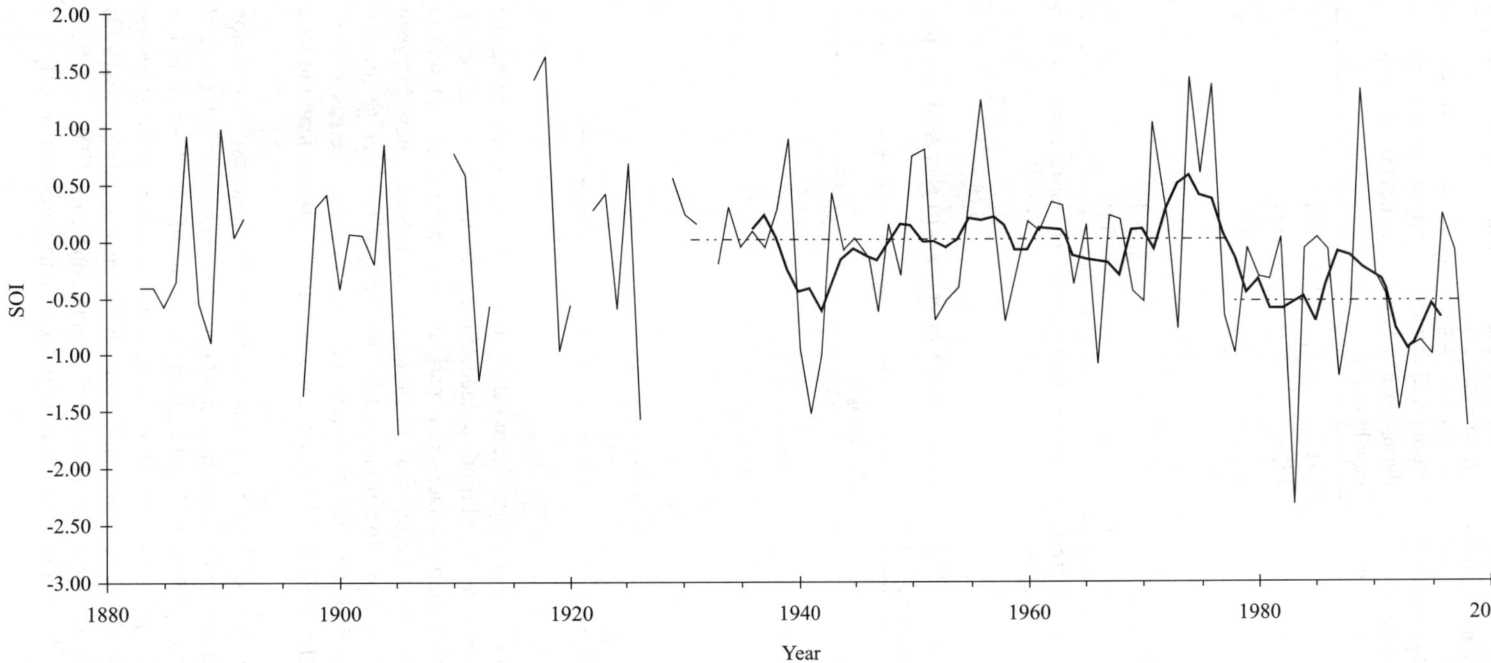

Figure 5. Recent trends in the state of the SOI (Southern Oscillation Index) for the years 1882 to 1998. Thin line represents hydrological year annual mean (July to June). Heavy line represents five year running mean. Also shown is the period mean for the years 1930 to 1977 (+0.02) and 1978 to 1998 (–0.53). Data source: NOAA Climate Diagnostic Bulletin.

aerosols, an increase in the abundance of greenhouse gasses within the atmosphere, or even a significant regime change in the global climate system (Wang, 1995; Webster & Palmer, 1997; Self et al., 1997).

4.2 *Longer term climate change and the Little Ice Age*

Marked late Holocene climatic variability in the Andes of southern Peru and northern Bolivia has been identified from a number of palaeoenvironmental records. These include the Quelccaya ice core from southern Peru (latitude 13°56'S) (Thompson et al., 1984, 1985, 1986; Thompson, 1995) and lake level data from Lake Titicaca southern Peru/northern Bolivia (latitude 15°-16°S) (Street-Perrott & Harrison, 1985). Both sites are situated within the high altitude Andean Altiplano above 3500 m. The 1500-year record of inter-annual climate change from the Quelccaya ice core was obtained by integrating a number of parameters which record the physical and chemical nature of the atmosphere. Identifiable annual laminations within the ice core have allowed precise dating and demonstrate both annual and decadal/century scale climatic variability. Clear correlations between lake level, ice core variables and instrumented climatic parameters have been recognised for the twentieth century (Thompson et al., 1986). The two strong ENSO events that occurred between 1975 and 1984 were associated with substantially reduced net snow accumulation, elevated conductivity levels and higher $\delta^{18}O$ values at Quelccaya. These indicators imply a marked reduction in precipitation and an increase in the atmospheric dust content during the two ENSO events. Lake level data prior to about 1970 show reduced lake levels coeval with drought conditions at Quelccaya, with a lag of one year representing the length of time it takes for meltwater to reach Lake Titicaca. Previous El Niño droughts, in particular those occurring during the years 1891, 1941-42, 1957 and 1972 were concurrent with low lake-levels (Caviedes, 1984). Significantly, both these palaeoclimate records correlate well with instrumented precipitation data from El Alto near La Paz in northern Bolivia for the period 1915 to 1984.

The Quelccaya ice core record of climate change is especially valuable as it encompasses the period of climatic deterioration associated with the Little Ice Age (c. 1450 to 1850, Grove, 1988). Climatic signatures within the ice core show marked differences between the Little Ice Age period and that of the twentieth century. Climate conditions prior to 1880 appear to have been characterised by cooler temperatures (lower value $\delta^{18}O$), lower precipitation (lower net accumulation), and increased atmospheric dustiness (greater particulate concentration and liquid conductivity). Thompson et al. (1986) suggest that the period encompassing the Little Ice Age between 1530 and 1880 was characterised by reduced $\delta^{18}O$ values and demonstrate that decadally averaged $\delta^{18}O$ values closely match Northern Hemisphere decadally averaged temperatures from 1580 to 1980 (Landsberg, 1985). Furthermore, historical observations from the region immediately adjacent to the Quelccaya ice cap are consistent with a cooler and drier climate with the altitudinal limit for cultivation 150 years ago 200 m lower than at present. Since about 1880 the ice core data indicate higher temperatures and increased precipitation (Thompson et al., 1986).

Although these records of climatic variability have been obtained from sources that are between 600 km and 900 km north of the Rio Alizos study area, they depict consistent climatic trends for the central Andes which are in phase and appear to correlate well with those observed from records within the Tarija region and the Alizos catchment. This is important as the evidence for climate change at Quelccaya and Lake Titicaca may also reflect climate variations in southern Bolivia. Against this background the following section presents the long term flood data from the Rio Alizos and compares this record to the climate records discussed above.

5 FLOOD FREQUENCY AND CHANNEL CHANGE

River terraces, boulder berms and splays within the five study reaches were surveyed to classify the sediment units into relative and absolute age groups according to lichen thalli size distribution. On each recognisable terrace surface the largest five lichens were measured. A Kruskal-Wallis one-way analysis of variance test was carried out on the thalli size frequency distributions to identify a number of flood periods. This identified 17 distinct clusters, or periods of flooding which were classed from A to Q (Table 5) using the lichen age-size relationship to establish absolute ages. Each of these has a P value > 0.05 (significant at the 95% confidence level) and the age classes have a range reflecting the oldest and youngest lichens found within each flood unit group.

The flood units grouped into age classes (A to Q) are likely to represent one or several floods, but because of the uncertainties in lichen growth rates, cannot be assigned to a single flood event. It is clear from Figure 6 that the oldest flood units in the Alizos catchment are present furthest upstream within reaches 1 and 2. However at reach 5, furthest from the mountain front, only flood deposits post-dating the late 1800s are found. This pattern

Table 5. Determination of flood unit age classes based upon lichen thalli size frequency distribution.

Age class	Number of alluvial units per class	Class age spread	Class P value (P > 0.05)	Class versus class P value (P < 0.05)
A	6	1694-1738	0.365	A – B (0.016)
B	7	1777-1803	0.461	B – C (0.001)
C	8	1809-1828	0.601	C – D (0.012)
D	3	1832-1837	0.681	D – E (0.048)
E	7	1845-1855	0.960	E – F (0.013)
F	5	1859-1868	0.563	F – G (0.029)
G	6	1874-1879	0.617	G – H (0.001)
H	4	1885-1888	0.979	H – I (0.001)
I	4	1896-1905	0.341	I – J (0.001)
J	14	1909-1919	0.474	J – K (0.006)
K	4	1925-1938	0.248	K – L (0.043)
L	10	1944-1956	0.552	L – M (0.023)
M	1	1960	-	M – N (0.001)
N	7	1964-1970	0.155	N – O (0.001)
O	3	1973-1976	0.278	O – P (0.001)
P	4	1983-1988	0.593	-
Q	53	1989-1995	Lichens absent	-

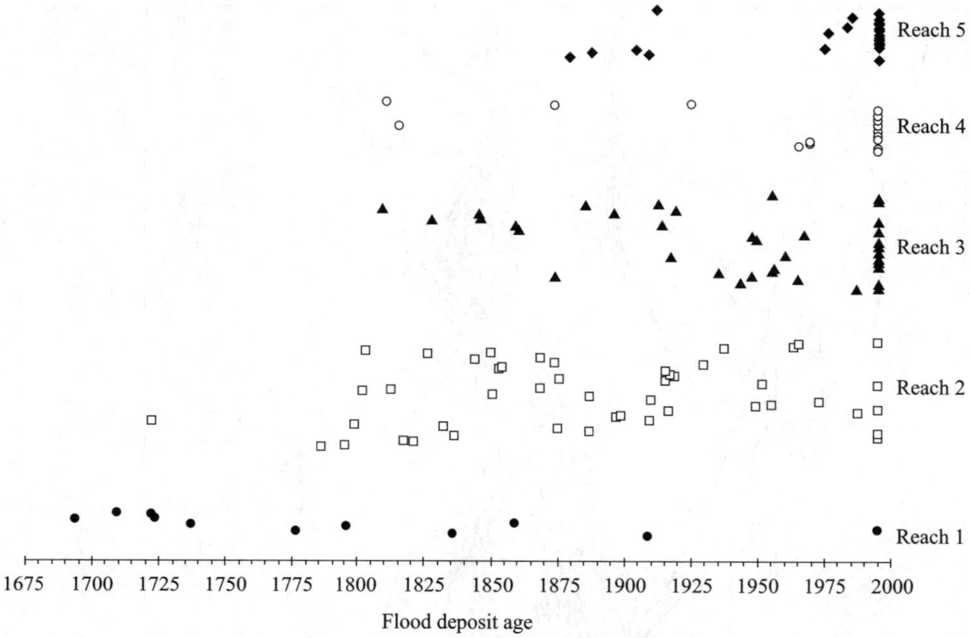

Figure 6. Age estimates of individual alluvial flood units within each of the five study reaches.

is likely to reflect the greater preservation potential for older deposits within the now more entrenched upper reaches (Figs. 7 and 8), compared to the stable or aggrading lower reaches (Figs. 9 to 11) where rates of valley floor sediment reworking are much higher. A ^{14}C date of 660 ± 90 BP (cal AD 1220 to 1435, Beta-122920) was obtained from a bone within a shallow pit on the highest alluvial fan surface in reach 1 and probably associated with the building of historic floodwater irrigation structures on the fan surface. The dates obtained indicate major fan incision is likely to have commenced sometime after the early-thirteenth to early-fifteenth century. The channel within reach 3 (Fig. 9), however, has migrated from the southern to the northern side of the valley floor since the mid-nineteenth century and incised its bed producing a stepped sequence of alluvial fills. This may have been in response to a reduction in the supply of sediment from upstream reaches. The valley floor at reach 5 (Fig. 11) has a relatively low relief with alluvial deposits dated to two main periods, the late nineteenth to early twentieth century, which are dissected by recent headcuts, followed by sedimentation since the 1970s which continues through to the present.

6 CLIMATIC CONTROLS OF FLOOD FREQUENCY

The age classes H to P covering the years 1885 to 1988 (vertical grey bands) and the lichen-based age estimates of individual alluvial units within the five study reaches (solid black dots) can be compared with the SOI record for the period 1882 to 1998 (Fig. 12). The upper and lower histograms show the number of individual wet season months (between

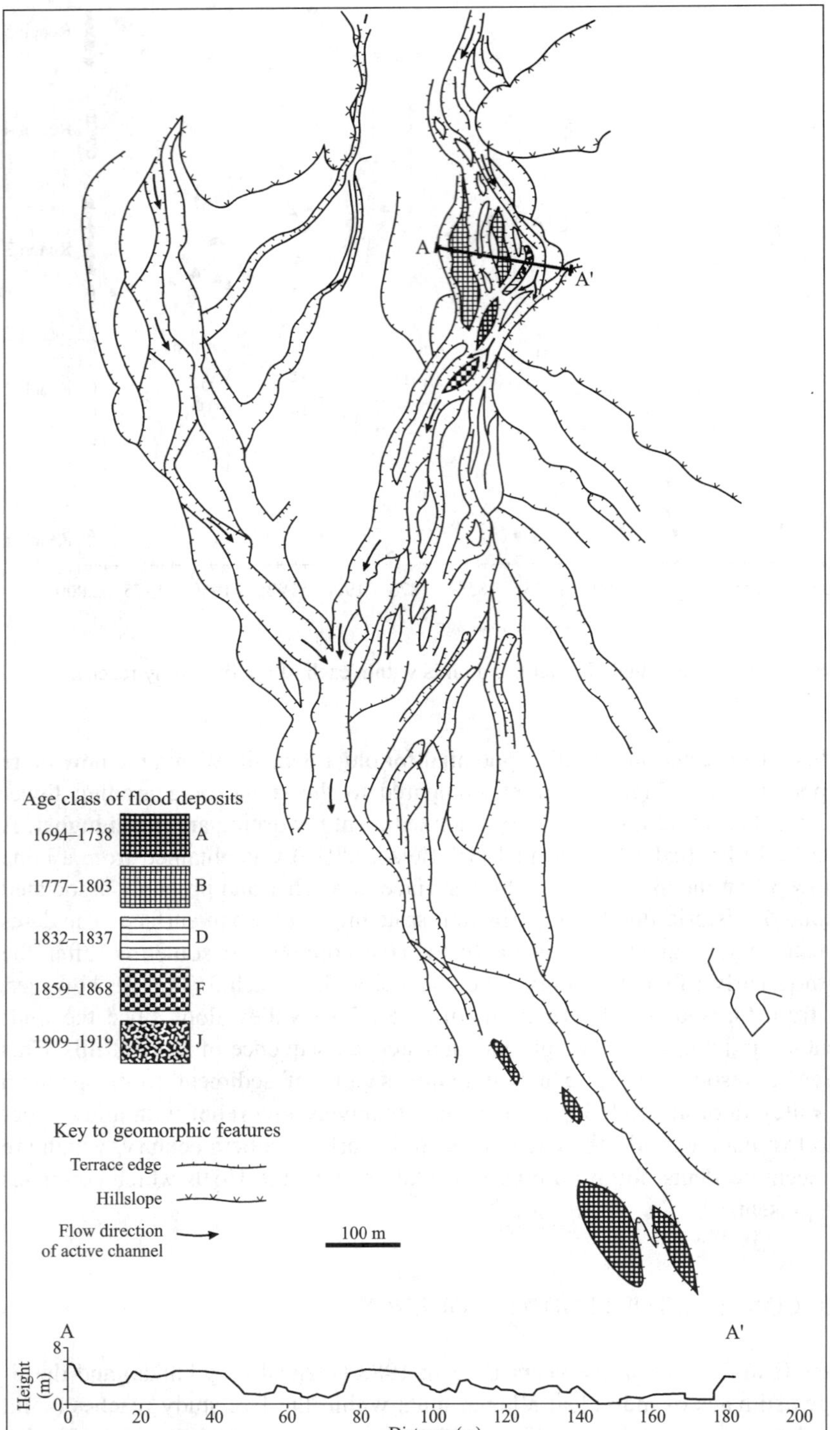

Figure 7. Geomorphological map of reach 1, showing age classes of valley floor alluvial units and valley floor cross-section.

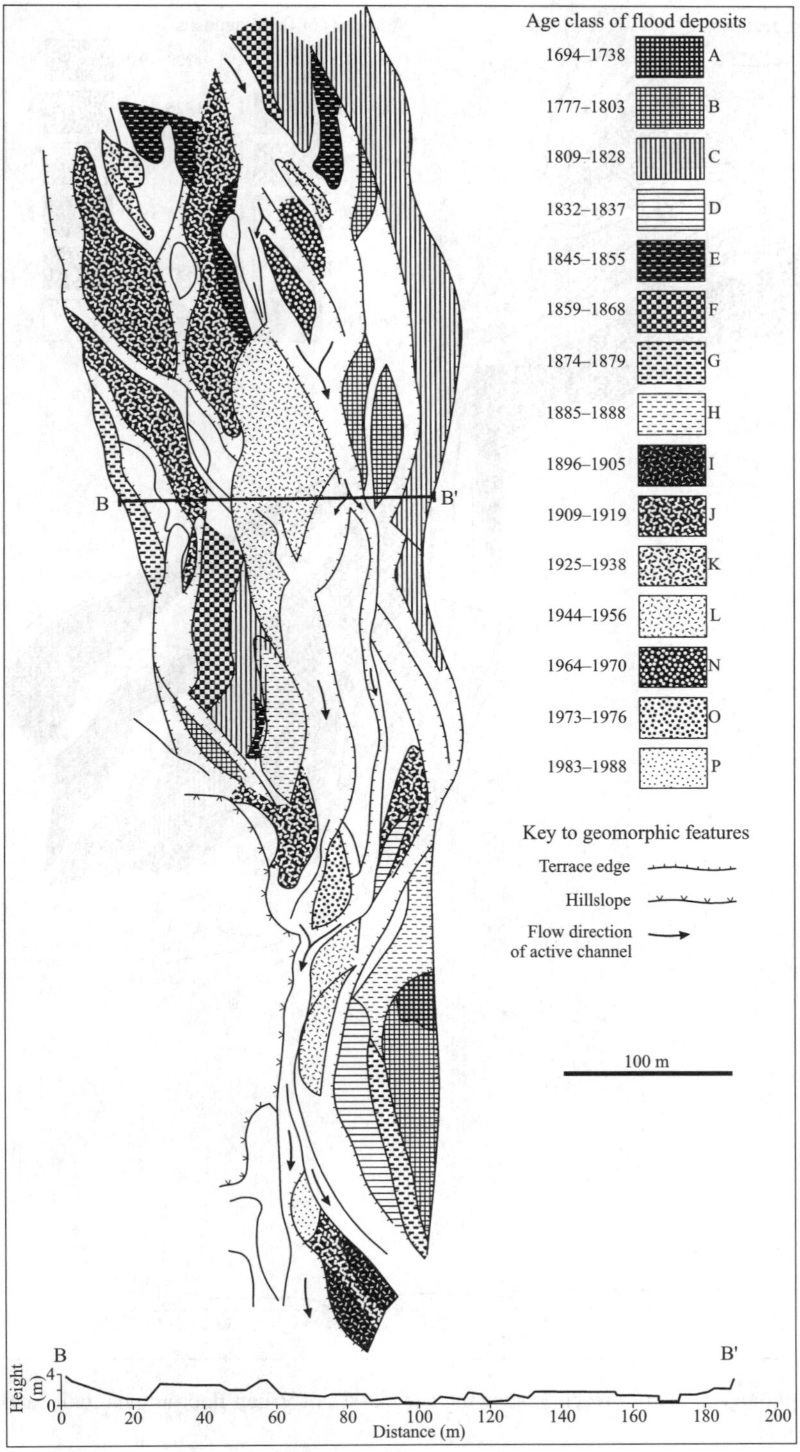

Figure 8. Geomorphological map of reach 2, showing age classes of valley floor alluvial units and valley floor cross-section.

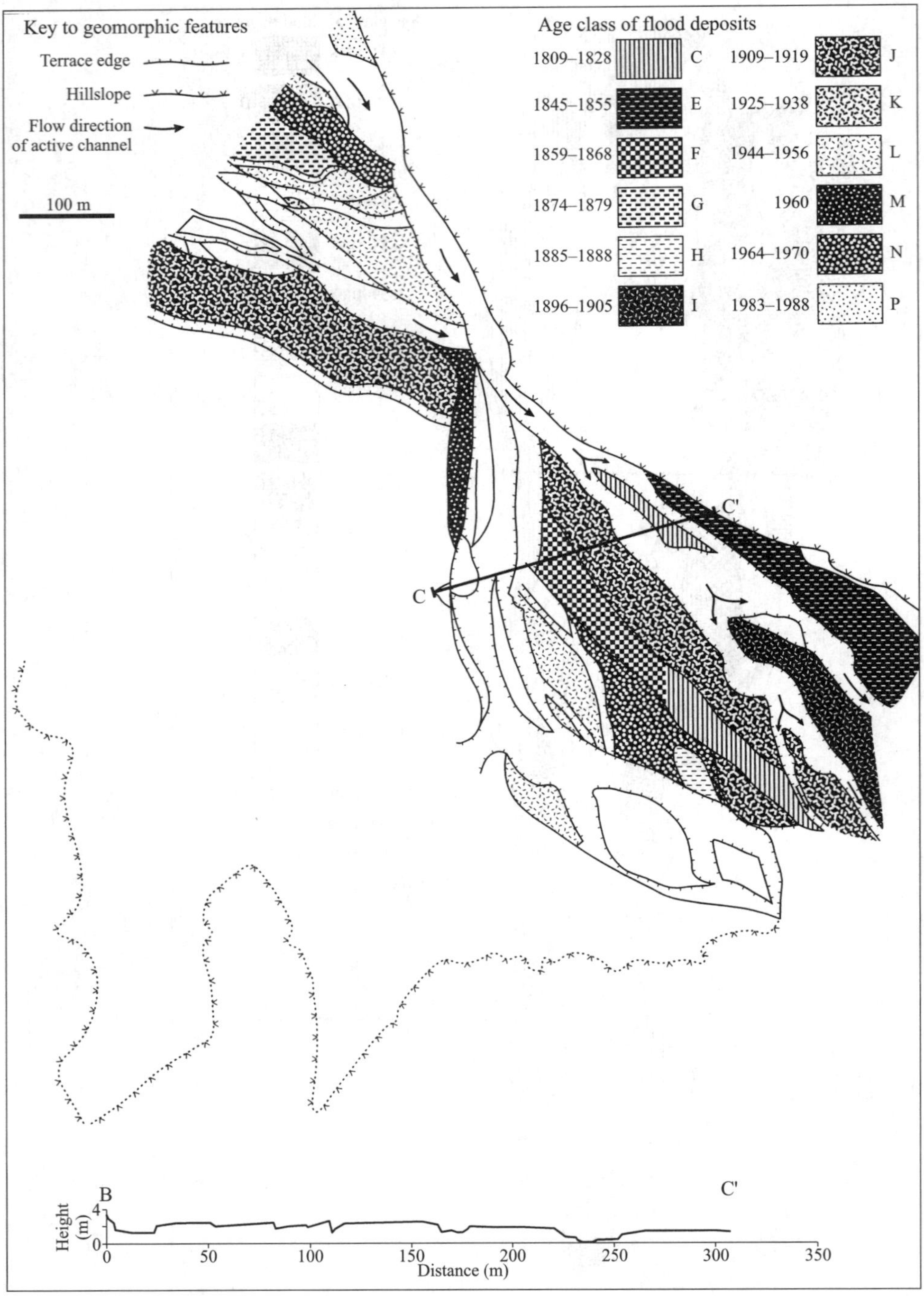

Figure 9. Geomorphological map of reach 3, showing age classes of valley floor alluvial units and valley floor cross-section.

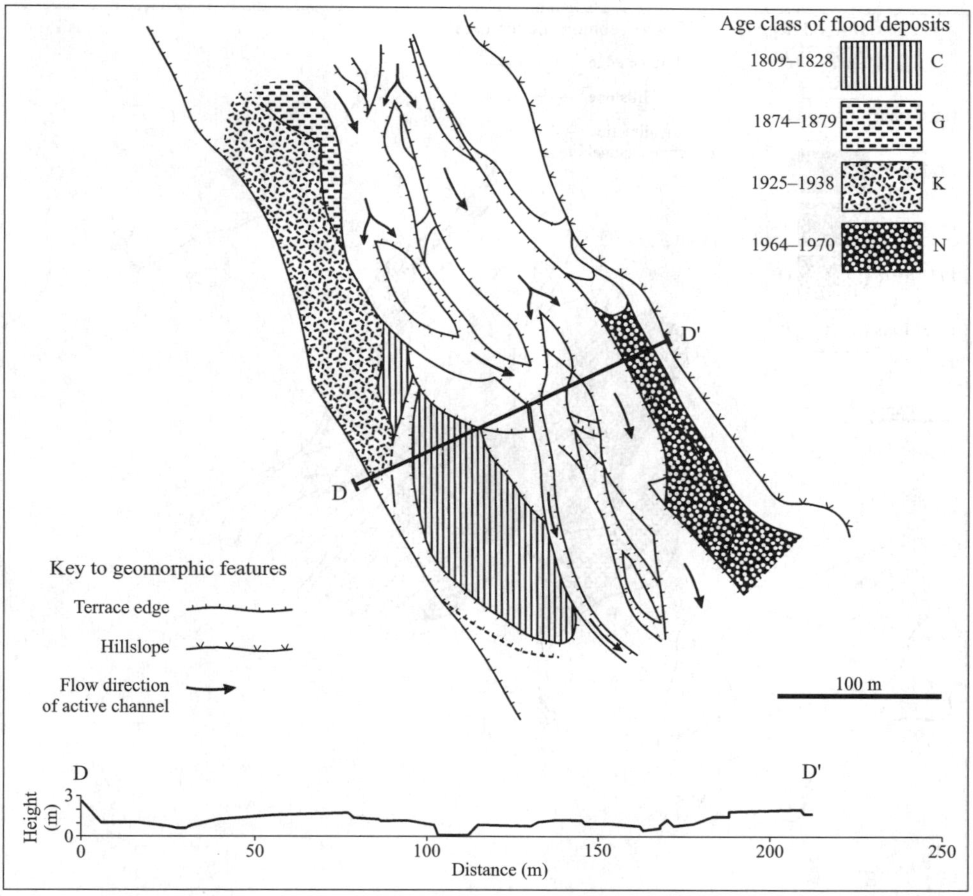

Figure 10. Geomorphological map of reach 4, showing age classes of valley floor alluvial units and valley floor cross-section.

October and April) per hydrological year with a SOI less than –1.5 (El Niño) and greater than +1.5 (La Niña), respectively. These thresholds were chosen as they represent the most extreme phases of the SOI and as such are likely to indicate possible extremes in local climatic conditions. The strong El Niño years of 1940-41, 1982-83 and the early 1990s are clearly evident, as are the strong La Niña events of 1902-03 and 1973-75. The number of months with a positive and negative SOI that were concurrent with the lichen-based age classes were counted and expressed as a percentage of the total number of positive and negative months. 51% of the months with a negative (El Niño) index fell within lichen-based age classes, while 77% of months with a positive (La Niña) index fell into these same age classes. This suggests that there is a greater likelihood that periods of flooding and coarse gravel deposition tend to be associated with La Niña-type phases of the Southern Oscillation, whereas El Niño-type conditions are equally as likely to be coeval with depositional as with non-depositional phases. Although there is by no means a strong relationship between the state of the SOI and the frequency of flood depositional events, it appears that during the last 100 years flooding seems to be concentrated in wet (La Niña)

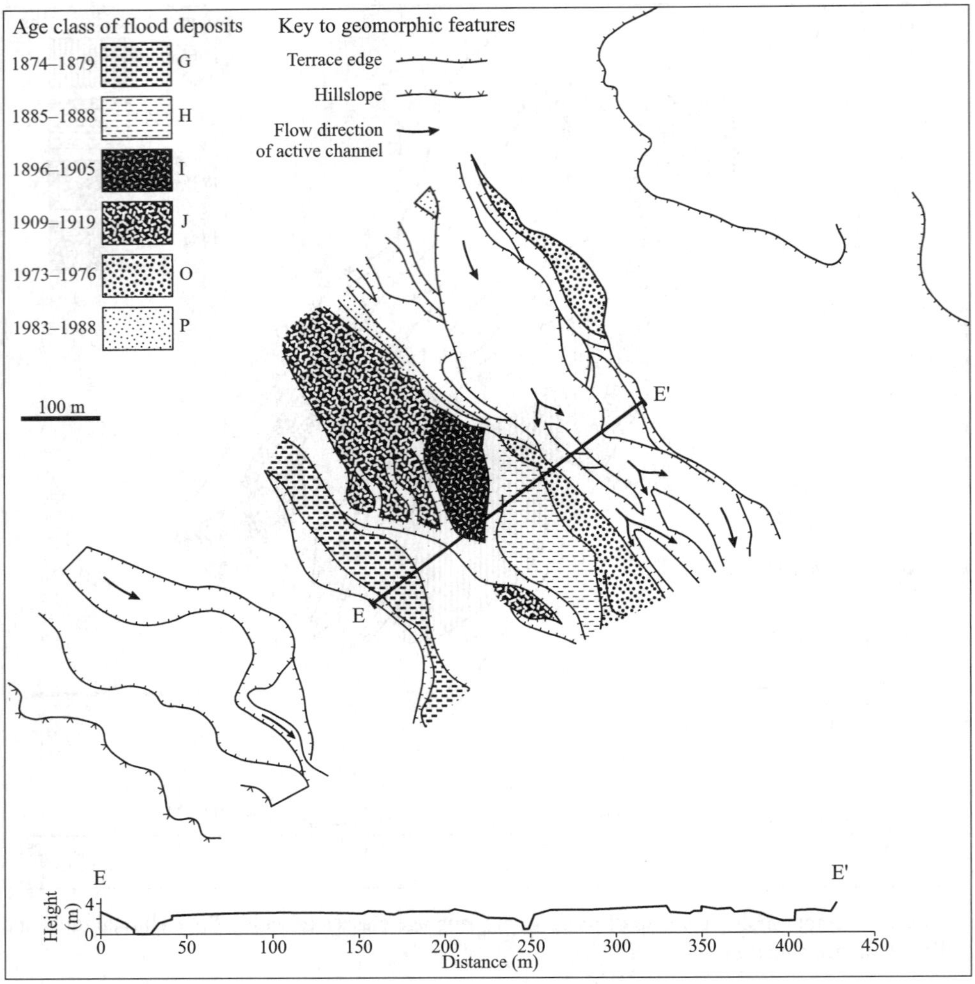

Figure 11. Geomorphological map of reach 5, showing age classes of valley floor alluvial units and valley floor cross-section.

periods. For example, during the strong El Niño years of 1940-41 and 1982-83, we have no record of major flood events within the Rio Alizos and the earlier of these two periods coincides with a ten year drought that affected the major part of the eastern Andes Altiplano from southern Peru to northern Bolivia between 1935 and 1945. The periods 1973-76 and 1897-1906, on the other hand, exemplify phases of alluviation concurrent with months characterised by a positive SOI (age classes I and O). In some periods, however, these patterns were reversed, most notably between 1983-88 (age class P) when the Southern Oscillation was dominated by negative indices.

The record of monthly precipitation data from Tarija Airport, which constitutes the longest available series (1951 to 1994), can be used to analyse these associations further. The dates of the fifteen wettest months from this series have also been plotted in Figure 12.

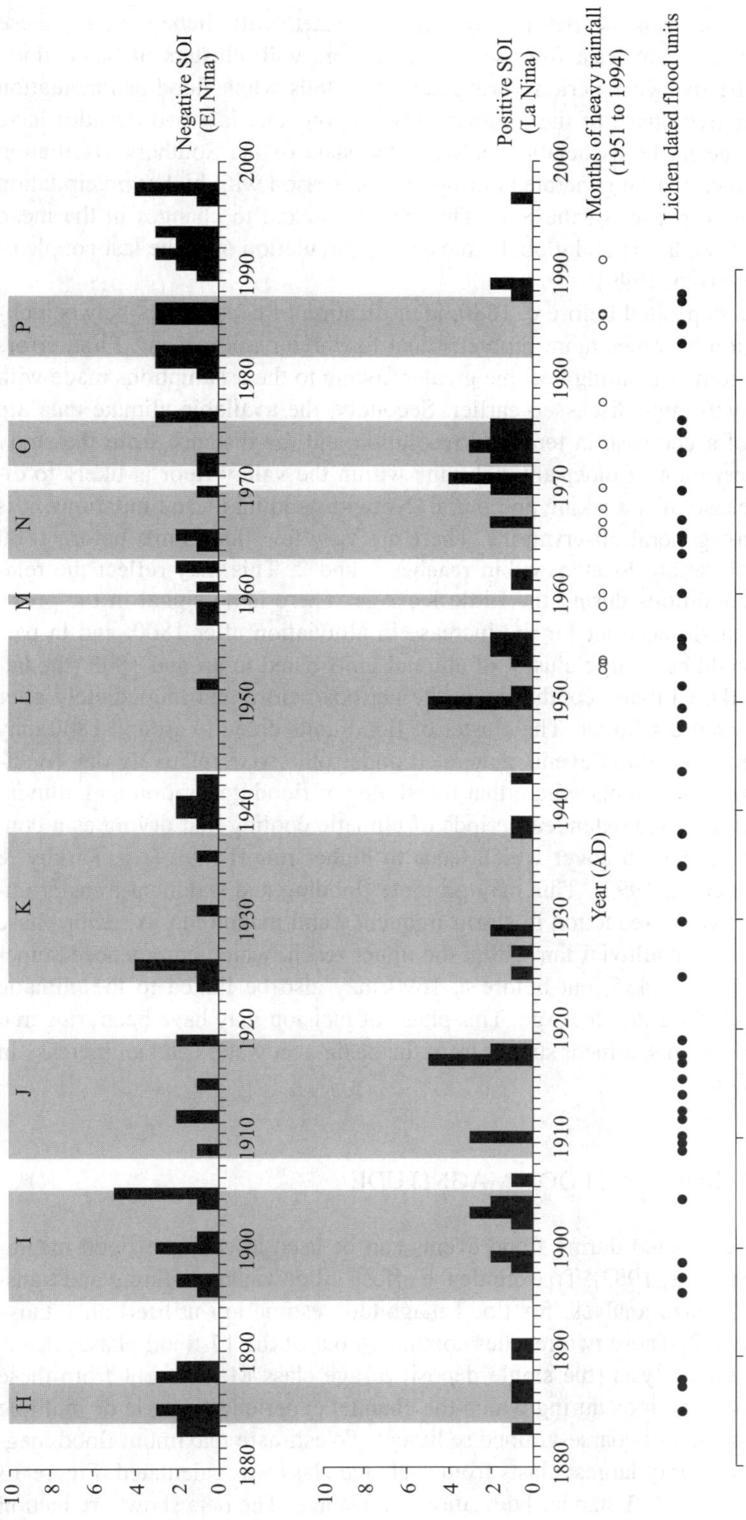

Figure 12. Lichen-based age classes H to P (vertical grey bands) and individual dated alluvial units (solid black dots) for the years 1880 to 1990 shown against monthly frequency of negative and positive states of the Southern Oscillation Index (top and bottom histograms respectively) and the occurrence of extreme monthly rainfall at Aasana (1951 to 1994).

Taking into account the possible dating vagaries associated with lichenometry, these months appear to correlate, at least from visual inspection, with clusters of dated flood units and highlight the five year period during the mid-1980s when flood sedimentation coincided with a negative phase of the Southern Oscillation. The last two decades have therefore seen a change in the association between the state of the Southern Oscillation and the quantity of precipitation generated during the wet season with higher precipitation associated with negative phases of the SOI. This may be linked to changes in the mean state of the Southern Oscillation and global atmospheric circulation over the last couple of decades (Allan & D'Arrigo, 1999).

For the flood units deposited before c. 1880, identification of causal links between climate and sedimentation becomes more problematical for three main reasons. First, errors associated with lichenometric dating become greater, owing to the assumptions made with respect to lichen growth rates discussed earlier. Secondly, the available climate data are less useful because of a decrease in temporal resolution and the distance from the study site. Lastly, the preservation of older alluvial units within the valley floor is likely to diminish over time because of reworking and burial. Notwithstanding these limitations, it is possible to make some general observations. There are very few flood units before 1800 and the majority of these are located within reaches 1 and 2. This may reflect the relatively dry and cold conditions during the Little Ice Age. There is no signal in the proxy climate records that might account for the increase in alluviation after 1800, and in particular, why there should be a large cluster of alluvial units dated to around 1800 (the latter half of age class B). Climatic conditions in the periods before and immediately after 1800 were apparently not dissimilar. The cluster of flood units dated to around 1800 may represent single or multiple flood events generated under otherwise relatively dry conditions. It has been demonstrated elsewhere that thresholds of flood generation and alluviation tend to be lowered during extended periods of climatic cooling and drying as a consequence of reduced vegetation cover which leads to higher runoff rates (e.g. Kirkby & Neale, 1987; Prentice et al., 1992). This may promote flooding and sediment transfer under a climatic regime with a reduction in storm frequency and magnitude. A major phase of channel incision into the alluvial fan within the upper reach, which commenced sometime after cal AD 1220 to 1435, but before c. 1690 may also be linked to the climatic cooling associated with the Little Ice Age. This phase of incision may have been triggered by a significant reduction in sediment supply from the basin headwaters and an increase in effective stream erosion.

7 CLIMATIC CONTROLS OF FLOOD MAGNITUDE

The size of boulders deposited during flood events can be used to estimate flood magnitude (Costa, 1983; Williams, 1983). To exclude the effects of downstream fining and transmission losses, boulder size analysis for flood magnitude estimation utilised only those data from reaches 1 and 2. These two reaches contain 16 out of the 17 flood phases determined from lichen data analysis (the single deposit of age class M is absent from these reaches) which represent periods during which the channel experienced single or multiple flood events and deposition of coarse-grained sediment. To estimate maximum flood magnitude the mean of the twenty largest clasts from each age class was calculated. Figure 13 plots mean boulder size to +/– 1 standard deviation against age. The data show a reduction

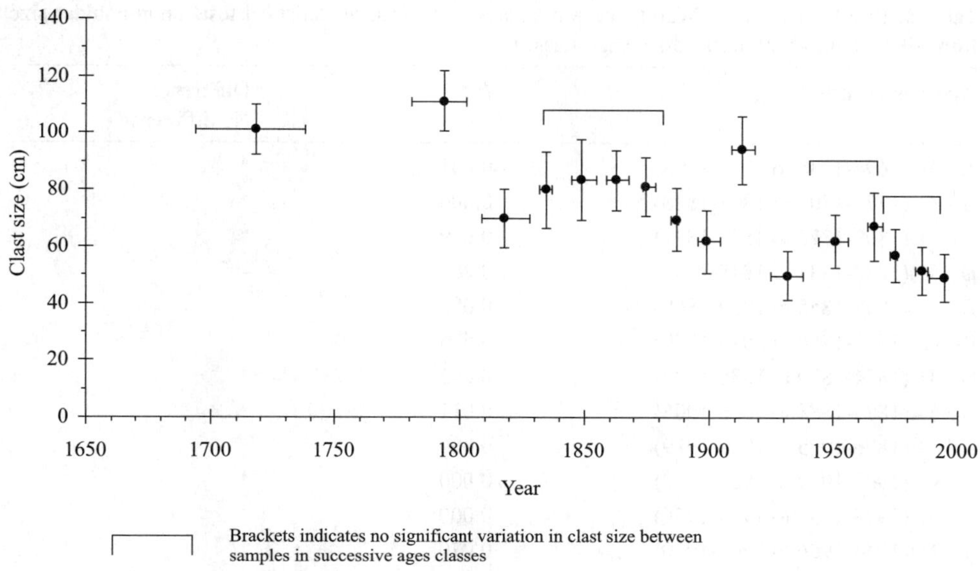

Figure 13. Flood magnitude proxy estimates based upon the mean of the twenty largest clasts (length of intermediate axes) from flood deposits within each age class (reach 1 and 2).

in flood magnitude (based on boulder size and assuming no change in gradient or flood channel width) from the early 1700s through to the late 1980s and early 1990s with the intermediate axes of boulders deposited in recent years being about half the size of boulders deposited during the early-eighteenth century. There is, however, an increase in boulder size between the years 1777-1803, 1909-1919 and 1964-1970, while flood events during the periods 1809-1828 and 1925-1938 are represented by relatively small boulders. A Kruskal-Wallis one-way analysis of variance statistical test was applied to the boulder size data to determine whether the size of boulders deposited during one particular period was significantly different from those deposited during the immediate preceding or following period. Table 6 shows the results of these tests which utilised the twenty largest clasts from deposits within each age class. *P* values of less than 0.05 indicate significant (at the 95% confidence level) differences between two successive age classes, whereas *P* values greater than 0.05 indicate the two samples may have been drawn from the same population and therefore show no significant difference in clast size.

The largest boulders deposited within the four periods (A to D) during the early 1700s to the early 1800s show statistically significant variations in size between age classes (Fig. 13). Clast size then remained relatively constant between 1832 and 1879 before decreasing towards the 1930s except for a brief period (1909-1919) during which deposited boulders were almost twice the size (approximately 90 cm) of those deposited during the preceding and following depositional periods (I and K). After 1944 there are two distinct periods showing negligible variation in clast size. The first 1944 to 1970 (age classes L and N) where clasts have intermediate axes of around 65 cm, followed by a period from 1973 to 1995 during which clast size reduced to about 50 cm. This most recent 22 year period, however, is characterised by a general reduction in clast size towards the present.

Table 6. Results of Kruskal-Wallis one-way analysis of variance statistical tests upon boulder size from alluvial units within successive age classes.

Age classes tested	*P* value	Difference * No difference -
A – B (1694-1738 to 1777-1803)	0.001	*
B – C (1777-1803 to 1809-1828)	0.000	*
C – D (1809-1828 to 1832-1837)	0.009	*
D – E (1832-1837 to 1845-1855)	0.203	-
E – F (1845-1855 to 1859-1868)	0.685	-
F – G (1859-1868 to 1874-1879)	0.570	-
G – H (1874-1879 to 1885-1888)	0.002	*
H – I (1885-1888 to 1896-1905)	0.011	*
I – J (1896-1905 to 1909-1919)	0.000	*
J – K (1909-1919 to 1925-1938)	0.000	*
K – L (1925-1938 to 1944-1956)	0.000	*
L – N (1944-1956 to 1964-1970)	0.070	-
N – O (1964-1970 to 1973-1976)	0.005	*
O – P (1973-1976 to 1983-1988)	0.126	-
P – Q (1983-1988 to 1989-1995)	0.225	-

The long term reduction in boulder size may be in response to a decrease in flood magnitude and/or a reduction in the size of clasts available for transportation. This 300 year period encompasses the latter stages of the Little Ice Age for which palaeoclimate records indicate temperatures and precipitation less than the present day average (Thompson et al., 1986). Although the climate for much of the period between about 1530 and 1880 was drier and cooler than the present century, there are a number of records of extreme rainfall and flooding events during this period. For example, the course of the Rio Salado in northwestern Argentina shifted in response to exceptionally wet months during 1703 and 1709 (Prieto, 1997) prompted by a strong La Niña cycle. The increased susceptibility of hillslopes to erosion during the cold and dry Little Ice Age and accelerated rates of mechanical breakdown is likely to have lead to a greater proportion of large calibre clasts available for entrainment during storm events. Therefore, larger alluvial clasts which date to the Little Ice Age may partly reflect changes in sediment supply and sediment calibre rather than solely flood magnitude. Indeed, when the composite decadally-averaged snow accumulation record from the Quelccaya ice-cap, Peru is compared to the boulder size record from dated flood deposits within reaches 1 and 2 (Fig. 14) it is clear that the largest clasts were deposited during periods when rates of snow accumulation were lower than average. The most prominent of these periods (1730s, 1770 to 1790, 1830 to 1850 and 1908 to 1912) are likely to have witnessed particularly cold and dry conditions leading to a reduction of vegetation cover on the hillslopes and an increase in runoff during storm events.

The apparent trend towards reduced flood magnitudes over time may also reflect an exhaustion of large calibre sediment source areas such as screes and gullies that were more active during the seventeenth to nineteenth centuries. Short periods where boulder

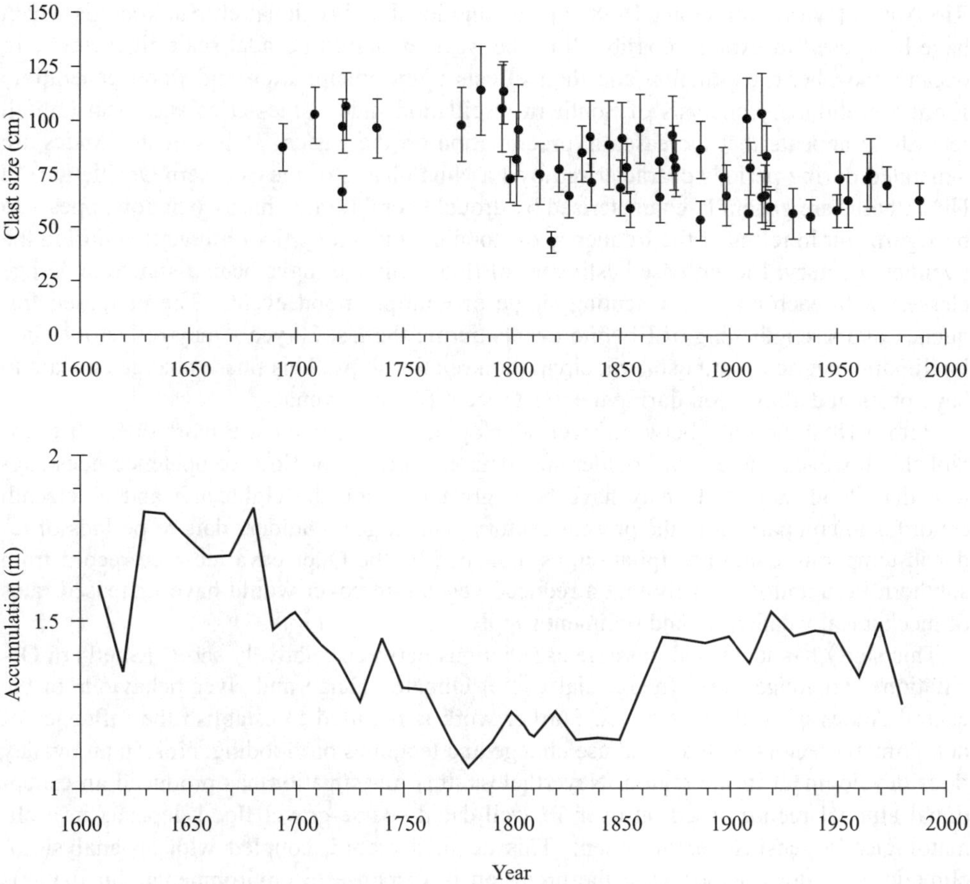

Figure 14. Mean of the five largest boulders (+/– 1 standard deviation) from lichen dated flood deposits within reach 1 and 2 compared with decadally averaged net snow accumulation (c. 1600-1980) from a composite of two cores from the Quelccaya ice cap, Peru (after Thompson, 1995).

sizes increased, such as between 1909 and 1919 may reflect a re-coupling of hillslope sediment sources to the active channel. This pattern of sediment exhaustion has been observed in comparable seasonally-dry, steepland environments in Crete in the eastern Mediterranean where Little Ice Age climates appear to have promoted an increase in the volume and calibre of hillslope material, followed by a reduction during the following period of climatic amelioration (Maas et al., 1998).

8 CONCLUSION

Five valley floor reaches within the Rio Alizos of southern Bolivia have been investigated to identify major controls of river development and flooding during the last c. 300 years. A series of coarse boulder-cobble deposits have been dated by lichenometry to establish a detailed flood history. This investigation has extended previous work undertaken in the

Rio Alizos (Warburton et al., 1998) by utilising local and regional climate records which have been used to explore possible links between annual to decadal scale fluctuations in ocean/atmospheric circulation and their effects upon precipitation and flood generation. From a preliminary analysis of Southern Oscillation Index time-series and local rainfall records, it appears that increases in precipitation on the eastern slopes of the Andes are generated during periods characterised by La Niña phases of the Southern Oscillation. El Niño phases are generally characterised by drought conditions. This association appears to be significant in terms of the frequency of flooding and valley floor alluviation during the twentieth century. Lichen-based estimates of flood unit ages have been assigned to 17 age classes, with each class representing single or multiple flood events. The increased frequency and strengthening of El Niño events during the last 20 years may reflect a switch in climatic and ocean/atmospheric circulation relationships. This phase change appears to have promoted alluviation during the most recent El Niño events.

Before 1880 the links between river development and climate are more difficult to establish. However, the use of boulder size data as a proxy for flow competence does suggest that flood magnitude may have been greater during the eighteenth and nineteenth centuries in comparison to the present century. The largest boulders date to periods of reduced temperature and precipitation, as indicated by the Quelccaya ice core record from southern Peru and it is likely that a reduced vegetation cover would have increased rates of mechanical weathering and sediment supply.

This study has identified possible associations between relatively short- (Southern Oscillation) and longer-term (neoglacial cycle) climate change and river behaviour in the central Andes of southern Bolivia. Further work is required to establish the influence of non-climatic factors such as land use change and tectonics on flooding, erosion and valley floor development in the region. Nevertheless, this investigation has produced an exceptional alluvial record based on over 17 well-dated coarse-gravel flood deposits in a climatologically sensitive environment. This detailed record, coupled with an analysis of climatic data, does demonstrate the precision in response to environmental fluctuations that alluvial sediments in mountain environments can display.

ACKNOWLEDGEMENTS

The authors are extremely grateful to Iain Gulley and Anthony Smith for their help in the completion of the geomorphological and location maps and also to Prof. John Lewin whose comments improved this chapter. Two anonymous reviewers provided additional constructive comments on the manuscript. This research was funded by DGXII, Commission of the European Communities.

REFERENCES

Abbott, J.T. & Valastro, Jnr. S. 1995. The Holocene alluvial records of the chorai of Metapontum, Basilicata and Croton, Calabria, Italy. In: Lewin, J., Macklin, M.G. & Woodward, J.C. (eds), *Mediterranean Quaternary River Environments*. Balkema, Rotterdam, 195-205.

Aceituno, P. 1988. On the functioning of the Southern Oscillation in the South American sector. Part I: surface climate, *Monthly Weather Review*, 116: 505-524.

Aceituno, M.B. & Montecinos, A. 1993. Circulation anomalies associated with dry and wet periods in the South American Altiplano. In: *Fourth International Conference on Southern Hemisphere Meteorology and Oceanography, American Meteorological Society,* 330-331.

Allan, R.J. & D'Arrigo, R.D. 1999. 'Persistent' ENSO sequences: how unusual was the 1990-1995 El Niño?, *The Holocene*, 9: 101-118.

Allan, R.J., Lindesay, J. & Parker, D. 1996. *El Niño Southern Oscillation and climatic variability*. CSIRO, Australia.

Barker, G.W. & Hunt, C.O. 1995. Quaternary valley floor erosion and alluviation in the Biferno Valley, Molise, Italy: the role of tectonics, climate sea-level change, and human activity. In: Lewin, J., Macklin, M.G. & Woodward, J.C. (eds), *Mediterranean Quaternary River Environments*. Balkema, Rotterdam, 145-157.

Benedict, J.B. 1988. Techniques in lichenometry: identifying the yellow rhizocarpons, *Arctic and Alpine Research*, 20: 285-291.

Benito, G., Machado, M. J. & Pérez-González, A. 1996. Climate change and flood sensitivity in Spain. In: Branson, J., Brown, A.G. & Gregory, K.J. (eds), *Global Continental Changes: the Context of Palaeohydrology, Geological Society Special Publications,* 115: 85-98.

Benito, G., Baker, V.R. & Gregory, K.J. 1998. (eds), *Palaeohydrology and Environmental Change*. Chichester, John Wiley.

Bradley, R.S. 1985. *Quaternary Paleoclimatology*. Allen and Unwin, Winchester, Mass, USA.

Bradley, R.S. & Jones, P.D. (eds), 1995. *Climate Since AD 1500*. Routledge, London.

Caviedes, C.N. 1984. El Niño 1982-83. *Geographical Review*, 74: 267-290.

Costa, J.E. 1983. Paleohydraulic reconstruction of flash-flood peaks from boulder deposits in the Colorado Front Range. *Geological Society of America Bulletin,* 94: 986-1004.

Craig, A.K. & Shimada, I. 1986. El Niño flood deposits at Batán Grande, northern Peru. *Geoarchaeology*, 1: 29-38.

DeGraff, J.V. 1994. The geomorphology of some debris flows in the southern Sierra Nevada, California. *Geomorphology*, 10: 231-252.

Easton, R.M. 1994. Lichens and rocks: a review. *Geoscience Canada*, 21: 59-76.

Enfield, D.B. & Cid, L.S. 1991. Low-frequency changes in El Niño-Southern Oscillation, *Journal of Climate*, 4: 1137-1146.

Gallart, F. & Clotet-Perarnau, N. 1988. Some aspects of the geomorphic processes triggered by an extreme rainfall event: the November 1982 flood in the eastern Pyrenees. *Catena Supplement*, 13: 79-95.

Gutiérrez Elorza, M. & Peña Monné, J.L. 1990. Upper Holocene climatic change and geomorphological processes on slopes and infilled valleys from archaeological dating (N.E. Spain). In: Imeson, A.C. & De Groot, R.S. (eds), *Landscape Ecological Impact of Climatic Change on the Mediterranean Region.* Universities of Wegeningen, Utrecht and Amsterdam, 1-18.

Harvey, A.M., Alexander, R.W. & James, P.A. 1984. Lichens, soil development and the age of Holocene valley floor landforms: Howgill Fells, Cumbria. *Geografiska Annaler*, 66A: 353-366.

Heine, K. 1987. Anthropogenic sedimentological changes during the Holocene in Mexico and Central America. In: Starkel, L., (ed.) *Anthropogenic Sedimentological Changes During the Holocene. Striae*, 26: 51-63.

Innes, J.L. 1983. Lichenometric dating of debris-flow deposits in the Scottish Highlands. *Earth Surface Processes and Landforms*, 8: 579-588.

Kirkby, M.J. & Neale, P.H. 1987. A soil erosion model incorporating seasonal factors. In: V. Gardiner. (ed.), *International Geomorphology* 1986, Part II, Chichester: John Wiley, 189-210.

Kousky, V E., Kagano, M.T. & Cavalcanti, I.F.A. 1984. A review of the Southern Oscillation: oceanic-atmospheric circulation changes and related rainfall anomalies, *Tellus*, 36A: 490-502.

Landsberg, H.E. 1985. Historic weather data and early meteorological observations. In: Hecht, A.D. (ed.), *Palaeoclimate Analysis and Modeling*, New York: Wiley, 27-70.

Lewin, J., Macklin, M.G. & Woodward, J.C. (eds) 1995. *Mediterranean Quaternary River Environments*. Balkema, Rotterdam.

Lockwood, J.G. 1984. The Southern Oscillation and El Niño, *Progress in Physical Geography*, 8: 102-110.

Longfield, S.A. & Macklin, M.G. 1999. The influence of recent environmental change on flooding and sediment fluxes in the Yorkshire Ouse basin, *Hydrological Processes*, 13: 1051-1066.

Maas, G.S., Macklin, M.G. & Kirkby, M.J. 1998. Late Pleistocene and Holocene river development in Mediterranean steepland environments, south-west Crete, Greece. In: Benito, G., Baker, V.R. & Gregory, K.J. (eds), *Palaeohydrology and Environmental Change*. Chichester, John Wiley, 153-166.

Macklin, M.G., Rumsby, B.T. & Heap, T. 1992. Flood alluviation and entrenchment: Holocene valley floor development and transformation in the British uplands, *Geological Society of America Bulletin*, 104: 631-643.

Macklin, M.G., Lewin, J. & Woodward, J.C. 1995. Quaternary fluvial systems in the Mediterranean basin. In: Lewin, J., Macklin, M.G. & Woodward, J.C. (eds), *Mediterranean Quaternary River Environments*, Balkema, Rotterdam, 1-25.

Merrett, S.P. & Macklin, M.G. 1999. Historic river response to extreme flooding in the Yorkshire Dales, northern England. In: Brown, A.G. & Quine, T.A. (eds), *Fluvial Processes and Environmental Change*, Chichester, John Wiley, 345-360.

Milne, A. 1986. *Floodshock: the drowning of planet earth*. Alan Sutton Publishing, Gloucester.

Molina Sempere, C.M., Vidal-Abarca, M. & Suárez, M.L. 1994. Floods in south-east Spanish areas: a historical and environmental review. In: Rossi, G., Harmancioglu, N. & Yevjevich, V. (eds), *Coping with Floods*, Kluwer Academic Press, Dordrecht, 271-278.

Poesen, J.W.A. & Hooke, J.M. 1997. Erosion, flooding and channel management in Mediterranean environments of southern Europe. *Progress in Physical Geography*, 21(2): 157-199.

Pope, K. & van Andel, T.H. 1984. Late Quaternary alluviation and soil formation in the southern Argolid: Its history, causes and archaeological implications. *Journal of Archaeological Science*, 11: 281-306.

Prentice, C.I., Guiot, J. & Harrison, S.P. 1992. Mediterranean vegetation, lake levels and palaeoclimate at the Last Glacial Maximum. *Nature*, 360: 658-660.

Preston, D.A., Macklin, M.G. & Warburton, J. 1997. Fewer people, less erosion: the twentieth century in southern Bolivia, *Geographical Journal*, 163: 198-205.

Prieto, M.R. 1997. Variaciones climáticas en el NOA durante el período colonial. In: Reboratti, C. (ed.), De hombres y tierras. Una historia ambiental del Noroeste argentino. *Proyecto Desarrollo agroforestal en comunidades rurales del Noroeste Argentino*, Salta, 60-75.

Prosser, I.P., Chappell, J. & Gillespie, R. 1994. Holocene valley aggradation and gully erosion in headwater catchments, south-eastern highlands of Australia. *Earth Surface Processes and Landforms*, 19: 465-480.

Ropelewski, C.F. & Halpert, M.S. 1987. Global and regional scale precipitation patterns associated with the El Niño/Southern Oscillation, *Monthly Weather Review*, 115: 1606-1626.

Rumsby, B.T. & Macklin, M.G. 1994. Channel and floodplain response to recent abrupt climate change: the Tyne basin, northern England. *Earth Surface Processes and Landforms*, 19: 499-515.

Rumsby, B.T. & Macklin, M.G. 1996. River response to the last neoglacial (the 'Little Ice Age') in northern, western and central Europe. In: Branson, J., Brown, A.G. & Gregory, K.J. (eds), *Global Continental Changes: the Context of Palaeohydrology*. Geological Society Special Publication, 115: 217-233.

Selby, M.J. 1993. *Hillslope Materials and Processes*. 2nd Edition, Oxford University Press, Oxford.

Self, S., Rampino, M.R., Zhao, J. & Katz, M.G. 1997. Volcanic aerosol perturbations and strong El Niño events: no general correlation, *Geophysical Research Letters*, 24: 1247-1250.

Stone, P.B. 1992. *The State of the World's Mountains: A global report*. Zed Books Ltd, London and New Jersey.

Street-Perrott, F.A. & Harrison, S.P. 1985. Lake levels and climate reconstruction. In: Hecht, A.D. (ed.), *Palaeoclimate Analysis and Modelling*, New York: John Wiley, 291-340.

Tapley Jnr., T.D. & Waylen, P.R. 1990. Spatial variability of annual precipitation and ENSO events in western Peru, *Hydrological Sciences*, 35: 429-446.

Thompson, L.G., Mosley-Thompson, E. & Arnao, B. 1984. El Niño-Southern Oscillation events recorded in the stratigraphy of the tropical Quelccaya ice cap, Peru, *Science*, 226: 50-53.

Thompson, L.G., Mosley-Thompson, E., Bolzan, J.F. & Koci, B.R. 1985. A 1500-year record of tropical precipitation in ice cores from the Quelccaya Ice Cap, Peru. *Science*, 229: 971-973.

Thompson, L.G., Mosley-Thompson, E., Dansgaard, W. & Grootes, P.M. 1986. The Little Ice Age as recorded in the stratigraphy of the tropical Quelccaya Ice Cap, *Science*, 234: 361-364.

Thompson, L.G. 1995. Ice core evidence from Peru and China. In: Bradley, R.S. & Jones, P.D. (eds), *Climate Since AD 1500*, Routledge, London, 517-548.

van Andel, T.H., Zangger, E. & Demitrack, A., 1990. Land use and soil erosion in Prehistoric and Historical Greece. *Journal of Field Archaeology*, 17: 379-396.

Wang, B. 1995. Interdecadal changes in El Niño onset in the last four decades, *Journal of Climate*, 8: 267-285.

Warburton, J., Macklin, M. & Preston, D.A. 1998. Fluvial hazards in a steepland mountain environment, southern Bolivia. In: *Geomorphological Hazards in High Mountain Environments. GeoJournal*, Kluwer Academic Publishing, 229-243.

Webster, P. J. & Palmer, T. N. 1997. The past and future of El Niño, *Nature*, 390: 562-564.

Wells, L. E. 1987. An alluvial record of El Niño events from northern coastal Peru. *Journal of Geophysical Research*, 92: 14463-14470.

Wells, L.E. 1990. Holocene history of the El Niño phenomenon as recorded in flood sediments of northern coastal Peru. *Geology*, 18: 1134-1137.

Wells, S.G. & Harvey, A.M. 1987. Sedimentologic and geomorphic variations in storm-generated alluvial fans, Howgill Fells, northwest England. *Geological Society of America Bulletin*, 98: 182-198.

Williams, G.P. 1983. Palaeohydrological methods and some examples from Swedish fluvial environments: I Cobble and boulder deposits. *Geographiska Annaler*, 65A: 227-243.

Woodward, J.C. 1995. Patterns of erosion and suspended sediment yield in Mediterranean river basins. In: Foster, I.D.L., Gurnell, A.M. & Webb, B.W. (eds), *Sediment and Water Quality in River Catchments*, Chichester, John Wiley, 365-389.

Part 4: Geoarchaeological perspectives and the human impact

11. The Holocene fluvial sedimentary record and alluvial geoarchaeology in the Nile Valley of northern Sudan

JAMIE C. WOODWARD
School of Geography, University of Leeds, UK

MARK G. MACKLIN
Institute of Geography and Earth Sciences, University of Wales, Aberystwyth, UK

DEREK WELSBY
The Sudan Archaeological Research Society, The British Museum, London, UK

1 INTRODUCTION

It is now well known that Nile basin hydrology is closely linked to the intensity of the African monsoon and large fluctuations in discharge and sediment transfer during the Quaternary Period have been driven by changes in global climate (Williams & Adamson, 1980; Hassan, 1981; Rossignol-Strick et al., 1982; Hulme, 1994). These changes are recorded in pollen and lake level data from across Africa (e.g. Gasse et al., 1980; Ritchie et al., 1985), and these palaeoclimate records have been integrated to establish the duration and spatial extent of wet and dry phases during the Late Quaternary (Street & Grove, 1979; Street Perrot et al., 1989; Hassan, 1996). In the headwaters of the Blue and White Niles, marked changes in precipitation and runoff regime have taken place throughout the Quaternary as a result of changes in the Earth's orbit (Milankovitch Cycles), global ocean temperature anomalies, and migrations of the Inter-Tropical Convergence Zone (ITCZ) (Hulme, 1994). As the nature of Late Quaternary fluvial sedimentation and erosion in the Nile Valley downstream of Khartoum is intimately related to the flood hydrology and sediment yield of the major headwater catchments, the vicissitudes of Pleistocene and Holocene climate have produced a varied and complex geomorphological and stratigraphic record in the Saharan Nile (Butzer & Hansen, 1968; Adamson et al., 1980; Williams & Adamson, 1982).

Long term fluctuations in Nile basin runoff are also recorded in the Nile Delta. Here, heavy mineral suites in sediment cores signal the changing importance of headwater basin sediment source areas during the Late Pleistocene and Holocene because the Blue and White Niles drain lithologically distinctive terrains (Foucalt & Stanley, 1989). However, in contrast to the Nile Delta records and these intensively studied areas upstream (Butzer & Hansen, 1968; Williams & Adamson, 1982), there is little well-dated geomorphological and stratigraphical information on long-term river behaviour for much of the Sudanese Saharan Nile downstream of Khartoum. In this part of the Nile basin the nature, age and spatial extent of the alluvial sedimentary record is not well known (Butzer, 1980). Furthermore, in many of these reaches, detailed archaeological surveys of the valley floor have not been carried out.

Previous investigations of long-term fluvial processes in the Nile Valley have commonly been hampered by the scarcity of suitable samples for radiocarbon dating and the age of many of the sedimentary sequences is poorly constrained (e.g. Butzer & Hansen, 1968). Good geochronological control is essential for any attempt to elucidate the nature of Late Pleistocene and Holocene river behaviour and to compare alluvial histories with

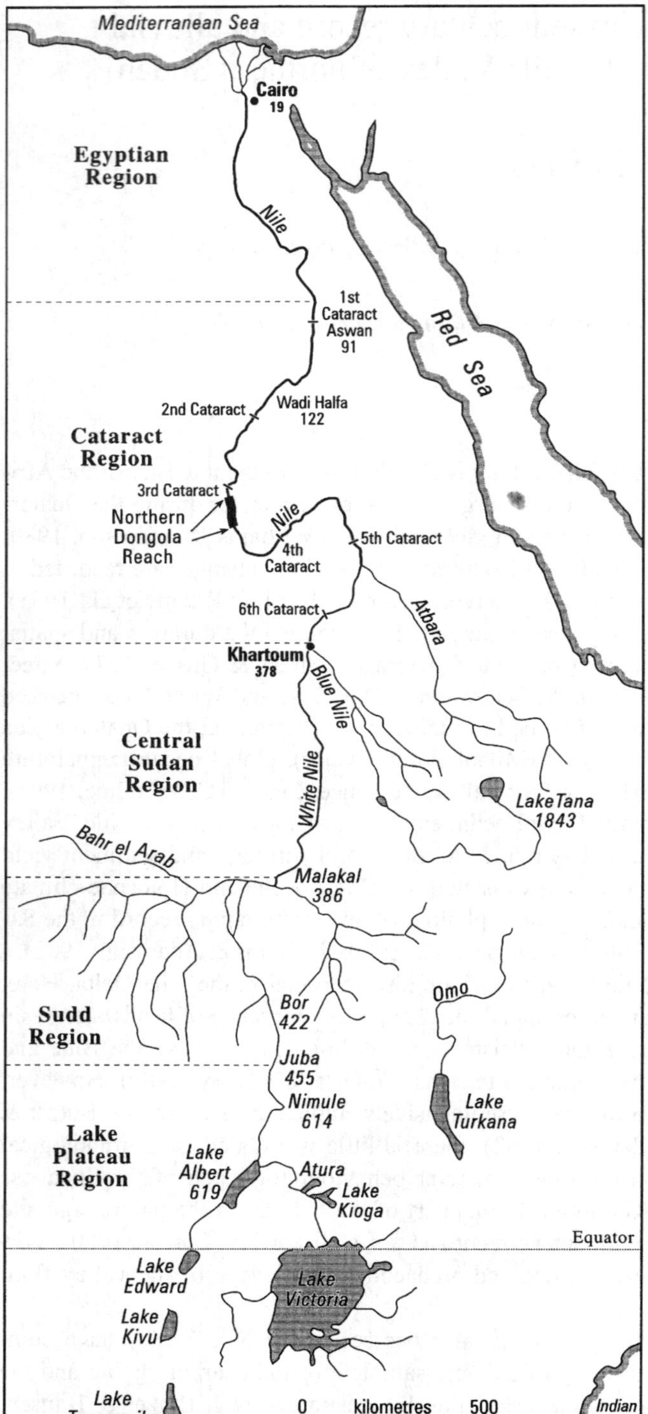

Figure 1. Map of the Nile basin showing the major tributaries and the main locations and sites mentioned in the text (after Said, 1994).

proxy climate records and archaeological data. This paper reports recent geomorphological research into Holocene river development in the Northern Dongola Reach of Sudanese Nubia (Fig. 1). Optically stimulated luminescence (OSL) dating techniques have been applied to develop a timescale for Holocene river activity in the region. This work was conducted in association with a large-scale archaeological survey of the valley floor by a team from the Sudan Archaeological Research Society (SARS) of the British Museum (Welsby, 1995). The aims and objectives of the archaeological survey and our geomorphological work are outlined below.

2 ALLUVIAL GEOARCHAEOLOGY AND THE NORTHERN DONGOLA REACH SURVEY

Variability in the source, seasonality and volume of discharge has exerted a major influence on the nature and distribution of human activity in the Nile corridor. Indeed, flood magnitude and resource availability have exerted a profound influence on the riparian landscapes and peopling of the Nile Valley since Lower Palaeolithic times (Wendorf & Schild, 1976; Butzer, 1980; Said, 1993). Most research into the impact of Nile flood dynamics on riparian land use and the archaeological record has been focused on the Egyptian Nile Valley (Butzer, 1981; Hassan, 1997). Thus, while many of the prominent archaeological sites in the Sudan – such as Kawa, Meroe and Kerma – have been excavated at various times during the present century (Fig. 2), large sectors of the Sudanese Nile Valley have not yet been systematically surveyed. In addition, many archaeological sites in the Sudanese Nile Valley are currently under threat from agricultural development (Welsby, 1995).

A SARS team has recently completed a detailed archaeological field survey in the Northern Dongola Reach. The survey area includes the well preserved site of Kawa that was first excavated in the 1920s and 1930s (Fig. 2). This site includes a temple built by Tutankhamun around 1350 BC and a large Kushite town (Welsby, 1998). Kawa is probably the most important site surviving in the Sudanese Nile Valley for the study of the period from the 18th Dynasty to the 4th century AD and is currently being surveyed and excavated by a new SARS project (Welsby, 1998). The surrounding area was not surveyed during the early investigations and much of the Northern Dongola Reach remained unexplored prior to the Northern Dongola Reach Survey (NDRS) which began in January 1993.

The NDRS focused on the nature and distribution of archaeological sites from the prehistoric period up to recent times. One particularly striking outcome of this work is the close correspondence between sites dating to the Kerma Period (c. 2500-1500 BC) and the margins of former Nile channel belts, some of which are located more than 15 km away from the present channel. As this pattern emerged, it became increasingly clear that a full interpretation of the cultural sequence in the Northern Dongola Reach would not be possible without an understanding of the Holocene alluvial record – including the age and mode of formation of the palaeochannel belts in the region. This paper outlines the archaeological survey data and presents the results of geomorphological and stratigraphical work on the Holocene fluvial record in the Northern Dongola Reach. This geoarchaeological work is ongoing and has three broad aims:

1. To conduct a geomorphological and stratigraphical evaluation of Holocene river behaviour in the Northern Dongola Reach.

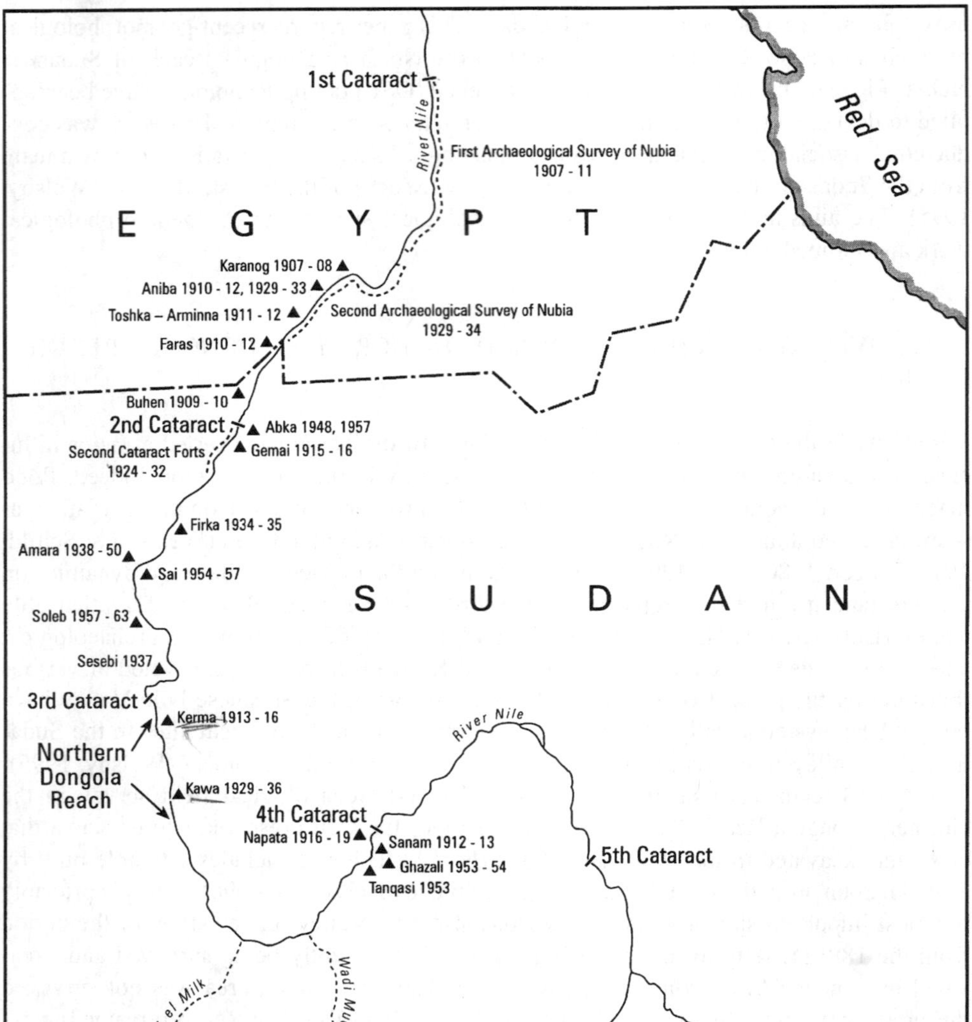

Figure 2. The main archaeological sites in the Sudanese Nile Valley and the Northern Dongola Reach Survey concession. Dates are given for the main phases of survey and excavation at each site (after Adams, 1977).

2. To establish the age of the alluvial sedimentary sequence and the palaeochannel belts using OSL and radiocarbon dating techniques.
3. To elucidate the temporal and spatial relationship between archaeological sites and the palaeochannel belts and assess the roles of climate change and flooding in long-term river behaviour and riparian land use.

3 THE NILE BASIN SEDIMENT SYSTEM

The Nile is the world's longest river (c. 6670 km) draining around one tenth of the African continent (Fig. 1). The present average contribution to main channel flows from each of the three main tributary basins of the Blue Nile, White Nile and Atbara is shown in Figure 3. Runoff from the Ethiopian highlands via the Blue Nile and Atbara accounts for roughly 70% of the annual water discharge and > 95% of the suspended sediment load – emphasising the importance of the East African monsoon (Foucalt & Stanley, 1989; Hulme, 1994). In contrast, the White Nile contributes less than one third of the discharge and its sediment load is small in comparison to the other two main tributaries – accounting for only 3% of the total load (Foucalt & Stanley, 1989). The mean annual suspended sediment load of the modern Nile (upstream of the High Aswan Dam) is estimated to be around 120 million tonnes per year (Milliman & Syvitski, 1992), but the magnitude of the conveyance losses to the channel and floodplain zone is not known.

The Blue Nile drains the mountains of Ethiopia and produces torrential flows during the summer monsoon (July to October) and very low flows during the dry season. In fact, the ratio of maximum to minimum flows can exceed 20:1 and from July to August, the ratio of the Blue Nile flows to those of the White Nile can be of the order of 97:1 (Andah & Siccardi, 1991). Thus, the White Nile, which drains the equatorial lakes of the central African plateau, is crucial for maintaining flows in the Saharan Nile from November to June (Figs 1 and 3B). The gauged flood record of the 20th century reveals considerable inter-decadal and inter-annual variability in Nile discharge (Evans, 1994). For example, 1916 recorded the maximum annual water yield of 120 billion cubic metres in contrast to the minimum of only 42 billion cubic metres for 1984 (Hulme, 1994).

4 LATE QUATERNARY HISTORY OF THE NILE

The dimensions of the Nile catchment have not been constant during the Quaternary, with periodic severing of the White Nile headwaters serving to enhance the seasonality of flows. Until the late Middle Pleistocene, the geomorphological evolution of the Egyptian Nile Valley was strongly influenced by more local runoff with major east bank wadis supplying the bulk of the sediment and water. Butzer (1980) has argued that almost all of the Pleistocene gravel-sand units in Upper Egypt can be attributed to regional runoff from the Red Sea Hills.

The importance of the Blue Nile sediment system has been outlined above and Williams & Adamson (1980) have proposed a model for the behaviour of this basin during the Last Glacial Maximum and the early Holocene which is illustrated in Figure 4. In very general terms, the Late Quaternary history of the Nile can be summarised as follows:

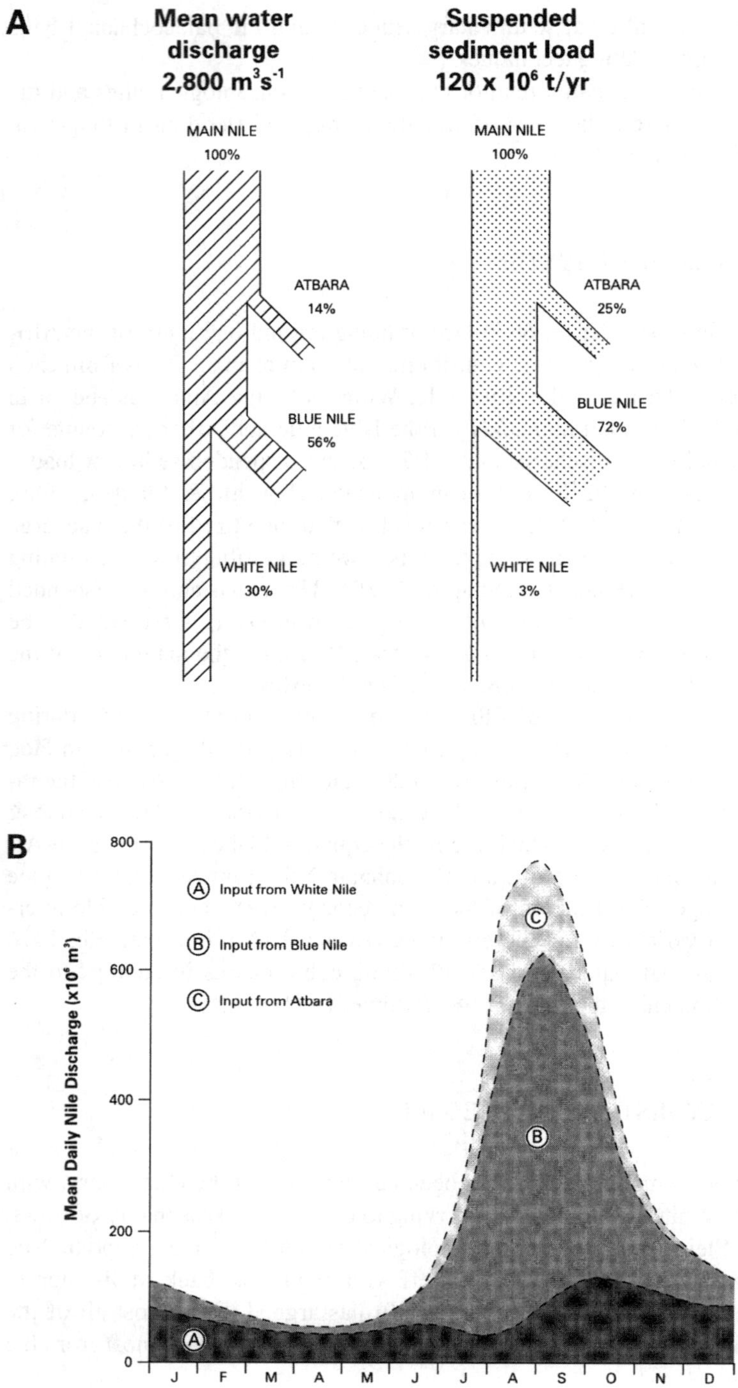

Figure 3. A) The water and suspended sediment budget of the present Nile basin (after Foucalt & Stanley, 1989). Water yield and sediment load data after UNESCO and Milliman & Syvitski (1992) respectively. B) The seasonal pattern of discharge of the main Nile and the contributions from the Blue Nile, White Nile and Atbara (after Hurst, 1952).

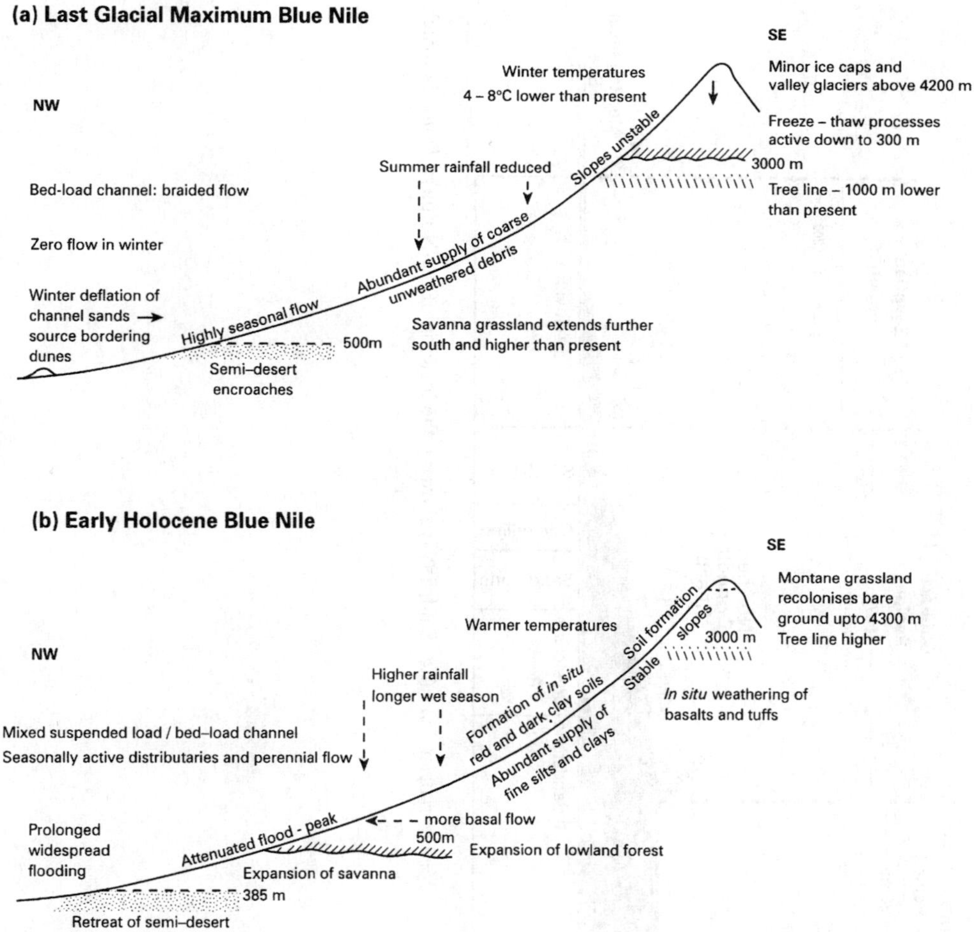

Figure 4. The Blue Nile during: a) The Last Glacial Maximum and b) The early Holocene (after Williams & Adamson, 1980).

A last glacial dry phase from c. 20 to 12.5 ka BP associated with cold conditions in the Ethiopian Highlands produced rapid erosion from bare hillslopes as the tree line stood about 1000 m lower than today. Throughout this period, large volumes of coarse and fine sediments were delivered to the fluvial system and net aggradation took place in many reaches of the main channel downstream (Williams et al., 1998) (Fig. 4). At the global scale, these conditions were produced by the highly reflective ice sheets, the cool oceans, and the equatorward-extended sea-ice borders (which displaced the polar front and mid-latitude westerlies towards the equator) and the reduction in the extension and intensity of the African monsoon (Hulme, 1994).

Towards the end of the last cold stage, more humid conditions developed as the increase in summer insolation in the Northern Hemisphere enhanced thermal contrasts between land and ocean and African monsoon intensity increased. Thus, both African lake levels and Nile flows were generally higher between c. 12.5 and 5 ka BP (Street & Grove,

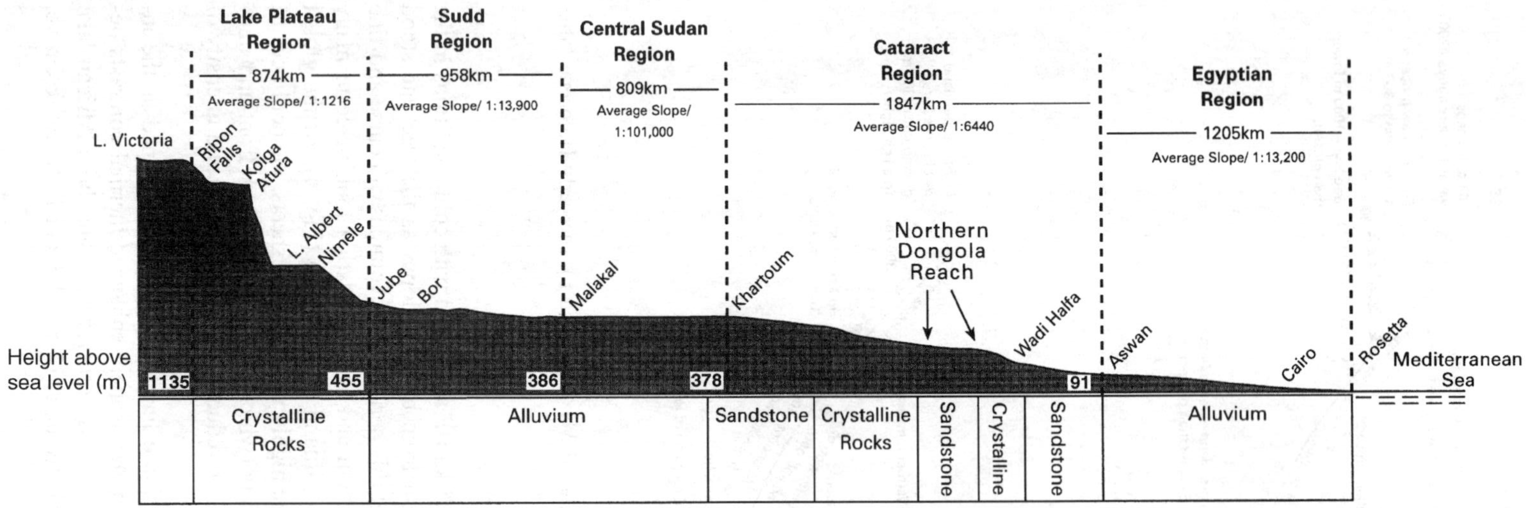

Figure 5. The long profile of the Nile from Lake Victoria to the Mediterranean Sea (after Said, 1994) showing the location of the Northern Dongola Reach in the cataract region.

1979; Williams et al., 1998), although this period was punctuated by several short-lived dry periods (Gasse & van Campo, 1994; Hassan, 1996). The main channel of the Nile shifted from the cold stage, sediment-charged braided system to a predominantly single-channel as the Ethiopian uplands were stabilised by vegetation and sediment yield declined (Fig. 4). During this early Holocene humid phase the northern front of monsoonal rain extended about 700 km northward from its present position and recharged groundwater stores, enhanced fluvial runoff and created lakes across the Sahara (Ritchie et al., 1985; Hassan, 1996). In contrast, the last five thousand years or so has been characterised by a progressive increase in aridity (Williams et al., 1998) with no significant runoff inputs downstream of the Atbara (Fig. 1).

5 THE NORTHERN DONGOLA REACH

Figure 5 shows the long profile of the Nile from Lake Victoria (1135 m above sea level) to the Mediterranean Sea. The cataract region of the main Nile Valley lies between Khartoum (378 m a.s.l.) and Aswan (91 m a.s.l.) where the channel falls over 280 m over a distance of some 1847 km. The Northern Dongola Reach is a relatively low gradient stretch of the Nile Valley which lies upstream of the Third Cataract. The wide alluvial plain of the Dongola Reach extends for a distance of about 313 km between the Fourth and Third Cataracts and the channel has a low average gradient of 0.083 m km^{-1}. This part of the valley is underlain by Nubian Sandstone and the lower part of this stretch includes all of the Northern Dongola Reach where our archaeological and geomorphological investigations have taken place (Figs 1 and 5). Upstream of Dongola the Nile Basin covers an area of some 1,610,000 km^2 and has a mean annual water yield of 89 billion cubic metres.

Plate 1 is a SPOT satellite image of the Northern Dongola Reach which shows the palaeochannel belts and the other major geomorphological features in the region. The absence of prominent river terrace features and the low relief of the terrain (excluding the aeolian sand dunes) from the modern Nile to the bedrock plateau in the east, indicates that the alluvial valley floor has filled relatively uniformly in the recent geological past (Fig. 6 and Plate 1). The Dongola Reach is an extensive alluvial basin containing a series of major palaeochannel belts and smaller palaeochannel features where localisation of sedimentation along river channels has created subtle elevation differences between upstanding levées and lower-lying inter-channel areas. Large-scale avulsion may be initiated by floods that breach levées, shift channel position or behead channels through channel incision (Richards et al., 1993).

6 THE KERMA PERIOD AND ARCHAEOLOGICAL SURVEY IN THE NORTHERN DONGOLA REACH

6.1 *The Kerma Period*

The Egyptian retreat from Nubia under the Thirteenth Dynasty, late in the 18th century BC, correlates with the rise of a rich culture at Kerma in the most fertile part of Sudanese Nubia (Phillipson, 1993). Kerma is an ancient city on the Nile in central Nubia located approximately 40 km upstream of the 3rd cataract region and approximately 55 km downstream

Figure 6. The Nubian Sandstone bedrock plateau which marks the eastern edge of the alluvial valley floor in the Northern Dongola Reach and the eastern margin of the archaeological survey (see SPOT image).

of Kawa (Fig. 2). Kerma was an important strategic location at the southern limit of Egyptian control and commanded an extensive network of trading routes (Bonnet, 1992). The Kerma Period spanned a thousand years from c. 2500 to 1500 BC and great wealth and a remarkable level of craftsmanship – particularly in pottery – were attained. Archaeological research in the Kerma necropolis of Sai (about 100 km to the north of Kerma) lead to the identification of four phases in the development of the Kerma civilization (see Bonnet, 1992) and these are shown in Table 1. This sub-division is based on several criteria including contrasts in pottery styles, funerary ritual and tomb morphology.

6.2 *The Northern Dongola Reach Survey*

The Sudan Archaeological Research Society held a concession for survey and trial excavation for an area 80 km north-south by a maximum of 18 km east-west on the east bank of the Nile (Welsby, 1995). Under the direction of Derek Welsby, the Northern Dongola Reach Survey began in January 1993 and involved systematic field survey along east-west transects at 5 minutes of latitude intervals (approximately 9.25 km), and the recording of the more visible sites throughout the survey area (Welsby, 1995). The area was divided into 24 grid squares (each 5′ of latitude by 5′ of longitude) and extended from the modern settlements of Eimani in the north to Mulwad opposite el Khandaq in the south (Fig. 7). The eastern limit of the valley floor is marked by a Nubian sandstone bedrock scarp which lies approximately 10 m above the valley floor (Fig. 6).

Table 1. The four chronological periods of the Kerma culture (after Bonnet, 1992) and the earlier and later cultures in the Northern Dongola Reach.

Period	Dates
Neolithic	Before 3500 BC
Pre-Kerma	3500 to 2500 BC
Early Kerma	2500 to 2050 BC
Middle Kerma	2050 to 1750 BC
Classic Kerma	1750 to 1580 BC
Final Kerma	1580 to 1500 BC
New Kingdom	1500 – c. 1070 BC
Kushite: Napatan	9th to 4th Century BC
Meroitic	4th Century BC to 4th Century AD
Post-Meroitic	4th Century AD to 6th Century AD
Medieval	6th Century AD to 15th Century AD

A four-season survey recorded over 450 new sites and many of these date to the Neolithic and Kerma periods (Fig. 7 and Table 1). Two main palaeochannel belts to the east of the modern Nile were identified during the 1993/1994 season. The eastern palaeochannel belt which runs roughly north-south (traversing grid squares I, M and P, Fig. 7) close to the bedrock scarp was named the Alfreda Nile and this turns westwards to join the central palaeochannel belt (in the southeastern corner of grid square R) which was named the Hawawiya Nile (Welsby, 1995). Downstream of this confluence zone the palaeochannel belt flows close to the Seleim Basin (Plate 1 and Fig 7) and was named the Seleim Nile by Welsby (1995).

In this part of the Nile Valley the present course of the river is an incised, predominantly single channel system (Plate 1) and is referred to here as the Dongola Nile. Sites dating to the Neolithic and Kerma Periods were found on the margins of the Alfreda, Hawawiya and Seleim palaeochannels and by the Dongola Nile. Neolithic sites are found throughout most of the survey area although they are more dispersed than the Kerma sites and a significant number are not closely associated with the major palaeochannels (Fig. 8). The Neolithic inhabitants utilised large mounds for their burials and such features are often found several kilometres apart – mainly between the Alfreda and Hawawiya palaeochannel belts and well to the east of the Seleim Nile (Welsby, 1995). A particularly important feature of the survey data is the close association of the Kerma sites with the major palaeochannel belts (Fig. 8). The Kerma sites in the Northern Dongola Reach include numerous cemeteries, occupation areas indicated by spreads of pottery, and prominent raised mounds with some isolated buildings (Fig. 9). Apart from Kawa itself, later sites dating to the Pharaonic and Napatan periods were rare or non-existent throughout the survey area (Welsby, 1995). Indeed, all the evidence for Meroitic, Medieval and later settlement (Table 1) was located along the banks of the present course of the Dongola Nile (Welsby, 1995). Elucidating the relationship between river behaviour and riparian settlement and land use during the Kerma and later periods is a particular focus of this project.

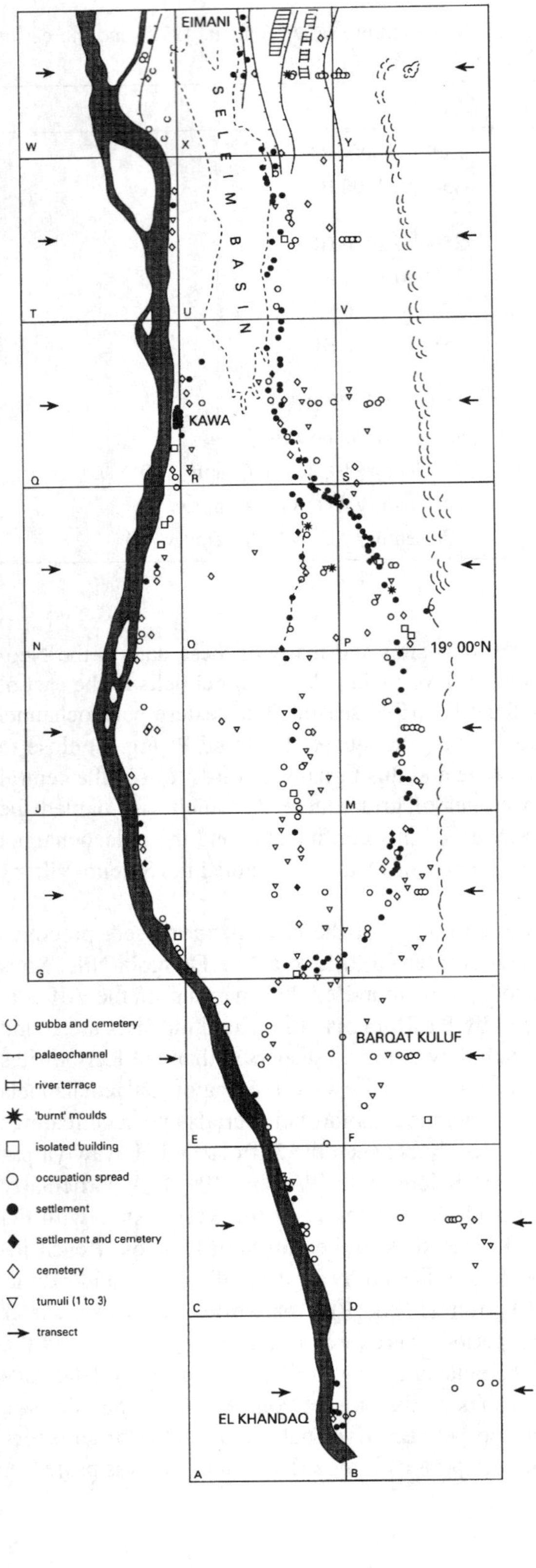

Figure 7. The distribution of archaeological sites of various ages in the Northern Dongola Reach recorded by the British Museum SARS survey (after Welsby, 1995).

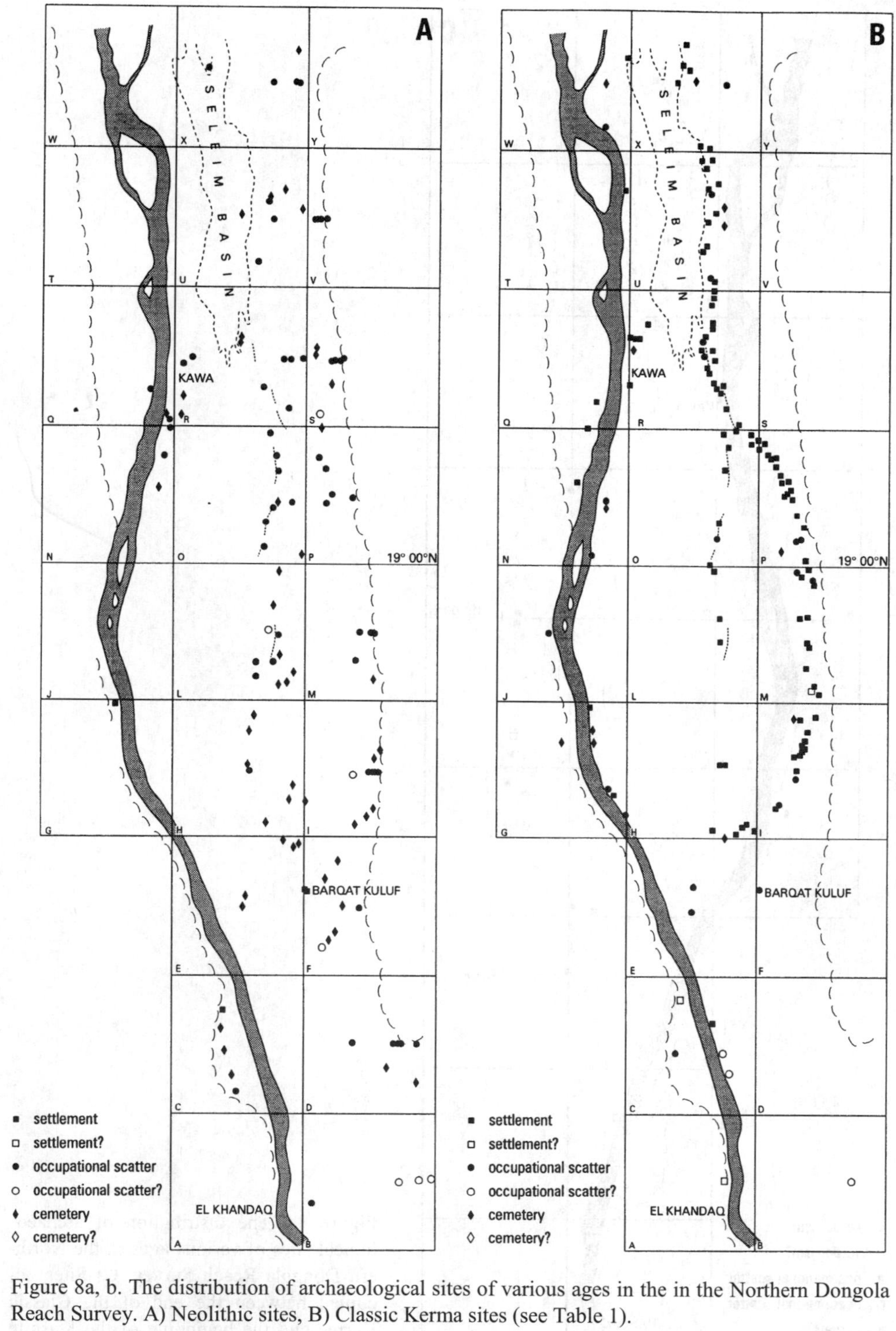

Figure 8a, b. The distribution of archaeological sites of various ages in the in the Northern Dongola Reach Survey. A) Neolithic sites, B) Classic Kerma sites (see Table 1).

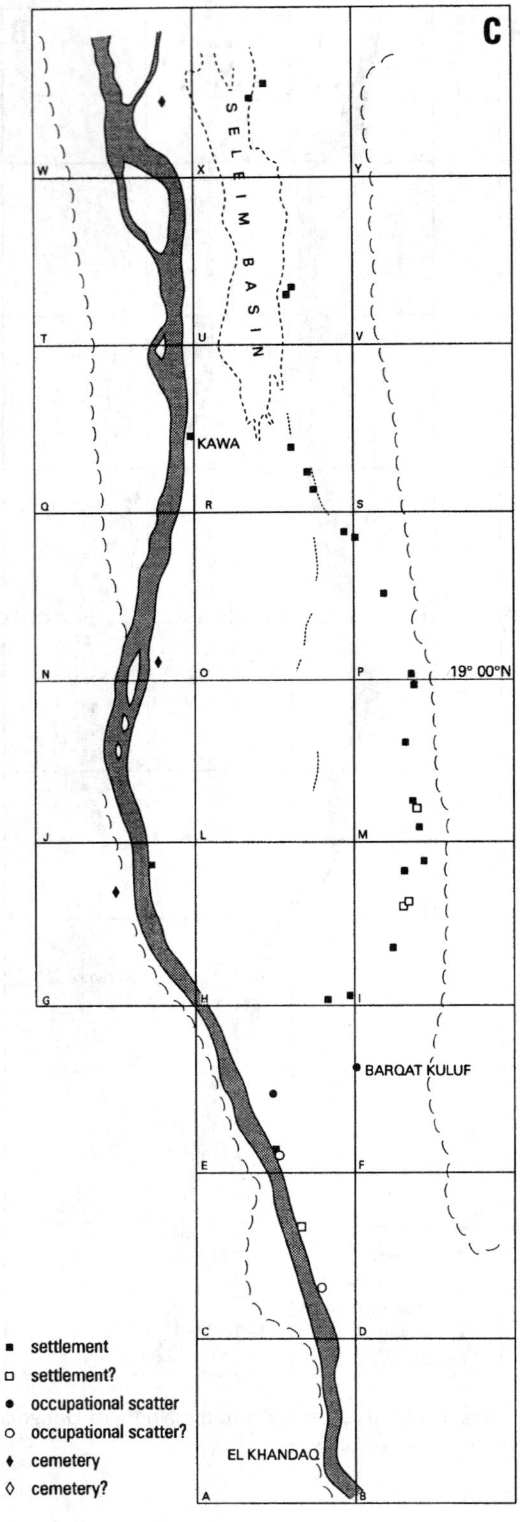

Figure 8c. The distribution of archaeological sites of various ages in the Northern Dongola Reach Survey. C) Sites occupied between the end of the Classic Kerma and the beginning of the Kushite periods (see Table 1).

Plate 1. A SPOT satellite image of the Northern Dongola Reach. This image shows the present channel of the Dongola Nile and the archaeological survey area to the east. The bedrock plateau at the eastern margin of the Holocene alluvial valley floor is clearly marked and the major palaeochannel belts can be identified in the upper half of the image. Note the presence of several large aeolian dune fields in the central and western parts of the survey area. The Seleim basin is at the top of this image to the east of the modern Nile (compare to Figure 7). This image covers an area of 60 km × 60 km.

Figure 9a. A Kerma settlement mound (Site P5) within the Alfreda Palaeochannel belt.

Figure 9b. The Kerma building under excavation at P4 (see Welsby, 1998) close to Pit 14 (see Fig. 11).

Figure 9c. Kerma pottery collected during the survey.

Figure 9d. Excavation of a Kerma cemetery in the Northern Dongola Reach.

7 ALLUVIAL SEDIMENTS AND LANDFORMS IN THE NORTHERN DONGOLA REACH

7.1 *Fluvial Stratigraphy and Geomorphology*

Geomorphological fieldwork in the Northern Dongola Reach was undertaken during the 1995/1996 and 1996/1997 field seasons in order to provide geomorphological and palaeoenvironmental context for the changing pattern of settlement revealed by the archaeological survey data (Macklin & Woodward, 1997, 1998). Agricultural development based on the exploitation of groundwater for irrigation is currently taking place in many parts of the archaeological survey concession. Large areas of the palaeochannel belts have been divided into small holdings of a few hectares and each is served by a series of groundwater pumps housed at the bottom of a pit. These square pits have been excavated into the alluvium to depths which typically range from 3 to 6 m. They are normally 3 to 4 m across with straight, stable sides and they provide excellent 3D exposures in the alluvial sequence – occasionally down to Nubian sandstone bedrock (Fig. 10). These pits are present throughout the survey area and these exposures have provided most of our stratigraphical and sedimentological data.

Figure 10. Holocene alluvial sediments and Nubian Sandstone bedrock exposed in a shallow (now disused) groundwater pump pit section in the Northern Dongola Reach. A thin (20 to 30 cm) layer of sandy gravel overlies the bedrock. In general terms, the rest of the section comprises two massive units of silt with fine sand separated by a thin (c. 10 to 15 cm) unit of well sorted fine sand at the same level as JCW's hat. The top of the sequence is covered by recent blown sand.

The stratigraphic and sedimentological record has been recorded in detail at 31 pits and their locations are shown in Figure 11. They extend throughout the survey area stretching from Pit 29 on the eastern margin of the Seleim Basin in the north, to Pit 24 in the southern portion of the survey area near Barqat Kuluf less than 5 km from the modern Nile. Pits 4, 7, 8, 9 and 16 (and 12 and 13 in the south) are close to the bedrock plateau in the east of the survey area (Fig. 11) and demonstrate that the entire area is underlain by Nile alluvium which ranges in thickness from < 2 m to > 7 m. Pits 5, 22, 26 and 27 are located within the Hawawiya Nile palaeochannel belt and Pits 1, 2, 18, 19 and 29 are located within the Seleim Nile palaeochannel belt. Most of the other pits are located within or close to the Alfreda Nile palaeochannel belt (Fig. 11). The comparatively small number of pit sections on the Hawawiya palaeochannel belt is due to the presence of large dune systems in this part of the reach and less extensive agricultural development (see Plate 1).

7.2 *Geochronological Dating*

Optically stimulated luminescence (OSL) and radiocarbon techniques have been used to date the alluvial sediments and four of the dated sections are described below. Luminescence dating was carried out by Dr Mark Bateman at the University of Sheffield following the procedures outlined in Bateman & Catt (1996). Samples were collected in the field from well sorted fine sand units (exposed in groundwater pump pits) by hammering a section of steel tubing (4 cm ID × 20 cm) horizontally into the section until it was flush with the section face. The sample tube was then excavated and sealed at both ends. In the laboratory all organics, carbonates and all minerals except quartz were removed from the sediment samples. Dosimetry was undertaken by ICP analysis of the elemental concentrations of uranium, thorium and potassium. Dose rates were attenuated for the grain sizes used (90-125 μm) and for a palaeomoisture level based on the present with 5% errors. Sample analysis followed a multiple aliquot additive dose protocol (MAAD) with stimulation using a 75w Halogen lamp filtered with a GG420 filter and OSL signal monitored through a U-340 filter (Bateman & Catt, 1996). All measurements were undertaken in a Riso automated luminescence reader. Inter-aliquot scatter was normalised using the total equivalent dose procedure. All the Northern Dongola Reach samples generated consistent results, growing linearly with dose with very low inter-aliquot scatter. The resultant palaeodose estimates (ED) could therefore be determined with confidence and a good level of precision (Mark Bateman, personal communication). We believe that the very effective zeroing of these sediments prior to burial by later flood sediments may be partly due to aeolian reworking during the dry season. Material suitable for radiocarbon dating was extremely rare; however, two charcoal samples were collected from exposures in Pits 12 and 26. These were both associated with archaeological features (pottery and burnt earth, respectively) exposed in section.

1.3 *Seleim Nile Palaeochannel Belt*

1.3.1 *Pit 18 (N19°07’26.0” E30°33’09.5”)*

This pit lies close to the southern edge of the Seleim Basin approximately 7 km due east of the modern Nile (Fig. 11). This site is located on the right bank of the Seleim Nile palaeochannel system about 4 km downstream of the confluence of the Alfreda and Hawawiya

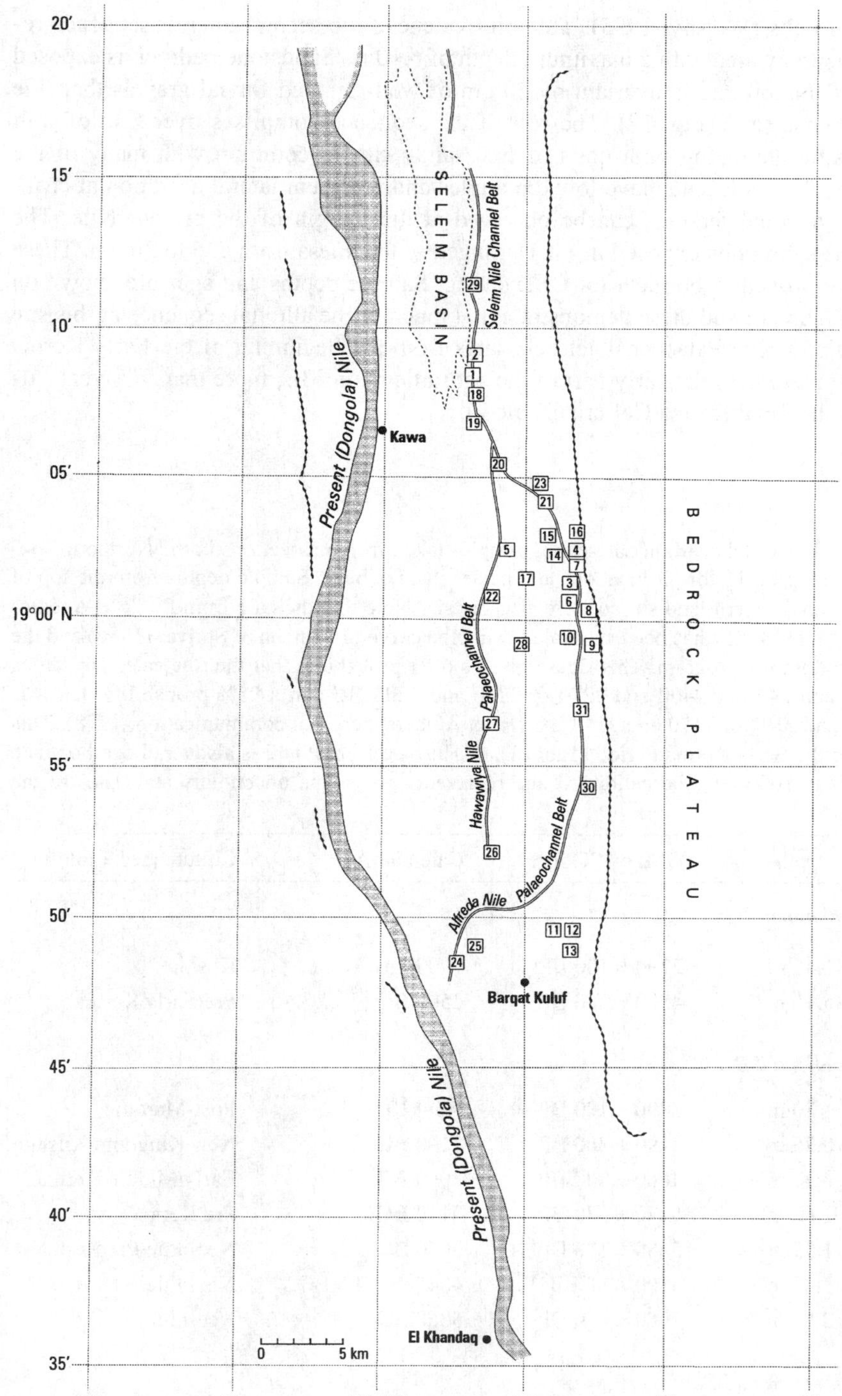

Figure 11. The distribution of groundwater pump pits selected for detailed study where the alluvial sequences have been logged. The major palaeochannel belts in the Northern Dongola Reach and the bedrock plateau marking the eastern extent of the survey area are also shown. Compare to SPOT image.

palaeochannel belts (Welsby, 1995). This pit evidenced the second longest sequence recorded in the survey area with a maximum depth of 6.10 m. Sandstone bedrock is exposed at the base of this pit and is overlain by 20 cm of well-rounded fluvial gravels that fine upwards to coarse sand (Fig. 12). The rest of the sequence comprises over 5 m of well sorted fine sand alternating with units of fine sandy silt. In common with many of the other sections, these silt units have low dip angles and represent lateral accretion at channel margins. Identical features can be observed at the margin of the present Nile. The prominent fine sand units exposed in Pit 18 ranged in thickness from c. 5 to 30 cm. These materials have proved to be ideal for OSL dating. Sample depths and ages are shown on Table 2 and Figure 12 and these demonstrate that much of the alluvial sequence at this site within the Seleim Nile Palaeochannel belt dates from the beginning of the Early Kerma Period until sometime in the early part of the 1st millennium BC, more than 700 years after the end of the Final Kerma Culture (Table 2).

Table 2. The OSL and radiocarbon dates from samples taken from pit sections in the Northern Dongola Reach. See Figure 11 for pit locations and palaeochannel belts. Sample depths from the top of the logged section (modern land surface) are also given. *Note that the date from Pit 26 is a radiocarbon date (BM-3128) that has been calibrated using the curve of Pearson & Stuiver (1986) and the OxCal v2.18 calibration program. This figure gives a 68% probability that the true calendar date is between 2460 and 2420 or 2400 and 2270 or 2250 and 2200 BC and a 95% probability that it is between 2460 and 2190 or 2170 and 2140 BC (Janet Ambers, personal communication, 1998). This sample dates to the Early Kerma Period (Table 1). **The date from Pit 12 is also a radiocarbon date on charcoal (Beta 100605). The calibrated age represents a 2 sigma uncertainty and dates to the Neolithic Period.

Pit No.	Depth	OSL or ^{14}C Age	Calendar Age	Culture (see Table 1)
Seleim Palaeochannel Belt:				
18	1.45 m	2790 ± 100 BP	790 BC	Kushite
18	5.15 m	4500 ± 300 BP	2500 BC	Pre/Early Kerma
Alfreda Palaeochannel Belt:				
23	1.30 m	1490 ± 100 BP	AD 510	Post-Meroitic
14	0.75 m	3190 ± 300 BP	1190 BC	New Kingdom/Kushite
14	4.40 m	4060 ± 300 BP	2060 BC	Early/Middle Kerma
7	1.60 m	5170 ± 530 BP	3170 BC	Pre Kerma
24	1.55 m	5680 ± 300 BP	3680 BC	Neolithic/Pre Kerma
12	1.83 m	5100 ± 80 BP**	4045 to 3705 BC	Neolithic
4	2.55 m	7060 ± 430 BP	5060 BC	Neolithic
Hawawiya Palaeochannel Belt:				
26	1.2 m	3830 ± 50 BP*	2460 to 2140 BC	Early Kerma
5	2.30 m	7100 ± 1090 BP	5100 BC	Neolithic
		7490 ± 1120 BP	5490 BC	Neolithic

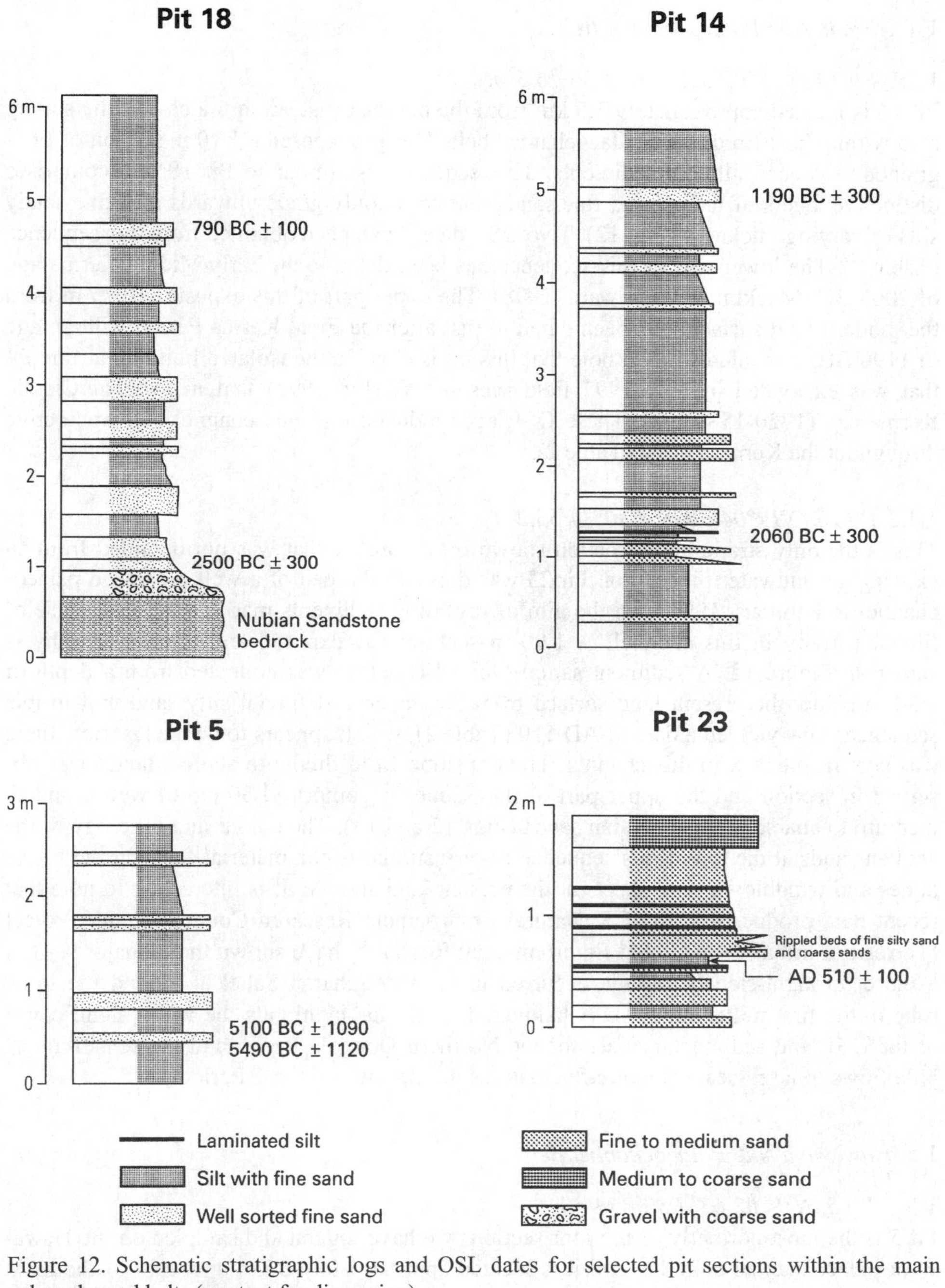

Figure 12. Schematic stratigraphic logs and OSL dates for selected pit sections within the main palaeochannel belts (see text for discussion).

1.4 *Alfreda Nile Palaeochannel Belt*

1.4.1 *Pit 14 (N19°02'31.4" E30°36'16.5"):*
Pit 14 is located approximately 3.5 km from the bedrock plateau in the east of the survey area within the Alfreda Nile palaeochannel belt. This pit exposed a 5.70 m section of fine-grained Holocene alluvial sediments. This sequence is similar to Pit 18 and comprises distinctive layers of well sorted fine sands that commonly grade upwards into fine sandy silts of varying thickness (Fig. 12). Two OSL dates have been obtained from this sequence (Table 2). The lower part of this sequence has been dated to an Early/Middle Kerma age of 2060 BC (Macklin & Woodward, 1998). The upper part of this exposure (0.75 m from the modern land surface) has been dated to just after the Final Kerma Period with an age of 1190 BC. It is interesting to note that this pit is close to the isolated building at Site P4 that was excavated in 1996/1997 field season (Welsby, 1997) that may be of Classic Kerma age (1750-1580 BC). These OSL ages indicate that this channel belt was active throughout the Kerma Period (Table 2).

1.4.2 *Pit 23 (N19°04'38.0" E30°34'53.2")*
This is the only stratigraphic section shown on Figure 11 that was not obtained from an existing groundwater pit section. Pit 23 was dug into the bed of a well preserved palaeochannel in February 1997 with the aim of revealing sediments marking the last phase of fluvial activity in this channel. A 1.90 m section was exposed and the stratigraphy is shown in Figure 12. A sediment sample for OSL dating was collected from a depth of 1.30 m below the present land surface from the uppermost fluvial silty sand unit in this sequence. This yielded a date of AD 510 (Table 2), which appears to be the last time there was significant flow in this channel. The transition from fluvial to aeolian facies was observed in section and the upper part of this sequence comprised 30 cm of well-rounded, medium to coarse grained aeolian sands (Figs 12 and 13). The colour and lithology of the aeolian sands at the top of this sequence is very similar to the material forming the active dunes and windblown sand sheets at the present land surface. It is interesting to note that recent data produced by the UK Natural Environment Research Council Tigger Project (Terrestrial Initiative in Global Environmental Research) have shown that a major shift to a more 'drought-ridden' regime occurred in the sub-Saharan Sahel at around the same time in the first millennium AD (Chaloner, 1997). This highlights the wider significance of the OSL and sedimentary data for the Northern Dongola Reach and the sensitivity of Nile flows to large-scale climatic fluctuations during the Holocene Period.

1.5 *Hawawiya Nile Palaeochannel Belt*

1.5.1 *Pit 5 (N19°02.62' E30°33.94')*
Pit 5 is the most northerly of the four sections we have logged and sampled on the Hawawiya palaeochannel complex and it is located approximately 6 km upstream of the main palaeochannel belt confluence with the Alfreda Nile (Fig. 11). The alluvial sediments exposed at this site reach a maximum depth of 2.8 m below the modern land surface (Fig. 12). Two OSL dates were obtained from a sample collected from a well sorted fine sand unit at a depth of 2.3 m and this gave ages of 5100 BC and 5490 BC (Table 2). Pit 5 is a comparatively shallow pit and the depth of the alluvial sequence in this part of the

Figure 13. The sediments exposed in Pit 23. This section was excavated in February 1997 within the centre of the palaeochannel form shown in Figures 14 and 15. The sediments above the trowel are coarse-grained aeolian sands. Below the trowel the section comprises grey Nile silts interbedded with fine- to medium-grained alluvial sands (see Fig. 12).

Hawawiya palaeochannel complex is not known. Nevertheless, these are the oldest dates so far obtained for alluvial materials in the Northern Dongola Reach and they demonstrate that this channel complex was active in the early Holocene. It is significant that a number of Neolithic sites have been recorded in this part of the Hawawiya Nile (Fig. 8) (Welsby, 1995).

8 PALAEOCHANNEL MORPHOLOGY AND PLANFORM

At several locations in the Northern Dongola Reach the morphology of the Alfreda and Seleim Nile palaeochannel belts is well preserved at the present land surface. In contrast, the channels of the Hawawiya Nile are not so well preserved due to aeolian erosion (Welsby, 1995) and perhaps also their greater antiquity. Two extended transects have been surveyed across typical landform associations in the study area. Cross section 1 passed immediately to the north of Pit 14 (Fig. 11) across part of the Alfreda Nile palaeochannel belt. It covered a distance of 2.1 km extending westwards from the bedrock plateau that demarcates the eastern edge of the survey area. This transect includes a gently sloping alluvial fan complex (that drains from a small catchment in the bedrock plateau) as well as a prominent Kerma settlement mound (Site P5) (Fig. 9a). Approximately 50 m to the north of the settlement mound at P5, a well preserved Kerma building (Site P4) was excavated during the 1996/1997 field season (Fig. 9b). Around 700 m of low relief terrain lies between the Kerma settlement mound at P5 and the bedrock plateau forming a palaeochannel complex beginning at the base of the lower fan (Fig. 14). This feature is mantled by a thin veneer of very gently sloping blown sand which overlies fine-grained alluvial sediments.

The second transect covered a distance of almost 1.4 km and was surveyed across the palaeochannel feature where Pit 23 was excavated (Fig. 14). This channel has a relief of at least 2 m from channel bed to levée top and is approximately 140 m wide. This profile highlights the convex form of some of these alluvial systems with well developed levées bordered by low relief, inter-channel flood basins (Fig. 15) and it is likely that this channel was part of a larger system of anabranching channels (Macklin & Woodward, 2001) (see Plate 1).

9 DISCUSSION AND CONCLUSIONS

The ten OSL and two radiocarbon dates reported here cover a time range of about 6000 years from the sixth millennium BC to the first millennium AD (Table 2). The Kerma Period spans only a small part of this period (2500 to 1500 BC, Table 1) and these dates show that the palaeochannel belts of the Northern Dongola Reach were regularly – if not permanently – inundated for much of the Holocene, before, during and after the major phase of Kerma Period occupation. The youngest OSL dates indicate that the Alfreda and Seleim Nile Palaeochannels were active until at least 800 BC and, perhaps episodically, up to about AD 500 (Macklin & Woodward, 1998).

During the 1994/1995 season, an almost uninterrupted spread of occupation material of Neolithic age was surveyed on either side of a 'wide depression' within the Hawawiya palaeochannel belt. The Kerma sites on the Hawawiya palaeochannel do not contain pottery

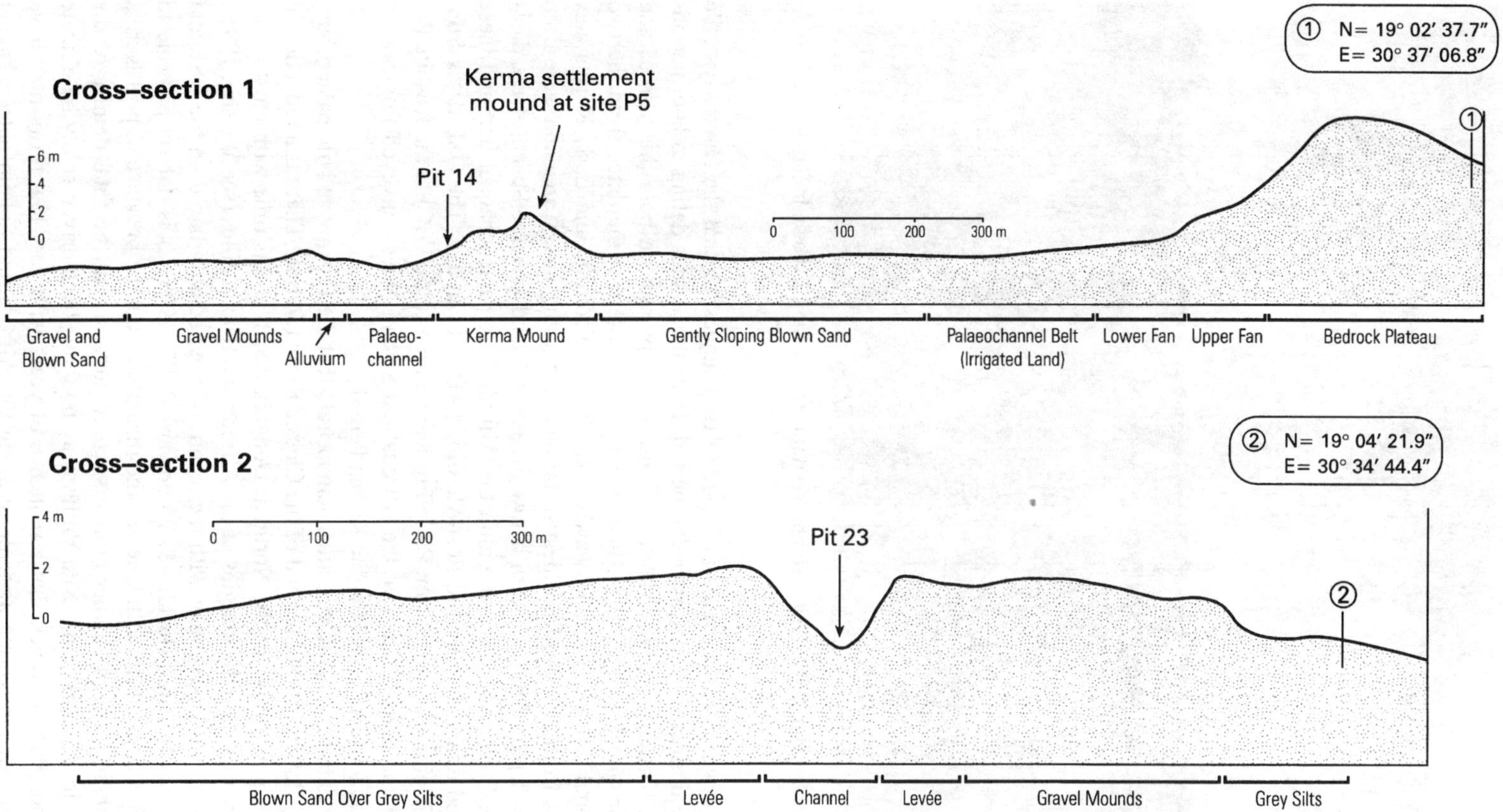

Figure 14. Cross section 1 that passed to the north of Pit 14 and cross section 2 showing the palaeochannel where Pit 23 was excavated.

Figure 15. The palaeochannel and levée complex within cross section 2 shown in Figure 14.

later than the Classic Kerma Period (1750-1580 BC) suggesting that the Hawawiya Nile ceased to flow during or shortly after that period. In contrast, sites dating to later periods (Final Kerma, Kushite and Post-Meroitic, Table 1) are present on both the Alfreda and Seleim palaeochannel belts (Fig. 8). It is interesting to note that the dates from the two dated Pits (5 and 26) within the Hawawiya Nile are of Neolithic and Early Kerma age. The OSL dates from Pit 5 are the oldest so far obtained in the survey area and were from sediments at a depth of only 2.30 m below the present land surface. Seven of the dates listed in Table 2 are from six pits within the Alfreda Nile palaeochannel belt and these range from Neolithic (5060 BC) to Post-Meroitic (AD 510) in age. The OSL dates from Pit 18 on the Seleim Nile range from Early/Pre-Kerma times (2500 BC) to the Kushite Period (790 BC) and demonstrate that the Alfreda and Seleim Niles conveyed floodwaters for at least two thousand years after the Kerma Period.

The pottery data from the Kerma sites demonstrate that the Hawawiya Nile palaeochannel belt ceased flowing before the end of the Classic Kerma Period. There is a reduction in settlement across the whole of the Northern Dongola Reach at this time with most of the later sites located close to the margins of the Alfreda and Seleim Niles (see Welsby, 1995). This pattern is in good agreement with the record of river activity based on the OSL and ^{14}C dates obtained to date (Macklin & Woodward, 1997, 1998). It is not yet possible to comment upon changes in flood frequency and magnitude, but the presence of post-Kerma settlements, albeit in much smaller numbers, could suggest that the final demise of the Kerma culture in this part of the Nile Valley may not have been driven solely by climate change. It is possible, however, that a sequence of years with very low flows may have devastated much of the Kerma agriculture along the Alfreda and Seleim palaeochannel belts and, after the Kerma Period, available water resources were only sufficient to support a much smaller population. As Hassan (1998, p. 37) has stated:

Contrary to the traditional wisdom of Greek visitors in classical antiquity who consorted with priests rather than farmers, Nile floods are capricious. One in every five floods is harmful: over-flooding the fields, too low to irrigate the higher edge of the floodplain, unseasonably early or late, staying too long or falling too quickly. In addition, the dynamics of the river – changing its course, breaking levées, or over-silting basins – can change the fortunes of villages and homesteads.

Establishing the sequence and timing of palaeochannel abandonment in the Northern Dongola Reach is the first step in attempting to identify the cause of these cultural changes. At this point it is instructive to consider some of the previous work on the early and middle Holocene palaeochannels on the Gezira Plain in the lower reaches of the Blue Nile. Williams & Adamson (1980) state that a major phase of Blue Nile incision may have started in the early Holocene, but its effects only became significant by middle Holocene times where former distributary channels that flowed across the Gezira Plain were beheaded and dried out – only transmitting flows during exceptional floods. This phase of incision seemed to coincide with a shift from a coarse- to fine-grained sediment load, and with an overall decline in flow volumes after the mid-Holocene as flood flows became concentrated in a single main channel (Williams & Adamson, 1980). In the case of the palaeochannels of the Gezira Plain, Adamson et al. (1982, p. 187) have observed that:

'progressively less and less of the flood flow would be diverted into them [palaeochannels] as the main channel incised further and their own channels became lined with silt.'

It is interesting to compare this scenario with the sedimentary sequence recorded in Pit 23 on the youngest dated sediments of the Alfreda Nile palaeochannel where the thin bands of fluvial silt are finally replaced by coarse aeolian sands representing arid conditions (Fig. 12). The deposition of these silt units may represent exceptionally high flows from an incising Dongola Nile that periodically inundated the Alfreda Nile channels with shallow, silt-laden floodwaters. Subsequent channel flows were too low to inundate the palaeochannel belts and this process could have resulted from a decrease in flood magnitude as monsoon intensity waned in the first millennium AD.

Further work is required in the Northern Dongola Reach to establish the altitudinal relationships between the major palaeochannel belts and the modern Dongola Nile. For example, long-term sediment deposition in the Hawawiya Nile may have elevated this channel belt above the surrounding flood basins, so that subsequent flows were conveyed by the Alfreda Nile to the east. These survey data will form an important part of any climatic, tectonic or other geomorphological explanation for the abandonment of the main palaeochannel belts and for the concentration of flood flows in the Dongola Nile. Detailed analysis and stratification of the Neolithic, Kerma and later pottery from across the survey area is still in progress and these chronological data will allow the relationship between river behaviour and human activity in the Northern Dongola Reach to be further refined. At the same time, it is interesting to note that several large palaeochannels within the Northern Dongola Reach identified on the SPOT satellite imagery are not associated with Kerma or later sites and further fieldwork is required to document the stratigraphy and establish the age of these features.

ACKNOWLEDGEMENTS

MGM and JCW are grateful for the invitation to participate in the Northern Dongola Reach Survey and they thank the Sudan Archaeological Research Society of the British Museum for financial and logistical support. We thank Isabella Welsby Sjöström who has analysed the pottery from the NDRS and this work contributed to the development of Figure 8. We also thank David Appleyard and Lois Wright of the Graphics Unit in the School of Geography at Leeds for drawing the figures and John Dodds for assistance with image processing. We are also grateful to Dr Mark Bateman (University of Sheffield) who carried out the OSL analyses and the referees who provided helpful comments on this paper. This paper is dedicated to the memory of Kevin Jenkins who worked on this project for a few short weeks in the Autumn of 1997.

REFERENCES

Adams, W.Y. 1977. *Nubia: Corridor to Africa.* Allen Lane, London and Princeton.

Adamson, D.A., Gasse, F., Street, F.A. & Williams, M.A.J. 1980. Late Quaternary History of the Nile. *Nature,* 288: 50-55.

Adamson, D.A., Williams, M.A.J. & Gillespie, R. 1982. Palaeogeography of the Gezira and of the lower Blue and White Nile valleys. In: Williams, M.A.J. & Adamson, D.A. (eds) *A Land Between Two Niles: Quaternary Geology and Biology of the Central Sudan,* Balkema, Rotterdam, 165-219.

Andah, K. & Siccardi, F. 1991. Prediction of hydrometeorological extremes in the Sudanese Nile region: a need for international cooperation. In: *Hydrology for the Water Management of Large River Basins,* IAHS Publication No. 201: 3-12.

Bateman, M. D. & Catt, J.A. 1986. An absolute chronology for the raised beach and associated deposits at Sewerby, East Yorkshire, England. *Journal of Quaternary Science,* 11: 389-395.

Bonnet, C. 1992. Excavations at the Nubian royal town of Kerma: 1975-91. *Antiquity,* 66: 611-625.

Butzer, K.W. 1980. Pleistocene History of the Nile Valley in Egypt and Lower Nubia. In: Williams, M.A.J. & Faure, H. (eds) *The Sahara and The Nile.* Balkema, Rotterdam, 248-276.

Butzer, K.W. 1981. Long-term Nile flood variation and political discontinuities in Pharaonic Egypt. In: Clark, J.D. & Brandt, S. (eds) *The Causes and Consequences of Food Production in Africa.* Berkeley, University of California Press,

Butzer, K.W. & Hansen, C.L. 1968. *Desert and River in Nubia: Geomorphology and Prehistoric Environments at the Aswan Reservoir.* University of Wisconsin Press, Madison.

Chaloner, W. 1997. Global change: the time dimension. *NERC News,* Autumn 1997, 14-15.

Evans, T. 1994. History of Nile Flows. In: Howell, P.P. & Allan, J.A. (eds) *The Nile: Sharing a Scarce Resource.* Cambridge University Press, Cambridge, 27-63.

Foucault, A. & Stanley, D.J. 1989. Late Quaternary palaeoclimatic oscillations in East Africa recorded by heavy minerals in the Nile delta. *Nature,* 339: 44-46.

Gasse, F., Rognon, P. & Street, F.A. 1980. Quaternary history of the Afar and Ethiopian Rift Lakes. In: Williams, M.A.J. & Faure, H. (eds) *The Sahara and The Nile.* Balkema, Rotterdam, 361-400.

Gasse, F. & van Campo, E. 1994. Abrupt post-glacial climatic events in West Asia and North Africa monsoon domains. *Earth and Planetary Science Reviews,* 1256: 435-456.

Hassan, F.A. 1981. Historical Nile Floods and their implications for climatic change. *Science,* 212: 1142-1145.

Hassan, F.A. 1996. Abrupt Holocene climatic events in Africa. In: Pwiti, G. & Soper, R. (eds) *Aspects of African Archaeology,* Papers from the 10th Congress of the PanAfrican Association for Prehistory and Related Studies, University of Zimbabwe Publications, Harare, 83-89.

Hassan, F.A. 1997. The dynamics of a riverine civilization: a geoarchaeological perspective on the Nile Valley, Egypt. *World Archaeology,* 29: 51-74.

Hassan, F.A. 1998. Climatic change, Nile floods and civilization. *Nature and Resources,* 34: 34-40.

Hulme, M. 1994. Global climate change and the Nile basin. In: Howell, P.P. & Allan, J.A. (eds) *The Nile: Sharing a Scarce Resource.* Cambridge University Press, Cambridge, 139-162.

Hulme, M.A. 1994. Global climate change and the Nile basin. In: Howell, P.P. & Allan, J.A. (eds) *The Nile: Sharing a Scarce Resource.* Cambridge University Press, Cambridge, 139-162.

Hurst, H.E. 1952. *The Nile: A general account of the river and the utilization of its waters.* London, Constable (Second Edition).

Macklin, M.G. & Woodward, J.C. 1997. Holocene Alluvial History in the Northern Dongola Reach of the Nile: the 1996/1997 field season. *Sudan and Nubia,* 1: 13-15.

Macklin, M.G. & Woodward, J.C. 1998. Alluvial architecture and luminescence dating of Holocene palaeochannels in the Northern Dongola Reach of the Nile. *Sudan and Nubia,* 2: 21-25.

Macklin, M.G. & Woodward, J.C. 2001. Holocene alluvial history and the palaeochannels of the Northern Dongola Reach of the Nile: A preliminary report. In: Welsby, D.A. (ed.) *Survey and Excavation in The Northern Dongola Reach.* SARS Monograph (in press).

Milliman & Syvitski, J.P.M. 1992. Geomorphic/Tectonic control of sediment discharge to the ocean: the importance of small mountainous rivers. *The Journal of Geology*, 100: 525-544.

Pearson, G.W. & Stuiver, M. 1986. High precision calibration of the radiocarbon time scale, 500-2500 BC. *Radiocarbon*, 29: 839-861.

Phillipson, D.W. 1993. *African Archaeology*. Cambridge University Press, Cambridge.

Richards, K.S., Chandra, S. & Friend, P. 1993. Avulsive channel systems: characteristics and examples. In: Best, J.L. & Bristow, C.S. (eds) *Braided Rivers,* Geological Society Special Publication, No.75: 195-203.

Ritchie, J.C., Eyles, C.H. & Haynes, C.V. 1985. Sediment and pollen evidence for an early to mid-Holocene humid period in the eastern Sahara. *Nature,* 314: 352-355.

Rossignol-Strick, M., Nesteroff, W., Olive, P. & Vergnaud-Grazzini, C. 1982. After the Deluge: Mediterranean Stagnation and Sapropel Formation. *Nature,* 295: 105-110.

Said, R. 1993. *The River Nile: Geology, Hydrology and Utilization*, Pergamon Press, Oxford.

Said, R. 1994. Origin and Evolution of the Nile. In: Howell, P.P. & Allan, J.A. (eds) *The Nile: Sharing a Scarce Resource.* Cambridge University Press, Cambridge, 17-26.

Street, F.A. & Grove, A.T. 1979. Global maps of lake-level fluctuations since 30,000 years BP. *Quaternary Research*, 10: 83-118.

Street-Perrot, F.A., Marchand, D.S., Roberts, C.N. & Harrison, S.P. 1989. *Global Lake-Level Variations from 180,000 to 0 Years Ago: A Palaeoclimatic Analysis*. US Department of Energy Technical Reports (TRO46), 1-213.

Welsby, D.A. 1995. The Northern Dongola Reach Survey, the 1994/95 season. a: The Survey. *SARS Newsletter,* 8: 2-7.

Welsby, D.A. 1997. The Northern Dongola Reach Survey, the 1996/97 season: Excavation at Sites O16, P1, P4 and P37. *Sudan and Nubia,* 1: 2-10.

Welsby D.A. 1998. Survey and excavation at Kawa, the 1997/98 season. *Sudan and Nubia*, 2: 15-20.

Wendorf, F. & Schild, R. 1976. *Prehistory of the Nile Valley*. New York, Academic Press.

Williams, M.A.J. & Adamson, D.A. 1980. Late Quaternary depositional history of the Blue and White Nile rivers in central Sudan. In: Williams, M.A.J. & Faure, H. (eds) *The Sahara and the Nile: Quaternary Environments and Prehistoric Occupation in Northern Africa,* Balkema, Rotterdam, 281-304.

Williams, M.A.J. & Adamson, D.A. 1982. (eds) *A Land Between Two Niles: Quaternary Geology and Biology of the Central Sudan*, Balkema, Rotterdam.

Williams, M.A.J., Dunkerley, D.L., De Deckker, P., Kershaw, A.P. & Chappell, J. 1998. *Quaternary Environments.* 2nd ed., Arnold, London.

12. Lateglacial and Holocene fluvial records from the central part of the Paris Basin (France)

JEAN-FRANÇOIS PASTRE, NICOLE LIMONDIN-LOZOUET & ANNE GEBHARDT
Laboratoire de Géographie Physique, Cedex, France

CHANTAL LEROYER
Centre National de Préhistoire, Périgueux, France

MICHEL FONTUGNE
Laboratoire des Sciences du Climat et de l'Environnement, Cedex, France

VINCENT KRIER
AFAN, Paris, France

1 INTRODUCTION

Studies of Lateglacial and Holocene alluvial sedimentary sequences from the Paris Basin (Middle Seine river basin) have recently been stimulated by the increase in rescue archaeology over the last ten years (Krier et al., 1991; Pastre et al., 1991, 1997; Leroyer, 1997). The aim of this paper is to give a synthetic view of the most significant environmental features offered by this area. The environmental records were obtained by simultaneous studies of geomorphological, sedimentological, palynological and malacological data from the same sections or cores radiocarbon dated. Extensive information was recovered from archaeological excavations, quarries and machine-dug trenches. Auger-drillings and cores were also largely used to determine the stratigraphy and the geometry of deposits in wet zones with thick sedimentation (5 to 10 m), otherwise impossible to observe.

The main tributary rivers of the River Seine (Oise, Marne, Yonne) have a mean discharge of less than 100 m^3/sec^{-1} with moderate winter and spring floods. They rise in the Jurassic belt, flow down through Cretaceous chalk and incise large Tertiary plateaus of the central part of the Paris Basin (Fig. 1). Tributaries include several small rivers with mean discharges of only a few m^3/sec^{-1}. Floodplains are wide (0.5 to 4 km for the main valleys; 0.1 to 1 km for small catchment) and flat (1 to 2.5‰ slope for the central part of the basin) and offer good conditions for fluvial or local organic sedimentation. Holocene polyphased sequences of peat, organic and detritic silts are often 6 to 10 m thick and data from main and small valleys give complementary information. Alluvial records from the centre of the basin around Paris (< 100 km from the town) are generally more complete. Alluvial sequences in headwater areas underlain by chalk show more hiatuses, because of a steeper gradient, and more active channels. Slopes and plateaus are covered with Pleistocene loess (mainly from the Last Glacial: Lautridou, 1985) overlain by Holocene brown soils which are susceptible to erosion.

Main valleys have large floodplains with only a restricted number of Middle Pleistocene river terraces. Floodplains are underlain by a lower alluvial unit of Last Glacial fluvial sand and gravels (e.g. Fig. 2) buried by Lateglacial and Holocene silts. Lateglacial sequences include large channel forms and flat banks. Small valleys show different records owing to variability of the geological, hydrological and edaphic context. Lateglacial deposits here are generally poorly developed, but sometimes yield interstadial channel deposits.

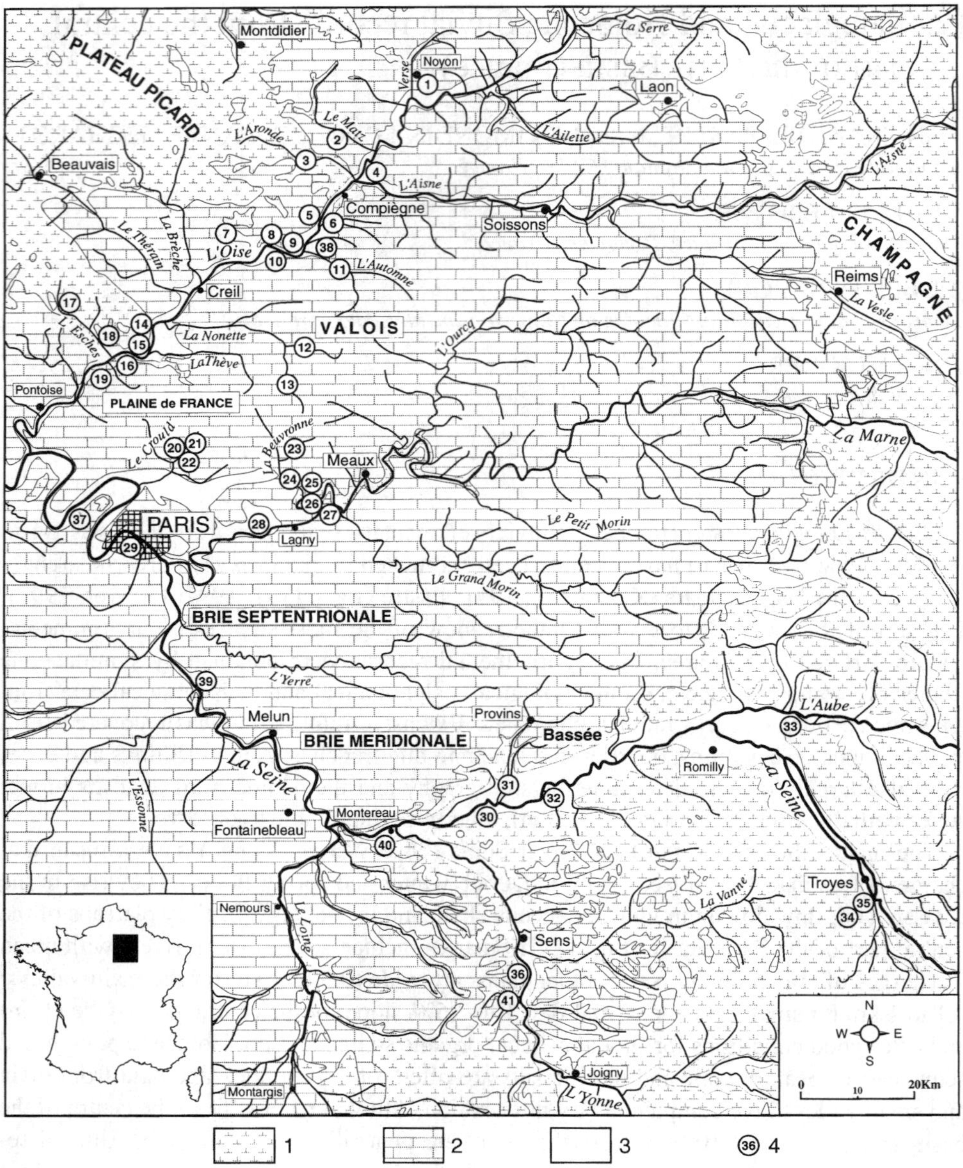

Figure 1. Location map of sites studied in the Paris Basin. 1. Cretaceous chalk – 2. Tertiary limestones, sands and clays – 3. Quaternary alluvia – 4. Study sites: *1. Beaumont-Les-Noyon – 2. Vandélicourt – 3. Montmartin – 4. Choisy-au-Bac – 5. Armancourt – 6. La Croix-Saint-Ouen – 7. Sacy-le-Grand – 8. Houdancourt – 9. Longueuil-Sainte-Marie – 10. Rhuis – 11. Béthisy-Saint-Martin – 12. Baron – 13. Ver-sur-Launette – 14. Boran-sur-Oise – 15. Bruyères-sur-Oise – 16. Beaumont-sur-Oise – 17. Amblainville – 18. Belle-Eglise – 19. Mours – 20. Goussainville (Les Aulnaies-du Pont)- 21. Goussainville (Le Crould)- 22. Gonesse – 23. Nantouillet – 24. Claye-Souilly – 25. Fresnes-sur-Marne – 26. Annet-sur-Marne – 27. Lesches – 28. Neuilly-sur-Marne – 29. Paris – 30. Bazoches-les-Bray – 31. Les Ormes-sur-Voulzie – 32. Noyen-sur-Seine – 33. Boulages – 34. Saint-Pouange – 35. Saint-Léger – 36. Etigny. – 37. Rueil-Malmaison* – Upper Palaeolithic Magdalenian sites: *38.Verberie, 39. Etiolles, 40. Pincevent, 41. Marsangy.*

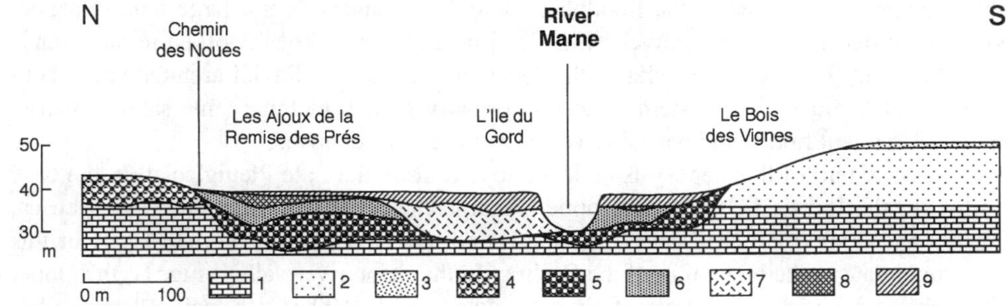

Figure 2. Cross-section of the Marne valley at Fresnes-sur-Marne (site 25, Fig. 1). 1. Upper Lutetian marls and limestones – 2. Bartonian sands – 3. Alluvial terrace 1 (Middle Pleistocene) – 4. Alluvial terrace 2 (Middle Pleistocene, Saalian) – 5. Alluvial apron 3 (Upper Pleistocene, Weichselian): sand and gravels – 6. Late Pleniglacial to Lateglacial fine sands and silts – 7. Holocene (Preboreal to early Subboreal): organic silts and tufas – 8. Holocene (Preboreal to Subboreal): peat – 9. Holocene (Subboreal – Subatlantic): clayey silts.

Holocene deposits are, however, well represented. In the Oise river basin, for example, there are peat deposits between 3 to 8 m thick (alkaline peat-bog), while at north (Plaine-de-France) and east of Paris (Brie), stratified sequences are generally composite with peat, organic and detritic silts, and local tufas.

The 36 sites investigated have used different field-work approaches. In main valleys floodplains, profiles were directly observed, whereas near river channels, information was mainly obtained from drilling and cores (10 to 20 m resolution). In small valleys, most of peaty stratigraphies, situated under the water table, were studied by drilling and cores (10 to 50 m resolution), using a small drilling-machine. Palynological sampling was performed at 5 cm intervals and malacological samples taken at 10 to 20 cm intervals. Descriptive sedimentological study, was undertaken by granulometric, organic carbon, carbonate and mineralogical analyses. ^{14}C datings, were mainly obtained by ß-counting and more rarely by AMS in Gif radiocarbon laboratory. Substages of the Lateglacial and the Holocene are used in chronostratigraphical sense following Mangerud et al. (1974). All ^{14}C dates were converted in years calibrated BP, using the Stuiver & Reimer (1993) programme.

2 LATEGLACIAL FLUVIAL RECORDS AND THEIR CLIMATIC CONTROL

During the Lateglacial, major modifications of the river regime occurred which significantly influenced valley floor erosion and sedimentation processes. The pattern of fluvial and environmental change during this period is now reasonably well known (e.g. Bohncke et al., 1987; Walker et al., 1994; Vandenberghe, 1993; Vandenberghe et al., 1994; Rose, 1995) and, similar to other parts of northwest Europe, river development in the Middle Seine river basin was controlled by climate and vegetation dynamics.

2.1 *Transition from the Pleniglacial to the Lateglacial (18,000-15,500 cal. BP) and Bølling (15,500-14,000 cal. BP)*

The main valleys of the Paris Basin show a sandy-silty alluvial unit which is discontinuous, but is well-preserved in the middle Oise river valley. This deposit, 2 to 3 meters thick, extends over most of the floodplain (1 to 3 km) and covers a large palaeochannel system cut into Pleniglacial gravels (Fig. 3). This unit consists of reworked tertiary sands and clays and (mainly Weichselian) Pleistocene loess and its alluvial architecture is consistent with braided river system. Stacked normally graded bedding (fine sand to marls) indicate frequent floods and partial reworking of previous deposits.

The lower part of this deposit is believed to date from the Late Pleniglacial. In the Oise valley, in the loessic belt area, the upper Pleniglacial loess deposits, dated between ca. 22,000 and 20,000 BP (Haesaerts, 1984a), do not cover this unit. The latest part of this unit could date to the beginning of the Bølling. In the Seine river valley (Fig. 1), the Upper Palaeolithic Magdalenian levels of Etiolles, dated ca. 15,400-15,100 cal. BP (Valladas, 1994), are interbedded in sand and silts (Roblin-Jouve, 1994) which indicate significant seasonal floodings. In the Oise river valley, the Magdalenian levels of Verberie (Fig. 1), dated ca. 14,600 cal. BP by ^{14}C (Audouze et al., 1991) and ca. 13,000 BP by TL (Valladas, 1981, 1994), postdate fine bedded sand deposits which are typical of this unit.

Associated environmental bio-indicators are scarce. Malacofaunas are poorly preserved and locally reworked. In the lower Marne river valley, they include assemblages with *Pupilla muscorum*, *Vallonia costata*, *V. pulchella* and *Trichia hispida*. This fauna is comparable to those of Great Britain (Kerney, 1963; Preece & Bridgland, 1998), and of the Somme river valley of the same age (Limondin, 1995; Limondin-Lozouet, 1997, 1998). Palynological spectras are poor and contain numerous intruded taxas.

The alluvial succession is similar to many other rivers of northwest Europe, such as the Maas (Vandenberghe et al., 1994; Kasse et al., 1995) or the Vistula valley (Starkel, 1994) where an aggradation following an incision is recorded towards the end of this period.

Climatic amelioration in the Bølling led to a reduction in sediment supply and a major modification of hydrodynamic style and regime. In the main valleys, flow was concentrated in a low sinuosity meandering channel frequently with several secondary channels (1 to 3 in the middle Oise valley). Soon after, flow in these secondary channels ceased and was concentrated in the main channel, where incision continued. This major incision phase has been recognized in numerous valleys of northwest Europe, including Belgium (Munaut & Paulissen, 1973; Haesaerts, 1984a and b; Cleveringa et al., 1988; Kiden, 1991), Middle Seine basin and Somme, France (Pastre et al., 1991, 1997; Roblin-Jouve, 1994; Antoine, 1997a-c), Maas and Mark, Netherlands (Bohncke & Vandenberghe, 1987; Bohncke et al., 1987; Vandenberghe et al., 1987; Vandenberghe, 1993), Vistula and Warta, Poland (Kozarski, 1983, 1991; Kalicki, 1991; Starkel, 1994; Vandenberghe et al., 1994). In the middle Oise basin, the effects of this major phase of incision were particulary pronounced at the end of the first half of the Bølling (Fig. 3). This episode probably lasted just a few centuries and at La Croix-Saint-Ouen (Fig. 1) it is dated by wood fragments which gave a ^{14}C AMS age of 14,957-14,111 cal. BP (Pastre et al., 1997).

In the second part of the Bølling, channel stabilisation occured. In the Oise valley, the main river channel appears to have been more than 200 m wide and cut several meters into Pleniglacial gravels. In the Seine valley, the Magdalenian levels of Pincevent, dated between ca. 14,500 and 14,000 cal. BP (Valladas, 1994), are interbedded with overbank silt deposits.

As shown on Figure 4 (zone 1), vegetation during this period was characterized by a steppic flora with *Artemisia*, some *Betula* and *Juniperus* (Van Zeist & Van Der Spoel-Walvius, 1980; Leroyer, 1987). In abandoned channels and flood basins of Bølling age, some discrete organic layers with hygrophyts are found (Pastre et al., in press), but local peat formation, as in the Somme river basin (Antoine, 1997a-c), has not yet been recorded.

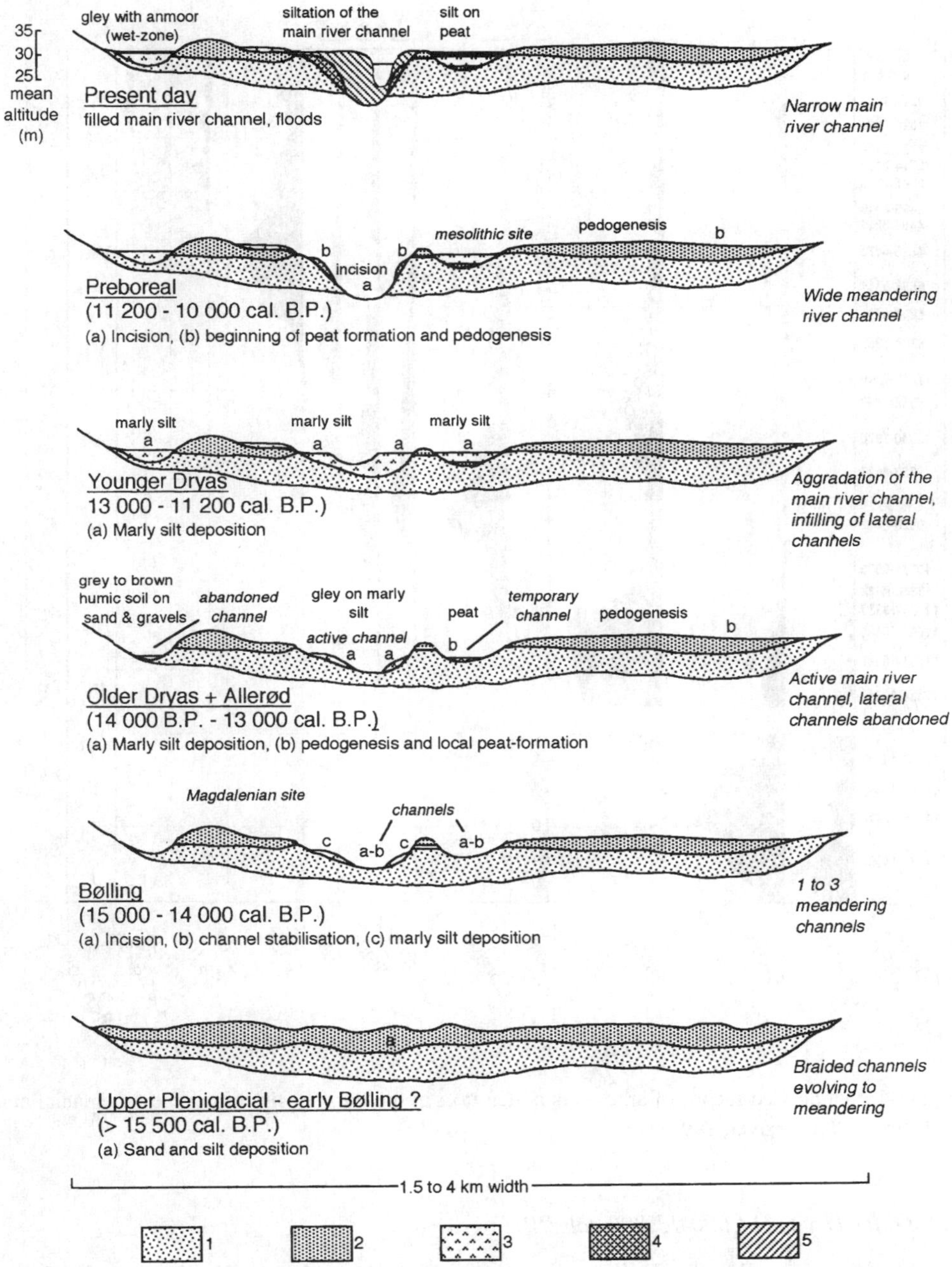

Figure 3. Schematic evolution of the Middle river Oise valley from Upper Pleniglacial to Present. 1. Sands and gravels – 2. Bedded sands and silts – 3. Marly silts – 4. Peat – 5. Argillaceous silts.

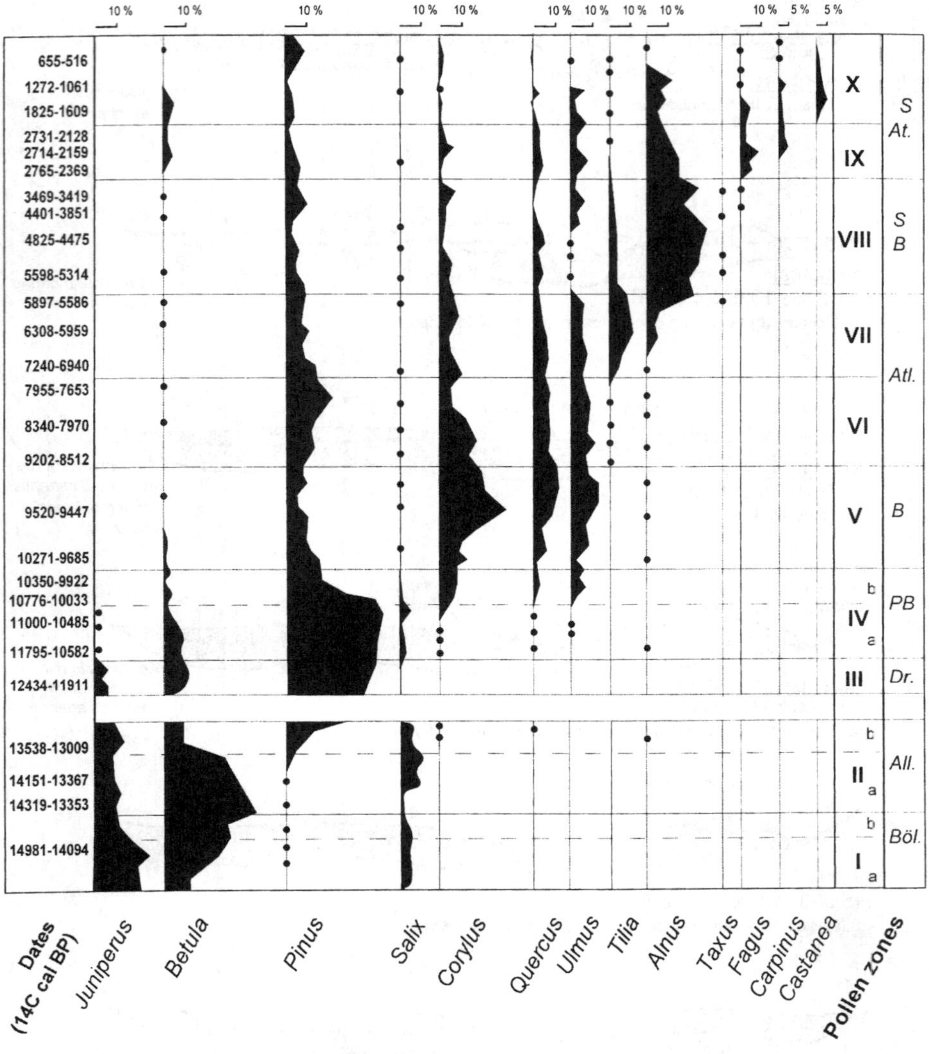

Figure 4. Synthesised record of arboreous pollen taxa in the Paris Basin during the Lateglacial and Holocene (after Leroyer, 1997).

2.2 *Older Dryas (14,000-13,800 cal. BP)*

In the Oise river valley, the Bølling river channel is covered by marly silts (Fig. 3). These undated, chalk-rich sediments, predating the Allerød suggest sedimentation rates were higher in the Older Dryas, as shown by Antoine (1997a-c) in the Selle river valley to the north of the Paris Basin. A sedimentary record found at Rueil-Malmaison in the Seine valley, near Paris (Epipalaeolithic site of Le Closeau, Bodu, 1995), might also correspond to this short cold phase. At this site, two Azilian levels were dated from 14,455-13,795 and 11,299-12,258 cal. BP (Bodu & Valentin, 1997). They were found on a sand and gravel bar located on the bank of a large channel separated by 60 cm of silt.

2.3 *Allerød (13,800-13,000 cal. BP)*

The Allerød began with a short period of incision, followed by a decrease in fluvial activity and the formation of a thin soil. On river banks, this soil is grey to light brown in colour. On the edge of the main river channel, a gley soil is found which laterally passes into fine-grained fluvial deposits, which are slightly organic in the channel. This palaeosol, unknown in the area until recently, is considered to be a regional marker-bed and has been recognized in many localities along the Oise valley. It is also known in the Seine valley at Rueil, in the Bassée area, and has been recently recognized in the Aube and the Yonne rivers valleys (Boulages, Etigny, Fig. 1). At Rueil this soil was formed before and during a Middle Azilian level dated ca. 13,100 cal. BP and ceased developing at ca. 12,800 cal. BP. It is correlated with the 'Belloy soil' in the Somme river basin (Fagnart, 1997; Antoine, 1997a-c) and the Allerød soil from the Netherlands (Van Geel et al., 1989) and England (Preece, 1994).

Pedogenesis is contemporaneous with local peat-formation in palaeochannels with high water levels. In the Automne valley (core-site of Bethisy-Saint-Martin, Fig. 1), the upper part of some organic beds (Fig. 7) give an age of 13,284-12,833 cal. BP (Pastre et al., 1997). In the Oise river valley, at Houdancourt (Fig. 1), a peat in a small palaeochannel, has been dated between ca. 14,000 and 12,800 cal. BP. Soil formation and organic sedimentation both occur during the second half of the Allerød. There is, however, no clear morpho-sedimentary impact related to the intra-Allerød cold period (Lehman & Keigwin, 1992).

Expansion of vegetation, recognized in Picardie (Van Zeist & Van der Spoel-Walvius, 1980) and the Seine River basin (Leroyer, 1997), show two successive phases. In the first phase (Fig. 4, IIa), birch dominates and then is replaced by pine after ca. 13,200 cal. BP (Fig. 4, IIb).

Increasing vegetation cover appears to have resulted in reduced slope erosion with sedimentation rates decreasing during the second half of the Allerød in the main water channel of the Oise river.

The malacogical response in the Oise and Seine valleys was very consistent. Associations from the grey soil are rich and moderately diverse (Fig. 5). They are dominated by open-land taxa *Vallonia pulchella*, *V. costata*, *Pupilla muscorum* and *Trichia hispida*, accompanied by several mesophilous molluscs and a very characteristic xerophilous group. This last group contains two species of modern central-European distribution *Helicopsis striata* and *Trochoidea geyeri*. After the Allerød, both these taxa decrease and finally disappear from the Paris Basin in the early Holocene. The environment characterised by these malacofaunas is stable and indicative of interstadial climatic conditions (Limondin, 1995; Limondin-Lozouet, 1998; Preece and Bridgland, 1998).

2.4 *Younger Dryas (13,000-11,200 cal. BP)*

Cold conditions during the Younger Dryas resulted in renewed river erosion and sedimentation throughout northwest Europe (e.g. Rose et al., 1980; Bohncke et al., 1987; Brown et al., 1994; Vandenberghe et al., 1994; Kasse et al., 1995; Rose, 1995; Collins et al., 1996). Recent results obtained from the Paris Basin, also show marked changes in river behaviour at this time. ^{14}C dates of around 12,800 cal. BP, obtained from Houdancourt and Rueil show that sedimentation probably began around this time, which soon covered most

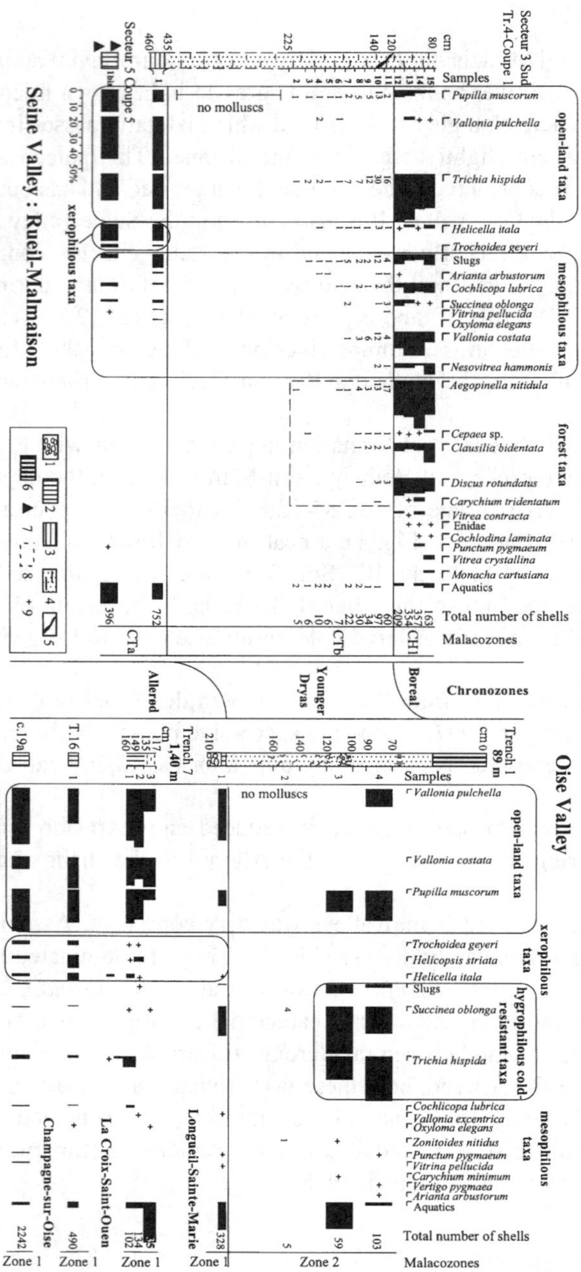

Figure 5. Lateglacial molluscan diagrams from the Seine and Oise valleys. Some assemblages from the Younger Dryas deposits appear very poorly developed, and consequently absolute values of taxa occurrences are given rather than percentages. 1. Fluvial gravels – 2. Sandy soil – 3. Grey clay – 4. Calcareous silts – 5. Gravel bed – 6. Modern soil – 7. Azilian industries: the lower assemblage was radiocarbon dated 14,455-13,795 cal. BP, the upper to 13,361-12,961 cal. BP – 8. Holocene intrusive taxa – 9. Single shell.

of lower parts of the major valleys (e.g. Fig. 3). Part of the main river channel and lateral channels were filled with marly silts up to 4 meters thick (thickness depending on the previous level of incision). Overbank deposition took place on lower bank with secondary channels infilled by marls. Coarser sandy calcareous silts deposition occured in the main channel. Similar to the Somme basin (Antoine, 1997a-c), the high $CaCO_3$ content of these sediments (40 to 70%) is mainly due to erosion of the chalk in upper part of the basin. It indicates microgelivation of slopes underlain by chalk, with probably seasonal rill wash processes. The cessation of minerogenic input is not yet well-dated as chronological markers are lacking in these alluvial sequences.

Reworked organic matter in basal silts reflect the erosion of the Allerød soil from the slopes. In soft sediments with high water content, cryoturbations structures were observed in the lower part of the deposit and through Allerød beds below. The best example is found at the channel of Houdancourt in the Oise valley where involutions warp Allerød peat and silts (Pastre et al., in press). In the Oise and the Seine valleys, earlier Younger Dryas deposits have no shell material. This breakup of malacological populations reflects a strong environmental shift and might be related with climatic impact. Later Younger Dryas deposits (Fig. 5), yield impoverished faunas dominated by cold-resistant hygrophilous taxa, *Succinea oblonga*, *T. hispida* and open-land molluscs (*P. muscorum*, *V. pulchella*). Conditions were slightly less severe allowing the re-development of malacological populations, but mesophilous and xerophilous species, which occur in the Allerød assemblages do not reappear.

The palynological record is also poorly preserved, principally because of the minerogenic character of these sequences (Leroyer, 1997), though some sections show steppic plants (*Artemisia*).

3 HOLOCENE FLUVIAL RECORD: CLIMATIC AND HUMAN IMPACT

3.1 *Preboreal (11,200-10,000 cal. BP)*

In the main valleys, incision at the beginning of the Holocene seems to have been very rapid, probably just a few centuries, according to the dates obtained from the first organic deposits of the main river channel around 11,000 cal. BP. Incision was restricted to the main Lateglacial river-channel, indicating that this channel was still active during the Younger Dryas. Incision created a wide meandering channel of intermediate sinuosity, more than twice large as that of the present day. In small valleys, a single channel cutting 1 to 2 m through Pleniglacial and Lateglacial sediments is more commonly found (e.g. in the Nonette valley, Fig. 1: Pastre et al., 1991, 1997).

During the second half of the Preboreal, detrital input significantly decreased. In the main valleys, dark brown organic clays are mixed with fine calcareous clay and silt fluvial deposit. In smaller valleys, black organic soils and gyttjas occur on the bank (e.g in the Nonette and Aronde valleys, Fig. 1) and peat formation began. These deposits give a palynological succession of *Pinus*, *Betula* and *Salix* (Fig. 4, zone IVa), followed by a *Pinus* and *Corylus* phase (Fig. 4, zone IVb); (Van Zeist & Van Der Spoel-Walvius, 1980; Gauthier, 1995; Leroyer, 1997). From recent palynological investigations (Leroyer, 1997), the transition is dated to around ca. 10,250 cal. BP (Fig. 4).

3.2 *Boreal and Lower Atlantic (10,000-8800-6800 cal. BP)*

The Boreal and the lower part of the Atlantic appear to have been periods of environmental stability. On margin of the main river channel in major valleys, fine and highly organic sedimentation occurred resulting in a reduction in both channel depth and width. In the small valleys northeast of Paris, from where most of the data have been obtained, peat-formation occurred. In the northern area (Aronde, Automne, Matz, Nonette, Verse; Fig. 1), alkaline peat-bogs with *hypnum* moss and hygrophyts expanded across the whole part of the floodplain. They range in thickness between 1.5 and 3.5 m and have thin beds with *Lymnaea* and *Planorbis* locally associated with short flood periods. In the Plaine-de-France (Fig. 1), peat layers are interbedded with thin beds (1 to 20 cm thick) of fine calcareous tufa. The most extensive beds with oncolites (1 to 5 m thick) occur in the Crould valley (Gonesse, Fig. 1), but also in the Marne valley (Fresnes, Fig. 1) where they are linked to Ca-rich springs rising from Tertiary limestones (Lutetian).

The palynological data obtained for these periods (Fig. 4), show the dominance of *Corylus* during the regional zone V situated after ca. 9900 cal. BP. During the zone VI, dated ca. 8950-7450 cal. BP, *Corylus* is still dominant over *Quercus* and *Ulmus*, though *Pinus* increases, and *Tilia* appears. The zone VII is characterized by the maximum of *Tilia*, the appearance of *Taxus* and the increase of *Alnus* prior to ca. 5500 cal. BP.

3.3 *Upper Atlantic and Lower Subboreal (6800-4500 cal. BP)*

During the Boreal and Lower Atlantic periods, the ecological impact of Mesolithic and Lower Neolithic peoples was very limited. However, some expansion of ruderal plants did occur which could be related to human activity (Leroyer, 1997). At the end of the Atlantic, the impact of the Middle Neolithic cultures is clear. Palynological data obtained from the Middle Seine (Châtenay and Noyen in the Bassée area, Fig. 1) and lower Marne (Annet, Fresnes, Fig. 1) river basins show significant amounts of ruderal plants and cereals (Leroyer, 1997; Leroyer et al., 1997), though clearance episodes seems to have been localised. In the Marne and Oise valleys, some sedimentological changes also occur at this time. In main river channels, deposition of organic-rich silts was followed by organo-detritic silts with channels slowly filling up. Lateral erosion processes and silt and sand deposits, are recorded around 5500 cal. BP (e.g. Annet for the river Marne: 5596, 5314 cal. BP; Armancourt for the river Oise: 5721, 5052 cal. BP), which have still little impact on channel dimensions and geometry.

3.4 *Upper part of the Subboreal (4500-2800 cal. BP)*

The upper part of the Subboreal was a phase of major environmental and hydrodynamic change. In the middle Oise river valley, several Upper Neolithic sites situated along the banks are buried by clayey silts which cover the previous organic-rich sediments. The archaeological site of La Croix-Saint-Ouen, near Compiègne (Talon et al., 1995), is typical in this respect (Fig. 6). Here, peaty dark-brown silts of the main river channel contain several Late Neolithic (Seine-Oise-Marne culture) artefacts with wood dated to 2824 BC by dendrochronology (about 4800 cal. BP). These are followed by light-brown clayey silts containing remains of the Late Neolithic Gord culture, followed by Lower Bronze Age material. Both levels are dated between ca. 4500-3800 cal. BP. These beds characterized a

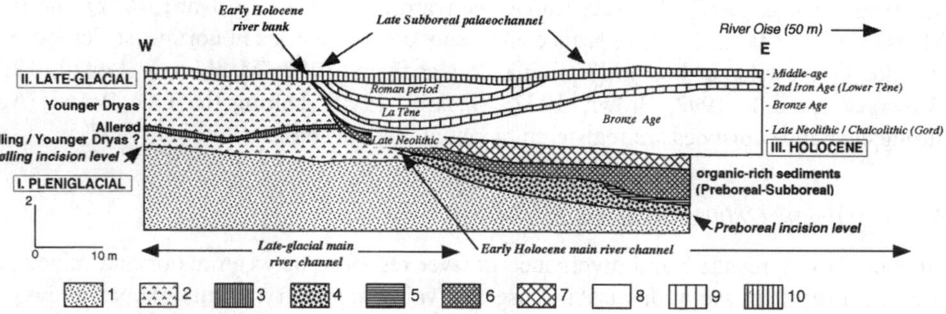

Figure 6. Cross section of the Oise river bank at La Croix-Saint-Ouen (Oise). 1. Sand and gravels (Weichselian) – 2. Marly silts – 3. Poorly organic silt – 4. Gravels – 5. Organic-rich clay – 6. Peat – 7. Organic clayey silt – 8. Clayey silts – 9. Humic silts – 10. Present day cultural soil.

major change of the sedimentological record with peaty silts (Total Organic Carbon > 5%) being replaced by detrital clayey silts (TOC < 0.6%) which principally come from the brown soils covering the loessic plateaus and slopes. Fine grained sedimentation filled the margin of main river channel and in the lower Marne valley it reached side channels by about between ca. 4400-3700 BP when peat-formation ceased (Leroyer et al., 1997). Thin interbedded soils indicate temporary phases of stabilization. Malacofaunas are mainly aquatic. Species such as *Anisus leucostoma* and *Lymnaea truncatula* indicate a quiet, vegetation-rich fluvial environment, though, they are able to survive seasonal draining of river banks. During this period the extension of *Alnus* was very significant (Fig. 4, zone VIII), as also observed along the other rivers of N-W Europe (e.g. Huntley & Birks, 1983; Bennett & Birks, 1990).

The increase of river activity during the Bronze Age is well documented in the middle Oise valley. Channel movement and incision occur frequently around 4500-3500 cal. BP. In La Croix-Saint-Ouen, a small channel was cut (Fig. 6) and was filled during the second Iron Age Period (La Tène culture). Downstream, in Longueil-Sainte-Marie, a similar channel cut-off is documented before deposition of an organic silt dated 4418-3927 cal. BP. In Boran and Champagne (Fig. 1), lateral channel erosion is noticeable between ca. 4500-4000 cal. BP and is succeeded by a loamy sedimentation during middle and late Bronze Age. In small river valleys, there is also a marked change in river dynamics at this time. At Houdancourt (Fig. 1), a small tributary of the Oise river, incision was followed by a coarse sedimentation with wood dated from 3830-3495 and 3450-3486 cal. BP. The change from organic peaty silts to light brown detritic silts could be interpreted as a consequence of Upper Neolithic and Lower Bronze Age agriculture. However, by comparison with the Middle Neolithic, neither palynological nor archaeological data show a significant increase in human settlements during these periods. Although ruderals appear relatively frequent, cereals are scarce on palynological diagrams (Leroyer, 1997). The effects of a higher rainfall or cultural practices such as burning for land clearance, may have been important but are difficult to demonstrate. During the Late Subboreal (3800-2800 cal. BP), pronounced channel incision suggest an increase in flow. The re-activation of Lateglacial lateral channels in several places along the Oise river valley clearly illustrates this process. An increase in river activity during the Late Subboreal resulting in higher sedimentation rates or incision has also been found in the upper part of the Middle Seine valley (Krier et al., 1991; Dzana, 1997), the river Rhône (Bravard et al., 1991; Bravard, 1992), the river Meuse (Lefèvre et al., 1993) in France and many other countries in northwest Europe (e.g. Poland: Godlowska et al., 1987; England: Needham, 1992; Macklin & Lewin, 1993; Brown & Keough, 1992, Brown, 1997; Macklin, 1999). During the Late Bronze Age, strong clearance episodes are registered by palynological studies (Leroyer, 1997).

3.5 *Subatlantic (2800-present cal. BP)*

Human activity resulted in a divergence in river response between major and minor valleys, as a result of variations in land use and valley geometry. In the major valleys, the silty aggradation continued with sedimentation in river channels resulting in accelerated bank erosion and reworking of older deposits. During the Second Iron Age (La Tène culture), an increase in farming activity is shown at many archaeological sites (Buchenschutz & Méniel, 1994; Malrain et al., 1996). Archaeological remains are frequently found in the fine-grained deposits of side channels such as those at La Croix-Saint-Ouen (Fig. 6). They could indicate a decrease in fluvial activity and could be related to land-clearing. In the Oise river valley, open-land malacofaunas dominated by *Vallonia pulchella* are found as would be expected with the clearance of floodplain vegetation. This continued during the Roman times and resulted in the complete filling of lateral channels with main river channels being reduced to half, or even one third, of their original early Holocene width sometimes with the development of islands.

The response to environmental change by the small tributaries in the Paris Basin was rather different. Renewed peat-formation during the first part of the Subatlantic is observed in some rivers where it replaced minerogenic sedimentation during the Subboreal in rivers such as Nonette (Fig. 7) or Esquillons (Houdancourt), tributaries of the river Oise (Fig. 1). The most significant peat formation and organic silt deposition occurred in first and second order streams where there had been no previous Holocene sedimentation. This can be seen, for example, in the small Aulnaies-du-Pont valley, near Roissy in the Plaine-de-France area (Figs 1 and 7), where peat deposition starts at 2720-2366 cal. BP or in the Guyonne river valley (archaeological site from Neauphle-le-Vieux, 35 km W from Paris, not shown on Fig. 1) after the Roman conquest as shown by the presence of *Juglans* (Leroyer & Allenet, 1997). This may be related to clearance causing increased run-off. These organic deposits directly overlie coarse Upper Pleistocene sediments and suggest very little activity in these small valleys until the Late Holocene. Extensive loam deposits (often more than 2 m thick) overlay organic sediments, and generally cover the whole floodplain. Archaeological, palynological and radiocarbon dates indicate that these units are of post-medieval age. In the small Aulnaies-du-Pont valley (Goussainville, Fig. 1), the peat formation ceased at 646-516 cal. BP (Pastre et al., 1997) and in the same Plaine-de-France region similar deposits overlay Middle Age levels from the Carolingian period (Bahain, 1997). In the river Esches valley to the northwest (Amblainville, Belle-Eglise, Fig. 1), clay and silt deposits bury Roman buildings and Upper Middle-age artefacts. The close stratigraphical position of these archaeological remains together with the organic-rich nature of the deposits suggest a significant decrease in erosion during the early Middle-Ages period, as seen elsewhere in mainland Europe at this time (Brown, 1997). Accelerated alluviation and colluviation observed during and after the fifteenth century can be related to the increase in population and the Little Ice Age climatic deterioration between 400-150 cal. BP (Lamb, 1982; Grove, 1988). The homogeneous nature of the sediment and the lack

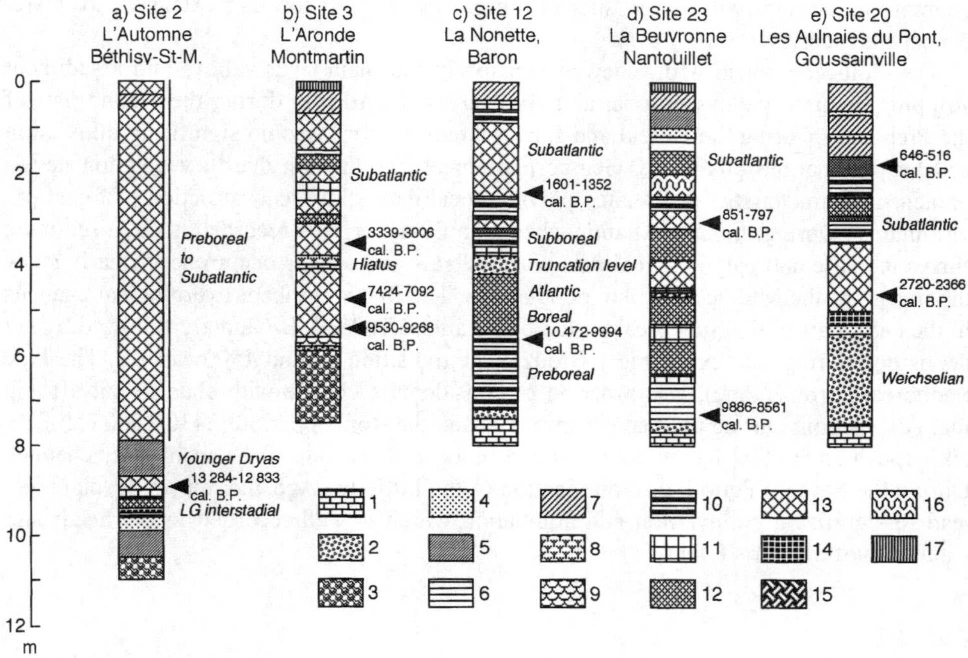

Figure 7. Typical sedimentary sequences (cores) from small valleys north of Paris (Plaine-de-France, Valois and Picardie areas). 1. Bedrock (chalk or limestones) – 2. Silt and gravels (poorly rounded) – 3. Sand, silt and gravels (well rounded) – 4. Sand – 5. Silty sand – 6. Silt – 7. Clayey silt (upper facies) – 8. Calcareous silt – 9. Marl – 10. Organo-mineral silts – 11. Peaty loam – 12. Silty peat – 13. Peat – 14. Humic peat – 15. Tufaceous peat – 16. Silt with reduced Fe – 17. Soil.

of interbedded soils suggest rapid sedimentation probably linked to arable cultivation. Soil development or peat formation noticed at the top of these sequences (according to catchment or local conditions), suggest a decrease in erosion, following the conversion of the floodplain to pasture or wetland forest, river channelisation and better land use practices.

4 CONCLUSIONS

The Lateglacial and Holocene floodplain record of the central part of the Paris Basin though comparable to other rivers valleys in northwest Europe, does show some unique features. During the first half of the Bølling, prior to 14,500 BP cal., significant incision occurred, sedimentation decreased, and there was a change from a braided to a meandering channel system. The transition from Bølling to Allerød was a period of instability and local sedimentation, which could be related to the climatic deterioration of the Older Dryas. The latter part of the Allerød shows the extensive development of a thin soil and local peat-formation in channels. Related malacofaunas with typical species *Helicopsis striata* and *Trochoidea geyeri* indicate an open-ground environment. Although the Allerød appears as a relatively quiet hydrologic period, by contrast, the sedimentary impact of the Younger Dryas is one of the most impressive for northwest Europe. Significant geomorphic activity with marly silts infilling channels and forming extensive overbank deposits is recorded.

The Holocene began with renewed incision in the main river valleys and a sediment stripping in small valleys. Fluvial activity decreased markedly during the second part of the Preboreal. During the Boreal and Lower Atlantic, there was no significant alluviation and peat formation took place. Geomorphological stabilization due to vegetation development is characteristic. Associated palynological data allow reconstruction of forest environments. During the Late Atlantic, change in fluvial style is recorded, though sediment fluxes increase and palynological data with ruderals and cereals occurrences clearly show the impact of the Middle Neolithic populations. The first significant hydrological changes in the early part of the Subboreal period occur after 5,700 cal. BP and are followed by silt deposition during Late Neolithic – Bronze Age transition around 4500 cal. BP. The Late Subboreal (Bronze Age), was a period of considerable change with channel cut-off and loam deposition. During the Second Iron Age and the Roman period (2430-1500 cal. BP), silt deposition infilled lateral channels and reduced the width of the main river channel. During the Modern Period, the combination of the Little Ice Age and human development lead to significant colluviation and alluviation which is well recorded in the headwater tributaries of the Paris Basin.

REFERENCES

Antoine, P. 1997a. Modifications des systèmes fluviatiles à la transition Pléniglaciaire-Tardiglaciaire et à l'Holocène: l'exemple du bassin de la Somme (nord de la France). *Géographie physique et Quaternaire*, 1997, vol. 51, no 1: 93-106.

Antoine, P. 1997b. Evolution tardiglaciaire et début Holocène des vallées de la France septentrionale: nouveaux résultats. *Comptes-rendus de l'Académie des Sciences, Paris, Sciences de la Terre et des Planètes*, 325: 35-42.

Antoine, P. 1997c. Evolution tardiglaciaire et holocène de la moyenne vallée de la Somme In: Fagnart J.P et Thévenin A. (eds) *Le Tardiglaciaire en Europe du Nord-Ouest*, Editions du CTHS, Paris, 13-26.

Audouze, F., Cahen, D., Keeley, L.H. & Schmider, B. 1991. Le site magdalénien du Buisson Campin à Verberie (Oise). *Gallia Préhistoire, Paris,* 24: 99-143.

Bahain, 1997. *Louvres, 'Le Bois d'Orville', DFS de sauvetage urgent SRA Ile-de-France*. Unpublished report.

Bennett, K.D. & Birks, H.J. 1990. Postglacial history of alder (*Alnus glutinosa*) in the British Isles. *Journal of Quaternary Science*, 5: 123-133.

Bodu, P. 1995. Un gisement à Federmesser sur les bords de Seine: 'le Closeau' à Rueil-Malmaison (Hauts-de-Seine). *Bulletin de la Société Préhistorique française*, 92, no 4: 451-455.

Bodu, P. & Valentin, P. 1997. Groupes à Federmesser ou Aziliens dans le sud et l'ouest du Bassin Parisien. Propositions pour un nouveau modèle d'évolution. *Bulletin de la Société Préhistorique française*, 94, no 3: 341-347.

Bohncke, S. & Vandenberghe, J. 1987. Palaeohydrological development in the southern Netherlands during the last 15,000 years In: Starkel L., Gregory K.J., Thornes J.B. (eds). *Temperate Palaeohydrology*. Chichester, John Wiley, 253-281.

Bohncke, S., Vandenberghe, J., Coope, R. & Reiling, R. 1987. Geomorphology and palaeoecology of the Mark Valley (southern Netherlands): Palaeoecology, palaeohydrology and climate during the Weichselian Late Glacial. *Boreas*, 16: 69-85.

Bravard, J.P., Peiry, J.L. et Gadiolet, P. 1991. La formation de la plaine alluviale holocène du Rhône à l'amont de Lyon. *Physio-Géo*, 22-23: 167-172.

Bravard, J.P. 1992. Les rythmes d'évolution morphologique des vallées françaises au Tardiglaciaire et à l'Holocène. *Bulletin de l'Association des Géographes français, Paris*, 3: 207-226.
Brown, A.G. 1997. *Alluvial geoarchaeology: floodplain archaeology and environmental change*, Cambridge, Cambridge University Press.
Brown, A.G. & Keough, M. 1992. Holocene floodplain metamorphosis in the East Midlands, United Kingdom. *Geomorphology*, 4: 433-45.
Brown, A.G., Keough, M.K. & Rice, R.J. 1994. Floodplain evolution in the East Midlands, United Kingdom: the Lateglacial and Flandrian alluvial records from the Soar and Nene valleys. *Philosophical Transactions of the Royal Society of London*, Series A, 348: 261-293.
Buchenschutz, O. & Méniel, P. 1994. Les installations agricoles de l'Âge du Fer en Ile-de-France. Actes du colloque de Paris. *Etudes d'Histoire et d'Archéologie*, 4.
Cleveringa, P., De Gans, W., Huybrechts, W. & Verbruggen, C. 1988. Outline of river adjustments in small river basins in Belgium and the Netherlands since the Upper Pleniglacial. In: Lang, G. & Schlüchter, C. (eds) *Lake, Mire and River environments*, Balkema, Rotterdam, 123-132.
Collins, P.E.F., Fenwick I.M., Keith-Lucas, D.M. & Worsley, P. 1996. Late Devensian river and floodplain dynamics and related environmental change in northwest Europe, with particular reference to a site at Woolhampton, Berkshire, England. *Journal of Quaternary Science*, 11(5): 357-375.
Dzana, J.G. 1997. *Le lit de la Seine de Bar à Montereau*. Unpublished Thesis, Université de Paris I.
Fagnart, J.P. 1997. *La fin des temps glaciaires dans le Nord de la France*. Mémoires de la Société Préhistorique Française, 24, Paris.
Gauthier, A. 1995. Résultats palynologiques de séquences holocènes du Bassin parisien: histoire de la végétation et action de l'Homme. *Palynosciences*, 3: 3-17.
Godlowska, M., Koslowski, J., Starkel, L. & Wasilikowa, K. 1987. Neolithic settlement at Pleszow and changes in the natural environment in the Vistula valley. *Przeglad Archeologiczny,* 34: 133-159.
Grove, J. 1988. *The Little Ice Age*. Methuen, London.
Haesaerts, P. 1984a. Aspects de l'évolution du paysage et de l'environnement en Belgique au Quaternaire In: Cahen D. et Haesaerts P. (eds) Peuples chasseurs de la Belgique dans leurs cadre naturel. *Institut royal des Sciences naturelles de Belgique*, Bruxelles, 28-39.
Haesaerts, P. 1984b. Les formations fluviatiles pléistocènes du bassin de la Haine (Belgique). *Bulletin de l'Association Française pour l'Etude du Quaternaire*, 21: 19-26.
Huntley, B. & Birks, H.J.B. 1983. *An Atlas of Past and Present Pollen maps for Europe: 0-13,000 BP years ago*, Cambridge, Cambridge University Press.
Kalicki, T. 1991. The evolution of the Vistula River Valley between Cracow and Niepolomice in Late Vistulian and Holocene times In: Starkel L. (ed.) Evolution of the Vistula River Valley during the last 15,000 years. *Polish Academy of Sciences*, Wroclaw, 11-37.
Kasse, K., Vandenberghe, J. & Bohncke, S. 1995. Climatic changes and fluvial dynamics of the Maas during the late Weichselian and early Holocene In: Frenzel, B., Vandenberghe, J., Kasse, K., Bohncke, S. & Gläser, B. (eds) *European river activity and climatic change during the Lateglacial and early Holocene*, *Paläoklimaforschung*, 14: 124-150.
Kerney, M.P. 1963. Lateglacial deposits on the chalk of south-east England. *Philosophical Transactions of the Royal Society of London*, Series B 246: 203-254.
Kiden, P. 1991. The Lateglacial and Holocene evolution of the Middle and Lower River Scheldt, Belgium In: Starkel, L., Gregory, K.J. & Thornes, J.B. (eds), *Temperate Palaeohydrology*. Chichester, John Wiley, 283-299.
Kozarski, S. 1983. River channel adjustment to climatic change in west central Poland In: Gregory (ed.). *Background to Palaeohydrology*. Chichester, John Wiley, 355-374.
Kozarski, S. 1991. Warta-a case study of a lowland river. In: Starkel, L., Gregory, K. & Thornes, J. (eds) *Temperate Palaeohydrology*. Chichester, John Wiley, 189-195.
Krier, V., Leroyer, C. & Limondin, N. 1991. Approche pluridisciplinaire de l'environnement du site holocène de Verrières (vallée de la Seine, Aube). *116e Congr. nat. des Soc. sav., Chambéry, Pré-protohistoire*, 47-59.

Lamb, H.H. 1982. *Climate, History and the Modern World*. Methuen, London.

Lautridou, J.-P. 1985. *Le cycle pléniglaciaire en Europe du nord-ouest et plus particulièrement en Normandie*. Thèse d'Etat en Géographie, Université de Caen.

Lefèvre, D., Heim, J., Gilot, J. & Mouthon, J. 1993. Evolution des environnements sédimentaires et biologiques à l'Holocène dans la plaine alluviale de la Meuse (Ardennes, France): premiers résultats. *Quaternaire*, 4(1): 17-30.

Lehman, S.J. & Keigwin, L.D. 1992. Sudden changes in North Atlantic circulation during the last deglaciation. *Nature*, 356: 757-762.

Leroyer, C. 1997. *Homme, climat, végétation au Tardi- et Postglaciaire dans le Bassin parisien: apports de l'étude palynologique des fonds de vallée*. Unpublished thesis, Université de Paris I, vol. 2.

Leroyer, C. & Allenet G. 1997. Analyse pollinique In: Giligny, F. (ed.). *Les occupations pré- et protohistoriques du vallon de la Guyonne*, DFS de sauvetage urgent, AFAN-SRA Ile-de-France, unpublished report, 67-86.

Leroyer, C., Pastre, J.-F., Fontugne, M. & Limondin, N. 1997. Le Tardiglaciaire et le début de l'Holocène dans le bassin aval de la Marne (Seine-et-Marne, France): chronostratigraphie et environnement des occupations humaines In: Fagnart, J.P et Thévenin, A. (eds) *Le Tardiglaciaire en Europe du Nord-Ouest:* 151-164, Editions du CTHS, Paris.

Limondin, N. 1995. Lateglacial and Holocene malacofaunas from archaeological sites in the Somme Valley (North France). *Journal of Archaeological Science*, 22: 683-698.

Limondin-Lozouet, N. 1997. Les successions malacologiques du Tardiglaciaire et du début de l'Holocène dans la vallée de la Somme In: Fagnart, J.P et Thévenin, A. (eds) *Le Tardiglaciaire en Europe du Nord-Ouest:* 39-46, Editions du CTHS, Paris.

Limondin-Lozouet, N. 1998. Successions malacologiques du Tardiglaciaire weichsélien: corrélations entre séries du Nord de la France et du Sud-Est de la Grande-Bretagne. *Quaternaire*, 9(3): 217-225.

Macklin, M.G. 1999. Holocene River environments in prehistoric Britain: human interaction and impact. *Quaternary Proceedings*, 7: 521-530.

Macklin, M.G. & Lewin, J. 1993. Holocene river alluviation in Britain. *Zeitschrift für Geomorphologie (Supplement)*, 88: 109-122.

Malrain, F., Maréchal, D. & Pinard, E. 1996. Occupation des sols et parcellaires dans la moyenne vallée de l'Oise du IVe avant au XIVe siècle après J.-C., In: Chouquer (ed.). Les formes du paysage, T2, *Actes du 3e colloque d'Ager, Orléans*, 21-42.

Mangerud, J., Andersen, S.T., Berglund, B.E. & Donner, J.J. 1974. Quaternary stratigraphy of Norden, a proposal for terminology and classification. *Boreas*, 3: 109-128

Munaut, A.V. & Paulissen, E. 1973. Evolution et paléoécologie de la vallée de la Petite Nèthe au cours du Post-Würm (Belgique). *Annales de la Société géologique de Belgique*, 96(II): 301-346.

Needham, S. 1992. Holocene alluviation and interstratified settlement evidence in the Thames valley at Runnymede Bridge In: Needham S. & Macklin M.G. (eds), *Archaeology Under Alluvium*. Oxbow Books, Oxford, 249-260.

Needham, S. & Macklin, M.G. (eds), 1992. *Archaeology Under Alluvium*. Oxbow Books, Oxford.

Pastre, J.F., Cecchini, M., Dietrich, A., Fontugne, M., Gauthier, A., Kuzucuoglu, C., Leroyer, C. & Limondin, N. 1991. L'évolution holocène des fonds de vallées au nord-est de la région parisienne (France): premiers résultats. *Physio-géo*, 22-23, 109-115.

Pastre, J.F., Fontugne, M., Kuzucuoglu, C., Leroyer, C., Limondin-Lozouet, N., Talon, M. & Tisnérat, N. 1997. L'évolution tardi- et postglaciaire des lits fluviaux au nord-est de Paris (France). Relations avec les données paléoenvironnementales et l'impact anthropique sur les versants. *Géomorphologie: relief, processus, environnement*, 4: 291-312.

Pastre, J.F., Leroyer, C., Limondin-Lozouet, N., Chaussé, C., Fontugne, M., Gebhardt, A., Hatté, C. & Krier, V. in press. Le Tardiglaciaire des fonds de vallée du Bassin Parisien (France). *Quaternaire*.

Preece, R.C. 1994. Radiocarbon dates from the 'Allerød soil' in Kent. *Proceedings of the Geologists's Association*, 105: 111-123.

Preece, R.C. & Bridgland D.R. 1998. *Late Quaternary environmental change in North-West Europe. Excavations at Holywell Coombe, South-East England.* Chapman and Hall, London.

Roblin-Jouve, A. 1994. Le milieu physique In: Taborin Y. (ed.). *Environnements et habitats magdaléniens dans le centre du Bassin parisien.* Documents d'Archéologie Française, Paris, no 43: 12-35.

Rose, J. 1995. Lateglacial and early Holocene river activity in lowland Britain. *Paläoklimaforschung*, 14: 51-74.

Rose, J., Turner, C., Coope, G.R. & Bryan, M.D. 1980. Channel changes in a lowland river catchment over the last 13,000 years In: Cullingford, R.A., Davidson, D.A. & Lewin, J. (eds), *Timescales in Geomorphology*. Chichester, John Wiley, 159-175

Starkel, L. 1994. Reflection of the glacial-interglacial cycle in the evolution of the Vistula river Basin, Poland. *Terra Nova*, 6: 486-494.

Stuiver, M & Reimer, P.J. 1993. Extended ^{14}C database and revised CALIB 3.0 ^{14}C age calibration program. *Radiocarbon,* 35: 215-230.

Talon, M., Bernard, V., Billand, G., Cottiaux, R., Pastre, J.-F., Pernaud, J.-M. & Prost, D. 1995. Le niveau organique Néolithique récent du site stratifié de la station d'épuration à La Croix-Saint-Ouen (Oise). Premiers résultats. *Revue archéologique de Picardie*, no spécial 9: 83-103.

Valladas, H. 1981. Datation par thermoluminescence de grès brulés de foyers de quatre gisements du Magdalénien final du Bassin Parisien. *Comptes-rendus de l'Académie des Sciences, Paris*, t. 292 série II, 355-358.

Valladas, H. 1994. Chronologie des sites du Magdalénien final du Bassin parisien In: Taborin Y. (ed.). *Environnements et habitats magdaléniens dans le centre du Bassin parisien*. Documents d'Archéologie Française, Paris, 65-68.

Vandenberghe, J. 1993. Changing fluvial processes under changing periglacial conditions. *Zeitschrift für Geomorphologie*, Suppl. Bd 88: 17-18.

Vandenberghe, J., Bohncke, S., Lammers W. & Zilverberg, L. 1987. Geomorphology and palaeoecology of the Mark valley (southern Netherlands): Geomorphological valley development during the Weichseliean and Holocene. *Boreas*, 16: 55-67.

Vandenberghe, J., Kasse, C., Bohncke, S. & Kozarski, S. 1994. Climate-related river activity at the Weichselian-Holocene transition: a comparative study of the Warta and Maas rivers. *Terra nova*, 6: 476-485.

Van Geel, B., Coope, R.G. & Van der Hammen, T. 1989. Palaeoecology and stratigraphy of the Late Glacial type section at Usselo (The Netherlands). *Reviews of Palaeobotany and Palynology*, 31: 359-448.

Van Zeist, W. & Van Der Spoel-Walvius, M.R. 1980. A palynological study of the Lateglacial and the Post-glacial in the Paris basin. *Palaeohistoria*, XXII: 68-109.

Walker, M.J.C., Bohncke, S.J.P., Coope G.R., O'Connel M., Usinger H. & Verbruggen C. 1994. The Devensian/Weichselian Lateglacial in nortwest Europe (Ireland, Britain, north Belgium, the Netherlands, northwest Germany): IGCP-253. *Journal of Quaternary Science*, 9: 109-118.

13. Historic and possible prehistoric human impacts on flo plain sedimentation, North Branch of the Susquehanna River, Pennsylvania, USA

DONALD M. THIEME
Honesdale, USA

1 INTRODUCTION

The role of human activity in shaping the channels, banks, and floodplains of the Earth's rivers is a subject of growing concern among both geographers and geologists (Brown, 1996; Butzer, 1976; Knox, 1977; Lewin & Macklin, 1987; Starkel, 1991; Macklin & Lewin, 1989). Rivers in the eastern United States have been particularly well studied in this regard (Wolman & Schick, 1967; Wolman, 1967; Trimble, 1974; Costa, 1975; Lindner, 1987) due to the prominent imprint of slightly more than four centuries of Euroamerican settlement, including at least two centuries of urbanisation and modern industry. Human impacts identified in my study of the North Branch of the Susquehanna River Valley in northeastern Pennsylvania (Fig. 1) conform to certain patterns noted in these previous studies. A parallel and not at all conflicting objective of my study has been to look at the alluvial sediment itself as an archive of the unique historical sequence of human impacts on the local channel alignments, sediment yield, and floodplain sedimentation (Thieme, 1999, 2000; Thieme & Schuldenrein, 1998).

The present paper begins with a review of types of human impact on rivers and floodplains which have been documented in previous studies. Patterns specific to humid temperate midlatitude regions such as eastern North America are emphasised. A summary of the geological and glacial history of the catchment is then presented, followed by a discussion of meteorological conditions associated with gauged floods and other floods of recorded history. Impacts of 19th century industrial development are then discussed in some detail. Activities which impacted the river in the reaches studied include logging of the dense pine and hardwood forests, mining of anthracite coal, and construction of a canal system for transporting the coal. Several remnants of the canal system can still be found within the river channel and floodplain.

The paper concludes with a brief summary of Holocene alluvial terrace remnants, several of which contain deposits which appear to result from the introduction of maize agriculture after AD 800 by the late prehistoric Clemson Island, Owasco, and Wyoming Valley cultures (Thieme, 1999; Thieme & Schuldenrein, 1998). An earlier and more regionally uniform episode of valley alluviation also appears to have been associated with disturbance of the vegetation within the catchment. This occurred at approximately 4500 BC, when hemlock trees were stricken by an insect pest (Davis, 1981; Filion & Quinty, 1993; Watts, 1979). Analogy with historic floods of equivalent magnitude suggests that the transported hillslope sediments were deposited in the valley flats by late winter to spring floods with a heavy snowmelt component. Prehistoric people rapidly colonized these new floodplain

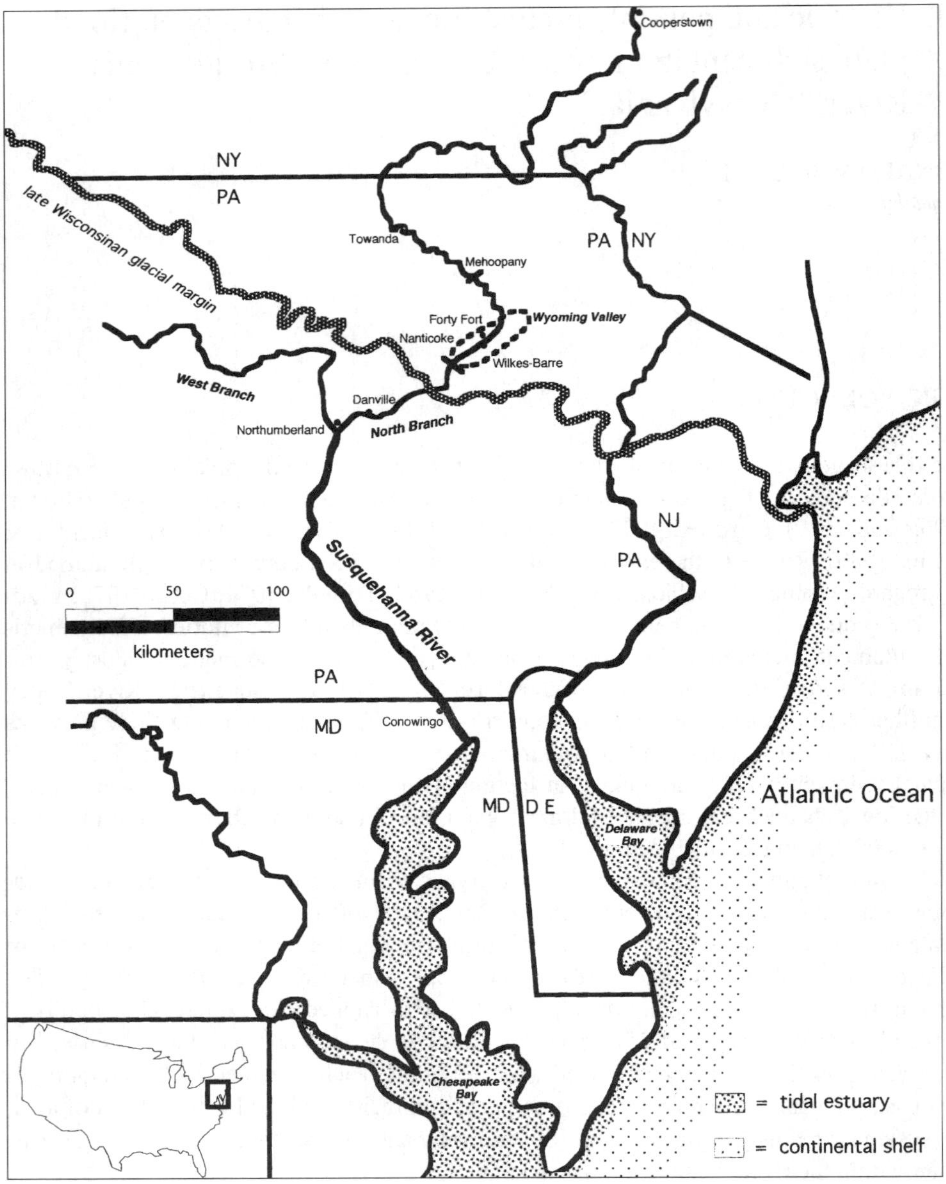

Figure 1. North Branch of the Susquehanna River Valley, Pennsylvania, USA.

surfaces with numerous large settlements dating between 3000 BC and 50 cal. BC (Custer, 1996, p. 163-216; Funk, 1993a, p. 284-288). Predominantly climatic as opposed to anthropogenic controls on the earlier alluvial stratigraphy produced packages of sediment and soil-forming intervals which can be correlated between terrace remnants despite the fact that these remnants are rarely mappable for more than two kilometers in a given reach (Dineen, 1993; Engel et al., 1996; Pazzaglia & Gardner, 1993; Scully & Arnold, 1981; Thieme, 2000; Thieme & Schuldenrein, 1998; Vento et al., 1999, p. 776-777).

2 PREVIOUS HUMAN IMPACT STUDIES AND REGIONAL PATTERNS

If we distinguish at the outset between *direct* ('channel-phase') and *indirect* ('land-phase') impacts of human activity (Knighton, 1998, p. 308) then most examples of the former in the eastern United States are associated with engineering features for which we have some type of historical documentation. Knowledge of such impacts is relevant to planning future construction of levees, dams, canals, bridges, or channelization projects. Studies of submerged or abandoned structures and floodplain deposits frequently complement or fill in gaps in the written or graphical documentation. Such studies may also serve as a basis for interpreting material remains and sediments associated with non-literate societies or societies whose texts and drawings have been lost or cannot be completely deciphered (Brown, 1997, p. 257-260; Clay, 1992; Hebda et al., 1991; Howard, 1993; Rohn, 1963; Salisbury, 1988, 1991, 1992; Siemens, 1998; Turner, 1983; Vivian, 1992; White et al., 1999; Wilshusen et al., 1997; Zangger, 1998).

General tendencies of river response to certain artificial structures are relatively well understood. Artificial levees, for example, encourage bedload deposition upstream (Knighton, 1998, p. 125) and continual rebuilding of levees along rivers such as the Huang Ho in China resulted in their becoming perched above their surrounding floodplain (Stoddart, 1978). River reaches downstream of dams typically experience a reduction in the magnitude of peak floods and a decrease in sediment load (Meade & Parker, 1985; Williams & Wolman, 1984). According to Petts (1979) the channel cross-section will decrease over 50 percent in area. However, Williams & Wolman (1984) found instances of increasing, decreasing, and constant width below dams in the western United States.

Effects of artificial channelisation have been reviewed by Brookes (1985, 1988). Confining flow within channel banks which would otherwise spill onto the river floodplain will tend to increase the mean annual discharge. A 50 percent increase in discharge caused the Cheslotta River in Canada to widen from 5 m to 75-100 m and entrench 10-15 m below its former floodplain over a period of 20 years following flow diversion (Kellerhals et al., 1979). Artificial cutoffs designed to improve navigation depths and reduce flood heights along the lower Mississippi River increased the gradient by 12 percent (Winkley, 1982), inducing greater bed-material mobility. Development of aggrading bars has resulted in navigation channels which are actually much shallower than they were 90 years ago.

Indirect or 'land-phase' impacts can be further subdivided into: 1.) those associated with mining and other specific extractive industries; and 2.) those resulting from more general 'land use' changes within the river catchment. G. K. Gilbert's study of the impacts of hydraulic gold mining on rivers in the Sierra Nevada (Gilbert, 1917) was one of the earliest geological studies to address impacts of *mining* on river floodplains. Use of pressurized jets of water for hydraulic sluicing produced immense volumes of poorly sorted sediment which far exceeded the channel capacity of rivers such as the Yuba and the Bear (James, 1991). Rapid bed aggradation and storage of the excess sediment on floodplains and in backflooded tributary valleys resulted in higher flood stages and loss of land suitable for cultivation. Similar impacts on a smaller scale resulted from brief experiments with hydraulic gold mining in the Blue Ridge mountains of the eastern United States (Leigh, 1994, 1997).

The *active transformation* of river floodplains with techniques such as hydraulic sluicing is less common than *passive dispersal* of mine wastes by runoff and stream discharge driven by atmospheric precipitation (Lewin & Macklin, 1987). Passively dispersed mine wastes tend to be smaller in volume, less uniformly distributed through a given reach, and are often better sorted due to a longer transport history. They consequently result in less significant aggradation of the channel bed, although channel migration may be promoted by lateral accumulation of mining waste (Knighton, 1998, p. 324).

Channel-bed aggradation and accelerated deposition on floodplains also result from *deforestation*, which appears to have begun considerably earlier in Europe than in other parts of the world. Distinct phases of accelerated valley-floor sedimentation during the Roman and Medieval periods have been reconstructed by Starkel (1991) based on alluvial stratigraphy in the Vistula River valley. More limited impacts probably resulted from Neolithic forest clearance but these have proven difficult to differentiate from the effects of Holocene climate change.

Because European farming practices were introduced relatively recently and many landscape changes were carefully documented, much can be learned about the impacts of deforestation from sequences in the United States (Knighton, 1998, p. 317). While much of the forest had been cleared for farmland in New York and New England by AD 1700, virgin forests lasted until at least AD 1800 in Pennsylvania (Stranahan, 1993, p. 83) and Wisconsin (Knox, 1977). Clearcutting by commercial loggers began in the middle to late 19th century, and some particularly drastic impacts occurred in remote forests such as the Great Smoky Mountains (Whittaker, 1956) and the redwood forests of northern California (Madej & Ozaki, 1996).

Loss of vegetation cover is particularly significant on steep slopes, and this was a crucial problem affecting intensive row-crop agriculture in many parts of the southeastern United States during the 19th century (Trimble, 1974). Much sediment eroded from former tobacco and cotton plantations is still stored, either as colluvial-sheetwash deposits on hillslopes or as alluvium in floodplains and channels (Meade, 1976, 1982; Phillips, 1993).

Impacts of *urbanisation* have been investigated by many researchers following the pioneering study by Wolman (1967) on rivers in Maryland. Growth of cities increases the area of paved surface and the efficiency of drains in routing storm water to natural channels (Hollis, 1975; Leopold, 1968). This results in a flashier runoff regime with shorter lag times and higher peak discharges. As shown by Wolman (1967), there is a large initial increase in sediment yield when soil is exposed to runoff on construction sites followed by a sharp decline since sediments are sealed beneath newly paved surfaces.

Interactions between human beings and river floodplains commonly appear to have been *cyclical*, since temporary economic crises gave rivers and vegetation communities brief periods to recover between impacts from specific technologies. While the most drastic or 'direct' impacts tend to occur latest in those valleys which have been studied so far, there probably are many exceptions which would be worthy of scientific study. Of the four centuries during which Euroamericans have been interacting with the rivers of the eastern United States, the 19th century represents the climax for certain extractive and manufacturing industries in areas which have yet to be urbanised. Anthracite coal reserves made certain reaches of the North Branch of the Susquehanna River particularly central to 19th century industrial development, with consequences for the river floodplain which are explored below.

3 GEOLOGICAL, GLACIAL, AND GEOMORPHOLOGICAL SETTING

The Susquehanna River is the largest river on the Atlantic seaboard in the northeastern United States, both in terms of channel length and drainage basin area. The river's ancient bedrock valley is incised through folded sedimentary rocks and filled with coarse gravel outwash. The active channel falls a total of 365 m as it traverses the 720 kilometers from the river headwaters near Cooperstown, New York to the Chesapeake Bay (see Fig. 1). The upstream limit of sea-level control on river stage during the Holocene is defined by a 30 m falls at Conowingo, Maryland with a slope of 0.0015 (Pazzaglia & Gardner, 1993; Thompson, 1990). The average slope for the remaining longitudinal profile is 0.0005, which falls on the threshold of meandering to braided streams according to Leopold & Wolman (1957). In certain of the lower reaches, the channel is in fact braided around numerous elongate islands composed primarily of Holocene alluvium.

The North Branch of the Susquehanna River is joined by the West Branch at Northumberland. The North Branch drainage basin upstream of Northumberland totals more than 30,000 km^2. The channel is typically less than six meters deep and channel widths range from 150 m to 450 m. Widths are partly controlled by the shape of the bedrock valley and the extensive inset deposits of glacial outwash. The widest reaches occur in synclinal troughs of the strongly folded Silurian through Pennsylvanian sedimentary lithologies (Berg et al., 1980) of the Ridge and Valley physiographic province. Synclinal valleys with inset terraces of outwash and alluvium occur in the reaches from Danville to Northumberland and through the 'Wyoming Valley' at Wilkes-Barre. Much narrower reaches occur where the meandering valley is incised straight through resistant sandstone and conglomerate beds as it crosses the anticlines between these valleys. On the Appalachian Plateau between Towanda and Wilkes-Barre, the river follows the grain of more subtle folds and abuts bluffs of over 100 m on the outside of many large-amplitude meanders (Leopold et al., 1964, p. 313).

Terraces of alluvium or coarse outwash flank the channel in most reaches and stand above the arbitrary 'bankfull' channel definable on the basis of 20th century discharge records (Table 1). While some silty or clayey sediments are present, these are mostly restricted to thin (< 50 cm) overbank veneers or relict backswamps within the synclinal valleys. The sediment in the modern channel perimeter is mostly bedload, with a median diameter of 5 cm or greater (Pazzaglia & Gardner, 1993, p. 85; Thieme, 2000). This accords with the wide, shallow channel shape indexed by the dimensionless ratio (*w/d*) of channel width to depth. Values for *w/d* of 50.8 for Towanda, 35.5 for Wilkes-Barre, and 69.1 for Danville typify rivers with less than 10% silt and clay in the channel perimeter (Schumm, 1960).

Table 1. North Branch of the Susquehanna River Bankfull Discharges calculated as the 1.5 yr R.I. for 1908-1971.

Gauging Station	Drainage Basin Area (km^2)	Gauge Elevation (m MSL)	Channel Width (m)	Bankfull Depth (m)	Q_{bkf} – Bankfull Discharge (m^3s^{-1})
Towanda	20,194	211.7	244	4.8	2,591
Wilkes-Barre	25,796	156.1	213	6.0	3,029
Danville	28,800	131.5	366	5.3	3,228

Large, arcuate meanders in many reaches of the bedrock valley are less typical of 'bed-load' rivers with a wide, shallow channel cross-section (Schumm, 1977, p. 153-157). This tends to support the hypothesis presented by Sevon (1986) that these meanders were carved by a somewhat more powerful stream with a much deeper channel cross-section. Meander wavelengths of approximately 6 km are found in the reaches between Towanda and the Wyoming Valley, for example, where wavelengths of less than 3 km are predicted from the bankfull discharges using the equations of Dury (1965) or Ackers & Charlton (1970). Regardless of their origin, the spacing of these arcuate meanders should influence the development of regions of turbulent flow (Yalin, 1976). Assuming that such regions are maintained or become even more pronounced as discharge increases beyond bankfull stage, meander spacing probably affects the distribution of river energy during flood events. The bedrock valley meanders may hence be partly responsible for the discontinuous distribution of terrace remnants, particularly those consisting of Holocene alluvium.

Approximately two-thirds of the North Branch valley upstream of Northumberland in Pennsylvania was glaciated during the late Wisconsinan, or oxygen isotope Stage 2 (~35-10 ka). There are no local radiocarbon dates for deglaciation but a recent regional synthesis by Cotter et al. (1986) suggests an age of ~18 ka for the initial retreat from the Wisconsinan terminal moraine. The moraine was incised following the retreat of the ice, and outwash sand and gravel were deposited at sucessively lower terrace levels. Peltier (1949) identified these levels with several 'stillstands' of the ice sheet where it built recessional moraines. The 'Valley Heads' position at the southern end of the Finger Lakes in New York State has been tightly dated to ~14 ka (Muller, 1965; Muller & Calkin, 1993) but intervening positions are poorly constrained.

Peltier also mapped a 'Mankato' alluvial terrace on the 'first bottom,' approximately 5 m above mean low water, which he thought was constructed during a periglacial climate (Peltier, 1949, p. 79-80). Due in part to the subsequent invention of ^{14}C dating and in part to extensive archaeological investigations, these 'terracettes' or terrace remnants are now known to be typically Holocene (Pazzaglia & Gardner, 1993; Thieme & Schuldenrein, 1998). Relatively few ^{14}C dates have been obtained on the basal sediments, however, and three ages greater than 9 ka are reported later in this paper (Table 5). Periods of particularly rapid alluvial deposition have been attributed to tropical storm events or other general changes in climate and atmospheric circulation (Vento & Rollins, 1989; Vento et al., 1999). Certain activities of the prehistoric indigenous people of the North Branch valley also appear to have played a role (Thieme,1999, 2000) in addition to the more recent settlements and industries of Euroamericans (Lottick, 1992; Petrillo, 1986; Stranahan, 1993, p. 75-115).

4 HISTORICAL FLOOD FREQUENCY AND METEOROLOGY

Severe weather events play some role in nearly all sediment transport within river basins. Some obvious exceptions would be the sort of active transformation of river floodplains discussed with regard to hydraulic mining, certain direct impacts of river engineering such as dam bursts, and perhaps some transport associated with tectonic disturbance or tidal incursion. Floods vary, however, both in magnitude and in their meteorological origins. Changes in climate, vegetation, or land use may influence not only the magnitude and frequency of floods in general but that of the various categories for floods in a given catchment.

Floods on the North Branch of the Susquehanna River were observed by Euroamerican settlers as early as the late 18th century (Hoyt & Langbein, 1955, p. 429) and continuous records of river discharge were being kept at several gauging stations by the late 19th century. Use of rating curves based on the period of continuous gauging to assess relative flood magnitudes may be inappropriate given that the 'stationarity' of the gauging records is compromised by the very impacts from both human activity and climate change which are being documented in this paper (Brown, 1996, p. 57; Gordon et al., 1992, p. 250; Knox, 1984, 1985). Nonetheless, recurrence intervals projected on the basis of the years 1908-1971 (Thieme, 2000) are being used in the present paper as arbitrary indices. The years from 1972 on were omitted from the flood frequency analyses due to the anomalous magnitude of the June 1972 event triggered by Tropical Storm Agnes, as seen in the annual flood series for the Wilkes-Barre gauging station (Fig. 2).

Early fall to late spring floods should be most typical based on catchment hydrology, since both runoff and groundwater recharge are at a maxiumum during these months of water surplus (Thornthwaite Associates, 1964, p. 573). The earliest recorded flood in Pennsylvania occurred on October 5, 1786 and is known as the 'pumpkin' flood because of its destruction and entrainment of that agricultural crop (Hoyt & Langbein, 1955, p. 429). A discharge of 5,351 m^3s^{-1} at Wilkes-Barre has been estimated by the U.S.G.S. (1999), which would only be a '25 year flood' using the frequency curve for 1908-1971. Of four other early floods whose heights were recorded prior to the first systematic gauging, by far the most significant occurred on March 17th (St. Patrick's Day) of 1865. The depth of 10.1 m at Wilkes-Barre would represent a discharge of 6569 m^3s^{-1} using the modern rating curve and a '50 year flood' using the frequency curve for 1908-1971.

While the peak annual floods within the period of gauging are quite consistent between the three principal gauging stations, they are clearly a 'mixed population' in terms of meteorological cause. One method for better estimating the true probability that a given type

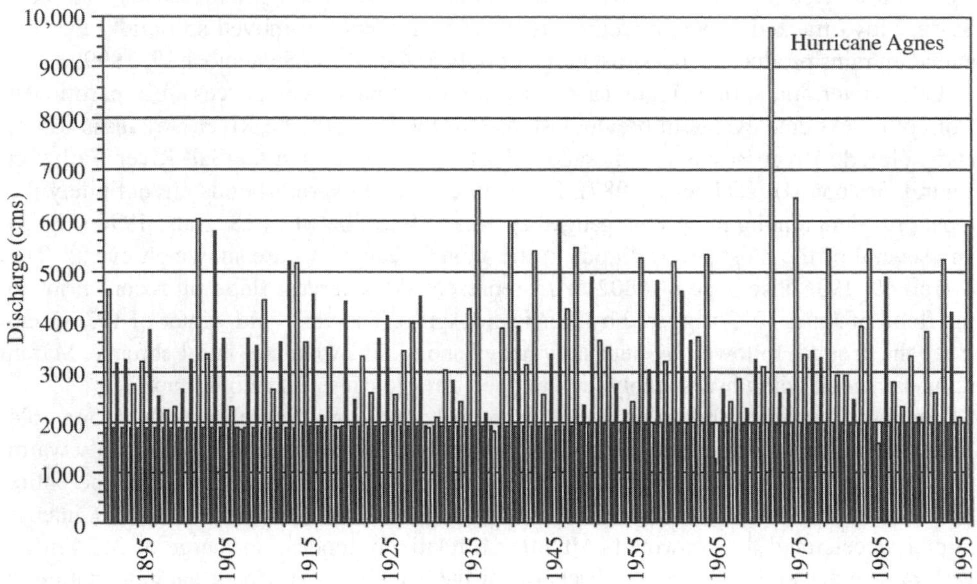

Figure 2. Annual Flood Series for Wilkes-Barre, Pennsylvania.

of event may recur or may have occurred with a given frequency in the deep past is to partition those events by hydrometeorological cause (e.g. Brown, 1991, 1996; Bevan, 1993; Hirschboeck, 1988). While the flood of June 1972 is clearly an anomalous outlier in the annual flood series for Wilkes-Barre (Fig. 2), for example, causes for that event diagnosed by Namias (1973) and Hirschboeck (1988, p. 35-37) are probably common to most if not all of the other *tropical storm* related events in the series.

Tropical Storm Agnes was born in the waters off Mexico's Yucatan Peninsula, growing to become the first hurricane of the season by June 18, 1972. Thereafter it weakened so that it was no longer a hurricane and became embedded between a ridge and a trough in the upper air flow, effectively constituting an 'extratropical cyclone' (Djuric, 1994, p. 167-168). This 'steering' of the storm's path took it across the populous and industrially developed mid-Atlantic coast to north-central Pennsylvania, where it stalled for almost 24 hours and was fed by moist air flowing from the abnormally warm western Atlantic. The 'meridional gradient' in sea-surface temperatures (SSTs) was particularly steep during 1972 (Namias, 1973), with colder than normal water in northerly latitudes of the North Atlantic coinciding with the especially warm water off the eastern seaboard of the United States.

The June 1972 floods on the Susquehanna were among the most destructive in United States history, costing more than 48 deaths and 1.5 billion dollars in Pennsylvania, and 117 deaths and 3.1 billion dollars in the United States as a whole (Bailey et al., 1975, p. 83). The measured discharges of 9060 m^3s^{-1} at Towanda, 9768 m^3s^{-1} at Wilkes-Barre, and 10,277 m^3s^{-1} at Danville are around three times larger than the mean annual flood for 1908-1971 but are nonetheless within 'maximum likelihood' limits defined from previous extreme floods for drainage basins of this size (Hoyt & Langbein, 1955, p. 59-61). At least twelve other annual flood peaks at Wilkes-Barre between 1891 and 1999 are related to cyclones or extratropical cyclones generated along the polar front during late summer and fall months (Elsner & Kara, 1999; NOAA, 1999). Particularly significant were the flood of November 17, 1926 with a discharge of 3426 m^3s^{-1} gauged at Wilkes-Barre and the flood of October 21, 1927 with a discharge of 3992 m^3s^{-1}. Most recently, Tropical Storm Floyd tracked almost directly across the catchment but moved so rapidly that the attendant rains produced a peak discharge of only 554 m^3s^{-1} on September 19, 1999.

Late winter and spring floods caused by *snowmelt* have been successfully partitioned from other extreme events in previous studies of the South Platte River, Arkansas River, and Colorado River Basins in Colorado (Elliott et al., 1982) and the Salt River Valley in central Arizona (Hirschboeck, 1987). Late winter to early spring floods are definitely the most prevalent among the events gauged at Wilkes-Barre between 1891 and 1996 (Fig. 3) and several of the most severe floods in the annual peak series are snowmelt events. The March 20, 1936 discharge of 6002 m^3s^{-1} represented the largest flood on record prior to the flood of June, 1972 triggered by Hurricane Agnes. The very cold winter of 1935-1936 froze the ground, followed by unusually heavy snowfalls in January and February. March rains coincided with unseasonably high temperatures to cause very high runoff.

Another significant snowmelt flood began on January 18, 1996 ahead of a strong cold front. In the day leading up to the floods, strong southerly winds and unseasonably warm temperatures created ideal conditions to melt the snowpack that was covering the entire Mid-Atlantic region. As the cold front moved through the area on January 19 and 20, intense rainfall accelerated the snowmelt. Although a relatively modest discharge of 3624 m^3s^{-1} was gauged at Wilkes-Barre, the discharge at the Tunkhannock Creek gauging station is more than twice the mean annual flood (Table 2). In fact, the annual flood series for the

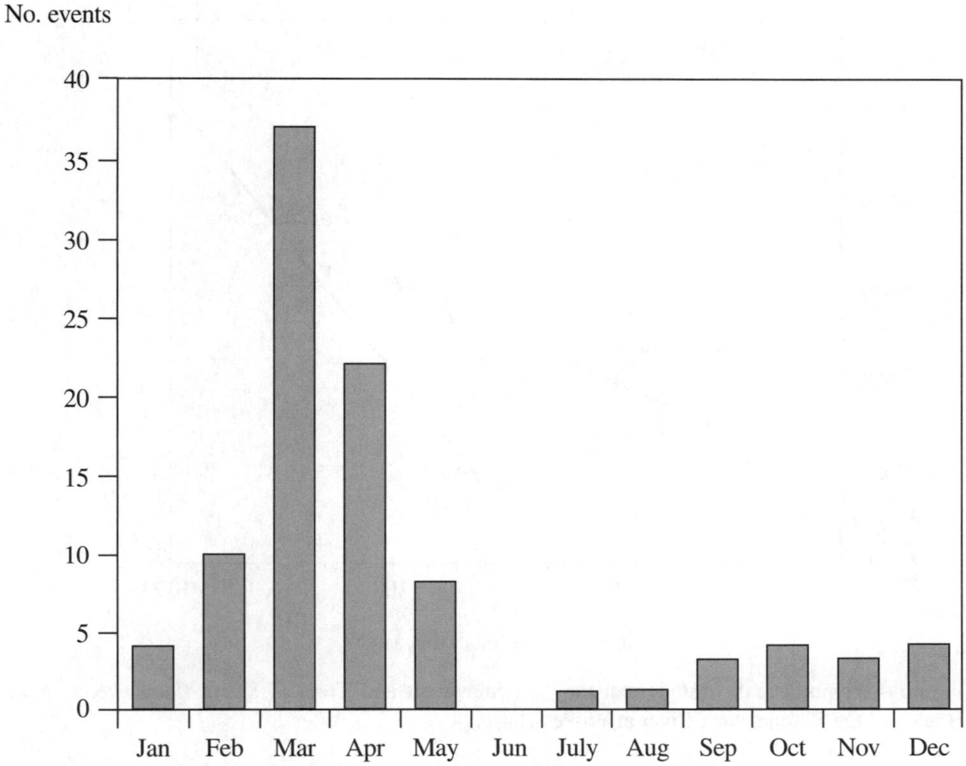

Figure 3. Distribution within the calendar year of peak annual floods recorded at Wilkes-Barre between 1891 and 1996.

Table 2. Snowmelt Flood Discharges recorded at the Tunkhannock Creek and Wilkes-Barre Gauging Stations.

	Tunkhannock Creek	Wilkes-Barre
March 1, 1964	669	5323
April 16, 1983	643	3907
March 16, 1986	750	2774
April 2, 1993	617	5238
January 20, 1996	858	3624
Q_{maf}	401	3570
Q_{bkf}	316	3029

All values are in m^3s^{-1}.

gauging station on this small (992 km^2) catchment records almost exclusively snowmelt events and was therefore used to help partition the series for the catchment as a whole.

Figure 4 is based upon nine extreme events whose genesis has been detailed above. Events in which tropical storms track through the catchment probably recur no more

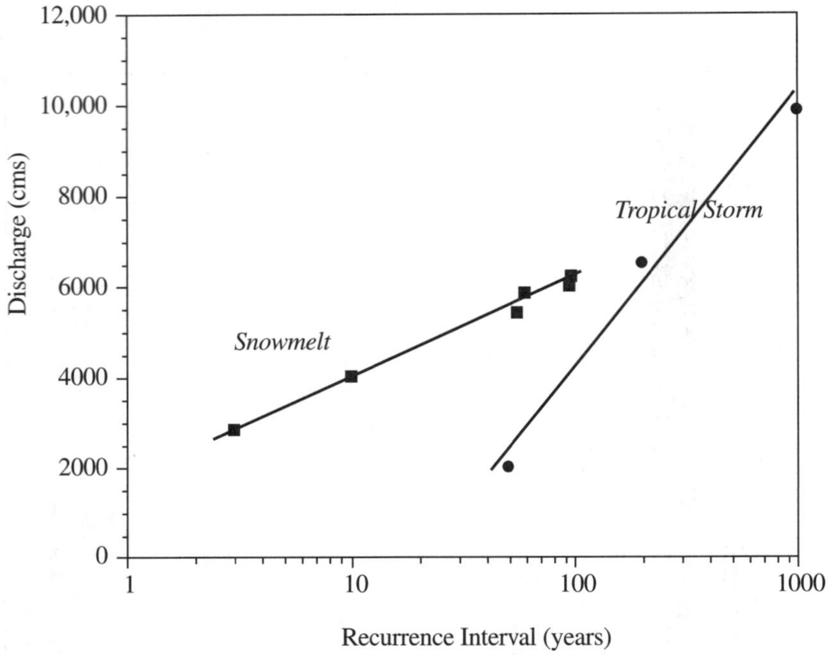

Figure 4. Hypothetical 'nested frequency' of Snowmelt and Tropical Storm flood events, North Branch of the Susquehanna River at Wilkes-Barre.

frequently than once every 100 years. Tropical storms do play a role, however, in the meteorological conditions responsible for other frontal and extratropical cyclonic precipitation events.While the largest snowmelt floods on record do not exceed 8000 m^3s^{-1} these are the most common event, exceeding the bankfull discharge once every two years on average. Snowmelt also plays a crucial role in flooding along high gradient tributary streams such as Tunkhannock Creek, and snowmelt floods are therefore a likely mechanism to deliver sediment into the trunk stream valley as well as to deposit it overbank on natural levee surfaces.

Late winter and spring flood events are linked to the movement of warm, moist air masses, through precipitation inputs to the local hydrological cycle and through rapid melting of the snowpack. This meteorological linkage is corroborated for the winters of 1865, 1936, and 1996 by large negative values for the 'North Atlantic Oscillation (NAO)' index, associated with weakened zonal flow in the upper atmosphere across the North Atlantic basin (Jones et al., 1997; Walker & Bliss, 1932). The winter of 1996 was also particularly severe in much of Europe, with record cold temperatures and heavy snowfalls (Marshall & Kushnir, 1997, p. 9). Snowmelt floods during the years 1830-1865 are discussed below in connection with 19th century canal engineering, and these also generally correlate with events described for the end of the 'Little Ice Age' in Europe (Grove, 1988; Lamb, 1982). Proximity to the Atlantic Ocean itself clearly plays some role in driving these synchronous events (Rossi, 1999) although the linkages with Atlantic Ocean sea surface temperatures (SSTs) are not straightforward and cannot be extrapolated prior to the period of instrumental record (Marshall & Kushnir, 1997).

5 HISTORIC HUMAN IMPACTS AND THE NORTH BRANCH CANAL

A number of direct modifications as well as more indirect impacts of human activity have contributed to the present channel alignments and particularly to floodplain sedimentation within the North Branch of the Susquehanna River valley. Flood control projects were first contemplated following floods in 1901 and 1902, and by 1917 the US Army Corps of Engineers had resolved that 'the only practicable way to protect the communities (along the river was) by means of … levees and embankments (Stranahan, 1993, p. 125)'. Money for Corps projects was not appropriated, however, until after the flood of March, 1936, and the work was actually delayed until after another severe flood struck in April, 1940. A 20 km long system of levees was eventually built on both banks of the river through the Wyoming Valley. The levees were designed to contain flows 11 m deep, approximately a meter deeper than the 1936 event. They were nonetheless overtopped by the flood of June, 1972. Toe stabilization was added immediately thereafter, and the levees have recently been raised 1-1.5 m.

The most direct and extensive modification, however, extending through nearly the entire catchment, is a canal constructed during the 19th century to transport Wyoming Valley anthracite coal (Inners, 1988; Petrillo, 1986; Shank, 1982, p. 40). The construction was driven by competition between Philadelphia, New York City, and other markets for the anthracite (Petrillo, 1986, p. 4). Plans for canal construction were not finalized until 1826, nearly a decade after work began on the Erie Canal in 1817. The first incarnation ran completely along the northwest bank of the river between the juncture with the West Branch at Northumberland and Nanticoke (Fig. 5). It featured seven wooden lift locks, which were simply earthern chambers lined with planks. Its dimensions were 8.5 m wide at the bottom, 12 m wide at the top, and 1.2 m deep. Five aqueducts were constructed to carry the canal across major streams as well as five stone culverts, 24 wooden culverts, four waste weirs, eight lock houses, and numerous roads and bridges for a finished cost of over a million dollars. Although this was more than 100 percent over budget, two boats did successfully carry coal a distance of 36 km from Nanticoke to Berwick in October of 1830 (Petrillo, 1986, p. 53).

Upstream of Nanticoke the canal boats were poled through slackwater impounded behind a crib dam. The river fell 2 m over the breast of the dam, which stood 3.7 m above the river bed. The slackwater extended upstream to the mouth of Solomon Creek on the southeast bank, and terrace remnants consisting of sediments which settled out of the slackwater are present at approximately the same distance upstream on the northwest bank. The section illustrated by Figure 6 is less than 200 m downstream of the mouth of Wadham Creek. The column is characterized by thin laminae with coal silt defining increments which resemble annual varves. Close to two meters of sediment accumulated in less than forty years prior to 1870, implying a sedimentation rate of > 5 cm yr^{-1}.

An extension of the canal through the Wyoming Valley on the southeast bank behind Wilkes-Barre opened in 1834, fed by a dam on the Lackawanna River. Canal construction was hampered by several floods and other geomorphic events chronicled in Table 3. Most of these occurred in winter or spring months and appear to be related to cold regional climate at the end of the Little Ice Age (Grove, 1988; Lamb, 1982). Flood records for the Connecticut River at Hartford and the Ohio River at Pittsburg (Hoyt & Langbein, 1955) also show peak floods in the interval 1830-1865. The North Branch Canal was eventually extended upstream to join with a canal along the Chemung River in New York State.

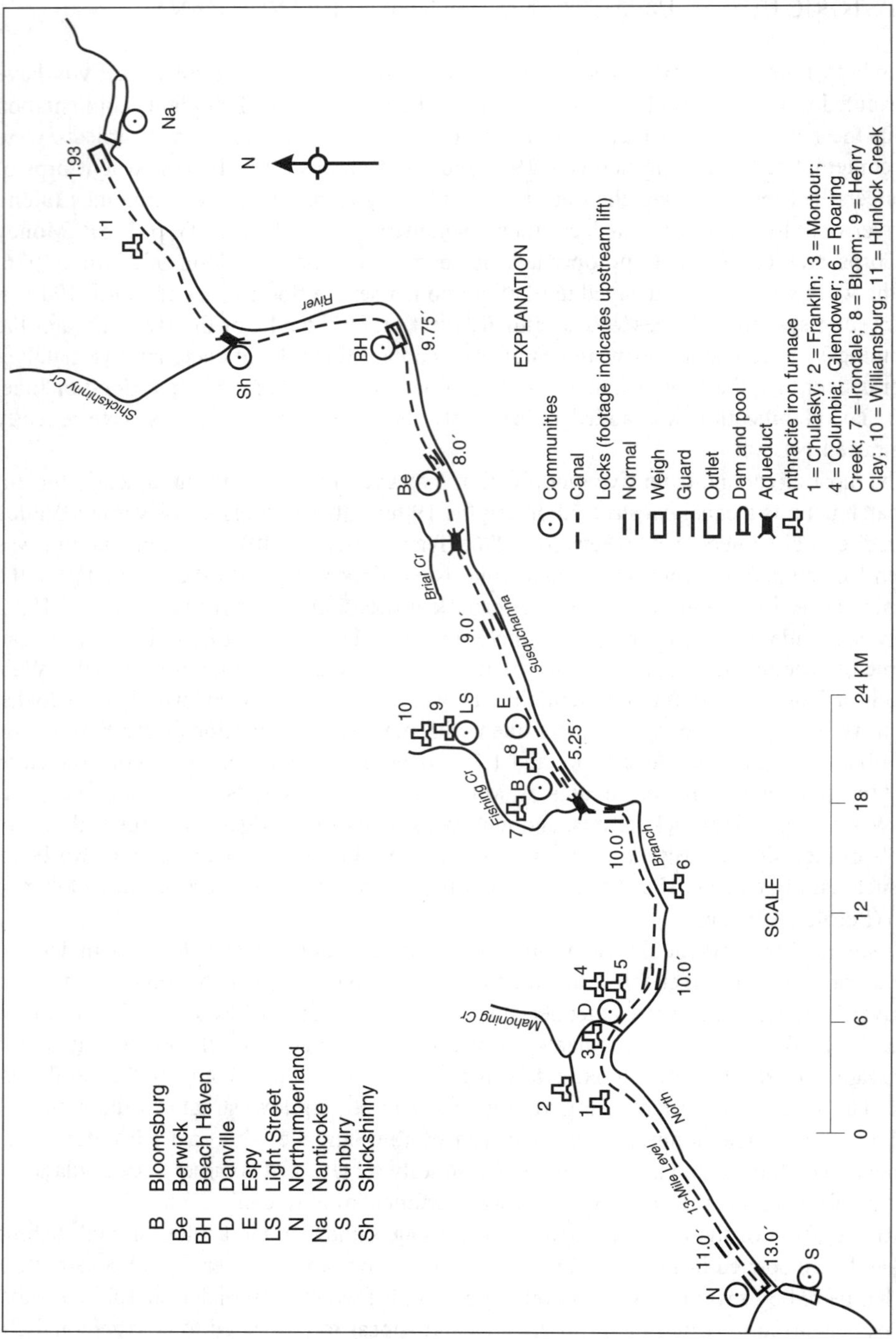

Figure 5. The North Branch Canal from Northumberland to Naticoke, Pennsylvania.

Figure 6. Terrace remnant constructed by slackwater in the pool behind a crib dam at Nanticoke, Pennsylvania for the North Branch Canal.

Table 3. Chronology of floods and other events impacting the North Branch Canal, 1830-1865 (as chronicled by Petrillo, 1986).

Year	Events
1832	January ice jam at Shickshinny, dam at Nanticoke undermined. Spring snowmelt flood shattered the western side of the dam and carried away the guard lock.
1833	Spring snowmelt flood caused another breach of the Nanticoke crib dam and carried away the lockhouse.
1839	April 13 snowmelt flood carried away the dam across the Lackawanna and the Mill Creek aqueduct in the Wyoming Valley.
1854	Heavy rains in May caused a slide along the canal in Wilkes-Barre. June 27 floods caused a landslide at Shickshinny.
1857	November 11 flood on the Chemung River closed that canal.
1864-65	River choked with ice from December 20 to March.
1865	March 17-18 was the worst flood in Wilkes-Barre in previous history. Covered the canal from Nanticoke Dam to Northumberland.

Remnants of a crib dam at Mehoopany are preserved in the middle of this stretch (Fig. 7), which was not completed until the 1850s. The crib dam was intended to fill the canal downstream to the Lackawanna River. Maintaining sufficient water proved a challenge, however, because the water percolated through the coarse gravelly sediment and found its natural level in the river channel. Beds of glacial lake clay eventually had to be mined and transported to provide a lining or fill at least a meter thick (Petrillo, 1986, p. 109).

The episodic 'environmental' events chronicled in Table 3 forced the canal engineers to make many costly repairs which ran them over budget but also probably led to a general improvement in canal design. In addition to the aforementioned relationship with regional climate, timber harvest throughout the catchment also probably affected the magnitude of floods and their sediment load. Logging in the dense pine and hardwood forests of northern Pennsylvania first became a commercial enterprise in the 1830s and 1840s, by which point most of the virgin growth had been cut from the forests in New York and New England (Stranahan, 1993, p. 83).

Although the primary focus of the lumbermen was on the tributary valleys of the West Branch in northcentral Pennsylvania, spring snowmelt was also used on most of the tributaries of the North Branch of the Susquehanna River to transport logs of pine, hemlock, oak, and hickory down to river landings. Crude, shallow, flat-bottomed 'arks' were then typically built with these logs in contrast to the much larger rafts of lumber from the West Branch headwaters (Stranahan, 1993, p. 80). The arks were navigated all the way downriver to the Chesapeake Bay, and this traffic was accomodated within the canal system by features such as a 'chute' around the crib dam at Nanticoke (Petrillo, 1986, p. 178-181). In addition to hoop-poles, barrel-staves, spokes, hubs, and other lumber products, transportation of salt and gypsum are mentioned.

While sediment yield within the catchment was clearly much higher in the 19th century than prior to Euroamerican settlement and industry, it may never be possible to quantify the role played by climate, logging, agricultural runoff, or runoff from the massive culm piles which accumulated near the Wyoming Valley anthracite mines. The silt and sand from the culm piles can be recognized and used as a tracer, however, and accumulations

Figure 7. Remnant of the Horse Race crib dam at Mehoopany, Pennsylvania.

are sufficiently deep in the lower Susquehanna Valley that coal has been excavated using steamboats and barges and sold to utilities (Stranahan, 1993, p. 148-149). Sediment yields in the Wyoming Valley and other basins affected by mining still depend on the amount of coal processed and the methods used to dispose of the fine material (Williams & Reed, 1972).

In the absence of reliable long-term records of sediment yield capable of demonstrating the trends associated with the various environmental changes in the catchment, the contemporary data on tributary valleys which were not affected by mining can be used for 'space-time substitution' of the sort pioneered by Wolman (1967). The sample of 14 tributary valleys from the Ridge and Valley physiographic province (Table 4) shows a moderate trend of decrease in annual sediment yield with increase in forest cover (r = –69, r^2 = .48).

Substituting time for space (Walling, 1996, p. 50-51), we can hypothesize that sediment yields tripled from approximately 30-50 t km^{-2} yr^{-1} to more than 100 t km^{-2} yr^{-1} as a result of forest clearance and other impacts from Euroamericans. These figures are quite conservative when compared to those compiled worldwide by Morgan (1986) and Walling (1995, 1996). Of approximately 1.5 × 10^6 t of sediment entering the catchment each year, less than a tenth of this is currently allocated to storage within floodplains in the Wyoming Valley and other synclinal basins (Williams & Reed, 1972). Artificial structures for flood control built during the present century probably reduce the amount of sediment storage, whereas certain of the 19th century structures of the North Branch canal system apparently resulted in sedimentation at levels above the present bankfull stage.

Table 4. Relationship between Percent Forest Cover and Annual Sediment Yield for 14 Tributary Valleys, Susquehanna River Drainage Basin (Data from Williams & Reed, 1972).

Station	Drainage Basin Area (km^2)	Percent Forest Cover	Sediment Yield ($t\ yr^{-1}$)	Sediment Yield ($t\ km^{-2}\ yr^{-1}$)
Fishing Creek	710	54	24,036	39.0
Spring Creek	226	33	8707	38.5
Bald Eagle Creek	878	44	24,490	27.9
Marsh Creek	114	59	3628	31.8
Penns Creek	780	65	20,862	26.7
E. Mahantango Creek	420	33	40,816	97.2
Juniata River	2113	71	46,259	21.9
Dunning Creek	445	65	9070	20.4
Raystown Branch	1958	69	61,678	31.5
Kichacoquillas Creek	425	59	9070	21.3
Bixler Run	39	52	907	23.3
Shermans Creek	518	57	15,419	29.8
Yellow Breeches Creek	559	31	25,397	45.4
Swatara Creek	873	35	66,213	75.8
Sample Mean		51.9		37.6
Sample s.d.		13.7		21.6

Linkages between sediment yield and vegetation or land use changes are relevant to the interpretation of flood events and floodplain stratigraphy because river transport of fine sediment is highly dependent on supply (Knighton, 1998, p. 120). Recent floods such as the June, 1972 event triggered by Tropical Storm Agnes have generated traceable sediment packages precisely because their high discharges occurred in periods of abundant supply. In the Wyoming Valley, Agnes flood deposits consisting of fine sand and silt derived primarily from historic alluvium 'protected' by the system of levees were mapped at elevations of up to 170 m AMSL (Flippo & Lenfest, 1973). By the time it reached this elevation the channel was nearly 2 km wide, a tenfold increase over the 213 m 'natural' channel width reported in Table 1.

Not only the historic floodplain but all known Holocene terrace remnants and many of the older gravel terraces were inundated for as long as four days, from June 22-25, 1972. Despite the anomalous magnitude of this flood, its deposits were still patchy or discontinuous in distribution (Bailey et al., 1975, p. 76-78). Judging from this example, single large-magnitude events are unlikely to account for valley-wide depositional units in the Holocene stratigraphic record. Such units more likely represent multiple discharges greater than the bankfull level resulting from common causes operative during periods of abundant sediment supply (Brackenridge, 1988; Knox, 1983, 1995; Ritter et al., 1973; Schumm & Brackenridge, 1987). As noted by both Lewin (1978, p. 419-421) and Brackenridge (1988), lateral migration of the channel and associated stream energies within the floodplain should also be a factor in periods of aggradation as well as degradation.

6 PREHISTORIC FLOODPLAINS AND ARCHAEOLOGICAL CONTEXTS

Given the discontinuous sedimentation associated with large magnitude historic floods, it is not surprising that former floodplains of the North Branch of the Susquehanna River are preserved in discontinuous terrace remnants. A nearly continuous historic floodplain (T-0) is present in wider reaches such as the synclinal Wyoming Valley (Hollowell, 1973, p. 17) but even this surface may be missing or confined to one side of the channel in reaches where the bedrock valley narrows. Within the Wyoming Valley, the system of levees discussed above protects much of what had been the active floodplain as well as the former floodplain surfaces of the Holocene alluvial terraces. This formerly active, 'historic' floodplain is typically flanked by a low first terrace surface (T-1a) 1-3 m above mean low water (Thieme & Schuldenrein, 1998).

Archaeological contexts from the early agricultural Clemson Island, Owasco, and Wyoming Valley cultures are found within the Wyoming Valley T-1a deposits on distinct, laterally traceable former floodplain surfaces which are commonly stacked on top of one another in stratigraphic section (Garrahan, 1990; Smith, 1973; Thieme, 1999; Thieme & Schuldenrein, 1998). Figure 8 illustrates terrace and floodplain stratigraphy at Forty Fort, the location of a small private airfield surrounded by late prehistoric sites referred to by their excavators as Airport I (36LU44) and Airport II (36LU77). Four house patterns encircled by an oval stockade were identified at the Airport II site (Garrahan, 1990), and small fragments of charred wood from a deep cultural pit, Feature 41, were radiocarbon dated to approximately AD 1029 in calendar years (see Table 5).

Riverward of the Airport II site, three buried soils were observed within the upper meter of T-1a deposits exposed in a borrow pit (see Fig. 9).Clear smooth contacts between Ab and overlying C horizons imply episodic deposition or *stratic* rather than *cumulic* soil profile development as defined by Gerrard (1992, p. 114-116) or Birkeland (1984, p. 184-189). Cultural refuse was concentrated in the darker horizons and may have added thickness and lowered the Munsell color value. A calibrated date of AD 1008 on loose fine charcoal from the 4Ab (see Table 5) precedes the date for Feature 41 at the Airport II site by less than the lifespan of a single human generation. Two further radiocarbon determinations, from an ossuary associated with the Airport II site (Herbstritt et al., 1997), fall in between these bracketing ages. Isotopic analyses of the ossuary skeletal samples provide further indications of cultigens in the diet of these individuals, as previously reported for a contemporaneous, culturally related skeletal population by Vogel & van der Merwe (1977).

Stratic profiles with 'stacked A-horizons' have previously been described by Vento & Rollins (1989) for the Susquehanna River valley and by Stewart (1990) and Wall & Stewart (1996, p. 202-220) for the Delaware River valley. While these authors ascribe their genesis to late Holocene climate fluctuations, the sediment yield data summarized above tend to favor an alternative explanation of human impact through cyclical deforestation associated with maize agriculture (Thieme, 1999). A similar pattern of landscape evolution associated with an early agricultural village has been proposed for the vicinity of the Grand Banks site in Canada (Crawford et al., 1998; Walker et al., 1997).

Slightly higher Holocene alluvial terrace remnants at elevations 3-5 m above mean low water were referred to as the T-1b terrace by Thieme & Schuldenrein (1998). One such surface is shown on the south bank opposite Forty Fort in Figure 8. Although relatively rare within the Wyoming Valley and other wider reaches of the bedrock valley, these terrace remnants occur throughout the Susquehanna River valley (Dineen, 1993; Engel et al.,

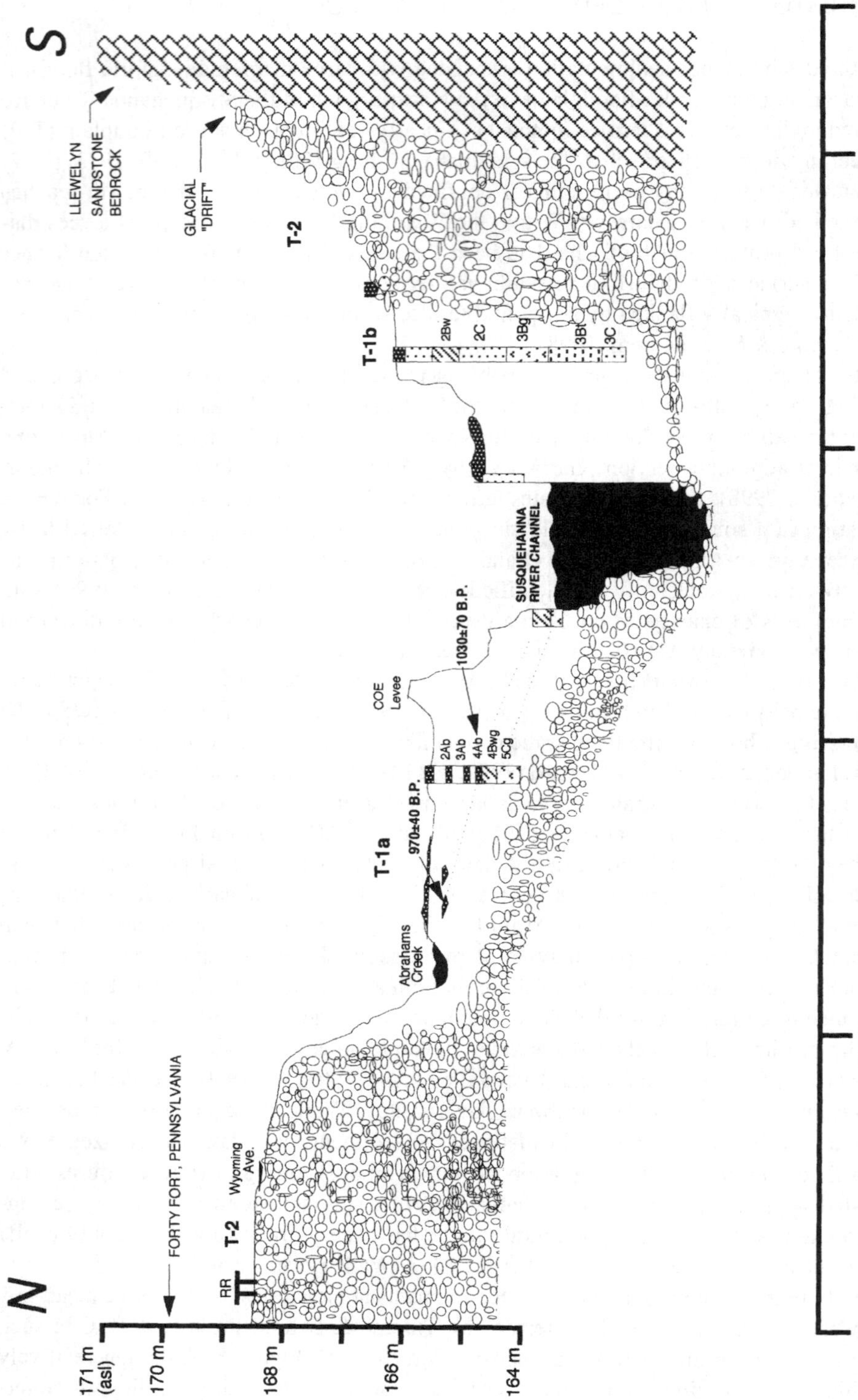

Figure 8. Cross-section of the North Branch of the Susquehanna River Valley in the vicinity of Forty Fort, Pennsylvania.

Table 5. Representative Radiocarbon Dates from Prehistoric Floodplains and Archaeological Contexts.

Site Filing Number	River km (upstream from mouth)	Depth bls (cm)	Context and material dated	Laboratory Number	^{14}C years BP	Calibrated Calendar Year Intercept(s)*
36LU77	306	50-100	Charred wood fragments from Feature 41, a deep pit	PITT-392	970 ± 40	AD 1029
36LU77	306	15	Charred wood fragments from burial cluster VII	Beta-27646	990 ± 90	AD 1023
36LU77	306	27	Charred wood fragments from burial clusters XI and XII	Beta-27648	1010 ± 120	AD 1020
36LU44	306	60-70	Bulk sediment from 4Ab soil horizon in Borrow Pit Section 1	Beta-82386	1030 ± 70	AD 1008
36LU60	299	50	Charred wood from shallow pit	Beta-82385	1170 ± 70	AD 887
36LU128	294	< 50	Charred wood from midden	Beta-105632	1170 ± 50	AD 887
NA	309	500-520	Bulk sediment from T-1b River Cutbank	Beta-82384	5310 ± 70	4246-4000 BC
NA	309	300	Twigs and leaves preserved in slackwater clay	UGa-7816	9520 ± 90	9080 BC
36LU105	265	150	Bulk sediment overlying bedload gravels	Beta-24447	5990 ± 110	4999-4724 BC
36LU90	265	200	Bulk sediment overlying bedload gravels	Beta-43796	8150 ± 140	7133 BC
36CO17	246	275	Charred wood from bedload gravels	Beta-37856	9060 ± 100	8269 BC
36NB117	200	260	Charred wood from cultural pit feature	A-10053	9165 ± 210	8402 BC

*All radiocarbon dates calibrated using Stuiver et al., 1998.

1996; Pazzaglia & Gardner, 1993; Scully & Arnold, 1981; Vento & Rollins, 1989) and represent much of what was mapped as the 'Mankato' terrace by Peltier (1949). Cumulic soil profiles with multiple buried cambic (Bw) horizons are typical (Engel et al., 1996, p. 287-288; Vento et al., 1999, p. 776-777) although more strongly developed Bt or Btx horizons have been described by both Cremeens et al. (1998) and Thieme & Schuldenrein (1998).

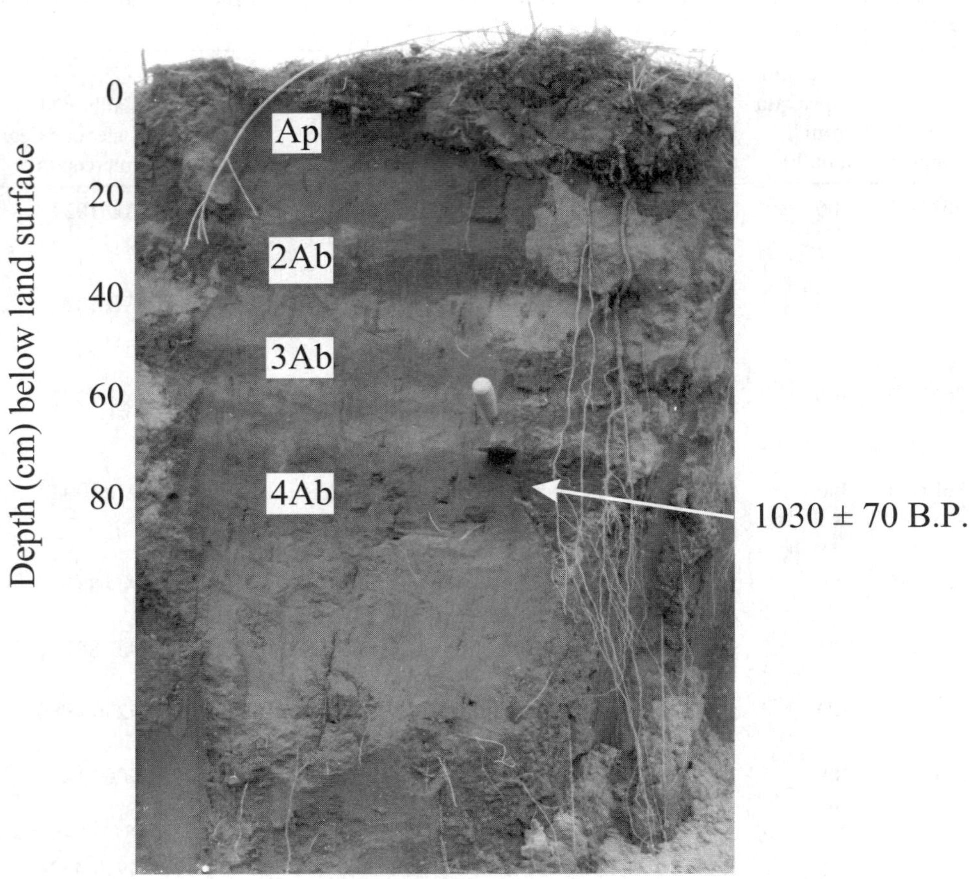

Figure 9. Section 1 in the Borrow Pit at Forty Fort.

Virtually all deep stratified archaeological sites in the North Branch of the Susquehanna River T-1b deposits excavated to date register an interval of soil formation and apparent stability of channel alignments and floodplain landforms lasting from approximately 3000 until 50 cal. BC (4.2-2.1 ka BP). This is evident from the extensive array of radiocarbon dates both on bulk sediment and on charcoal from prehistoric pit features. Six of the twelve samples from the Mifflinville Bridge site (36CO17) date to this interval (Wall et al., 1990) as do two of the three samples from the Susquehanna Steam and Electric Station sites (Hayes et al., 1981). Fourteen of the 24 radiocarbon dates for the Gould Island site (36LU105) and twelve of the thirteen for the Jacobs site (36LU90) are in the interval 3000-50 cal. BC (Weed & Wenstrom, 1992) as well as five of the six radiocarbon dates for site 36LU132 (Miller, 1995, 1998). Even sites with significant late occupations by early agricultural villagers may register this soil-forming interval and associated hunter-gatherer floodplain settlement. Seven of 34 samples submitted from the Harding Flats site (36WO55) at the mouth of Tunkhannock Creek, for example, dated between 3000 and 50 cal. BC (East, 1998).

Archaeological contexts older than 3000 cal. BC are much less common in alluvial strata throughout eastern North America, and prehistoric hunter-gatherers were probably less numerous and/or distributed differently across the landscape (Funk, 1993b, p. 173-188 Custer, 1996, p. 133-162; Stewart, 1991). Detailed investigation of river basin sediments also suggests, however, that many apparent gaps in the archaeological record coincide with intervals of rapid vertical accretion and/or extensive lateral reworking in major river valleys (Artz, 1995; Bettis & Hajic, 1995; Butzer, 1978; Schuldenrein, 1994). All five of the sites in the North Branch of the Susquehanna River valley from which basal Holocene radiocarbon dates have been obtained have sedimentation rates greater than 1 mm yr^{-1} during the interval immediately preceding 3000 cal. BC. Furthermore, of six basal dates compiled in Table 5, two of the samples resting directly on bedload gravels are younger than 5000 cal. BC. Scouring of the surfaces which people inhabited by high energy floods and rapid burial of the scatters associated with their short-term occupations help to account for the record that we do have of the Early Archaic and Middle Archaic cultural periods.

In many of the T-1b sections, beds of particularly coarse sand are found immediately beneath the soils which formed after 3000 cal. BC. The interval 7000-3000 cal. BC (8-4 ka BP) is considered to have had the warmest climate since the end of the Pleistocene, with mean July temperatures up to 2°C warmer than at present in the northern hemisphere (COHMAP members, 1988, p. 1048; Webb et al., 1993, p. 458-465). Pollen of thermophilous tree species is more abundant in the 'oak zone' (C) of Deevey (1939), as is also true for the time-equivalent 'Atlantic' period in England (Godwin, 1975, p. 455-472) and continental Europe (Sernander, 1908). There thus appear to be circum-Atlantic controls on the timing if not necessarily the direction or magnitude of Holocene climate change.

One indirect consequence of the post-Pleistocene warming trend in eastern North America appears to have been the spread of an insect pest (Bhiry & Filion, 1996; Davis, 1981; Filion & Quinty, 1993) which attacked the eastern hemlock (*Tsuga canadensis*). Locally, Barnosky et al. (1988, p. 178) dated the relative decline in *Tsuga* pollen to approximately 4550 cal. BC (5700 BP). Loss of the hemlock canopy from valley side slopes probably increased the sediment yield considerably, and some effects on alluvial deposition were proposed previously for the upper reaches of the North Branch of the Susquehanna River valley in New York state (Scully & Arnold, 1981, p. 341). Analogy with historic floods of equivalent magnitude on tributaries such as Tunkhannock Creek suggests that the transported hillslope sediments were deposited in the valley flats by late winter to spring floods with a heavy snowmelt component.

Sedimentological analyses hold much promise as researchers continue to examine the dynamics of floodplain development and human occupation of floodplain surfaces following deglaciation. The coarse sand beds of the early to middle Holocene are particularly significant given that suspended sediment has a fairly consistent composition within the entire Susquehanna River basin of 10 percent sand, 50 percent silt, and 40 percent clay (Williams & Reed, 1972). Significantly greater stream power during peak floods would be inferred assuming a stable channel configuration (Ritter et al., 1995, p. 22). In many reaches, however, the channel has always had a tendency to braid and deposit longitudinal bars which constrict subsequent flows.

Extant islands within the North Branch of the Susquehanna River valley are clearly only a sample of the many which were constructed as longitudinal bars and, in many cases, attached as T-1b landforms to the valley margin. I recently obtained a radiocarbon date of approximately 9520 ± 90 BP (9080 cal. BC) on twigs and stems from 300 cm bls in an elongate depression running between a T-1b terrace and the riser of an outwash terrace 4 km upstream of Forty Fort in the Wyoming Valley. Overlying clayey settling deposits register a backswamp setting with some episodic channeling events or crevasse splays represented by sandy interbeds. Pollen analyses are planned to document local as well as regional vegetation changes associated with the changing floodplain.

7 CONCLUSIONS

Correlation of sediment packages and soil-forming intervals between the Holocene alluvial terrace remnants in the North Branch of the Susquehanna River valley remains a work in progress. Much has already been learned, however, through the investigation of river responses and floodplain dynamics during the more than ten centuries within which humans appear to have been directly or indirectly impacting the river. Even the most direct actions to control or exploit the river's energy have had effects which are both more complex and more rapid than might be suspected based on their stratigraphic record. While particular floods have produced significant changes in the floodplains, many changes are more directly related to the supply of sediment, controlled in part by the rate of runoff and erosion within the entire catchment.

The 'major discontinuities' at 8500-8000 BP (7550-7050 cal. BC) and 5000-4500 BP (3800-3300 cal. BC) found by Knox (1983, 1995) in alluvial chronologies from throughout North America bracket the primary interval during which the floodplains were established on which people subsequently settled during the Late Archaic and Woodland culture periods. Within this interval there was an abrupt loss of hemlock trees within the forest canopy, and this was suggested above to have resulted in considerable erosion of valley side slopes and a supply of coarser sediment which the river redistributed into islands, longitudinal bars, and ultimately a terrace 3-5 m above present mean low water.

While a coherent story is emerging for most of the major river valleys in eastern North America, we have yet to conduct many detailed studies of alluvial deposits in tributary valleys. Different magnitudes for floods recorded at tributary gauging stations suggest that these deposits may represent different episodes in the Holocene stratigraphy or at the very least represent the same episodes differently. Studies of colluvial and alluvial fan sequences are also needed and these should either corroborate or contradict the effects of the hemlock decline suggested above. In northern Vermont, Bierman et al. (1997) found that soil profiles developed during an interval which includes the hemlock decline while more sediment was eroding off hillsides prior to 6000 BP.

Evidence from the T-1a deposits in the Wyoming Valley suggests that prehistoric agriculture did impact floodplain sedimentation, although these impacts were clearly much less significant than those associated with Euroamerican settlement during the 18th and 19th centuries. Maximum settlement yields occurred sometime during the middle to late 19th century in conjunction with canal construction and anthracite mining, as evidenced by fine sand and coal silt deposited at > 5 cm yr^{-1} from the slackwater impounded behind the crib dam at Nanticoke.

The North Branch of the Susquehanna River has a unique combination of *direct* human impacts associated with river transportation and *indirect* impacts associated with extractive industries which makes it quite similar to many rivers in England and continental

Europe. Sediment yields are consequently anomalously high during the 'farming period' defined from a sample of smaller river valleys by Wolman (1967). Because much of the catchment has yet to be urbanised, however, 20th century sedimentation has been controlled as much by climate as by land use change. The frequency and magnitude of both snowmelt floods and tropical storm events is related in ways that are only now being understood to the interaction of regional air masses with the currents and sea surface temperatures of the adjacent Atlantic Ocean. Undoubtedly these regional controls have a great deal to do with the specific manner in which Knox's (1983) global discontinuities are manifest in the alluvial sedimentary sequences of river basins in the eastern United States.

REFERENCES

Ackers, P. & Charlton, F.B. 1970. Dimensional analysis of alluvial channels with special reference to meander length. *Journal of Hydraulics Research*, 8: 287-316.

Artz, J.A. 1995. Geological contexts of the early and middle Holocene archaeological record in North Dakota and adjoining areas of the northern Plains. In: Bettis, E.A. III (ed.), *Archaeological Geology of the Archaic Period in North America*, Boulder: Geological Society of America, 67-86.

Bailey, T.F., Patterson, J.C., & Paulus, J.L.H., 1975. *Hurricane Agnes rainfall and floods, June-July, 1972*. Washington, D.C.: US Geol. Survey Professional Paper, no. 924: 1-398.

Barnosky, A.D., Barnosky, C.W., Nickman, R.J., Ashworth, A.C., Schwart, D.P., & Lantz, S.W. 1989. Late Quaternary Paleoecology at the Newton site, Bradford Co., northeastern Pennsylvania: *Mammuthus columbi*, palynology, and fossil insects. In: Laub, R.S., Steadman, D.W. & Miller, N. (eds), *Proceedings of the Smith Symposium on Late Pleistocene and Early Holocene Paleoecology and Archaeology of the Eastern Great Lakes Region*. Buffalo: Buffalo Society of Natural Sciences.

Berg, T.M., Edmunds, W.E., Geyer, R.E., Glover, H.D., Hoskins, D.M., MacLachlan, D.B., Root, S.I., Sevon, W.D. & Socolow, A.A., 1980. *Geologic Map of Pennsylvania*: Harrisburg: Pennnsylvania Geological Survey.

Bettis, E.A. III, & Hajic, E.R. 1995. Landscape development and the location of evidence of Archaic cultures in the upper Midwest. In: Bettis, E.A. III (ed.), *Archaeological Geology of the Archaic Period in North America*, Boulder: Geological Society of America, 87-113.

Bevan, K., 1993. Riverine flooding in a warmer Britain. *Geographical Journal*, 159: 157-161.

Bhiry, N. & Filion, L. 1996. Mid-Holocene hemlock decline in eastern North America linked with phytophagous insect activity. *Quaternary Research*, 45: 312-330.

Bierman, P., Lini, A., Zehfuss, P., Church, A., Davis, P.T., Southon, J. & Baldwin, L. 1997. Postglacial ponds and alluvial fans: recorders of Holocene landscape history. *GSA Today*, v. 7, no. 10: 1-8.

Birkeland, P.W. 1984. *Soils and Geomorphology*. New York: Oxford University Press.

Brackenridge, G.R. 1988. River flood regime and floodplain stratigraphy. In: Baker, V.R., Kochel, R.C. & Patton, P.C. (eds), *Flood Geomorphology*, New York: Wiley-Interscience, 139-156.

Brookes, A. 1985. River channelisation: traditional engineering methods, physcial consequences, and alternative practices. *Progress in Physical Geography*, 9: 44-73.

Brookes, A. 1988. Channelised rivers: perspectives for environmental management. Chichester.

Brown, A.G. 1991. Hydrogeomorphological changes in the Severn basin during the last 15,000 years: orders of change in a maritime climate. In: Starkel, L., Gregory, K.J., & Thornes, J.B. (eds), *Temperate paleohydrology*. Chichester, John Wiley, 57-72.

Brown, A.G. 1996. Human dimensions of palaeohydrological change. In: Branson, J., Brown, A.G. & Gregory, K.J. (eds), *Global Continental Changes: the Context of Palaeohydrology*, London: Geological Society Special Publication No. 115, 57-72.

Brown, A.G. 1997. *Alluvial Geoarchaeology*. Cambridge: Cambridge University Press.

Butzer, K.W. 1976. *Early Hydraulic Civilization in Egypt: A Study in Cultural Ecology*. Chicago: University of Chicago Press.

Butzer, K.W. 1978. Changing Holocene environments at the Koster site: a geo-archaeological perspective. *American Antiquity*, 43(3): 408-413.

Clay, P. 1992. A Norman mill dam at Hemington Fields, Castle Donnington, Leicestershire. In: Needham, S. & Macklin, M.G. (eds), *Alluvial Archaeology in Britain*, Oxford: Oxbow, 163-168.

COHMAP Members 1988. Climatic changes of the last 18,000 years: observations and model simulations. *Science*, 241: 1043-1052.

Costa, J.E. 1975. Effects of agriculture on erosion and sedimentation in Piedmont province, Maryland. *Bulletin of the Geological Society of America*, 86: 1281-1286.

Cotter, J.F.P., Ridge, J.H., Evenson, E.B. Sevon, W.D., Sirkin, L. & Stuckenrath, R. 1986. The Wisconsinan history of the Great Valley, Pennsylvania and New Jersey, and the age of the 'Terminal Moraine.' In: Cadwell, D.H. (ed.), *The Wisconsinan Stage of the First Geological District, Eastern New York*, New York State Museum, Albany, New York, 22-49.

Crawford, G.W., Smith, D.G., Desloges, J.R. & Davis, A.M. 1998. Floodplains and agricultural origins: a case study in south-central Ontario, Canada. *Journal of Field Archaeology*, 25: 123-137.

Cremeens, D.L., Hart, J.P. & Darmody, R.G. 1998. Complex pedostratigraphy of a terrace fragipan at the Memorial Park site, central Pennsylvania. *Geoarchaeology*, 13(4): 339-359.

Custer, J.F. 1996. *Prehistoric Cultures of Eastern Pennsylvania*. Harrisburg: Pennsylvania Historical and Museum Commission.

Davis, M.B. 1981. Outbreaks of forest pathogens in Quaternary history. *Proceedings of the Fourth International Palynological Conference, Lucknow, India (1976-1977)*, 216-227.

Deevey, E.S., Jr. 1939. Studies on Connecticut lake sediments. 1. A postglacial climatic chronology for southern New England. *American Journal of Science*, 237: 691-723.

Dineen, R.J. 1993. Postglacial terrace formation, lateral river migration, and prehistoric settlement patterns. In: Funk, R. F. (ed.), *Archaeological Investigations in the Upper Susquehanna Valley, New York State*, Buffalo: Persimmon Press, 100-114.

Djuric, D. 1994. *Weather Analysis*. Englewood Cliffs: Prentice-Hall.

Dury, G.H. 1965. *Theoretical implications of underfit streams*. US Geol. Survey Professional Paper, no. 452-C.

Dwyer, T.R., Mullins, H.T. & Good, S.C. 1996. Paleoclimatic implications of Holocene lake-level fluctuations, Owasco Lake, New York. *Geology*, 24(6): 519-522.

East, T.C. 1998. The Harding Flats site (36WO55), a multicomponent site in northeastern Pennsylvania. *Eastern States Archaeological Federation Abstracts*, 3.

Elliott, J.G., Jarrett, R.D. & Ebling, J.L. 1982. *Annual snowmelt and rainfall peak-flow data on selected foothills region streams, South Platte River, Arkansas, River, and Colorado River Basins, Colorado*. US Geol. Survey Open File Report, no. 82-426.

Elsner, J.B. & Kara, A.B. 1999. *Hurricanes of the North Atlantic, Climate and Society*. New York: Oxford University Press.

Engel, S.A., Gardner, T.W. & Ciolkosz, E.J. 1996. Quaternary soil chronosequences on terraces of the Susquehanna River, Pennsylvania. *Geomorphology*, 17: 273-294.

Filion, L. & Quinty, F. 1993. Macrofossil and tree-ring evidence for a long-term forest succession and Mid-Holocene hemlock decline. *Quaternary Research*, 40: 89-97.

Flippo, H.N. & Lenfest, L.W. 1973. *Flood of June 1972 in Wilkes-Barre Area, Pennsylvania*. Washington, D.C.: US Geological Survey.

Funk, R.F. 1993a. Subsistence, settlement, and seasonality. In: Funk, R.F. (ed.), *Archaeological Investigations in the Upper Susquehanna Valley, New York State*, Buffalo: Persimmon Press, 245-312.

Funk, R.F. 1993b. The Upper Susquehanna sequence and chronology. In: Funk, R.F. (ed.), *Archaeological Investigations in the Upper Susquehanna Valley, New York State*, Buffalo: Persimmon Press, 141-213.

Garrahan, F.D. 1990. Airport II site: a Clemson Island/Owasco settlement on the North Branch of the Susquehanna River. *Pennsylvania Archaeologist*, 60(1): 1-31.

Gerrard, J. 1992. *Soil geomorphology: an integration of pedology and geomorphology*. London: Chapman and Hall.

Gilbert, G.K. 1917. *Hydraulic mining debris in the Sierra Nevada*. US Geol. Survey Professional Paper, no. 104.

Godwin, H. 1975. *The History of the British Flora, a factual basis for phytogeography*. Cambridge: Cambridge University Press.

Gordon, N.D., McMahon, T.A. & Finlayson, B.L. 1992. *Stream Hydrology: An Introduction for Ecologists*. New York: John Wiley and Sons.

Grove, J. 1988. *The Little Ice Age*. London: Methuen.

Hayes, D.R., Roper, D.C., Schuldenrein, J. & Stinson, W.R. 1981. *Archaeological Investigations at the Susquehanna SES: the Susquehanna SES floodplain*. Jackson, Michigan: Commonwealth Associates.

Hebda, R.J., Siemens, A.H. & Robertson, A. 1991. Stratigraphy, depositional environment, and cultural significance of Holocene sediments in patterned wetlands of central Veracruz, Mexico. *Geoarchaeology*, 6(1): 16-84.

Herbstritt, J.T., Garrahan, F.D., Adovasio, J.M., Andrews, R.C., Dirkmaat, D.C., Hyland, D.C., McVay, W.F. & Pedlar, D.R. 1997. The bone pit at Fort Fort: A 10th century Owasco ossuary. Paper presented at the 68th Annual Meeting of the Society for Pennsylvania Archaeology, Wilkes-Barre.

Herbstritt, J.T., 1988. A reference for Pennsylvania radiocarbon dates. *Pennsylvania Archaeologist*, 58(2): 427-469.

Hirschboeck, K.K. 1987. Hydroclimatically-defined mixed distributions in partial duration flood series. In: Singh, V.P. (ed.), *Hydrologic frequency modeling*, Boston: D. Reidel, 192-205.

Hirschboeck, K.K. 1988. Flood hydroclimatology. In: Baker, V.R., Kochel, R.C. & Patton, P.C. (eds), *Flood geomorphology*, New York: Wiley-Interscience, 27-49.

Hollis, G.E. 1975. The effect of urbanisation on floods of different recurrence intervals. *Water Resources Research*, 11: 431-435.

Hollowell, J.R. 1973. *Hydrology of the Pleistocene sediments in the Wyoming Valley, Luzerne County, Pennsylvania*. Harrisburg: Pennsylvania Geological Survey.

Howard, J.B. 1993. A paleohydraulic approach to examining agricultural intensification in Hohokam irrigation systems. In: Scarborough, V.L. & Isaac, B.L. (eds), *Economic Aspects of Water Management in the Prehistoric New World*, Greenwich: JAI Press, 263-264.

Hoyt, W.G. & Langbein, W.B. 1955. *Floods*. Princeton, New Jersey: Princeton University Press.

Inners, J.D. 1988. Susquehanna Steam Electric (Nuclear) Station of PP&L and North Branch Canal. In: Inners, J. (ed.), *Bedrock and glacial geology of the North Branch Susquehanna lowland and the Eastern Middle Anthracite field, northeastern Pennsylvania*. Harrisburg: Field Conference of Pennsylvania Geologists.

James, L.A. 1991. Incision and morphologic evolution of an alluvial channel recovering from hydraulic mining sediment. *Bulletin of the Geological Society of America*, 103: 723-736.

Jones, P.D., Jonsson, T. & Wheeler, D. 1997. Extension to the North Atlantic Oscillation using early instrumental pressure observations from Gibraltar and South-West Iceland. *International Journal of Climatology*, 17: 1433-1450.

Kellerhals, R., Church, M. & Davies, L.P. 1979. Morphological effects of interbasin river diversions. *Canadian Journal of Civil Engineering*, 6: 18-31.

Knighton, D. 1998. *Fluvial Forms and Processes*. London: Arnold.

Knox, J.C. 1977. Human impact on Wisconsin stream channels. *Annals of the Association of American Geographers*, 67: 323-342.

Knox, J.C. 1983. Responses of river systems to Holocene climates. In: Wright, H.E., Jr. (ed.), *Late Quaternary environments of the United States, Volume 2: The Holocene*, Minneapolis: University of Minnesota Press, 26-41.

Knox, J.C. 1984. Fluvial responses to small scale climate changes. In: Costa, J.E. & Fleisher, P.J. (eds), *Developments and Applications of Geomorphology*. Berlin: Springer-Verlag.

Knox, J.C. 1985. Responses of floods to Holocene climatic change in the Upper Mississippi Valley. *Quaternary Research*, 23: 287-300.

Knox, J.C. 1995. Fluvial systems since 20,000 years B.P. In: Gregory, K.J., Starkel, L. & Baker, V.R. (eds), *Global Continental Palaeohydrology*. Chichester, John Wiley, 87-108.

Lamb, H.H. 1982. *Climate, History, and the Modern World*. London: Methuen.

Leigh, D.S. 1994. Mercury contamination and floodplain sedimentation from former gold mines in north Georgia. *Water Resources Bulletin*, 30(4): 739-748.

Leigh, D.S. 1997. Mercury-tainted overbank sediment from past gold mining in north Georgia, USA. *Environmental Geology*, 30(3/4): 244-245.

Leopold, L.B. 1968. *Hydrology for urban planning*. US Geol. Survey Circular, no. 259.

Leopold, L.B. 1994. *A View of the River*. Cambridge, Massachusetts: Harvard University Press.

Leopold, L.B. & Wolman, M.G. 1957. River channel patterns – braided, meandering, and straight. Washington, D.C.: US Geological Survey Professional Paper, 282B: 39-85.

Leopold, L.B., Wolman, M.G. & Miller, J.P. 1964. *Fluvial Processes in Geomorphology*. San Francisco: W.H. Freeman Co.

Lewin, J. 1978. Floodplain geomorphology. *Progress in Physical Geography*, 7: 408-437.

Lewin, J. & Macklin, M.G. 1987. Metal mining and floodplain sedimentation in Britain. In: Gardiner, V. (ed.), *International geomorphology 1986, part I.*, Chichester: John Wiley, 1009-1027.

Lindner, C.R. 1987. Geoarchaeology of Culturally Induced Flood Impacts: Schoharie Valley, Eastern New York. Ph.D. dissertation, Department of Anthropology, State University of New York at Albany.

Lottick, S.T. 1992. *Bridging Change: A Wyoming Valley Sketchbook*. Wilkes-Barre: Wyoming Valley Historical and Geological Society.

Macklin, M.G. & Lewin, J. 1989. Sediment transfer and transformation of an alluvial valley floor: the River South Tyne, Northumbria, UK. *Earth Surface Processes and Landforms*, 14: 233-246.

Madej, M.A. & Ozaki, V. 1996. Channel response to sediment wave propagation and movement, Redwood Creek, California, USA. *Earth Surface Processes and Landforms*, 21: 911-927.

Marshall, J. & Kushnir, Y. 1997. A 'white paper' on Atlantic climate variability. Published on the World-Wide Web at http://geoid.mit.edu/accp/avehtml.html.

Meade, R.H. 1976. Sediment problems in the Savannah River basin. In: Dillman, B.I. & Stepp, J.M. (eds), *The future of the Savannah River*, Clemson: Water Resources Research Institute, 105-129.

Meade, R.H. 1982. Sources, sinks, and storage of river sediment in the Atlantic drainage of the United States. *Journal of Geology*, 90: 235-252.

Meade, R.H. & Parker, R.S. 1985. Sediment in rivers of the United States. In: *National Water Summary 1984*, US Geol. Survey Water-Supply Paper, 2275: 49-60.

Miller, P.E. 1995. Archaeological data recovery, the Skvarek site (36LU132), proposed Shickshinny/Mocanaqua bridge replacement (S.R. 0239), Salem and Conyngham Townships, Luzerne County, Pennsylvania. Wilkes-Barre: Pennsylvania Department of Transportation.

Miller, P.E. 1998. Lithic projectile point technology and raw material use in the Susquehanna River valley. In: *The Archaic Period in Pennsylvania, hunter-gatherers of the early and middle Holocene*, Harrisburg: Pennsylvania Historical and Museum Commission, 91-119.

Morgan, R.P.C. 1986. *Soil Erosion and Conservation*. Harlow: Longman.

Muller, E.H. 1965. Quaternary geology of New York. In: Wright, H.E. & Frey, D.G. (eds), *The Quaternary of the United States*, Princeton: Princeton University Press, 99-112.

Muller, E.H. & Calkin, P.E. 1993. Timing of Pleistocene glacial events in NY State. *Canadian Journal of Earth Sciences*, 30: 1829-1849.

Namias, J. 1973. Hurricane Agnes – an event shaped by large-scale air-sea systems generated during antecedent months. *Quarterly Journal of the Royal Meteorological Society*, 99: 506-51.

NOAA (National Oceanic and Atmospheric Administration), 1999, Information obtained through the World-Wide Web at http://wesley.wwb.noaa.gov, http://sgi62.wwb.noaa.gov, and http://ngdc.noaa.gov.

Pazzaglia, F.J. & Gardner, T.W. 1993. Fluvial terraces of the lower Susquehanna River. *Geomorphology*, 8: 83-113.

Peltier, L.C. 1949. *Pleistocene terraces of the Susquehanna River*, Pennsylvania: Harrisburg, Pennsylvania Geological Survey.

Petrillo, F.C. 1986. *Anthracite and Slackwater: The North Branch Canal 1828-1901*. Easton, Pennsylvania: Center for Canal History and Technology.

Petts, G.E. 1979. Complex response of river channel morphology to reservoir construction. *Progress in Physical Geography*, 3: 329-362.

Phillips, J.D. 1993. Pre- and post-colonial sediment sources and storage in the Lower Neuse basin, North Carolina. *Physical Geography*, 14: 272-284.

Ritter, D.F., Kinsey, W.F. III. & Kauffmann, M.E. 1973. Overbank sedimentation in the Delaware river valley during the last 6000 years. *Science*, 179: 174-175.

Ritter, D.F., Kochel, R.C. & Miller, J.R. 1995. *Process Geomorphology*. Dubuque: Wm. C. Brown.

Rohn, A.H. 1963. Prehistoric soil and water conservation on Chapin Mesa, southwestern Colorado. *American Antiquity*, 28: 441-455.

Rossi, T. 1999. Climate. In: Schultz, C.H. (ed.), *The Geology of Pennsylvania*, Harrisburg: Pennsylvania Geological Survey, 658-665.

Salisbury, C.R. 1988. A Saxon and Norman fish weir at Colwick, Nottinghamshire. In: Aston, M. (ed.), *Medieval Fish, Fisheries, and Fishponds in England*, Oxford: British Archaeological Reports, 329-351.

Salisbury, C.R. 1991. Primitive British fishweirs. In: Good, G.L., Jones, R.H. & Ponsford, M.W. (eds), *Waterfront Archaeology: Proceedings of the Third International Conference, 1988*, London: Council for British Archaeology Research Report 74, 76-87.

Salisbury 1992. The archaeological evidence for paleochannels in the Trent valley. In: Needham, S. & Macklin, M.G. (eds), *Alluvial Archaeology in Britain*, Oxford: Oxbow, 155-162.

Schuldenrein, J. 1994. Alluvial site geoarchaeology of the Middle Delaware valley: a fluvial systems paradigm. *Journal of Middle Atlantic Archaeology*, 10: 1-21.

Schumm, S.A. 1960. The shape of alluvial channels in relation to sediment type: US Geol. Survey Professional Paper, 352B: 17-30.

Schumm, S.A. 1977. *The Fluvial System*: Wiley-Interscience, New York, 17-30.

Schumm S.A. & Brackenridge, G.R. 1987. River responses. In: Ruddiman, W.F. & Wright, H.E. Jr. (eds), *North America and adjacent oceans during the last deglaciation*, Boulder, Colorado: Geological Society of America, 221-240.

Scully, R.W. & Arnold, R.W. 1981. Holocene alluvial stratigraphy in the Upper Susquehanna River Basin. *Quaternary Research*, 15(3): 327-344.

Sernander, R. 1908. On the evidence of postglacial changes of climate furnished by the peatmosses of northern Europe. *Geologiska Föreningens Stockholm Förhandlingar*, 30: 465-478.

Sevon, W.D. 1986. A sesquicentennial story, Susquehanna River water gaps: many years of speculation. In: *Celebrating 150 years of the Pennsylvania Geological Survey*. Harrisburg: Pennsylvania Geological Survey.

Shank, W.H. 1982. *Towpaths to Tugboats, A History of American Canal Engineering*. York, Pennsylvania: The American Canal and Transportation Center.

Siemens, A.H. 1998. *A Favored Place: San Juan River Wetlands, Central Veracruz, AD 500 to the Present*. University of Texas Press, Austin.

Smith, I.F. 1973. The Parker site: a manifestation of the Wyoming Valley culture. *Pennsylvania Archaeologist*, 43(3-4): 1-56.

Starkel, L. 1991. The Vistula River valley: a case study for central Europe. In: Starkel, L., Gregory, K.J. & Thornes, J.B. (eds), *Temperate palaeohydrology*, Chichester: John Wiley, 473-495.

Stevens, M.A., Simons, D.B. & Richardson, E.V. 1975. Nonequilibrium river form: *Journal of the Hydraulics Division, American Society of Civil Engineers*, HY5: 557-566.

Stewart, R.M. 1990. *Archaeology, sedimentary sequences, and environmental change in the Delaware River basin*. Harrisburg: Pennsylvania Historical and Museum Commission.

Stewart, R.M. 1991. A Middle Archaic period sampler: introduction. *Journal of Middle Atlantic Archaeology*, 7: 1-5.

Stoddart, D.R. 1978. Geomorphology in China: *Progress in Physical Geography*, 2: 187-236.

Stranahan, S.Q. 1993. *Susquehanna: River of Dreams*. Baltimore: The Johns Hopkins Press.

Thieme, D.M. 1999. Hands washed in a muddy stream: corrections and further thoughts on 'Wyoming Valley landscape evolution…': *Pennsylvania Archaeologist*, 69(2): 45-52.

Thieme, D.M. 2000. *A Stratigraphic and Chronometric Investigation of the Alluvial Deposits of the North Branch of the Susquehanna River*. Ph.D. dissertation, Department of Geology, University of Georgia.

Thieme, D.M. & Schuldenrein, J. 1998. Wyoming Valley landscape evolution and the emergence of the Wyoming Valley culture complex. *Pennsylvania Archaeologist*, 68(2): 1-17.

Thompson, G.H., Jr. 1990. Geomorphology of the Lower Susquehanna River Gorge. In: *Carbonates, Schists, and Geomorphology in the Vicinity of the Lower Reaches of the Susquehanna River*, Harrisburg: Guidebook for the 55th Annual Field Conference of Pennsylvania Geologists, 86-106.

Thornthwaite Associates 1964. *Average Climatic Water Balance Data of the Continents, Part VII, United States*. Centerton, New Jersey: Laboratory of Climatology.

Trimble, S.W. 1974. *Man-Induced Soil Erosion on the Southern Piedmont, 1700-1970*. Ankenny, Iowa: Soil Conservation Society of America.

Turner, B.L.II 1983. The excavations of raised and channelized fields at Pultrouser Swamp. In: Turner, B.L.II. & Harrison, P.D., (eds), *Pultrouser Swamp: Ancient Maya Habitat, Agriculture, and Settlement in Northern Belize*, Austin: University of Texas Press, 30-51.

U.S.G.S. (United States Geological Survey), 1999. Information obtained through the World-Wide Web at http://www.rvares.er.usgs.gov/hcdn-report and http://waterdata.usgs.gov/nwis-w/PA.

Vento, F.J. & Rollins, H.B. 1989. *Development of a Late Pleistocene-Holocene genetic stratigraphic framework as it relates to atmospheric circulation and climatic change in the upper and central Susquehanna River drainage basin*. Harrisburg: Pennsylvania Historical and Museum Commission.

Vento, F.J., Donahue, J. & Adovasio, J.M. 1999. Geoarchaeology. In: Schultz, C.H. (ed.), *The Geology of Pennsylvania*, Harrisburg: Pennsylvania Geological Survey, 770-777.

Vivian, R.G. 1992. Chacoan water use and managerial decision making. In: Doyel, D.E. (ed.), *Anasazi Regional Decision Making*, Albuquerque: Maxwell Museum of Anthropology, 45-57.

Vogel, J.C. & van der Merwe, N.J. 1977. Isotopic evidence for early maize cultivation in New York State. *American Antiquity*, 42: 238-242.

Walker, G.T. & Bliss, E.W. 1932. World Weather V. *Memoirs of the Royal Meteorological Society*, 4: 53-84.

Walker, J., Desloges, J.R., Crawford, G.W. & Smith, D.G. 1997. Floodplain formation processes and archaeological implications at the Grand Banks site, lower Grand River, southern Ontario. *Geoarchaeology*, 12(8): 865-887.

Wall, R.D. & Stewart, R.M. 1996. *Sturgeon Pond site (28ME114) data recovery*. Trenton: New Jersey Department of Transportation.

Wall, R.D., Stewart, R.M. & Koldehoff, B. 1990. Phase II Archaeological Investigations for the Mifflinville Bridge Replacement Project, S. R. 2028, Section 004, Columbia County, Pennsylvania. East Orange, New Jersey: Louis Berger and Associates, Inc. 53-84.

Walling, D.E. 1995. Suspended sediment yields in a changing environment. In: Gurnell, A. & Petts, G. (eds), *Changing River Channels*. Chichester, John Wiley, 149-176.

Walling, D.E. 1996. Erosion and sediment yield in a changing environment. In: Branson, J., Brown, A.G. & Gregory, K.J. (eds), *Global continental changes: the context of palaeohydrology*, London: The Geological Society, 43-56.

Watts, W.A. 1979. Late Quaternary vegetation of central Appalachia and the New Jersey Coastal Plain. *Ecological Monographs*, 49: 427-469.

Webb, T. III, Bartlein, P.J., Harrison, S.P. & Anderson, K.H. 1993. Vegetation, lake levels, and climate in eastern North America for the past 18,000 years. In: Wright, H.E., Kutzbach, J.E., Webb, T. III, Ruddiman, W.T., Street-Perrott, F.A. & Bartlein, P.J. (eds), *Global climates since the last glacial maximum*, Minneapolis: University of Minnesota Press, 415-467.

Weed, C.S. & Wenstrom, W.P. 1992. *Cultural resources investigations of 36LU90 (Jacobs site) and 36LU105 (Gould Island site), Luzerne County, Pennsylvania*. New Orleans: New World Research, Inc.

White, N.M., Knetsch, J. & Jones, C.B. 1999. Archaeology, history, fluvial geomorphology, and the mystery mounds of northwest Florida. *Southeastern Archaeology*, 18(2): 142-156.

Whittaker, R.H. 1956. Vegetation of the Great Smoky Mountains. *Ecological Monographs*, 26: 1-80.

Williams, G.P. & Wolman, M.G. 1984. *Downstream effects of dams on alluvial rivers*. Washington, D.C.: US Geol. Survey Professional Paper, no. 1286.

Williams, K.F. & Reed, L.A. 1972. *Appraisal of Stream Sedimentation in the Susquehanna River Basin*. Washington, D. C.: US Geological Survey Water Supply Paper 1532-F.

Wilshusen, R.H., Churchill, M.J. & Potter, J.M. 1997. Prehistoric reservoirs and water basins in the Mesa Verde region: intensification of water collection strategies during the Great Pueblo Period. *American Antiquity*, 62(4): 664-681.

Winkley, B.R. 1982. Response of the Lower Mississippi to river training and realignment. In: Hey, R.D., Bathurst, J.C. & Thorne, C.R. (eds), *Gravel bed rivers*. Chichester, John Wiley, 659-680.

Wolman, M.G. 1967. A cycle of sedimentation and erosion in urban river channels. *Geografiska Annaler*, 49-A, 385-395.

Wolman, M.G. & Schick, A. 1967. Effects of construction on fluvial sediment, urban and suburban areas of Maryland. *Water Resources Research*, 3(2): 451-464.

Yalin, M.S. 1976. *Mechanics of sediment transport*. Oxford: Pergamon Press.

Zangger, E. 1998. The Port of Nestor. In: Davis, J.L. (ed.), *Sandy Pylos: An Archaeological History from Nestor to Navarino*, Austin: University of Texas Press, 69-74.

14. Archaeological resources, preservation and prospection in the Trent Valley: The application of Geographical Information Systems to Holocene fluvial environments

ANDY J. HOWARD
School of Geography, University of Leeds, UK

KEITH CHALLIS
Trent and Peak Archaeological Unit, University Park, Nottingham, UK

MARK G. MACKLIN
Institute of Geography and Earth Sciences, University of Wales, Aberystwyth, UK

1 INTRODUCTION

The archaeological record preserved within alluvial systems in Great Britain reflects the attraction of river valleys to human communities since earliest prehistoric times and the suitability of such landscapes to preserve high-resolution cultural and proxy environmental histories (Bell & Neumann, 1997; Macklin, 1999; Needham & Macklin, 1992; Passmore & Macklin, 1997). However, the spatial and temporal evolution of fluvial systems is complex with in-channel and floodplain dynamics influencing both archaeological site location and the quality of preservation (Lewin, 1992; Macklin et al., 1992; Passmore & Macklin, 1997). In a recent review, Howard & Macklin (1999) proposed an empirically-based model of Holocene river valley development in Britain, which explicitly considered the impact of fluvial processes on the archaeological record. Whilst this work provides an important conceptual framework, further systematic dovetailing of archaeological and geomorphological data sets are required to develop integrated models of fluvial and archaeological landscape evolution. Once constructed, these can be rigorously interrogated to predict the location, level of preservation, and the potential for archaeological remains in any area of river valley. This is particularly important given the threats to the resource by construction, aggregate extraction and drainage of valley floors resulting in the lowering of water tables and oxidation of alluvium (Darvill & Fulton, 1998). Whilst the last decade has seen increasing interdisciplinary collaboration, perhaps the greatest problem has been an inability to integrate large data sets – particularly those generated by the archaeological community – within a geomorphological framework. Geographical Information Systems (GIS) provide a potential solution to this problem and this paper provides three examples of reach-scale integration of geomorphological and archaeological data from the Trent Valley of Nottinghamshire, UK (Fig. 1). The reaches chosen have contrasting fluvial styles from piedmont through to lowland as defined by Howard & Macklin (1999).

2 BACKGROUND TO THE TRENT VALLEY

The Trent is one of the major river systems of Great Britain (Fig. 1). In terms of main channel length and catchment area (Ward, 1981) it is ranked fifth (149 km) and second (7490 km^2) respectively in Britain, and it has the second highest mean annual discharge (82.21 m^3 s^{-1}). The catchment is formed in mudstones and siltstones of the Triassic Mercia

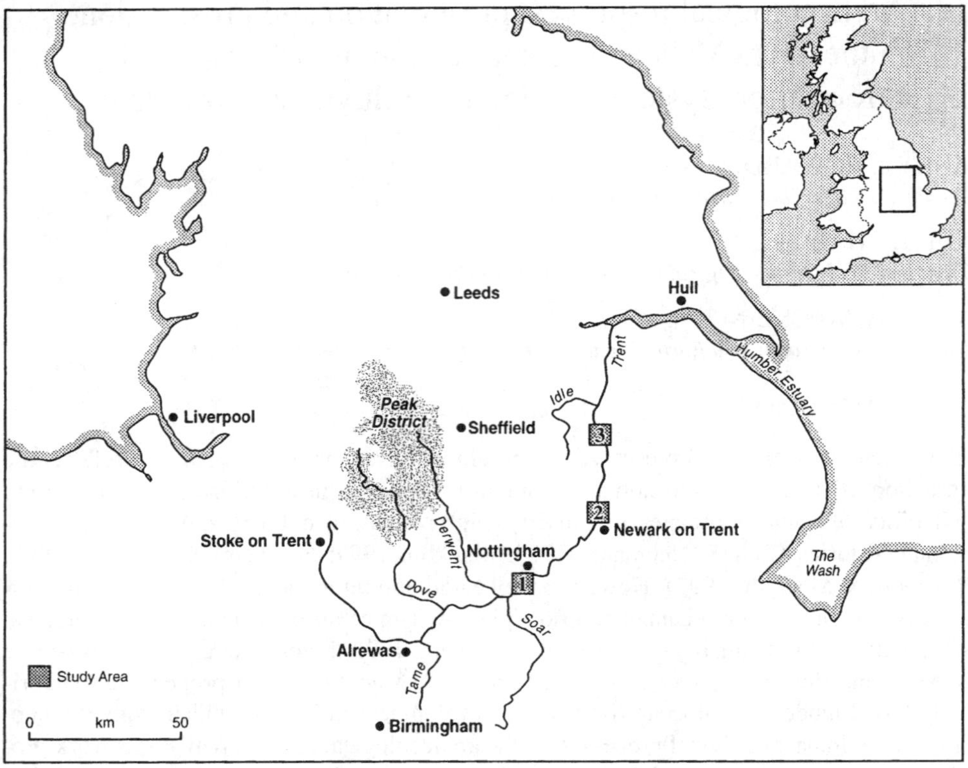

Figure 1. Location map showing the River Trent and its main tributaries. The Nottinghamshire survey area is situated between the confluences of the Rivers Soar and Idle. Boxed areas mark the three study reaches of Barton in Fabis (1), Cromwell (2) and Walkeringham to West Burton (3).

Mudstone Group. It rises on the moorlands of Staffordshire and flows southeast to Alrewas, before flowing northeast to Newark on Trent, and subsequently north to the Humber Estuary. During the last Pleistocene cold stage, the Devensian, the Trent lay beyond the limits of glaciation, but was fed by meltwater, resulting in the deposition of coarse sands and gravels across the valley floor. Since c. 5000 BP, these coarse sands and gravels have been periodically reworked by a laterally mobile River Trent (Howard et al., 1999; Salisbury, 1992; Salisbury et al., 1984). This high energy activity in the later Holocene contrasts with the low energy alluviation that has prevailed in other major British river basins, such as the Thames (Shotton, 1978; Needham, 1992), and Severn (Brown, 1987), or smaller catchments such as the Nene and Soar (Brown et al., 1994). Work by Brown (1998) on enhanced channel mobility of the Trent during the Medieval Warm Period and Little Ice Age suggests that its mobility is a response to increased discharge from the Derwent and Dove tributaries. These rivers drain the uplands of the Peak District and their confluences with the Trent are located approximately 40 km apart in the middle reaches of the catchment (Fig. 1).

The Trent Valley has a long history of archaeological discovery both within the coarse-grained alluvial deposits and upon the terrace surfaces of the valley floor (Knight &

Howard, 1995; Riley, 1987). The lateral mobility of the river may have resulted in both the destruction and/or burial of cultural remains including fishweirs (Salisbury, 1981), bridges (Cooper & Ripper, 1997; Salisbury, 1995) and mill dams (Clay, 1992). Fine-grained alluviation, possibly commencing in the Romano-British period (Buckland & Sadler, 1985), may also have been important for the burial of archaeology away from high energy processes. In addition to the cultural remains, the potential of organic sediments preserved in palaeochannels and other low-lying depressions on the valley floor for providing proxy records of climate, vegetation and land-use (Dinnin, 1997) has been increasingly recognised.

3 GIS METHODOLOGY

GIS provide a methodology for manipulating, analysing and presenting geographic data in a flexible, computer based model (Burrough, 1986; Worboys, 1995). Generic GIS software provide a range of methods to represent collected data, principally based on raster and vector data models (Dangermond, 1990). Some real world phenomena such as terrain, land-cover, or remotely sensed data, are most effectively represented by continuous raster or cell-based modelling methods. Other phenomena, such as stream courses, boundaries and site locations, are more suited to representation as discrete entities using a vector (line and node) model. The complementary raster and vector data models provide the basic GIS representation of reality and each offers access to different suites of modelling tools.

The use of GIS in the study of fluvial geomorphology has thus far focused largely on the collection, integration and analysis of geomorphological data sets at a reach scale. Studies examining river channel planform change have utilised the mapping and analysis facilities offered by GIS to build time-slice models of river valleys, based on integration of historic mapping, air-photographs and field survey, to show channel movement and past channel behaviour (Gurnell et al., 1994; Gurnell et al., 1999; Leys & Werritty, 1999). Applications linking archaeology and geomorphology have largely focused on the relationship between patterns of human activity and the physical landscape, usually viewed in an environmentally deterministic framework (e.g. Gillings, 1995).

The GIS model of the Trent Valley comprises a number of components representing aspects of the natural and human modified environment, and was implemented using ESRI's ArcView 3.1 GIS. This software package runs effectively on a desktop PC and provides tools to integrate raster and vector data sources and point-based information derived from databases. A comprehensive suite of data integration and analysis tools is provided within the package (ESRI, 1998a).

4 COMPONENTS OF THE GIS

Surface terrain is represented by a Digital Terrain Model (DTM), a continuous surface constructed from a series of discrete observations by means of a suitable interpolation algorithm (Weibel & Heller, 1991). Two major problems associated with terrain modelling are the accuracy and reliability of the data source for the DTM and the suitability of the interpolation algorithm used to generate the model. For example, the Ordnance Survey

(OS) 1:50,000 Panorama terrain data, which formed the basis of the DTM in this study, provides a ±3 m contour height accuracy with a contour interval of 10 m (Ordnance Survey, 1997). In a geoarchaeological context, variations in relief corresponding to a terrace surface or palaeochannel may fall within the error envelope and will not be represented. In addition, sampling bias introduced by the regular, 10 m contour interval can itself have a disproportionate effect on the resulting DTM, particularly in areas of low relief such as floodplains (Wood, 1996). Whilst the OS 1:10,000 Profile terrain data offer a higher degree of accuracy (±1.5 m vertical accuracy with a 5 m contour interval), it proved prohibitively expensive for the Trent Valley study.

Data for high-resolution DTMs must be obtained by field survey or from remotely sensed sources, both of which are beyond the scope of the present study. Traditional manual field survey techniques are time-consuming and costly; however, the use of newer, more automated survey techniques may provide a cost-effective and efficient methodology for gathering dense, high-accuracy data for DTMs. The use of Global Positioning Systems (GPS) in an archaeological and geomorphological context is a relatively recent innovation. Standard differential GPS offer a compromise between speed and ease of data collection, affordability of equipment and accuracy degradation, typically providing decimetre levels of accuracy (Higgitt & Warburton, 1999). The increasing availability and affordability of millimetre-accurate carrier wave differential GPS provide a rapid and effective means of gathering the large volumes of data needed for accurate DTMs (Barratt et al., 2000). Remote sensing has traditionally relied upon manual photogrammetric techniques (Lane et al., 1993). However, new remotely sensed data sources such as LiDAR (Jaafar et al., 1999) – while at present only available for small areas – offer the promise of regional centimetre-accurate DTM data in due course.

The use of an appropriate interpolation algorithm when generating a DTM is critical. For most applications, modelling the land surface using some form of triangulated irregular network (TIN) based interpolation algorithm will produce the best result. TIN-based methods mirror the action of a human cartographer by constructing a network of triangles linking all data points, as a guide to interpolation. TINs honour every data point directly, since these form the vertices of the triangles used in modelling the terrain. Logical strings representing break lines, stream channels and other landscape features may be introduced into the triangulation to allow accurate representation of topographical structures (McCullagh, 1988). Triangulation also has the advantage of being computationally less intensive than many other interpolation algorithms, allowing faster generation of complex terrain models from large data sets. The DTM for the study area was built from OS 1:50,000 Panorama data, which comprises contours, spot-heights, break lines and stream courses. The model was generated using a triangulation-based algorithm (the TIN option within ArcView 3D Analyst; ESRI 1998b) to ensure that the resulting DTM included all of the topographic detail within the original data set.

Surface geology and geomorphological information was manually digistised from 1:50,000 British Geological Survey (BGS) maps to produce a vector data set showing the surface distribution of geological deposits. As well as the issues of accuracy associated with digitising from documentary sources (Dunn et al., 1990; Gurnell et al., 1994; Downward, 1995; Leys & Werrity, 1998), the accuracy of geological mapping may vary, and within the Holocene floodplain, these data were enhanced by incorporating observations of palaeochannels from black and white aerial photographs (1:10,000 scale). The photographs were computer rectified (using the Rastools image rectification package within the

MapInfo GIS) and the detail of palaeochannels interactively digitised on-screen from the rectified images. Examination of a number of sequences of aerial photographs ranging in date from the 1940s to the 1970s, proved necessary to adequately map all palaeochannels, many of which were only intermittently visible from the air. The resulting vector data set showing the location, extent and form of palaeochannels, was imported into ArcView. In addition, the contemporary course of the River Trent was captured from the OS 1:2,500 Landline digital mapping, which also supplied other relevant details of modern topography such as roads and urban areas.

Where sufficient records existed, sub-surface deposits were modelled from publically accessible borehole data provided by the BGS. These data are infrequently and irregularly distributed across the floodplain and valley floor and presented a number of modelling problems, particularly in the choice of an appropriate interpolation algorithm to produce a continuous surface from the discrete data. Kriging, a variogram-based interpolation algorithm, is particularly suited to modelling such data (Oliver & Webster, 1990) and produces an estimate of error in the interpolated surface, which may assist in gauging certainty when modelling hidden sub-surface relief from scant data sets. Observations from borehole records were tabulated within a spreadsheet before importing into the Surfer surface modelling software package (Golden Software Inc.). A Kriging interpolation algorithm was used to generate continuous sub-surface models for the base of the fine-grained alluvium and base of the valley floor sand and gravel. These DTMs were imported into ArcView and further modified by incorporating the boundary of alluvium (as mapped by the BGS) as a constraint on the extent of the sub-surface model.

Archaeological site distribution was modelled from two discrete, and to some extent complementary data sets; the Nottinghamshire County Council Sites and Monuments Record and the Royal Commission on Historic Monuments for England (RCHME) Monarch database. Both data sources provide a set of point-based, geo-referenced records of site locations and descriptive observations. The data sources were combined and reconciled within a proprietary database where duplicated and contradictory records were identified and removed. The resulting 'cleaned' data set was modified to use standardised fields and descriptive thesauri (RCHME, 1993; RCHME, 1998) before importing into ArcView.

The query capabilities of ArcView allow the production of maps showing by period and type, the distribution of archaeological sites across the valley floor. However, archaeological site data are not easily represented as point data which may be divorced from any indication of their spatial extent (Harris & Lock, 1990). The most cogent archaeological site types for this study are those represented by cropmarks. Clearly, simple point-based distributions of such sites do not indicate their spatial extent or morphological character and hence much vital information is lost. Within the study this shortcoming has been mitigated to some extent by incorporation of digitised raster versions of the RCHME's National Mapping Programme maps of archaeological features which are derived from comprehensive digitising of cropmark and other evidence from aerial photographs (Bewley, 1995). These maps, while comprehensive, are by no means complete, and are at best an interim solution. In addition, the uncertainly inherent in archaeological observations is particularly difficult to represent in a GIS. For example, the record of a single find of a Romano-British *fibula* (brooch) might represent a number of scenarios from a truly isolated artefact, indicating accidental loss in antiquity, through to an extensive buried site, perhaps a cemetery or a settlement for which no other indicator exists. Since the extent and character are unknown, it is impossible to model.

5 RESULTS

Examples of the outputs from the GIS model for the Trent Valley are illustrated for the three study reaches (Figs 2a, b & c; 3a & b; and Plate 1). Figures 2a-c show the distribution of archaeology, identified by single geo-referenced points, with respect to the DTM, mapped geology and palaeochannels. Reach one, around Barton in Fabis (Fig. 2a) is situated to the east of the Trent-Derwent-Soar confluence. In this reach, the western side of the floodplain between Toton and Beeston has been extensively quarried and therefore many of the palaeochannels and archaeological sites referred to on Figure 2a have already been destroyed. The contemporary channel is single thread and meandering, although palaeochannels with dimensions similar to the present river are identified in a narrow belt adjacent to the contemporary channel. Archaeological sites are scattered widely across the fine grained alluvium, and at several points, are in close proximity to abandoned channels.

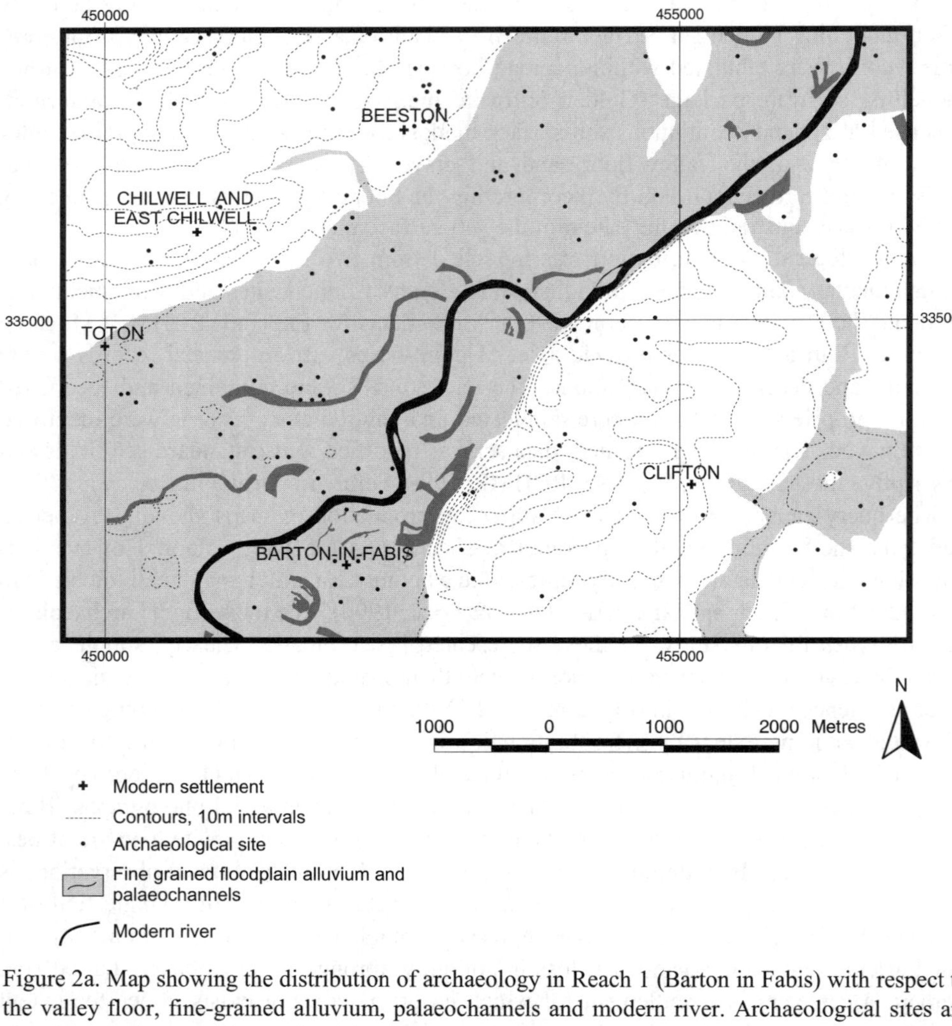

Figure 2a. Map showing the distribution of archaeology in Reach 1 (Barton in Fabis) with respect to the valley floor, fine-grained alluvium, palaeochannels and modern river. Archaeological sites are shown as single geo-referenced points.

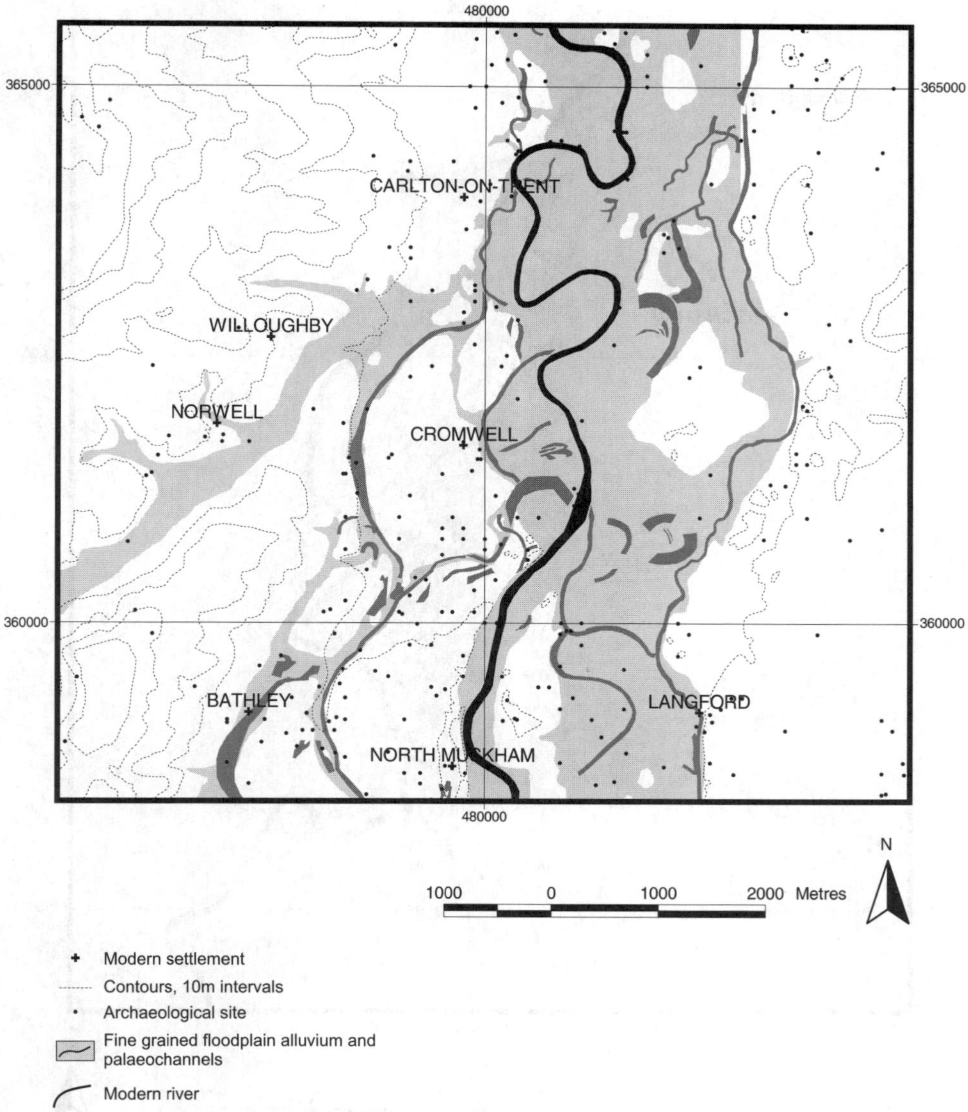

Figure 2b. Map showing the distribution of archaeology in Reach 2 (Cromwell) with respect to the valley floor, fine-grained alluvium, palaeochannels and modern river. Archaeological sites are shown as single geo-referenced points.

In addition, archaeological sites appear to form a sinuous pattern north of the river between Toton and Barton in Fabis, and to the southeast of Beeston where they may indicate human activity adjacent to palaeochannels unrecognised through geological and geomorphological surveys. No gravel islands, which may have proved attractive to settlement, are recorded within this reach of the floodplain.

Reach two (Fig. 2b) is located to the north and south of Cromwell village, c. 5 km downstream of Newark-on-Trent. Extensive gravel extraction operations occur in this area

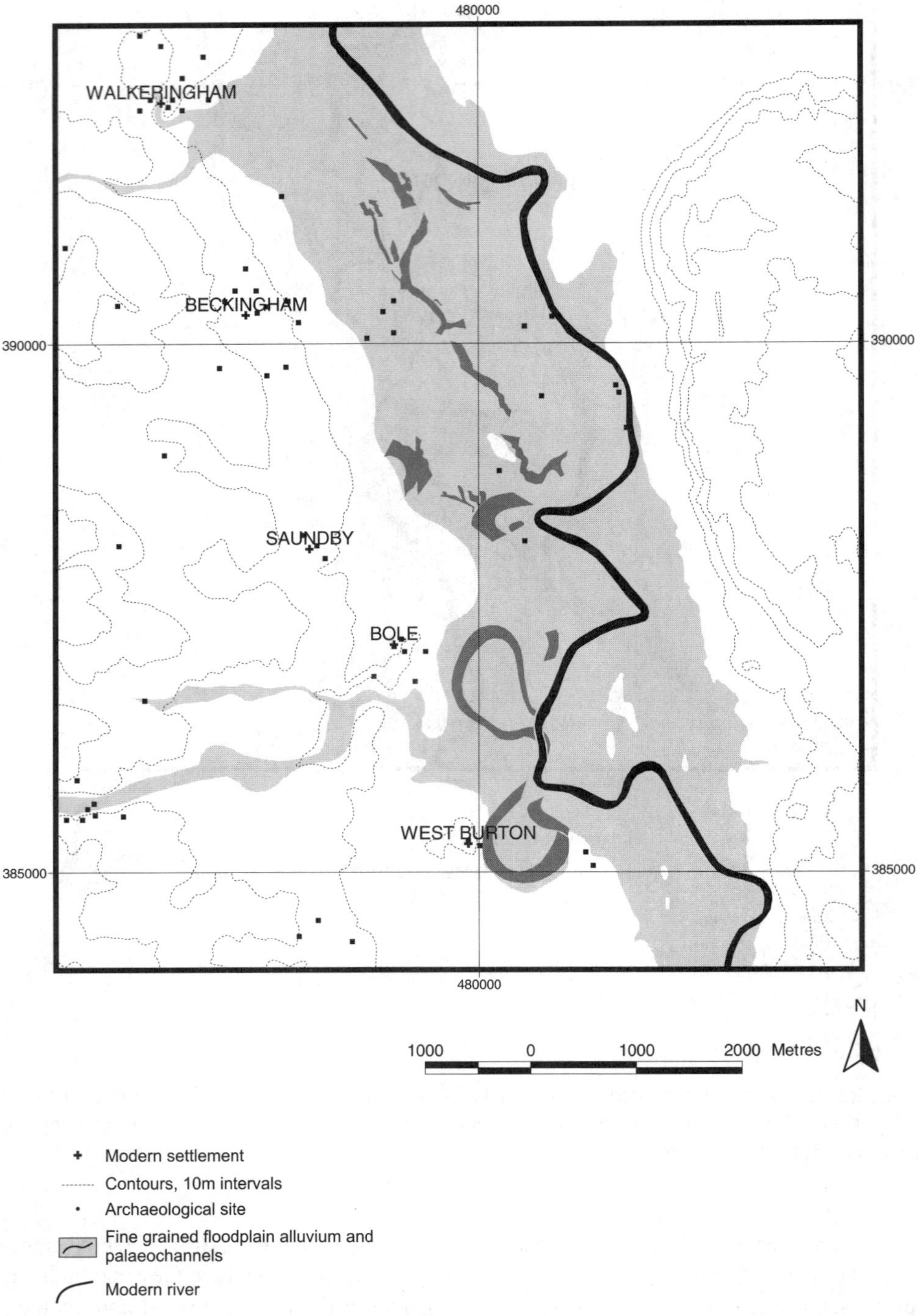

Figure 2c. Map showing the distribution of archaeology in Reach 3 (Walkeringham to West Burton) with respect to the valley floor, fine-grained alluvium, palaeochannels and modern river. Archaeological sites are shown as single geo-referenced points.

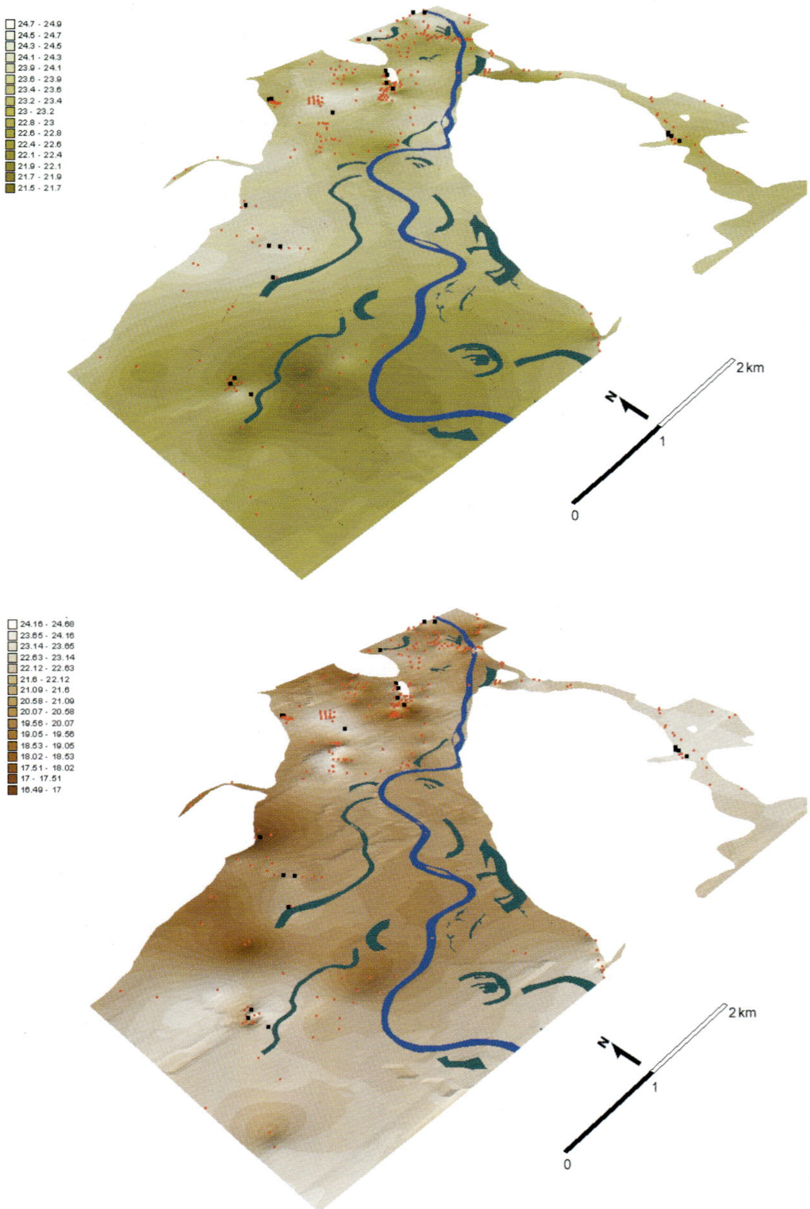

Plate 1. Three dimensional modelling of valley floor sediments in Reach 1 (Barton in Fabis). Borehole localities are marked by red dots and organic sediments recorded in boreholes by black squares. The upper model illustrates the basal relief of fine grained alluvial deposits, whilst the lower model illustrates the basal relief of the sands and gravels. Palaeochannels and the contemporary river are shown and altitude is recorded in metres OD.

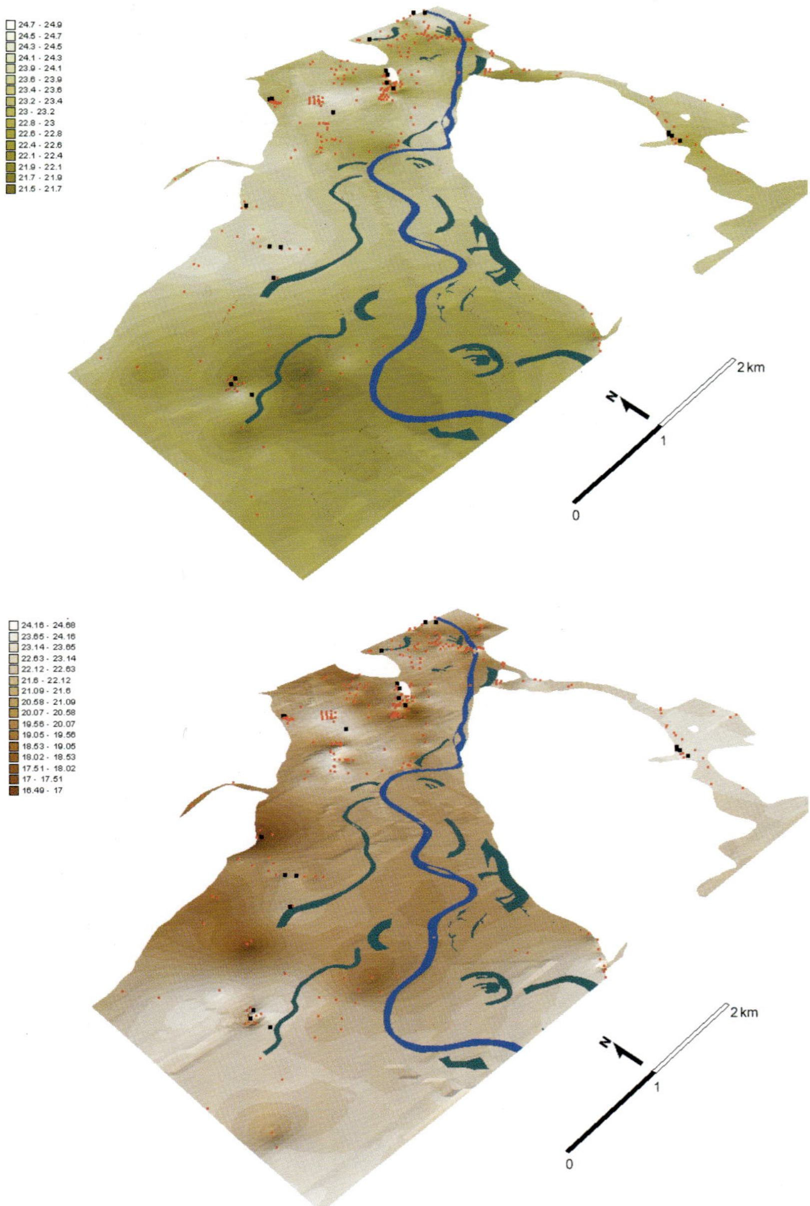

Plate 1. Three dimensional modelling of valley floor sediments in Reach 1 (Barton in Fabis). Borehole localities are marked by red dots and organic sediments recorded in boreholes by black squares. The upper model illustrates the basal relief of fine grained alluvial deposits, whilst the lower model illustrates the basal relief of the sands and gravels. Palaeochannels and the contemporary river are shown and altitude is recorded in metres OD.

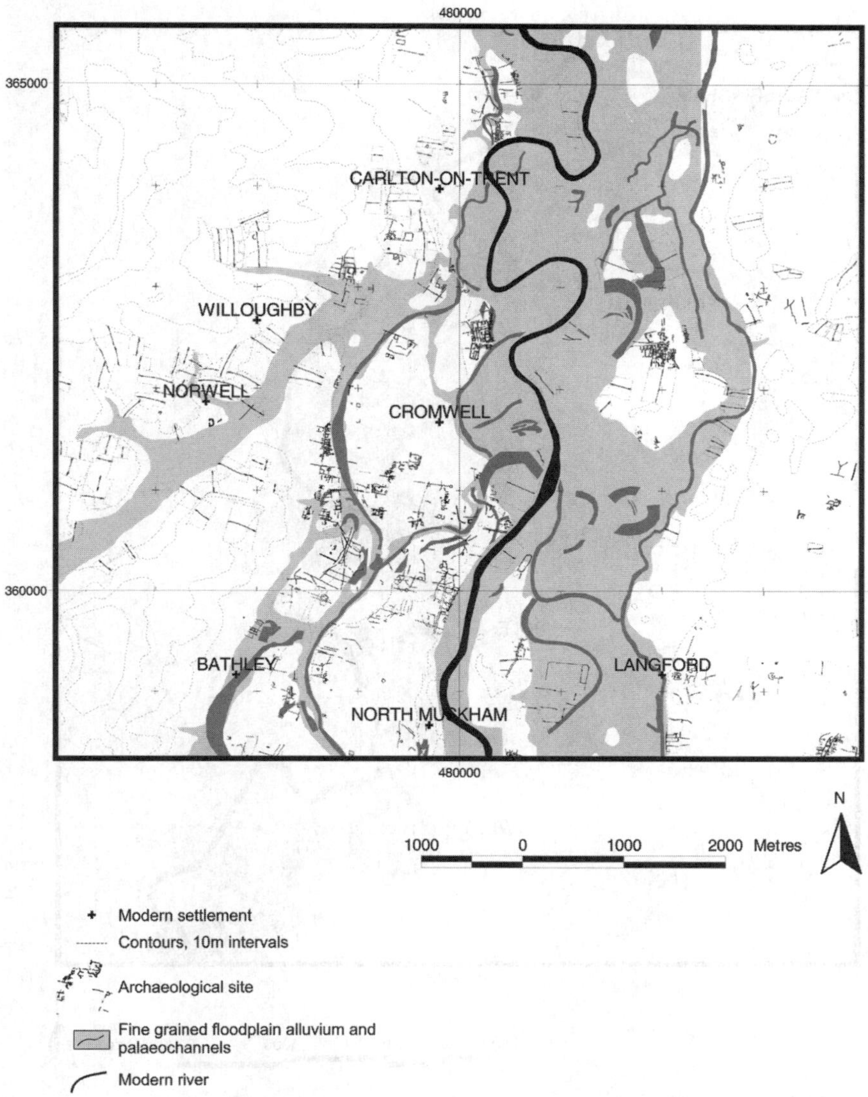

Figure 3a. Map showing the distribution of archaeology in Reach 2 (Cromwell) with respect to the valley floor, fine-grained alluvium, palaeochannels and modern river. Archaeological sites are shown as digitized raster versions of archaeological features.

on both sides of the floodplain. The contemporary channel is single thread and sinuous, but with a well-developed series of palaeochannels occupying the Holocene valley floor. In contrast to reach one, fewer archaeological sites are recorded upon or within the fine grained alluvium, although there does appear to be a close association with palaeochannels. The majority of the sites recorded are clustered on well-drained gravel islands rising

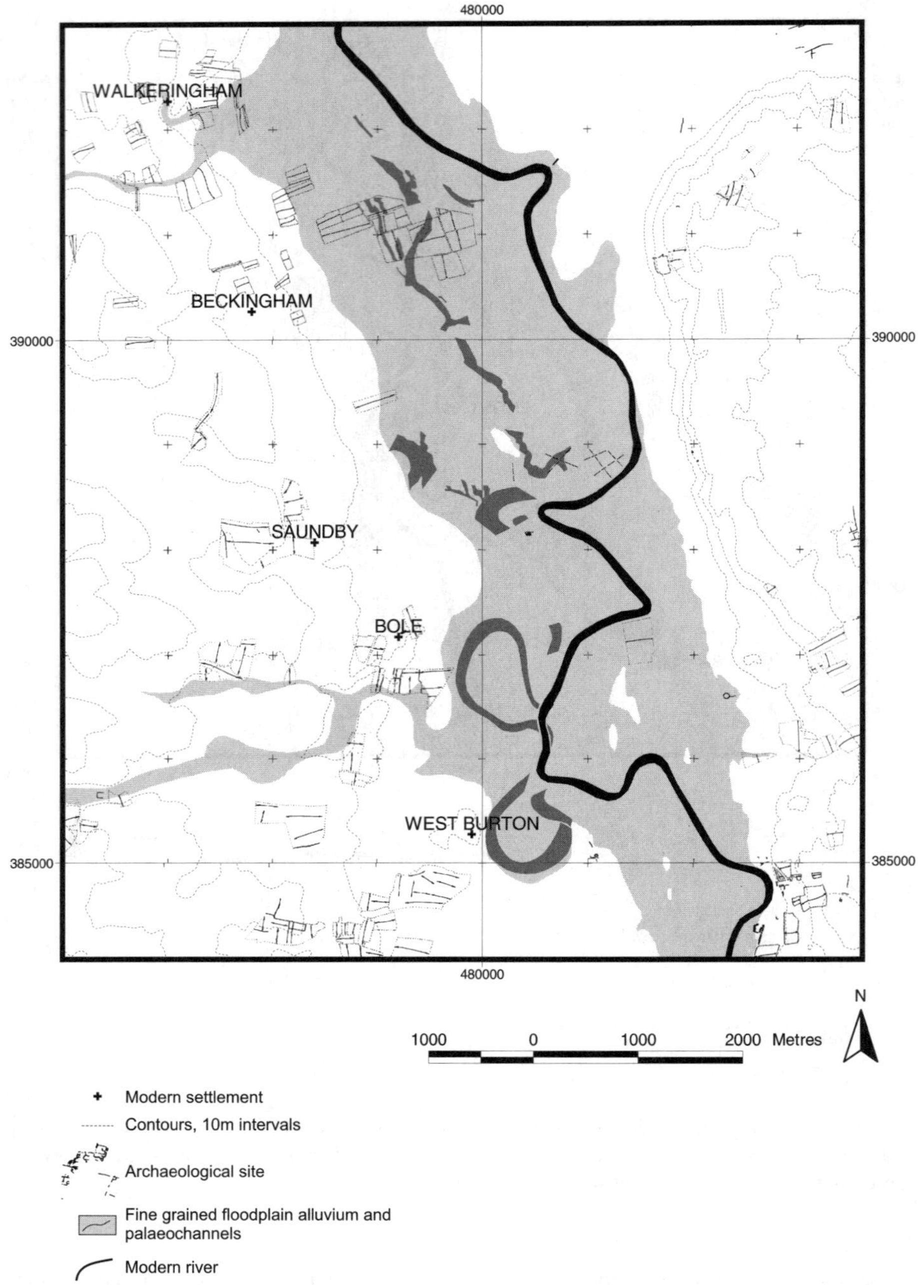

Figure 3b. Map showing the distribution of archaeology in Reach 3 (Walkeringham to West Burton) with respect to the valley floor, fine-grained alluvium, palaeochannels and modern river. Archaeological sites are shown as digitized raster versions of archaeological features.

above the alluvium such as around Cromwell and North Muskham, or on the higher Pleistocene terraces that flank the valley floor (north and east of Langford). On many of the gravel islands, archaeological sites are located close to the mapped edges. However, on the basis of single point data, the relationship of the archaeology on the gravel islands to the surrounding fine grained alluvium is unclear. An important question is whether archaeological sites extend down into (and are buried by) fine-grained alluvium or are truncated by erosion at the gravel island edge.

The issue of archaeological site visibility and the potential for burial of structural and artefactual remains in areas of fine grained alluvium is exemplified in reach 3 (Fig. 2c). Located c. 27 km downstream of Newark on Trent, between Walkeringham and West Burton, the contemporary channel is single thread and slightly sinuous, with a well-developed belt of palaeochannels situated across the valley floor, including the prominent abandoned meanders of Bole Ferry and Bole Round. In this reach of the valley floor very few archaeological sites are visible on the fine grained alluvium or gravel islands. This may reflect the burial of former landsurfaces and archaeology, including gravel islands through high rates of vertical accretion (see Howard & Macklin, 1999), or differing land use strategies and practices in a landscape zone which is regularly waterlogged. However, with single point data, the relationships between geology and archaeology are difficult to interpret. These difficulties can, in part, be overcome by the incoporation of digitised raster versions of archaeological features recorded from cropmarks on aerial photographs. Figures 3a and 3b illustrate overlays of cropmarks for reaches 2 and 3 and should be compared with Figures 2b and 2c. The cropmark plots clearly indicate the significant spatial extent of many of the archaeological records located on gravel islands, terraces and fine grained alluvium and the complex relationships that exist between archaeology and geomorphic processes at the interface of different landscape zones. Cropmarks regularly disappear at the interface of palaeochannels and the fine-grained alluvium at the edges of gravel islands and valley side terraces. The GIS overlay maps allow the identification of such zones of complexity, which can be studied through further fieldwork to provide an understanding of whether archaeological features have been eroded or buried by fluvial processes.

All the reaches show evidence of abandoned channels across the valley floors and may indicate a previously braided or anastomosing style of sedimentation (cf. Brown, 1998). Plate 1 illustrates an attempt to model the three dimensional stratigraphy of reach 1 to the base of the fine-grained alluvium and base of the valley floor sands and gravels. Although the distribution of borehole data is uneven, the thickest sequences of fine grained alluvial deposits are located in the southern part of the reach, probably associated with the infilling of a former channel belt. The basal surface of the sand and gravel is highly variable across the reach with numerous depressions, possibly the result of fluvial scouring or solution of the underlying gypsum-bearing Mercia Mudstone. Organic sediments are recorded in boreholes across the valley floor, some associated with these deeper depressions. However, no organic remains are recorded in the numerous palaeochannels and this reflects the interests of the commissioning organisations, primarily the mineral extraction and construction industries. These palaeochannels are without doubt of Holocene age and have the potential to provide detailed proxy records of climate and land-use. Had the boreholes been commissioned with the intention of identifying sediments with palaeoenvironmental potential, it is likely that the number of organic records from abandoned channels would be significant.

6 CONCLUSIONS

This study illustrates the potential of GIS for the integration and analysis of large geomorphological and archaeological data sets. However, the usefulness of GIS is dependent on the quality of the primary information. Data generated from geological maps, aerial photographs, borehole and archaeological records all require interpretation and 'cleaning' prior to incorporation into a GIS. In addition, the absence of data such as organic records from palaeochannels indicates the potential short-falls of the approach.

The results presented in this paper nevertheless indicate the usefulness of GIS and analysis of the spatial distribution of archaeology within the landscape allows relationships to be clearly identified in reaches with rather different fluvial characteristics. The apparent decrease of archaeological sites upon the valley floor downstream, may reflect changing fluvial dynamics from a system dominated by episodic aggradation and incision resulting in the development of river terraces attractive for settlement, to one dominated by the vertical accretion of fine-grained material in which archaeology is buried and preserved at depth. GIS clearly identifies interface zones where the relationship of archaeology to geomorphic processes are complex and which need be elucidated through small-scale geoarchaeological investigations (e.g. test pitting, trial trenching). The modelling of sub-surface valley floor morphology and identification of distinct stratigraphic levels provide opportunities for prospection, as well as following recorded cropmarks into deeply alluviated areas. The recognition of organic remains on the valley floor, especially where close to records of human activity, provide the opportunity for the integration of cultural and environmental data sets, particularly if water tables are high. In addition, although beyond the scope of the Trent Valley Survey funded to date, it is possible and certainly desirable to analyse the distribution of archaeology by type and period, allowing an assessment of changing human use of the landscape through time. The identification of channel belts and stable terrace zones within the valley floor will allow the identification of areas of *in situ* and reworked archaeological remains. As pressure increases on natural resources, the integration of geomorphological and archaeological data sets is essential to understand fully, the management of this diminishing resource.

ACKNOWLEDGEMENTS

The development of a GIS for the Trent Valley was funded by English Heritage, Nottinghamshire County Council, and the Environment Agency. Work on the project was undertaken by Andy Howard (Leeds), Keith Challis, Steve Malone and Gill Robson (Trent and Peak Archaeological Unit) and Lucie Dingwall and Vince Gaffney (Birmingham University Field Archaeology Unit). AJH and MGM wish to thank the School of Geography, University of Leeds and the Leverhulme Trust for providing on-going financial support for geoarchaeological research. The comments of two anonymous referees improved this manuscript and are gratefully acknowledged.

REFERENCES

Barratt, G., Gaffney, V., Goodchild, H. & Wilkes, S. 2000. Survey at Wroxeter using carrier phase differential GPS surveying techniques. *Archaeological Prospection*, 7: 133-143.

Bell, M. & Neumann, H. 1997. Prehistoric intertidal archaeology and environments in the Severn Estuary, Wales. *World Archaeology,* 29: 95-113.

Bewley, R.H. 1995. A national mapping programme for England. Luftbildarchaologie in Ost – und Mitteleleuropa. *Forschungen zur Archaologie im Land Brandenburg*, 3: 83-92.

Brown, A.G. 1998. Fluvial evidence of the Medieval Warm Period and the Late Medieval Climatic deterioration in Europe. In: Benito, G., Baker, V.R. & Gregory, K.J. (eds) *Palaeohydrology and Environmental Change*. Chichester, John Wiley & Sons, 43-52.

Brown, A.G. 1987. Holocene floodplain sedimentation and channel response of the lower Severn, United Kingdom. *Zeitschrift fur Geomorphologie,* 31: 293-310.

Brown, A.G., Keough, M. & Rice, R.J. 1994. Floodplain evolution in the East Midlands, United Kingdom: the Lateglacial and Flandrian alluvial record from the Soar and Nene valleys. *Philosophical Transactions of the Royal Society of London,* A348: 261-93.

Buckland, P.C. & Sadler, J. 1985. The nature of late Flandrain alluviation in the Humberhead Levels. *East Midland Geographer,* 8: 239-51.

Burrough, P.A. 1986. *Principles of Geographical Information Systems for Land Resource Assessment*. London: Clarendon Press.

Clay, P. 1992. A Norman mill dam at Hemington Fields, Castle Donington, Leicestershire. In: Needham, S. & Macklin, M.G. (eds) *Alluvial Archaeology in Britain*. Oxford: Oxbow Monograph 27: 163-68.

Cooper, L. & Ripper, S. (eds) 1997. *The Hemington Bridges*. University of Leicester Archaeological Services 97/27.

Dangermond, J. 1990. A classification of software components commonly used in geographic information systems. In: Pequet, D.J. & Marble, D.F. (eds) *Introductory Readings In Geographic Information Systems*. London: Taylor & Francis. 30-51.

Darvill, T. & Fulton, A.K. 1998. MARS: *The Monuments at Risk Survey in England 1995. Main Report*. Bournemouth and London. Bournemouth University and English Heritage.

Dinnin, M. 1997. Holocene beetle assemblages from the Lower Trent floodplain at Bole Ings, Nottinghamshire, UK. In: Ashworth, A.C., Buckland P.C. & Sadler, J.P. (eds) Studies in Quaternary Entomology. An Inordinate Fondness For Insects. London: Quaternary Research Association. *Quaternary Proceedings* 5: 83-104.

Downward, S.R. 1995. Information from topographic survey. In: Gurnell, A. & Petts, G. (eds) *Changing River Channels*. Chichester, John Wiley & Sons, 303-23.

Dunn, R., Harrison, A.R. & White, J.C. 1990. Positional accuracy and measurement of error in digital databases of land use: an empirical study. *International Journal Geographical Information Systems,* 4: 385-98.

ESRI. 1998a. *What's new in ArcView GIS Version 3.1*. ESRI White Paper, July 1998. Redlands, California.

ESRI. 1998b. *ArcView 3D analyst features*. ESRI White Paper, December 1998. Redlands, California.

Gillings, M. 1995. Flood dynamics and settlement in the Tisza valley of north-east Hungary: GIS and the Upper Tisza project. In: Lock, G. & Stancic, Z. (eds) *Archaeology and Geographical Information Systems: A European Perspective*. Taylor and Francis, London, 67-84.

Gurnell, A.M., Downward, S.R. & Jones, R. 1994. Channel planform change on the River Dee meanders 1876-1992. *Regulated Rivers: Research and Management,* 9: 187-204.

Gurnell, A.M., Bickerton, M., Angold, P., Bell, D., Morrissey, I., Petts, G.E. & Sadler, J. 1998. Morphological and ecological change on a meander bend: the role of hydrological processes and the application of GIS. *Hydrological Processes,* 12: 981-93.

Harris, T.M. & Lock, G.R. 1990. The diffusion of a new technology: a perspective on the adoption of geographic information systems within UK archaeology. In: Allen, K.M.S., Green, S.W., Ezra, B.W. & Zubrow, B.W. (eds) *Interpreting Space: GIS and Archaeology*. London: Taylor & Francis, 33-53.

Higgitt, D.L. & Warburton, J. 1999. Applications of differential GPS in upland fluvial geomorphology. *Geomorphology,* 29: 121-34.

Howard, A.J. & Macklin, M.G. 1999. A generic geomorphological approach to archaeological interpretation and prospection in British river valleys: a guide for archaeologists investigating Holocene landscapes. *Antiquity,* 73(281): 527-41.

Howard, A.J., Smith, D.N., Garton, D., Hillam, J. & Pearce, M. 1999. Middle to Late Holocene environments in the Middle to Lower Trent Valley. In: Brown, A.G. & Quine, T.A. (eds), *Fluvial Processes and Environmental Change*. Chichester, John Wiley, 165-78.

Jaafar, J., Priestnall, G., Mather, P.M. & Viera, C.A. 1999. Construction of DEM from LiDAR DSM using Morphological Filtering, Conventional Statistical Approaches and Neural Networks. In: Pan, P. & Barnsley, M. (eds) Earth Observation: From Data to Information. *Proceedings of the 25th Annual Conference and Exhibition of the Remote Sensing Society*, 299-306.

Knight, D. & Howard, A.J. 1995. *Archaeology and alluvium in the Trent Valley: An archaeological assessment of the floodplain and gravel terraces*. Unpublished report, Trent & Peak Archaeological Trust, Nottingham.

Lane, S.N., Richard, K.S. & Chandler, J.H. 1993. Developments in photogrammetry: the geomorphological potential. *Progress in Physical Geography,* 17(3): 306-28.

Lewin, J. 1992. Alluvial sedimentation style and archaeological sites: the Lower Vyrnwy, Wales. In: Needham, S. & Macklin, M.G. (eds) *Alluvial Archaeology in Britain*. Oxford: Oxbow Monograph, 27: 103-10.

Leys, K.F. & Werritty, A. 1999. River channel planform change: software for historical analysis. *Geomorphology,* 29: 107-20.

Macklin, M.G. 1999. Holocene river environments in prehistoric Britain: Human Interaction and Impact. *Quaternary Proceedings*, 7: 521-30.

Macklin, M.G., Passmore, D.G. & Rumsby, B.T. 1992. Climatic and cultural signals in Holocene alluvial sequences: the Tyne basin, northern England. In: Needham, S. & Macklin, M.G. (eds) *Alluvial archaeology in Britain.* Oxford: Oxbow Monograph 27: 123-39.

McCullagh, M.J. 1988. Terrain and surface modelling systems: theory and practice. *Photogrammetric Record* 12(72): 747-99.

Needham, S. 1992. Holocene alluviation and interstratified settlement evidence in the Thames Valley at Runnymede Bridge. In: Needham, S. & Macklin, M.G. (eds) *Alluvial Archaeology in Britain.* Oxford: Oxbow Monograph 27: 249-60.

Needham, S. & Macklin, M.G. (eds) 1992. *Alluvial Archaeology in Britain*. Oxford: Oxbow Monograph 27.

Oliver, M.A. & Webster, R. 1990. Kriging: a method of interpolation for geographical information systems. *International Journal of Geographical Information Systems*, 4(3): 313-32.

ORDNANCE SURVEY 1997. *Land-Form PANORAMA Users Guide.* Southampton: Ordnance Survey.

Passmore, D.G. & Macklin, M.G. 1997. Geoarchaeology of the Tyne Basin: Holocene River Valley Environments and the Archaeological Record. In: Tolan-Smith C. (ed.) Landscape Archaeology in Tynedale. University of Newcastle upon Tyne: Tyne Solway Ancient and Historic Landscapes *Research Programme Monograph*, 1: 11-27.

Riley, D.N. 1987. *Air Photography and Archaeology*. London: Duckworth.

Royal Commission for the Historic Monuments of England. 1993. *Recording England's past. A data standard for the extended national archaeological record.* London: Royal Commission for the Historic Monuments of England.

Royal Commission for the Historic Monuments of England. 1998. *MIDAS A manual and data standard for monument inventories*. Swindon: Royal Commission for the Historic Monuments of England.

Salisbury, C.R. 1995. An 8th Century Mercian bridge over the Trent at Cromwell, Nottinghamshire, England. *Antiquity*, 69: 1015-18.

Salisbury, C.R. 1992. The archaeological evidence for palaeochannels in the Trent Valley. In: Needham, S. & Macklin, M.G. (eds) *Alluvial Archaeology in Britain*. Oxbow: Oxbow Monograph, 27: 155-62.

Salisbury, C.R. 1981. An Anglo-Saxon weir at Colwick, Nottinghamshire. *Transactions of the Thoroton Society of Nottinghamshire*, 84: 26-36.

Salisbury, C.R., Whitley, P.J., Litton, C.D. & Fox, J.L. 1984. Flandrian courses of the River Trent at Colwick, Nottingham. *Mercian Geologist*, 9: 189-207.

Shotton, F.W. 1978. Archaeological inferences from the study of alluvium in the lower Severn-Avon valleys. In: Limbrey, S. & Evans, J.G. (eds) *The effect of man on the landscape: the lowland zone. Council for British Archaeology Research Report,* 21: 27-32.

Ward, R.C. 1981. River systems and river regimes. In: J. Lewin (ed.) *British Rivers*. London: Allen & Unwin, 1-33.

Weibel, R. & Heller, M. 1991. Digital terrain modelling. In: Maguire, D.J. Goodchild, M.F. & Rhind, D.W. (eds) *Geographical information systems: principles and applications.* Harlow: Longman Scientific and Technical, 269-97.

Wood, J. 1996. *The geomorphological characterisation of digital elevation models*. Unpublished PhD Thesis, University of Leicester.

Worboys, M.F. 1995. *GIS: A Computing Perspective.* London: Taylor & Francis.

Part 5: Modelling and monitoring of sediment fluxes and river channel dynamics

15. Holocene sediment budgets in an upland gravel bed river: the River South Tyne, northern England

DAVID G. PASSMORE
Department of Geography, University of Newcastle, Newcastle upon Tyne, UK
MARK G. MACKLIN
Institute of Geography and Earth Sciences, University College of Wales, Aberystwyth, UK

1 INTRODUCTION

Sediment budgeting techniques have emerged as a useful tool for investigating spatial and temporal discontinuities in river basin sediment erosion, transport and deposition, and for gauging the relative contribution of differing catchment sediment sources to basin yields (e.g. Trimble, 1983, 1995). Particular attention has focused on the role of sediment storage in recognition of its importance in determining the response of fluvial systems to environmental change (Phillips, 1991). This has prompted budgeting of sediment storage at scales ranging from individual river bends (Neill, 1983) and bar complexes (Passmore et al., 1993; Lane et al., 1994, 1995) through discrete reaches (Martin & Church, 1995; Ashmore & Church, 1998; Huisink, 1999) to entire drainage basins (Trimble, 1983; Phillips, 1991; Campo & Desloges, 1994). There remains, however, a lack of detailed understanding of spatial and temporal variability in sub-basin scale flux and storage of sediment, or what Trimble (1995) terms 'distributed' budgets.

In upland and piedmont valley reaches in the UK, despite the increasing interest in Holocene valley floor development and environmental change in these environments (see Macklin, 1999 for a recent review), there is little information on long-term (10^2-10^4 years) residence time of stored sediment and the rates and patterns of sediment transfer (Merrett & Macklin, 1999). Yet river systems in these settings are liable to be at their most sensitive to external environmental change (Ferguson, 1981; Brown & Quine, 1999). Sediment budgeting techniques therefore offer a means of enhancing our understanding of fluvial response to climate and land-use changes of differing scale and character beyond those of the instrumental and documentary record.

Holocene sediment budgeting studies that are based on field evidence, however, are faced with problems of fragmentary survival of sedimentary sequences and establishing dating control for periods of incision and alluviation. Nevertheless, recent work by Huisink (1999) has demonstrated that morphological approaches to sediment budgeting at extended reach scales can establish upper and lower bound rates of fluvial sediment fluxes over timescales of the order of 10^4 years.

This paper employs similar techniques to assess changes in Holocene fluvial sediment storage in a 10.5 km long reach of the South Tyne, a gravel-bed river in northern England. Holocene alluvial histories have been established at a number of sites in the South Tyne basin (Macklin et al., 1992a and b; Passmore et al., 1993; Macklin et al., 1998; Passmore & Macklin, 2000) and this study builds on detailed investigations of Holocene fluvial terraces in a 2.75 km sub-reach of the valley between Lambley and Featherstone Castle

(Passmore & Macklin, 2000). Cartographic and aerial photograph sources have also been used to quantify channel and bar erosion and deposition since the mid-nineteenth century within the study reach (Macklin & Lewin, 1989; Passmore et al., 1993; Macklin et al., 1998), but budgeting of patterns and rates of sediment transfer in the South Tyne have hitherto been confined to short-term (1990-1993) analyses of recent morphological changes in the active channel environment at Lambley-Featherstone (Sohag, 1993; Macklin et al., 1998).

This study has developed a digitised database that combines the results of geomorphological mapping, topographic survey and geochronological analyses of late Pleistocene and Holocene fluvial terraces. Here we use the database to (i) quantify the volume of Holocene and late Pleistocene fluvial sediment storage presently infilling the valley floor, (ii) estimate rates of sediment reworking and export from fluvial storage over discrete periods defined by Holocene incision and deposition cycles on the basis of quantified morphological changes and (iii) to examine the relationship between valley boundary conditions and long-term patterns of sediment transfer and storage.

2 BACKGROUND TO THE STUDY REACH

The study reach extends over 10.5 km of the South Tyne valley between Eals Bridge (NY 682553) and the confluence with Haltwhistle Burn (NY 716637) (Fig. 1). Here the northerly-flowing river emerges from the North Pennine massif and turns east towards Tynedale and the North Sea. The catchment area of the South Tyne at the upper and lower study reach limits is 254 km^2 and 414 km^2, respectively. Underlying solid geology comprises Carboniferous limestone, sandstone and, at Lambley village (Fig. 1), a local intrusive dolorite outcrop of the Whin Sill. During the Late Devensian Dimlington Stadial, middle and lower reaches of the South Tyne, including the study reach downstream of Lambley (Fig. 1), were overrun by easterly flowing ice originating from the Lake District and Solway (Raistrick, 1931; Beaumont, 1968; Passmore, 1994). Upland reaches of the South Tyne, however, lay beneath a local ice cap centred on Cross Fell (Catt, 1991a and b).

The study reach is characterised by alluvial basins up to 600 m wide that are separated by narrow reaches confined by steep till-mantled and bedrock bluffs (Fig. 1). Overall reach gradients of the present river average 0.006 m m^{-1}, although steeper gradients (up to 0.01 m m^{-1}) are evident at the upstream limits of the study reach (valley km 0-1) and below Coanwood Viaduct (valley km 3.5-5) immediately downstream of the outcropping Whin Sill (Fig. 2). The river is currently incising into tills and in some parts of the valley is flowing on bedrock. Present river planforms are characterised by confined meandering channels that locally divide around overlapping active gravel bars and occasionally larger, relatively stable vegetated islands. This planform morphology is typical of upland northern British rivers and has been likened to wandering gravel rivers described by Neill (1973) and Church (1983) (Macklin & Lewin, 1989; Passmore et al., 1993; Passmore & Macklin, 2000).

Previous investigations in the South Tyne basin have demonstrated that sub-reaches of the study area at Eals, Coanwood, Lambley, Wydon and Haltwhistle (Fig. 1) have exhibited episodic channel braiding and lateral instability since the mid-nineteenth century (Macklin & Lewin, 1989; Passmore et al., 1993; Macklin et al., 1998). The architecture of late Holocene alluvial fills at Lambley-Featherstone indicates floodplain construction for the past 3000 years has also been dominated by coarse-grained migratory bed and bar

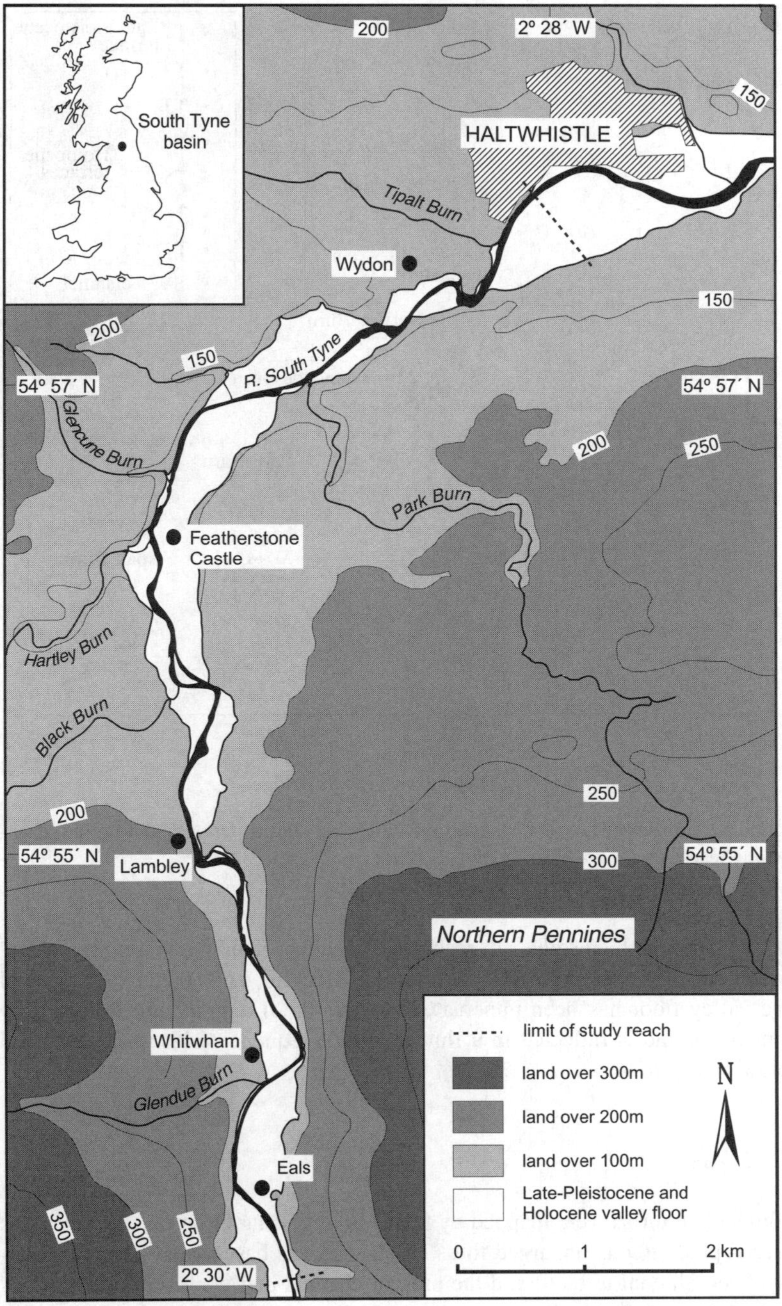

Figure 1. Map of study reach showing relief, area of valley floor (including late Pleistocene and Holocene fluvial terraces and alluvial fans), present river channel and location of South Tyne basin.

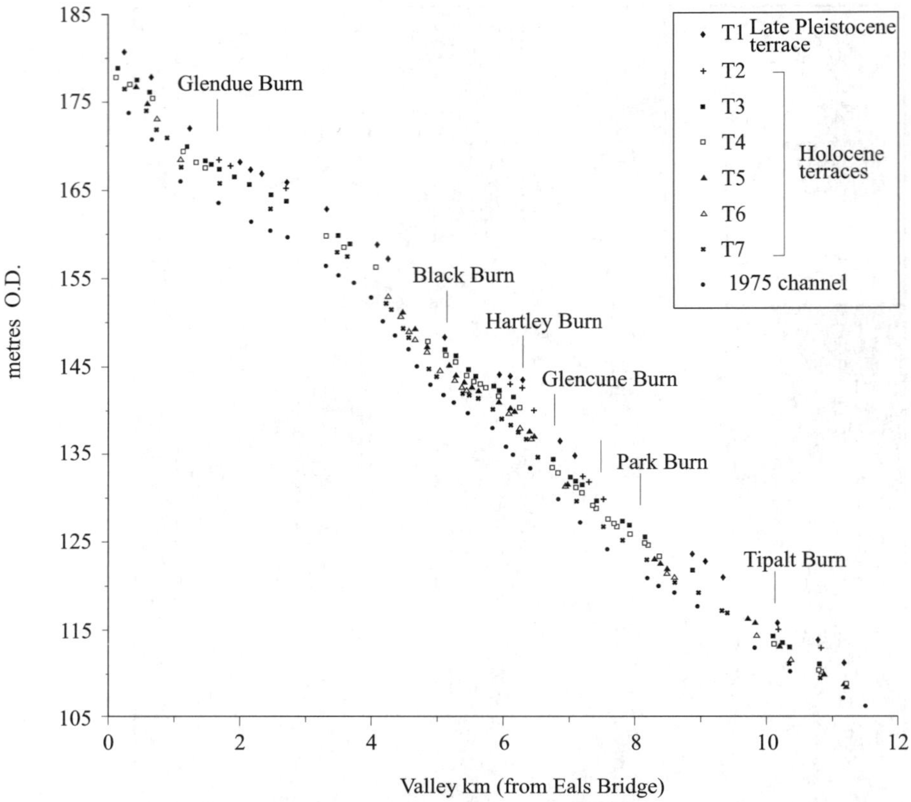

Figure 2. Longitudinal profiles of Late-Pleistocene and Holocene alluvial terraces and present channel bed in the study reach.

assemblages, associated with episodic sub-reach scale channel braiding and bank erosion, with thin veneers of overbank sediment (Passmore & Macklin, 2000). Here, postglacial incision of the valley floor has been punctuated by periods of aggradation followed by channel entrenchment and is reflected in a fluvial terrace sequence with younger fluvial units inset below older surfaces.

3 METHODOLOGY

Valley floor fluvial landforms were mapped by field walking using Ordnance Survey maps and vertical aerial photographs (enlarged to c.1:3000 scale) as base maps. Over two hundred spot heights on alluvial terraces and the present channel bed were surveyed to O.D. benchmarks. Additional data on alluvial terrace and channel bed elevations were available from surveyed cross-profiles at Lambley-Featherstone (Passmore & Macklin, 2000) and Haltwhistle (Bishop et al., 1992; Passmore & Macklin, 1997) (Fig. 1). Longitudinal profiles of alluvial terrace surfaces and channel thalweg elevations were then plotted in relation to the valley centre-line (Fig. 2). The thickness and lithostratigraphy of alluvial terraces have been primarily determined from eroding river bank sections logged in the vicinity of Eals, Whitwham, Lambley-Featherstone and Wydon (Fig. 1). Additional data has been obtained from machine trenching and sediment coring of terrace and palaeochannel fills at Lambley-Featherstone (Passmore & Macklin, 2000) and Haltwhistle (Bishop et al., 1992; Macklin & Passmore, 1993; Passmore & Macklin, 1997).

Dating control for Holocene alluvial units have been provided by detailed investigations at Lambley-Featherstone and Haltwhistle; these have yielded five ^{14}C assays on wood fragments recovered from alluvial fills, supplemented by lichenometric techniques, trace metal geochemistry of fine-grained alluvium and palaeomagnetic dating of an inorganic fine-grained palaeochannel fill (Passmore & Macklin, 1997, 2000). Historic maps and aerial photographs document channel and floodplain development throughout the study reach since c. AD 1865 (Macklin et al., 1998; Passmore & Macklin, 2000). All ^{14}C dates quoted here have been calibrated according to calibration curves after Stuiver et al. (1993) using the programme OxCal (v.2.18, Oxford Radiocarbon Accelerator Unit).

Valley floor and alluvial fan margins, terrace scarps and active channel and bar margins on historic maps were digitised to National Grid co-ordinates using ARC/INFO. Transformation of pre-National Grid maps was accomplished with reference to farm buildings and field boundaries with known coordinates. The study reach was subdivided into 0.5 km sub-reaches with boundaries established normal to the valley centre-line. Map coverages were then interrogated for terrace and active channel zone areas for the period c.1865 to present. This information was merged in a database with values of alluvial unit thickness determined either directly from lithostratigraphic analyses or, where unavailable, extrapolated from the nearest adjacent sub-reach. Estimates of sediment volume of each terrace and for the post-1860 active channel zone were calculated for each sub-reach. Details of sediment budgeting procedures are set out below after a summary of Holocene valley floor development and total sediment storage in the study reach.

4 FLUVIAL TERRACE SEQUENCE

Fluvial terraces are preserved as discontinuous elements between 2-8 m above the present channel bed and are paired in wider parts of the valley (Figs 2-5). Terrace correlations in the study reach have been based on dated terrace sequences established at Lambley-Featherstone, where six fluvial terraces (designated T1-6) that predate the mid-nineteenth century AD have been identified (Passmore & Macklin, 2000, Fig. 4), and in the lower part of the study reach at Haltwhistle (Fig. 5). Maximum terrace elevations of 8 (T1), 7 (T2), 6 (T3), 5 (T4), 4 (T5) and 3 m (T6) above present river bed levels occur upstream of valley km 6.5 (Figs 2-4), although locally (e.g. between valley km 4-5) these terrace surfaces are 1-2 m lower as the result of differential channel bed incision. Downstream of valley km 6.5, terrace elevations generally lie no more than 6 m above the present channel bed and at Haltwhistle (valley km 10-10.5) terraces T3-6 converge (Fig. 2).

Investigations at Lambley-Featherstone have not yielded direct dating controls for terraces T1-T3. However, T1 is believed to be equivalent to similarly high-elevation, late Pleistocene gravel terraces recorded in river valleys elsewhere in northeast (Passmore & Macklin, 1997) and northwest England (e.g. Harvey, 1985; Tipping, 1995) that most probably formed as outwash deposits during de-glaciation. Deposition of alluvial unit T3

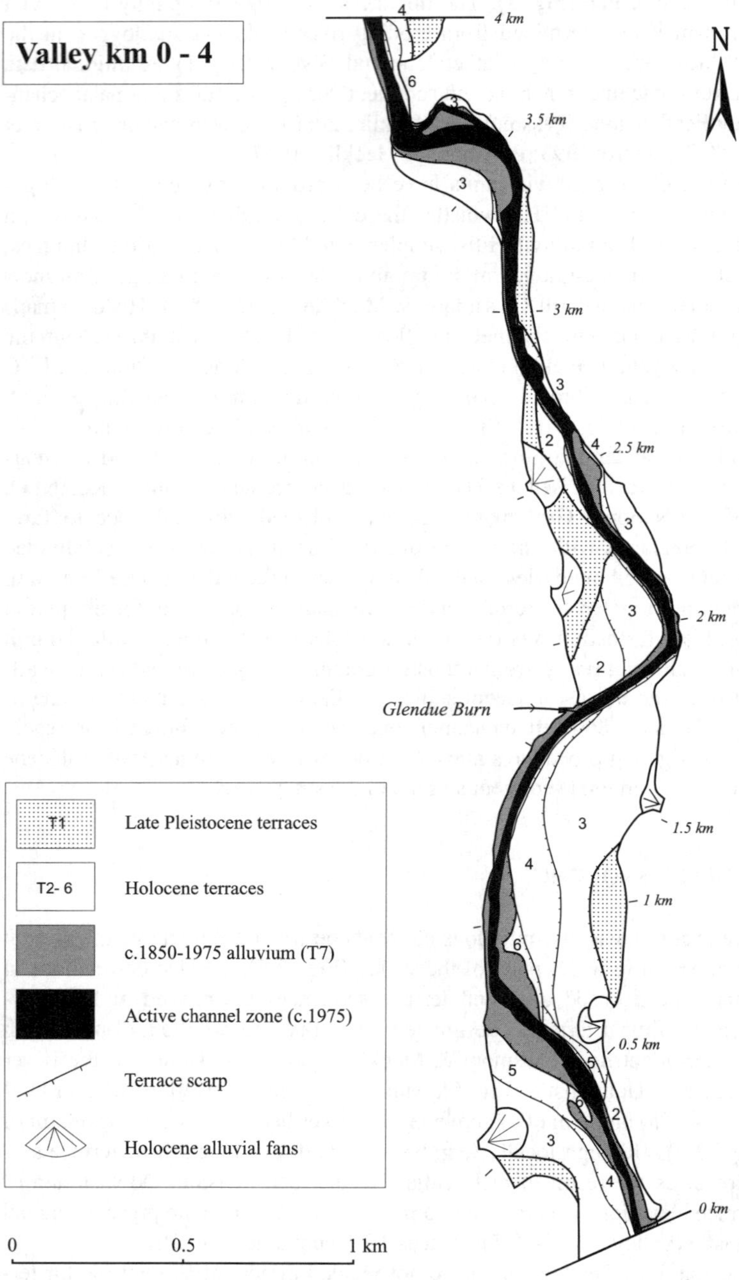

Figure 3. Geomorphological map of valley km 0-4 showing late Pleistocene and Holocene alluvial terraces, alluvial fans, the active valley floor between c. AD 1850-1975 and the active channel zone c. AD 1975.

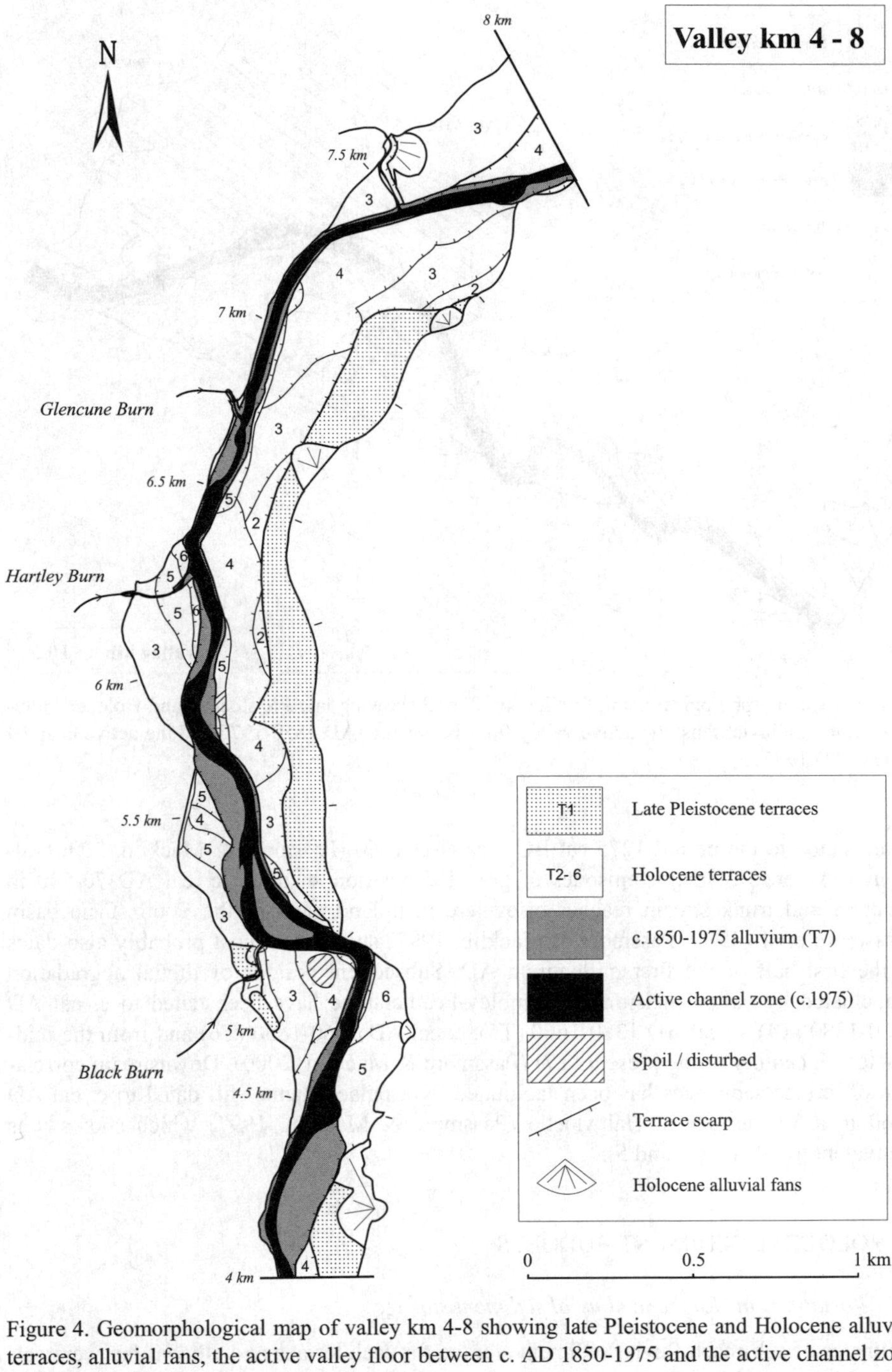

Figure 4. Geomorphological map of valley km 4-8 showing late Pleistocene and Holocene alluvial terraces, alluvial fans, the active valley floor between c. AD 1850-1975 and the active channel zone c. AD 1975.

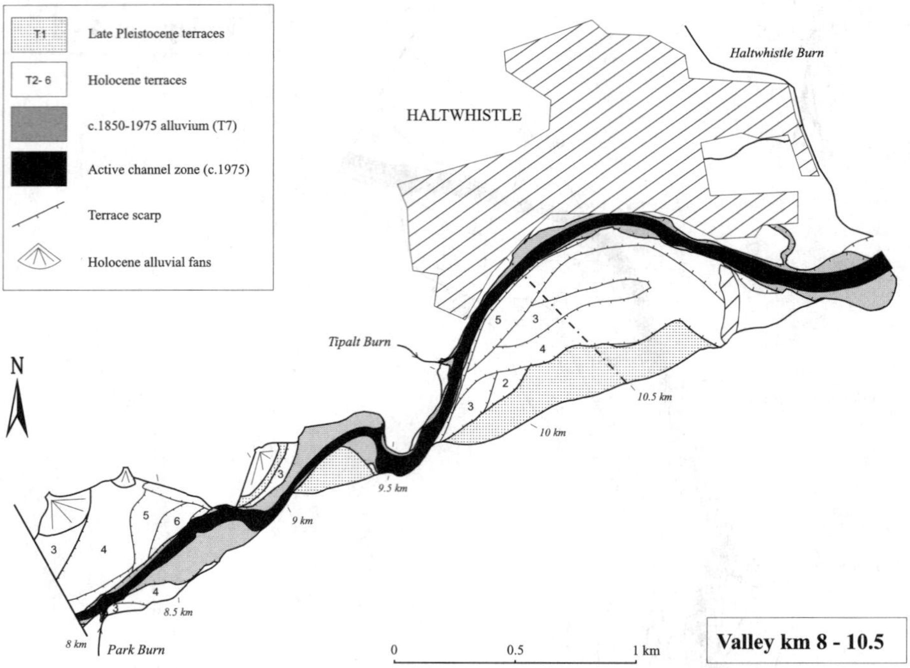

Figure 5. Geomorphological map of valley km 8-10.5 showing late Pleistocene and Holocene alluvial terraces, alluvial fans, the active valley floor between c. AD 1850-1975 and the active channel zone c. AD 1975.

is bracketed to the period 1275 cal BC – cal AD 1160 (Passmore & Macklin, 2000), although evidence of major episodes of gravel deposition spanning c. cal AD 70-540 in tributary and trunk stream reaches elsewhere in upland parts of the South Tyne basin (Passmore et al., 1993; Passmore & Macklin, 1997) suggests it most probably also dates to the first half of the first millennium AD. Subsequent periods of fluvial aggradation (punctuated by valley incision) at Lambley-Featherstone have been dated to c. cal AD 1160-1380 (T4), c. cal AD 1380-1660 (T5), c. cal AD 1660-1850 (T6) and from the mid-nineteenth century to the present (T7) (Passmore & Macklin, 2000). Downstream correlation of terrace sequences has been facilitated by a palaeochannel fill dated to c. cal AD 1260 in a 3 m terrace at Haltwhistle (Passmore & Macklin, 1997) which shows it is equivalent to T4 (Figs 2 and 5).

5 HOLOCENE SEDIMENT BUDGETS

5.1 *Variations in Holocene alluvial sediment storage*

Estimates of sub-reach fluvial sediment storage for each terrace and alluvial fan were calculated by multiplying terrace/fan surface area by alluvial unit thickness. For T5 and later alluvial units which are partially inset within till and (or) bedrock, thickness was calculated as the difference between terrace surface elevations and the present river bed. Thickness values for T1-4 have been calibrated by reference to averaged unit depths recorded in cut-bank exposures at Lambley-Featherstone. Thickness values are assumed to be uniform throughout each fluvial terrace. While this assumption is supported by observations of extended cut-bank exposures at Lambley-Featherstone (Passmore & Macklin, 2000), it is likely to over-estimate sediment storage volumes and hence values recorded here should be regarded as upper-bound estimates. Thickness values for alluvial fans were obtained by geometric means assuming a representative triangular cross-section based on survey points at the fan apex and distal margin; unit depths were then multiplied by planform areas.

Table 1 shows total present alluvial sediment storage in the study reach amounts to 7.8 million m^3. The majority of sediment is stored in the relatively wide valley floors in the vicinity of Eals (valley km 0-2) and Featherstone Castle-Wydon (valley km 5.5-8.5), which respectively accommodate c.17% and c.48% of the total store (Fig. 6). Late Pleistocene sediments account for 37% of total storage, and exceed 50% of storage in some sub-reaches, while only valley km 7.5-8.5 lack sediment of this age (Fig. 6). Holocene sediments account for 63% of stored alluvium and are dominated (22% of the total sediment volume) by terrace T3 deposits. Only 4% of stored Holocene alluvium dates to before the 1st millennium BC (T2), while alluvial fans comprise 5% of the total Holocene sediment store (Table 1; Fig. 7).

To estimate net sediment export from each sub-reach during the Holocene we assume (i) that surface elevations of T1, representing the maximum aggradation level of late Pleistocene alluvial deposits, were uniform across the valley floor of each sub-reach at the beginning of the Holocene (although it is very likely that the South Tyne valley floor was entrenched, to some degree, shortly after de-glaciation), and (ii) that valley floor margins have not changed significantly during the Holocene. Net Holocene sediment export is calculated by multiplying the surface area of Holocene terrace units and the post-1850 channel belt by the height difference between respective units and T1, and then summing these values for each sub-reach. Height differences for T5 and later fluvial units have been reduced by 1 m to compensate for incision of these units into tills and(or) bedrock (on the basis of detailed studies at Lambley-Featherstone; see Passmore & Macklin, 2000).

Net Holocene fluvial sediment losses for each sub-reach are plotted against present fluvial storage volumes in Figure 7. This shows that overall net sediment export by fluvial processes during the Holocene is broadly comparable to that remaining in storage within the study reach (the mean ratio of these values being 0.99, with a range between 0.45-1.72; Fig. 7) with nearly 8 million m^3 of sediment having been lost from the study reach.

5.2 *Rates of Holocene alluvial sediment reworking*

Estimates of rates of alluvial sediment reworking in the study reach are based on sediment transfers accomplished between active channels and terrace stores within each sub-reach. We have no information on inputs of sediment delivered to the upstream part of the reach, nor for sediment passing through the reach, and hence no attempt is made here to develop sediment budgets that quantify *total* reach sediment inputs and outputs. We also make no attempt to differentiate between coarse and fine sediment mobilised from fluvial stores. To facilitate comparison of rates of sediment reworking and export, all values discussed below have been standardised to per annum rates using the timespans for fluvial aggradation and incision cycles defined in the previous section.

Table 1. Volumetric estimates of fluvial sediment storage and % of total sub-reach sediment for Late Pleistocene (T1), Holocene (T2-T6) and post-1850 (T7) fluvial terraces throughout the study reach.

Reach	Alluvial fans, Late Pleistocene (T1), Holocene (T2-T6) and post-1850 (T7) fluvial terraces: Sediment volume *(% of total sub-reach alluvium)*																
(valley km)	Alluvial fans		T1		T2		T3		T4		T5		T6		Sum (T1-T6)	post-1850		Sum (all)	
0.5	9,300	3	78,900	26	26,100	9	124,400	41	21,800	7	9,100	3	0	0	269,600	34,600	12	304,200	4
1	36,200	7	252,900	47	0	0	106,700	20	34,900	7	42,000	8	15,600	3	488,300	45,700	9	534,000	7
1.5	1,200	0	57,000	17	0	0	189,100	55	67,100	20	400	0	0	0	314,800	28,800	8	343,600	4
2	10,100	6	29,000	18	72,200	44	26,200	16	6,300	4	0	0	0	0	143,800	20,700	13	164,500	2
2.5	30,900	8	203,800	52	17,500	4	97,200	25	5,200	1	0	0	0	0	354,600	33,900	9	388,500	5
3	15,500	9	87,500	51	34,600	20	3,400	2	0	0	0	0	0	0	141,000	31,100	20	172,100	2
3.5	0	0	58,600	51	0	0	16,100	14	4,700	4	0	0	0	0	79,400	35,300	31	114,700	1
4	0	0	46,100	25	0	0	50,400	27	12,800	7	7,300	4	0	0	116,600	71,200	38	187,800	2
4.5	56,400	13	252,300	59	0	0	0	0	6,800	2	15,800	4	9,500	2	340,800	87,200	23	428,000	5
5	7,600	2	27,500	9	0	0	57,700	18	41,200	13	67,700	21	64,600	20	266,300	51,400	17	317,700	4
5.5	21,600	4	261,700	43	0	0	151,200	25	27,300	5	13,500	2	24,000	4	499,300	106,300	18	605,600	8
6	0	0	352,800	52	10,800	2	94,100	14	100,400	15	25,600	4	32,500	5	616,200	58,100	9	674,300	9
6.5	0	0	173,800	36	39,700	8	92,400	19	77,500	16	44,100	9	14,400	3	441,900	44,400	9	486,300	6
7	21,600	4	269,900	52	21,800	4	128,900	25	24,700	5	19,500	4	0	0	486,400	31,800	6	518,200	7
7.5	10,300	2	142,500	26	84,300	15	134,100	24	151,700	28	0	0	0	0	522,900	27,600	5	550,500	7
8	36,900	8	0	0	0	0	306,100	69	72,900	16	0	0	0	0	415,900	28,600	7	444,500	6
8.5	69,300	17	0	0	0	0	68,200	17	201,700	50	22,100	5	7,800	2	369,100	34,700	10	403,800	5
9	18,500	10	48,800	25	0	0	11,200	6	11,500	6	2,400	1	17,400	9	109,800	83,400	48	193,200	2
9.5	20,400	7	182,300	61	0	0	25,300	8	0	0	0	0	0	0	228,000	73,100	26	301,100	4
10	0	0	96,900	54	2,400	1	23,300	13	1,700	1	0	0	6,700	4	131,000	50,200	28	181,200	2
10.5	0	0	278,500	58	27,800	6	39,800	8	51,100	11	22,400	5	14,800	3	434,400	46,700	10	481,100	6
Totals *(% total alluvium)*	365,800	5	2,900,800	37	337,200	4	1,745,800	22	921,300	12	291,900	4	207,300	3	6,770,100	1,024,800	13	7,794,900	100

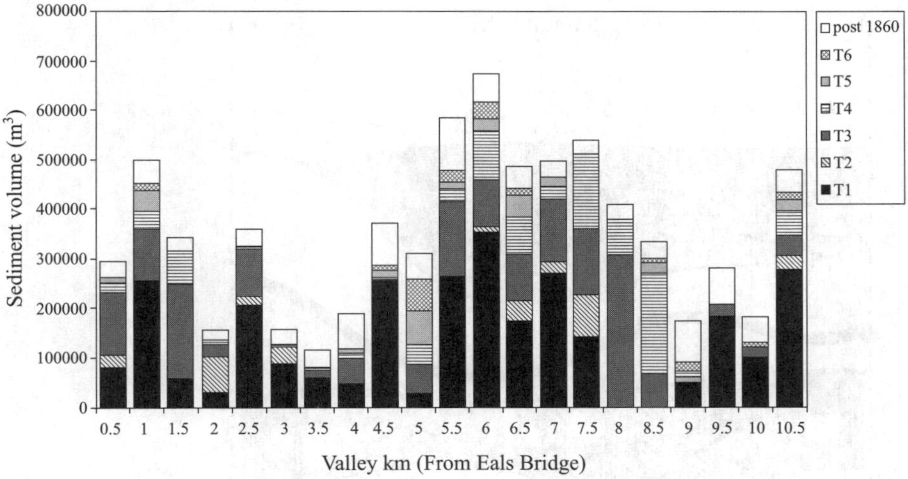

Figure 6. Volume (m^3) of late Pleistocene and Holocene alluvial terraces for each 0.5 km sub-reach of the South Tyne valley between Eals and Haltwhistle.

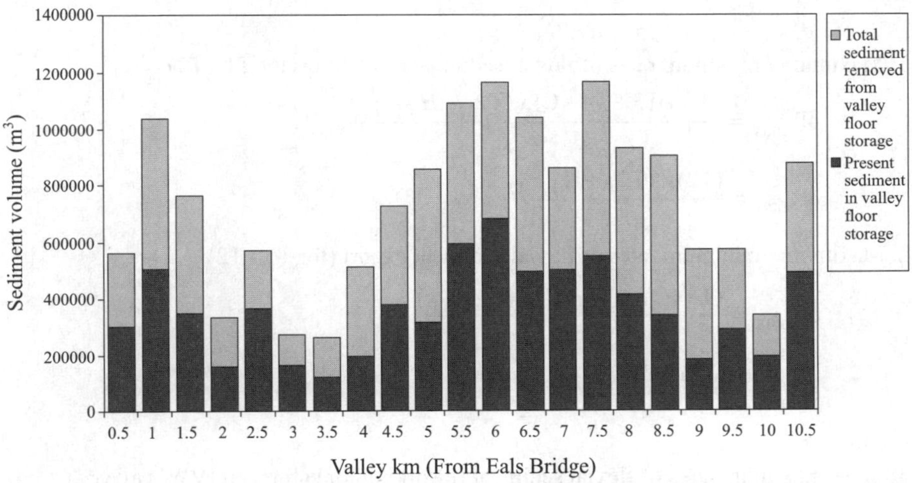

Figure 7. Volumes (m^3) of net Holocene fluvial sediment export and present fluvial sediment storage (including alluvial fans) for each 0.5 km sub-reach of the South Tyne valley between Eals and Haltwhistle.

Figure 8 shows the calculations used to estimate maximum and minimum rates of Holocene valley floor alluvial sediment reworking. Thus, for T2 (dating to between c.9550-1275 cal BC) for example, maximum rates of sediment reworked on the valley floor are obtained by summing the area of T2, all younger terraces and the modern channel area, and multiplying this value by the thickness of T2 and the height difference between T1 and T2 (Fig. 8). Using the same calculation, but restricting the area value to that of T2

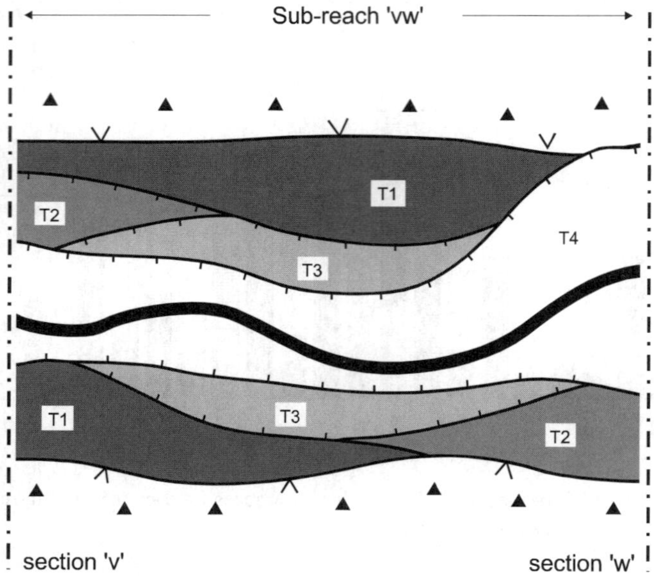

1. Maximum / minimum rates of fluvial sediment reworking (for T1 - T2)

$$\delta R_{max} = \frac{(T2 + T3 + T4 + C)\, x\, (T2_t + H_{T1-T2})}{t}$$

$$\delta R_{min} = \frac{(T2)\, x\, (T2_t + H_{T1-T2})}{t}$$

2. Maximum / minimum rates of fluvial sediment export (for T1 - T2)

$$\delta E_{max} = \frac{(T2 + T3 + T4 + C)\, x\, (H_{T1-T2})}{t}$$

$$\delta E_{min} = \frac{(T2)\, x\, (H_{T1-T2})}{t}$$

where:

δR_{max} = maximum rates of fluvial sediment reworking for sub-reach 'VW' ($m^3\,yr^{-1}$)
δR_{min} = minimum rates of fluvial sediment reworking for sub-reach 'VW' ($m^3\,yr^{-1}$)
δE_{max} = maximum rates of fluvial sediment export for sub-reach 'VW' ($m^3\,yr^{-1}$)
δE_{min} = minimum rates of fluvial sediment export for sub-reach 'VW' ($m^3\,yr^{-1}$)
T_n = area of alluvial terrace
C = area of modern channel (m^2)
$T2_t$ = depositional thickness of terrace T2 (m)
H_{T1-T2} = height difference between surface of T1 and T2 (m)
t = timespan for deposition of T2 (yrs)

Figure 8. Calculations of sediment reworking and net sediment export for a hypothetical Holocene fluvial terrace sequence.

Table 2. Volumetric estimates of fluvial sediment reworked and net sediment export from fluvial storage for the entire study reach over timespans defined by fluvial aggradation and incision cycles (see text for details).

Period	Sediment reworked ($m^3 yr^{-1}$)			Net sediment export ($m^3 yr^{-1}$)		
	min	max	mean	min	max	mean
c.9550-1275 BC (T1-T2)	40	990	520	20	410	210
c.1275 BC – AD 1160 (T2-T3)	690	2330	1510	230	760	490
c. AD 1160-1380 (T3-T4)	4230	12450	8340	2430	6590	4510
c. AD 1380-1660 (T4-T5)	920	4530	2720	170	1080	620
c. AD 1660-1850 (T5-T6)	1090	10670	5880	350	2740	1550
c. AD 1850-1975 (T6-T7)	8460	14540	11500	1610	2800	2210

only, gives an estimate of minimum rates of sediment reworked on the valley floor for the same period. For sub-reaches where T2 is not present, maximum rates are based on the area of all younger terraces and the modern channel, while minimum rates become zero.

Maximum and minimum rates of sediment reworking for each incision-deposition phase are shown for each sub-reach in Figure 9 and are summarised for the entire study reach in Table 2. Highest rates of valley floor fluvial reworking are recorded between c. cal AD 1160-1380 (T4) and c. cal AD 1850-1975 (T7) with annual minimum and maximum values for the entire study reach estimated at 4230-12,450 m^3 yr^{-1} and 8460-14,540 m^3 yr^{-1}, respectively (Table 2). Enhanced values are also evident between c. cal AD 1660-1850 (1090-10,760 m^3 yr^{-1}). Lower rates of sediment transfer are characteristic for the period c. cal AD 1380-1660 (920-4530 m^3 yr^{-1}) and the period before c. AD 1160 (Table 2), although the latter most probably reflect, to some degree, the relatively poor chronological resolution for terraces deposited before the second millennium AD.

All periods show spatial variability in rates of sediment reworking. For periods before the late-fourteenth century AD, higher values are evident in sub-reaches with relatively wide valley floors (Figs 3-5, 9). Peak values are recorded for the period between c. cal AD 1160-1380 (T3-T4) between valley km 8-8.5, at the confluence with Park Burn, where minimum and maximum rates are estimated at 920-1400 m^3 yr^{-1} respectively. After c. cal AD 1380, however, rates of activity are reduced at, and upstream of, Park Burn through valley km 6-9.5, and coincide with high estimated rates of reworking (between 90-850 m^3 per annum for cal AD 1660-1860) in the relatively confined reach between valley km 8.5-9.5 (Figs 3-5, 9).

5.3 *Rates of Holocene sediment export*

Estimates of maximum and minimum net loss of fluvial sediment for each sub-reach are made on similar basis to calculations of valley floor reworking, but are adjusted to exclude the volume of sediment remaining in terrace storage (Fig. 8). For example, maximum rates of sediment export for the period c. 9550-1275 cal BC are obtained by summing the area of T2 and all younger terraces and the modern channel area, and multiplying this value by the height difference between T1 and T2 (Fig. 8). These calculations also give a lower-bound estimate of sediment transfer through the study reach.

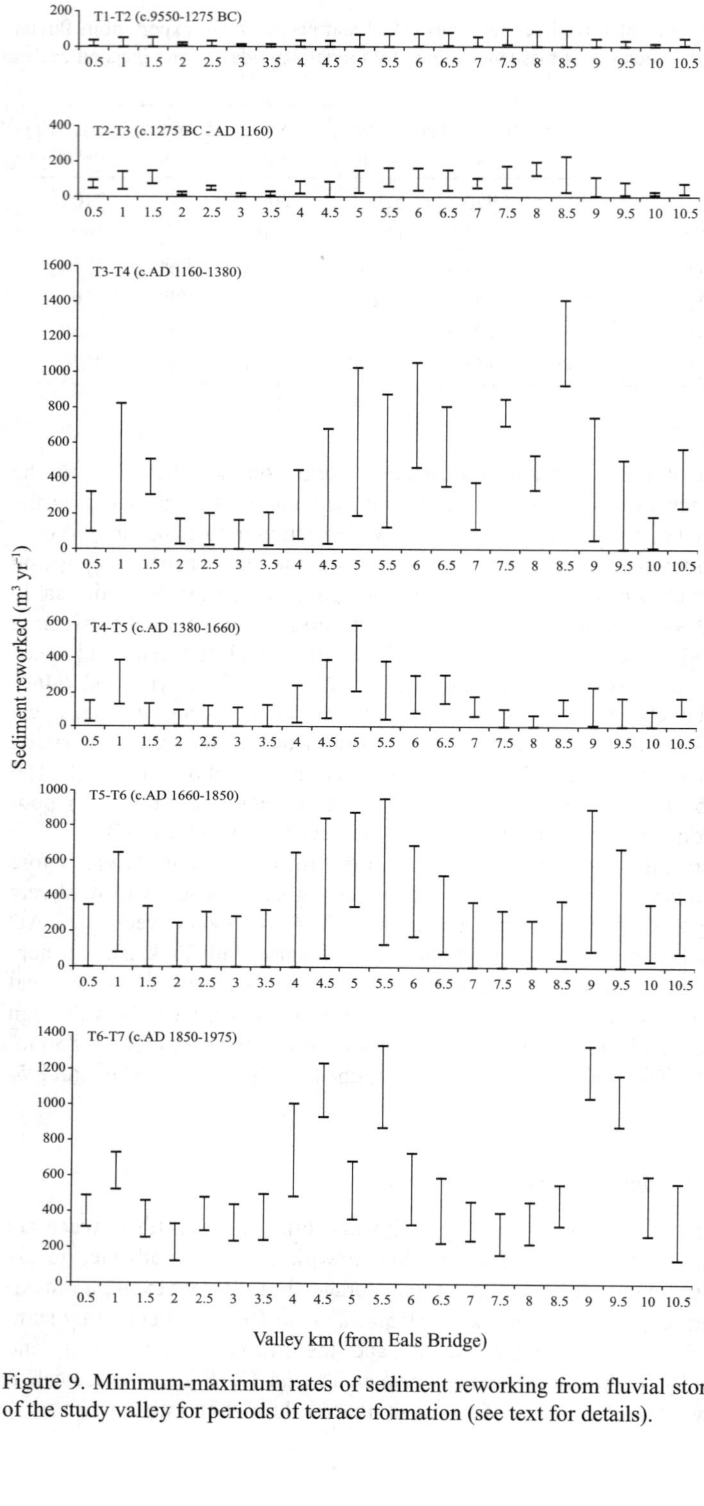

Figure 9. Minimum-maximum rates of sediment reworking from fluvial storage for each sub-reach of the study valley for periods of terrace formation (see text for details).

Spatial patterns of sediment export mirror those for valley floor reworking and show that relatively high rates of sediment transfer are likely to have occurred in wider parts of the valley floor (Fig. 10, Table 2). Values of net sediment export from the study reach also underline the accelerated fluvial activity between c. cal AD 1160-1380; minimum and maximum rates of sediment export over this period are 2430 and 6590 m^3 yr^{-1}, respectively, and are the highest estimated values recorded for the Holocene (Table 2). Valley km 8-8.5 also exhibits the peak range of export values for a discrete sub-reach with minimum and maximum rates estimated at 70 and 940 m^3 yr^{-1}, respectively (Fig. 10).

6 DISCUSSION

6.1 *Valley floor morphology and sediment sources*

Pleistocene glacial and glaciofluvial drift deposits are recognised as constituting the most widespread and important source of coarse sediment to upland British rivers (e.g. Harvey, 1985; Macklin & Lewin, 1986; 1993; Brown & Quine, 1999). In the South Tyne valley these are represented by T1 terraces that are present in nineteen of twenty-one 500 m sub-reaches (Fig. 6), and account for around two-fifths of fluvial sediment stored in the study reach. Although these terraced alluvial units are presently largely decoupled from the modern active channel by inset Holocene fluvial terraces (Figs 3-5), it seems likely that for much of the Holocene period late Pleistocene deposits contributed a significant proportion of fluvial sediment yields. Similar observations in modern and formerly glaciated river basins have been considered as evidence for a paraglacial cycle of upland geomorphic response (after Church & Ryder, 1972) that extends well into post-glacial timescales (e.g. Church & Slaymaker, 1989; Ashmore, 1993; Warburton, 1999).

Valley floor morphology inherited from Pleistocene glaciation and controlled by bedrock geology has also conditioned rates and patterns of Holocene fluvial sediment transfer. Wider alluvial basins form the largest fluvial sediment stores in the study reach and have experienced the highest rates of sediment reworking and export (Figs 9 and 10). Relatively narrow valley reaches constricted by steep bedrock and till bluffs at Lambley-Whitwham and Wydon (Figs 3-7), by contrast, offer comparatively little accommodation space for fluvial sediment storage and have functioned as zones of sediment transfer over the Holocene (Passmore & Macklin, 2000). However, valley floor 'bottlenecks' at Lambley (valley km 4, located where the valley is incised into the dolerite Whin Sill) and Wydon (valley km 9.5) (Figs 3-5) may have promoted channel instability in reaches lying immediately upstream.

High rates of sediment transfer in the narrow Wydon reach (valley km 9.5) and downstream of Lambley village (valley km 5.5) may also reflect tributary sediment input from Park Burn (the largest tributary to join the study reach) and Black Burn (Figs 1, 3-5). Elsewhere in the study reach tributary inputs are either decoupled from the trunk stream by fluvial terraces or appear to have little direct relationship with recent historic channel and floodplain instability. This may be because many tributary streams join where the trunk channel is entrenched and is competent to flush tributary sediment input. Over Holocene timescales, however, these confluence zones may have been the focus of episodic local instability.

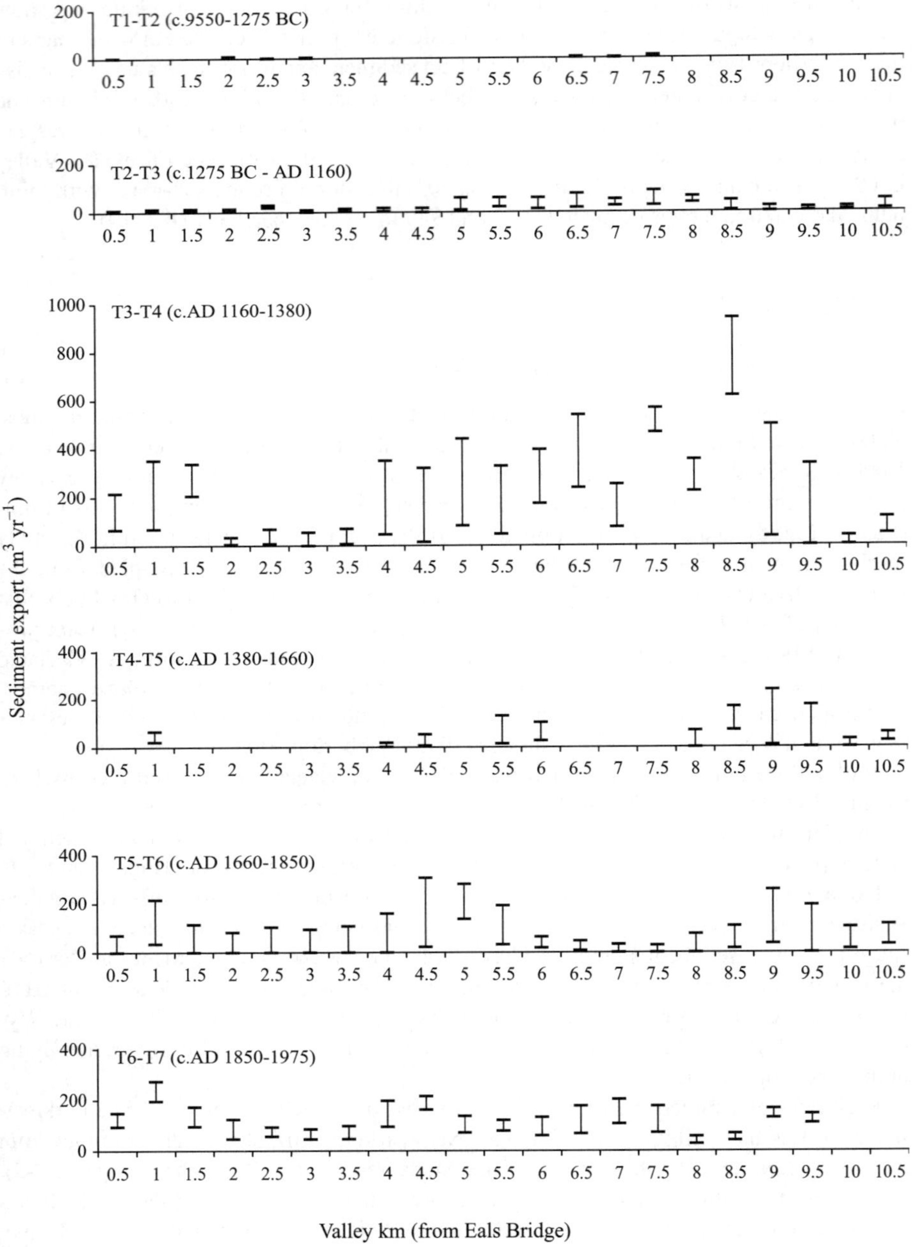

Figure 10. Minimum-maximum rates of net sediment export from fluvial storage for each sub-reach of the study valley for periods of terrace formation (see text for details).

Rates of valley floor reworking and fluvial sediment transfer also show marked temporal variability over Holocene timescales (Figs 9 and 10), although it is accepted that these estimates are conditioned by available dating controls. Low annual rates of fluvial activity during the prehistoric and early historic period (T1-T3) are, in particular, likely to reflect the long intervals between inferred and actual dates. Indeed, based on dated alluvial sequences elsewhere in the South Tyne catchment, T3 sediments are perhaps more likely to have been deposited during the mid-first millennium AD when many valley floors in the catchment were transformed through the effects of climatic deterioration on fluvial systems sensitised by unprecedented rates of catchment disturbance by human activity (Passmore & Macklin, 1997; Passmore & Macklin, 2000). Here we note that T3 terraces are developed at a scale (accounting for 23% of the total fluvial sediment storage, Table 1) consistent with such a transformation.

The highest rates of valley floor reworking and sediment transfer over the past 1000 years occurred between c. cal AD 1160-1380 (T4) and are also enhanced from c. cal AD 1660 to the present. Fluvial activity during these periods is likely to have been exacerbated by catchment disturbance associated with medieval and later agricultural activity (Passmore & Macklin, 1997), while channel instability after the mid-seventeenth century onset of North Pennine base-metal mining will also have been promoted by impeded rates of vegetation colonisation on bar and floodplain surfaces contaminated by metalliferous fine sediment (Macklin & Lewin, 1989; Passmore et al., 1993; Passmore & Macklin, 2000). However, these periods also coincide with well-documented climatic transitions at the end of the Medieval Warm Period and the latter stages of the Little Ice Age neoglacial. Both periods are likely to have witnessed a higher incidence of large flood events (Rumsby & Macklin, 1994, 1996; Brown, 1998), while documentary evidence testifies to a period of higher flood frequency and magnitude during the mid-late nineteenth century (Passmore et al., 1993; Rumsby & Macklin, 1994)

By contrast, comparatively low potential rates of valley floor reworking and sediment transfer occurred between c. cal AD 1380-1660 (Table 2) and coincide with a relatively mild and drier climate before cooling at the end of fifteenth century. Lower rates of river activity during this period are consistent with fluvial records from many other British and European catchments that show a lower frequency of flooding (Rumsby & Macklin, 1996).

Estimates of late Holocene alluvial sediment export from the study reach (Table 2) may be compared to an estimate of modern sediment yield of c.1300 m^3 yr^{-1} for an upland catchment of 414 km^2 (equivalent to the downstream limit of the study reach) obtained using an empirical formula for British river basins developed by Newson (1986). This value is less than one-third of the mean total estimate of annual sediment export from fluvial storage in the study reach for the period c. cal AD 1160-1380 (Table 2), and is also exceeded by equivalent estimates for the periods c. cal AD 1660-1850 and 1850-1975.

Comparison of the scale and rate of late Holocene fluvial activity in the South Tyne with other British rivers may also be made by estimating bank erosion rates (in mm yr^{-1}) for periods spanning incision and fluvial deposition cycles in the study reach. These have been obtained by considering the maximum and minimum areas of reworking for each period and dividing by 500 to achieve a unit area of erosion for each 0.5 km sub reach. Values are then standardised to per annum bank erosion rates. Results are shown in Table 3 together with rates of bank erosion for various periods over the past 150 years reported from British rivers with catchment areas of between 85 and 620 km^2. For the periods since the twelfth century AD in the South Tyne, mean estimated rates of bank erosion in the

Table 3. Bank erosion rates for (i) selected UK river catchments for various periods during the nineteenth and twentieth century, and (ii) sub-reaches of the study valley (expressed as minimum, maximum and mean values) for periods defined by Holocene terrace formation (see text for details).

Reference	Catchment	Catchment area (km^2)	Bank erosion rate ($mm\ yr^{-1}$)
Knighton (1973)	Bollin-Dean, Cheshire	c.260	230
McGreal and Gardiner (1977)	Lagan, Northern Ireland	85	10-80
Hooke (1979)	Various Devon rivers	10-620	80-1180
Lewin (1987)	Upper Severn	164-211	25-1000
Gardiner (1983)	Lagan, Northern Ireland	85	76-139
Ashbridge (1995)	Culm, Devon	276	227-329
Passmore et al. (1993)	South Tyne	800	226-2000
This study	South Tyne	414	*min-max (mean)*
c.9550-1275 BC (T1-T2)	"	"	5-50 (25)
c.1275 BC – AD 1160 (T2-T3)	"	"	10-150 (75)
c. AD 1160-1380 (T3-T4)	"	"	220-2010 (850)
c. AD 1380-1660 (T4-T5)	"	"	130-1320 (395)
c. AD 1660-1850 (T5-T6)	"	"	265-2100 (650)

study valley are comparable to values recorded in reaches of the Upper Severn, South Tyne and some larger Devon rivers, and are significantly higher than rates typical of lower relief British catchments.

Sediment budgeting of late Holocene valley floor reworking and fluvial sediment export in this reach of the South Tyne underlines, therefore, the propensity of upland and piedmont zones in formerly glaciated catchments to experience some of the highest rates of fluvial activity in British river valleys (Brown & Quine, 1999). Rates of fluvial activity, in common with many other upland British catchments (e.g. Harvey & Renwick, 1987; Tipping, 1994; Merrett & Macklin, 1999), appear to have increased over the past 1000 years. This tendency may, at least in part, be a function of the greater preservation potential of younger alluvial units and their higher chronological resolution; however, it is also likely to reflect the combination of (i) progressive entrenchment of the valley floor leading to confined and deepened flood flows and (ii) contamination of the active river corridor by metalliferous fine sediment by seventeenth century and later mining operations (Passmore & Macklin, 2000).

Enhanced rates of valley floor reworking and sediment transfer since the twelfth century AD are therefore likely to reflect the increasing sensitivity of the South Tyne to changes in late Holocene flood regimes, and have most probably been associated with episodically higher sediment yields than those of present rivers. It is also noted, furthermore, that rates of sediment transfer and export reported here represent an approximation of upper and lower bound values, and that the averaging of rates of change between bracketing dates is liable to underestimate actual rates of transfer experienced during large flood events that are 'geomorphologically effective' (e.g. Newson & Macklin, 1990; Coulthard et al., 1999).

7 CONCLUSIONS

Sediment budgeting using morphological techniques has established that in this upland reach of the South Tyne valley postglacial river erosion has removed some 50% of late Pleistocene alluvial sediment deposited during regional de-glaciation. This has left a total of c.7.9 million m^3 of sediment stored within the study reach in the form of late Pleistocene and Holocene fluvial terraces, and a further 0.37 million m^3 as Holocene alluvial fans. Postglacial reworking of local late Pleistocene and Holocene fluvial sediment stores appears to have contributed a significant, and perhaps dominant, proportion of sediment transferred through this part of the South Tyne catchment.

Variability in the rate of valley floor reworking and sediment transfer is evident over Holocene timescales, and particularly over the past 1000 years during which the South Tyne has become increasingly entrenched and sensitive to changes in flood regimes. Maximum estimated rates of sediment flux occur between c. AD 1160-1380, c. AD 1660-1850 and c. AD 1850-1975 and coincide with episodes of increased flood frequency and magnitude and catchment land use changes. Rates of sediment export from local valley floor storage during these periods exceed best estimates for modern sediment yields.

Spatial patterns of valley floor reworking and sediment transfer, however, are governed by valley floor morphology inherited from Pleistocene glaciation and controlled by bedrock geology. Maximum rates of Holocene fluvial activity occur in wider alluvial basins whereas relatively narrow reaches, which are confined by bedrock and till bluffs, have stored less sediment and have acted as zones of Holocene sediment transfer. It is concluded that sediment budgeting techniques offer insights into the scale and rates of Holocene valley floor development that will complement advances in modelling of Holocene fluvial systems, as well as directly informing management frameworks for river valley environments.

ACKNOWLEDGEMENTS

The authors would like to thank the two referees for their helpful comments, and NERC (Grant No.GR3/7617A) for funding the research. Our thanks are also due to Andrew Donnelly and M.A. Sohag for assisting with fieldwork.

REFERENCES

Ashbridge, D. 1995. Processes of river bank erosion and their contribution to the suspended sediment load of the River Culm, Devon. In: Foster, I.D.L., Gurnell, A.M. & Webb, B.W. (eds) *Sediment and Water Quality in River Catchment Systems*. Chichester, John Wiley, 229-245.

Ashmore, P.E. 1993. Contemporary erosion of the Canadian landscape. *Progress in Physical Geography,* 17(2): 190-204.

Ashmore, P.E. & Church, M.A. 1998. Sediment Transport and River Morphology: A Paradigm for Study. In: Klingeman, P.C. et al. (eds) *Gravel-Bed Rivers in the Environment*. Colorado, Water Resources Publications, LLC, 115-148.

Beaumont, P. 1968. *A history of glacial research in northern England.* Durham, University of Durham, Department of Geography Occasional Paper 9.

Bishop, M.C., Passmore, D.G. & Ryder, P.F. 1992. *Haltwhistle Bypass Archaeological Assessment.* University of Newcastle upon Tyne Archaeological Practice, Unpublished Report to Ove Arup & Partners.

Brown, A.G. 1998. Fluvial Evidence of the Medieval Warm Period and the Late Medieval Climatic Deterioration in Europe. In: Benito, G., Baker, V.R. & Gregory, K.J. (eds) *Palaeohydrology and Environmental Change.* Chichester, John Wiley, 43-52.

Brown, A.G. & Quine, T.A. 1999. Fluvial Processes and Environmental Change: An Overview. In: Brown, A.G. & Quine, T.A. (eds) *Fluvial processes and Environmental Change.* Chichester, John Wiley, 1-28.

Campo, S.H. & Desloges, J.R. 1994. Sediment yield conditioned by glaciation in a rural agricultural basin of southern Ontario, Canada. *Physical Geography*, 15(6): 495-515.

Catt, J.A. 1991a. Late Devensian glacial deposits and glaciations in eastern England and the adjoining offshore region. In: Ehlers, J., Gibbard, P. & Rose, J. (ed.) *Glacial Deposits of Great Britain and Ireland.* Rotterdam, Balkema, 61-68.

Catt, J.A. 1991b. The Quaternary history and glacial deposits of east Yorkshire. In: J. Ehlers, P. Gibbard, & J. Rose (ed.) *Glacial Deposits of Great Britain and Ireland.* Rotterdam, Balkema, 185-191.

Church, M. 1983. Pattern of instability in a wandering gravel bed channel. In: Collinson, J.D. & Lewin, J. (eds), *Modern and Ancient Fluvial Systems. International Association of Sedimentologists Special Publication*, 6: 169-180.

Church, M. & Ryder, J.M. 1972. Paraglacial sedimentation; a consideration of fluvial processes conditioned by glaciation. *Geological Society of America Bulletin,* 83: 3059-3072.

Church, M. & Slaymaker, O. 1989. Disequilibrium of Holocene sediment yield in glaciated British Columbia. *Nature*, 337: 452-454.

Coulthard, T.J., Kirkby, M.J. & Macklin, M.G. 1997. Modelling hydraulic, sediment transport and slope processes, at a catchment scale, using a cellular automaton approach. In: Abrahart, R.J. (ed.) *Proceedings of the First International Conference on GeoComputation*, Volume 1. School of Geography, University of Leeds, 248-281.

Coulthard, T.J., Kirkby, M.J. & Macklin, M.G. 1999. Modelling the impacts of Holocene environmental change in an upland river catchment using a cellular automaton approach. In: Brown, A.G. & Quine, T.A. (eds) *Fluvial processes and Environmental Change*. Chichester, John Wiley, 31-46.

Ferguson, R.I. 1981. Channel forms and channel change. In: J. Lewin (ed.) *British Rivers.* London, Allen and Unwin, 90-125.

Gardiner, T. 1983. Some factors promoting channel bank erosion, River Lagan, County Down. *Journal of Earth Science (Royal Dublin Society),* 5: 231-239.

Harvey, A.M. 1985. The river systems of North-west England. In: R.H. Johnson (ed.) *The Geomorphology of North-West England.* Manchester, Manchester University Press, 122-142.

Harvey, A.M. & Renwick, W.H. 1987. Holocene alluvial fan and terrace formation in the Bowland Fells, Northwest England. *Earth Surface Processes and Landforms*, 12: 249-57.

Hooke, J.M. 1979. An analysis of the processes of river bank erosion. *Journal of Hydrology*, 42: 39-62.

Huisink, M. 1999. Lateglacial river sediment budgets in the Maas valley, the Netherlands. *Earth Surface Processes and Landforms*, 24: 93-109.

Knighton, A.D. 1973. Riverbank erosion in relation to streamflow conditions, River Bollin-Dean, Cheshire. *East Midlands Geographer*, 5(8): 416-426.

Lane, S.N., Chandler, J.H. & Richards, K.S. 1994. Developments in monitoring and modelling small-scale river bed topography. *Earth Surface Processes and Landforms*, 19: 349-368.

Lane, S.N., Richards, K.S. & Chandler, J.H. 1995. Within-Reach Spatial Patterns of Process and Channel Adjustment. In: Hickin, E.J. (ed.) *River Geomorphology.* Chichester, John Wiley, 105-130.

Lewin, J. 1987. Historical channel changes. In: Gregory, K.J., Lewin, J. & Thornes, J.B. (eds), *Palaeohydrology in Practice*. Chichester, John Wiley, 161-175.

Macklin, M.G. 1999. Holocene river environments in prehistoric Britain: human interaction and impact. *Journal of Quaternary Science*, 14(6): 521-530.

Macklin, M.G. & Lewin, J. 1989. Sediment transfer and transformation of an alluvial valley floor: The River South Tyne, Northumbria, UK. *Earth Surface Processes and Landforms*, 14: 233-246.

Macklin, M.G. & Lewin, J. 1993. Holocene river alluviation in Britain. *Zeitschrift fur Geomorphologie (Supplement)* 88: 109-122.

Macklin, M.G., Passmore, D.G. & Rumsby, B.T. 1992. Climatic and cultural signals in Holocene alluvial sequences: the Tyne Basin, Northern England. In: S. Needham & M.G. Macklin (eds) *Alluvial Archaeology in Britain*. Oxford, Oxbow Monograph, 27: 123-140.

Macklin, M.G., Rumsby, B.T., Heap, T. & Passmore, D.G. 1994. Thinhope Burn, Northumberland. In: J. Boardman & J. Walden (ed.) *Cumbria Field Guide*. Oxford, Quaternary Research Association Field Guide, 50-58.

Macklin, M.G., Passmore, D.G. & Newson, M.D. (1998) Controls of short and long term river instability: processes and patterns in gravel-bed rivers, the Tyne basin, northern England. In: Klingeman, P.C. et al. (eds) *Gravel-Bed Rivers in the Environment*. Colorado, Water Resources Publications, LLC, 257-278.

Martin, Y. & Church, M. 1995. Bed-material transport estimated from channel surveys: Vedder River, British Columbia. *Earth Surface Processes and Landforms*, 20: 347-361.

Mauquoy, D. & Barber, K. 1999. A replicated 3000 yr proxy-climate record from Coom Rigg Moss and Felecia Moss, the Border Mires, northern England. *Journal of Quaternary Science*, 14(3): 263-275.

McGreal, W.S. & Gardiner, T. 1977. Short-term measurements of erosion from a marine and fluvial environment in County Down, Northern Ireland. *Area*, 9: 285-289.

Merrett, S.P. & Macklin, M.G. 1999. Historic River Response to Extreme Flooding in the Yorkshire Dales, Northern England. In: Brown, A.G. & Quine, T.A. (eds) *Fluvial Processes and Environmental Change*. Chichester, John Wiley, 345-360.

Neill, C.R. 1983. Bank erosion versus bed load transport in a gravel river. In: *River meandering; Proceedings of the Rivers '83 Conference*, New Orleans, Louisiana. American Society of Civil Engineers Special Publication, 204-221.

Newson, M.D. 1986. River basin engineering – fluvial geomorphology. *Journal of the Institute of Water Engineers Science*, 40: 307-324.

Newson, M.D. & Macklin, M.G. 1990. The geomorphologically effective flood and vertical instability in river channels: a feedback mechanism in the flood series for gravel bed rivers. In: White, W.R. (ed.) *International Conference on River Flood Hydraulics*. Chichester, John Wiley, 123-141.

Passmore, D.G. 1994. *River response to Holocene environmental change: the Tyne basin, northern England*. Unpublished PhD Thesis, University of Newcastle upon Tyne.

Passmore, D.G., Macklin, M.G., Brewer, P.A., Lewin, J., Rumsby, B.T. & Newson, M.D. 1993. Variability of Late Holocene braiding in Britain. In: J.L. Best & C.S. Bristow (ed.) *Braided Rivers*. London, Geological Society, 205-232.

Passmore, D.G. & Macklin, M.G. 1997. Geoarchaeology of the Tyne Basin: Holocene river environments and the archaeological record. In: Tolan-Smith, C. (ed.) *Landscape Archaeology in Tynedale*. University of Newcastle upon Tyne, Tyne-Solway Monograph No.1: 11-27.

Passmore, D.G. & Macklin, M.G. 2000. Late Holocene channel and floodplain development in a wandering gravel-bed river: The River South Tyne at Lambley, northern England. *Earth Surface Processes and Landforms*, 25: 1237-1256.

Patton, P.C. & Schumm, S.A. 1975. Gully erosion, northwestern Colorado: a threshold phenomenon. *Geology*, 3: 88-90.

Phillips, J.D. 1991. Fluvial sediment budgets in the North Carolina Piedmont. *Geomorphology*, 4: 231-241.

Raistrick, A. 1931. The glaciation of Northumberland and Durham. *Proceedings of the Geologists Association*, 42: 281-291.

Raistrick, A. & Jennings, B. 1965. *A History of Lead Mining in the Pennines.* London, Longmans.

Rumsby, B.T. & Macklin, M.G. 1994. Channel and floodplain response to recent abrupt climate change; the Tyne basin, northern England. *Earth Surface Processes and Landforms,* 19: 499-516.

Rumsby, B.T. & Macklin, M.G. 1996. River response to the last neoglacial (the 'Little Ice Age') in northern, western and central Europe. In: Branson, J., Brown, A.G. & Gregory, K.J. (eds), *Global Continental Changes: the Context of Palaeohydrology*, Geological Society Special Publication, 115: 217-233.

Schumm, S.A. 1973. Geomorphic thresholds and complex response of drainage systems. In: M. Morisawa (ed.) *Fluvial Geomorphology.* SUNY, Binghampton Publications in Geomorphology, 299-309.

Sohag, M.A. 1993. *Sediment tracing, bed structure and morphological approaches to sediment transport estimates in a gravel-bed river: the River South Tyne, Northumberland, UK.* Unpublished PhD thesis, University of Newcastle upon Tyne.

Stuiver, M., Long, A. & Kra, R.S. (eds) 1993. 1993 Calibration issue. *Radiocarbon,* 35(1).

Tipping, R. 1994. Fluvial chronology and valley floor evolution of the Upper Bowmont Valley, Borders Region, Scotland. *Earth Surface Processes and Landforms*, 21: 77-91.

Tipping, R. 1995. Holocene evolution of a lowland Scottish landscape: Kirkpatrick Fleming. Part III, fluvial history. *The Holocene,* 5: 184-195.

Trimble, S.W. 1983. A sediment budget for Coon Creek basin in the Driftless area, Wisconsin, 1853-1977. *American Journal of Science*, 283: 454-474.

Trimble, S.W. 1995. Catchment Sediment Budgets and Change. In: Gurnell, A. & Petts, G. (eds) *Changing River Channels*. Chichester, John Wiley, 201-215.

Warburton, J. 1999. Environmental Change and Sediment Yield from Glacierised Basins: The role of Fluvial Processes and Sediment Storage. In: Brown, A.G. & Quine, T.A. (eds) *Fluvial Processes and Environmental Change*. Chichester, John Wiley, 385-408.

16. The potential for high resolution fluvial archives in braided rivers: quantifying historic reach-scale channel and floodplain development in the River Feshie, Scotland

BARBARA RUMSBY & ROSS MCVEY
Department of Geography, The University of Hull, UK

JAMES BRASINGTON
Department of Geography, Cambridge University, UK

1 INTRODUCTION

Fluvial response to rapid environmental change, especially abrupt, short-term climatic fluctuations, is a major current focus of research in fluvial systems (cf. FLAG Focus 3 and 4), reflecting the wider concern over the implications of recent, and potential future, global climate change for earth surface systems. Studies of past fluvial activity have much to offer as analogues of channel and floodplain response, although their utility as indicators of rapid environmental change is contingent on the retrieval of high resolution information from the alluvial archive. Recent studies in a wide range of basins have provided mounting evidence of the sensitivity of upland fluvial systems to short-lived climate change in the last millennium (Rumsby & Macklin, 1996 and references therein). Rumsby & Macklin (1996), for example, have shown that upland basins in north, west and central Europe experienced enhanced fluvial activity in response to episodes of increased flooding coincident with Little Ice Age climate changes. Within the British uplands, the responsiveness of fluvial systems in north-west and north-east England to late Holocene climate and land-use changes has also been well-established over the past decade (e.g. Harvey et al., 1984; Macklin et al., 1992a). However, to date there have been few such studies in Scottish river basins. Despite growing evidence that hillslopes in the Scottish Highlands were highly sensitive to late Holocene climate changes, with enhanced debris cone accumulation and debris flow activity (Ballantyne, 1991), relatively little is known of the degree and timing of coupling of alluvial fans and debris cones with fluvial systems, and the susceptibility and morphological response of channels and floodplains themselves.

The majority of contemporary upland fluvial systems in Britain comprise single thread, wandering gravel-bed rivers, with occasional divisions around mid-channel bars. Extensive braiding in Britain is, however, relatively rare, there being only one major active braided system, the River Spey draining the Cairngorm Mountains in Scotland (Ferguson, 1981; Passmore et al., 1993). A number of studies have nonetheless indicated that braiding was more widespread in the past, especially in the eighteenth and nineteenth centuries and also during earlier periods of the Holocene. Passmore et al. (1993) examined the timing of late Holocene braiding in two major British rivers, the River Tyne and the Upper Severn. While at present braiding in these systems is limited to a few sites, they identified significant increases in the occurrence of braiding during the eighteenth-nineteenth centuries, the thirteenth-fourteenth centuries and the late Roman period. Channel transformation and aggradation during these periods was attributed to increased rates of coarse sediment supply, associated with tributary and trunk stream incision in response to enhanced flood frequency and magnitude. Further evidence for widespread braiding in the late nineteenth century is provided by Werritty & Ferguson's (1980) study of planform changes on the Feshie, Scotland (see below).

In dynamic braided rivers the stratigraphic record is likely to comprise a complex sequence of spatially and temporally discontinuous units. Research on the longer-term evolution of braided systems has often tended to underestimate the heterogeneity of such alluvial sequences, correlating deposits over large areas from a typically small number of samples and with limited consideration of the preservation potential of individual units. In such dynamic systems, the longer-term preservation of alluvial units is therefore a key question, with implications for interpretation of the deposits associated with the more extensive distribution of braided rivers in Britain and NW Europe earlier in the Quaternary (cf. FLAG Focus 3). Investigation of active and recently active braided systems provides a basis for understanding patterns and rates of channel change and the complexity of the alluvial sequence arising from a range of climatic and environmental conditions. In this paper we undertake an investigation of the dynamics of the braided River Feshie, Scotland, with the aim of evaluating the temporal and spatial variability and preservation of the sedimentary record. This study is based on an interpretation of high-resolution geomorphological maps derived from a combination of aerial photography and cartography with topographic survey of contemporary active channel and bar morphology. A spatially and temporally nested approach is adopted, with investigations focusing on three overlapping reaches (10^0-10^3 m) and with a timeframe varying from the late historic period to the present (10^0-10^2 a). The focus on the later historical period, with abundant documentary evidence, was chosen as it is arguably the best period to test the potential for, and preservation of, high resolution alluvial archives in braided systems. This study takes advantage of recent developments in survey technology and computing power, through the use of the GPS (Global Positioning System) to survey modern channel topography and a GIS environment to store, manage and analyze geomorphological maps.

2 STUDY AREA

The River Feshie is a tributary of the Spey, draining a total area of 235 km^2 at the western edge of the Cairngorm Mountains, Scotland (Fig. 1). The catchment is underlain primarily by Moinian schists, which dominate the contemporary coarse sediment load, and Cairngorm granites in the north and east of the basin. The Feshie lies within the limits of the late Devensian ice advance and headwater tributaries, particularly those draining from the east, are also likely to have been affected by Loch Lomond nival activity (Werritty & Ferguson, 1980). The present form of the valley strongly reflects the glacial legacy, with wide, U-shaped reaches alternating with narrower sections where the valley is laterally confined by bedrock (former glacial breaches) or outwash terraces. Extensive Holocene terrace development is restricted to the wider reaches, comprising a series of unpaired and discontinuous units inset within the high, fluvio-glacial terrace surface (Robertson-Rintoul, 1986). At the present day, the Feshie is actively re-working Pleistocene and Holocene valley floor sediments. Abundant availability of coarse sediment combined with steep slopes, a very flashy flow regime and high stream power, results in widespread braiding which is particularly well developed within three active 'sedimentation zones' (*sensu*

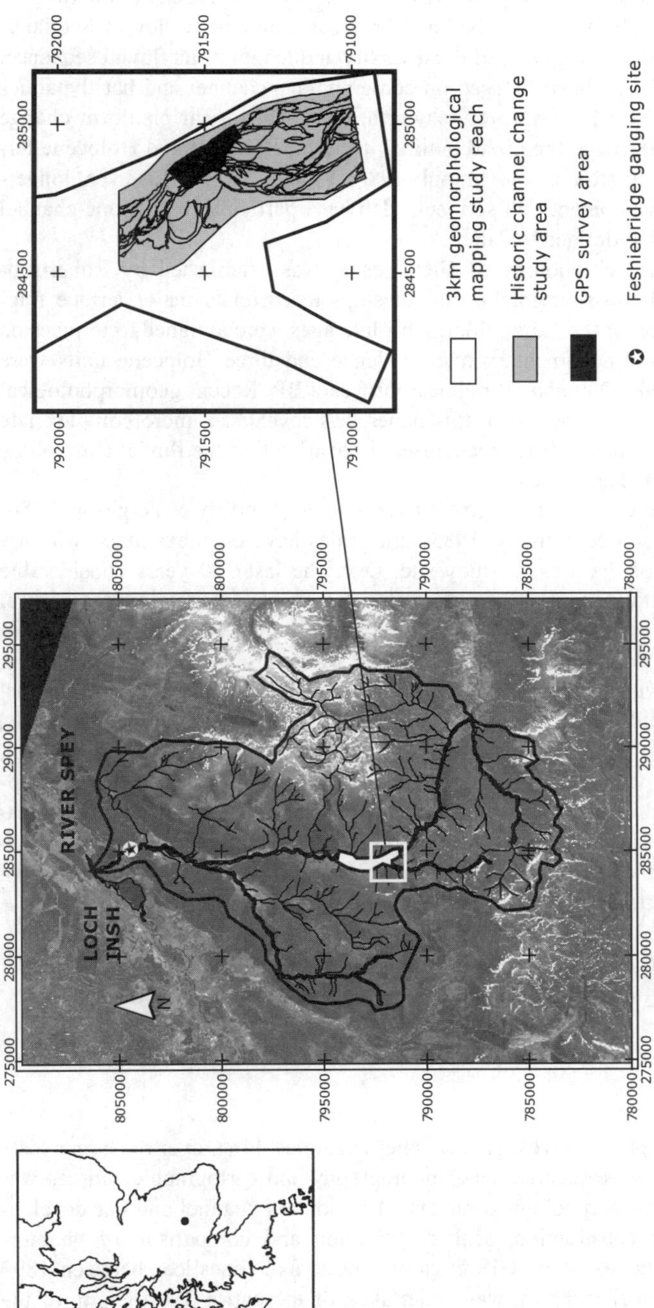

Figure 1. River Feshie location map. Inset shows location of the three, nested study reaches in the Upper Feshie.

Church & Jones, 1982; Desloges & Church, 1989) located within wider reaches of the valley floor. The present study concentrates on the most southerly of these active braided reaches, upstream of Glenfeshie Lodge (NN 842933), with a drainage area of ~80 km^2.

The Feshie has previously been described as 'the most important valley in Scotland' (Werritty & McEwen, 1997) on the basis of the diversity and length of its fluvial sequence. To date, studies on the Feshie have focused on contemporary channel and bar dynamics (Ferguson & Werritty, 1983; Ferguson & Ashworth, 1992), historical planform change (Werritty & Ferguson, 1980) and the broad outline of the Late Glacial and Holocene terrace sequence in the valley (Robertson-Rintoul, 1986; Young, 1976). However, longer-term channel and bar development and sediment delivery, particularly hillslope-channel linkages, have not been well defined to date.

The Post-Glacial terrace chronology of Glen Feshie was established by Robertson-Rintoul (1986) who used soil-stratigraphic relationships to correlate major terrace fragments over a 12 km section of the valley floor. Absolute ages were assigned to terraces on the basis of ^{14}C dating and documentary map evidence and three Holocene units were identified, dating to ~3600, 1000 and 80 radiocarbon years BP. Recent geomorphological mapping in the upper Feshie, reported in this paper, has revealed a more complex late Holocene terrace sequence than hitherto recognised, indicating that the fluvial chronology has yet to be fully elucidated and dated.

Channel pattern change and bar development over historic (Werritty & Ferguson, 1980) and contemporary (Ferguson & Werritty, 1983) timescales have been examined within a 3 km reach downstream of the present study site. Over the last 200 years considerable variations in channel location and planform characteristics were identified, with a general decrease in sinuosity and an increase in braiding particularly marked since the late nineteenth century (Werritty & Ferguson *ibid.*). The role of floods was seen as central in channel pattern change, with high magnitude events responsible for changing the channel pattern to an unstable, braided one. Subsequent lower flows rework bar and bed materials, resulting in rationalisation and reorganisation into a simpler single thread main channel, with mid-channel bars. Investigation of short term channel dynamics led to the development of a tentative model of braided river development which linked episodic progradation of alternate diagonal bars with bank erosion opposite accreting bar margins and avulsion around and across the bar. This model was supported by Ferguson & Ashworth (1992) who surveyed patterns of velocity, shear stress and bedload transport within a ~75 m length reach of the Upper Feshie on a single day (2 May 1986) during a diurnally-varying snowmelt flow (peaking at around bankfull). Their results suggested that bar growth proceeds with the transfer of sediment from pool head to pool tail and bar front.

3 METHODS

Two approaches were adopted to investigate channel dynamics. First, over the larger scale and longer-term, analysis of sequential aerial photography and cartographic sources was used to infer the recent geomorphological history of braidplain channel and bar development. This involved the construction, scale-rectification and comparison of geomorphological maps within the ArcView GIS environment at five timeslices between 1899 and 1997. Geomorphological mapping was undertaken of the entire 3 km length of the Upper Feshie braidplain, although detailed temporal analysis was restricted to a smaller

(880 × 350 m), nested reach at the upstream end of the full reach. While this approach provides an important insight into the two-dimensional pattern of channel change, it fails to account for vertical adjustments and sheds little light on longer-term sediment budgets. In order to begin to address this methodological shortcoming, a second approach was employed to monitor the three-dimensional pattern of channel change in a 220 × 70 m subsection of the main upstream reach. A detailed topographic survey of contemporary bed and bar morphology provided data which was used to derive a Digital Terrain Model (DTM) of the reach. Here the results from a survey in 1998 are presented which will be used as the baseline to assess patterns of three-dimensional channel change by directly differencing DTMs developed from future annual surveys. This survey was undertaken using a differential GPS-based survey system with the potential to acquire high precision (±1-5 cm) digital topographic data, without the limitations of 'line-of-sight' and post-processing required of traditional tacheometric survey methods (Van Sickle, 1996).

3.1 *Geomorphological mapping*

A series of historical geomorphological maps were constructed for the upstream subsection from aerial photographs flown in 1946, 1955, 1964 and 1989 at a range of scales and from the Second Edition OS 6' map dated to 1899 (Table 1). In addition, the entire 3 km reach was mapped from colour vertical APs flown in 1997 at ~1:10,000 scale. The use and limitations of sequential aerial photography for deriving information on channel planform change are well-established (Hooke & Kain, 1982; Hooke, 1997). A noted shortcoming is that the resolution and precision of information derived from APs decreases as scale increases, making it difficult to obtain more than the broad channel plan and major gravel units for the relatively small streams of upland Britain on scales smaller than 1:10,000 (Hooke & Kain, 1982). A second problem is related to the geo-referencing and overlaying of APs at different scales. In this study these problems are somewhat mitigated through the use of image analysis software and GIS to rectify and digitally analyze photographic data. Each AP was scanned to give digital images at a common resolution of ~2.4 pixels per m (0.4 m per pixel) and georeferenced within Erdas Imagine. A polynomial transformation was used to register images to the British National Grid. Table 1 displays the root mean squared error (metres) of x and y coordinates for control points on each image used in the study. While there is some variation between dates, the results indicate a mean, error

Table 1. Aerial photograph and map details and control point root mean squared errors associated with geo-referencing to British National Grid.

Year	Day & month	Contact scale	Format	Control point errors (m) X	Y	Total
1997	5 August	1:10,000	Colour	0.19	0.29	0.40
1989	7 May	1:24,000	B & W	0.27	0.34	0.44
1964	31 August	1:27,000	B & W	1.35	2.97	3.26
1955	22 July	1:10,000	B & W	1.70	1.99	2.62
1946	8 May	1:10,000	B & W	1.29	0.89	1.57
1899	'Winter'	1:10,560	B & W	0.91	1.56	1.81

in x and y of less than 3.00 m and maximum total error of 3.26 m. This implies a limit of detection of differences between paired images, both of which contain errors, of ~4.5 m.

Once registered to a common projection images were interpreted and geomorphological features digitized on-screen within ArcView GIS. The active zone was classified into three categories, channel, exposed gravel and vegetated gravel, and the inactive (palaeo-) zone into a further three, palaeochannels, bars and terraces edges. The key advantage of on-screen digitizing over using a standard tablet is the facility to vary the scale, particularly to enlarge and zoom-in on features of interest. Within ArcView, areas occupied by each type of geomorphological feature in each map date were calculated and locational differences between maps identified using map overlay functions. The potential magnitude of errors in areal estimates associated with digitizing boundaries of features was assessed through repeat digitising of the outline of a medium-sized gravel bar within the active zone on the 1997 AP. The bar outline was digitised 30 times and bar area calculated for each run. This gave a mean area of 2168.5 m^2 and standard deviation of 11.3 m^2, implying a 95% confidence interval of 22.7 m^2 (2σ). Thus, the potential error associated with areal estimates derived from digitizing is in the order of 1% of the total area.

3.2 *Topographic mapping*

The advantages of a three-dimensional, as opposed to planimetric, evaluation of channel change have been discussed earlier. Nonetheless, until recently, there have been relatively few attempts to monitor and interpret the patterns and processes involved in channel adjustment in a fully three-dimensional manner. Indeed, most attempts to monitor vertical channel dynamics have been based on repeat survey of coarsely spaced cross-sections with consequent emphasis on lateral as opposed to longitudinal patterns and processes. In large part these methodological shortcomings can be considered a response to the difficulty of acquiring morphological data at a resolution sufficient to allow a full description of channel form. Within the last decade, however, developments in survey technology have enhanced the potential to capture terrain information at scales and levels of precision previously unattainable with traditional tacheometric-based survey methods. Lane et al. (1993) for example, have demonstrated the possibility to rapidly obtain detailed topographic representation of exposed channel bars using digital terrestrial photogrammetry. A key advantage of such highly distributed topographic data is the ability to derive digital terrain models (DTMs) of channel form. Through a combination of repeat survey and terrain modelling, fully three-dimensional patterns of channel change, highlighting areas of net scour or aggradation can then be identified by directly differencing DTMs derived from surveys at different epochs (Lane, 1998; Stojic et al., 1998).

While the photogrammetric approach has much to offer in terms of monitoring exposed channel areas, and even submerged bedforms under clear water, this method is unsuitable for bed survey in rivers with highly coloured or turbid waters such as the Feshie. The recent adaptation of the Global Positioning System for terrestrial surveying offers a versatile alternative method to acquire rapid, weather independent, high precision information on exposed and submerged topography. Here, a detailed topographic survey of the 220 × 70 m sub-section of the upper study reach was carried out in summer 1998 using Geotronics Geotracer 2000 GPS survey equipment. This technology involves the identification of an unknown ground position by radio tracking the distances to at least 4 from a constellation of 24 satellites and resolving the ground position by trilateration. Rapid 'stop and go' survey with two GPS receivers (one static and one roving) was used to acquire digital British National Grid coordinates in real time (1-3 seconds) without post-processing, allowing the collection of 2000+ survey points per day over complex terrain. A quasi-systematic survey method was adopted, in which individual bed and bar units were surveyed in tightly spaced transects, while important topographic features (e.g. breaks of slope) were later infilled. At a small number of locations the channel was too deep (> 1 m) to allow access with the GPS equipment and topography was determined using traditional levelling techniques.

In addition to the survey of the main reach, 19 control points were established which were reoccupied 2-3 times daily to assess the survey precision and nature of GPS errors (Table 2).

Results from the control point survey reveal that planform precision is superior to measurements in the vertical, although the results imply 95% confidence intervals of approximately 4 cm in x and y and just 5 cm in the z. The distribution of observations was found to be approximately normal. Individual observations may deviate more markedly from the mean, but even so the maximum range of only 8.5 cm in z in a total of 399 measurements is acceptable for most geomorphological applications. A 0.25 m grid-based DTM

Table 2. GPS survey control point standard deviations and ranges.

Point	N	Standard deviation (m)			Maximum range (m)		
		Easting	Northing	Elevation	Easting	Northing	Elevation
1	19	0.017	0.015	0.026	0.068	0.086	0.075
2	19	0.017	0.028	0.024	0.060	0.081	0.088
3	19	0.023	0.031	0.026	0.080	0.119	0.085
4	19	0.019	0.021	0.024	0.081	0.068	0.074
5	19	0.017	0.018	0.023	0.073	0.073	0.067
6	19	0.018	0.018	0.026	0.073	0.081	0.097
7	19	0.019	0.021	0.023	0.060	0.081	0.068
8	19	0.018	0.026	0.026	0.078	0.106	0.082
9	19	0.019	0.000	0.026	0.071	0.073	0.101
10	19	0.014	0.023	0.023	0.056	0.072	0.075
11	19	0.016	0.023	0.031	0.050	0.075	0.100
12	19	0.014	0.028	0.028	0.050	0.092	0.102
13	19	0.022	0.028	0.028	0.083	0.147	0.079
14	19	0.023	0.021	0.024	0.082	0.060	0.076
15	19	0.026	0.015	0.026	0.087	0.086	0.088
16	19	0.019	0.023	0.027	0.078	0.080	0.085
17	19	0.017	0.010	0.027	0.064	0.086	0.080
18	19	0.027	0.023	0.031	0.100	0.087	0.099
19	19	0.023	0.023	0.027	0.081	0.071	0.083
20	19	0.024	0.018	0.026	0.088	0.070	0.091
21	19	0.016	0.015	0.027	0.059	0.086	0.097
	Mean	**0.019**	**0.020**	**0.026**	**0.072**	**0.085**	**0.085**

of the reach was derived by quintic resampling of a TIN (triangular irregular network) constructed by Delauney triangulation of the GPS survey points. Data analysis was conducted using the Arc/Info GIS.

4 GEOMORPHOLOGICAL MAPS

The geomorphological map covering the 3 km study area represents at least four phases of fluvial activity (Fig. 2). The current active zone (delimited by the solid black line) is highly braided, with extensive development of mid-channel and lateral gravel bars. Much of the gravel is unvegetated or sparsely vegetated reflecting the frequent reworking and transfer of sediment in the reach. Adjacent to, and up to ~1 m above, the active zone are a series of palaeochannels and bar fragments, including the large, well-vegetated island close to the junction between the Upper Feshie and Allt Lorgaidh. Historic and OS map evidence (see below) suggests that these were active in the nineteenth-early twentieth centuries. Palaeochannel planforms associated with this unit appear to be more sinuous than at present and less highly braided, with fewer mid-channel bars and flow concentrated in several well-defined channels to the west and east of the present active zone. However, caution is needed when interpreting this pattern due to the varying preservation potential of mid channel bars and bends depending on the mechanism of channel change. For example, where chute cut-offs occur, preservation of the outermost cut banks of curved channels may dominate and give a misleading impression of total sinuosity. The back edge of this terrace is marked by a major break of slope up to 2 m in height that is particularly well defined on the west side of the valley floor. At least three higher fluvial terraces are present in the study area at elevations of up to 4m above the present channel bed. This suggests a somewhat more complex sequence of Holocene valley floor development than that indicated by Robertson-Rintoul (1986) who identified only two terraces above the level of the nineteenth century deposits, tentatively dated to 3600 and 1000 uncal. BP.

4.1 *Summary of changes, 1899-1997*

The detailed historical geomorphological maps of the 880 m study reach are displayed in Figure 3 and the areal extent of each feature compared in Table 3. Active channel and bar development is discussed in detail later and general trends outlined below.

The greatest channel extents are shown on the 1899 and May 1989 maps, at 12.6% and 9.3% of the total area of ~280,000 m^2, with a minimum of 5.1% in July 1955. Peak extent in both exposed and vegetated gravel area is shown in August 1964, with minima in May 1989 and 1899, indicating an inverse relationship with estimated flow stage. This notwithstanding, there do seem to be some general trends between successive maps, with an increase in gravel area 1899-1946-1955-1964, a marked decline between 1964 and 1989 and a moderate increase 1989-1997. The effects of stage differences between the maps are reduced through comparison of total active zone area, combining channel, exposed gravel and vegetated gravel categories. This indicates a key distinction between the two most recent maps and the four older ones, with the former exhibiting much reduced active zone extent. Palaeochannel extent increases between successive maps, 1946-1997, reflecting the overall reduction in active zone extent and rationalisation of the channel pattern in the

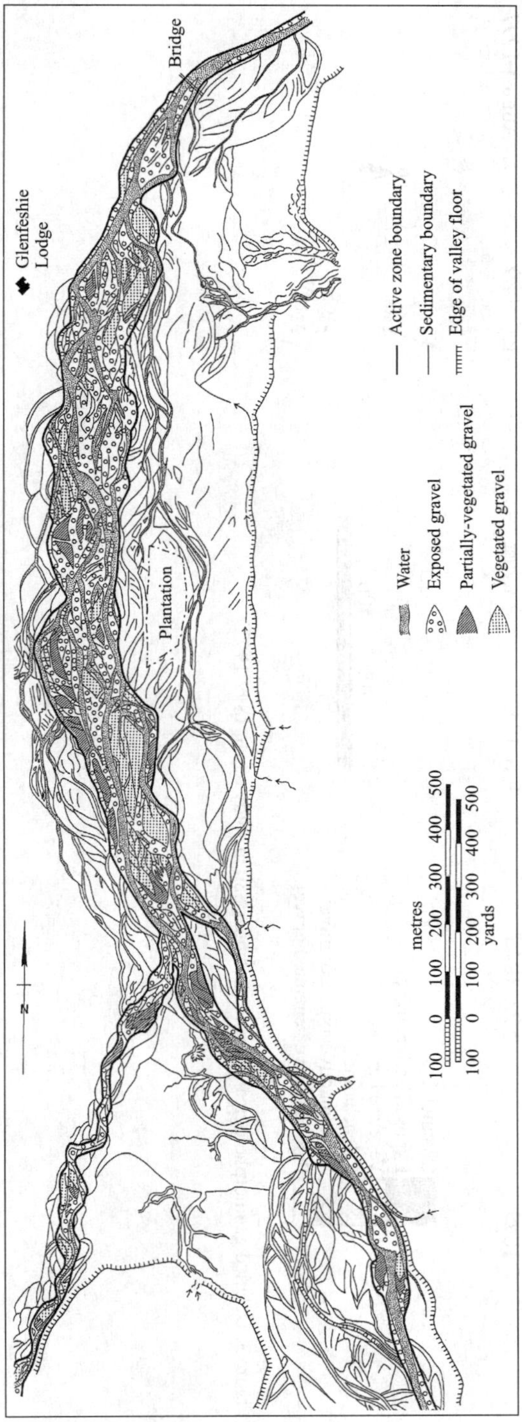

Figure 2. Geomorphological map of the 3 km study reach.

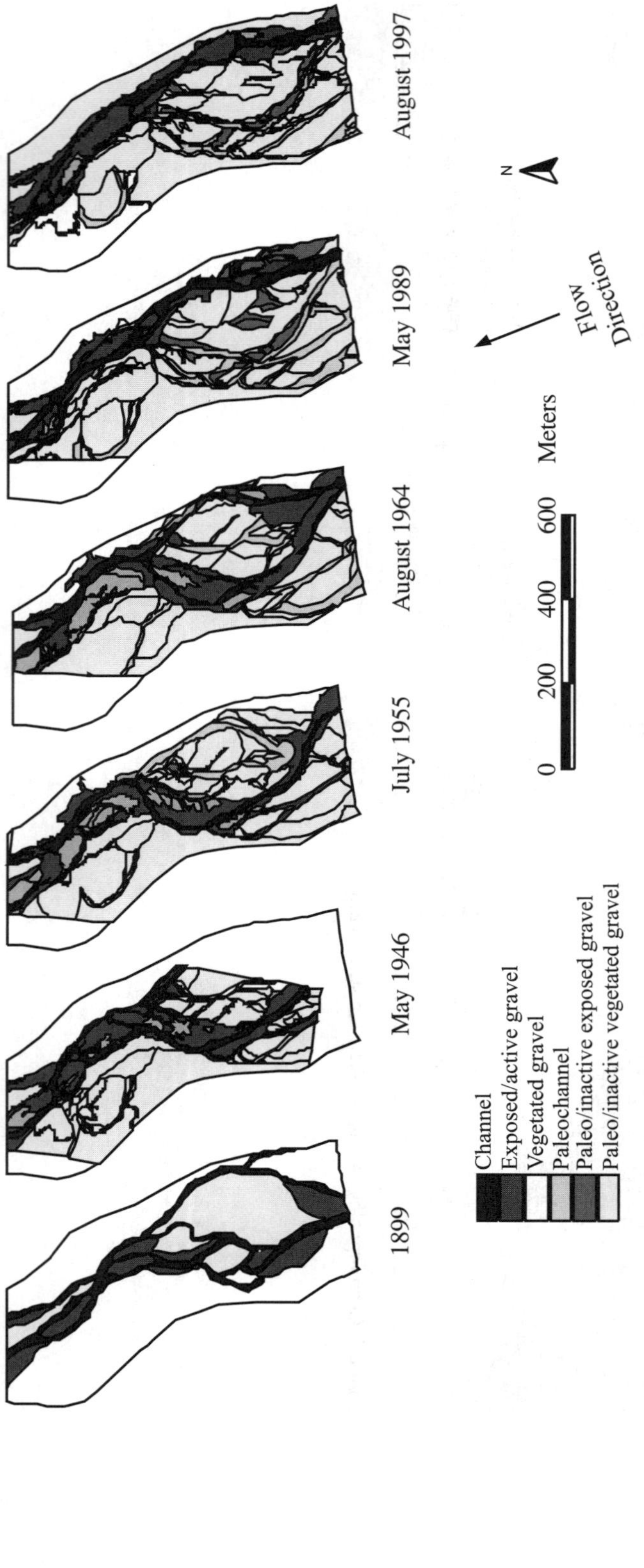

Figure 3. Detailed geomorphological maps of the upper 880 m study reach, 1899-1997.

Table 3. Areal extent of features identified on Upper Feshie geomorphological maps.

Area (Percent of total area)	1899	May-46	Jul-55	Aug-64	May-89	Aug-97
Channel	12.6	6.8	5.1	7.4	9.3	6.3
Exposed gravel	12.7	14.0	15.5	18.5	8.1	13.8
Vegetated gravel	0	2.5	6.9	7.6	0.4	1.3
Total active zone (C+EG+VG)	25.3	23.3	27.4	33.5	17.7	21.4
Palaeochannel	0	2.9	5.2	6.5	9.0	9.0
Palaeo-exposed gravel	0	0	0.3	0	7.9	3.5
Palaeo-vegetated gravel	19.8	38.2	49.8	41.9	47.3	55.3
Unclassified	54.9	35.6	17.3	18.1	18.0	10.7
Total area (m^2)	281,084	281,205	281,205	281,205	280,645	279,952
Active zone remaining in 1997 (m^2)	19,475	3154	8722	49,547	4624	59,914
Active zone remaining in 1997 (% of original active area)	27.4	4.8	11.3	52.6	9.3	100

upper half of the reach (see below), particularly since the 1960s. The pattern of changes in palaeo-exposed and vegetated gravel is similar to the previous category, with an overall increase from 38.2% in 1946 to 58.8% in 1997. The exception to this pattern is the decrease in area shown on the 1964 map. This is related to channel avulsion in the upper half of the reach between 1955 and 1964, with re-occupation of the eastern anabranch along with continued maintenance of flow in the western branch. Thus, several areas shown as palaeo- or inactive prior to 1964 were being actively re-worked by 1964.

There are a number of caveats that complicate a straightforward interpretation of the changes identified above. First, the 1946 map is incomplete as the AP did not provide full cover of the study area, truncating the upper, SE section of channel. Second, there are differences in the flow stage represented in each image. The APs were flown at different times of the year: 1946 and 1989 were taken in May, 1964 and 1997 in August and 1955 in July. The precise survey date for the 1899 OS map is not stated, however, OS 1:10,560 maps usually represent channel boundary at 'normal winter water level' (Hooke & Kain, 1982). Thus, the a proportion of the increase in channel area and reduced extent of exposed gravel on the 1899 map, and to a lesser extent on the 1946 and 1989 maps, may be stage related. Representation of palaeo-features (e.g. terrace edges) is less likely to be affected, although it is possible that palaeochannels may be inundated during higher flows.

A third caveat relates to the criteria used for definition of features on maps and APs. All APs were interpreted and digitized by the same individual, thus there is internal consistency in the identification and delimitation of features. However, it is not known what criteria the surveyor and cartographer of the 1899 OS map used to define gravel bars. Active zone boundaries as delineated on APs incorporate three types of feature: channel (area under water at time of survey), exposed gravel and vegetated gravel. These are active in the sense that they are subject to inundation and erosional and depositional modification by flows equivalent to the mean annual flood. Palaeo-exposed and vegetated gravel is more stable and not inundated in such flows. The distinction between active and stable gravel bars will always be artificial and, to some extent arbitrary, and is clearly stage dependent. Furthermore, delimitation of vegetated gravel boundaries is not always straightforward as vegetation cover may be subject to seasonal changes in density and extent. Hence, discussion of preservation potential below focuses on total active zone extent, grouping channel, exposed gravel and vegetated gravel as one unit (see Table 3). Comparison of palaeo-features is restricted to maps derived from AP sources (1946-1997) as palaeochannels and relict gravel/terrace edges are not explicitly depicted on the 1899 map.

Caution is essential in interpolating change between successive maps, as they only provide snapshots (Hooke, 1997). Inferred patterns and rates of change are dependent on the temporal spacing and must be considered minimum estimates. Extrapolation of trends identified in the study section to the entire 3 km braided reach is also precarious due to the inherent spatial heterogeneity of response in such a dynamic system. It would be expected that changes propagate downstream in some way, but with sensitive dependence on upstream conditions.

5 HISTORIC CHANNEL AND BAR DEVELOPMENT

Extraction of active channel and bar features from the geomorphological maps allow the major changes in channel and bar configuration between each map to be examined in greater detail (Fig. 4), although subject to the qualifications connected with definition of active gravel outlined in caveat three above.

5.1 *1899-1946*

Channel pattern on the 1899 map displays a significant difference in the style and spatial scale of channel division between the upper and lower half of the study area that is maintained for over 65 years (until at least 1964). The upper reach is anastomosed, exhibiting large-scale division of two main channel branches around a vegetated island (400 × 170 m in extent). The western channel belt is wider and more complex with several mid-channel and lateral bars and vegetated islands. The eastern branch is narrower, single thread and contains little exposed gravel. The lower half of the reach, below the convergence of the two channels, is divided on a smaller scale, with mid-channel and attached gravel bars that are relatively narrow (20 m) and elongated parallel to flow direction. In the lowermost section of the reach flow diverges either side of a vegetated island which remains a key feature throughout the whole study period through to 1999, with periodic extension and trimming of its upstream margins. The 1946 image is incomplete and changes in the SE of the reach cannot be determined. However, the map suggests maintenance of the anastomosing pattern upstream, with straightening and rationalisation in the western channel which appears to be single thread. This is accompanied by westward migration, and possibly avulsion, of the main channel and amalgamation of three mid-channel bars to form a major lateral bar. Downstream, there appears to be an increase in the width of the active zone, with eastward migration of the eastern channel. Adjacent to the vegetated island at the bottom of the reach, the main channel has avulsed by ~50 m, switching to a more westerly route around a large lateral bar. The major changes between 1899 and 1946 comprise a reduction in braiding intensity and increase in the extent of exposed gravel. An increase in the size of individual bars appears to have been accomplished through downstream migration and amalgamation of mid-channel bars. Attachment of mid-channel bars in the upper half and lowermost sections of the reach is associated with channel avulsion and switching.

5.2 *1946-1955*

There is continued simplification of the channel pattern in the upper half of the reach and cessation of flow through the eastern channel by 1955. The western channel has increased in sinuosity, with downstream migration of the bend apex and development of a second apex upstream to produce a double-headed meander. The bar configuration in this zone is broadly similar to 1946 although there appears to be an increase in the vegetation cover on bar surfaces, possibly indicating relative stability and lack of frequent inundation, although changes associated with seasonal factors cannot be discounted. Downstream channel division appears to be less extensive, although this may, in part, be stage-related (with lower flows in July 1955 than in May 1946) and three mid-channel bars have joined to form a complex lateral bar. This has been accompanied by channel switching to the eastern edge of the valley floor, the development of the right-angled meander bend upstream and deflection of flow eastwards, dissecting the lower margin of the large lateral bar. This easterly branch appears to take most of the flow in the lower part of the reach in 1955, with narrowing of the main western channel and apparent severing of its connection with upstream flow. There is some trimming and accretion at the upstream margins of the lower island and switching of main channel flow to the east. The 9 year period between 1946 and 1955, therefore, was characterised by a reduction in channel division at a large and small-scale, and increase in channel sinuosity, with amalgamation and attachment of bars.

5.3 *1955-1964*

In contrast to the above period, significant changes have occurred during the 9 years between July 1955 and August 1964, particularly in the upper part of the reach. Here major channel avulsion has begun resulting in the re-occupation of the easterly channel which now appears to be taking the highest proportion of flow. There are considerable new expanses of exposed gravel, including a large triangular bar immediately downstream of the channel divergence, suggesting significant sediment supply to the reach. Downstream change is less obvious, the eastern channel is still dominant although there has been switching to the west of the lowermost vegetated island and re-occupation of a former channel route. The long, attached bar complex is now well vegetated and, apart from accretion of gravel upstream and some trimming to its eastern edge, appears to be stable. The major changes upstream suggest an influx of sediment to the reach, most probably associated with a large, geomorphologically-effective flood in September 1961, with a peak recorded discharge of 200 m^3s^{-1} at Feshiebridge (Werritty & Ferguson, 1980).

5.4 *1964-1989*

There have again been significant changes between the two maps, although over a longer period of time (25 years). The western channel zone in the upper half of the reach is no longer active and all flow is directed along the eastern channel which has migrated

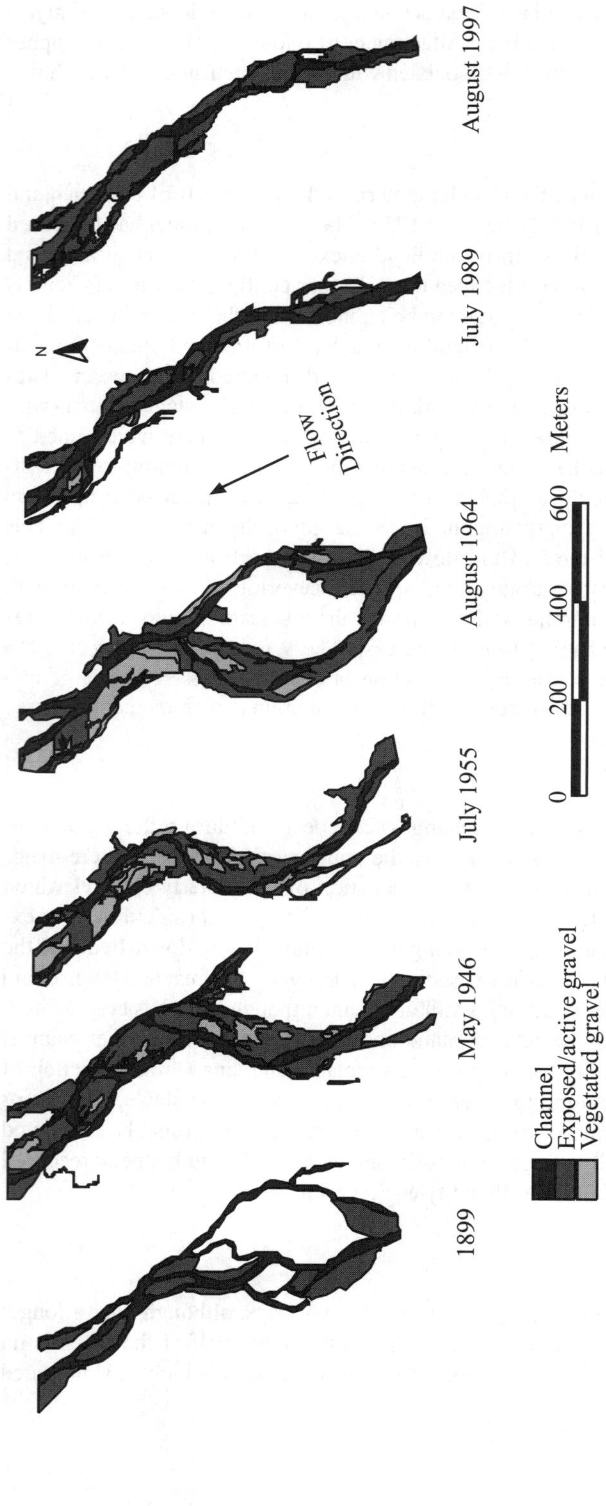

Figure 4. Channel and bar development, 1899-1997.

slightly further to the east. There is a marked reduction in the extent of exposed gravel throughout the study reach, although it is particularly evident in the upper section. This may be stage-related, with higher flows in May 1989 than August 1964. However, it is possible that the main channel has incised slightly, possibly in response to a reduction in sediment supply from immediately upstream. Despite the reduction in exposed gravel area, the number of mid-channel bars in the lower section has increased due to channel avulsion and bar dissection. This is readily apparent adjacent to the downstream vegetated island. The lack of vegetation of bar surfaces suggests instability of the gravel substrate and relatively recent reworking. In summary, there has been an overall eastward movement of the active channel zone during this period, accompanied by an increase in the number of mid-channel bars and a reduction in exposed gravel area and the size of individual bars.

5.5 *1989-1997*

No major changes have taken place in the location and configuration of the active channel zone in the eight years between 1989 and 1997, although there is a small reduction in channel sinuosity, an increase in the size of, and downstream migration and progradation of, gravel bars. This is especially prominent in the lower half of the reach where a large mid-channel bar has migrated downstream and connected with a lateral bar, forcing a switch in channel location to the west and re-occupation of a palaeochannel last active 42+ years ago (shown as connected to main flow on 1946 map). Accretion of gravel to the upstream margin of the lower vegetated island is linked with the westwards migration of the adjacent channel.

6 BAR TYPES, BAR DEVELOPMENT AND SEDIMENT SUPPLY

Braid bars are notoriously difficult to classify as their morphology and planform is stage dependant and changes over time. There have been numerous definitions and typologies published in the literature (see Bridge, 1993 for a review), and Ferguson (1993) counsels 'unambiguous classification may be impossible'. In the present study, bar morphology and plan are compared between maps derived from APs taken at different times of the year, at varied, low to moderate flow stages. Chute channels may divide bars at higher stages but may be inactive at lower flows. Most of the bars in the study area are not simple, as is apparent in the above discussion many are compound features, with periodic accretion and erosion, so that their form is variable over time and complex in origin. These uncertainties notwithstanding, it is possible to identify two broad types of active gravel bar in the study area, mid-channel and lateral (or attached), that accord well with Ferguson & Werritty's (1983) notation. In addition, there are a number of more permanent, vegetated 'islands' which are themselves complex assemblages of former braid bar and channel units. Channel division around the 'island' in the upper part of the reach is sufficiently large-scale to be classed as anastomosing as 'the length of channel segments exceeds the length of first order channels around individual first order bars' (*sensu* Bridge, 1993: 21).

There are significant differences in the size, shape and stability of bars as depicted on successive APs, a marked discontinuity in braiding style between 1964 and 1989 and an increase in total active gravel area between 1899 and 1964. As mentioned above, changes in active gravel extent between 1899 and 1946 should be viewed with a certain amount of caution due to probable (but unknown) differences in the way in which active gravel was delineated on the map compared with the APs. However, between 1946 and 1964 there was a corresponding increase in the number and size of attached bars and a decline in the number of mid-channel bars and, hence, braiding intensity. The mechanism for this appears to be the downstream migration of mid-channel bars, overlapping with lateral bars and amalgamation of gravel units. Hence, the majority of active gravel bars in the study reach are compound units, comprising a number of differently aged gravel units.

Notwithstanding the difficulties of inferring events between discrete timeframes, it is evident that some bars persist and survive events for extended periods of time, although their margins undergo a complex history of erosional and depositional modification, and they may eventually be incorporated into the floodplain and preserved in the alluvial record. A key element in the stabilisation and longer-term preservation potential of gravel bars is the degree to which they become vegetated. Vegetation helps stabilise bar surfaces by increasing tensile and shear strength and, hence, resistance to erosion and encouraging the trapping and accumulation of fines. The major reduction in active gravel area in the lower half of the study reach between 1964 and 1989 is associated with vegetation, stabilisation and incorporation of lateral bars into the floodplain (e.g. the progressive increase in vegetation cover on the surface of the long, lateral bar attached to the western bank between 1946 and 1964). This process of floodplain accretion leads to the development of a mosaic of multi-aged sediment units and a high probability that laterally adjacent floodplain units will vary significantly in age, complicating interpretation of the alluvial record.

With caution, it is possible to make some inferences about sediment supply to the study reach on the basis of changes in the extent of active gravel bars over time. This necessitates distinguishing between autogenic (local) supply, generated as a result of feedbacks within the active channel zone, and influx of sediment as a consequence of extrinsic factors. The former is strongly linked with deflection and divergence of flow associated with downstream bar migration and subsequent bank erosion and has been well documented for individual events further downstream (Ferguson & Werritty, 1983). The reduction in active gravel extent in the post-1989 active channel belt is probably a consequence of the cut-off of the western channel belt in the upper reach sometime in the 1960s, and isolation of a major sediment source. Over the longer-term, extrinsic factors controlling sediment supply from upstream are likely to be influential. Werritty & Ferguson's (1980) study of longer-term channel development downstream of the present Upper Feshie study reach, suggests the current, extensively braided pattern was already present in the late nineteenth century. There is mounting evidence from the UK (Gilvear, 1993; Passmore et al., 1993) and north west Europe (Rumsby & Macklin, 1996) of enhanced fluvial instability in upland river basins during the late nineteenth century, including increased intensity and extent of braiding. This appears to be linked with cooler and wetter climatic conditions, enhanced flood frequency and magnitude and increased hillslope and tributary sediment supply (e.g. Macklin et al., 1992b; Rumsby & Macklin, 1996). The changes in braiding intensity and active gravel extent in the study area are consistent with an influx of sediment to the reach, triggering a cycle of aggradation and degradation as it is transported through. In the case of the Upper Feshie, a significant volume of sediment has not been transferred downstream, but currently remains in storage in the 1964 cut-off channel belt.

7 BRAIDING AND CHANNEL AVULSION

Channel division in the study reach occurs across the full range of the hierarchy, from channel belt (anastamosing), through to dominant main channels and chute channels associated with bar dissection. Throughout much of the time period covered by the study, up until 1964, a distinction in the style of channel division is maintained in the upper and lower half of the reach. The upper reach is dominated by large scale channel belt division around a stable, vegetated island and is thus classed as anastomosed. The lower reach, however, comprises a single, braided channel belt. After 1964 the contrast between the two zones is terminated with the cut-off of the upper western channel belt and the whole reach is characterised by a low sinuosity, relatively narrow active channel zone, with main channel division. The presence of chute channels, associated with lateral and diagonal bar dissection, varies with flow stage but they are present throughout the study period.

The time interval between geomorphological maps varies from 8 to 47 years, hence, it is not possible to infer channel change mechanisms in detail, however, some general trends are apparent. Avulsion appears to dominate over progressive bank migration, although the latter would be more difficult to detect from the long time intervals between successive images. This is true at scales ranging from anabranch switching around the upper vegetated island through to main channel segments. It is less clear in the case of smaller-scale chute channels and bar dissection features where bar erosion and accretion processes are likely to be important (cf. Ferguson & Werritty, 1983). There seem to be certain preferred channel paths and re-occupation of inactive channels and channel belts is common. This supports the supposition that topography is a key control on channel migration in braided gravel-bed rivers (Warburton et al., 1993; Werritty & Ferguson, 1980). Re-occupation of former channels suggests their maintenance as topographic lows, indicating that infilling is limited over the time period studied (i.e. 100 years). Furthermore, the periodic re-occupation and scouring of channels implies the time span over which such channels are active may be underestimated as only the channel-bed morphology associated with the final occupation phase prior to avulsion is likely to be preserved in the alluvial record.

8 ACTIVE ZONE REWORKING AND PRESERVATION POTENTIAL

With the same provisos about inferring rates of change between successive maps, the extent of reworking of active zone units identified on images gives an indication of the longer-term preservation potential of channel and bars morphologies in dynamic upland rivers. Total active zone areas (channel, exposed gravel and vegetated gravel) for each map date are overlain in Figure 5, which also shows the percentage of active zone deposits remaining in 1997. The highest survival rates are associated with the 1964 active zone, reflecting the major anabranch avulsion in the upper part of the reach. Low survival of deposits associated with the 1946, 1955 and 1989 maps highlights the rapid re-working of material within the active zone.

In their study of bar development and channel changes on the Upper Feshie over a 9 year period, Ferguson & Werritty (1983) distinguished between normal development of braided channels, involving episodic progradation of longitudinal bars, and interruptions

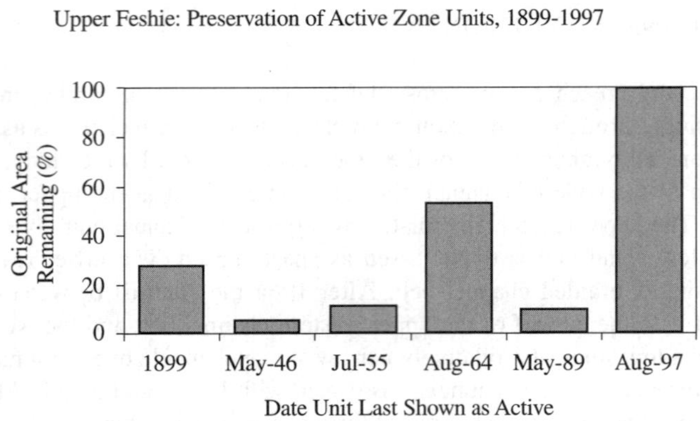

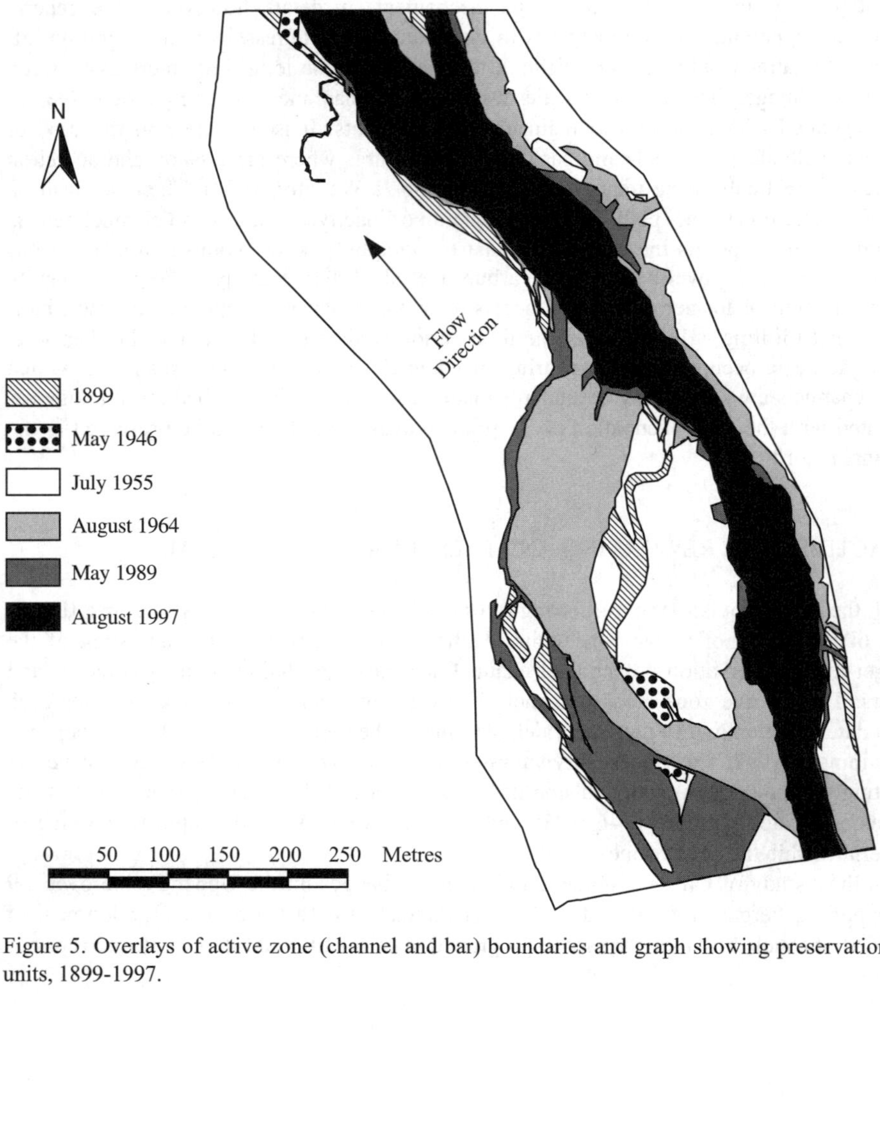

Figure 5. Overlays of active zone (channel and bar) boundaries and graph showing preservation of units, 1899-1997.

to development by chute formation or avulsion. The present study suggests interruptions may dominate in the longer term. Over the 100 year period, channel belt avulsion has contributed larger amounts of sediment storage to the floodplain compared with bar accretion. Examination of older terraces on the geomorphological map (Fig. 2) shows that channel belt deposits continue to be a key element.

9 GPS SURVEY AND TERRAIN MODELLING

GPS-based survey of the sub-section of the upper reach was undertaken in the summer of 1998 over a seven day period under low flow conditions (~2 m^2s^{-1}). Over 9500 points were acquired over the 13.92 ha area of the reach giving a mean intensity of 0.63 points m^{-2} (Fig. 6a, b). The sampling resolution was varied in line with the topographic complexity. Greatest attention was given to sensitive areas (riffles, diffluences, confluences) and important breaks of slope (bar and bank toes and tops) while some of the deeper, less easily navigable areas are less well controlled (e.g. the north east right channel pool, Fig. 6c).

The digital terrain model derived from the survey (Fig. 6c) clearly demonstrates the continuity of bedforms in the reach and the stage dependent nature of any attempt to classify and delimit morphological units. The overall topographic variability is significant in such a small reach, with up to 4.6m difference between the topographic highs and lows. Flow can be visualized to enter from upstream through a single thread channel which bifurcates into two riffles around a major mid-channel bar and again in the left channel around a smaller triangular bar. However, in addition to the major bedforms, some subtle features are also revealed which cannot be easily observed in a standard analysis of APs. For example, significant topographic variability is evident on the main mid-channel bar which is partially dissected by a number of right-left trending chutes as well as variations in the main channel thalweg and roughness.

A comparison with the 1997 aerial photography for the reach reveals little change in the broad structure of the reach, except for the now clear exposure of the small triangular mid-channel bar in the left channel below the main diffluence. Repeat surveys of the reach were undertaken in June 1999 & July 2000, and a direct comparison of the reach between these two surveys is a current focus of work.

10 CONCLUSIONS

This study has demonstrated the application of two approaches which provide insight into the two- and three-dimensional patterns of channel and bar development on the braided River Feshie over the late historical period. Lateral channel adjustment since 1899 has been investigated using historic maps and APs. Image analysis software and a GIS have been used to geo-reference, rectify, overlay and digitally-analyze geomorphological maps to provide a quantitative assessment of channel pattern changes and the survival potential of the morphologies and sediments associated with former active channel zones. The second approach uses high precision GPS survey of contemporary channel and bed topography to derive elevation data and enable construction of a DTM of the study reach. The

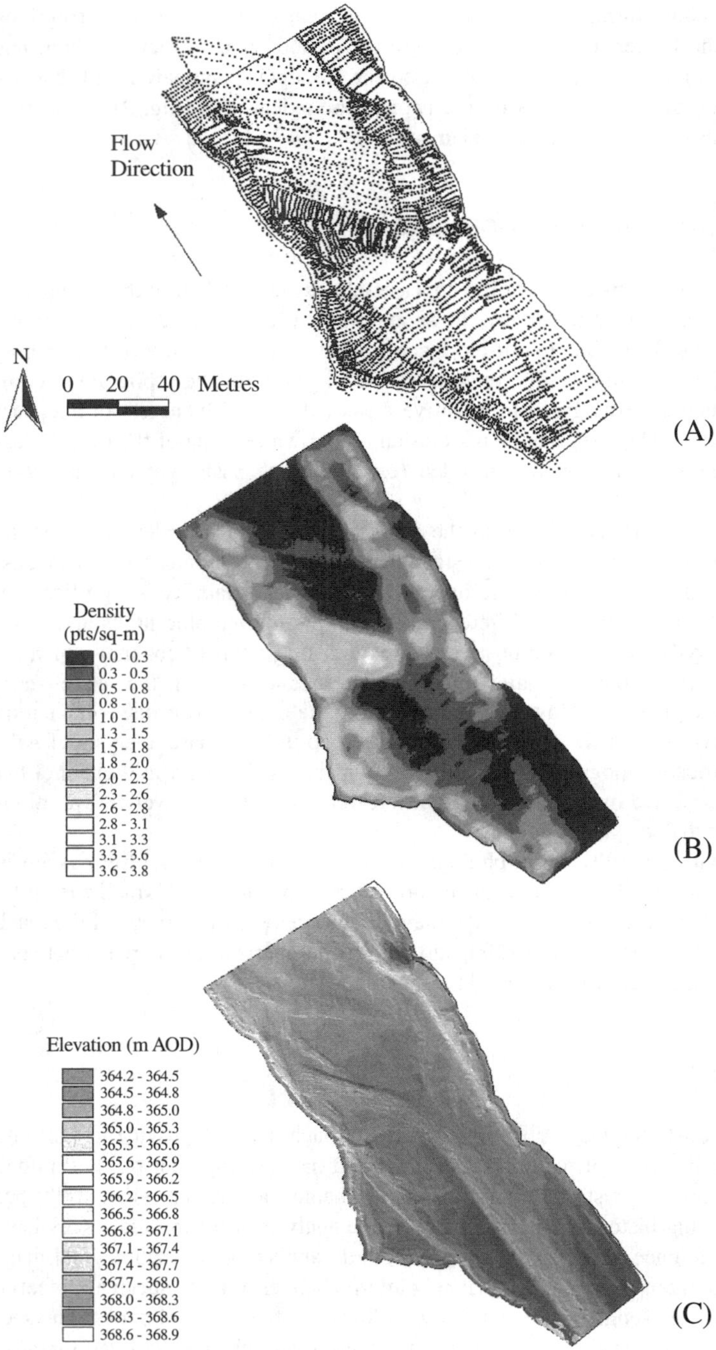

Figure 6. GPS survey results showing (A) survey point distribution, (B) survey point density, (C) 0.25 × 0.25 m DTM.

distributed topographic surface of the model highlights the continuity of morphological units within braided systems and emphasises the difficulty of defining, and comparing, channel and bar boundaries which are so stage dependent.

The temporal and spatial complexity of the sedimentary and morphological record in dynamic braided systems has been well demonstrated in this study by the lateral and vertical discontinuity of alluvial units. Over the 100 year study period, bar development has taken place through downstream migration, dissection and amalgamation and most bars are clearly compound features. Across the full hierarchy of channels, switching and avulsion appear to be the key mechanism of change as opposed to progressive lateral migration. Valley floor topography is a primary control on the avulsion process, with periodic re-occupation of former channels which remain as depressions, hence, channel belts are also likely to be multi-aged. Over the last century a high proportion of the active valley floor of the Upper Feshie study area has been re-worked laterally, with limited preservation of deposits and landforms associated with the 1899 system. The highest degree of preservation is associated with the western channel belt which was abandoned due to avulsion in the 1960s, putting a large volume of gravel into storage. Preservation of this channel belt in the longer term is contingent on its retention as a topographic low. Although sediment supply at the upstream diffluence is a critical factor, the present active zone is now locked against the eastern valley wall and a tendency towards incision may constrain the potential for future lateral re-working within this reach.

Braided and near-braided rivers are dynamic systems, with rapidly evolving morphologies and high sensitivity to variations in flood frequency and sediment supply associated with climatic and land cover changes. However, this study has demonstrated that the inherent responsiveness of systems such as the Feshie constrains their utility as high resolution fluvial archives, due to the high probability of valley floor reworking and low preservation potential of sediments and landforms in the alluvial record. This raises important questions about what exactly the alluvial record tells us in such systems and under what conditions preservation potential is enhanced. The approach described herein provides a means for beginning to address these questions as it allows the spatial and temporal heterogeneity of the fluvial archive to be examined and interpreted. However, to fully explain the alluvial record a more complete understanding of form-process interactions than hitherto obtainable is required.

Ongoing and future investigations on the Feshie aim to help elucidate these issues in several respects. The first is the quantification of contemporary morphological change and its relationship with hydrological events and series of events. This involves direct monitoring of channel topography by GPS survey, construction and differencing of DEMS for different epochs, as discussed above. Monitoring of individual events is difficult where the flow regime is highly variable and time to peak of the flood hydrograph is rapid. Additional constraining factors are the time taken to complete the survey and the need for sufficiently low flows to allow access the channel. Hence, the study focuses on the time averaged response to series of events over a flood season. More frequent survey may be possible by combining analytical photogrammetry for exposed morphology with GPS survey for sub-aqueous topography. Although not fully developed at present, airborne multi- and hyperspectral remote sensing techniques such as ATM and CASI offer great potential for extending high resolution topographic survey over large areas (e.g. Winterbottom & Gilvear, 1997). The second focus of work is to piece together the fragmentary and discontinuous Holocene fluvial record of the Upper Feshie through high resolution topographic survey, stratigraphic investigation and a dating programme. Combined with the investigations of contemporary channel evolution, we can then start to disentangle autogenic and exogenic factors and assess the extent and nature of fluvial response to Holocene environmental changes.

ACKNOWLEDGEMENTS

We would like to thank the following: West Highland Estates and Scottish Natural Heritage for permission to undertake research in Glenfeshie; The University of Hull and BGRG Research and Publication Fund for contributing to fieldwork expenses; Alan Werritty and Rob Ferguson for advice, and access to an unpublished report; Derek Fraser at SEPA for providing flow data. We also acknowledge the help of two anonymous referees for constructive comments on the original manuscript.

REFERENCES

Ballantyne, C.K. 1991. Holocene geomorphic activity in the Scottish Highlands. *Scottish Geographical Magazine*, 107: 84-98.

Brazier, V. & Ballantyne, C.K. 1989. Late Holocene debris cone evolution in Glen Feshie, western Cairngorm Mountains, Scotland. *Transactions of the Royal Society of Edinburgh: Earth Sciences*, 80: 17-24.

Bridge, J.S. 1993. The interaction between channel geometry, water flow, sediment transport and deposition in braided rivers. In: Best, J.L. & Bristow, C.S. (eds), *Braided Rivers*, Geological Society Special Publication No. 75, 13-72.

Church, M. & Jones, D. 1982. Channel bars in gravel-bed rivers. In: Hey, R.D., Bathurst, J.C. & Thorne, C.R. (eds), *Gravel Bed Rivers*. Chichester, John Wiley, 291-338.

Desloges, J.R. & Church, M. 1989. Wandering gravel bed rivers, *Canadian Geographer*, 33: 360-4.

Ferguson, R.F. 1993. Understanding braiding processes in gravel-bed rivers: progress and unsolved problems. In: Best, J.L. and Bristow, C.S. (eds), *Braided Rivers*, Geological Society Special Publication No. 75, 73-88.

Ferguson, R.F. 1981. Channel forms and channel changes. In: Lewin, J. (ed.), *British Rivers*, Allen and Unwin, London, 90-125.

Ferguson, R.I. & Ashworth, P.J. 1992. Spatial patterns of Bedload transport and channel change in braided and near-braided rivers. In: Billi, P. et al (eds*), Dynamics of Gravel-Bed Rivers*. Chichester, John Wiley, 477-496.

Ferguson, R.I. & Werritty, A. 1982. Bar development and channel changes in the gravelly River Feshie, Scotland. *Special Publication International Association of Sedimentologists*, 6: 181-193.

Gilvear, D. J. 1993. River managment and conservation issues on formerly braied river systems: the case of the River Tay, Scotland. In: Best, J.L. & Bristow, C.S. (eds), *Braided Rivers*, Geological Society Special Publication No. 75, 231-240.

Gurnell, A.M. 1997. Channel change on the river Dee meanders, 1946-1992, from the analysis of air photographs. *Regulated Rivers: Research and Management*, 13: 13-26.

Harvey, A.M, Alexander, R.W. & James, P.A. 1984. Lichens, soil development and age of Holocene valley floor landforms, Howgill Fells, Cumbria. *Geografiska Annaler*, 66A: 353-66.

Hooke, J.M. 1997. Styles of Channel Change. In: Thorne, C.R., Hey, R.D. & Newson, M.D. (eds), *Applied Fluvial Geomorphology for River Engineering and Management*. Chichester, John Wiley, 237-268.

Hooke, J.M. & Kain, R.J.P. 1982. *Historical Changes in the Physical Environment*. Butterworths, London.

Lane, S.N. 1998. The use of digital terrain modelling in the understanding of dynamic river channel systems. In: Lane, S.N., Richards, K.S. & Chandler, J. (eds), *Landform Monitoring, Modelling and Analysis*. Chichester, John Wiley, 311-342.

Lane, S.N., Richards, K.S. & Chandler, J.H. 1993. Developments in photogrammetry; the geomorphological potential. *Progress in Physical Geography*, 17: 306-328.

Macklin, M.G. & Lewin, J. 1989. Sediment transfer and transformation of an alluvial valley floor: the River South Tyne, Northumbria, UK. *Earth Surface Processes and Landforms*, 14: 223-46.

Macklin, M.G., Passmore, D.G. & Rumsby, B.T. 1992a. Climatic and cultural signals in Holocene alluvial sequences: the Tyne basin. In: Needham, S. & Macklin, M.G. (eds), *Alluvial Archaeology in Britain*, Oxbow Monograph 27, Oxbow Press, Oxford, 123-39.

Macklin, M.G., Rumsby, B.T. & Heap, T. 1992b. Flood alluviation and entrenchment: Holocene valley floor development and transformation in the British uplands. *Geological Society of America, Bulletin*, 104: 631-643.

Nicholas, A.P., Ashworth, P.J., Kirkby, M.J., Macklin, M.G. & Murray, T. 1995. Sediment slugs: large-scale fluctuations in fluvial sediment transport rates and storage volumes. *Progress in Physical Geography*, 19: 500-519.

Passmore, D.G., Macklin, M.G., Brewer, P.A., Lewin, J., Rumsby, B.T. & Newson, M.D. 1993. Variability of late Holocene braiding in Britain. In: Best, J.L. & Bristow, C.S. (eds), *Braided Rivers*, Geological Society Special Publication No. 75, 205-229.

Stojic, M., Chandler, J., Ashmore, P. & Luce, J. 1998. The assessment of sediment transport rates by automated digital photogrammetry. *Photogrammetric Engineering and Remote Sensing*, 64: 387-395.

Robertson-Rintoul, M.S.E. 1986. A quantitative soil-stratigraphic approach to the correlation and dating of postglacial river terraces in Glen Feshie, south-west Cairngorms. *Earth Surface Processes and Landforms*, 11: 605-17.

Rumsby, B.T. & Macklin, M.G. 1996. River response to the last neoglacial (the 'Little Ice Age') in northern, western and central Europe. In: Branson, J., Brown, A.G. & Gregory, K.J. (eds) *Global Continental Changes: the Context of Palaeohydrology*, Geological Society Special Publication No. 115, 217-233.

Van Sickle, J. 1996. *GPS for Land Surveyors*. Ann Arbor Press, Michigan.

Warburton, J., Davies, T.R.H. & Mandl, M.G. 1993. A meso-scale field investigation of channel change and floodplain characteristics in an upland braided gravel-bed river, New Zealand. In: Best, J.L. & Bristow, C.S. (eds), *Braided Rivers*, Geological Society Special Publication No. 75, 241-257.

Werritty, A. & Ferguson, R.I. 1980. Pattern changes in a Scottish braided river over 1, 30 and 200 years. In: Cullingford, R.A., Davidson, D.A. & Lewin, J. (eds), *Timescales in Geomorphology*. Chichester, John Wiley, 53-68.

Werritty, A. & McEwen, L.J. 1997. Fluvial landforms and processes in Scotland. In: Gregory, K.J. (ed.), *Fluvial Geomorphology of Great Britain*, GCR Series, Chapman & Hall, London, 21-32.

Winterbottom, S.J. & Gilvear, D. 1997. Quantification of channel bed morphology in gravel-bed rivers using airborne multispectral imagery and aerial photography, *Regulated Rivers*, 13: 489-499.

Young, 1976. The terraces of Glen Feshie, Inverness-shire. *Transactions of the Royal Society of Edinburgh*, 69: 501-12.

[illegible]

[illegible] 1996 [illegible]

[illegible]

[illegible] 1993 [illegible]

[illegible]

[illegible]

[illegible]

[illegible]

[illegible] 1992 [illegible]

17. Exploring the possibilities and limitations of modelling Quaternary fluvial dynamics: a case study of the River Meuse

L.A. TEBBENS & A. VELDKAMP
Laboratory of Soil Science and Geology, Wageningen University, the Netherlands

1 INTRODUCTION

Within sedimentology and geomorphology, the use of numerical fluvial process models has increased in recent years (Snow & Slingerland, 1990; Koltermann & Gorelick, 1992; Howard et al., 1994; Mackey & Bridge, 1995; Veldkamp & Van Dijke, 1998). The main advantage of fluvial process models is that they provide a quantitative description of river response to internal and external forcing within the fluvial system. In this way, they can be of great benefit, providing insights into non-linear fluvial response over longer time-spans or into complex-response dynamics in the fluvial system (Schumm, 1977; Snow & Slingerland, 1989). Numerical fluvial process models can test theoretical concepts of fluvial response and morphological evolution that are based on field data reconstructions (cf. Merritts et al., 1994). They provide better conception of how processes acting on large spatio-temporal scales influence local river morphology and sedimentary records. Therefore, such models can be of great value to field geomorphologists who aim to understand their research results for single river reaches within the context of longer timescales, or who wish to integrate them within the spatial scale of the fluvial drainage basin.

As geomorphological and sedimentological models have mostly been validated for small scales only, they are not however, necessarily appropriate when applied to larger spatio-temporal scales (Howes & Anderson, 1988). Non-linear processes complicate the extrapolation of fluvial dynamics from small spatio-temporal scales to large spatio-temporal scales (Snow & Slingerland, 1989; Rigon et al., 1994; Vandenberghe, 1995). Moreover, many fluvial systems tend to self-organisation, for example, they might respond to base-level change by internally adjusting their channel dimensions or patterns (Schumm, 1993). A model should thus give an open-system description, which preferably allows for all possible (linear and non-linear) feed-back mechanisms resulting in self-organisation. Another essential aspect of model development is testing the validity of the model-input relationships for different spatio-temporal scales than those on which the input relationships were initially based. A systematic scale analysis can address this aspect. This scale analysis is based on the assumption that the mathematical equations in a model can only be applied if they refer to the same spatio-temporal scale as the original observations from which they were derived. A scale analysis can be carried out in a similar way as the commonly-applied unit analysis, except that not only the units should tally, but also their magnitudes e.g. years, centuries, millennia etc. An example with respect to temporal resolution is the reconstruction of a regional mean uplift rate from river valley age and depth. A 1-Ma old valley of 100 m depth will yield a net uplift rate estimate of 100 m/Ma. If this valley contains a series of dated terraces, a more detailed mean uplift rate can be reconstructed on a 100 ka resolution. Van den Berg (1994, 1996) gives a clear example of uplift rates that are dependent on temporal scale by showing that these recalculated uplift rates deviate from the Ma average. Examples with respect to spatial scale are spatial variability in the river system and the resolution on which calculations are made. The size of the fluvial system will largely determine its stream power and consequently the response to external forcing. Natural fluvial systems in humid-temperate climatic settings have increasing discharges in downstream direction, have small and large tributaries, may experience considerable loss or gain of drainage area following tributary capture, demonstrate lateral channel-belt shifts due to avulsions and are influenced by sea-level change in their lower reaches. However, most of these factors are often not addressed in modelling efforts at the spatial scale of a flume experiment. Furthermore, different river reaches along the fluvial longitudinal profile will mutually influence each other, because erosion in upstream reaches might cause deposition in downstream reaches (Schumm, 1993). Finally, Schoorl et al. (2000) exemplify how different spatial resolutions of input digital elevation models and varying pathways of water distribution over the gridcells might influence calculations of sediment generation and redistribution in the modelled landscape.

Numerical fluvial models have been developed for various spatial scales, ranging from experimental flumes to large-scale models describing entire river basins (cf. Willgoose et al., 1991). Because flume experiments yield a wealth of data from measurable variables, they have been the starting-point for many models. This led to very detailed and highly complex models able to simulate sediment transport in great detail, sometimes at the spatial scale of a single grain and at single-second time resolution. One example of such an almost purely physically based model is the SEDSIM dynamic process model (Tetzlaff & Harbaugh, 1989), which mainly focuses on sediment transport, erosion and deposition at the scale of individual grains. The Navier-Stokes force-momentum law lies at the basis of the model, which delimits its use to small scales only. Nevertheless, SEDSIM has been applied at large spatial and temporal scales. Koltermann & Gorelick (1992) linked SEDSIM calculations to large-scale processes such as tectonics, sea-level dynamics and climatic change. From a temporal scale point of view, these links seem rather conflicting with the basic model assumptions. For example, they use only one or two daily discharges for each simulated year and recalculate sea-level and crustal movements from millions of years to annual rates. Therefore, it is most likely that the application of Koltermann & Gorelick (1992) is far beyond SEDSIM's original validity and sensitivity domain. Their apparent realistic results most probably stem from extensive calibration exercises, as indicated by the six different calibration parameters, and the absence of an independent model validation. It seems that an extensive model sensitivity analysis could have prevented this misapplication.

In general, numerical models simulate large-scale processes for long timespans by using scale factors, probabilities or cleverly designed computing schemes, to scale up physical laws and results from data collected in short-term and small drainage system experiments, on which whole theories are based (Hugget, 1985). These three techniques to cope with the larger temporal and spatial scales all have disadvantages. Schumm (1977) and Schumm et al. (1987) have clearly demonstrated that although small-scale experiments have a strong qualitative resemblance with large-scale processes, they can not be simply scaled up with a scale factor to reliably represent large-scale systems. The introduction of stochastic elements (probabilities) in sedimentary models (cf. Bridge & Leeder, 1979) originates from the opinion that sedimentation can be envisaged as a spatio-temporal, random phenomenon (Crowley, 1984; Allen, 1978). This approach excludes dynamic complex-response processes, which are known to dominate the fluvial system (Schumm, 1977; Bull, 1991). The validation to use specially designed computing schemes (like SEDSIM) originates from the observation that the actual deposition time needed to build up sedimentary records usually comprises only a small fraction of the time-span that such records represent. Such reasoning neglects the fact that the formation of a net-depositional record could well be the result of many alternating erosional and depositional events in time, instead of a series of stacked, short depositional events with prolonged quiet periods in between.

Another strategy for long-term modelling is to use generalised process relationships, which are also valid for longer timespans. This approach aims at a generalised description of system behaviour, such as quantifying sediment fluxes in a fluvial system without bothering about channel patterns or grain-size distribution. The strongest simplification is usually found in limiting the amount of spatial dimensions. Examples are the 2-D simulations of longitudinal-profile development of fluvial systems using forward-modelling techniques (cf. Snow & Slingerland, 1990; Howard et al., 1994; Veldkamp & Van Dijke, 1998). These models usually describe the dynamics in a fluvial system with a set of non-linear relationships between river valley gradient, discharge and sediment supply. Apart from strongly simplified 2-D models, one can also abandon the goal to make a full numerical (mathematical) model, as reliable quantitative knowledge for that purpose is still too limited and unreliable. Instead, a model can be constructed using qualitative descriptions, for which sufficient knowledge is available. This type of heuristic modelling is done with finite-state modelling, resulting in coarse and discrete conceptual models (e.g. Veldkamp & Vermeulen, 1989).

Independent of the chosen modelling technique, one should always attempt to make a detailed model analysis. Modellers of fluvial behaviour try to capture the usually descriptive and qualitative hypotheses of river response and translate them into quantitative model concepts. Therefore, a thorough model analysis is highly relevant to test the sensitivity and validity of initial model assumptions and indirectly the validity of the modellers' initial hypotheses. However, even for a firmly theoretically based model significant derivations from reality can be experienced (Anderson, 1988; Oreskes et al., 1994). Physically based computer simulation models appear especially sensitive to parameter and input values, or to the numerical methods which are utilised (Hugget, 1985, Schoorl et al., 2000). Model assessment usually consists of performing a sensitivity analysis in order to clarify the crucial aspects of the model and to analyse the uncertainty of the model outputs. Such a sensitivity analysis comprises the study of the influence of variations in model parameters, inputs etc. on calculated model outputs and starts with a determination of sensitivity sources. Such sources can be found in model structure, input variability, initial and/or boundary conditions, model parameters and the applied computational scheme or model operation. Unfortunately, a standard approach does not exist to demonstrate model sensitivity for numerical and finite state models dealing with grid-based landscape simulations.

Validation of numerical models in general has been a topic of debate (Oreskes et al., 1994). The authors share their view that a real verification or validation of earth scientific models is not possible, because we are dealing with fundamentally open systems which can never be accurately described by simplified subsystem descriptions. Thus, the primary value of numerical models is mainly heuristic. They are representations at best, useful for guiding further research and to challenge existing formulations, rather than pretending to be an exact, verified copy of the real world (Oreskes et al., 1994). Therefore, what is generally referred to as validation is in reality a plausibility study of the conceptual model assumptions and boundary conditions. Nevertheless, numerical modelling of highly simplified stream systems is able to demonstrate various types of non-linear behaviour (Snow & Slingerland, 1990). In this paper, we will discuss the sensitivity and plausibility of the FLUVER2 model set (Veldkamp & Van Dijke, 1998, 2000). The model set consists of a 2-D numerical model simulating the development of the longitudinal fluvial profile and it provides a semi-3D finite-state model, simulating local river valley development. FLUVER2 is designed for Quaternary timespans and consequently has a coarse temporal (10^3-10^6 years) and spatial resolution (10^2-10^6 m).

2 FLUVER2 MODEL SET

2.1 *Model description*

On a time scale of 10^3 to 10^6 years, the fluctuations in the external forcing parameters (tectonic uplift rate, sea level and climate) perturb the balance between the sediment supply to the fluvial system and its sediment transport capacity. The longitudinal-profile model describes these long-term fluvial dynamics on a basin-wide scale with equations that relate the sediment supply to sediment transport capacity of the fluvial system (Veldkamp & Van Dijke, 1998, 2000). The 2D model includes both length and vertical adjustments of the river profile and allows for the different sizes of tributary catchments as well as climatic forcing of hillslope processes. However, channel-specific parameters like width, depth and morphological pattern are not included, and neither are lateral channel belt shifts. The semi-3D finite-state model simulates local river valley development, i.e. for a given section along the longitudinal profile. It includes lateral erosion processes too and thus enables the assessment of preservation potential of a fluvial sedimentary record within the modelled river-valley section.

The 2D model requires preferably well-dated input data for tectonics, climate and eustatic sea-level changes on a time-scale of 10^3-10^6 years to simulate the development of the fluvial longitudinal profile in response to external forcing. For example, the relevant climatic and sea-level input for the north-west European setting might include normalised $\delta^{18}O$ data from the GRIP ice-core or ODP records (Dansgaard et al., 1993; Funnel, 1995; Raymo et al., 1998). The model links the trunk river stream power (river valley gradient · discharge) to the sediment supply from the upstream drainage area to quantify the sediment fluxes in the fluvial system. Assuming mass continuity and dynamic equilibrium, the model then calculates the sediment flux resulting from the difference in sediment detachment rate and settlement rate from node to node for 1 km segments along the longitudinal river profile. A surplus of sediment supply over sediment transport capacity at a given position along the profile raises the height (*h*) of the inter-nodal profile segment and thus mimics a sedimentation event (positive *dh*-value, in m). Net removal of sediment lowers the height of the inter-nodal profile segment and represents the fluvial erosion process (negative *dh*-value). The simulated fluvial dynamics are visualised in the form of a Profile Evolution Map (PEM; Veldkamp & Van Dijke, 1998). The PEM shows the temporal evolution of *dh*-values along the longitudinal profile. For example, consider that for every time-step of 20 years a *dh*-value is calculated for every segment along the longitudinal profile, which is plotted after every period of 1000 years. If 250 of these 1000-year time-slices are successively plotted in one figure (like in Plate 1), then the resulting PEM will visualise how the longitudinal profile has evolved during the last 250 ka. It indicates where along the longitudinal profile erosion (negative *dh*) or deposition (positive *dh*) occurred and when. Next, the timing, allocation and intensity of these erosional and depositional events can be compared with fieldwork data from selected study areas to evaluate the likelihood of model assumptions and – even more interesting – of model predictions.

2.2 *Calibration and sensitivity analysis longitudinal profile model*

The FLUVER2 model set has been applied to the Allier/Loire, the Meuse, the Lower Rhine and the Aller (upstream part of the Weser) fluvial systems (Veldkamp, 1992; Tebbens & Veldkamp, in press; Veldkamp & Van Dijke, 1998, 2000; Stemerdink, 1999). As the application for the River Meuse fluvial system for the last 250 ka is the most elaborated and detailed one, we will discuss the model sensitivity and plausibility for various simulations of River Meuse fluvial dynamics.

The 250-0 ka BP calibration of the FLUVER2 model set for the River Meuse required input data on the tectonic uplift and subsidence rates along the longitudinal profile, sea-level changes and climatic parameters. Tectonic uplift rates were taken from Van den Berg (1994, 1996). The sea-level input for 124-18 ka BP was based on the eustatic sea-level curve of Chappell & Shackleton (1986) and Chappell et al. (1996), who derived this curve from a section of uplifted Huon coral terraces in Papua-New Guinea. A more detailed record for 18-0 ka BP was based on Fairbanks (1989). For the period 250-124 ka BP, we fitted the detailed sea-level record of Fairbanks (covering a range of 121 m sea-level rise) to the astronomically tuned and well-dated SPECMAP normalised $/^{18}$O-isotope stratigraphy of Martinson et al. (1987). Climatic fluctuations were simulated by normalising the GRIP ice-core isotope data (Dansgaard et al., 1993) and linking the youngest (Holocene) normalised δ^{18}O value to measured present-day mean annual Meuse discharges (Berger & Mugie, 1994). For detailed information we refer to Tebbens (1999) and Tebbens & Veldkamp (in press).

The longitudinal-profile model includes three main variables to tune and stabilise the computational procedure and/or to improve the performance of the simulations. They are the bedrock/alluvium *erodibility (k_sed,* in m^{-2}), the *sediment travel distance* (*d_tra*, in km) and a *hillslope erosion variable (difco*, in yr^{-1}*)*. These variables interfere in the calculation of sediment detachment rates and sediment settlement rates as follows. *K_sed* (erodibility) quantifies the ease with which the rocks or sediments in a longitudinal profile segment are detached. When multiplied with the discharge, it determines the stream power of the river to erode the bedrock or alluvium and, as such, it directly determines the sediment detachment rate. *D_tra* or travel distance is the downstream distance over which a certain amount of sediment that is already in transport is dispersed along the longitudinal profile. Dividing the incoming sediment flux by this sediment travel distance yields a settlement rate for the sediment load in any inter-nodal segment. The hillslope erosion variable (*difco*) only becomes operative in fluvial reaches with local relief, i.e. where the altitude of the watershed differs from that of the river valley. If the hillslope erosion variable is multiplied with the amount of local relief, then hillslope erosion processes are simulated, generating an additional sediment flux into the fluvial system in sloping areas. The *difco*-variable is proportionally scaled with relevant climatic input data (in this case the normalised $\delta^{18}O$-values from the GRIP ice-core), to induce highest hillslope supply during coldest climatic conditions. Anthropogenic land-use changes that might cause locally increased hillslope erosion are not accounted for in this long-term model.

To tune the model, it was run with different settings for the *k_sed* and *d_tra* variables, while external-forcing input and hillslope erosion dynamics were not changed. Next, the simulated fluvial dynamics as visualised in the Profile Evolution Maps (PEMs) were compared with real-world and geomorphological data to assess the performance of the model simulations. Such an assessment resulted in Figure 1, showing calibration stability fields for 250 ka model runs at several combinations of *k_sed* and *d_tra*. The zero-field (0) represents numerical model instability, occurring at values of *k_sed* > $36 \cdot 10^{-10}$ m^{-2}. Below this value, the model is numerically stable. Several fill- and cut-terraces clearly indicate Late Weichselian to Holocene degradation (incision) in the middle and lower Meuse reaches (> 600 km; Tebbens et al., 1999). The minus-field (–) in Figure 1 represent those model runs that did not generate this incisional phase: the PEMs corresponding to this field do not show negative *dh*-values around 10 ka BP, nor around 125 ka BP. Additionally, the PEM representing the lower part of this stability field (Plate 1 PEM 'A') shows various artefacts related to the persistence of several initial gradient knickpoints.

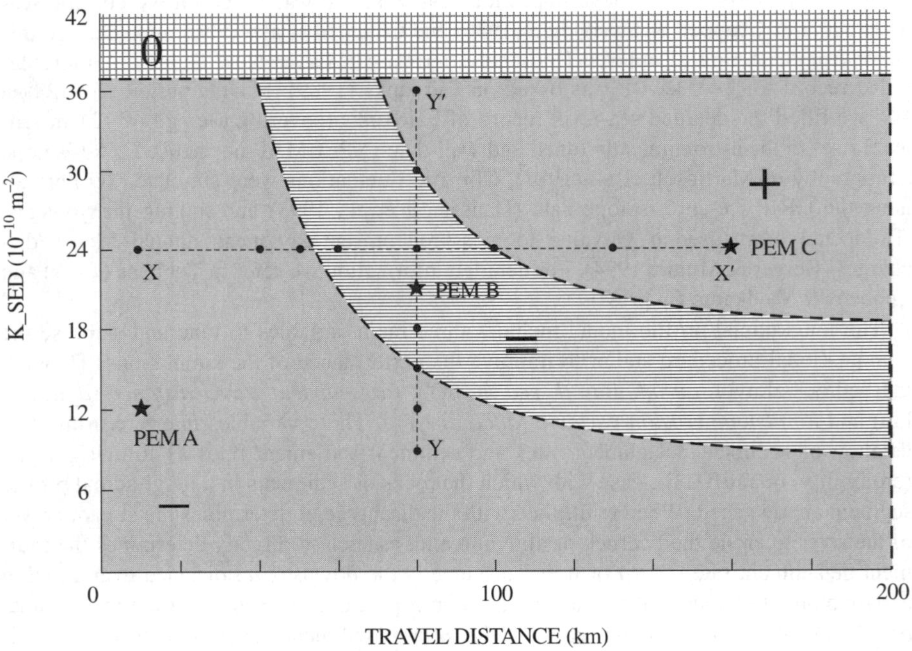

Figure 1. Stability fields for a 250-ka BP calibration of the FLUVER2 longitudinal-profile model for the River Meuse fluvial system at several combinations of *k_sed* and *d_tra*. The modelled fluvial dynamics for a model run in each stability field are visualised in Profile Evolution Maps A, B and C respectively (PEM 'A', 'B' and 'C' in this figure correspond to those in Plate 1). The X-X' and Y-Y' transects correspond to Figures 3 and 4 respectively.

This illustrates that the input values for *k_sed* and *d_tra* determining net fluvial erosion apparently were too low. The model runs in the plus-field (Fig. 1: +) yielded too much fluvial erosion. Their final longitudinal profiles showed such extreme erosion in the middle reaches, that the PEMs corresponding to this stability field (e.g. Plate 1: PEM 'C') indicated the position of the Holocene terrace intersection unrealistically far inland. In the PEM, the position of the terrace intersection in the Meuse lower reach can be found at the location where upstream erosion (negative *dh*-values) changes into sedimentation related to the Eemian and Holocene sea-level highstands (large positive *dh*-values at ~125 ka and ~10-0 ka BP). PEM C in Plate 1 shows that the position of the terrace intersection shifts upstream with time within the 700-1000 km reach, but eventually comes too far inland (~710 km) with respect to the real-world and present-day terrace intersection at ~765 km. This leaves a considerable, but arbitrary, central field (Fig. 1: =) to include those combinations of *k_sed* and *d_tra* that yield the most realistic simulations of longitudinal profile development. PEM 'B' in Plate 1 is the result of such a simulation. It corresponds to a model run for which the position of the modelled coastline, the terrace intersection and the spatial distribution of fill-terraces and cut-terraces largely agree with the actual geomorphological field setting.

Transect X-X' (Figs 1 and 2) visualises a sensitivity test for different values of the sediment travel-distance (*d_tra*). Interglacial sea-level highstands cause increased sedimentation and profile aggradation related to the process of gradient backfilling, which may reach up to hundreds of kilometres inland (Schumm, 1993; Törnqvist, 1998). The downstream location where lower-reach fluvial degradation (negative *dh*-values) changes into backfilling-related aggradation (positive *dh*-values) corresponds to the position of the terrace intersection. Figure 2 clearly illustrates that with increasing values of *d_tra* at a constant erodibility (*k_sed* = $24 \cdot 10^{-10}$ m^{-2}), the locations of the modelled coastline and the

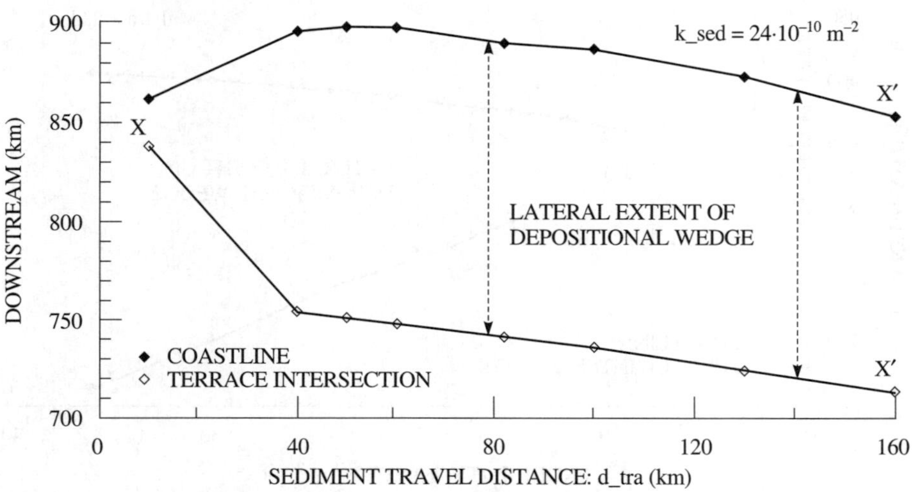

Figure 2. The downstream locations of the modelled coastline and terrace intersection at 0-ka BP (recent Holocene) and the lateral extent of the depositional wedge (see text) as a function of the sediment travel-distance parameter (*d_tra*) at a constant *k_sed*-value of $24 \cdot 10^{-10}$ m^{-2}. The X-X' transect is indicated in Figure 1.

terrace intersection will shift inland. A low sediment travel distance of 10 km exports too little sediment from the supplying areas (see also Plate 1: PEM 'A'), leading to dominant profile aggradation and a modelled coastline too far inland at 860 km downstream of the river source. The present-day coastline is found at 874 km (Berger & Mugie, 1994). At a sediment travel distance of ~50 km, the modelled coastline reaches a maximum extension at ~895 km. At *d_tra*-values higher than ~100 km, too much sediment is exported to the sea. This induces less aggradation and lowering of the river profile in the subsiding lower reach and consequently causes both the coastline and terrace intersection to be located further inland again. The spatial difference between the modelled downstream locations of the terrace intersection and the coastline corresponds to the lateral extension of the depositional wedge or coastal prism (Talling, 1998; Törnqvist, 1998). Figure 2 shows that at a constant erodibility factor of $24 \cdot 10^{-10}$ m^{-2} and for sediment travel distances > 40 km, the depositional wedge for the simulated system amounts some 150 km. This lateral extension seems to be independent from the sediment travel distance.

If we now keep the sediment travel distance constant at 82 km (see below) and increase the erodibility factor (Transect Y-Y' in Figs 1 and 3), then the modelled coastline steadily extends in a seaward direction. High erodibility values represent high sediment supply to the fluvial system and lead to profile aggradation if fluvial transport capacity does not suffice to transport the extra amount of sediment. Apparently, the resulting aggradation is able to outpace long-term subsidence in the North Sea Basin area, which leads inevitably to delta progradation. Thus, increasing erodibility values will lead to profile aggradation and delta progradation, which flatten the river gradient progressively. Consequently, the effects of an interglacial sea-level highstand on sedimentation should be noticed further inland. Indeed, the lateral extension of the depositional wedge increases from only 32 km for the lowest *k_sed*-value to a maximum 240 km for the highest *k_sed*-value (Figure 3).

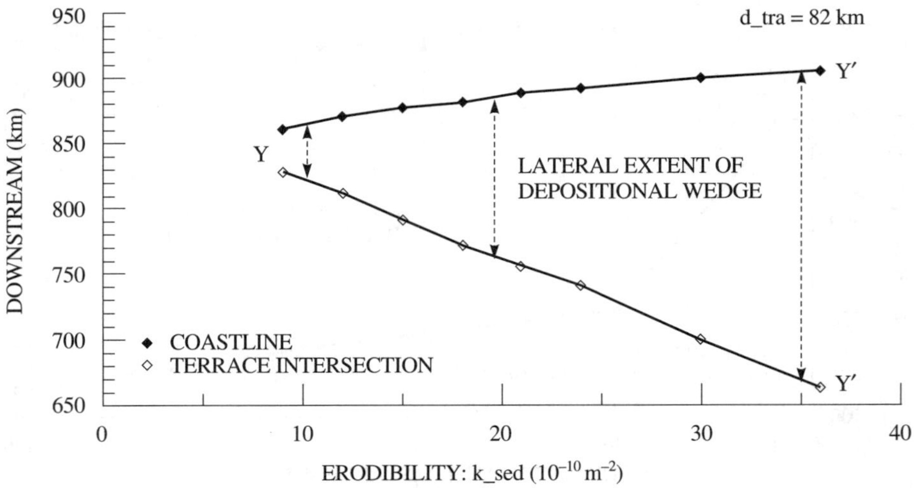

Figure 3. The downstream locations of the modelled coastline and terrace intersection at 0-ka BP (recent Holocene) and the lateral extent of the depositional wedge (see text) as a function of the erodibility parameter (*k_sed*) at a constant *d_tra*-value of 82 km. The Y-Y' transect is indicated in Figure 1.

In order to calibrate to real-world data we chose to base the sediment travel distance (*d_tra*) on the present-day longitudinal profile (for which we also have discharge characteristics) and kept this value constant during the whole simulation. The process of large-scale grain-size sorting yields the most accurate indicative value for the sediment travel distance. Field data from the Meuse case study indicate that gravels grade into sands and silts at ~82 km downstream of the middle reach low-mountain range (Ardennes). Within the stability field of most realistic model runs (=-field, horizontally hatched), a set of plausible *k_sed*-values can most likely be obtained at *d_tra* = 82 km.

As a final test, all aforementioned factors were kept constant and hillslope erosion rates were reduced by 50% to assess their influence on fluvial development. The resulting simulations (not shown) were compared with the model runs presented above. Notably, aggradation events in the Ardennes area appeared less severe and because less sediment was supplied to the system, the downstream location of the coastline and delta progradation were reduced too. In the stability plot (Fig. 1), this produced a roughly parallel shift of boundary lines to the left, but the performance of the Profile Evolution Maps essentially remained the same. Only the intensity and therefore the net amount of degradation and aggradation decreased.

Stemerdink (1999) applied the FLUVER2 model to reconstruct the Early Quaternary dynamics of the Lower Rhine system. It is well-known that the Rhine extended its drainage system in the course of time owing to the process of upstream river capture and/or headward erosion (e.g. Quitzow, 1974; Gibbard, 1988). Consequently, the initial segments of the longitudinal profile had to be situated at the most probable Late Pliocene position in the current middle reach, instead of at the present-day Alpine source of the Rhine. Moreover, the drainage-system extensions had to be simulated with instantaneously increasing upstream discharges. These instantaneous increases in river discharges tended to destabilise the model calculations, because it proved difficult to back them up with a simultaneous and appropriate change of sediment load at the moment of drainage-system extension. Therefore, another stability plot (Fig. 4) had to be made to check the combined effects of the initial sediment load within the first segments of the longitudinal profile (Qs[0]) and the erodibility factor *k_sed* on the model performance. Due to these initialisation problems, only semi-stable model states were obtained, while the stability field with most reliable outcomes was very small. This type of model instability is not caused by a model limitation as such, but by the lack of reliable quantitative data on fluvial dynamics due to stream capture.

3 SEMI-3D VALLEY DEVELOPMENT MODEL

Statements about the preservation potential of sedimentary records in certain river valley reaches based on results of the 2D longitudinal profile model will be ambiguous, because the latter does not incorporate the effects of lateral erosion. A simulation of semi-3D river valley development can provide insights into this issue. Veldkamp & Vermeulen (1989) developed a finite-state model to simulate the formation of fluvial terraces at the spatial scale of a regular river valley. This model provides insight into the preservation potential of fluvial sediments within terraces and basins or grabens as determined by the interplay of fluvial processes and tectonics. Effects of local slope processes and tributaries are not taken into account. This means that modelled terrace scarps will not be prone to long-term

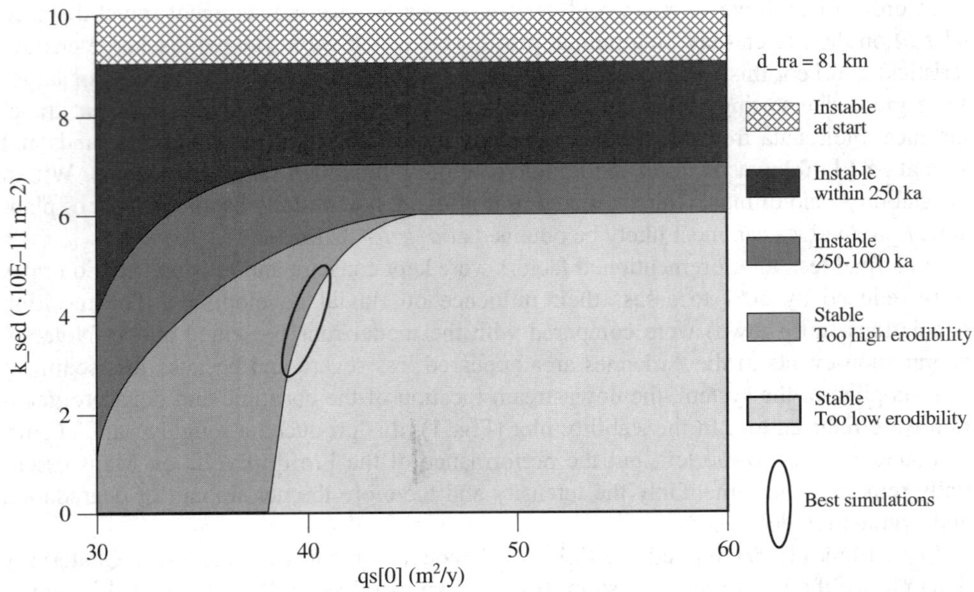

Figure 4. Stability fields for a 2.4-1.2 Ma BP calibration of the FLUVER2 longitudinal profile model for the River Rhine system at several combinations of Qs[0] and *k_sed.*

slope instability and thus these features will be more conspicuous in the modelled landscape than in the field. Furthermore, local tributary alluvial fans will not be represented in the modelled topography, whereas real topography might include them. The basis for the finite-state model is a three-dimensional grid. The model changes the topography of the simulated landscape by removing or adding grid-cells in this basic grid. The extent of the selected Meuse valley reach is arbitrarily set at 10 km · 10 km · 0.5 km (Plate 2). The semi-3D river valley model uses the following three output variables from the longitudinal profile model: valley gradient (m/m), erosion/sedimentation (m) and discharge (m^3/yr). These variables require a partial translation into variables suitable for a 3-D description of local valley development. Valley gradient is straightforward and does not require a transformation. Discharge is used to calculate floodplain width using empirical relationships based on Schumm (1977, p. 115) and Chorley et al. (1984, p. 310). With the longitudinal and cross-sectional dimensions known, the calculated erosion-sedimentation thickness (m) can be converted into a valley volume (m^3).

To allow grid-cell changes, each direction in the simulated valley has to be subdivided into individual grid cells having their own specific width, length and height properties. Plate 2 shows how the three calculated internal variables are related to the X-, Y- and Z-grid directions and how a terraced landscape develops during 1-Ma of uplift. River floodplain width is directly related to the X-axis, the longitudinal river valley gradient to the Y-axis and the volume of erosion or deposition with the combined X-, Y- and Z-axes. The length, width and height properties of the grid cells will determine the precision of the outputs. For example, in case the sizes of the grid cells are set at 200·200·1 m, the grid will have 50, 50 and 500 cells in the X-, Y- and Z-directions respectively. However, this chosen grid size might be too coarse to visualise the relief changes that arise from the calculated erosion and deposition events, resulting in an unwanted loss of model sensitivity. Therefore, it is necessary to check the sensitivity effects of the sizes of selected grid cells on the simulated landscape changes.

3.1 *Sensitivity of semi-3D valley development model*

We used the finite-state model version that was calibrated for a terraced section of the River Meuse valley near Maastricht (Veldkamp & Van den Berg, 1993) to perform a sensitivity analysis. The standard methodology to perform such an analysis is using Monte Carlo simulations (Janssen et al., 1993). Monte Carlo simulations estimate the variability of the used model input parameters assuming a normal distribution. About 100 different randomly selected input combinations were used to run the model. Two different aspects were evaluated: 1) whether any sensitivity was lost by the conversion of variables into grid cells; 2) the overall model performance.

1. By the conversion of valley gradient, floodplain width and erosion/sedimentation into grid cells, relief changes that fall within the range of twice their standard deviations are considered as insignificant and need not necessarily be displayed by the grid cells. The grid should thus at least be able to visualise relief changes in the simulated 3D landscape that are the result of changes in the internal model inputs equal or larger than twice their standard deviations.
   A comparison of the maximum allowed grid size (calculated from the input variability) with the used grid size (200·200·1 m, Table 1) indicates that the selected grid cell properties are approximately half of the maximum allowed grid size for all grid directions. This indicates that the selected landscape grid size of 200·200·1 m is sufficiently detailed to prevent a reduction in model sensitivity by the conversion of calculated variables into grid-cell properties. Moreover, it is large enough to suggest a realistic resolution. Because the chosen grid size is within the range of calculated standard deviations of the input variables, model output terraces of only one grid unit must be considered as too small to have any realistic value. A modelled terrace-like feature should thus be at least two grid units wide and long to be classified as a terrace within the model.
2. None of the sampled input combinations caused a model crash or instability, indicating robustness of the finite-state model. All simulations yielded a river valley with fill terraces (*sensu* Bull, 1991). These results indicate that terrace formation, and related sediment preservation in the simulated valley segment, is not a model property that needs careful or endless calibration under the given conditions. When higher or lower uplift rates are used as model input, none or only a few terraces develop. The semi-3D

Table 1. Comparison of the selected grid size and calculated maximum grid size based on two times the standard deviation of the internal variables.

AXIS	Selected grid size (m)	Maximum grid-size allowed (m)
X	200	395.6
Y	200	454.5
Z	1	2.4

model thus gives insight into the preservation of sediment bodies, because not every deposition-erosion sequence will automatically yield a preserved terrace (Veldkamp & Van den Berg, 1993).

Statistical analysis was applied to check whether input parameters have a strong contribution to the model output variability. We included the amount of terraces, their surface age, river valley depth and width. Only terrace surface age was weakly positively correlated to sediment flux magnitude ($r = 0.36$). This flux is related to the hillslope erosion processes that dominate during a glacial. Multivariate statistics indicated that there is no dominant unidirectional drive of one of the input variables to output variability. Most inputs display a small ($r < 0.1$) but significant contribution to model output variability. This indicates that none of the input variables is superfluous and each contributes to the simulated system dynamics.

The sensitivity analysis indicated that the semi-3D valley development model is not susceptible to sensitivity loss due to the translation of input variables into grid-cell properties. Furthermore, all input variables contributed significantly to output variability indicating that no variable is superfluous. It has to be noted that we only evaluated model behaviour under the assumption that the used heuristic rules and principles are correct. It is not possible to evaluate this aspect in a systematic way. This means that a perceived unrealistic model performance can stem from unrealistic expert rules.

4 PLAUSIBILITY OF FLUVER2 SIMULATIONS

The sensitivity analysis of the FLUVER2 model as calibrated for the Meuse system was done with both stability diagrams and Monte Carlo simulations. Both methods were used to examine the effect of variations in model input and parameter values upon model behaviour and outputs. The analysis demonstrated that the model responds in a theoretically realistic manner to representative variations of model input and parameter values. It also illustrated FLUVER2 to be sufficiently selective and sensitive to cope with realistic variations in the simulated system and that the used model parameters and inputs all seem to contribute to overall model sensitivity. Thus, it can be concluded that the model set simulates the development of the 2-D longitudinal profile and local valley development sensitive enough to test hypotheses about long-term fluvial dynamics.

The remaining question is how plausible the obtained model results are. The best way to check this is to compare model results with independent field data that were not directly used in the modelling and calibration exercises. As was already done above, these data might include the location of the terrace intersection, the location of the coastline, the net amount of incision and the amount and thickness of terrace bodies. In the case of the longitudinal-profile model, a more or less independent check could be established by relating measured bulk-geochemical sediment composition (Tebbens et al., 1998) to calculated sediment fluxes. During the calibration, no attention was given to the sediment-flux composition and the modelled geochemical records. Figure 5 demonstrates that a more than casual match exists between measured and simulated K/Al-ratios for fine-grained Meuse sediments deposited during the last 15 ka. This match suggests that the model describes sediment-flux generation and mixing in a plausible way. Unfortunately, detailed bulk-geochemical records do not exist for the simulated Meuse terrace sequences. This excludes a similar comparison of independent model and real-world properties. The local

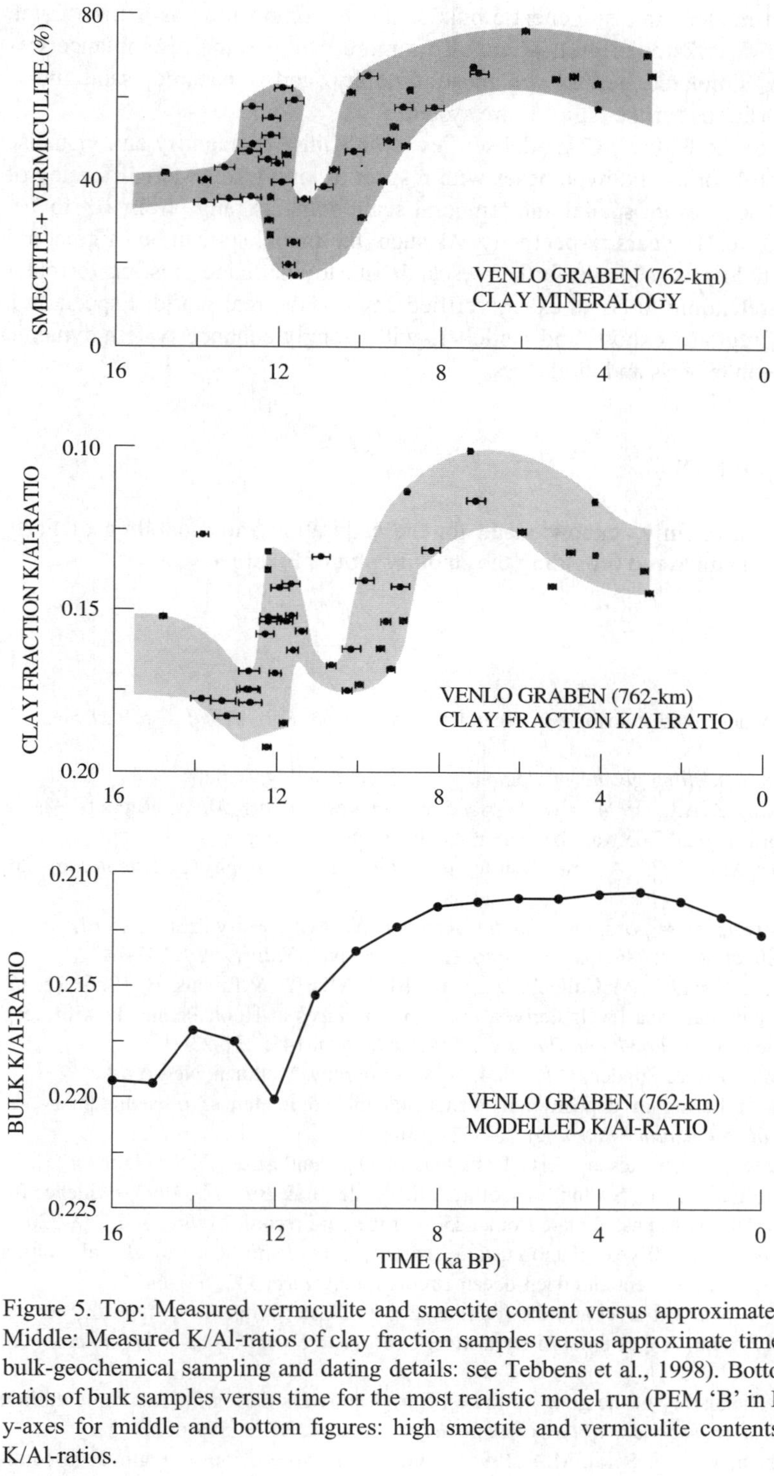

Figure 5. Top: Measured vermiculite and smectite content versus approximate time of deposition. Middle: Measured K/Al-ratios of clay fraction samples versus approximate time of deposition. (For bulk-geochemical sampling and dating details: see Tebbens et al., 1998). Bottom: Modelled K/Al-ratios of bulk samples versus time for the most realistic model run (PEM 'B' in Fig. 2). Note reverse y-axes for middle and bottom figures: high smectite and vermiculite contents coincide with low K/Al-ratios.

valley development model can also generate bulk-sediment composition as demonstrated for the Allier basin (Veldkamp, 1992), where also a more than passing resemblance was observed between simulated terrace chemostratigraphy and measured sand bulk-geochemical compositions for the Allier/Loire system.

The main value of the FLUVER2 model set lies in its ability to quantify and visualise the consequences of theories and hypotheses with respect to long-term external forcing of fluvial dynamics. The relevant spatial and temporal scale domains range from 10^2 to 10^6 metres and from 10^3 to 10^6 years respectively. As such, the model set can be of great aid to fluvial geomorphologists to guide further research and to challenge existing formulations, rather than pretending to be an exact, verified copy of the real world. Especially a close interaction of regional experts and modellers will strongly enhance system dynamic insight for both region experts and modellers.

ACKNOWLEDGEMENTS

Chris Stemerdink is gratefully acknowledged for the calibration and modelling of Early Quaternary Rhine dynamics and providing the stability plot of Figure 4.

REFERENCES

Allen, J.R.L. 1978. A neglected innovator in meander studies. *Can. Soc. Petrol. Geol. Memoir*, 5: 199-209.

Anderson, M.G. 1988. *Modelling geomorphological systems*. John Wiley, London.

Berger, H.J E. & Mugie, A.L. 1994. *Hydologische systeembeschrijving Maas*. Rapport 94.022 Rijksinstituut voor Integraal Zoetwaterbeheer en Afvalwaterbehandeling.

Bridge, J.S. & Leeder, M.R. 1979. A simulation model of alluvial stratigraphy. *Sedimentology*, 26: 617-644.

Bull, W.B. 1991. *Geomorphic responses to climatic change*. Oxford University Press, Oxford.

Chappell, J. & Shackleton, N.J. 1986. Oxygen isotopes and sea level. *Nature*, 324: 137-140.

Chappell, J., Omura, A., Esat, T., McCulloch, M., Pandolfi, J., Ota, Y. & Pillans, B. 1996. Reconciliation of late Quaternary sea levels derived from coral terraces at Huon Peninsula with deep sea oxygen isotope records. *Earth and Planetary Science Letters*, 141: 227-236.

Chorley, R.J., Schumm, S.A. & Sugden, D.F. 1984. *Geomorphology*. Methuen, New York.

Crowley, K.D. 1984. Filtering of depositional events and the completeness of sedimentary sequences. *Journal of Sedimentary Petrology*, 54: 127-136.

Dansgaard, W., Johnsen, S.J., Clausen, H.B., Dahl-Jensen, D., Gundestrup, N.S., Hammer, C.U., Hvidberg, C.S., Steffensen, J.P., Sveinbjörnsdottir, A.E., Jouzel, J. & Bond, G. 1993. Evidence for general climatic instability of past climate from a 250-kyr ice-core record. *Nature*, 364: 218-220.

Fairbanks, R.G., 1989. A 17,000-year glacio-eustatic sea level record: influence of glacial melting rates on the Younger Dryas event and deep-ocean circulation. *Nature*, 342: 637-642.

Funnell, B.M., 1995. Global sea-level and the (pen)insularity of Late Cenozoic Britain. In: Preece, R. C. (ed.), Island Britain: A Quaternary Perspective. *Geological Society Special Publication*, 96: 3-13.

Gibbard, P.L. 1988. The history of the great northwest European rivers during the past three million years. *Philosophical Transactions of the Royal Society of London*, B138: 559-602.

Howard, A.D., Dietrich, W.E. & Seidl, M.A. 1994. Modeling fluvial erosion on regional to continental scales. *Journal of Geophysical Research*, 99: 13971-13986.

Howes, S. & Anderson, M.G. 1988. Computer simulation in geomorphology. In: Anderson, M.G. (ed.), *Modelling geomorphological systems*. John Wiley, London.

Hugget, R.J. 1985. *Earth surface systems*, Springer-Verlag, Berlin, Heidelberg.

Janssen, P.H.M., Heuberger, P.S.C. & Sanders, R. 1993. *UNCSAM software package for uncertainty and sensitivity analysis using Monte Carlo sampling*. RIVM Bilthoven. The Netherlands.

Koltermann, C.E. & Gorelick, S.M. 1992. Paleoclimatic signature in terrestrial flood deposits. *Science*, 256: 1775-1782.

Mackey, S.D. & Bridge, J.S. 1995. Three-dimensional model of alluvial stratigraphy: theory and application, *Journal of Sedimentary Research*, B65: 7-31.

Martinson, D.G., Pisias, N.G., Hays, J.D., Imbrie, J., Moore, T.C. & Shackleton, N.J. 1987. Age dating and the orbital theory of the Ice Ages: Development of a high-resolution 0 to 300,000-year chronostratigraphy. *Quaternary Research*, 27: 1-29.

Merritts, D.J., Vincent, K.R. & Wohl, E.E. 1994. Long river profiles, tectonism, and Eustasy: A guide to interpreting fluvial terraces. *Journal of Geophysical Research*, 99: 14031-14050.

Oreskes, N., Shrader-Frechette, K. & Belitz, K. 1994. Verification, validation and confirmation of numerical models in the Earth Sciences. *Science*, 263: 641-644.

Quitzow, H.W. 1974. Das Rheintal und seine Entstehung. Bestandsaufnahme und Versuch einer Synthese. In: Macar, P. (ed.): *L'évolution Quaternaire des bassins fluviaux de la Mer du Nord méridionale*, 53-104.

Raymo, M.E. Ganley, K., Carter, S., Oppo, D.W., McManus, J. 1998. Millennial-scale climate instability during the early Pleistocene epoch. *Nature*, 392: 699-702.

Rigon, R., Rinaldo, A. & Rodriguez-Iturbe, I. 1994. On landscape self-organization. *Journal of Geophysical Research*, 99: 11,971-11,993.

Schoorl, J.M., Sonneveld, M.P.W. & Veldkamp, A. 2000. 3D Landscape process modelling: the effect of DEM resolution. *Earth Surface Processes and Landforms*, 25, 1025-1034.

Schumm, S.A. 1977. *The fluvial system*, John Wiley, New York.

Schumm, S.A, Mosley, M.P. & Weaver, W.E. 1987. *Experimental fluvial geomorphology*. John Wiley, New York.

Schumm, S.A. 1993. River response to baselevel change: implications for sequence stratigraphy. *Journal of Geology*, 101: 279-294.

Snow, R.S. & Slingerland, R.L. 1990. Stream profile adjustment to crustal warping: nonlinear results from a simple model. *Journal of Geology*, 98: 699-708.

Stemerdink, C. 1999. *Longitudinal profile development of the Early Quaternary Rhine: A 2-D modelling exercise with special emphasis on the Tiglian sedimentary cycles in the southern Netherlands*. Unpublished MSc thesis Wageningen University, Laboratory of Soil Science and Geology.

Talling, P.J. 1998. How and where do incised valleys form if sea level remains above the shelf edge? *Geology*, 26: 87-90.

Tebbens, L.A., Veldkamp, A. & Kroonenberg, S.B. 1998. The impact of climate change on the bulk and clay geochemistry of fluvial residual channel infillings: the Late Weichselian and Early Holocene River Meuse sediments, the Netherlands. *Journal of Quaternary Science*, 13: 345-356.

Tebbens, L.A. 1999. *Late Quaternary evolution of the Meuse fluvial system and its sediment composition. A reconstruction based on bulk sample geochemistry and forward modelling*. PhD thesis Wageningen University, Wageningen, the Netherlands.

Tebbens, L.A., Veldkamp, A., Westerhoff, W. & Kroonenberg, S.B. 1999. Fluvial incision and channel downcutting as a response to Lateglacial and Early Holocene climate change: the lower reach of the River Meuse (Maas), the Netherlands. *Journal of Quaternary Science*, 14, 59-75.

Tebbens, L.A. & Veldkamp, A. (in press). Modelling longitudinal profile development in response to Late Quaternary tectonics, climate and sea-level change: the River Meuse. *Global and Planetary Change*.

Tetzlaff, D.M. & Harbaugh, J.W. 1989. Simulating clastic sedimentation. *Computer methods in the geosciences*. Van Nostrand Reinhold.

Törnqvist, T.E. 1998. Longitudinal profile evolution of the Rhine-Meuse system during the last deglaciation: interplay of climate change and glacio-eustasy? *Terra Nova*, 10: 11-15.

Van den Berg, M.W. 1994. Neotectonics of the Roer Valley rift system. Style and rate of crustal deformation inferred from syn-tectonic sedimentation. *Geologie en Mijnbouw*, 73: 143-156.

Van den Berg, M.W. 1996. *Fluvial sequences of the Maas. A 10 Ma record of neotectonics and climate change at various time-scales*. PhD thesis Wageningen Agricultural University, Wageningen.

Vandenberghe, J. 1995. Time scales, climate and river development. *Quaternary Science Reviews*, 14: 631-638.

Veldkamp, A. & Vermeulen, S.E.J.W. 1989. River terrace formation, modelling and 3-D graphical simulation. *Earth Surface Processes and Landforms*, 14: 641-654.

Veldkamp, A. 1992. A 3-D model of Quaternary terrace development, simulations of terrace stratigraphy and valley asymmetry, a case study in the Allier terraces (Limagne, France). *Earth Surface Processes and Landforms*, 17: 487-500.

Veldkamp, A. & Van den Berg, M.W. 1993. Three-dimensional modelling of Quaternary fluvial dynamics in a climo-tectonic dependent system. A case study of the Maas record (Maastricht, the Netherlands). *Global and Planetary Change*, 8: 203-218.

Veldkamp, A. & Van Dijke, J.J. 1998. Modelling long-term erosion and sedimentation processes in fluvial systems: A case study for the Allier/Loire system. In: Benito, G., Baker, V.R. & Gregory, K.J. (eds) *Palaeohydrology and environmental change*. John Wiley, 53-66.

Veldkamp, A. & Van Dijke, J.J. 2000. Simulating internal and external controls on fluvial terrace stratigraphy: a qualitative comparison with the Maas record. *Geomorphology*, 33, 225-236.

Willgoose, G., Bras, R.L. & Rodriguez-Iturbe, I. 1991. Results from a new model of river basin evolution. *Earth Surface Processes and Landforms*, 16: 237-254.

18. Modelling the impacts of different flood magnitudes and frequencies on catchment evolution

TOM J. COULTHARD, MARK G. MACKLIN
Institute of Geography and Earth Sciences, University of Wales, Aberystwyth, UK

MIKE J. KIRKBY
School of Geography, University of Leeds, UK

1 INTRODUCTION

Within fluvial geomorphology, considerable uncertainty surrounds the role of large infrequent floods (catastrophism) relative to the cumulative effect of more frequent lower magnitude floods (gradualism) in catchment evolution. As the largest floods often cause extensive damage to land and property and their deposits are usually well preserved, it is tempting to conclude that these extreme events are dominant. Yet equilibrium and dominant discharge concepts suggest that channel geometry is adjusted to relatively frequent floods of moderate magnitude, as there are fairly strong links between channel geometry and channel discharge statistics (Wolman & Miller, 1960).

Part of this uncertainty concerning the role of large, infrequent floods may originate from the nature of observation. For example, the effects of a single large event can be relatively easy to quantify as there is often a dramatic change from the previous landscape to the new one. Therefore, measurements of volumes of sediment moved can be calculated with reasonable certainty. However, assessing the more gradual impact of smaller flood events is hampered by the long time periods over which change occurs and the relatively minor alterations associated with each event. This means that areas have to be monitored over long periods of time and the changes noted may be small and especially susceptible to measurement error.

When examining longer time-scales (greater than 100 years), some workers have tried to link alluvial stratigraphies to single large floods or periods of enhanced flooding (e.g. Macklin & Lewin, 1993). Whilst this approach has proved useful and revealed many insights into the roles of climate change and land use on alluvial systems, the temporal resolution of both climate and stratigraphic record is commonly too coarse to establish direct links between individual floods and alluviation/aggradation chronologies.

Computer modelling may offer a solution to these problems, and there have been many successful modelling schemes focussed on the river channel system (e.g. Nicholas & Walling, 1997; Bates et al., 1997). However, these have not incorporated important catchment-wide processes that influence sediment supply or hydrology, such as vegetation, lithology, sediment size and transport (Merrett & Macklin, 1999), slope channel coupling and the shape and flashiness of the hydrograph (Knighton, 1998).

To incorporate these processes and address these issues, a cellular automaton model is used to model a small upland catchment in the Yorkshire Dales. This model has been successfully applied to investigate the long-term effects of climate and land use change on catchment morphology and sediment discharge (Coulthard et al., 2000) and the role of non-linear processes in catchment dynamics (Coulthard et al., 1998). Here this model will be used to compare two groups of simulations. The first represents periods of sustained moderate flooding and the second represents a single extreme flood event.

2 METHOD

The cellular model used is conceptually simple, yet complex in operation and a full description is given in the following section. Within the model, the catchment is represented by an array of uniform square grid cells (2 × 2 m). Each grid cell is assigned initial values for elevation, water discharge, water depth, drainage area and grain-size fractions. For each timestep or iteration, these values are updated in relation to the immediate neighbours according to rules applied to every cell. These rules fall into five groups covering hydrological, flow routing, fluvial erosion and deposition, slope processes and vegetation modelling. For every iteration the model steps through these five groups sequentially, applying them (where appropriate) to every cell. These rules are discussed in the following sections. All units, unless otherwise specified, are in metres and seconds.

2.1 *Hydrological Model*

For every minute of a model run, a simple hydrological mode calculates the soil saturation for each cell (J_t). The saturation for the next time-step (T, here 60 seconds) is then calculated (J_{t+1}), but for this an additional parameter j_t is carried over, which before each calculation is set to the previous iterations j_{t+1}. Then, if the rainfall rate (r in m hr^{-1}) equals 0, J_{t+1} is calculated according to Equation 1.

$$j_{t+1} = \frac{j_t}{1 + \left(\frac{j\,T}{m}\right)} \qquad J_{t+1} = \frac{m}{T}\log\left(1 + \left(\frac{j_t T}{m}\right)\right) \tag{1}$$

If rainfall is not equal to 0, Equation 2 is used.

$$j_{t+1} = \frac{r}{\left(\frac{r - j_t}{j_t}\exp\left(\left(\frac{(0-r)T}{m}\right)+1\right)\right)} \qquad J_{t+1} = \frac{m}{T}\log\left(\frac{(r - j_t) + j_t \exp\left(\frac{rT}{m}\right)}{r}\right) \tag{2}$$

Within these expressions, m is the key variable, controlling the rise and fall of the soil moisture deficit, as per the exponential soil water parameter m in TOPMODEL (Beven &

Kirkby, 1979). If the cell deficit is required between these two time steps a simple linear interpolation is used.

2.2 *Flow Routing*

For each grid cell, a runoff threshold is calculated (Equation 3) which is based upon the amount of water that will infiltrate into the soil, a balance of the hydraulic conductivity (*K*), the slope (*S*) and the horizontal spacing (*Dx*).

$$Threshold = KS(Dx)^2 \tag{3}$$

This is then subtracted from the soil saturation value produced from Equations 1 and 2 and the proportion above is treated as run off, that below as subsurface flow. This subsurface flow is routed using a multiple flow algorithm as described by Desmet & Govers (1996) (Equation 4).

$$Q_i = Q_o \frac{S_{in}^x}{\sum S_i^x} \tag{4}$$

Here Q_i is the fraction of discharge delivered to the neighbouring cell $_i$ from the total cell discharge (Q_o) in m^3s^{-1}, according to the slope *S* between the cell and its relative neighbours *i*, numbering from 1 – *x* (*x* ranging from 3 to 8 depending on the number of neighbours). With surface flow, the depth is calculated using Manning's equation (Equation 5).

$$Q = \frac{A\left(R^{0.67} S^{0.5}\right)}{n} \tag{5}$$

Where *A* is cross sectional area, *R* is hydraulic radius, *S* is slope and *n* is Manning's coefficient. In the case of a cell 1 m wide, this can be re-arranged to give Equation 6, with width (*w*) as 1, leaving depth (*d*). Where cells larger than 1 m are used, the model divides the discharge of the cell by the width then calculates depth for the whole cell as if it were 1 m.

$$Q = d\left(\frac{R^{0.67} S^{0.5}}{n}\right) \tag{6}$$

However, in order to calculate the depth, the hydraulic radius has to be resolved. This is a difficult approximation as it is not easy to estimate the hydraulic radius for a grid cell or part of a channel 1 m wide. However, for a rectangular/trapezoidal channel the hydraulic radius can be simplified as in Equation 7.

$$R = \frac{A}{p} = \frac{dw}{w + 2d} \tag{7}$$

If *w* is greater than *d*, then the 2*d* in the denominator can be ignored leaving *d*, as in Equation 8. This simplification is only valid where the channel width is much greater than the depth. Within this model an individual cell depth can be greater than its width, but it is assumed that this only occurs when the cell is part of a much wider channel.

$$R \approx \frac{dw}{w} = d \tag{8}$$

This means that Equation 6 can be re-written to calculate depth as Equation 9.

$$d = \left(\frac{Qn}{S^{0.5}} \right)^{3/5} \tag{9}$$

Excessively low slopes can result in excessive depths being calculated. To account for this when the slope is less than 0.005, the depth is set to the same value as discharge. Three different methods of calculating the slope were tested:

– the average slope of the neighbours
– the greatest slope of all the neighbours
– the average of all positive slopes

By trial and error the latter was found to be most stable, but it must be acknowledged that the program shows some sensitivity to the method of slope calculation.

Water is then routed according to Equation 10 where the depth of water as well as cell elevation is considered.

$$Q_i = Q_o \frac{[(e+d)-e_i]^x}{\sum[(e+d)-e_i]^x} \tag{10}$$

Here Q_i is the discharge routed to cell i, Q_o the total discharge in the cell, e is the elevation and d depth of water (in metres) for each neighbouring cell i. In both these expressions, differences in slope between diagonal neighbours are accounted for by dividing by $\sqrt{(2[Dx^2])}$. The calculation of depth is an important approximation as it allows discharges to be routed over, as well as around, obstacles.

Both the hydrological and hydraulic sections rely heavily on the use of a scanning algorithm, developed especially for this model. Many conventional DEM-based hydrological models are forced to sort the altitudes to enable the routing of flow from the highest to lowest point. Sorting is a computationally expensive task, and to 'bubble sort' 4 million points takes over 2 hours on a Sun Ultra Sparc workstation. Even more refined sorting algorithms such as 'QuickSort' may take several minutes. In hydrological models this may not be a problem, as sorting only needs to be performed once. However, as this model erodes and deposits, the path of the water from the highest to lowest grid cell constantly changes. This means that for each cycle of erosion and deposition, the cells have to be re-sorted, which with conventional algorithms would take an unacceptably long time.

The 'scanning' algorithm sorts the altitudes by ascribing each cell the output from the hydrological model. This value is then pushed, like water across the map, from left to right, right to left, up and down (Fig. 1).

For each push or scan, the value is routed only to the three cells immediately in front. If the three cells are all higher than the contributing cell, but the combined water depth and elevation of the contributing cell is higher, water is retained in the contributing cell up to the depth of the obstruction whilst the is rest routed on. This process has the effect of filling hollows that the model creates with water. These hollows or sinks frequently have to be removed from DEM data (Goodchild & Mark, 1987; Hutchinson, 1989).

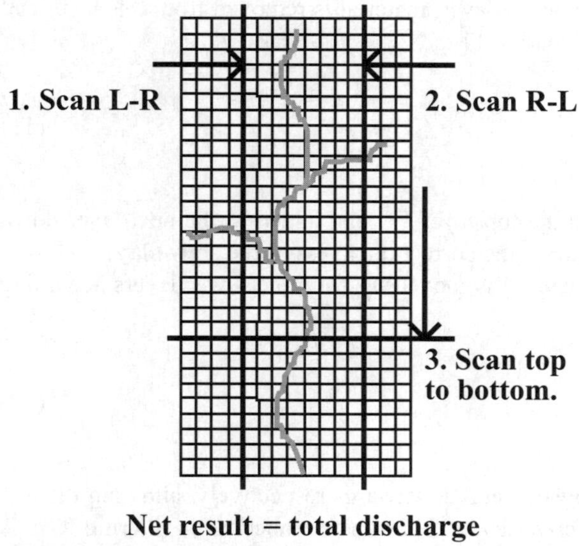

Figure 1. Diagram depicting the scanning across of the catchment area.

The maximum depth calculated from each of these four scans is recorded and used to drive fluvial erosion and deposition. This routing procedure allows flow trapped after one iteration to be incorporated in the next iteration's scans. This storage in corners and bends maintains a flow in the channel even around complex channel patterns and meanders. For example, with a meander sequence, water may be routed around the first corner, but be trapped by the second. However, in the next iteration, this water is still there, to be released in the next scan, and replaced by more water from upstream, allowing the continuum of flow.

Another way to visualise the whole process is to imagine covering a 3D map surface with water droplets. If these are squeezed across with a giant squeegee – as in each scan – some water will remain in depressions or in the channel. If the same amount of water droplets is applied again, and scanned, the remaining water will have acted as a store in the hollows and bends, the same amount applied will be removed at the base. This method gives very similar results to that of the conventional method – even in complex areas such as confluences – in a fraction of a second, thus generating a huge time saving.

2.3 *Fluvial Erosion and Deposition*

To represent the erosion and deposition of different grain sizes and the development of an armoured surface layer, an active layer system is used, similar to that of Parker (1990), Hoey & Ferguson (1994) and Toro-Escobar et al. (1996). This model, however, uses twelve active layers. One for bedload, one for the surface active layer and ten further subsurface layers. The depth of the surface active layer is defined as $2D_{90}$ with the ten layers below at $4D_{90}$. Nine grain sizes are represented from 0.004 to 1.024 m in whole phi classes (-2ϕ to -10ϕ). Furthermore, the surface active layer has two additional categories representing a protective surface vegetative mat and bedrock.

When material is added to the top active layer, material is removed from this layer and added to the next layer down, as in Equation 11.

$$E_i = \left(\frac{F_i^x}{\sum F_{i-n}^x} \right) \left(\sum F_{i-n}^x - A \right) \tag{11}$$

Here E_i is the amount removed from the top layer F^x and added to the next layer down F^{x+1} of grain-size fraction $_i$. A represents the correct thickness of the active layer ($2D_{90}$ or $4D_{90}$). For erosion or degradation, material is moved up from the lower layers according to Equation 12.

$$E_i = \left(\frac{F_i^{x+1}}{\sum F_{i-n}^{x+1}} \right) \left(A - \sum F_{i-n}^x \right) \tag{12}$$

Both these expressions propagate upwards and downwards respectively, allowing the displacement of material through the active layers. No transfer function or filtering term is used (Hoey & Ferguson, 1994; Toro-Escobar et al., 1996), as such a term has no temporal scaling. This scaling is necessary given the variable time step used in the model's operation as discussed later. This representation of the river bed allows the development of an armoured surface, and the storage of deposited sediment in the stratigraphy of the other ten active layers.

The amount eroded by fluvial action from cell to cell is determined using the Einstein-Brown formulation (Einstein, 1950). This was chosen as much information is available about the local hydraulic conditions (Gomez & Church, 1989) and the total load is calculated from the sum of fractions eroded. This is well suited to the nine grain-size classes and active layer system used. The formula used here takes the form of Equation 13.

$$\psi = \frac{(\rho_s - \rho) D}{\rho d S} \tag{13}$$

Where ψ is the balance between the forces moving and restraining the particle, $\rho_s - \rho$ is the relative density of the submerged sediment, D is the grain size (metres), d is the flow depth and S is the energy slope. A dimensionless bedload transport rate ϕ is then calculated.

$$\phi = q_s \sqrt{\frac{\rho}{(\rho_s - \rho) g D^3}} \tag{14}$$

ϕ is then related to ψ by the relationship plotted by Einstein (1950),

$$\phi = 40 (1/\psi)^3 \tag{15}$$

A rearrangement of Equations 14 and 15 then allows q_s (the volumetric sediment load in $m^3\ s^{-1}$) to be calculated. For each grid cell, the amount in each grain-size class which can be eroded is calculated, and removed from the active layer of the cell in question, and deposited to the active layer of the downstream cell.

The grass layer is treated as a boulder of diameter 0.26 m. This was calculated from field shear stress measurements carried out by Prosser (1996). Bedrock is similarly treated as being an immovable boulder 100 m in diameter.

2.4 *Slope Processes*

Mass movement is simply represented as an instantaneous removal process. When the slope between adjacent cells exceeds a threshold (currently 0.5), material is moved from the up-slope cell to the one below until the angle is lower than the threshold. As a small slide in a cell at the base of a slope may trigger more movement upslope, the model continues to check the adjacent cell's rows until there is no further movement. Where the cells border the main channel, material is transferred from active layer to active layer and the layers' grain size proportions are updated accordingly. If the amount removed from the upslope cell is greater than the active layer thickness, material from the subsurface is also added.

Soil creep is calculated between each cell for every month of model time according to Equation 16

$$Creep\left(yr^{-1}\right)=\frac{S0.01}{Dx} \tag{16}$$

New cell values are updated simultaneously and where cells border the channel, material is transferred to the active layer of the receiving cells. As this model is designed and applied to small, vegetated temperate river catchments, rill growth, rainsplash and sheet wash processes are not significant and are omitted. However, the these could easily be added if the model was applied to a semi-arid catchment for example.

2.5 *Vegetation growth*

Vegetation re-growth is simply represented in the model. It has no interactions with the hydrological model as its purpose is simply to allow a protective turf mat to develop over flood deposits. This process was considered important in the preservation of alluvial flood units. An extra fraction is added to the surface active layer to represent this turf cover. A simple linear growth model adds to this layer in monthly time steps if the cell is not under water and after ten years of uninterrupted growth, full cover will develop. If this layer is eroded, material is removed from the grass fraction and treated as if washed out of the catchment, instead of re-depositing. The addition or removal of vegetation has no effect upon the elevation of the cell, but partial removal will allow material from beneath to become exposed and vulnerable to erosion.

For all functions except creep and vegetation growth, a variable time-step is used. This is adjusted so the maximum amount that can be removed or deposited from one cell to another is a small proportion of the average slope (< 10%). A maximum time step of 120 seconds is introduced allowing significantly rapid progress in low flow situations, yet not too long as to miss any part of a storm. The boundary conditions were fixed so that none of the edge cells can be altered in elevation, but material can pass over them, similar to the conditions at the outflow of a flume. The net result of these sub-models is that for the relatively simple inputs of a topography, vegetation cover and rainfall, a sequence of erosion and deposition in a small catchment can be simulated.

3 THE STUDY CATCHMENT

For this study, the model was applied to Cam Gill Beck, a small (4.2 km^2) upland tributary of the River Wharfe, North Yorkshire, UK (Fig. 2). This catchment is steep, dropping from 700 m to 230 m over 3 km and the underlying lithology is predominantly Carboniferous limestone with grit and sandstones above 500 m (Raistrick & Illingworth, 1949). For the model, catchment topography was digitised from 1:10,000 scale Ordnance Survey maps combined with extra valley floor detail from an EDM survey. This data was processed using the TOPOGRID command in ARC-INFO to create a 2 m by 2 m resolution DEM containing approximately 1 million cells (Fig. 3).

4 FLOOD SIMULATIONS

Two groups of simulations were carried out. Firstly, to model periods of moderate sustained flooding, a ten year, hourly rainfall data set (1985 to 1995) from Church Fenton,

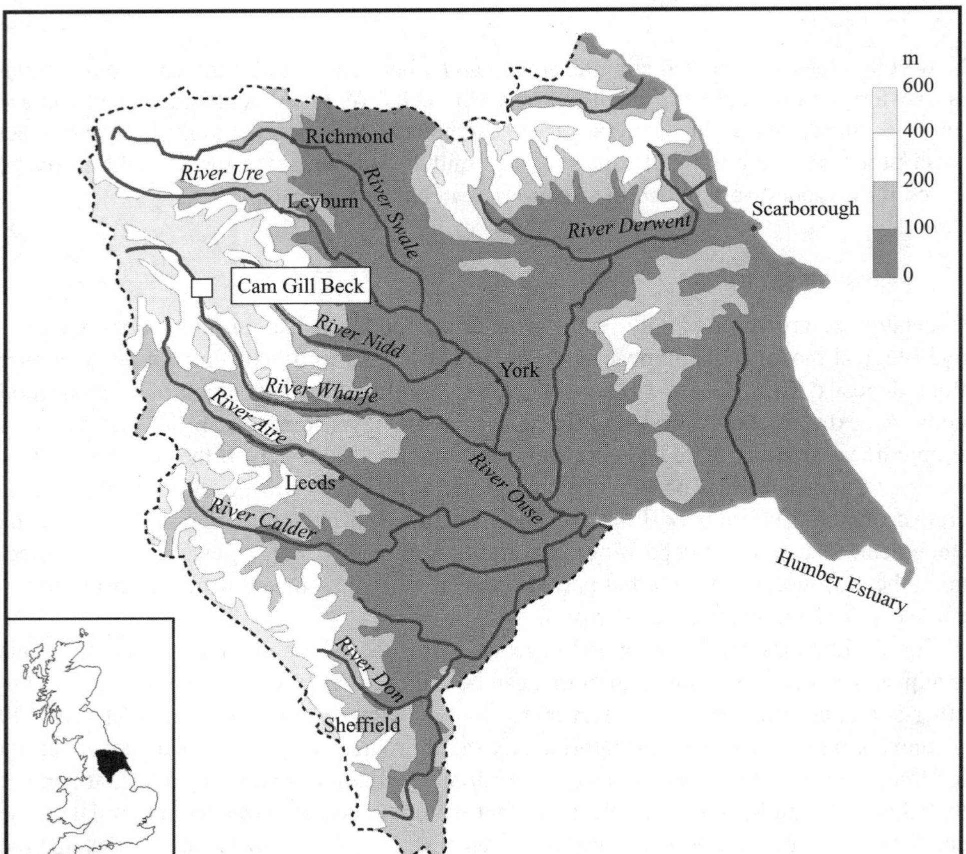

Figure 2. Location and relief of Cam Gill Beck in Yorkshire, UK.

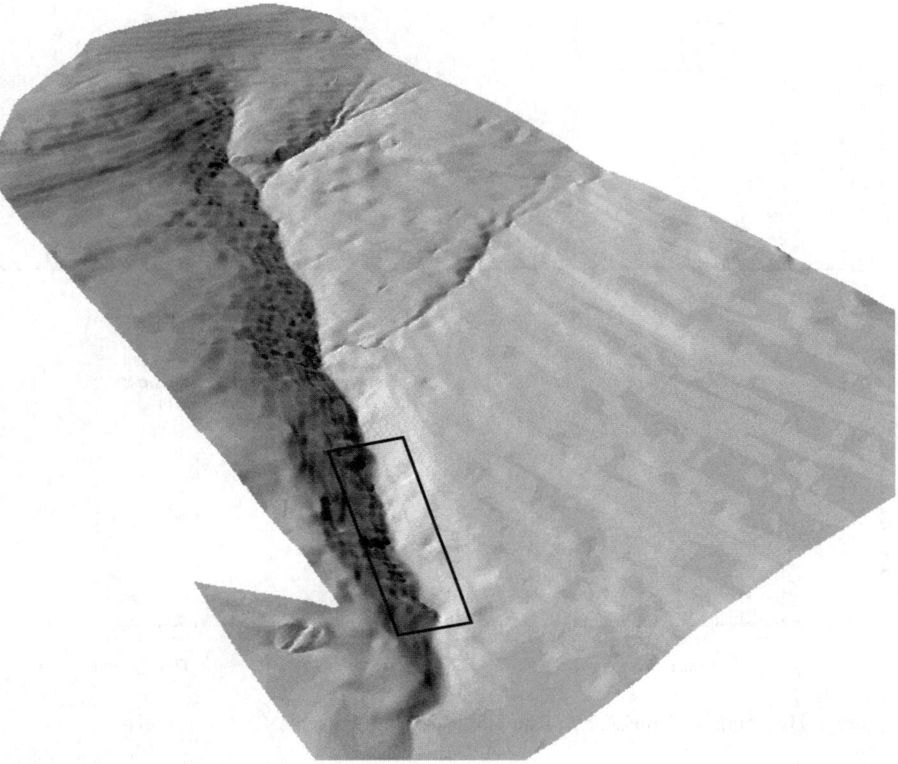

Figure 3. 3d projection of Cam Gill Beck DEM, viewed from the south. Scale 1400 by 2800 m. The boxed area is detailed in Figures 7 and 8.

North Yorkshire (SE 515 370) was used to drive the model. This was multiplied by factors of 1, 1.5 and 2 (*Run 1, Run 1.5 and Run 2*), to bracket a range of potential climates and the hydrographs produced by these sequences are detailed in Figure 4. Comparison to a 1995-1996 rainfall record at nearby Coverdale (SE 014 787, altitude 240 m) reveals that the *Run 1.5* scenario best represents present day conditions. This is supported by a series of 16 long term simulations also carried out on Cam Gill Beck covering a much wider range of climatic scenarios (Coulthard et al., 2000). After the first ten years of simulation, the sequence was repeated a further nine times, using the surface left from the previous ten years to simulate a total of 100 years erosion and deposition. Secondly, to simulate an extreme flood event, the same rainfall sequence was edited to include two consecutive hours of extreme rainfall of 100 mm hr^{-1} and 80 mm hr^{-1} respectively. These amounts were chosen as they were similar to the value of 192 mm in 2 hours recorded by Evans (1996) prior to an extreme flood in Wycoller Beck in the Central Pennines. The initial surface was taken from the end of *Run 1* detailed above.

At the end of every 10 year simulation run, the elevations were saved and subtracted from the initial conditions to determine the sediment yield. Because of the size of the DEM used, it was impractical to define initial conditions for every grid cell. Therefore, a 'spin up' time is required to allow these initial conditions to evolve. The catchment begins

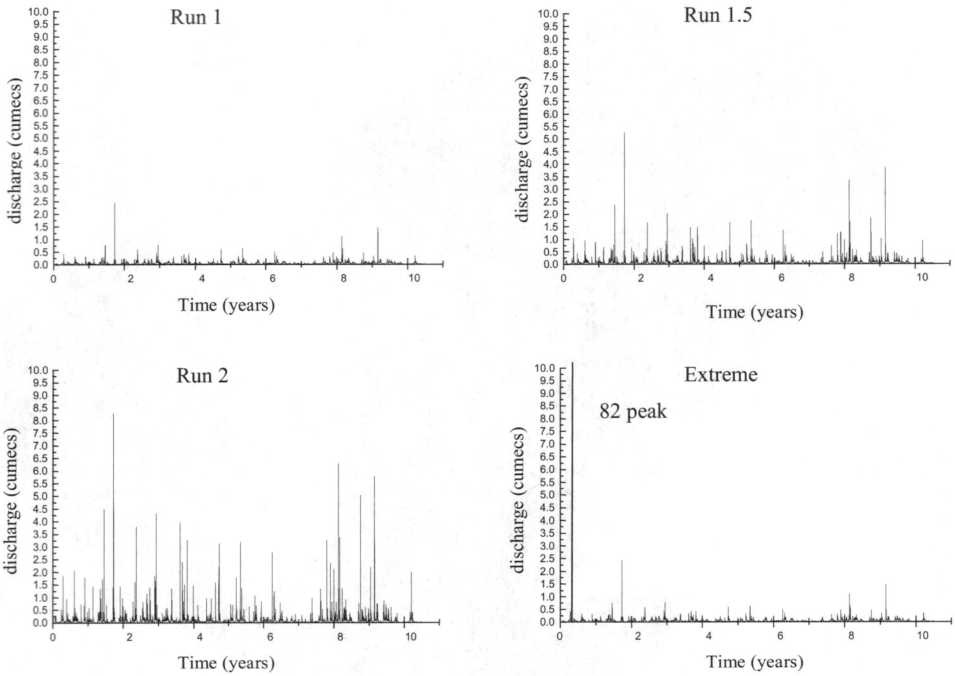

Figure 4. Hydrographs from Run 1, Run 1.5, Run 2 and Extreme.

with equal distributions of grain-size in every cell and, during this initialisation period, is subjected to small floods that allow a channel and bed armour layer to develop.

5 RESULTS AND MODEL VALIDATION

Figure 5 shows the ten-year bedload discharges from the three moderate flooding simulations. The initial peak is caused by the removal of fines from the initial conditions after which they stabilise producing a regular bedload discharge. This compares favourably with field measurements from other catchments (Fig. 6, Table 1). There is considerable scatter within Figure 6 that may reflect the physical differences between the catchments studied, and the wide variety of techniques used to measure bedload. However, the results of the three moderate flooding simulations lie within the values in Figure 6 and this indicates that the model is predicting sediment discharges within the correct range. Minor fluctuations in these bedload discharges after the initial peak may be caused by non-linear reactions within the model which develop because of the complexity of the interactions between the processes and forms within the catchment (Coulthard et al., 1998).

Figure 7 shows the extent of flood deposits left by the extreme flood simulation and compares them to those left by an extreme flood event documented in 1686. This major flood caused such damage that is was actually mentioned in Parliament at the time. One account details how on the night of the 18th February, 1686:

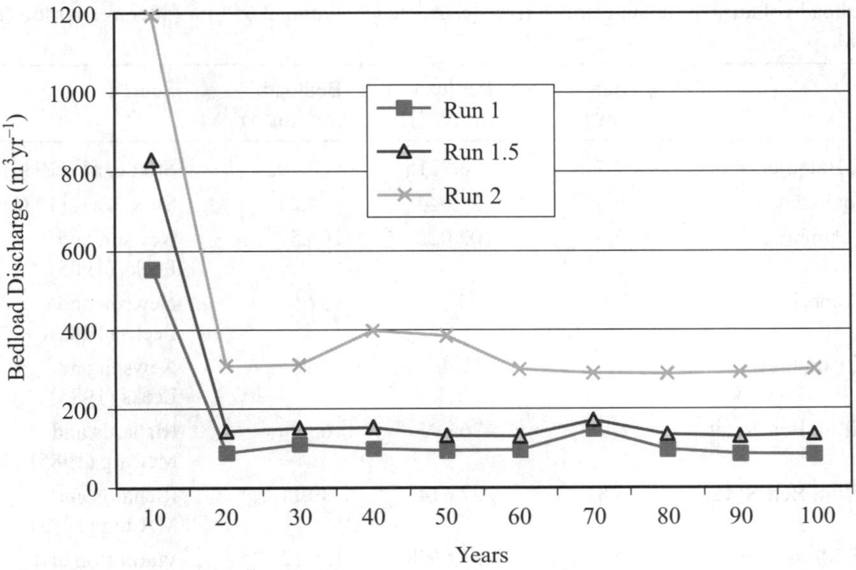

Figure 5. Bedload discharge for simulations Run 1, Run 1.5 and Run 2.

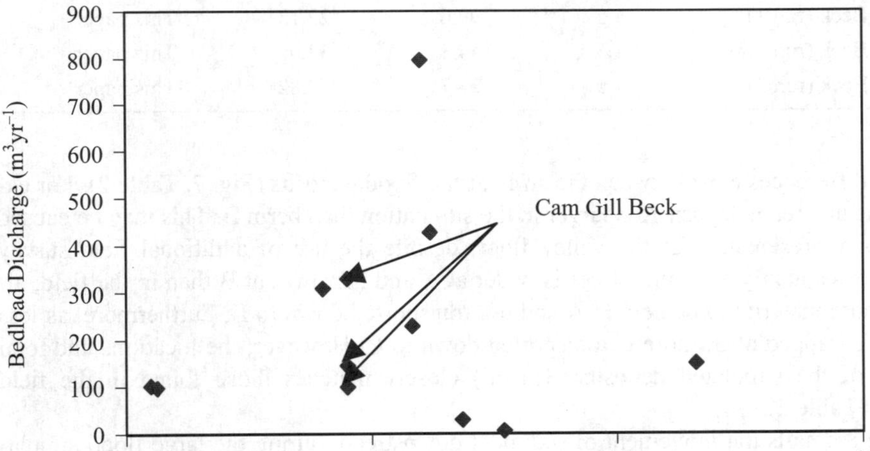

Figure 6. Bedload discharge against catchment area from Table 1.

> The whole of England was visited by a tempest, accompanied with thunder, which committed general devastation. The inhabitants of Kettlewell and Starbottom, in Craven, were almost all drowned in a violent flood. These villages are situate under a great hill, whence the rain descended with such violence for an hour and half, that the hill on one side opened, and casting up water into the air to the height of an ordinary church steeple, demolished several houses, and carried away the stones entirely. (Mayhall, 1860)

Table 1. Bedload volumes for upland British rivers calculated assuming 2.65 t m^3 (after Warburton & Evans, 1998).

Catchment	Area (km^2)	Bedload (m^3 yr^{-1})	Bedload (m^3 km^2 yr^{-1})	Source
Monachyle, Balquidder	7.7	6.1215	0.795	Stott et al. (1986)
Kirton, Balquidder	6.85	29.044	4.24	Stott et al. (1986)
Beckthorn, Cumbria	0.5	102.025	204.5	Newson and Leeks (1985)
Coledale, Cumbria	6.0	795	132.5	Newson and Leeks (1985)
Lanthwaite, Cumbria	4.0	307.4	76.85	Newson and Leeks (1985)
Allt a'Mhuillin, Ben Nevis	6.19	426.491	68.9	Richards and McCaig (1985)
Allt a'Mhuillin, Ben Nevis	5.82	227.614	39.11	Richards and McCaig (1985)
Rough Sike, Moor House	0.63	97.6658	155.02	Warburton and Evans (1998)
Trout Beck, Moor House	11.6	149.46	12.84	Warburton and Evans (1998)
Cam Gill Beck (run 1)	4.5	99.6	22.13	This paper
Cam Gill Beck (run 1.5)	4.5	144.3	32.06	This paper
Cam Gill Beck (run 2)	4.5	327.7	72.82	This paper

Some differences exist between the sizes of the flood deposits (Fig. 7, Table 2). For example, the upstream deposit A is larger in the simulation than berm B. This may be caused by a poor representation of the valley floor, despite the use of additional field survey points. Consequently the valley floor is wider at A and narrower at B than in the field, allowing more material to be held at A and not transported down to B. Furthermore, as less material is trapped at B, more is transported down to C. However, the locations and total volumes of the simulated deposits (331 m^3) closely matches those found in the field (350 m^3) (Table 2).

Figure 8 details the movement of sediment downstream during the large flood simulation with areas of deposition highlighted in white. The timing of these pictures is shown by the bars on the flood hydrograph in Figure 9, and a long profile of the same section in Figure 10. During the early part of the falling limb (Fig. 8, image 1), there is comparatively little deposition, with only the area upstream of *D* beginning to aggrade. By image 2, *D* has enlarged considerably and the deposits downstream have grown, with substantial quantities of sediment reaching *E*. By image 3, at the end of the flood wave, deposits upstream of *D* have been removed and *D* itself substantially reduced. This material has been re-deposited at *E* which has grown notably, especially when compared to image 1. Figure 10 shows that vertical changes in these deposits are substantial, with up to 2 m thickness of sediment being removed at *D*. These snapshots show that sediment may be moving through this reach in a pulse or wave, riding the falling limb of the hydrograph. This is confirmed by the long profile in Figure 10.

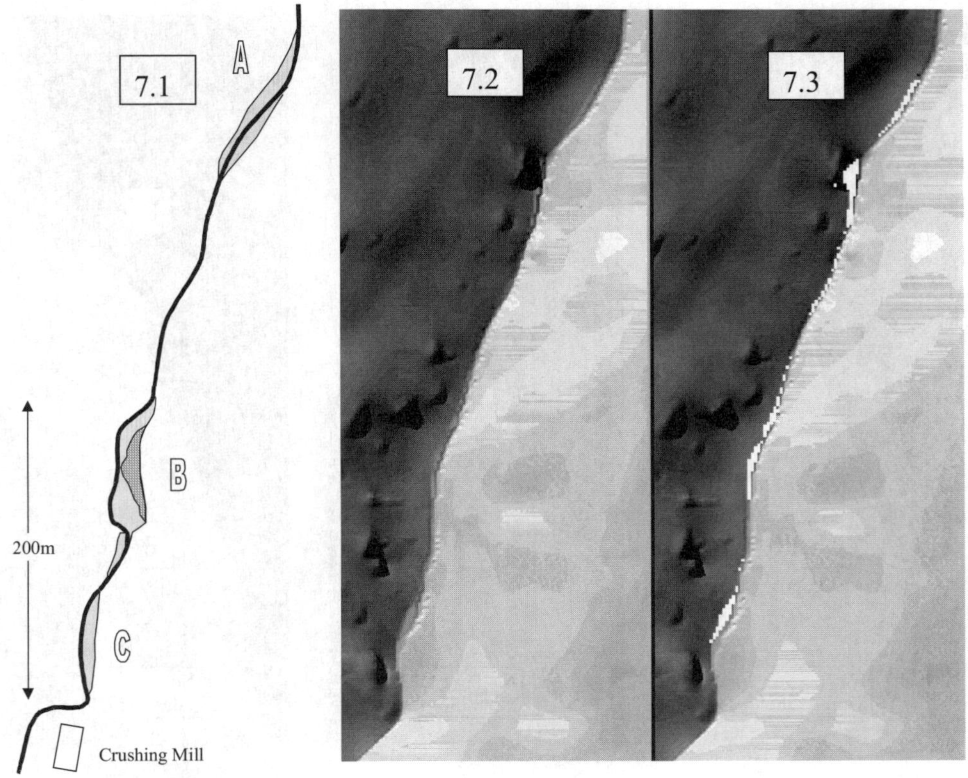

Figure 7. Plan view of section highlighted in Figure 1. (7.1) Map of flood deposits, (7.2) Shaded view of simulation results, (7.3) Shaded view with deposition from the simulated extreme flood highlighted in white.

Table 2. Comparison of sediment volumes from field measurements and simulation (All units in m^3).

Initial Conditions	Field	Extreme flood simulation
Volume eroded	N/A	5357
Volume deposited	N/A	2285
Balance (E-D)	N/A	3071
Volume Berm A	100	233.06
Volume Berm B	250	96.66

6 DISCUSSION AND CONCLUSIONS

These simulations provide new insights into the roles of extreme floods and periods of lower magnitude flood events in valley floor development. McEwan & Werritty (1988), examined the effects of flash floods in the Cairngorms and stated that:

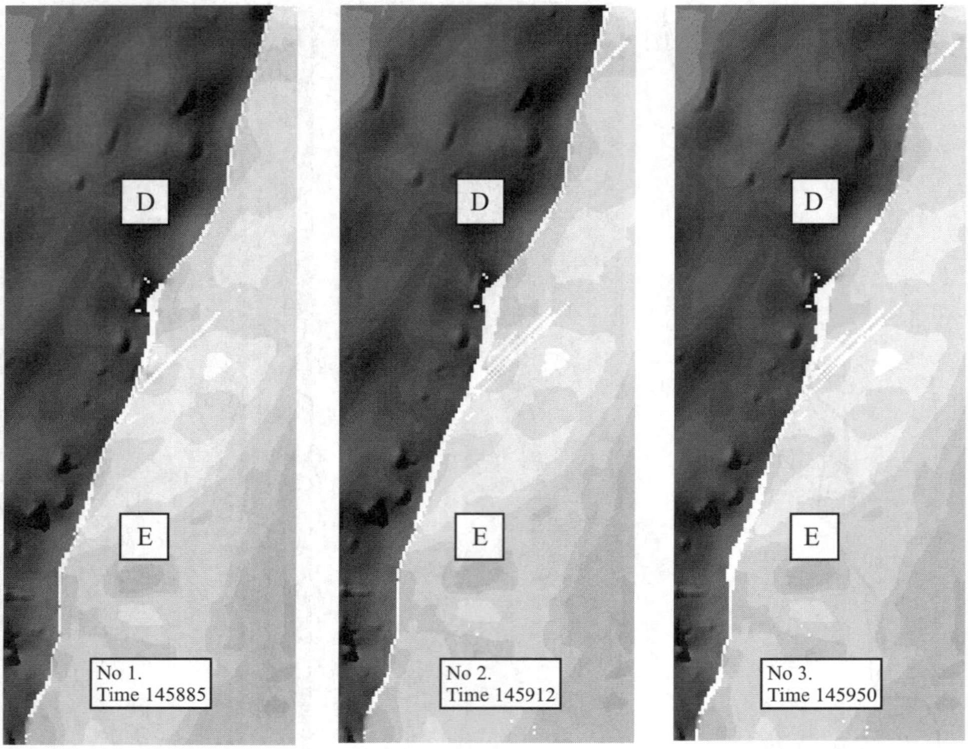

Figure 8. Plan views showing movement of sediment wave (highlighted in white).

'The long term evolution of boulder bed mountain torrents in upland Scotland must be regarded as being primarily controlled by the operation of catastrophic processes.'

In these simulations, the model predicts the bedload yield from the extreme flood to be 3000 m^3 compared to 3277, 1443 and 996 m^3 per 10 years for the more moderate floods (runs *medium 2*, *medium 1.5* and *medium 1*). Therefore, in the terms of work done removing sediment from the basin, the extreme flood is equivalent to 10 years at 150% greater rainfall magnitude, 20 years at current levels and 30 years at 75%. When modelled over 100 years, the more moderate storm regimes cause 350, 500 and 1000% more bedload discharge than the extreme event. It could be argued that the ten year data set used is unrepresentative and may contain extreme floods itself. Indeed, the sequence runs from 1985 to 1995, capturing a full spectrum of storms including Hurricane Charley, August 1986, the wettest day in the UK's record (Institute of Hydrology, 1988). However, the hydrographs (Fig. 4) show the peak discharges from runs 1, 1.5 and 2 ranging from 2.5, 5 and 8.5 $m^3 s^{-1}$ respectively, compared to 82 $m^3 s^{-1}$for the extreme flood. Thus, a ten year storm sequence giving a maximum flood of 2.5 $m^3 s^{-1}$ (*medium 1*) erodes more material in 30 years than one huge flood. Furthermore, the storms producing extreme events – especially those resulting from thunderstorms – are often very localised. For example, Evans (1996) estimates that the Wycoller Beck flood was caused by a small cell of rain approximately 500 m wide. The frontal weather systems associated with periods of sustained rainfall operate over regional scales of hundreds of square kilometres (Longfield & Macklin,

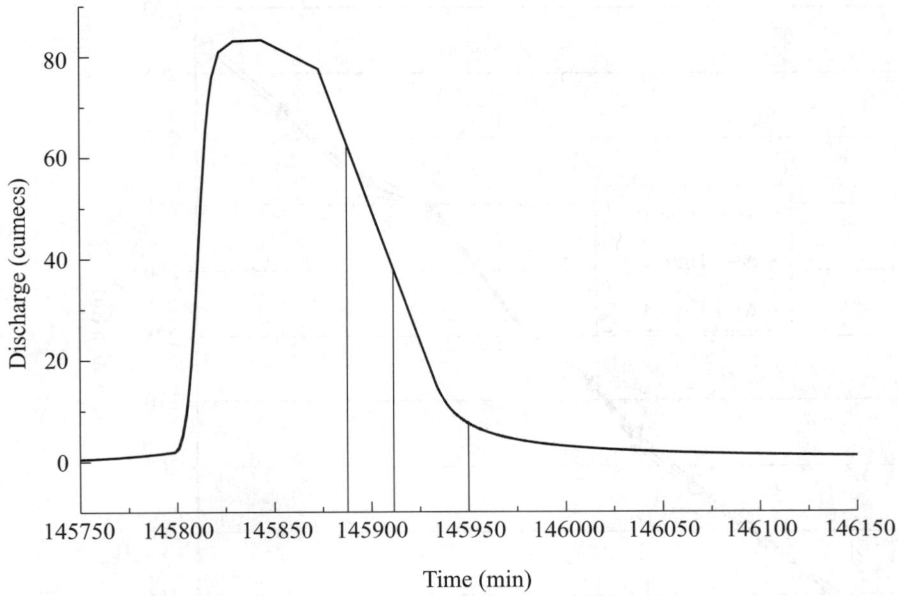

Figure 9. Hydrograph from simulated extreme flood. The vertical black lines indicate the timing of the views in Figure 8.

1999) and thus have the potential to cause flooding over a wider area. In view of this and the results from the model, in catchments like Cam Gill Beck long term periods of more moderate storm events will have a far greater impact on basin sediment yield and thus surface lowering.

Despite these indications that a single extreme event is not as effective at removing sediment when compared to long periods of moderate flooding, they can have a dramatic impact on catchments and may be a substantial hazard to human populations. In these simulations, the model shows they can have a large impact upon the valley floor morphology, leaving deposits *A B* and *C* (Fig. 7) standing up to 2 m above the main channel. It was expected that these deposits would be left exposed by the gradual incision of smaller flood events. However, the model showed that towards the end of the falling limb, the channel rapidly incised next to the main deposit *B*. Furthermore, other simulations illustrated that for up to two years after the extreme flood, large sections of this deposit *B* were removed by smaller floods of up to 6 m^3s^{-1}. However, this feature was preserved if these smaller floods were held back allowing a channel to incise next to the deposit and vegetation re-growth to stabilise it. Therefore, the timing and magnitude of subsequent flood events is vital for the preservation of such units, adding a further dimension of uncertainty when trying to interpret previous events from present morphologies.

Although smaller floods may have a greater cumulative role, large flood events can be important in the long term evolution of a catchment. A single large flood event may only be equivalent to ten years of moderate flooding, but if three events of such magnitude occur within ten years, then three times more erosion is carried out. Thus, when interpreting past regimes or climates from large flood deposits as in Macklin et al. (1992) and Merrett &

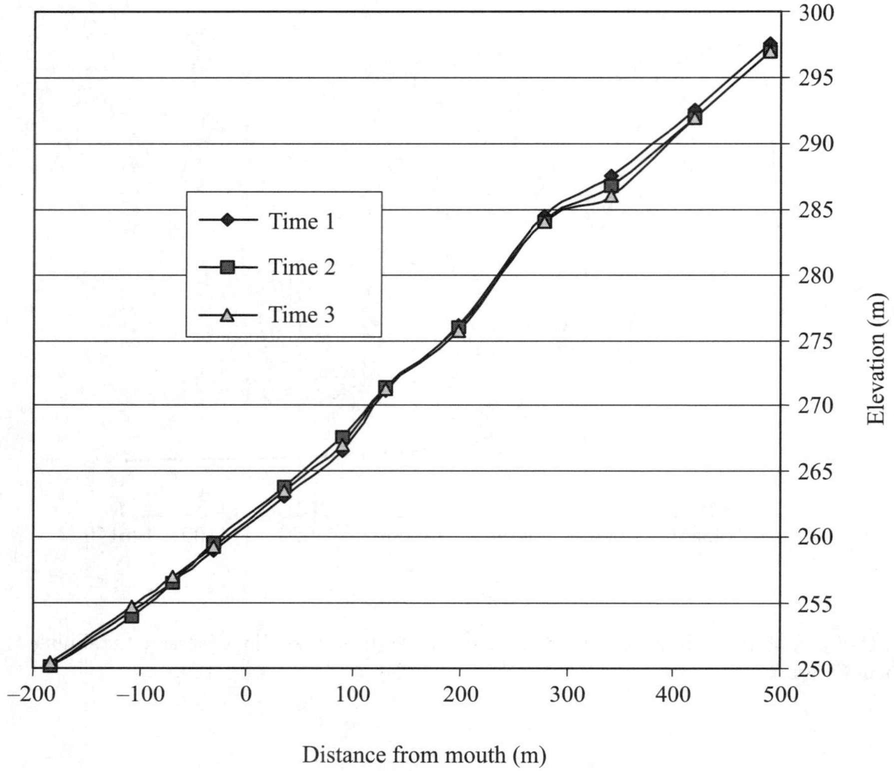

Figure 10. Long profile of section detailed in Figure 8.

Macklin (1999), we should be wary of inferences drawn from single or small clusters of flood events. But where there are large groups of large floods at, for example, a decadal resolution, this can be taken as a good proxy for periods of increased flood magnitude and thus catchment sediment mobility.

During the extreme flood simulation there is a downstream movement of a pulse, wave or slug of sediment (Nicholas et al., 1995), depositing in the early stages of the flood, then re-working and moving down the channel (Figs 8 and 10). As the long profile (Fig. 10) shows, this results in incision at higher points of the basin and deposition at lower. During the passage of the sediment wave, the model shows zones of sediment storage & transport similar to those identified by Macklin & Lewin (1989). Depositional areas are wider and have a lower slope (Figs 7 and 8), whereas there is little activity in the middle section as it is steeper and narrower with the sediment flushed through. Furthermore, as the slug passes through reaches *D* and *E*, there is a widening of the flood plain as the valley floor in-fills. This consequently narrows as the channel incises and removes the fill. The main sediment movement appears to occur during the falling limb of the hydrograph, suggesting that this part of the flood is most important.

This slug appears to be moving rapidly, travelling through the highlighted reaches in 65 minutes compared to rates varying form 0.1 to 5 km a year measured by other authors (Nicholas et al., 1995). The high slug celerity may be driven by the steep gradient of Cam

Gill Beck and the constrained valley floor, restricting substantial deposition. The limited extent of the flood deposits remaining from the event of 1686 supports this view. However, it must be remembered that, for validation purposes, there are very few measurements of large-scale sediment movement in upland catchments during extreme events. Yet this modelling approach shows great potential for studying the dynamics and effects of such large floods. If the spatial and temporal resolution were increased, the detail of the landforms, sediment movement and even the stratigraphies of the deposits could be simulated in much greater detail.

Notwithstanding these encouraging results, it must be remembered that this study is based on a single catchment. Figure 6 shows that considerable variance in bedload yield between catchments of different sizes and morphologies is to be expected. Furthermore, particular problems surround validation, which is hampered by three key factors:

1. As the model is attempting to simulate the past, we have no records of previous landforms or morphologies so the initial conditions for the model have to be estimated.
2. The use of the Wycoller Beck rainfall data is only an estimate of the 1686 flood. There is every possibility that deposits in Cam Gill Beck were formed by a flood of greater or smaller magnitude.
3. There are not enough dated deposits to validate the model. As only one date is recorded, all other validations are qualitative or based upon observation. Ironically, the model gives us far more information about the catchment morphology and grain-size than we are capable of measuring in the field.

However, this modelling technique offers great promise, and its design is generic allowing the easy application to any small catchment that can be digitised. The combination of hydrological, hydraulic and slope models captures many of the complex interrelationships within a catchment, at a scale rarely achieved. Whereas other modelling methods have concentrated upon modelling reaches during a single flood, this technique models the entire catchment over long-term sequences of flood events. Furthermore, for future studies, cellular automaton models respond well to the application of parallel programming techniques. Therefore, by dividing larger catchments up into smaller sections running on individual processors, with the margins linked together, the only limit to the spatial or temporal resolution of the study is the number of processors available.

ACKNOWLEDGEMENTS

The authors would like to thank Stephen Merrett and Dr Andy Howard for their advice, comments and assistance with surveying, Dr Joanna Schmidt of Leeds University Computing Services for help with the HPC facilities, and the Natural Environment Research Council for the provision of a research studentship (No. GT4/95/147/F) to TJC. Two anonymous referees provided valuable comments on the manuscript.

REFERENCES

Bates, P.D., Anderson, M.G., Hervouet, J.M. & Hawkes, J.C. 1997. Investigating the behaviour of two-dimensional finite element models of compound channel flow. *Earth Surface Processes and Landforms*, 22: 3-17.

Beven, K.J. & Kirkby, M.J. 1979. A physically based variable contributing-area model of catchment hydrology. *Hydrological Science Bulletin,* 24(1): 43-69.

Coulthard, T.J., Kirkby, M.J. & Macklin, M.G. 1998. Non-linearity and spatial resolution in a cellular automaton model of a small upland basin. *Hydrology and Earth System Sciences*, 2: 257-264.

Coulthard, T.J., Kirkby, M.J. & Macklin, M.G. (2000). Modelling geomorphic response to environmental change in an upland catchment. *Hydrological Processes,* 14: 2031-2045.

Desmet, P.J.J. & Govers. G. 1996. Comparison of routing algorithms for digital elevation models and their implications for predicting ephemeral gullies. *Int. J. Geographical Information Systems*, 10: 311-331.

Einstein, H.A. 1950. The bed-load function for sediment transport on open channel flows. *Tech. Bull. No. 1026, USDA, Soil Conservation Service.*

Evans, R. 1996. Hydrological impact of a high-magnitude rainfall event. *Advances in Hillslope Processes, Volume 1.* In: Anderson, M.G. & Brooks, S.M. (eds). Chichester, John Wiley, 98-127.

Goodchild, M.F. & Mark, D.M. 1987. The fractal nature of geographic phenomena. *Annals of Association of American Geographers,* 77: 265-278.

Gomez, B. & Church, M. 1989. An assessment of bed load sediment transport formulae for gravel bed rivers. *Water Resources Research*, 25(6): 1161-1186.

Hoey, T. & Ferguson, R. 1994. Numerical simulation of downstream fining by selective transport in gravel bed rivers: Model development and illustration. *Water Resources Research*, 30(7): 2251-2260.

Hutchinson, M.F. 1989. A new procedure for gridding elevation and stream line data with automatic removal of spurious pits. *Journal of Hydrology,* 106: 211-232.

Institute of Hydrology/British Geological Survey, 1998. *Hydrological Data United Kingdom, 1986 Year Book.* Natural Environment Research Council, Wallingford.

Knighton, D. 1998. *Fluvial Forms and Processes: A New Perspective*. Arnold, London.

Longfield, S.A. & Macklin, M.G. 1999. The influence of recent environmental change on flooding and sediment fluxes in the Yorkshire Ouse basin. *Hydrological Processes*, 7: 1051-1066.

Macklin, M.G. & Lewin, J. 1989. Sediment transfer and transformation of an alluvial valley floor: The river south Tyne, Northumbria, UK. *Earth Surface Processes and Landforms,* 14: 233-246.

Macklin, M.G., Rumsby, B.T. & Heap, T. 1992. Flood alluviation and entrenchment: Holocene valley floor development and transformation in the British uplands. *Geological Society of America, Bulletin*, 104: 631-643.

Macklin, M.G. & Lewin, J. 1993. Holocene river alluviation in Britain. *Zeitshrift für Geomorphologie* (Supplement), 88: 109-122.

Mayhall, J. 1860. *The Annals of Yorkshire; From the Earliest Period to the Present Time. Volume 1, 1856 BC to AD 1859*. H.C. Johnson, Leeds.

McEwan, L.J. & Werritty, A. 1988. The hydrology and long-term geomorphic significance of a flash flood in the Cairngorm Mountains, Scotland. *Catena*, 15: 361-377.

Merrett, S.P. & Macklin, M.G. 1999. Historic River Response to Extreme Flooding in the Yorkshire Dales, Northern England. In: Brown, A.G. & Quine, T.M. (eds), *Fluvial Processes and Environmental Change.* Chichester, John Wiley, 345-360.

Nicholas, A.P, Ashworth, P.J, Kirkby, M.J., Macklin, M.G. & Murray, T. 1995. Sediment slugs: large scale fluctuations in fluvial sediment transport rates and storage volumes. *Progress in Physical Geography,* 19(4): 500-519.

Nicholas, A.P. & Walling, D.E. 1997. Modelling flood hydraulics and overbank deposition on river floodplains. *Earth Surface Processes and Landforms*, 22: 59-77.

Parker, G. 1990. Surface based bedload transport relation for gravel rivers. *Journal of Hydraulic Research*, 28: 417-436.

Prosser, I.P. 1996. Thresholds of Channel Initiation in Historical and Holocene Times, South-Eastern Australia. In: Anderson, M.G. & Brooks, S.M. (eds) *Advances in Hillslope Processes,* Volume 2: 687-708 Chichester, John Wiley.

Raistrick, A. & Illingworth, J.L. 1949. *The Face of North-West Yorkshire*, Dalesman, Clapham.

Toro-Escobar, C.M. Parker, G. & Paola, C. 1996. Transfer function for the deposition of poorly sorted gravel in response to streambed aggradation. *Journal of Hydraulic Research*, 34(1): 35-51.

Wolman, M.G. & Miller, J.P. 1960. Magnitude and frequency of forces in geomorphic processes. *Journal of Geology*, 68: 54-74.

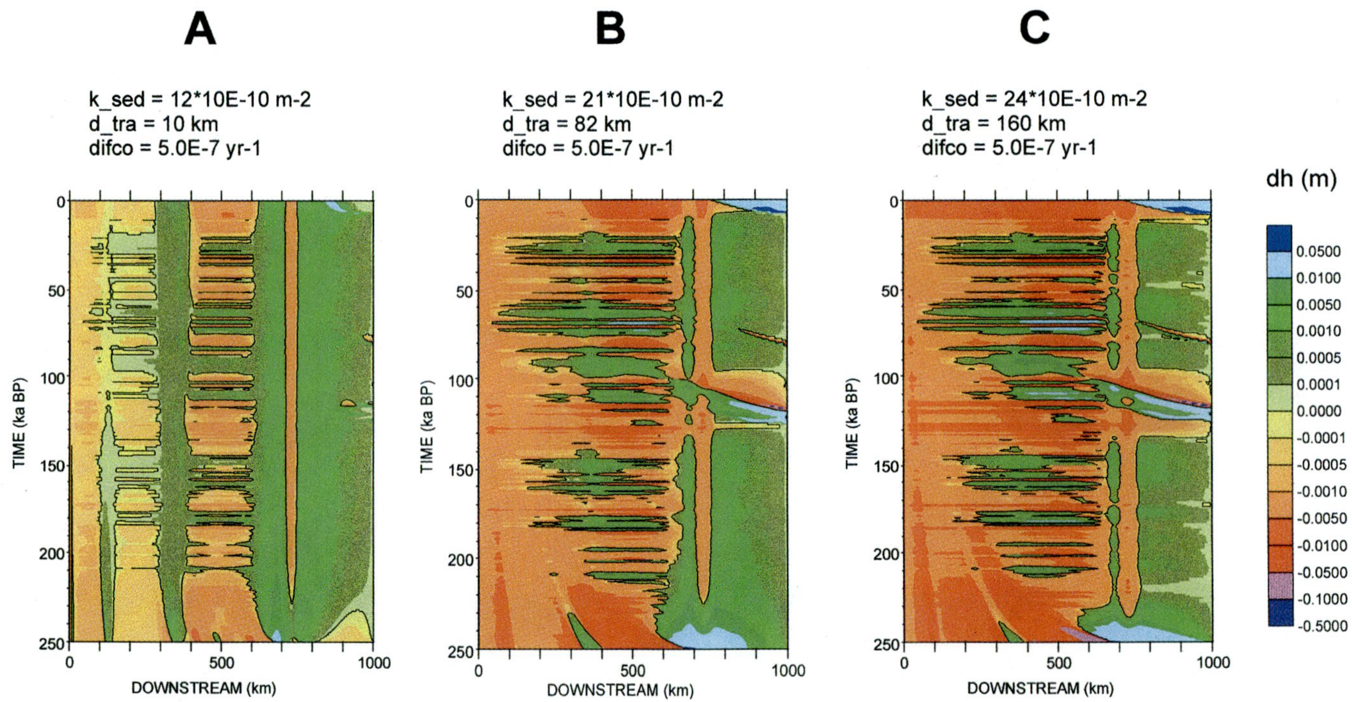

Plate 1. Profile Evolution Maps for FLUVER2 model runs with different settings for *k_sed* and *d_tra*, showing where along the Meuse longitudinal profile (x-axis) and when (y-axis) erosion (negative *dh*-values) or deposition (positive *dh*-values) events occur. PEM A: Note overall too low fluvial dynamics with knickpoint artefacts and lack of incisional phases in the lower reach (>600-km). PEM B: most realistic model run. Note incisional phases for >700-km reach at ~125-ka and ~10-ka BP, followed by strong backfilling-related aggradation (see text). The terrace intersection at 0-ka BP (recent Holocene) is located at ~760-km. PEM C: Note overall strong fluvial dynamics with erosion in middle reaches and incisional phases for the >600-km reach at ~125-ka and ~10-ka BP, followed by backfilling-related aggradation (see text). The terrace intersection at 0-ka BP (recent Holocene) is located irrealistically far inland at ~710-km.

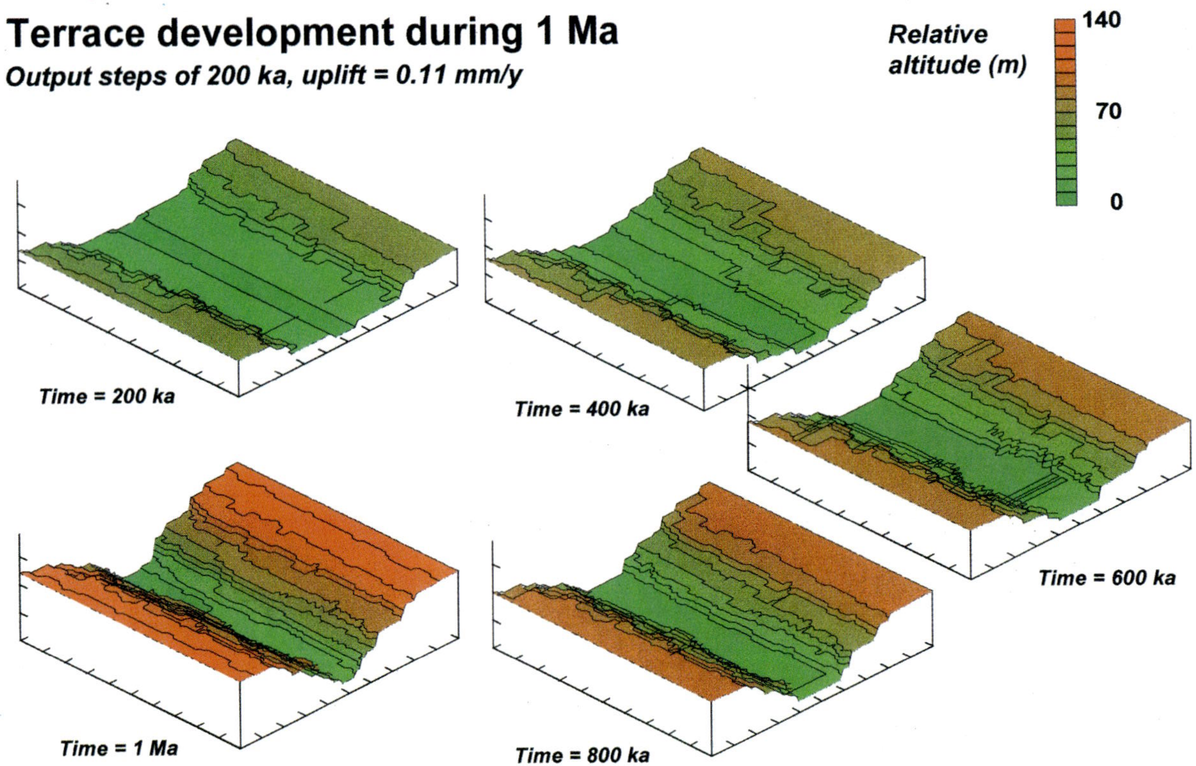

Plate 2. Finite-state 3-D model outputs to illustrate successive phases of local river valley and terrace development for a terraced valley section of the River Meuse (~622-632 km in PEM's A to C in Figure 2). Note the orientation of river-valley width in the x-direction (10-km), river-valley length in the y-direction (10-km) and river-valley/terrace altitude in the z-direction (150-m).

River Basin Sediment Systems: Archives of Environmental Change

List of Contributors

Mike D. Blum, Department of Geosciences, 214 Bessey Hall, University of Nebraska - Lincoln, Lincoln, NE 68588-0340, USA

Sjoerd Bohncke, Department of Quaternary Geology and Geomorphology, Faculty of Earth Sciences, Vrije Universiteit, De Boelelaan 1085, 1081 HV Amsterdam, The Netherlands

Olga Borisova, Institute of Geography, Academy of Sciences, Staromonetny 29, 109017, Moscow, Russia

James Brassington, Department of Geography, University of Cambridge, Downing Place, Cambridge, CB2 3EN, UK

David R. Bridgland, Department of Geography, University of Durham, South Road, Durham, DH1 3LE, UK

Keith Challis, Trent & Peak Archaeological Unit, University of Nottingham, University Park, Nottingham, NG7 2RD, UK

Tom Coulthard, Institute of Geography and Earth Sciences, University of Wales Aberystwyth, Llandinam Building, Aberystwyth, SY23 3DB, UK

Michel Fontugne, Laboratoire des Sciences du Climat et de l'Environement, Avenue de la Terrasse, 91118 Gif-sur-Yvette Cedex, France

Anne Gebhardt, Laboratoire d'Anthropologie, UMR 0153 CNRS, Campus de Beaulieu, 35042, Rennes Cedex, France

Peter Houben, Institute of Physical Geography, J.W. Goethe-University Frankfurt am Main, P.O. Box 11 19 32, D-60054 Frankfurt am Main, Germany

Andy Howard, School of Geography, University of Leeds, Leeds, LS2 9JT, UK

Kees Kasse, Department of Quaternary Geology and Geomorphology, Faculty of Earth Sciences, Vrije Universiteit, De Boelelaan 1085, 1081 HV Amsterdam, The Netherlands

Mike J. Kirkby, School of Geography, University of Leeds, Leeds, LS2 9JT, UK

Nikolai Kovalyukh, State Scientific Centre of Environmental Radiogeochemistry, Kyiv Radiocarbon Laboratory, 34 Palladin Avenue, 252680, Kyiv-142, Ukraine

Vincent Krier, AFAN, 7, rue de Madrid, 75008, Paris, France

Chantal Leroyer, Centre National de Préhistoire et UMR 9933, 38 avenue du 26ème R.I., 24000 Périgueux, France

John Lewin, Institute of Geography and Earth Sciences, University of Wales Aberystwyth, Llandinam Building, Aberystwyth, SY23 3DB, UK

Nicole Limondin-Lozouet, Laboratoire de Géographie Physique, UMR 8591 CNRS, 1 place Aristide Briand, 92195 Meudon Cedex, France

Mark G. Macklin, Institute of Geography and Earth Sciences, University of Wales Aberystwyth, Llandinam Building, Aberystwyth, SY23 3DB, UK

Darrel Maddy, Department of Geography, University of Newcastle, Newcastle, NE1 7RU, UK

Glenn Maas, Institute of Geography and Earth Sciences, University of Wales Aberystwyth, Llandinam Building, Aberystwyth, SY23 3DB, UK

Ross McVey, Department of Geography, The University of Hull, Hull, HU6 7RX, UK

Eilidh Meldrum, School of Geography, University of Leeds, Leeds, LS2 9JT, UK

Sabine Nolte, Am Grün 56A, D-35037 Marburg/Lahn, Germany

Andrey Panin, Geographical Faculty, Moscow State University, 119899 Moscow, Russia

David G. Passmore, Department of Geography, University of Newcastle, Newcastle, NE1 7RU, UK

Jean-François Pastre, Laboratoire de Géographie Physique, UMR 8591 CNRS, 1 place Aristide Briand, 92195 Meudon Cedex, France

Holger Rittweger, Institute of Physical Geography, J.W. Goethe-University Frankfurt am Main, P.O. Box 11 19 32 , D-60054 Frankfurt am Main, Germany

Barbara Rumsby, Department of Geography, The University of Hull, Hull, HU6 7RX, UK

Danielle Schreve, Department of Geography, Royal Holloway, University of London, Egham, Surrey, TW20 0EX, UK

Aleksey Sidorchuk, Geographical Faculty, Moscow State University, 119899 Moscow, Russia

Eric C. Straffin, Department of Geology, One University Plaza, Youngstown State University, Youngstown, OH 44555, USA

Leo A. Tebbens, Laboratory of Soil Science and Geology, Wageningen University, P.O. Box 37, 6700 AA Wageningen, The Netherlands

Donald M. Thieme, 203 7th Street, Apt.5, Honesdale, PA 18431, USA

Meindert Van den Berg, Netherlands Institute of Applied Geosciences, TNO-National Geological Survey, P.O Box 511, NL 8000 AM Zwolle, The Netherlands

Ton van Hoof, Department of Geophysics, Faculty of Earth Sciences, University of Utrecht, P.O. Box 80021, 3508TA Utrecht, The Netherlands

Antoni Veldkamp, Laboratory of Soil Science and Geology, Wageningen University, P.O. Box 37, 6700 AA Wageningen, The Netherlands

Jeff Warburton, Department of Geography, University of Durham, South Road, Durham, DH1 3LE, UK

Derek Welsby, Sudan Archaeological Research Society, The British Museum, London, WC1B 3DG, UK

Robert Westaway, 16 Neville Square, Durham, DH1 3PY, UK

Jamie C. Woodward, School of Geography, University of Leeds, Leeds, LS2 9JT, UK

Jürgen Wunderlich, Institute of Physical Geography, J.W. Goethe-University, Frankfurt am Main, P.O. Box 11 19 32, D-60054 Frankfurt am Main, Germany